AF291857

Ergebnisse der Mathematik
und ihrer Grenzgebiete

Volume 57

3. Folge

A Series of Modern Surveys
in Mathematics

For further volumes:
www.springer.com/series/728

Preface

In 1955, Chevalley [67] published constructions of simple groups of Lie type from Lie algebras. At that time, Dieudonné [119] had already published constructions of the classical groups and had started to focus on their geometric interpretation [120]. The universal geometric counterpart though was provided by Tits, whose approach and way of thinking by then began to transpire through lectures and papers such as [280]. However, the full extent of his constructions as well as the geometric classification of groups of Lie type (at least for rank at least 3) became available through [286]. This is his first comprehensive work on buildings, and deals primarily with the classification of spherical buildings of rank at least three. In later work, using joint work with Bruhat, Tits [290] took care of the classification of buildings of affine type and rank at least four (see [306] for an excellent account). Their automorphism groups are infinite groups, and we will only provide a simple example of the kind of geometry involved. One of the surprising aspects that came forward from Tits' work on buildings is the notion of a diagram. Diagrams prescribe what the geometries underlying the groups of Lie type look like locally. Here, a geometry is an abstract object, little more than a multipartite graph, and the local information alluded to concerns residues, that is, subgeometries induced on the set of vertices adjacent to a set of mutually adjacent vertices of that graph. In the case of bipartite graphs, the geometry is a so-called generalized polygon. As the word indicates, this is a very natural generalization of polygons. Generalized 3-gons, for instance, are projective planes. The ordinary 3-gon, a combinatorial triangle, is a generalized 3-gon and can be viewed as the smallest projective plane. Moreover, projective planes contain hoards of triangles.

In the classification of (non-Abelian) finite simple groups, the groups of Lie type play a crucial role, simply because, apart from these and the alternating groups, there are only 26 more—the so-called sporadic groups. Buekenhout launched the idea of employing the diagrammatic description of the geometries for groups of Lie type to a wider class of groups, preferably one that would lead to a classification of the finite simple groups by diagram geometry. By judiciously extending the classes of bipartite graphs allowed as residues, the right diagrams might occur that would fit all simple groups rather than just those of Lie type. The idea led to a flurry of activities, ranging from the construction of diagram geometries on the basis of known groups and a system of their subgroups to the classification of all geometries pertaining to a given diagram. Although quite a few diagrams have been found for finite simple groups and quite a few interesting classifications of geometries have seen the light, the classification of finite simple groups has been completed without a satisfactory framework offered by diagram geometry. Nevertheless diagram geometry structured several characterizations of individual sporadic groups, and provided tools that are useful for geometric alternatives to certain existing parts of the classification. Besides, a lot of finite group theory is of a very geometric nature, although the proofs are not always formulated in the associated terminology.

Incidence geometry, though, has beauty in its own right. This is not only reflected by quite intriguing diagrams for several simple groups, but also by striking axioms characterizing spaces related to classical geometries. Most impressive is the Buekenhout-Shult description of a polar space by means of the single condition that, for each line and each point off that line, either one or all points of the line are collinear with the point. From this axiom, together with some light nondegeneracy conditions, the full building belonging to any classical group distinct from a special linear group and of rank at least two (that is, having a subspace of dimension at least two in the natural representation space of the group on which the invariant defining form completely vanishes) can be reconstructed. Besides, whereas diagram geometry functions best in cases where ranks are finite, much of the polar space approach remains valid for spaces of arbitrary rank. The construction of geometries from spaces with few axioms is a major theme of diagram geometry. In this respect, the root filtration spaces form the counterpart of polar spaces. Their axioms are more involved, but examples exist for all finite groups of Lie type of rank at least two and distinct from $^2F_4$. In this book, these spaces are introduced and enough properties are derived so as to be able to characterize spaces related to projective spaces and to line Grassmannians of polar spaces.

This book provides a self-contained introduction to diagram geometry. The first three chapters are spent on the basic theory. The fourth chapter shows the tight connection with group theory; it deals with thin geometries, which are very close to quotients of Cayley graphs of Coxeter groups. These geometries are abundant in buildings, like the triangles in projective planes. We then treat projective and affine geometry in two chapters. These are geometries with a linear diagram and linear shadow spaces, which implies that they are matroids. This opens the door to variations of the geometries connected with buildings. We restrict ourselves to a limited number of variations, just enough to give the flavor of combinatorial structures like Steiner systems in the context of diagram geometry. The next four chapters are devoted to polar spaces. Their complete classification is found in Tits [286]. Here we use a different approach, starting with Veldkamp's method [297] of embedding the polar space in a projective space. There are exceptions, such as the polar spaces whose projective planes are not Desarguesian, to which we devote little attention. This reflects the idea that the book is primarily an introduction to diagram geometry and the associated synthetic treatment of fascinating geometric spaces. The intention is to give a flavor of the topic rather than an exhaustive treatment. The final chapter gives a brief introduction to the theory of buildings and shows that every spherical building of rank at least three is connected with a root filtration space. The references to the literature in the Notes sections are meant to enable the interested reader to further knowledge in several directions.

The switching of viewpoints within a single geometry by use of the diagram leads to axiom systems of various kinds for the classical geometries. As mentioned above, the root filtration spaces are special among these as every possible finite group of Lie type distinct from $^2F_4$ acts acts almost faithfully on such a space. A representative collection of these spaces is directly related to the Lie algebras introduced by Chevalley. It is the purpose of the second author to complete a second volume dedicated to these spaces and the non-classical geometries of spherical Coxeter type.

Preliminary versions of this book, including the intended chapters of the second volume, have been available on the Internet for over fifteen years. We are very grateful to comments received by enthusiastic readers and acknowledgments in the form of references to such a volatile site as the place where the individual chapters were to be found. We hope the readers will be pleased that—at least the first part of—the Internet version has finally been turned into a book.

We thank

(1) the person who has inspired and stimulated us most to write the book: Jacques Tits;
(2) three persons who have greatly contributed: Gábor Ivanyos, Ralf Köhl (né Gramlich), Antonio Pasini;
(3) the people who have helped in one way or another: Andries Brouwer, Philippe Cara, Hans Cuypers, Edmond Dony, Chris Fisher, Jonathan Hall, Simon Huggenberger, Cécile Huybrechts, William Kantor, Jan Willem Knopper, Dimitri Leemans, Shoumin Liu, Scott Murray, Jos in't panhuis, Erik Postma, Ernest Shult, Rudolf Tange, Hendrik Van Maldeghem, Jean-Pierre Tignol, David Wales.
(4) Joachim Heinze and Ute Motz at Springer-Verlag and professor Remmert for their continued interest and infinite patience over the years.

Brussels, Belgium   Francis Buekenhout
Eindhoven, The Netherlands   Arjeh M. Cohen

Leitfaden

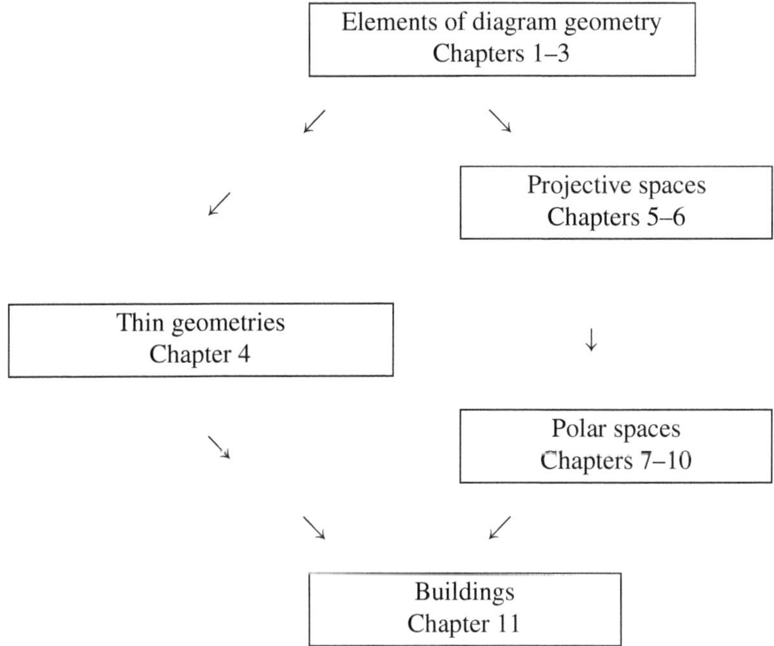

Each of the eleven chapters ends with a section of exercises and a section of notes. Exercises that are deemed hard are indicated by a ♮ at their beginning; those that are deemed computationally intensive by a ♯.

Contents

Chapter 1
Geometries

The fundamental structure in this book is a geometry. We look at a geometry as an incidence system: abstract objects that are related by means of incidence. The concept of a point lying on a line is carried over to the more abstract notion of an element of type `point` being incident to an element of type `line`. The basic ideas and definitions are given in this chapter. The usual related concepts like homomorphisms and subgeometries, and less general concepts such as connectedness and residues, are introduced. The important notion of residual connectedness is described in different but equivalent ways (Corollary 1.6.6).

Many of the examples we give display a lot of symmetry. In the later sections of this chapter, we show how the automorphism group of such a highly symmetric geometry can be used for a complete description of the geometry in terms of this group and some of its subgroups. Towards the end of the chapter, we describe how properties like residual connectedness can be expressed in term of these subgroups (Corollary 1.8.13).

1.1 The Concept of a Geometry

In this book, the word geometry is used in a technical sense, just as words like topology and algebra. It provides a generalization of the concept of incidence. In a broader context, it would be appropriate to speak of an *incidence geometry*, but in this book, there is no danger of confusion.

A geometry consists of elements of different types such as points, lines, planes (or vertices, edges, faces, cells, or subspaces of dimension i where i is an integer). In this context, the reader should momentarily abandon the usual physical viewpoint according to which a line is a set of points, an edge is a set of two vertices, and so on. The same status will be given to each of the different types of elements. Afterwards, we can assign the role of basic elements, traditionally played by points and lines, to any of these types.

F. Buekenhout, A.M. Cohen, *Diagram Geometry*,
Ergebnisse der Mathematik und ihrer Grenzgebiete. 3. Folge / A Series of
Modern Surveys in Mathematics 57, DOI 10.1007/978-3-642-34453-4_1,
© Springer-Verlag Berlin Heidelberg 2013

Fig. 1.1 The cube geometry

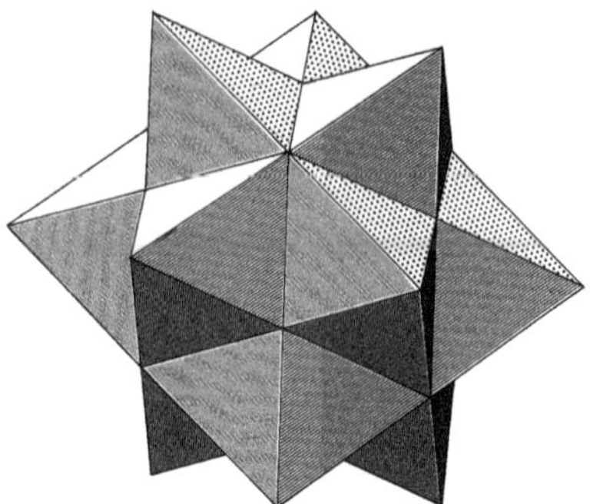

Example 1.1.1 See Fig. 1.1 for a picture of the cube. Let ε_1, ε_2, ε_3 be the standard basis of the Euclidean vector space $\mathbb{R}^3$ and consider the cube Γ whose 8 vertices are the vectors $\pm\varepsilon_1 \pm \varepsilon_2 \pm \varepsilon_3$. The edges (faces) of Γ can be viewed as pairs (respectively, quadruples) of vertices. By replacing each edge and face by its barycentric vector (up to a suitable scaling by a scalar multiple) in the physical cube, we can visualize Γ by 26 vectors: the 8 vectors corresponding to vertices, the 12 vectors $\pm 2\varepsilon_i \pm 2\varepsilon_j$ ($1 \leq i < j \leq 3$) corresponding to edges, and the 6 vectors $\pm 2\varepsilon_i$ ($1 \leq i \leq 3$) corresponding to faces. Incidence between elements of Γ can now be visualized as a line segment connecting two vectors. There are 72 line segments representing incident pairs of elements and 48 triangles representing incident triples.

In general, incidence is a symmetric, reflexive relation on the set of elements with no two elements of the same type incident. The rank of a geometry is the number of distinct types of elements. This number is almost always assumed to be finite. We use this word rather than dimension because the latter has several classical meanings which might not agree with this definition. We often use natural numbers for types; otherwise, types are written in a special `font`.

Example 1.1.2 The Euclidean affine plane $\mathbb{E}^2$ provides a rank two geometry with types `point` and `line` which we call the *real affine plane*. The Euclidean affine space $\mathbb{E}^3$ provides a rank three geometry with types `point`, `line`, and `plane` which we call the *real affine geometry of rank three*. The Euclidean distance plays no role here.

Example 1.1.3 A *polygon* in the Euclidean plane is loosely defined as a 2-dimensional simply connected shape whose boundary is made up of a finite number of points, called vertices, and line segments, called edges, such that each vertex is on exactly two line segments. Similarly, a *polyhedron* in the Euclidean space is a 3-dimensional simply connected shape whose boundary is made up of a finite number of vertices, edges (line segments again), and polygons in planes (called faces), such that each edge is on exactly two faces. It provides a rank three geometry whose types are `vertex`, `edge`, `face`. Well-known examples are pyramids, prisms, the tetrahedron, and the octahedron. Figure 1.2 illustrates two other famous examples:

Fig. 1.2 Icosahedron and dodecahedron

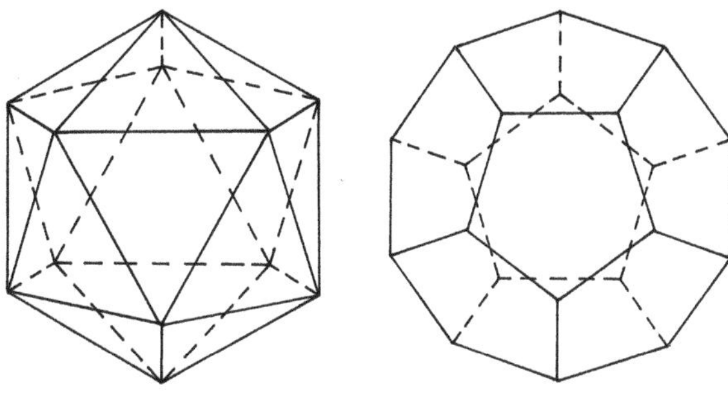

Fig. 1.3 A tiling of the Euclidean plane

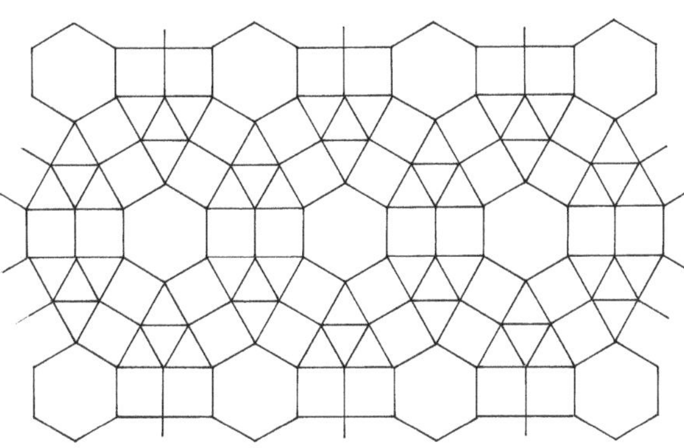

the icosahedron and the dodecahedron. These are dual to each other in the sense that, replacing the names `vertex`, `face` by `face`, `vertex`, respectively, we obtain each geometry from the other.

Example 1.1.4 An edge-to-edge tiling or *tessellation* of $\mathbb{E}^2$ by polygons gives a rank three geometry whose types are `vertex`, `edge`, `face`. The familiar pattern of a brick wall is not edge-to-edge. Well-known examples that we will come across later are the three regular tessellations with equilateral triangles, squares, and regular hexagons, respectively. Figure 1.3 gives one more tessellation.

Example 1.1.5 A face-to-face *tessellation* of $\mathbb{E}^3$ by polyhedra provides a rank four geometry whose types are `vertex`, `edge`, `face`, `cell`. The typical example is built up from cubes (as cells). Figure 1.4 depicts a less obvious example, in which all cells are copies of a so-called *truncated octahedron*.

Remark 1.1.6 There may be more than one way to view a familiar geometric object as a geometry. The cube, for instance, is a rank three geometry in the guise of a polyhedron, but a rank two geometry when the faces are disregarded. Also the polyhedron with 26 vertices (the vectors), 72 edges (the lines joining vectors representing incident elements of the cube), and 48 triangles of Fig. 1.1 can be interpreted as the rank three cube geometry.

Fig. 1.4 A tiling of
Euclidean space by truncated
octahedra

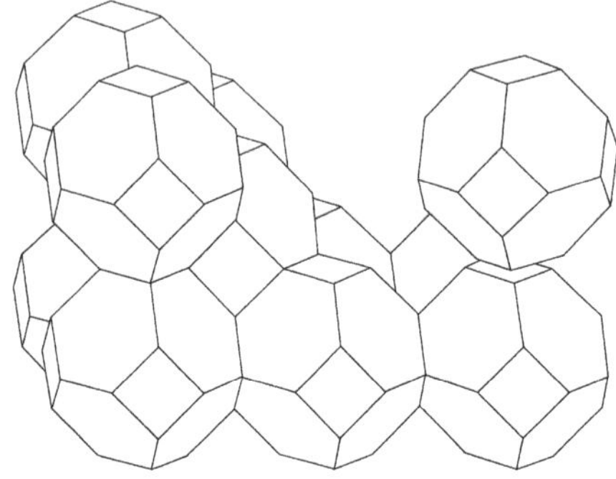

1.2 Incidence Systems and Geometries

In this section we give the formal definition of a geometry. It is close to the notion
of a multipartite graph.

For the duration of the section, I will be a set, called the *type set*. It need not be
finite. Its elements as well as its subsets are called *types*.

Definition 1.2.1 A *graph* is a set of *vertices* (singular: vertex) with a family of
(unordered) pairs of distinct vertices, called *edges*. Two vertices x, y are called
adjacent if $\{x, y\}$ is an edge; we also express this fact by saying that y is a *neighbor*
of x.

A graph is fully determined by the pair (V, E), where V is the vertex set and
E is the edge set, but also by $(V, \sim)$, where V is as before and $\sim$ is the adjacency
relation. For $\sim$ we use infix notation, so $x \sim y$ means the same as $\{x, y\} \in E$.

A graph is called *bipartite* if its vertex set can be partitioned into two non-empty
subsets, X_1 and X_2, such that there are no edges inside X_1 nor inside X_2. More
generally, a graph with vertex set X is called *multipartite* with partition $(X_i)_{i \in I}$ if
X is partitioned into non-empty subsets X_i $(i \in I)$ such that, for each $i \in I$, there
are no edges inside X_i.

A *complete graph* is a graph in which all pairs of vertices are edges. A *partial
subgraph* of a graph (V, E) is a graph (V', E') with $V' \subseteq V$ and $E' \subseteq E$. If A is a
subset of V, then the *subgraph* of (V, E) *induced* on A is the pair $(A, E \cap (A \times A))$.
If the subgraph is a complete graph, then A is often referred to as a *clique*.

Definition 1.2.2 A triple $\Gamma = (X, *, \tau)$ is called an *incidence system* over I if

(1) X is a set (its elements are also called *elements* of Γ);
(2) $*$ is a symmetric and reflexive relation on X; it is called the *incidence relation
of* Γ, usually written as an infix operator; so, for $x, y \in X$, incidence between
them is denoted $x * y$;
(3) τ is a map from X to I, called the *type map* of Γ, such that distinct elements
$x, y \in X$ with $x * y$ satisfy $\tau(x) \neq \tau(y)$; members of the pre-image $\tau^{-1}(i)$ are
called elements *of type i*, or *i-elements*.

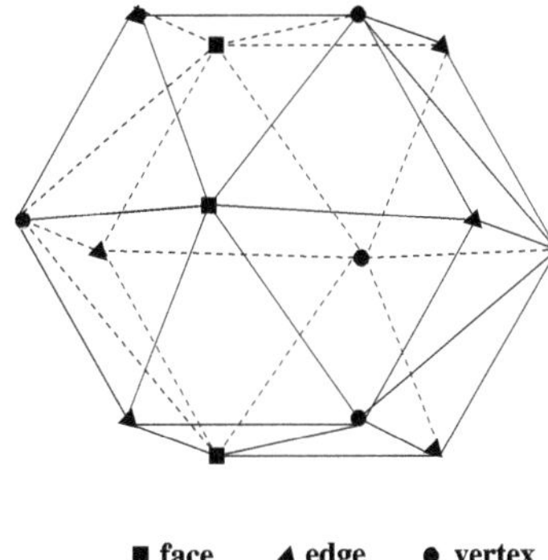

Fig. 1.5 The tetrahedron geometry of rank three pictured as a polyhedron in $\mathbb{E}^3$

In an incidence system $\Gamma = (X, *, \tau)$ over I, the set X is the disjoint union of the sets $X_i = \tau^{-1}(i)$, the inverse image of i under τ, for $i \in I$. Thus, $(X, *)$ is a multipartite graph with partitioning $(X_i)_{i \in \tau(X)}$. (The types in the codomain I of τ that are not in the image $\tau(X)$ play an insignificant role; we usually have $I = \tau(X)$.) The concept is illustrated by Fig. 1.5, where a picture of a geometry associated with the tetrahedron is drawn. Here, incidence is represented by edges; incidence of an element with itself is not drawn.

For an incidence system Γ over I, in which τ is surjective, we also write $((X_i)_{i \in I}, *)$ instead of $(X, *, \tau)$, suppressing τ. Observe that $X = \bigcup_{i \in I} X_i$ and that $\tau(X_i) = \{i\}$ for each $i \in I$, so that indeed Γ is uniquely determined by the pair $((X_i)_{i \in I}, *)$. The incidence relation $*$ on each X_i is the identity, so the new description of Γ is close to a multipartite graph but differs from it in two ways:

(1) Although the parts X_i are distinguished, the type map τ does not exist on a multipartite graph and is made explicit in Γ.
(2) The vertices of the multipartite graph are not adjacent to themselves but are incident in Γ.

Definition 1.2.3 Let $\Gamma = (X, *, \tau)$ be an incidence system over I. The set I is called the *type* of Γ and the cardinality of I is called the *rank* of Γ, and denoted by $\mathrm{rk}(\Gamma)$.

If $A \subseteq X$, we say that A is of *type* $\tau(A)$ and of *rank* $|\tau(A)|$, the cardinality of $\tau(A)$, denoted by $\mathrm{rk}(A)$. The *cotype* of A is $I \setminus \tau(A)$ and the *corank* of A is the cardinality of $I \setminus \tau(A)$.

Given $Y \subseteq X$, we write Y^* for the set of all elements incident with every member of Y. For $x \in X$, we sometimes write $x * Y$ instead of $x \in Y^*$. We also write x^* instead of $\{x\}^*$.

The graph with vertex set X whose edges are the unordered pairs $\{x, y\}$ of distinct vertices x, y with $x * y$ is called the *incidence graph* of Γ.

A *flag* of Γ is a set of mutually incident elements of Γ. Flags of Γ of type I are called *chambers*.

The type of the set X of elements of the incidence system Γ coincides with I, the type Γ, if and only if the type map τ is surjective.

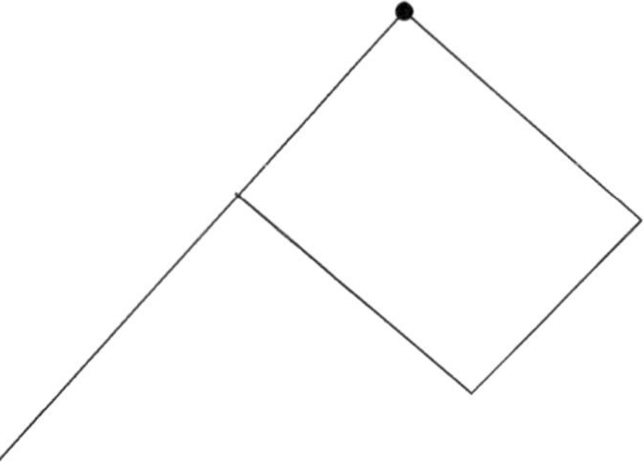

Fig. 1.6 A flag, consisting of an incident point, line, and plane

In an incidence system Γ over I, the restriction of the type map to every flag of Γ is an injection. For, a flag of Γ has at most one element of each type.

If $\sim$ denotes the relation on X of being distinct and incident, then, with the notation of Definition 1.2.1, the incidence graph is $(X, \sim)$. Moreover, flags are cliques.

The 48 chambers of the cube of Example 1.1.1 are displayed as the triangles in Fig. 1.1.

The word flag is inspired by the visualization in Fig. 1.6 of a flag in a projective geometry but can be based on the real affine space of Example 1.1.2 as well. The point of the flag is the ball at the top, the line of the flag is the pole with that top, and the plane is the cloth of the flag. A typical example of a flag in $\mathbb{E}^3$ is a triple consisting of a point, line, and plane such that the point is on the line and the line is on the plane (because then the point is also incident with the plane).

Remark 1.2.4 By Zorn's lemma, every flag is contained in at least one maximal flag, that is, a flag not properly contained in any other flag.

In an incidence system, chambers are maximal flags. In general, however, the converse does not hold: Exercise 1.9.3 gives examples of incidence systems over I in which maximal flags exist that do not have type I.

Definition 1.2.5 Let Γ be an incidence system over I. If every maximal flag of Γ is a chamber, then Γ is called a *geometry* over I.

In terms of the incidence graph, an incidence system is a geometry if and only if the image under the type map of every maximal clique is I. In this case, the type map is surjective.

Example 1.2.6 Figure 1.7 shows the usual picture of a tetrahedron in Euclidean affine space $\mathbb{E}^3$ and Fig. 1.5 represents the geometry by a polyhedron in $\mathbb{E}^3$ having $4 + 6 + 4 = 14$ vertices (elements) and $4 \times 3 \times 2 = 24$ triangular faces (chambers of the geometry).

As an abstract geometry, it is the special case $n = 4$ of the geometry $(X_1, \ldots, X_{n-1}, *)$ over $[n-1]$ that we define below for each $n > 1$. Here and elsewhere, $[n]$ denotes the set $\{1, \ldots, n\}$, and $\binom{X}{k}$, for a set X and a number k, the collection of

Fig. 1.7 The tetrahedron

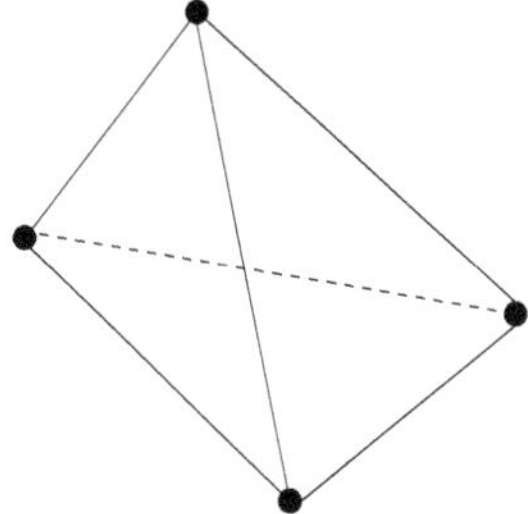

Fig. 1.8 An unfolding of the cube

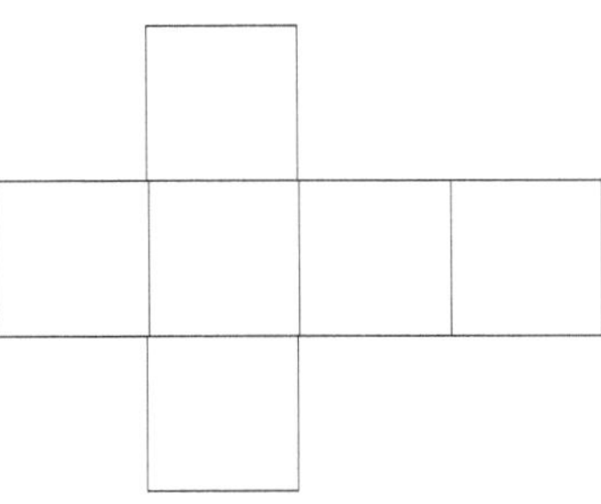

Fig. 1.9 Polygons viewed as geometries

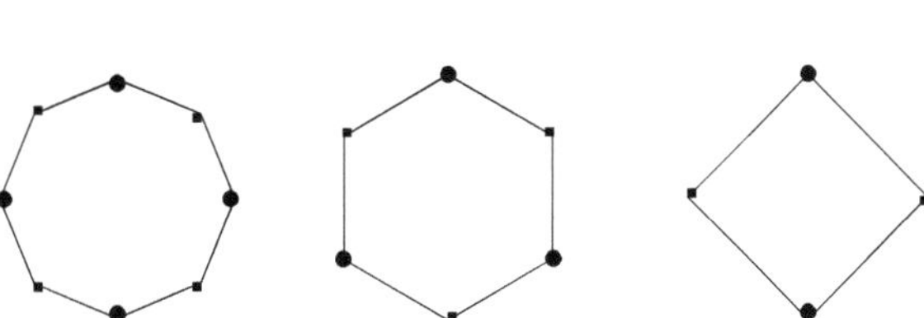

subsets of X of size k. Now $X_i = \binom{[n]}{i}$ for $i \in [n-1]$ and $*$ is defined by $U * W$ if and only if either $U \subseteq W$ or $W \subseteq U$. It has $2^n - 2$ elements and $n!$ chambers.

The condition defining the incidence relation $*$ is called *symmetrized inclusion*. More generally, the *symmetrized* version of a relation $\sim$ will be the relation $\sim'$ given by $x \sim' y$ if and only $x \sim y$ or $y \sim x$.

Definition 1.2.7 A geometry Γ is *firm* (respectively, *thick*) if every flag of type other than I is contained in at least two (respectively, three) distinct chambers of Γ. It is called *thin* if every flag of type $I \setminus \{i\}$ for some $i \in I$ is contained in exactly two chambers of Γ.

Example 1.2.8 A polygon in $\mathbb{E}^2$ is a familiar rank two geometry. It is thin. Its incidence graph is again a polygon, with twice as many vertices as the original; see Fig. 1.9. Let $n \in \mathbb{N}$, $n \geq 2$. The polyhedron in the Euclidean affine space $\mathbb{E}^3$ for $n = 3$, and, more generally, the *polytope* in $\mathbb{E}^n$, are thin geometries of rank n.

Remark 1.2.9 Although we almost always work with firm geometries, we give two cases where non-firm geometries naturally occur. First, the unfolding of a polyhedron in an elementary and practical view, as depicted in Fig. 1.8. This is a geometry with 14 vertices, 19 edges, and 6 faces, which is not firm. Second, a chamber, viewed as a subgeometry of a geometry (Definition 1.4.1), is clearly not firm.

1.3 Homomorphisms

In geometry, as in every structure theory, the concepts of homomorphism, isomorphism, and automorphism are essential. We also need homomorphisms of a more general kind, called weak homomorphisms. Fix a type set I and an incidence system $\Gamma = (X, *, \tau)$ over it.

Definition 1.3.1 Let $\Gamma' = (X', *', \tau')$ be an incidence system over I'. A *weak homomorphism* $\alpha : \Gamma \to \Gamma'$ is a map $\alpha : X \to X'$ such that, for all $x, y \in X$,

(1) $x * y$ implies $\alpha(x) *' \alpha(y)$;
(2) $\tau(x) = \tau(y)$ if and only if $\tau'(\alpha(x)) = \tau'(\alpha(y))$.

If, in addition, $I = I'$ and $\tau(x) = \tau'(\alpha(x))$ for all $x \in X$, then α is called a *homomorphism*.

Properties like injectivity applied to the weak homomorphism α will be understood to apply to the underlying map $X \to X'$. An injective homomorphism $\alpha : \Gamma \to \Gamma'$ of incidence systems is also called an *embedding* of Γ into Γ'.

A bijective weak homomorphism α whose inverse α^{-1} is also a weak homomorphism is called a *correlation*. If α is a homomorphism and a correlation, then we call α an *isomorphism* and write $\Gamma \cong \Gamma'$. In this case α^{-1} is also an isomorphism.

The correlations of Γ onto itself, called *auto-correlations*, form a group, denoted by $\mathrm{Cor}(\Gamma)$, whose multiplication is composition of maps. The isomorphisms of Γ onto itself, called *automorphisms*, also form a group, denoted $\mathrm{Aut}(\Gamma)$.

Remark 1.3.2 A weak homomorphism $\alpha : \Gamma \to \Gamma'$ preserves incidence and sends elements of the same type in I to elements of the same type in I'. If τ is surjective, then α induces a map $\alpha_I : I \to I'$. In case $\Gamma = \Gamma'$ and $I = I'$, the restriction of the map $\alpha \mapsto \alpha_I$ to $\mathrm{Cor}(\Gamma)$ is a homomorphism $\mathrm{Cor}(\Gamma) \to \mathrm{Sym}(I)$ of groups, with kernel $\mathrm{Aut}(\Gamma)$. Here, $\mathrm{Sym}(X)$ denotes the group of all permutations of a set X, where a *permutation* of X is a bijection $X \to X$. For, an auto-correlation α of Γ is an isomorphism if and only if it fixes types, that is, is the identity on I. In particular, the group $\mathrm{Aut}(\Gamma)$ is a normal subgroup of $\mathrm{Cor}(\Gamma)$.

Example 1.3.3 Let Δ be the rank two geometry consisting of the 20 vertices and 30 edges of the dodecahedron D, drawn in the right hand side of Fig. 1.2. Identifying opposite elements of D we obtain a natural homomorphism α from Δ onto the famous *Petersen graph* Pet with 10 vertices and 15 edges, depicted twice in Fig. 1.10.

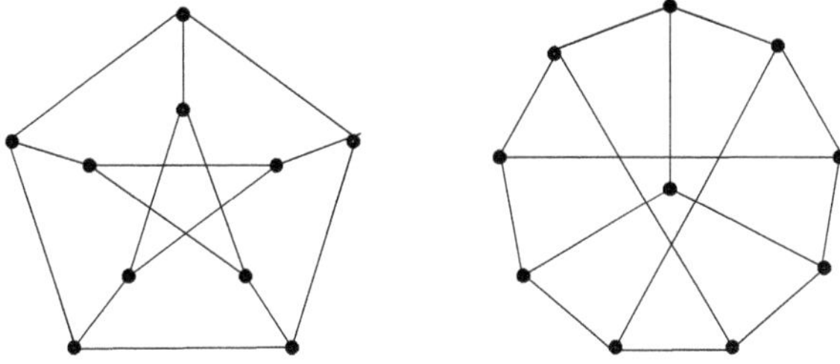

Fig. 1.10 The Petersen graph
in two guises

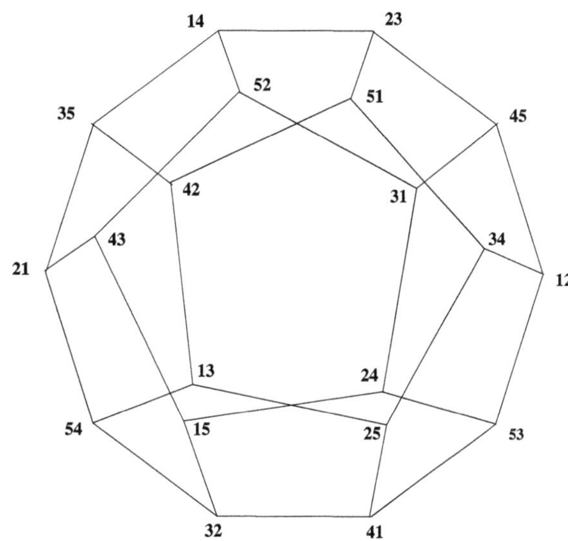

Fig. 1.11 Labelling the
vertices of a dodecahedron

This can be verified by labelling the vertices of D with ordered pairs (i, j) for distinct $i, j \in [5]$, in such a way that the vertices labelled (i, j) and (k, l) form an edge if and only if there is an even permutation of $[5]$ moving (i, j, k, l) to $(1, 2, 3, 4)$. This labeling is given in Fig. 1.11. The Petersen graph can be described as the vertex set $\{\{i, j\} \mid 1 \leq i < j \leq 5\}$ with $\{i, j\}$ and $\{k, l\}$ joined by an edge if and only if $\{i, j\} \cap \{k, l\} = \emptyset$. In this set-up, there is a homomorphism $\alpha : \Delta \to \mathrm{Pet}$ determined by $\alpha((i, j)) = \{i, j\}$.

The map interchanging (i, j) and (j, i) for all i, j clearly induces an automorphism z of order two of Δ. The group $\mathrm{Aut}(\Delta)$ is a direct product $C_2 \times \mathrm{Alt}_5$ of the cyclic group C_2 of order two generated by z (in general, C_n denotes the *cyclic group* (that is, a group generated by a single element) of order n) and the alternating group Alt_5 (that is, the group of all even permutations of $[5]$). The group $\mathrm{Aut}(\mathrm{Pet})$ is isomorphic to the symmetric group Sym_5 on five letters. The element z maps onto the identity on Pet, but new automorphisms (induced by odd permutations) appear.

Example 1.3.4 The homomorphism $\alpha : \Delta \to \mathrm{Pet}$ of Example 1.3.3 maps the 12 faces of D into six circuits of length 5. There are 12 circuits of length 5 on the Petersen graph falling into two classes: those coming from a face of D and the six

Fig. 1.12 The great stellated dodecahedron

others (each coming from a shortest path in D from a vertex to its opposite). Two members of the same class share a unique edge. Automorphisms coming from odd permutations in Sym_5 interchange the two classes, while those coming from even permutations (Alt_5) preserve them. The rank three geometry consisting of the vertices and edges of Pet and one of these classes of circuits of length 5 is called a *hemi-dodecahedron*. The map α provides a homomorphism from D onto the hemi-dodecahedron built from one of the two classes. But symmetry suggests that, for the other class of six circuits of length 5, there must be another rank three geometry isomorphic to D and a surjective weak homomorphism α' from it onto the other hemi-dodecahedron on Pet. This geometry is realized by the *great stellated dodecahedron* whose vertices are the same as those of the dodecahedron but whose faces are star-shaped pentagons (so-called *pentagrams*). See Fig. 1.12.

Definition 1.3.5 Let A be a *permutation group* on a set X, that is, a subgroup of the group $\mathrm{Sym}(X)$ of all permutations of X. A minimal A-invariant subset of X is called an *A-orbit* of X. The collection of all A-orbits is denoted by X/A.

Suppose that A is a group of automorphisms of Γ. The *quotient incidence system* of Γ by A is the incidence system $\Gamma/A = (X/A, */A, \tau/A)$ over I, where two orbits Ax and Ay of $x, y \in X$, respectively, are incident with respect to $*/A$ if there are $x' \in Ax$ and $y' \in Ay$ with $x' * y'$, and τ/A maps Ax to $\tau(x)$.

More generally, let E be an equivalence relation defined on X such that $(x, y) \in E$ implies $\tau(x) = \tau(y)$ whenever $x, y \in E$. The *quotient incidence system* of Γ by E is the incidence system $\Gamma/E = (X/E, */E, \tau/E)$ over I, where X/E is the set of equivalence classes of E, two classes e, f are incident with respect to $*/E$ if there are $x \in e$ and $y \in f$ with $x * y$, and τ/E maps e to $\tau(x)$ for $x \in e$.

Indeed, Γ/E is an incidence system over I and Γ/A is the special case of Γ/E, where E is defined by $(x, y) \in E$ if and only if the A-orbits of x and y coincide.

The map sending an element to its equivalence class under E defines a surjective homomorphism from Γ to Γ/E, and so this quotient is also the image of an incidence system homomorphism in the sense of Definition 1.3.1.

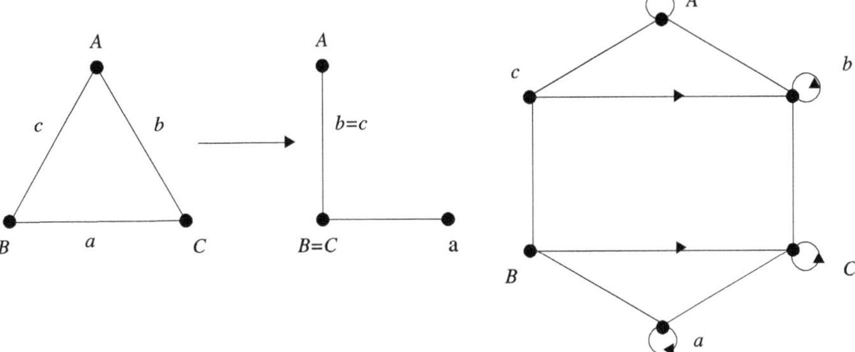

Fig. 1.13   The folding of a triangle

Example 1.3.6   The hemi-dodecahedron of Example 1.3.4 is the quotient of the do-decahedron (viewed as a rank three geometry) by the group of order two generated by z.

We will construct more examples of this kind, starting from the following thin geometry $\Gamma := (Y_1, Y_2, \ldots, Y_n, *)$, where $n \in \mathbb{N}$. For $j \in [n]$, an element of Y_j is a vector x of $\mathbb{R}^n$ of the form $x = \sum_{i \in J} \varepsilon_i x_i$, where $J \in \binom{[n]}{j}$ and $x_i \in \{\pm 1\}$ for all $i \in J$. For subsets $J \in \binom{[n]}{j}$ and $K \in \binom{[n]}{k}$ with $1 \le j \le k \le n$, the elements $x \in \sum_{i \in J} \pm \varepsilon_i$ and $y \in \sum_{i \in K} \pm \varepsilon_i$, are incident if $J \subseteq K$ and $y - x \in \sum_{i \in K \setminus J} \pm \varepsilon_i$; now, $*$ is the symmetrization of this relation. The set Y_j has size $2^j \binom{n}{j}$ and there are $2^n n!$ chambers. The special case $n = 3$ is the geometry of the cube of Example 1.1.1.

The group A generated by the scalar multiplication by -1 is a group of automorphisms of Γ of order two. The quotient incidence system Γ/A is again a geometry.

The group B of all sign changes in coordinates, is also a group of automorphisms of Γ of order 2^n. The quotient incidence system Γ/B has a single element of type n, which is incident with every element of the geometry.

Example 1.3.7   Start with a triangle as a rank two geometry. See Fig. 1.13. At the left hand side, the usual picture has been drawn and at the right hand side a picture that represents the incidence graph after removal of the directed edges. The points labelled A, B, C represent the vertices and the points labelled a, b, c represent the edges of the triangle. A *folding* of the triangle in the physical sense is depicted in the incidence graph by means of directed edges. The equivalence relation in which two elements are equivalent whenever the folding moves one to the other can be used to describe the physical folding as a surjective homomorphism. The folded geometry appears in the middle of Fig. 1.13.

Definition 1.3.8   In Remark 1.3.2 we saw that every auto-correlation α of Γ induces a permutation α_I of I. If α_I has order two, then α is called a *duality*. If both α_I and α have order two, then α is called a *polarity*. If both α_I and α have order 3, then α is called a *triality*.

Fig. 1.14 A regular
tessellation with labelled
vertices

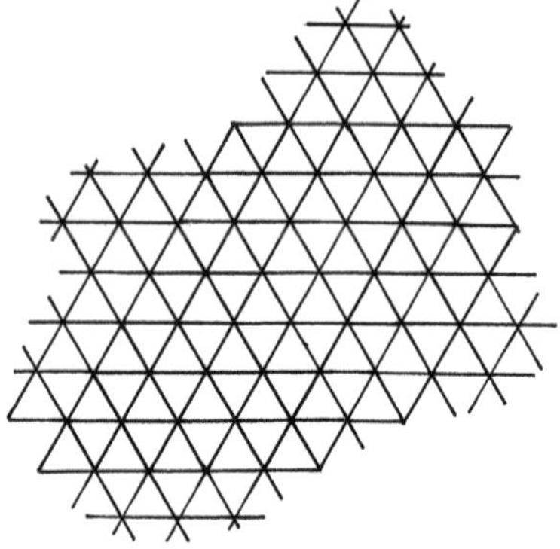

Example 1.3.9 Polarities and dualities in projective planes are classical examples of
auto-correlations. For those (as yet) unfamiliar with projective planes, we describe
the simplest case. Consider once more the triangle represented by the hexagon at
the right hand side of Fig. 1.13. A rotation over 60 degrees around the center results
in a correlation. Its square is a rotation over 120 degrees, which is a nontrivial automorphism. Therefore, the rotation over 60 degrees induces a duality that is not a
polarity. Rotation over 180 degrees leads a polarity.

Example 1.3.10 Consider the regular tessellation of $\mathbb{E}^2$ by triangles indicated in
Fig. 1.14. The vertices can be labelled by 1, 2, 3 in such a way that the labels of the
three vertices of each triangle are distinct. Let Γ be the geometry over [3] whose
elements of type i are the vertices labelled i for each $i \in$ [3] and in which two
elements are incident if they are vertices of a common triangle. It is easy to spot a
triality sending elements of types 1, 2, 3 onto elements of types 2, 3, 1, respectively.
There are also polarities, so $\mathrm{Cor}(\Gamma)/\mathrm{Aut}(\Gamma)$ is isomorphic to Sym_3.

Example 1.3.11 The *unit geometry* over I is the geometry $(I, *, \tau)$ over I where
$*$ is the 'complete relation' (every two elements are incident) and τ is the identity
map on I. For every geometry Γ over I the assignment of types is a homomorphism
from Γ onto the unit geometry over I. Similarly, every permutation of I determines
a correlation of the unit geometry over I and every injective map from I into a set I'
is a weak homomorphism of the unit geometry over I into the unit geometry over I'.

1.4 Subgeometries and Truncations

A polyhedron can be seen as a subgeometry of $\mathbb{E}^3$ consisting of a selection of the
points, lines, and planes, if no degeneracies like the one depicted in Fig. 1.15 occur.

In this section we formalize the notion of subgeometry as well as truncation,
which is the geometry that remains after the removal of all elements of certain des-
ignated types. We also introduce some of the geometries of our core interest, like

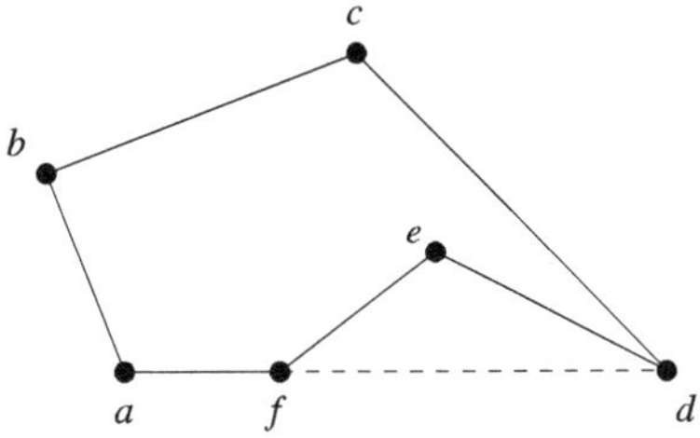

Fig. 1.15 A hexagon in the plane

projective and affine geometries related to a finite-dimensional vector space. Fix two sets of types, I and I', let $\Gamma = (X, *, \tau)$ be an incidence system over I, and let $\Gamma' = (X', *', \tau')$ be an incidence system over I'.

Definition 1.4.1 If $X' \subseteq X$, $*' \subseteq *$ (where both incidence relations are viewed as subsets of $X \times X$), $I' \subseteq I$, and $\tau' : X' \to I'$ is obtained by restricting τ to X', then Γ' is a *partial subsystem* of Γ over I'. In most (if not all) applications of this concept, we will deal with the particular case of a *subsystem*, where $*'$ is the restriction of $*$ to $X' \times X'$. In that case Γ' is determined by the choice of X' in X up to the choice of I' within I. If the choice is minimal, that is $I' = \tau(X')$, then $(X', *', I')$ is called the subsystem of Γ *induced* on X'.

A subsystem of Γ that is a geometry is called a *subgeometry* of Γ.

Remark 1.4.2 The hexagon in the Euclidean plane shown in Fig. 1.15 is a clear example of a partial subsystem that is not a subsystem: the incidence of point d with the Euclidean line af does not occur in the hexagon.

The subsystem of Γ induced on a flag of type F is isomorphic to the unit geometry over $\tau(F)$.

Exercise 1.9.6 shows that the property of being a subsystem can also be described in terms of an injective homomorphism.

Definition 1.4.3 Let A be a group of automorphisms of Γ. The *fixed subsystem* of Γ with respect to A is the subsystem induced on the set of elements of Γ that are fixed by A.

Example 1.4.4 Consider the rank two geometry of the triangle with points $\{a, b, c\}$ and edges $\{A, B, C\}$ such that $a, b \in C^*$, $b, c \in A^*$, and $a, c \in B^*$ take care of all incidences. The map α inducing $(b, c)(B, C)$ on the element set has fixed elements $\{a, A\}$, so the fixed subsystem with respect to the group $\{\mathrm{id}, \alpha\}$ has rank two. But neither $\{a\}$ nor $\{A\}$ extends to a flag of rank two, so the fixed subsystem is not a subgeometry of the triangle.

Example 1.4.5 The complex affine plane can be viewed as a rank two geometry in much the same way as the real affine plane of Example 1.1.2. Take A to be the group of order two consisting of complex conjugation and the identity, acting coordinate-wise on the complex affine plane. The fixed subsystem with respect to A is a subgeometry; it is the real affine plane $\mathbb{E}^2$ discussed in Example 1.1.2.

In Sect. 1.1, we mentioned that a cube can be seen both as a polyhedron and as a graph. Similarly, the Euclidean space $\mathbb{E}^3$ can be seen as a rank three geometry but also as a rank two geometry (for instance by disregarding its planes). These are instances of the general concept of truncation.

Definition 1.4.6 Let J be any subset of I. The *J-truncation* of Γ is the subsystem of Γ over J on the set $\tau^{-1}(J)$. It is denoted $_J\Gamma$.

Thus, an incidence system of rank n allows for 2^n truncations. If Γ is a geometry over I, and $J \subseteq I$, then $_J\Gamma$ is a geometry over J, so $_J\Gamma$ coincides with the subgeometry induced on $\tau^{-1}(J)$.

If, in Fig. 1.5, we delete the elements of type `face`, the picture of the {`vertex`, `edge`}-truncation of the tetrahedron arises.

Example 1.4.7 We construct examples of *division rings*, also known as skewfields. These are (associative but not necessarily commutative) rings with a unit in which every non-identity element has a multiplicative inverse. These division rings will be used as scalars for vector spaces, from which projective and affine geometries will emerge. Well-known non-commutative division rings are the quaternions; see Exercise 1.9.11(b). Here we introduce another series of division rings.

Let $\mathbb{K}$ be a field admitting an automorphism σ of order three. The fixed points of σ in $\mathbb{K}$ form a subfield $\mathbb{F}$ of $\mathbb{K}$ and $\mathbb{K}$ can be viewed as a vector space over $\mathbb{F}$. The dimension of $\mathbb{K}$ over $\mathbb{F}$ is equal to 3 (proving this fact is part of Exercise 1.9.12). Fix $a \in \mathbb{F}$. The associative algebra $\mathbb{D}$ over $\mathbb{K}$ is defined as the left vector space $\mathbb{D} = \mathbb{K} \oplus \mathbb{K}j \oplus \mathbb{K}j^2$ with the multiplication rules

$$jx = \sigma(x)j \quad \text{for } x \in \mathbb{K}, \quad \text{and} \quad j^3 = a.$$

As σ is not the identity, the multiplication is not commutative, so $\mathbb{D}$ cannot be a field. We show that, for suitable choices of a, the ring $\mathbb{D}$ is a division ring. We proceed in three steps, the first two of which are a general construction.

STEP 1. *The subfield $\mathbb{F}$ is the center of $\mathbb{D}$. For each $k \in \mathbb{K} \setminus \mathbb{F}$, its centralizer in $\mathbb{D}$ coincides with $\mathbb{K}$, that is, $\{x \in \mathbb{D} \mid xk = kx\} = \mathbb{K}$.*

For, if $x = x_0 + x_1 j + x_2 j^2 \in \mathbb{D}$ with $x_0, x_1, x_2 \in \mathbb{K}$, then $xk = kx$ implies

$$\left(k - \sigma(k)\right)x_1 = \left(k - \sigma^2(k)\right)x_2 = 0.$$

As $\sigma(k) = k$ or $\sigma^2(k) = k$ conflicts with $k \notin \mathbb{F}$, the left factors of x_1 and x_2 are nonzero, hence invertible, elements of $\mathbb{K}$, so we find $x_1 = x_2 = 0$. This proves $x = x_0 \in \mathbb{K}$, so the centralizer in $\mathbb{D}$ of k is contained in $\mathbb{K}$. As the other inclusion is obvious, Step 1 is complete.

We need the norm map $N : \mathbb{K} \to \mathbb{F}$ of $\mathbb{K}$ over $\mathbb{F}$, which is given by $N(x) = x\sigma(x)\sigma^2(x)$ for $x \in \mathbb{K}$. An important property of this map is the fact that it is multiplicative: $N(xy) = N(x)N(y)$ for all $x, y \in \mathbb{K}$. We are interested in the case where N is not surjective and, in fact, a is not a norm.

STEP 2. *If $a \in \mathbb{F} \setminus N(\mathbb{K})$, then $\mathbb{D}$ is a division ring.*

Assume that $\mathbb{D}$ is not a division algebra. This means there is a nonzero $x \in \mathbb{D}$ such that $x\mathbb{D} \neq \mathbb{D}$. Let

$$V = \{y \in \mathbb{D} \mid xy = 0\}.$$

This is the kernel of the linear map $\mathbb{D} \to \mathbb{D}$ given by $y \mapsto xy$, where $\mathbb{D}$ is viewed as a right vector space over $\mathbb{K}$. By standard linear algebra, we find $\dim_{\mathbb{K}}(V) + \dim_{\mathbb{K}}(x\mathbb{D}) = \dim_{\mathbb{K}}(\mathbb{D}) = 3$, so either $\dim_{\mathbb{K}}(V) = 1$ and $\dim_{\mathbb{K}}(x\mathbb{D}) = 2$, or $\dim_{\mathbb{K}}(V) = 2$ and $\dim_{\mathbb{K}}(x\mathbb{D}) = 1$. For each nonzero $y \in V$ we have $y\mathbb{D} \subseteq V$, hence $\dim_{\mathbb{K}}(y\mathbb{D}) = 1$ if $\dim_{\mathbb{K}}(x\mathbb{D}) = 2$. Substituting y for x if necessary, we may therefore assume $\dim_{\mathbb{K}}(x\mathbb{D}) = 1$, so $x\mathbb{D} = x\mathbb{K}$. This implies that there is some $k \in \mathbb{K}$ with $xj = xk$. The ensuing equation $x(j - k) = 0$ shows that right multiplication by $j - k$ is a singular linear map $\mathbb{D} \to \mathbb{D}$ of left vector spaces over $\mathbb{K}$. Its matrix with respect to the basis 1, j, j^2 is

$$\begin{pmatrix} -k & 1 & 0 \\ 0 & -\sigma(k) & 1 \\ a & 0 & -\sigma^2(k) \end{pmatrix}.$$

The matrix is singular and its determinant is $-N(k) + a$, so $a = N(k)$ is the norm of k.

STEP 3. *The setting of Step 2 is realized by the following choices.*

(1) $\mathbb{F} = \mathbb{Q}$.
(2) $\mathbb{K} = \mathbb{Q}(u)$, where $u^3 + u^2 - 2u - 1 = 0$ (*this equation is satisfied by the algebraic number* $2\cos(2\pi/7)$).
(3) $a = 2$.
(4) $\sigma \in \mathrm{Aut}(\mathbb{K})$ *determined by* $\sigma(u) = u^2 - 2$.

We verify the hard part of the proof that this is an example, namely that $a \notin N(\mathbb{K})$. Suppose that $x = x_0 + x_1 u + x_2 u^2 \in \mathbb{K}$, with $x_0, x_1, x_2 \in \mathbb{Q}$, satisfies $N(x) = 2$. Let g be the minimal positive integer with $gx_i \in \mathbb{Z}$ for each $i \in \{0, 1, 2\}$. A straightforward computation shows

$$N(x) = x_0^3 - x_0^2 x_1 + 5x_0^2 x_2 - 2x_0 x_1^2 + 6x_0 x_2^2$$
$$- x_0 x_1 x_2 + x_1^3 - x_1^2 x_2 - 2x_1 x_2^2 + x_2^3.$$

Applying this to gx, taking values modulo two, and expanding the right hand side of the equation below, we verify that

$$0 = g^3 N(x) \equiv 1 + (gx_0 + 1)(gx_1 + 1)(gx_2 + 1) \pmod{2}.$$

If some gx_i would be odd, then the product of three terms at the right hand side vanishes and the equation $0 \equiv 1 \pmod 2$ results, a contradiction. Therefore, each gx_i is even, so $gx_i = 2m_i$ for certain integers m_i $(i = 0, 1, 2)$. Moreover, by minimality of g, the number g is odd (if it were even, then the highest power of 2 dividing g would be equal to the highest power occurring in the denominator of some x_i and so gx_i would be odd, a contradiction). Now, writing $m = m_0 + m_1 u + m_2 u^2$, we have

$2g^3 = g^3 N(x) = N(gx) = N(2m) = 8N(m) \in 8\mathbb{Z}$, which contradicts that g is odd. Hence, there is no $x \in \mathbb{K}$ with $N(x) = 2$.

Remark 1.4.8 Let $\mathbb{D}$ be a division ring. Since $\mathbb{D}$ need not be commutative, it is necessary to specify, for a vector space over $\mathbb{D}$, whether it is a right or a left vector space, that is, whether we take the scalar multiplication as a right or a left action of $\mathbb{D}$ on V. We will mainly restrict ourselves to right vector spaces over $\mathbb{D}$. The reason is that most group actions in this book, including the action of the group $\mathrm{GL}(V)$ of linear transformations of V, are on the left, so that linearity can be nicely expressed as the associativity rule

$$g(v\lambda) = (gv)\lambda \quad \big(g \in \mathrm{GL}(V),\ v \in V,\ \lambda \in \mathbb{D}\big).$$

The mathematical consequences of this decision are minor, as the right scalar multiplication of $\mathbb{D}$ is equivalent to the left scalar multiplication of the opposite division ring, as explained in Exercise 1.9.11.

Example 1.4.9 Let $n \in \mathbb{N}$, $n \geq 1$, and let V be a vector space of finite dimension $n + 1$ over a division ring. The *projective geometry* $\mathrm{PG}(V)$ is defined as follows.

(1) The elements are all nonempty subspaces of V except $\{0\}$ and V itself.
(2) Subspaces U and W are incident if and only if either $U \subseteq W$ or $W \subseteq U$.
(3) The type of an element is its *affine dimension* (that is, its dimension as a vector space).

This incidence system is a geometry over $[n]$. Its elements of type 1 (respectively, 2) are usually called points (respectively, lines). The $[2]$-truncation of $\mathrm{PG}(V)$ is closely connected to the projective space on V, which will be introduced in Definition 5.2.1. These spaces will be studied extensively in Chaps. 5 and 6.

Here is a similar construction of affine geometries.

Example 1.4.10 Let $n \in \mathbb{N}$, $n \geq 1$, and let V be a vector space of finite dimension n over a division ring. The *affine geometry* $\mathrm{AG}(V)$ is defined as follows.

(1) The elements are all nonempty affine subspaces of V except V itself.
(2) Incidence is defined by symmetrized inclusion (see Example 1.4.9(3)).
(3) The type of an element of affine dimension i is equal to $i + 1$.

This incidence system is a geometry over $[n]$. Its elements of type 1 (respectively, 2) are usually called points (respectively, lines). The $[2]$-truncation of $\mathrm{AG}(V)$ is closely connected to the affine space on V, which will be introduced in Proposition 5.1.3. These spaces will be studied extensively in Chaps. 5 and 6.

Remark 1.4.11 Consider the case where $n = \infty$. We recall that if c is a cardinality with $c \leq n$, then $c + n = n$ and also that the class of all cardinalities is well ordered. Suppose we define $\mathrm{AG}(V)$ as above, with incidence being symmetrized inclusion. Letting the type be dimension would be a bad choice: all affine subspaces of finite codimension would have the same type. It seems more appropriate to define the type $\tau(S)$ as the pair (d, c) where $d = \dim S$, $c = \operatorname{codim} S$. Observe that $d + c = n$, hence at least one of d, c must be equal to n. Unfortunately, AG(V) as defined in this very natural way, is not even an incidence system because there exist elements with the same type $d = n = c$ which are contained in each other and which are therefore incident.

There is a natural total order $\leq$ on the set I of types: put $(d, c) \leq (d', c')$ if and only if $d \leq d'$ and $c \geq c'$. For, given two types (d, c), (d', c'), then $d + c = n = d' + c'$ and if $d < d'$, then $d < n$, whence $c = n$ and so $c \geq c'$. If we want to force a geometry AG(V) despite the preceding observations, we can truncate and keep as elements only those subspaces which have either finite dimension or finite codimension. This geometry is of denumerable rank.

The following definition regarding subgroups of $\operatorname{Cor}(\Gamma)$ is to be compared with Definition 1.4.3 for subgroups of $\operatorname{Aut}(\Gamma)$. It uses the action $\alpha \mapsto \alpha_I$ of $\operatorname{Cor}(\Gamma)$ introduced in Remark 1.3.2.

Definition 1.4.12 Let A be a group of auto-correlations of Γ. The *absolute* of Γ with respect to A is the incidence system $\Gamma_A = (X_A, *_A, \tau_A)$ over J, where

(1) the set J is the collection of all A-orbits K on I for which there are invariant flags of type K;
(2) the set X_A consists of all minimal (which is understood to imply non-empty) A-invariant flags of Γ;
(3) the relation $*_A$ on X_A is determined by $F *_A G$ if and only if $F \cup G$ is a flag of Γ;
(4) the function $\tau_A : X_A \to J$ is the map assigning to a minimal A-invariant flag F the set of A-orbits in $\tau(F)$.

Suppose that A is a subgroup of $\operatorname{Aut}(\Gamma)$. Let J be the set of all $j \in I$ for which there is an element of type j fixed by A. Replacing singletons of types by the actual type, we can view the absolute Γ_A as an incidence system over J. After a similar replacement of singletons by their elements in X_A, the absolute Γ_A will coincide with the fixed subsystem of Γ with respect to A.

Several important geometries to be met later on in this book, arise as absolutes with respect to a small group A of correlations of some bigger geometry Γ. They often have interesting simple groups of automorphisms which are subgroups of $\operatorname{Aut}(\Gamma)$ centralizing A.

Example 1.4.13 Let V be a vector space of finite dimension n over a field $\mathbb{F}$. A *bilinear form* on V is a map $f : V \times V \to \mathbb{F}$ such that, for each $v \in V$, both maps

$$f(v, \cdot) : V \to \mathbb{F}, \qquad f(\cdot, v) : V \to \mathbb{F}$$

are linear. A bilinear form f is called *symmetric* if, for all $v, w \in V$, we have $f(v, w) = f(w, v)$.

The *radical* of f is the following linear subspace of V.

$$V^\perp = \left\{ x \in V \mid f(x, y) = 0 \text{ for all } y \in V \right\}. \tag{1.1}$$

It is also denoted $\mathrm{Rad}(f)$.

Fix a basis $\varepsilon_1, \ldots, \varepsilon_n$ of V. A typical example of a symmetric bilinear form on V is $f_0(x, y) = \sum_{i=1}^n x_i y_i$ where $x = \sum_i x_i \varepsilon_i$, $y = \sum_i y_i \varepsilon_i$ with $x_i, y_i \in \mathbb{F}$.

For $x \in V$, we denote by $x^\perp$ the linear subspace of V consisting of all $y \in V$ with $f(x, y) = 0$. It is either a *hyperplane*, that is, a subspace of V of codimension 1, or all of V. If $x^\perp$ is a hyperplane for each nonzero vector in V, the form f is called *nondegenerate*. Observe that f is nondegenerate if and only if $\mathrm{Rad}(f) = \{0\}$. The example f_0 is nondegenerate. For an arbitrary subset X of V, set $X^\perp = \bigcap_{x \in X} x^\perp$. Clearly, $V^\perp = \mathrm{Rad}(f)$.

Suppose now that f is nondegenerate. If X is a k-dimensional subspace of V, then $X^\perp$ is an $(n - k)$-dimensional subspace and $(X^\perp)^\perp = X$. Hence, the map $\alpha : X \mapsto X^\perp$ is an auto-correlation of $\mathrm{PG}(V)$ (the projective geometry of Example 1.4.9) of order two; in other words, α is a polarity. We consider the absolute of $\mathrm{PG}(V)$ with respect to $A = \langle \alpha \rangle$. The A-orbit of a subspace X of V of dimension k consists of X and $X^\perp$. As $\dim(X^\perp) = n - k$ and α has order 2, we may interchange X and $X^\perp$ if needed, so as to force $\dim(X) \leq n/2$. Now $X * X^\perp$, where $*$ is the incidence relation of $\mathrm{PG}(V)$, means $X \subseteq X^\perp$, which is equivalent to X being *singular*, that is, $f(x, y) = 0$ for all $x, y \in X$. Therefore, by mapping the element $\{X, X^\perp\}$ of the absolute to the least dimensional member of this unordered pair, we can identify the absolute of $\mathrm{PG}(V)$ with respect to A with the set of all singular subspaces of V of affine dimension at most $n/2$. Incidence and type are as in $\mathrm{PG}(V)$: the former is given by symmetrized inclusion, whereas $\tau(X) = \dim(X)$. This geometry is a polar geometry, which is the subject of Chaps. 7–10.

1.5 Residues

The concept of a residue is a cornerstone of the theory of geometries. In the physical view, where elements are sets of points, this concept arises in different ways, such as subspaces (for the elements contained in a given subspace) and quotient spaces (for the elements containing a given subspace). In the study of polyhedra it appears as the *vertex figure*, that is, the collection of edges and faces containing a vertex together with their incidences. Residues unify and clarify these classical concepts. As usual, I will be a fixed set of types and $\Gamma = (X, *, \tau)$ an incidence system over I.

Definition 1.5.1 Let F be a flag of Γ. The *residue* of F in Γ is the subsystem Γ_F of Γ over $I \setminus \tau(F)$ on the point set $F^* \setminus F$.

For the sake of brevity, residues of flags of Γ are also referred to as residues of Γ. If $F = \{x\}$ for some $x \in X$, we also write Γ_x instead of Γ_F.

Example 1.5.2 Consider the cube as a geometry over {vertex, edge, face}. The residue of a face of the cube is a quadrangle. A little more thought yields that the residue of a vertex is a triangle over {edge, face}. The residue of an edge is a polygon with two vertices and two edges, which we call a *digon*. Other residues are less interesting. The residue of the empty flag is the full cube, the residue of a chamber is empty. The residue of a rank two flag consists of exactly two elements: the geometry is thin.

Similarly, consider the icosahedron (Fig. 1.2) as a geometry over {vertex, edge, face}. This geometry is also thin. The residue of a face is a triangle. The residue of a vertex is a pentagon and the residue of an edge is a digon as defined in Example 1.5.2. In each polyhedron or tessellation, the residue of an edge is a digon. Digons are very common indeed.

Proposition 1.5.3 *Suppose that F is a flag of Γ.*

(i) *A subset G of X_F is a flag of Γ_F if and only if $F \cup G$ is a flag of Γ.*
(ii) *If G is a flag of Γ_F, then $(\Gamma_F)_G = \Gamma_{F \cup G}$.*
(iii) *If Γ is a geometry, then its residue Γ_F is a geometry over $I \setminus \tau(F)$.*

Proof (i) Let G be a subset of X_F. Then G is a flag of Γ_F if and only if G is a flag of Γ and $G \subseteq F^*$, which in turn is equivalent to $G \cup F$ being a flag of Γ.

(ii) Suppose that G is a flag of Γ_F. By (i), $G \cup F$ is a flag of Γ. Now y is an element of $(\Gamma_F)_G$ if and only if $y \in (F^* \setminus F) \cap (G^* \setminus G) = (F \cup G)^* \setminus (F \cup G)$, that is, $y \in X_{F \cup G}$. This proves $(X_F)_G = X_{F \cup G}$. As incidence and the type map on $(X_F)_G$ are both obtained by repeated restrictions, we find $(\Gamma_F)_G = \Gamma_{F \cup G}$.

(iii) This is an immediate consequence of (ii).   □

For a subset J of I and a flag F of Γ, the notation $_J\Gamma_F$ can be interpreted as $(_J\Gamma)_F$ only if $\tau(F) \subseteq J$ and as $_J(\Gamma_F)$ only if $\tau(F) \subseteq I \setminus J$. So confusion between the residue of F in the J-truncation of Γ and the J-truncation of the residue of F in Γ due to absence of brackets will not arise.

Example 1.5.4 A face-to-face tessellation of $\mathbb{E}^3$ by cubes is a geometry over $I = $ {vertex, edge, face, cell}. The residue of a cell is a cube. The residue of a vertex consists of 6 edges, 12 faces, 8 cells, and is in fact an octahedron.

Example 1.5.5 Consider $\mathbb{E}^3$ as a geometry of rank three with points, lines, and planes. The residue of a plane is a plane, the real affine plane (of Example 1.1.2). In the residue of a point, any two lines are incident with a unique plane and any two planes are incident with a unique line. These are features of the *real projective plane*, that is $\mathrm{PG}(\mathbb{R}^3)$ (of Example 1.4.9). The residue of a line consists of all points and all planes on that line. Every such point is incident with every such plane. This looks much like the digon structure in that the incidence graph of this residue is a *complete bipartite graph*: each element is incident with each element of the other type. For this reason it is called a generalized digon; it will be properly introduced in Definition 2.1.1. We will write $\mathrm{K}_{m,n}$ for the complete bipartite graph whose parts have sizes m and n.

Example 1.5.6 Let V be a finite-dimensional vector space over the division ring $\mathbb{D}$ and let PG(V) be as in Example 1.4.9. Assume that U and W are incident subspaces of V with $\dim(W) = \dim(U) + 3$. For every flag F of PG(V) maximal with respect to containing U and W but no subspaces of dimension $\dim(U) + 1$ or $\dim(U) + 2$, the residue of F is PG(W/U), which is isomorphic to the projective plane PG($\mathbb{D}^3$).

Also, if L is a line of PG(V), so that $\dim(L) = 2$, then the residue of every flag maximal with respect to containing L but not a point, is a set of projective points, which is identified with L in classical projective geometry.

1.6   Connectedness

This section deals with connectivity, the extent to which elements of a geometry can be joined by chains of incident elements. Again I is a set of types.

Definition 1.6.1 Let Δ be a graph. If p and q are vertices of Δ, then a *path of length* n from p to q is a sequence $p = x_0, x_1, x_2, \ldots, x_n = q$ of vertices of Δ such that $\{x_i, x_{i+1}\}$ is an edge of Δ for $i = 0, \ldots, n-1$. We usually write

$$p = x_0, x_1, x_2, \ldots, x_n = q$$

to indicate the path. The minimal length of a path from p to q is called the *distance* between p and q and denoted by $d_\Delta(p, q)$, or just $d(p, q)$ if it is clear in which graph the distance is measured. If there is no such path, we say that the distance between p and q is ∞. If $d(p, q) = n$, we also say that p is *at distance* n from q. A path from p to q of length $d(p, q)$ is called a *minimal path* or a *geodesic* from p to q in Δ.

If we write $p \equiv q$ whenever there exists some path (always of finite length) from p to q, then $\equiv$ is clearly an equivalence relation on the vertex set of Δ. The classes of $\equiv$ are the *connected components* of Δ. A graph is called *connected* if it has exactly one connected component, i.e., if the vertex set is non-empty and every pair of vertices are joined by a path.

Let $\Gamma = (X, *, \tau)$ be an incidence system over I. Then Γ is said to be *connected* if its incidence graph is. A path in the incidence graph is called a *chain* in Γ. If, for $J \subseteq I$, all of its vertices, except possibly the endpoints, have types in J, then it is called a J-*chain*.

If Γ is a firm incidence system of rank one, then its incidence graph has at least two vertices and no edges, so it is not connected. However, we will be mostly concerned with firm geometries of rank at least two that are connected. It is easy to draw pictures of connected geometries with disconnected residues of rank two. Thus, connectedness leads to new restrictions on the geometries of our interest.

Definition 1.6.2 An incidence system Γ over I is called *residually connected* if each residue of Γ of rank at least two is connected.

Fig. 1.16 A chain from p to q

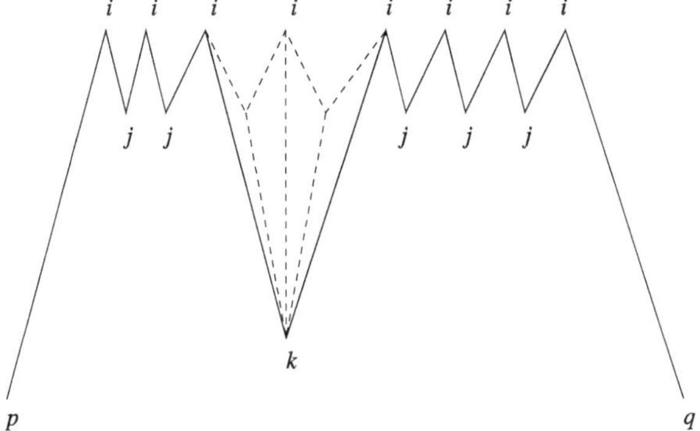

An incidence system over I is itself the residue of the empty flag. So, if $|I| \geq 2$ and it is residually connected, then it is connected.

Lemma 1.6.3 *Let I be finite and suppose that Γ is a residually connected incidence system over I. If i, j are distinct elements of I and p, q two elements of Γ, then there exists an $\{i, j\}$-chain from p to q.*

Proof We proceed by induction on the rank r of $\Gamma = (X, *, \tau)$. For $r = 2$ the property is obvious by connectedness of Γ. Assume $r \geq 3$. As Γ is connected, there is a chain from p to q in Γ, say

$$p = x_0, x_1, \ldots, x_n = q.$$

We want to find an $\{i, j\}$-chain from p to q. If $p * q$, we are done, so we may assume $n > 1$. Let a be the smallest index with $1 \leq a < n$ such that x_a is not of type i or j. We can assume that there is such an index, for otherwise the existing chain is as required. Let $k = \tau(x_a)$. The residue Γ_{x_a} is a geometry over $I \setminus \{k\}$ all of whose residues of rank at least two are connected. Therefore, the induction hypothesis on r applies and gives an $\{i, j\}$-chain from x_{a-1} to x_{a+1}, inside Γ_{x_a}. Figure 1.16 illustrates this argument.

If we replace x_a by this $\{i, j\}$-chain from x_{a-1} to x_{a+1} in the original chain, we find a new chain. Applying the same procedure to each element of the chain whose type is neither i nor j (there is one less of these in the new chain than in the original chain), we eventually arrive at an $\{i, j\}$-chain from p to q. $\square$

Exercise 1.9.18 gives an example showing that the finiteness requirement on I in the lemma above is needed.

The incidence system $(\{a, b\}, *, \tau)$ over $[3]$ with $\tau(a) = 1$ and $\tau(b) = 2$, in which all (three) pairs of elements are incident, is residually connected but not a geometry.

Lemma 1.6.4 *A residually connected incidence system is a geometry if no flag of corank one is maximal.*

Proof Let F be a maximal flag of a residually connected incidence system Γ over I (it exists by a straightforward application of Zorn's lemma). We need to show that F is a chamber. Assume it is not. By the hypotheses, $|I \setminus \tau(F)| \geq 2$, where τ is the type map of Γ. Now Definition 1.6.2 (recall that the empty residue is not connected) yields the existence of an element x of Γ_F, contradicting maximality of F. So, F must be a chamber. $\square$

For the case where I is finite, a weaker sufficient condition than the one of Lemma 1.6.4 is given in Exercise 1.9.17.

Theorem 1.6.5 *Suppose that Γ is a residually connected incidence system over I in which no flag of corank one is maximal. For each flag F of Γ, the residue Γ_F is a residually connected geometry.*

Proof Let F be a flag of Γ. By Lemma 1.6.4, Γ is a geometry and, by Proposition 1.5.3(iii), Γ_F is a geometry over $I \setminus \tau(F)$, where τ is the type map of Γ.

Let G be a flag of Γ_F with $|(I \setminus \tau(F)) \setminus \tau(G)| \geq 2$. As Γ is residually connected, Proposition 1.5.3(ii) gives that the residue $(\Gamma_F)_G = \Gamma_{F \cup G}$ is connected. Hence, Γ_F is residually connected. $\square$

Corollary 1.6.6 *Let Γ be a geometry over I and assume that I is finite. Then Γ is residually connected if and only if, for all distinct i, j in I and flags F of Γ having no elements of type i or j, the $\{i, j\}$-truncation ${}_{\{i,j\}}\Gamma_F$ is connected.*

Proof The 'if' part is obvious. As for the 'only if' part, suppose that Γ is residually connected. By Theorem 1.6.5, the properties of Γ carry over to Γ_F, so we can restrict ourselves to proving the result for $F = \emptyset$. This means that it suffices to show that, for every two distinct i, j in I, the subgraph of the incidence graph of Γ induced on $X_i \cup X_j$ is connected. For $\mathrm{rk}(\Gamma) \leq 1$ this is trivial and for $\mathrm{rk}(\Gamma) \geq 2$ it is a consequence of Lemma 1.6.3. $\square$

To end this section, we single out the geometries of our interest.

Definition 1.6.7 A geometry over I is called an *I-geometry* if it is firm and residually-connected.

1.7 Permutation Groups

In previous sections, projective spaces, affine spaces, regular polytopes, and regular tilings were associated with geometries having a lot of symmetry, that is, large groups of automorphisms. In this section, we explore how large a group of automorphisms of a geometry need be to fully describe the geometry in terms of the group. In fact, the main theorems are variations of the elementary but very basic

Theorem 1.7.5 below which describes the correspondence between subgroups and transitive representations of a given group.

Throughout this section, G denotes a group and X a set. The group $\mathrm{Sym}(X)$ of all permutations of X was introduced in Remark 1.3.2. We will let permutations act on the left. So, if $\sigma \in \mathrm{Sym}(X)$ and $x \in X$, then $\sigma(x)$ denotes the image of x under σ. For $n \in \mathbb{N}$, we often write Sym_n instead of $\mathrm{Sym}([n])$. According to Definition 1.3.5, a group is a permutation group on X if it is a subgroup of $\mathrm{Sym}(X)$. We use the following slightly more general notion.

Definition 1.7.1 A *representation* of G in X is a group homomorphism $\phi : G \to \mathrm{Sym}(X)$. In this case, X is referred to as a *G-set*. Such a representation is called *faithful* if ϕ is injective.

Let $x \in X$. The set $\{\phi(g)x \mid g \in G\}$ is called the *G-orbit* of x in X (compare Definition 1.3.5). The representation is called *transitive* if X is a single G-orbit. The *stabilizer* of x in G is the subgroup $\{g \in G \mid \phi(g)x = x\}$ of G; it is denoted by G_x.

As usual, we employ some abbreviations for the sake of readability. Instead of a permutation representation of G on X, we often speak of an *action* of G on X, and instead of $\phi(g)x$, where $g \in G$, $x \in X$, we also write gx if the representation ϕ is clear from the context. Similarly, for the orbit $\phi(G)x$ of $x \in X$ we often write Gx. Observe also that ϕ is involved in the definition of stabilizer but not visible in the corresponding notation.

If H is a subgroup of G, there is a standard way of constructing a transitive representation of G on the space $G/H = \{aH \mid a \in G\}$ of left cosets of H in G.

Definition 1.7.2 Let H be a subgroup of G. The map $\phi : G \to \mathrm{Sym}(G/H)$ given by $\phi(g)aH = gaH$ for all $g, a \in G$ is called the *representation* of G over H.

It is readily seen that this map ϕ is a transitive permutation representation. Its kernel is $\bigcap_{a \in G} aHa^{-1}$, the biggest normal subgroup of G contained in H.

For $H = 1$, this representation is faithful. Therefore, G can be viewed as a subgroup of $\mathrm{Sym}(G)$ by means of the embedding sending $g \in G$ to left multiplication by g on G.

Any transitive representation can be described as a representation over a subgroup. To be more precise, we need the notion of equivalence.

Definition 1.7.3 Let X' be a set. Two representations $\phi : G \to \mathrm{Sym}(X)$ and $\phi' : G \to \mathrm{Sym}(X')$ are said to be *equivalent* if there is a bijection $\beta : X \to X'$ such that $\beta\phi(g)\beta^{-1} = \phi'(g)$ whenever $g \in G$. In this case, we also say that X and X' are *isomorphic G-sets*.

For instance, any two sets of the same size n have natural actions of Sym_n and as such are isomorphic Sym_n-sets.

Example 1.7.4 If ϕ is the representation of a group G over a subgroup H and $a \in G$, then the stabilizer in G of aH is aHa^{-1}. Now ϕ is equivalent to the representation of G over aHa^{-1} by means of the map $\beta : G/H \to G/(aHa^{-1})$ given by $\beta(yH) = ya^{-1}(aHa^{-1})$. As a consequence, ϕ is determined up to equivalence by the conjugacy class of subgroups of G to which H belongs.

Theorem 1.7.5 *Let $\phi : G \to \mathrm{Sym}(X)$ be a transitive representation. For $x \in X$, the representation of G over G_x is equivalent to ϕ.*

Proof Take $\beta : X \to G/G_x$ to be the map given by

$$\beta(y) = gG_x \quad \text{if } g \in G \text{ satisfies } y = \phi(g)x.$$

Note that β is well defined. First, as X is a single G-orbit, for each $y \in X$, there is $g \in G$ with $y = \phi(g)x$. Second, if g and g' are elements of G, with $y = \phi(g)x$ and $y = \phi(g')x$, then $\phi(g^{-1}g')x = x$, so $g^{-1}g' \in G_x$, whence $gG_x = g'G_x$.

Also β is a bijection. For, if $y, y' \in X$ satisfy $\beta(y) = \beta(y')$, then $gG_x = g'G_x$ for $g, g' \in G$ with $\phi(g)x = y$ and $\phi(g')x = y'$. Taking $h, h' \in G_x$ with $gh = g'h'$, we find $y = \phi(g)x = \phi(gh)x = \phi(g'h')x = \phi(g')x = y'$, proving that β is injective indeed. Clearly, β is also a surjection, and so even a bijection.

Now write ϕ' for the representation of G over G_x, and let $g \in G$ and $z \in X$. Then, with $h \in G$ such that $z = \phi(h)x$, we find

$$\beta\phi(g)(z) = \beta\big(\phi(gh)x\big) = ghG_x = \phi'(g)(hG_x) = \phi'(g)\beta(z),$$

and so β is indeed an equivalence between the G-sets X and G/G_x. $\square$

Remark 1.7.6 Here is a more general and more sophisticated formulation of the theorem that incorporates intransitive actions. Let X be a G-set. Denote by X/G the collection of orbits of G on X (compare again Definition 1.3.5). So, if $x \in X$, then Gx belongs to X/G. Let $t : X/G \to X$ be a *transversal*, i.e., a map satisfying $Gt(y) = y$ for $y \in X/G$. Then there is a G-set isomorphism

$$\bigcup_{y \in X/G} G/G_{t(y)} \to X$$

given by $gG_{t(y)} \mapsto gt(y)$ ($g \in G$, $y \in X/G$). In the same vein, the variations of Theorem 1.7.5 for group actions on graphs and geometries obtained in this section and the next one can be formulated more generally.

Example 1.7.7 From a given representation $\phi : G \to \mathrm{Sym}(X)$, several others can be constructed. The homomorphism $\phi^{(2)} : G \to \mathrm{Sym}\binom{X}{2}$ given by

$$\phi^{(2)}(g)\{x, y\} = \big\{\phi(g)x, \phi(g)y\big\} \quad (x, y \in X, x \neq y)$$

is an example. The constructed representations are often clear once the underlying set (which is $\binom{X}{2}$ in the above example) is specified. We then refer to the representation as the action of G on this set *induced by* ϕ.

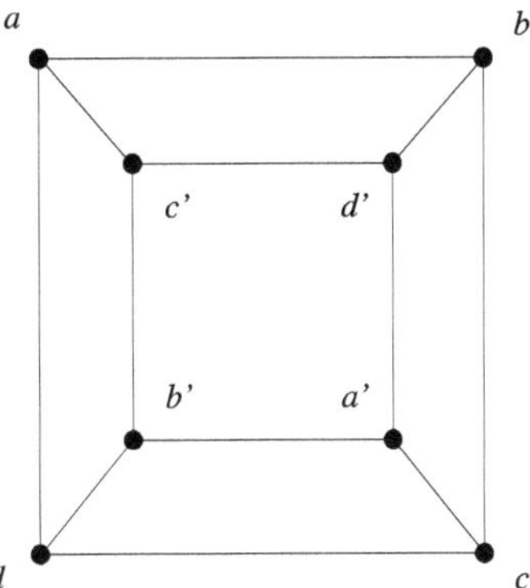

Fig. 1.17 The cube as a graph

Example 1.7.8 We study the subgroups of Sym_4. The definition of Sym_4 as the group of all permutations of the set [4] gives the following non-exhaustive list of transitive Sym_4-sets with indicated stabilizers.

(1) The set [4], with stabilizer isomorphic to Sym_3; example: $\langle(1,2),(2,3)\rangle$ is the stabilizer of 4.
(2) The set $\binom{[4]}{2}$ (of size six) of all 2-subsets of [4], with stabilizer isomorphic to $C_2 \times C_2$; example: $\langle(1,2),(3,4)\rangle$ is the stabilizer of $\{1,2\}$ (and also of $\{3,4\}$).
(3) The set of ordered pairs of distinct points of [4], with stabilizer isomorphic to C_2; example: $\langle(3,4)\rangle$ is the stabilizer of the ordered pair $(1,2) \in [4] \times [4]$.
(4) The set of partitions of [4] into two subsets of size two each, with stabilizer isomorphic to Dih_8, the dihedral group of order eight; example: $\langle(1,3),(1,2,3,4)\rangle$ is the stabilizer of $\{\{1,3\},\{2,4\}\}$.

It is also known that Sym_4 is the group of all rotations of the cube (see also Example 1.7.15). To see this, label the vertices of the cube as in Fig. 1.17.

There are four pairs of opposite vertices, namely $1 = \{a,a'\}$, $2 = \{b,b'\}$, $3 = \{c,c'\}$, and $4 = \{d,d'\}$. The element $z := (a,a')(b,b')(c,c')(d,d')$ is induced by the homothety $-\,\mathrm{id}$ on the Euclidean vector space $\mathbb{R}^3$ in which the cube embeds (with center of gravity at the origin). It is not a rotation. Each element of Sym_4 determines two automorphisms of the graph in Fig. 1.17, one of which is obtained from the other by multiplication by z. In particular, a unique automorphism corresponding to the element of Sym_4 is a rotation of the cube. For instance, $(3,4)$ corresponds to the automorphism $(c,d')(c',d)$ of the graph which is induced by a reflection of $\mathbb{R}^3$ and so maps to $(c,d')(c',d)z = (a,a')(b,b')(c,d)(c',d')$. This gives a faithful action of Sym_4 on the cube.

From this description we derive the following non-exhaustive list of transitive Sym_4-sets.

(5) The set of two tetrahedra in the cube, with stabilizer isomorphic to Alt_4.
(6) The set of faces, with stabilizer C_4.
(7) The vertex set, with stabilizer C_3.
(8) The set of flags of type {vertex, edge}, with stabilizer the trivial group.

Fig. 1.18 The subgroup
pattern of Sym_4

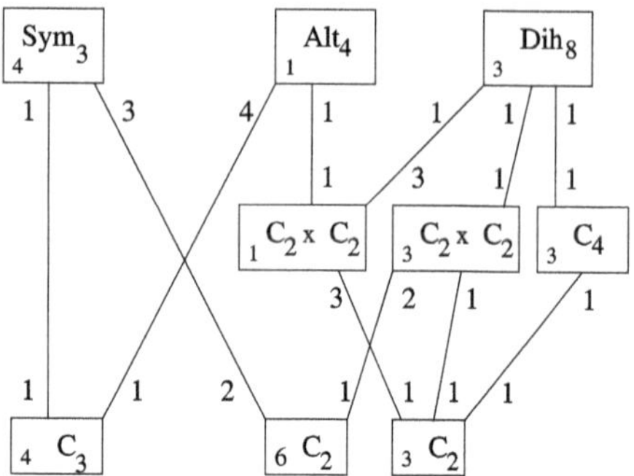

So far, we have found eight conjugacy classes of subgroups of Sym_4. There are
two more classes of proper subgroups, with representatives $\langle (1,2)(3,4) \rangle$ (isomorphic to C_2) and $\langle (1,2)(3,4), (1,3)(2,4) \rangle$ (isomorphic to $C_2 \times C_2$), respectively. In
particular, there are two conjugacy classes of groups isomorphic to C_2 in Sym_4.
Similarly for $C_2 \times C_2$ instead of C_2.

The nontrivial proper subgroups of Sym_4 obey the pattern displayed by Fig. 1.18.
Each box represents a conjugacy class of subgroups whose isomorphism type is as
inscribed. The cardinality of the conjugacy class can be found in the southwest
corner of the box. Inclusions are as indicated by the labelled connections between
boxes: the group at the bottom is a maximal subgroup of the one at the top (up to
conjugacy). For instance, each member of the class of Sym_3 contains three members
of the class of C_2 of size six and each of the latter is contained in two subgroups
isomorphic to Sym_3.

As a consequence, Sym_4 has eleven inequivalent transitive representations.

We are ready for variations of Theorem 1.7.5 to structures whose underlying set
admits a representation. The group of the representation should act on the structure
as a group of automorphisms.

Definition 1.7.9 If Δ denotes a structure (e.g., a vector space) and $\mathrm{Aut}(\Delta)$ the group
of all automorphisms of Δ, then we say that ϕ is a *representation* of G on Δ if it is
a group homomorphism $\phi : G \to \mathrm{Aut}(\Delta)$.

If Δ is a vector space, this means that ϕ is a so-called linear representation. If Δ
is just a set, then Definition 1.7.1 coincides with Definition 1.7.9.

As before, we also speak of a G-action on Δ when we mean a representation
in Δ.

In order to state the variations of Theorem 1.7.5 for other structures, we need the
notion of equivalence of representations of these structures.

Definition 1.7.10 Two representations ϕ, ϕ' of G on structures Δ, Δ' (of the same
kind), respectively, are called *equivalent* if there is an isomorphism $\beta : \Delta \to \Delta'$
establishing equivalence between the associated ordinary representations $\phi : G \to$
$\mathrm{Sym}(\Delta)$ and $\phi' : G \to \mathrm{Sym}(\Delta')$.

Here, $\mathrm{Sym}(\Delta)$ is the symmetric group on the natural set underlying Δ. For instance, the set of vectors if Δ is a vector space.

We now focus on graph structures.

Definition 1.7.11 Suppose that Δ is a graph and that G acts on Δ. A representation $\phi : G \to \mathrm{Aut}(\Delta)$ is called *edge transitive* if both ϕ and the induced action of ϕ on the set of edges of Δ are transitive.

Remark 1.7.12 We describe how to translate the data of a graph $\Delta = (X, \sim)$ with sufficient symmetry, namely a transitive representation $\phi : G \to \mathrm{Aut}(\Delta)$, into group information.

By Theorem 1.7.5, taking $x \in X$ and setting $H = G_x$, we may assume without loss of generality that $X = G/H$ and $\phi(g)aH = gaH$ $(a, g \in G)$. It remains to describe $\sim$ in terms of the groups G and H. For $a, b \in G$, we have $aH \sim bH$ if and only if $H \sim a^{-1}bH$, so that it suffices to identify the neighbors of the vertex H. To this end write $K = \{g \in G \mid H \sim gH\}$. Clearly, $K = hK = Kh$ for every $h \in H$, so K is a union of double H-cosets. Hence a graph on which G acts transitively is determined by the knowledge of a stabilizer H and a union K of double cosets of H in G.

As Δ is a graph, $H \sim rH$ implies $rH \sim H$, whence, by application of $\phi(r^{-1})$, also $H \sim r^{-1}H$. This shows that $r^{-1} \in K$, so $K = K^{-1}$. We now have a full description of $\sim$ terms of G, H, and K under the assumption that G is edge transitive; $aH \sim bH$ if and only if $a^{-1}b \in K$ (for $a, b \in G$).

Reversing the above, we start with a group G, a subgroup H and a union K of double cosets of H in G with $K = K^{-1}$. It is straightforward that $(X, \sim)$, where $X = G/H$ and $\sim$ is the relation on X given by $aH \sim bH$ if and only if $a^{-1}b \in K$ $(a, b \in G)$, is a graph admitting a transitive representation. This graph has a vertex whose stabilizer is H and whose neighbors are the vertices aH for $a \in K$.

Definition 1.7.13 Let H be a subgroup of G and M a subset of G. The graph with vertex set G/H and edges the unordered pairs $\{aH, bH\}$ $(a, b \in G)$ with $a^{-1}b \in \bigcup_{k \in M}(HkH \cup Hk^{-1}H)$ is called the *coset graph* on G/H determined by M. It is denoted $\Delta(G, H, M)$.

The group G acts transitively on such a graph $\Delta(G, H, M)$ by left multiplication on the vertex set. We focus on the simplest case, where K of Remark 1.7.12 consists of a single double coset, so we can take $M = \{r\}$ for a single $r \in G$ with $r^{-1} \in HrH$. This means that, if aH is adjacent to H for some $a \in G$, then $aH = brH$ for some $b \in H$. In other words, H, the stabilizer in G of the vertex H in X, is transitive on the set of the vertices adjacent to H. So G is edge transitive on the graph $\Delta(G, H, \{r\})$.

Proposition 1.7.14 *If $\phi : G \to \mathrm{Aut}(X, \sim)$ is an edge-transitive representation of G on a graph $(X, \sim)$, then, for $x \in X$, the representation of G on the coset graph $\Delta(G, G_x, \{r\})$, where $r \in G$ satisfies $x \sim \phi(r)x$, is equivalent to ϕ.*

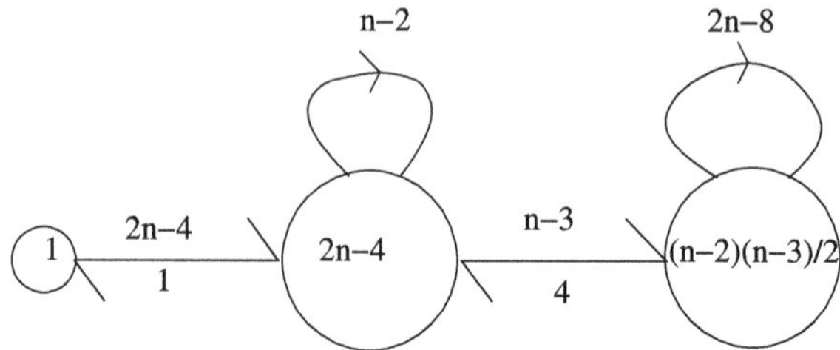

Fig. 1.19 The distribution diagram of Sym_n acting on $\binom{[n]}{2}$

Proof Fix $x \in X$ and take $\beta : G/G_x \to X$ to be the map sending gG_x to gx. Then β establishes the required isomorphism. $\qquad\square$

Example 1.7.15 Take $G = \mathrm{Sym}_4$ and $H = \langle(1,2,3)\rangle$. As we have seen in Example 1.7.8(7), the set G/H corresponds, as a G-set, to the vertex set of the cube, on which G acts. The graph Γ of the cube is equivalent to the coset graph $\Delta(G, H, \{(1,4)\})$. Each vertex of Γ has the same number of vertices adjacent to it; such a graph is called *regular* and the number is called the *valency* of Γ. Let us verify that the valency of this graph is three (as it should be). The valency is the number of H-cosets in $H(1,4)H$. As $H \cap (1,4)H(1,4)^{-1} = \langle(1,2,3)\rangle \cap \langle(4,2,3)\rangle = 1$, we have $|H(1,4)H| = |H(1,4)H(1,4)^{-1}| = |H| \cdot |H|/|H \cap (1,4)H(1,4)^{-1}| = 9$, and so the valency of Γ equals $9/|H| = 3$.

Example 1.7.16 Let $n \in \mathbb{N}$, $n \geq 2$, and let $G = \mathrm{Sym}_n$ act on $[n]$. Since G acts transitively on $\binom{[n]}{2}$ (see Example 1.7.8), the only graph structures on $[n]$ preserved by G are the empty graph (with empty edge set) and the complete graph (with edge set $\binom{[n]}{2}$). Next consider graphs with vertex set $V = \binom{[n]}{2}$. It is readily seen that G is transitive on the collection E_ε of ordered pairs $(x, y) \in V \times V$ with $\varepsilon = |x \cap y|$ taking values in $\{0, 1, 2\}$. As $|x \cap y| = |y \cap x|$, the relations E_ε are *symmetric*, that is, $(x, y) \in E_\varepsilon$ implies $(y, x) \in E_\varepsilon$. Now E_2 is the diagonal, $\{(x, x) \mid x \in V\}$. The other two G-orbits, E_0 and E_1, lead to graphs that are each other's complements (the complement of a graph (V, E) is the graph $(V, \binom{V}{2} \setminus E)$). The equivalent coset graphs are $\Delta(G, H, \{(2,3)\})$ and $\Delta(G, H, \{(1,3)(2,4)\})$, respectively, where $H = \langle(1,2)\rangle \times \mathrm{Sym}(\{3, \ldots, n\})$. The *distribution diagram* of (V, E_1) is Fig. 1.19. This diagram gives a summary of the division of the vertex set into H-orbits. The three H-orbits of sizes 1, $2n - 4$, and $(n - 2)(n - 3)/2$ are depicted by circles circumscribing these cardinalities. In addition, the number written near an arrow going from H-orbit A to H-orbit B denotes the number of vertices in B adjacent to a fixed vertex in A.

For $\varepsilon = 0$ and $n = 5$, we obtain the Petersen graph Pet of Example 1.3.3, as may become clear from Fig. 1.20.

For $\varepsilon = 0$ and $n = 6$, we obtain a graph connected with a geometry (a generalized quadrangle of order $(2, 2)$) that will appear in Example 2.2.8. To be more precise, it is the collinearity graph (see Definition 2.5.12) of this geometry. In general, it may be quite intricate to derive the distribution diagram from a given group and subgroup.

Fig. 1.20 The Petersen graph
Pet. At each node, the symbol
ij stands for $\{i, j\}$

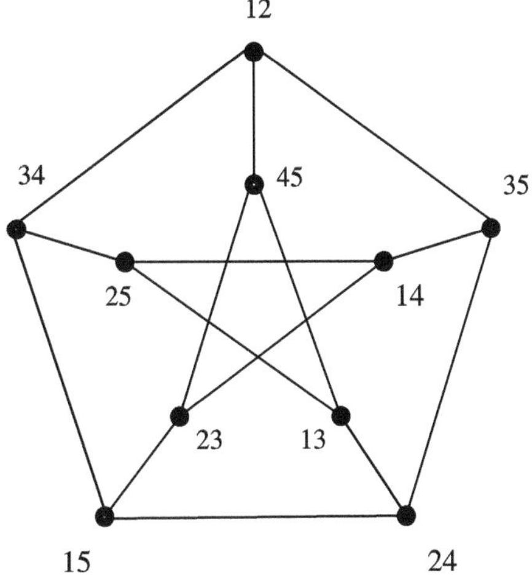

1.8 Groups and Geometries

In this section, we will give a description of geometries with sufficiently high symmetry in terms of groups. As usual, I denotes a type set. Furthermore, G will denote a group.

Let $\Gamma = (X, *, \tau)$ be an incidence system over I. As in Definition 1.7.9, a homomorphism $\phi : G \to \mathrm{Aut}(\Gamma)$ is a representation of G on Γ. Similarly, Definition 1.7.10 applies, so if $\Gamma' = (X', *', \tau')$ is also an incidence system over I, two representations $\phi : G \to \mathrm{Aut}(\Gamma)$ and $\phi' : G \to \mathrm{Aut}(\Gamma')$ are called equivalent if there is an isomorphism $\beta : \Gamma \to \Gamma'$ establishing equivalence between the ordinary representations $\phi : G \to \mathrm{Sym}(X)$ and $\phi' : G \to \mathrm{Sym}(X')$.

In contrast to permutation groups, it does not make sense to ask for transitivity of G on the element set X of Γ if the rank of Γ is at least two: $\mathrm{Aut}(\Gamma)$ preserves types, so each G-orbit consists of elements of the same type. Also, for (x, y) and $(x', y') \in X \times X$, the existence of $g \in G$ with $(\phi(g)x, \phi(g)y) = (x', y')$ implies that both (x, y) and (x', y') belong to $\tau^{-1}(i) \times \tau^{-1}(j)$ for certain $i, j \in I$. Thus, we are led to the following definition.

Definition 1.8.1 Let ϕ be a representation of G on an incidence system $\Gamma = (X, *, \tau)$ over I. We say that ϕ is *transitive* on Γ if, for each $i \in I$, the group G is transitive on the set $X_i = \tau^{-1}(i)$. We call ϕ *incidence transitive* on Γ if, for each $i, j \in I$, the induced representation

$$\phi_{ij} : G \to \mathrm{Sym}\big(\{(x_i, x_j) \in X_i \times X_j \mid x_i * x_j\}\big)$$

given by $\phi_{ij}(g)(x_i, x_j) = (\phi(g)x_i, \phi(g)x_j)$ is transitive. The map ϕ is called *flag transitive* if for each $J \subseteq I$, the group G is transitive on the set of all flags of type J.

Taking $i = j$ in the definition of incidence transitivity, we see that incidence transitivity implies transitivity.

Clearly, incidence transitivity is a consequence of flag transitivity. Moreover, if $|I| = 2$, the notions of flag transitivity and incidence transitivity coincide.

Lemma 1.8.2 *Suppose that* $\phi : G \to \mathrm{Aut}(\Gamma)$ *is a representation on the incidence system* $\Gamma = (X, *, \tau)$ *over* I. *If* ϕ *is incidence-transitive and* $\{x_i \mid i \in I\}$ *is a chamber of* Γ *with* $\tau(x_i) = i$, *then* Γ *is fully determined by* G, *and the system* $(G_i)_{i \in I}$, *where* G_i *is the stabilizer in* G *of* x_i, *in the following sense.*

(i) *For each* $i \in I$, *the map* $aG_i \mapsto \phi(a)x_i$ $(a \in G)$ *is an equivalence* $X_i \to G/G_i$ *of* G*-sets.*

(ii) *For each* $i, j \in I$ *and* $a, b \in G$, *we have* $\phi(a)x_i * \phi(b)x_j$ *if and only if* $aG_i \cap bG_j \neq \emptyset$.

(iii) $\tau(\phi(a)x_i) = i$.

Proof (i) is immediate from Theorem 1.7.5 and the fact that G is transitive on X_i.

(ii) If G is incidence transitive on Γ and $x_i * x_j$, then $\phi(a)x_i * \phi(b)x_j$ if and only if there exists $g \in G$ with $\phi(g)x_i = \phi(a)x_i$ and $\phi(g)x_j = \phi(b)x_j$. The latter is equivalent to $a^{-1}g \in G_i$ and $b^{-1}g \in G_j$ for some $g \in G$, so that $\phi(a)x_i * \phi(b)x_j$ can be reformulated as $aG_i \cap bG_j \neq \emptyset$.

(iii) As $\phi(G)$ is contained in $\mathrm{Aut}(\Gamma)$, it preserves types. $\square$

These observations lead to the following construction of incidence systems from groups.

Definition 1.8.3 Let $(G_i)_{i \in I}$ be a system of subgroups in G. The *coset incidence system* of G over $(G_i)_{i \in I}$, denoted $\Gamma(G, (G_i)_{i \in I})$, is the incidence system over I, whose elements of type i are the cosets of G_i in G and in which aG_i and bG_j are incident if and only if $aG_i \cap bG_j \neq \emptyset$. The group G_i is called the *standard parabolic subgroup* of G of type i (with respect to $\Gamma(G, (G_i)_{i \in I})$). Of course, if $\Gamma(G, (G_i)_{i \in I})$ is a geometry, it is also called a *coset geometry.*

The group G has a natural representation ϕ in this incidence system, referred to as the *geometric representation* of G over $(G_i)_{i \in I}$. It is given by $\phi(g)aG_i = gaG_i$ $(a, g \in G; i \in I)$.

Remark 1.8.4 (i) The notation G_i has been chosen so as to resemble the notation for the stabilizer in G of an element of type i. Indeed, G_i is the stabilizer of the element G_i of type i and $\{G_i \mid i \in I\}$ is a chamber of $\Gamma(G, (G_i)_{i \in I})$.

(ii) In Definition 1.8.3 we speak of a system $(G_i)_{i \in I}$ rather than a set to prevent confusion in case two subgroup G_j and G_k are the same for distinct $j, k \in I$. A coset of G_j coincides with a coset of G_k only if $G_j = G_k$. So, if the subgroups of the system are chosen to be mutually distinct, the union of G/G_i over all $i \in I$ is disjoint. Furthermore, an instance $G_j = G_k$ for distinct j and k does not provide an interesting geometry, as the $\{j, k\}$-truncation of $\Gamma(G, (G_i)_{i \in I})$ will then be a disjoint union of unit geometries.

(iii) Exercise 1.9.27 shows that a coset incidence system need not be a geometry.

(iv) Similarly to Example 1.7.4, conjugation by an element x of G leads to an isomorphism of coset incidence systems $\Gamma(G, (G_i)_{i \in I}) \to \Gamma(G, (xG_ix^{-1})_{i \in I})$.

Example 1.8.5 Let $G = \mathrm{Sym}_4$, and let G_1, G_2, G_3 be subgroups isomorphic to C_3, C_2 (having two fixed points in [4]), and C_4, respectively. The incidence system $\Gamma = \Gamma(G, (G_1, G_2, G_3))$ has the same number of elements of type i ($i \in [3]$) as the cube geometry. The given parabolic subgroups correspond to the transitive representations of Example 1.7.8(7), (3), and (6), respectively. However the cube is not the only such incidence system affording Sym_4. We use Remark 1.8.4(iv) to trim down the possibilities. The choice of G_1 is unique up to conjugacy; we fix it to be $G_1 = \langle (1, 2, 3) \rangle$. Once this choice has been made, the choice of G_3 is unique up to conjugacy as well, because G_1 acts transitively (by conjugation) on the set of three subgroups of G isomorphic to C_4. We fix it to be $G_3 = \langle (1, 3, 2, 4) \rangle$. This element maps to the graph automorphism $(a, c', b', d)(a', c, b, d')$ in the notation of Fig. 1.17.

In the conjugation on subgroups of G, the subgroup $\langle (1, 2) \rangle$ stabilizes G_1 and G_3 and has four orbits on the set of six conjugates of G_2, with representatives $\langle (1, 2) \rangle$, $\langle (1, 3) \rangle$, $\langle (1, 4) \rangle$, and $\langle (3, 4) \rangle$. So there are four distinct choices of G_2. In the case where, $G_2 = \langle (1, 4) \rangle$, the geometry Γ is necessarily the cube (compare with Example 1.7.15). Among the other three coset incidence systems, studied in Exercise 1.9.26, a geometry distinct from the cube appears.

Lemma 1.8.6 *Let $(G_i)_{i \in I}$ be a system of subgroups of G. The geometric representation of G over the coset incidence system $\Gamma(G, (G_i)_{i \in I})$ is incidence transitive.*

Proof Let $*$ denote incidence in $\Gamma(G, (G_i)_{i \in I})$. Suppose that $aG_i * bG_j$ holds for $a, b \in G$ and $i, j \in I$. We show that there is $g \in G$ such that $(gaG_i, gbG_j) = (G_i, G_j)$. Incidence of aG_i and bG_j means $a^{-1}b \in G_iG_j$, so there are $x \in G_i$, $y \in G_j$ such that $a^{-1}b = xy$. Now $g = x^{-1}a^{-1}$ satisfies $g(aG_i, bG_j) = (x^{-1}G_i, yG_j) = (G_i, G_j)$, as required. $\qquad\square$

Here is the converse.

Proposition 1.8.7 *Suppose that $\Gamma = (X, *, \tau)$ is an incidence system over I with a chamber $\{x_i \mid i \in I\}$ such that $\tau(x_i) = i$, and $\phi : G \to \mathrm{Aut}(\Gamma)$ is an incidence-transitive representation of G. The geometric representation of G over $(G_{x_i})_{i \in I}$ is equivalent to ϕ.*

Proof Direct from Lemma 1.8.2. $\qquad\square$

We will explore incidence-transitive representations somewhat further. The main goal is to translate properties like residual connectness of the incidence system into group-theoretical terms.

Definition 1.8.8 Given subgroups G_i ($i \in I$) of G, and $J \subseteq I$, we write G_J to denote $\bigcap_{j \in J} G_j$. Extending the Definition 1.8.3, we call this subgroup the *standard parabolic subgroup* of G of type J (so $G_{\{j\}} = G_j$ for each $j \in J$).

Lemma 1.8.9 *The coset incidence system* $\Gamma = \Gamma(G, (G_i)_{i \in I})$ *of* G *over* $(G_i)_{i \in I}$ *satisfies the following properties.*

 (i) *Γ is connected if and only if $G = \langle G_l \mid l \in I \rangle$.*
 (ii) *For $i, j, k \in I$, the group G is transitive on the set of flags of type $\{i, j, k\}$ if and only if $G_i G_j \cap G_i G_k = G_i(G_j \cap G_k)$.*
 (iii) *For each $J \subseteq I$, there is a natural injective homomorphism of incidence systems over $I \setminus J$*

$$\phi_J : \Gamma\big(G_J, (G_{J \cup \{i\}})_{i \in I \setminus J}\big) \to \Gamma_{\{G_j \mid j \in J\}}$$

 given by $\phi_J(aG_{J \cup \{i\}}) = aG_i$ ($a \in G_J, i \in I \setminus J$).
 (iv) *Given $J \subseteq I$, the homomorphism ϕ_J is surjective if and only if, for all $i \in I \setminus J$, we have $\bigcap_{j \in J}(G_j G_i) = G_J G_i$.*
 (v) *If ϕ_J is surjective for all $J \subseteq I$, then ϕ_J is an isomorphism for all $J \subseteq I$.*

Proof (i) Suppose that Γ is connected. Take $i \in I$. If $a \in G$, then there is a chain

$$G_i = a_0 G_{i_0}, a_1 G_{i_1}, a_2 G_{i_2}, \ldots, a_m G_{i_m} = a G_i$$

connecting the elements G_i and aG_i of Γ, with $a_0 = 1$ and $a_m = a$. Now $a_j G_{i_j} \cap a_{j+1} G_{i_{j+1}} \neq \emptyset$, so $a_j^{-1} a_{j+1} \in G_{i_j} G_{i_{j+1}}$ for $j = 0, \ldots, m-1$, whence

$$a = \big(a_0^{-1} a_1\big) \cdots \big(a_{m-2}^{-1} a_{m-1}\big)\big(a_{m-1}^{-1} a_m\big) \in G_{i_0} \cdots G_{i_{m-1}} G_{i_m},$$

and so $a \in \langle G_i \mid i \in I \rangle$. The converse is obtained by reversing the above argument.

(ii) Suppose that G is transitive on the set of all flags of type $\{i, j, k\}$. If $x \in G_i G_j \cap G_i G_k$, then $\{x^{-1} G_i, G_j, G_k\}$ is a flag of Γ, so that by the transitivity assumption there is $g \in G$ with $gG_i = x^{-1} G_i$, $gG_j = G_j$ and $gG_k = G_k$. This means that $x^{-1} \in gG_i$ and $g \in (G_j \cap G_k)$, so that $x \in G_i g^{-1} \subseteq G_i(G_j \cap G_k)$. Since obviously $G_i G_j \cap G_i G_k \supseteq G_i(G_j \cap G_k)$, we have established the 'only if' part of (ii).

As for the converse, consider a flag X of Γ of type $\{i, j, k\}$. We will establish that X lies in the G-orbit of $\{G_i, G_j, G_k\}$, the standard flag of type $\{i, j, k\}$. In view of incidence transitivity, we may assume (without loss of generality) $X = \{x^{-1} G_i, G_j, G_k\}$ for some $x \in G$. Thus, using incidences of the elements of this flag, we have $x \in G_i G_j \cap G_i G_k$. So, if $G_i G_j \cap G_i G_k = G_i(G_j \cap G_k)$, we have $x \in G_i x_1$ for some $x_1 \in G_j \cap G_k$, whence

$$\big\{x^{-1} G_i, G_j, G_k\big\} = \big\{x_1^{-1} G_i, x_1^{-1} G_j, x_1^{-1} G_k\big\} = x_1^{-1}\{G_i, G_j, G_k\}.$$

This shows that X is in the same G-orbit as the standard flag of type $\{i, j, k\}$. Hence (ii).

(iii) As $G_{J \cup \{i\}} \subseteq G_i$ and $a \in aG_i \cap G_j$ for all $i \in I \setminus J$, $j \in J$, $a \in G_J$, the map ϕ_J is well defined. Suppose $aG_{J \cup \{i\}} \cap bG_{J \cup \{k\}} \neq \emptyset$. Then also $aG_i \cap bG_k \neq \emptyset$, so ϕ_J is indeed a homomorphism. Suppose that $a, b \in G_J$ satisfy $\phi_J(aG_{J\cup\{i\}}) = \phi_J(bG_{J\cup\{i\}})$. Now $aG_i = bG_i$, so that $b^{-1}a \in G_i$. On the other hand, $b^{-1}a \in G_J$, so $b^{-1}a \in G_{J\cup\{i\}}$ whence $aG_{J\cup\{i\}} = bG_{J\cup\{i\}}$. This shows that ϕ_J is injective.

(iv) Suppose that ϕ_J is surjective. If $x \in \bigcap_{j\in J}(G_jG_i)$ for some $i \in I \setminus J$, then xG_i is an element of the residue of the flag $\{G_j \mid j \in J\}$. So we can find $x' \in G_J$ with $\phi_J(x'G_{J\cup\{i\}}) = xG_i$. Then $x'G_i = xG_i$, so $x \in x'G_i \subseteq G_JG_i$, proving $\bigcap_{j\in J}(G_jG_i) = G_JG_i$. The converse is equally straightforward.

(v) Suppose that ϕ is surjective for each $J \subseteq I$. Fix $K \subseteq I$. By (iii), ϕ_K is bijective. We need to show that ϕ_K^{-1} is a homomorphism. Let xG_i, yG_j, where $i, j \in I \setminus K$ and $x, y \in G_K$, be incident elements of the residue $\Gamma_{\{G_k \mid k\in K\}}$ in Γ of the flag $\{G_k \mid k \in K\}$. Then $y^{-1}x \in G_jG_i \cap G_K$. But the surjectivity of $\phi_{K\cup\{j\}}$ and (iv) yield $G_jG_i \cap G_K \subseteq G_jG_i \cap G_KG_i \subseteq G_{\{j\}\cup K}G_i$ whence $G_jG_i \cap G_K \subseteq (G_{\{j\}\cup K}G_i) \cap G_K = G_{\{i\}\cup K}G_{\{j\}\cup K}$ so that $y^{-1}x \in G_{\{j\}\cup K}G_{\{i\}\cup K}$, proving that $yG_{\{j\}\cup K}$ and $xG_{\{i\}\cup K}$ are incident elements of $\Gamma(G_K, (G_{\{i\}\cup K})_{i\in I\setminus K})$. Hence (v). $\qquad\square$

In 'good' geometries we expect that the homomorphisms ϕ_J of Lemma 1.8.9(iii) are isomorphisms. The next results gives equivalent criteria for this to hold. The first and last one characterize flag transitivity of G on Γ fully in terms of subgroups and therefore represent the goal we were after.

Theorem 1.8.10 *Let Γ be the coset incidence system of G over $(G_i)_{i\in I}$. If I is finite, then the following statements are equivalent.*

(i)   *G is flag transitive on Γ.*
(ii) *For each subset J of I, the homomorphism ϕ_J is an isomorphism.*
(iii) *For each subset J of I of size three, the group G is transitive on the set of flags of type J, and for each $i \in I$ the subgroup G_i is flag transitive on $\Gamma(G_i, (G_{\{i,j\}})_{j\in I\setminus\{i\}})$.*
(iv) *For each $J \subseteq I$ and each $i \in I \setminus J$, we have $G_JG_i = \bigcap_{j\in J}(G_jG_i)$.*

If one (whence all) of these properties hold, then Γ is a geometry.

Proof The last statement follows directly from the fact that in the flag-transitive case every flag can be transformed by an automorphism of Γ to a subset of the standard chamber $\{G_i \mid i \in I\}$.

(i) $\Rightarrow$ (ii) Let J be a subset of I and let aG_i be an element of the residue $\Gamma_{\{G_j \mid j\in J\}}$. The set $\{aG_i\} \cup \{G_j \mid j \in J\}$ is a flag of Γ, so by (i) there is $g \in G$ with $g^{-1}a \in G_i$ and $g \in G_J$, whence $a \in G_JG_i$. Taking $a_1 \in G_J$ such that $a \in a_1G_i$, we obtain $aG_i = \phi_J(a_1G_{J\cup\{i\}})$. Therefore, ϕ_J is surjective for all $J \subseteq I$. It follows from Lemma 1.8.9(v) that ϕ_J is an isomorphism.

(iii) $\Rightarrow$ (i) Let $J \subseteq I$ and let $X = \{x_jG_j \mid j \in J\}$ be a flag of type J. We want to establish that X is in the G-orbit of the standard flag $\{G_j \mid j \in J\}$. For $|J| \le 2$, this is true thanks to Lemma 1.8.6, and for $|J| = 3$, there is nothing to show, so we may assume $|J| \ge 4$. Fix $i \in J$. After applying x_i^{-1} to X, we may assume $x_i = 1$. For each $l \in J \setminus \{i\}$ we can replace x_l by a coset representative $x'_l \in G_i$ with $x_l G_l = x'_l G_l$. Let $j, k \in J \setminus \{i\}$. Since X is a flag, we have $x'^{-1}_k x'_j \in G_k G_j \cap G_i$. By transitivity of G on the set of flags of type $\{i, j, k\}$, this implies $x'^{-1}_k x'_j \in G_{\{i,k\}} G_{\{i,j\}}$ (cf. Lemma 1.8.9(ii)). Thus $x'_j G_{\{i,j\}}$ and $x'_k G_{\{i,k\}}$ are incident in $\Gamma(G_i, G_{\{i,l\}})_{l \in J \setminus \{i\}}$ and belong to its chamber $\{x'_l G_{\{i,l\}} \mid l \in J \setminus \{i\}\}$. By the assumption that G_i is flag transitive on the geometry $\Gamma(G_i, G_{\{i,l\}})_{l \in I \setminus \{i\}}$, we can find $h \in G_i$ with $x'_l G_{\{i,l\}} = h G_{\{i,l\}}$ for all $l \in J \setminus \{i\}$. Consequently,

$$X = \{G_i\} \cup \{x_l G_l \mid l \in J \setminus \{i\}\} = h\{G_l \mid l \in J\}$$

is in the same G_i-orbit as the standard flag, and we are done.

(ii) $\Rightarrow$ (iii) We proceed by induction on the rank $|I|$ of Γ. Clearly, the implication is trivial if $|I| \leq 2$, so assume $|I| \geq 3$.

Let $\{i, j, k\} \subseteq I$ be of size three. Observe that $\phi_{\{i,j\}}$ is surjective as it is an isomorphism, so that, by Lemma 1.8.9(iv), we have $(G_i G_k) \cap (G_j G_k) = G_{\{i,j\}} G_k$. According to Lemma 1.8.9(ii), this implies that G is transitive on the set of flags of type $\{i, j, k\}$.

Fix $i \in I$. We will consider the action of G_i on $\Gamma(G_i, (G_{\{i,j\}})_{j \in I \setminus \{i\}})$. The homomorphisms ϕ_J ($J \subseteq I \setminus \{i\}$) for the latter geometry are easily seen to be surjective: by Lemma 1.8.9(iv) it suffices to verify that, for each non-empty $J \subseteq I \setminus \{i\}$ and $k \in I \setminus (J \cup \{i\})$, the equality

$$\bigcap_{j \in J} (G_{\{j,i\}} G_{\{k,i\}}) = G_{J \cup \{i\}} G_{\{k,i\}}$$

holds. But this follows from

$$\bigcap_{j \in J} (G_{\{j,i\}} G_{\{k,i\}}) \subseteq \left(\bigcap_{j \in J \cup \{i\}} (G_j G_k) \right) \cap G_i = (G_{J \cup \{i\}} G_k) \cap G_i = G_{J \cup \{i\}} G_{\{k,i\}}$$

where the first equality is due to surjectivity of $\phi_{J \cup \{i\}}$.

By Lemma 1.8.9(v), the incidence system $\Gamma(G_i, (G_{\{i,j\}})_{j \in I \setminus \{i\}})$ satisfies (ii). Therefore, the induction hypothesis may be applied to it, giving that (iii) holds for $\Gamma(G_i, (G_{\{i,j\}})_{j \in I \setminus \{i\}})$. But then, by the part '(iii) $\Rightarrow$ (i)' already proven, G_i is flag transitive on this geometry. This establishes (iii).

(ii) $\Leftrightarrow$ (iv). This is straightforward from Lemma 1.8.9(iv) and (v). $\square$

Example 1.8.11 Let $G = \mathrm{Sym}_4$. Clearly, G cannot act flag transitively on the cube geometry, as the latter has 48 chambers and $|G| = 24$. But we can also derive this fact from Theorem 1.8.10 by viewing the cube as the coset geometry $\Gamma(G, (G_1, G_2, G_3))$ where $G_1 = \langle (1,2,3) \rangle$, $G_2 = \langle (1,4) \rangle$, and $G_3 = \langle (1,3,2,4) \rangle$ (see Example 1.8.5). Now $G_2 \cap G_1 G_3$ contains $(1,4)$, while $G_1 \cap G_2 = G_2 \cap G_3 = \{1\}$, so $G_1 G_3 \cap G_2 \neq (G_1 \cap G_2)(G_3 \cap G_2)$, which by Exercise 1.9.24, implies $G_1 G_2 \cap G_3 G_2 \neq (G_1 \cap G_3) G_2$. Alternatively, $G_1 G_3 \cap G_2 G_3 = G_3 \cup (1,4) G_3$. In both ways, it follows from Theorem 1.8.10 that G does not act flag transitively on the geometry.

Proposition 1.8.12 *Let $\Gamma = \Gamma(G, (G_i)_{i\in I})$ be a coset incidence system over I. The following two statements are equivalent.*

(i) *For each $J \subseteq I$ with $|I \setminus J| \geq 2$ we have $G_J = \langle G_{J\cup\{i\}} \mid i \in I \setminus J \rangle$.*
(ii) *For each $J \subseteq I$ and distinct $i, k \in I \setminus J$ we have $G_J = \langle G_{J\cup\{i\}}, G_{J\cup\{k\}} \rangle$.*

Proof Since (ii) $\Rightarrow$ (i) is obvious, we assume (i) and derive (ii). Let $J \subseteq I$ and $i, k \in I \setminus J$ be distinct. If $I = J \cup \{i, k\}$, then there is nothing to show, so assume that there is $r \in I \setminus (J \cup \{i, k\})$. Assertion (i) implies $G_{J\cup\{r\}} = \langle G_{J\cup\{r,l\}} \mid l \in (I \setminus J \cup \{r\}) \rangle$. Induction on the rank $|I|$ of Γ applied to the coset incidence system $\Gamma(G_{J\cup\{r\}}, (G_{J\cup\{r,l\}})_{l\in I\setminus(J\cup\{r\})})$ of rank $|I \setminus J| + 1 \leq |I| - 1$ gives $G_{J\cup\{r\}} = \langle G_{J\cup\{r,i\}}, G_{J\cup\{r,k\}} \rangle$. But then $G_{J\cup\{r\}} \subseteq \langle G_{J\cup\{i\}}, G_{J\cup\{k\}} \rangle$ for all $r \in I \setminus J$, whence

$$G_J = \langle G_{J\cup\{l\}} \mid l \in I \setminus J \rangle \subseteq \langle G_{J\cup\{i\}}, G_{J\cup\{k\}} \rangle. \qquad \square$$

A group-theoretic description of residual connectedness is easily derived from Proposition 1.8.12.

Corollary 1.8.13 *Suppose that I is finite and that $\Gamma = \Gamma(G, (G_i)_{i\in I})$ is a geometry over I on which G acts flag transitively. The following three statements are equivalent.*

(i) *Γ is a residually connected geometry.*
(ii) *$G_J = \langle G_{J\cup\{i\}} \mid i \in I \setminus J \rangle$ for each $J \subseteq I$ with $|I \setminus J| \geq 2$.*
(iii) *If $J \subseteq I$ and $i, k \in I \setminus J$ with $i \neq k$, then $G_J = \langle G_{J\cup\{i\}}, G_{J\cup\{k\}} \rangle$.*

Proof By Theorem 1.8.10, every flag-transitive coset incidence system is a geometry.

By Lemma 1.8.9(i), Γ is connected if and only if $G = \langle G_i \mid i \in I \rangle$. Application of this criterion to Γ_J for each subset J of I with $|I \setminus J| \geq 2$, which, by Theorem 1.8.10(ii) is isomorphic to $\Gamma(G_J, (G_{J\cup\{i\}})_{i\in I\setminus J})$, yields the equivalence of (i) and (ii). The equivalence of (ii) and (iii) is straightforward from Proposition 1.8.12. $\qquad \square$

Another consequence of Proposition 1.8.12 is the following set of sufficient conditions for an incidence system to be a flag-transitive geometry.

Corollary 1.8.14 *Suppose that $\Gamma = \Gamma(G, (G_i)_{i\in I})$ is a coset incidence system over the finite set I satisfying the following four conditions for all $J \subseteq I$ and all subsets $\{i, j, k\}$ of I of size three.*

(i) *$G_J = \langle G_{J\cup\{r\}} \mid r \in I \setminus J \rangle$.*
(ii) *G_i is flag transitive on $\Gamma(G_i, (G_{\{i,r\}})_{r\in I\setminus\{i\}})$.*
(iii) *$G \neq G_j G_k$ (i.e., G_j is not transitive on G/G_k).*
(iv) *$G_{\{i,j\}}$ has at most two orbits on the coset space $G_i / G_{\{i,k\}}$.*

Then G is flag transitive on Γ.

Proof Let $i, j, k \in I$ be mutually distinct. Due to (iv) there is $x \in G_i$ such that

$$G_i = G_{\{i,j\}}G_{\{i,k\}} \cup G_{\{i,j\}}xG_{\{i,k\}},$$

so that

$$G_iG_k = G_{\{i,j\}}G_k \cup G_{\{i,j\}}xG_k.$$

Therefore $G_iG_k \cap G_jG_k$ coincides with either $G_{\{i,j\}}G_k$ or G_iG_k. In the latter case, we have $G_iG_k = G_iG_k \cap G_jG_k \subseteq G_jG_k$, so $\langle G_i, G_j \rangle G_k = G_jG_k$. By (i) and Proposition 1.8.12, $\langle G_i, G_j \rangle = G$, so that $G = GG_k = \langle G_i, G_j \rangle G_k = G_jG_k$, contradicting (iii). Therefore, we must have

$$G_iG_k \cap G_jG_k = G_{\{i,j\}}G_k.$$

This is equivalent to $G_kG_i \cap G_kG_j = G_kG_{\{i,j\}}$, as can be seen by inversion of the elements in the sets at both sides. Hence, by Lemma 1.8.9(ii), G is transitive on the set of all flags of type $\{i, j, k\}$ and we can finish by (ii) and Theorem 1.8.10(iii). $\square$

Here is yet one more translation of an incidence system property to group theory.

Corollary 1.8.15 *Suppose that $\Gamma = \Gamma(G, (G_i)_{i \in I})$ is a coset incidence system over the finite set I on which G acts flag transitively. Then Γ is a firm geometry if and only if $G_{I \setminus \{j\}} \neq G_I$ for each $j \in I$.*

Proof By Theorem 1.8.10, Γ is a geometry. By flag transitivity, we only need show that, for each $j \in I$, the cardinality of the residue $\Gamma_{\{G_i \mid i \in I \setminus \{j\}\}}$ is at least two. Also by Theorem 1.8.10, such a residue is isomorphic to $\Gamma(G_{I \setminus \{j\}}, (G_I))$, and so has precisely $|G_{I \setminus \{j\}}/G_I|$ elements. $\square$

The above analysis of the coset incidence system $\Gamma(G, (G_i)_{i \in I})$ gives that it is an I-geometry if and only if

(1) $G_{I \setminus \{j\}} \neq G_I$ for each $j \in I$;
(2) $G_JG_i = \bigcap_{j \in J}(G_jG_i)$ for each $J \subseteq I$ and each $i \in I \setminus J$;
(3) $G_J = \langle G_{J \cup \{i\}} \mid i \in I \setminus J \rangle$ for each $J \subseteq I$ with $|I \setminus J| \geq 2$.

Example 1.8.16 We exhibit a group acting flag transitively on the projective geometries $\mathrm{PG}(V)$ of Example 1.4.9 for $V = \mathbb{D}^{n+1}$ where $\mathbb{D}$ is a division ring and $n \in \mathbb{N}$. We view V as a right vector space of column vectors over $\mathbb{D}$. We make use of the *general linear group* $\mathrm{GL}(V)$, consisting of all invertible linear transformations of V (introduced in Remark 1.4.8). With respect to the standard basis $\varepsilon_1, \ldots, \varepsilon_{n+1}$ of V, this group can be viewed as the set of invertible $(n+1) \times (n+1)$-matrices with entries in $\mathbb{D}$ and supplied with the usual matrix multiplication. The group $\mathrm{GL}(V)$ acts on V from the left by means of the usual multiplication of a matrix and a column vector.

It is well known from linear algebra that $\mathrm{GL}(V)$ coincides with $\mathrm{Aut}(V)$. As it maps subspaces of V to subspaces of the same dimension and preserves incidence, it induces an action on $PG(V)$. In general, this action is not faithful, as scalar matrices λI_{n+1} with $\lambda \in Z(\mathbb{D})$, the *center* of $\mathbb{D}$, induce the identity on the set of elements of $PG(V)$.

For $i \in [n]$, we define the subgroup G_i of $G = GL(V)$ as the stabilizer of the subspace V_i of $\mathbb{D}^{n+1}$ generated by all i first standard basis vectors $\varepsilon_1, \ldots, \varepsilon_i$ of V. This implies that G_i consists of all matrices in $GL(V)$ of the form

$$\begin{pmatrix} A_{i,i} & B_{i,n+1-i} \\ 0 & C_{n+1-i,n+1-i} \end{pmatrix},$$

where $X_{k,l}$ for X one of A, B, C, denotes a $k \times l$-matrix with entries in $\mathbb{D}$. We will establish that the map $\beta : \Gamma(G, (G_i)_{i \in [n]}) \to PG(V)$ given by $\beta(aG_i) = aV_i$ for $a \in G$ is an isomorphism of incidence systems over $[n]$. The set $c = \{V_1, V_2, \ldots, V_n\}$ is a flag of $PG(V)$, stabilized by $B = G_{[n]}$. Suppose $\{W_1, W_2, \ldots, W_t\}$ is a flag of $PG(V)$ with $\dim(W_i) < \dim(W_j)$ whenever $i < j$. Beginning with a basis of W_1 and extending it to a basis of the next W_i as we go along, we can find a basis $a_1, a_2, \ldots, a_{n+1}$ of V such that $a_1, a_2, \ldots, a_{\dim(W_i)}$ is a basis of W_i for each $i \in [t]$. The matrix m whose j-th column is a_j belongs to G and satisfies $m(V_{\dim W_i}) = W_i$ for each i, so $\{W_1, W_2, \ldots, W_t\}$ is the image under m of the subflag of c of type $\{\dim(W_1), \dim(W_2), \ldots, \dim(W_t)\}$. This establishes that G acts flag transitively on $PG(V)$. By Proposition 1.8.7, $\Gamma(G, (G_i)_{i \in [n]})$ is isomorphic to $PG(V)$, and the corresponding representations of G are equivalent.

If $\mathbb{D} = \mathbb{F}$ is a field, the determinant $\det : GL(V) \to \mathbb{F} \setminus \{0\}$ is a homomorphism of groups. Its kernel is the *special linear group* $SL(V)$. A slight adaptation to the above argument shows that $SL(V)$ also acts flag transitively on $PG(V)$.

The image of $GL(V)$ in $\mathrm{Aut}(PG(V))$ is denoted $PGL(V)$, and the image of $SL(V)$ in $\mathrm{Aut}(PG(V))$ is denoted $PSL(V)$.

Example 1.8.17 Adding translations to $GL(V)$, we obtain a group acting flag transitively on the affine geometries $AG(V)$ of Example 1.4.10 for $V = \mathbb{D}^n$ where $\mathbb{D}$ is a division ring and $n \in \mathbb{N}$, $n \geq 1$. Let $T(V)$ be the group of all *translations* of V, that is, the set of maps $t_a : V \to V$ for $a \in V$, determined by $t_a(x) = x + a$. The group $T(V)$ is isomorphic to the additive group underlying V by means of the group homomorphism $a \mapsto t_a$. It acts on $AG(V)$ and is transitive on V. Besides, it is normalized by $GL(V)$, so the two subgroups of $\mathrm{Aut}(AG(V))$ generate a semi-direct product. If $(c_i)_{i \in [n]}$ and $(d_i)_{i \in [n]}$ are two chambers of $AG(V)$, then the translation $t_{d_1 - c_1}$ of $T(V)$ will map c_1 to d_1, so, for proving that these chambers are in the same orbit it suffices to consider the case where $c_1 = d_1 = 0$. Subsequently, by Example 1.8.16, a suitable element of $GL(V)$ will map $(c_i)_{2 \leq i \leq n}$ to $(d_i)_{2 \leq i \leq n}$. Therefore, the semi-direct product of $T(V)$ and $GL(V)$ acts flag transitively on $AG(V)$.

1.9 Exercises

Section 1.1

Exercise 1.9.1 Let $n \in \mathbb{N}$, $n \geq 2$. The generalization to higher dimensions of the polygon in the Euclidean plane and the polyhedron in Euclidean space is the *polytope* in n-dimensional Euclidean affine space $\mathbb{E}^n$.

(a) Give a definition of polytope extending Example 1.1.3 and leading to a rank n geometry (in the intuitive sense; a formal definition is still lacking at this point).

(b) Extend the construction of Example 1.1.1 to the *hypercube* in the Euclidean vector space $\mathbb{R}^n$ whose vertex set consists of all points of the form $\pm\varepsilon_1 \pm \varepsilon_2 \pm \cdots \pm \varepsilon_n$.

Section 1.2

Exercise 1.9.2 The hexagon, viewed as a graph with six vertices and six edges, is bipartite.

(a) Show that the corresponding bipartition is unique.

(b) The hexagon is also tripartite, that is, its vertices can be partitioned into three subsets of size two such that no vertices from the same subset are adjacent. Verify that such a tripartition is not unique.

Exercise 1.9.3 (This exercise is used in Remark 1.2.4) Let $\Gamma = (X, *, \tau)$ be a geometry over I. Extend X to the disjoint union Y of X and a singleton $\{\infty\}$. Extend $*$ to the symmetric relation $\circ$ on $Y \times Y$ such that $\infty \circ y$ for $y \in Y$ if and only if $y = \infty$. Finally, pick $i \in I$ and extend τ to a map $\sigma : Y \to I$ by setting $\tau(\infty) = i$. Prove that if $|I| > 1$, then $(Y, \circ, \sigma)$ is an incidence system in which the type map is surjective that is not a geometry.

Section 1.3

Exercise 1.9.4 If A is a group of automorphisms of the geometry $\Gamma = (X, *, \tau)$ over I acting transitively on each $\tau^{-1}(i)$ ($i \in I$), then Γ/A is the unit geometry (cf. Example 1.3.11). True or false?

Exercise 1.9.5 Consider the graph Δ whose vertices are $\pm\{i, j\}$ for $i, j \in [5]$ with $i \neq j$, and in which, for $\varepsilon, \delta \in \{\pm\}$, there is an edge on $\varepsilon\{i, j\}$ and $\delta\{k, l\}$ if and only if $\{i, j\} \cap \{k, l\} = \emptyset$ and $\varepsilon\delta = -1$. Show that the map $\pm\{i, j\} \mapsto \{i, j\}$ yields a homomorphism from Δ to the Petersen graph. Verify that Δ is not isomorphic to the dodecahedral graph on 20 vertices described in Example 1.3.3.

Fig. 1.21 The Fano plane

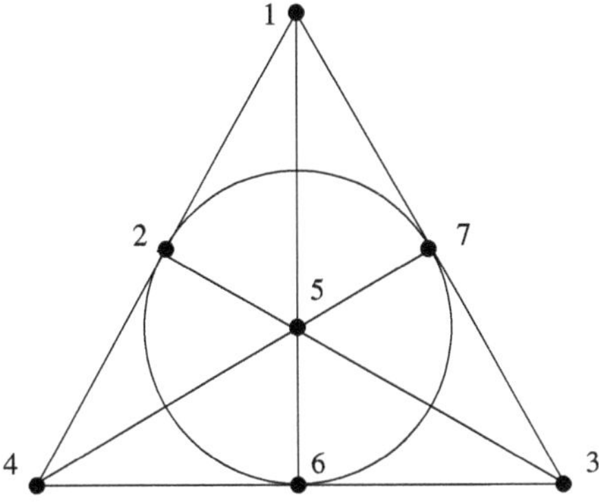

Section 1.4

Exercise 1.9.6 (This exercise is used in Remark 1.4.2) Let $\Gamma = (X, *, \tau)$ and $\Gamma' = (X', *', \tau')$ be incidence systems over I.

(a) Suppose that $\alpha : \Gamma' \to \Gamma$ is an injective homomorphism. Show that Γ' is isomorphic to a partial subsystem of Γ on $\alpha(X')$.
(b) Suppose that Γ' is a partial subsystem of Γ. Show that the identity map on X' is an injective homomorphism $\Gamma' \to \Gamma$.

Exercise 1.9.7 The rank two geometry Γ depicted in Fig. 1.21 is the projective plane of order 2, otherwise known as the *Fano plane*. Its point set is [7], and its lines are the sets $\{i, i + 1, i + 3\}$ (the numbers are interpreted modulo 7 with values in $\lfloor 7 \rfloor$).

(a) Describe a polarity of Γ.
(b) Determine the orders of $\mathrm{Aut}(\Gamma)$ and $\mathrm{Cor}(\Gamma)$.
   (*Hint:* Verify that $(1, 2, 3, 4, 5, 6, 7)$ is an automorphism, proving that $\mathrm{Aut}(\Gamma)$ is transitive on [7]. Use the picture to identify an automorphism that fixes the three points on a line.)
(c) Show that Γ is the projective geometry $\mathrm{PG}(\mathbb{F}_2^3)$ over the field $\mathbb{F}_2$ of order two as in Example 1.4.9.

Exercise 1.9.8 The choice of field is important to the existence of elements of a given type in the absolute geometry of Example 1.4.13. Let $\mathbb{F}$ be a field and consider the vector space $V = \mathbb{F}^n$ over $\mathbb{F}$, with standard basis $\varepsilon_1, \ldots, \varepsilon_n$. We define the bilinear form f on V by $f(x, y) = \sum_{i=1}^{n} x_i y_i$ where $x = \sum_i x_i \varepsilon_i$, $y = \sum_i y_i \varepsilon_i$.

(a) Show that the absolute geometry Γ with respect to f is empty if $\mathbb{F} = \mathbb{R}$.
(b) Show that Γ contains elements of each type in $[\lfloor n/2 \rfloor]$ if $\mathbb{F} = \mathbb{C}$. Here, for any integer m, we write $\lfloor m \rfloor$ to denote the largest integer less than or equal to m.
(c) Let $n = 2$ and $\mathbb{F} = \mathbb{F}_p$, the finite field of order p, for some odd prime p. Show that Γ contains elements of type 1 if and only if $p \equiv 1 \pmod 4$.

Exercise 1.9.9 Consider the geometry Γ over $I = [n-1]$ on the collection X of proper subsets of $[n]$ in which incidence is symmetrized inclusion, and the type of an element is its cardinality. Prove the following statements involving a permutation π of $[n]$.

(a) The map $\delta : X \to X$ given by $\delta(Y) = [n] \setminus \pi(Y)$ is a duality of Γ. It is a polarity if and only if π has order at most two.
(b) An element Y of Γ belongs to a flag fixed by δ if and only if $\pi^2(Y) = Y$ and $\pi(Y) \cap Y = \emptyset$.
(c) If $n = 2k$ and π is an *involution* (i.e., a bijection of order two) without fixed points, then the absolute of Γ with respect to $\langle \delta \rangle$ is a geometry over $[k]$.

Exercise 1.9.10 Let V be a vector space of infinite dimension. Show that the construction of the projective geometry $\mathrm{PG}(V)$ of Example 1.4.9 fails. Verify that, with symmetrized inclusion for incidence, a geometry over the positive integers can be constructed on the set of nontrivial finite-dimensional subspaces of V.

Exercise 1.9.11 Suppose that $\mathbb{D}$ is a division ring. Consider its *opposite* $\mathbb{D}^{\mathrm{op}}$, that is, the same set $\mathbb{D}$, but now supplied with the multiplication $\circ$ given by $x \circ y = yx$ for $x, y \in \mathbb{D}$.

(a) Prove that $\mathbb{D}^{\mathrm{op}}$ is also a division ring.
(b) Let $\mathbb{D} = \mathbb{H}$, the division ring of *rational quaternions* $\mathbb{Q} + \mathbb{Q}i + \mathbb{Q}j + \mathbb{Q}k$ with multiplication determined by $ij = -ji = k$ and $i^2 = j^2 = k^2 = -1$. Show that $\mathbb{D}$ is indeed a division ring.
(*Hint:* Define the norm map $N : \mathbb{D} \to \mathbb{Q}$ by $N(x_1 + x_2 i + x_3 j + x_4 k) = x_1^2 + x_2^2 + x_3^2 + x_4^2$ if $x_i \in \mathbb{Q}$ and use the identities $N(xy) = N(x)N(y)$ and $(x_1 - x_2 i - x_3 j - x_4 k)(x_1 + x_2 i + x_3 j + x_4 k) = N(x)$.)
(c) Prove that $\mathbb{H}$ is isomorphic to its opposite. Such an isomorphism is called an *anti-automorphism* of $\mathbb{H}$.
(*Hint:* Consider the homomorphism $\alpha : \mathbb{H} \to \mathbb{H}^{\mathrm{op}}$ determined by $\alpha(i) = j$, $\alpha(j) = i$, $\alpha(k) = k$.)
(d) Show that a left vector space V over $\mathbb{D}^{\mathrm{op}}$ is a right vector space over $\mathbb{D}$ via $v\lambda = \lambda v$ $(\lambda \in \mathbb{D}, v \in V)$.

There exist non-commutative division rings without anti-automorphisms (see Example 7.2.2) and division rings with anti-automorphisms but without anti-automorphisms of order two (see Example 7.2.16).

Exercise 1.9.12 Let $\mathbb{K}$ be a field admitting an automorphism σ of order three.

(a) Verify that the set $\mathbb{F}$ of elements of $\mathbb{K}$ fixed by σ is a subfield of $\mathbb{K}$.
(b) Show that there is a natural way to interpret $\mathbb{K}$ as a vector space over $\mathbb{F}$.
(c) Prove that the dimension of $\mathbb{K}$ over $\mathbb{F}$ is three.

Exercise 1.9.13 Consider a division ring $\mathbb{D}$ and its *opposite* $\mathbb{D}^{\mathrm{op}}$ in which $a \circ b = c$ if and only if $ba = c$ in $\mathbb{D}$ (cf. Exercise 1.9.11). Here, we denote multiplication in

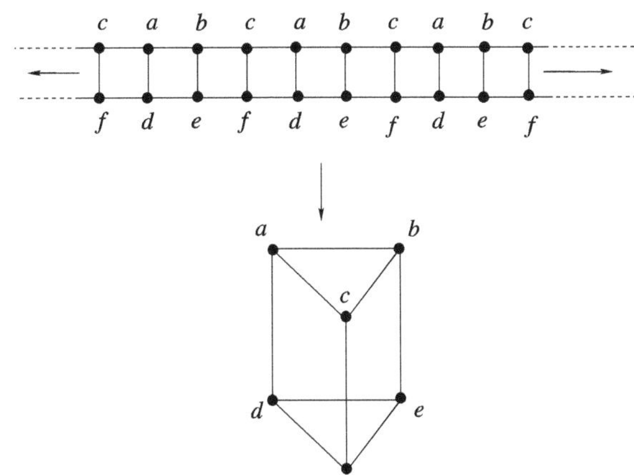

Fig. 1.22 An infinite covering of graphs. It can also be viewed as a covering of geometries of rank two, whose points are the vertices and whose lines are the edges of the corresponding graphs

$\mathbb{D}^{\mathrm{op}}$ by $\circ$ to prevent confusion with multiplication in $\mathbb{D}$. (Since $\mathbb{D}$ and $\mathbb{D}^{\mathrm{op}}$ have the same underlying set, it is easy to mix up the two multiplications.) Let V be a finite-dimensional right vector space over $\mathbb{D}$. The *dual vector space* $V^{\vee}$ consists of all $\mathbb{D}$-linear forms on V. It is a right vector space over $\mathbb{D}^{\mathrm{op}}$ by means of the scalar multiplication

$$(\phi\lambda)x = \lambda(\phi x) \quad \left(x \in V, \phi \in V^{\vee}, \lambda \in \mathbb{D}\right).$$

(a) Show that $\mathrm{PG}(V^{\vee})$ is weakly isomorphic to $\mathrm{PG}(V)$.
(b) Prove that if $\mathbb{D}$ is isomorphic to $\mathbb{D}^{\mathrm{op}}$, then $\mathrm{PG}(V^{\vee})$ is isomorphic to $\mathrm{PG}(V)$.

Section 1.5

Exercise 1.9.14 A homomorphism $\phi : \Gamma \to \Gamma'$ of incidence systems is called a *covering* if, for every element x of Γ, the restriction of ϕ to the residue Γ_x is an isomorphism onto $\Gamma'_{\phi(x)}$.

(a) Verify that the homomorphism suggested by Fig. 1.22 is a covering.
(b) Find a connected geometry on 16 vertices and 24 edges covering the rank two geometry on the vertices and edges of the cube.
(c) Prove that the composition of two coverings is again a covering.

Exercise 1.9.15 Let Γ be a geometry over I, let F be a flag, and J a subset of $I \setminus \tau(F)$. Prove $_J\Gamma_F = {_{J \cup \tau(F)}}\Gamma_F$.

Exercise 1.9.16 Let Γ be a geometry and A a subgroup of $\mathrm{Aut}(\Gamma)$ each nontrivial element of which maps every element of Γ to an element at distance at least four from it in the incidence graph of Γ. Show that the quotient map $\Gamma \to \Gamma/A$ is a covering (see Exercise 1.9.14).

Section 1.6

Exercise 1.9.17 Let I be finite and let Γ be a residually connected incidence system over I with type map τ. Prove that, if τ is surjective, then Γ is a geometry over I.

Exercise 1.9.18 Set $I = \{0, 0'\} \cup \{1/n \mid n \in \mathbb{N} \setminus \{0\}\}$. For each $i \in I \setminus \{0'\}$, denote by X_i the set of all closed intervals of length i in $\mathbb{R}$. Thus, X_0 is the set of singletons $\{a\}$ where $a \in \mathbb{R}$. Let $X_{0'} = \mathbb{R}$. It is a disjoint copy of X_0 with $a \in X_{0'}$ corresponding to $\{a\}$ in X_0. For every element $x \in \bigcup_{i \in I} X_i$, write $\phi(x)$ for the subset of $\mathbb{R}$ corresponding to x. So ϕ is the identity, except on $X_{0'}$, where, for $x \in X_{0'}$, we have $\phi(x) = \{x\}$. Let Γ be the incidence system determined by the multipartite graph with parts X_i $(i \in I)$, and incidence $*$ given by

$$x * y \quad \Leftrightarrow \quad \phi(x) \cap \phi(y) \neq \emptyset$$

for every $x \in X_i$, $y \in X_j$ $(i \neq j)$. Prove that Γ is a residually connected geometry over I for which the conclusion of Lemma 1.6.3 fails.

(*Hint:* To prove that Γ is a geometry, use the fact that the intersection of a collection of pairwise meeting closed intervals is a non-empty closed interval. To contradict the conclusion of the lemma, consider $\{0, 0'\}$-chains.)

Exercise 1.9.19 Let Γ be the incidence system defined in Exercise 1.9.18. Consider its subsystem Γ' obtained by removing 0 from $X_{0'}$.

(a) Verify that Γ' is residually connected but not a geometry.
(b) Show that the flag $F = \{[0, 1/n] \mid n \in \mathbb{N}\} \cup \{\{0\}\}$ of Γ' is maximal but not a chamber. Conclude that Lemma 1.6.4 does not hold if the condition that every flag of corank one be nonmaximal is removed from the hypotheses.

Exercise 1.9.20 (The Principle of Maximal Intersection) Consider a set P of points together with a family $\mathcal{B}$ of subsets, and a map $\tau : \mathcal{B} \to I$. We build an incidence system $\Gamma(P, \mathcal{B}, \tau) = (\mathcal{B}, *, \tau)$ over I. For blocks B_1 and B_2 with $\tau(B_1) \neq \tau(B_2)$, we let $B_1 * B_2$ if and only if $B_1 \cap B_2$ is maximal among all sets of the form $X \cap Y$ for $\tau(X) = \tau(B_1)$ and $\tau(Y) = \tau(B_2)$.

(a) Show that $\Gamma(P, \mathcal{B}, \tau)$ need not be a geometry.
(b) Consider the truncated octahedron depicted in Fig. 1.23. Let P be the set of its vertices, let $\mathcal{B}_1$ be the family of (six) sets of faces with four vertices, let $\mathcal{B}_2$ the family of (twelve) edges not contained in an element of $\mathcal{B}_1$, and let $\mathcal{B}_3$ be the family of (eight) faces with six vertices. Here, the faces and edges are viewed as vertex sets. Verify that $\Gamma(P, \mathcal{B}, \tau)$, where $\mathcal{B} = \mathcal{B}_1 \cup \mathcal{B}_2 \cup \mathcal{B}_3$ and $\tau : \mathcal{B} \to I$ takes value i on $\mathcal{B}_i$, is the cube geometry of rank three.

Exercise 1.9.21 Let Γ be the Fano plane (cf. Exercise 1.9.7). Show that there is a rank two geometry Γ' that is a tree (i.e., is connected and has no circuits) and admits a covering $\phi : \Gamma' \to \Gamma$ (see Exercise 1.9.14).

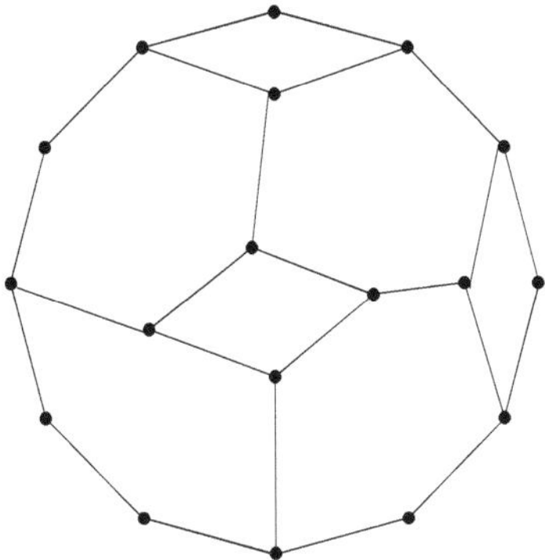

Fig. 1.23 A truncated octahedron (only those points and lines visible from the front are drawn)

Section 1.7

Exercise 1.9.22 (This exercise is used in Definition 4.2.5) Let H be a subgroup of a group G, and compare the representation α of G over H (cf. Definition 1.7.2) with the 'right' version $\beta : G \to \mathrm{Sym}(H \backslash\!\backslash G)$ defined by $\beta(g)Hk = Hkg^{-1}$ $(g, k \in G)$. Here, we write $H \backslash\!\backslash G = \{Hg \mid g \in G\}$ rather than $H \backslash G$ to avoid confusion with set exclusion. Show that α and β are equivalent.

Exercise 1.9.23 Prove that the icosahedron, viewed as a graph, has an edge-transitive group of automorphisms, and determine its distribution diagram.

Exercise 1.9.24 Let G_1, G_2, G_3 be subgroups of a group G. Prove that $G_2 G_1 \cap G_3 G_1 = (G_2 \cap G_3)G_1$ holds if and only if $(G_2 \cap G_1)(G_3 \cap G_1) = (G_2 G_3) \cap G_1$.

Section 1.8

Exercise 1.9.25 Consider the dihedral group G of order 12, generated by two involutions r_1 and r_2, so that $r_1^2 = r_2^2 = (r_1 r_2)^6 = 1$. Show that the incidence system $\Gamma(G, (G_1, G_2))$ where $G_1 = \langle r_1 \rangle$ and $G_2 = \langle r_2 \rangle$, is the hexagon geometry over [2], having six elements of each type.

Exercise 1.9.26 Consider $G = \mathrm{Sym}_4$ together with the subgroups $G_1 = \langle (1, 2, 3) \rangle$ and $G_3 = \langle (1, 3, 2, 4) \rangle$.

(a) Let $G_1 = \langle (1, 4) \rangle$. Show that the coset incidence system $\Gamma(G, (G_1, G_2, G_3))$ is the cube geometry.
(b) Determine, for each of the three other subgroup choices in Example 1.7.8, viz. $G_2 = \langle (1, 2) \rangle$, $\langle (1, 3) \rangle$, $\langle (3, 4) \rangle$, whether the coset incidence system $\Gamma(G, (G_1, G_2, G_3))$ is a geometry and whether it is flag transitive or not.

Exercise 1.9.27 Let $G = \mathrm{Sym}_4$. Consider the subgroups $G_1 = \langle(1,2,3)\rangle$, $G_2 = \langle(1,4)\rangle$, $G_3 = \langle(1,3,2,4)\rangle$ and $G_4 = \langle(2,3)\rangle$. Show that the incidence system $\Delta = \Gamma(G, (G_i)_{i\in[4]})$ over $[4]$ is not a geometry.

(*Hint:* Consider the flag $F = \{G_1, G_2, (1,4)G_3\}$. The only coset in G/G_4 incident with both G_1 and G_2 in Δ is G_4. But G_4 is not incident with $(1,4)G_3$, so the flag F is not contained in a chamber of Δ.)

Exercise 1.9.28 Let $\mathbb{D}$ be a division ring and V a vector space over $\mathbb{D}$ of dimension at least three.

(a) Show that $\mathrm{Cor}(\mathrm{PG}(V))/\mathrm{Aut}(\mathrm{PG}(V))$ has order two if and only if $\mathbb{D}$ has an anti-automorphism (cf. Exercise 1.9.11).
(*Hint:* Use Exercise 1.9.13.)
(b) Show that $\mathrm{Aut}(\mathrm{PG}(\mathbb{F}_2^3))$ is of order 168 (use Fig. 2.9) and isomorphic to $\mathrm{SL}(\mathbb{F}_2^3)$.
(c) Exhibit a *polarity* π of $\mathrm{PG}(\mathbb{F}_2^3)$ and find the absolute points of π (cf. Definition 1.4.3).
(d) Prove that the group $\mathrm{SL}(\mathbb{F}_2^3)$ is simple.
(*Hint:* Show that every normal subgroup is generated by the Sylow 7-subgroups of $\mathrm{SL}(\mathbb{F}_2^3)$.)

Exercise 1.9.29 Let G be the Frobenius group of order 21. This means that G has a normal subgroup $\langle c\rangle$ of order seven and a subgroup $\langle a\rangle$ of order three with $aca^{-1} = c^2$. Then $d := ca = c^{-1}ac$ is also an element of order three. Consider the coset incidence system $\Gamma = \Gamma(G, (\langle a\rangle, \langle d\rangle))$ over $[2]$ (so $\langle a\rangle$ is an element of type 1).

(a) Prove that Γ is isomorphic to $\mathrm{PG}(\mathbb{F}_2^3)$.
(b) Conclude that G acts flag transitively on $\mathrm{PG}(\mathbb{F}_2^3)$.

Exercise 1.9.30 (Cited in Proposition 4.4.4) Let $\phi : V \to \mathbb{D}$ be a nonzero linear form on the right vector space V of finite dimension at least two over the division ring $\mathbb{D}$ and let $a \in V \setminus \{0\}$.

(a) Prove that the map $r_{a,\phi} : V \to V$ defined by $r_{a,\phi}(v) = v - a\phi(v)$ for $v \in V$ belongs to $\mathrm{GL}(V)$ if and only if $\phi(a) \neq 1$, and compute its inverse.
(b) Show that, in its action on $\mathrm{PG}(V)$, the transformation $r_{a,\phi}$ with $\phi(a) \neq 1$ fixes the hyperplane $\phi^{-1}(0)$ of $\mathrm{PG}(V)$ point-wise as well as the point $a\mathbb{D}$ and all lines on $a\mathbb{D}$ of $\mathrm{PG}(V)$. An automorphism of $\mathrm{PG}(V)$ with these properties is called a *perspectivity* with *center* $a\mathbb{D}$ and *axis* $\phi^{-1}(0)$. If the center is on the axis, the automorphism is called a *transvection*; otherwise, it is called a *homology*.
(c) Show that $r_{a,\phi}$ induces a homology on $\mathrm{PG}(V)$ if and only if $\phi(a) \neq 0, 1$.
(d) Verify that every element of $\mathrm{GL}(V)$ inducing a homology on $\mathrm{PG}(V)$ can be written in the form $r_{a,\phi}$ with $\phi(a) \neq 0, 1$ up to an element of $Z(\mathrm{GL}(V))$ (the *center* of the group $\mathrm{GL}(V)$). In this case, the transformation $r_{a,\phi}$ itself is called a *pseudo-reflection*.

(e) A pseudo-reflection is called a *reflection* if and only if it has order two. Show that $r_{a,\phi}$ is a reflection if and only if $\phi(a) = 2 \neq 0$. The axis of this perspectivity is called its *mirror*.

(f) Prove that, if $\mathbb{D} = \mathbb{R}$, the group $\mathrm{GL}(V)$ is generated by all the pseudo-reflections in it, and that the group $\mathrm{PSL}(V)$ is generated by all the transvections in it.

Exercise 1.9.31 (Cited in Example 4.3.5 and Remark 4.5.4) Let V be a vector space over a field $\mathbb{F}$, let σ be an automorphism of $\mathbb{F}$ with $\sigma^2 = \mathrm{id}$, and let $f : V \times V \to \mathbb{F}$ be a *sesquilinear form* on V with respect to σ. This means that, for all $a, b, c, d \in V$ and all $\lambda, \mu \in \mathbb{F}$,

$$f(a + b, c + d) = f(a, c) + f(b, c) + f(a, d) + f(b, d),$$
$$f(a\lambda, b\mu) = \sigma(\lambda) f(a, b)\mu.$$

We assume in addition that f is *hermitian* with respect to σ, which means that $\sigma(f(x, y)) = f(y, x)$ for all $x, y \in V$. The subgroup of $\mathrm{GL}(V)$ of all linear transformations g preserving f (in the sense that $f(gx, gy) = f(x, y)$ for all $x, y \in V$) is denoted $\mathrm{U}(V, f)$ and called the *unitary group* on V with respect to f. A transformation in $\mathrm{GL}(V)$ is called *unitary* if it belongs to $\mathrm{U}(V, f)$. Prove that, for $r_{a,\phi}$ as in Exercise 1.9.30 to be a unitary reflection, it is necessary and sufficient that either $f(a, x) = 0$ for all $x \in V$ and $\phi(a) = 2 \neq 0$, or $f(a, a) \neq 0$ and $\phi(x) = 2f(a, a)^{-1} f(a, x)$ for all $x \in V$. In particular, if $f(a, a) \neq 0$, there is a unique unitary reflection with center $a\mathbb{F}$.

If $\sigma = \mathrm{id}$, then f is a symmetric bilinear form, in which case the group is denoted $\mathrm{O}(V, f)$, and called the *orthogonal group* with respect to f. Accordingly, a unitary reflection with respect to such a form f is called a *orthogonal reflection*.

1.10 Notes

Over the years 1880–1910 much work was done in the direction of abstract geometry. We out Moore [218] who made an interesting but hardly influential step towards abstract geometries. Hilbert's Foundations of Geometry [161] indicated another significant and influential step. But the key concepts of a diagram geometry, a weak homomorphism, and a flag residue as presented here are due to Tits [280, 282]. The present conceptual setting grew out of Buekenhout's work [44, 45]. Pasini's book [233] contains another description of the material presented here. The recent books [260, 292] also cover large parts of the topics of the current book.

Section 1.1

The cube, tetrahedron, octahedron, icosahedron, and dodecahedron are the five Platonic solids, which will reappear in Theorem 4.1.8. For more on tilings, see [144].

Section 1.2

For terminology on graphs, we mostly follow [36].

The word flag occurs in Borel's Bourbaki Seminar 121 in 1955, where he discusses Ehresmann's work. Usually, in algebraic geometry, the flag variety consists of all cosets of the Borel subgroup in a split reductive algebraic group (cf. [28, 168, 264]).

Section 1.3

The terminology for (weak) homomorphisms varies from author to author. In [233], for instance, a weak homomorphism is called a morphism and a homomorphism a type-preserving (or special) morphism, and in [287], a homomorphism is called a morphism (and so is understood to be type preserving).

The Petersen graph, introduced in Example 1.3.3, is named after Petersen because of a publication by Julius Petersen in which the graph appears, but in 1886 Kempe [193] had preceded him.

Section 1.4

Example 1.4.7 is the case of degree three of the well-known cyclic algebra construction; see for instance [197, Sect. 14]. These algebras lead to division rings isomorphic to their opposites but without anti-automorphisms of order two, which will be given in Example 7.2.16.

A good introduction to classical projective geometry is [155].

Section 1.5

Results liked Proposition 1.5.3 have their origin in [284] or, even earlier, in a construction of geometries from Lie groups in [280].

Section 1.6

Definition 1.6.2 generalizes the definition of residual connectedness in [287] from geometries to incidence systems. Lemma 1.6.4 shows that the classical definition applied to incidence systems would directly imply that each residually connected incidence system is a geometry.

Lemma 1.6.3 goes back to [280].

Section 1.7

The results in this section are fairly standard. The correspondence between transitive permutation groups and subgroups, described in Theorem 1.7.5, can be found in many textbooks. Among these, [235, 311] concentrate fully on permutation groups. A pleasant introductory textbook concerned with graphs and groups is [19]. A more elaborate theory, dealing also with covers of graphs, hinted at in Remark 1.7.6, is to be found in [112, 117, 252].

The earliest reference to coset geometries known to us is [284]. Lemma 1.8.9 provides the solution to an exercise in [286, p. 5].

More details on distribution diagrams, including those for the Johnson graphs, which generalize Example 1.7.16, can be found in [36].

The subgroup lattice of many more interesting finite groups than Sym_4, which was studied in Example 1.7.8, is known. For instance, for a sporadic group as large as ON (O'Nan's simple group; cf. Table 5.2), the result is reported on and used by Leemans in [200].

Section 1.8

Some of the material of this section, like Proposition 1.8.12, originates from [287].

In [5] the interplay between geometries and groups is dealt with from scratch. A more general version dealing with orbit spaces in the vein of Remark 1.7.6 can be found in [139].

Usually, a parabolic subgroup of a group G as in Definition 1.8.3 is understood to be a conjugate of a standard parabolic subgroup; but we do not need the notion in this book. Often, $N_G(G_i) = G_i$ holds for a standard parabolic subgroup G_i, so the representation of G on G/G_i is equivalent to the representation of G on the class of subgroups of G conjugate to G_i.

Section 1.9

The smallest thick projective plane, introduced in Exercise 1.9.7, is ascribed to Gino Fano [125], at the time when Pasch had already shown the incompleteness of Euclid's Elements.

There are many more notions of covering than the one introduced in Exercise 1.9.14. For instance, an m-covering is usually understood to be a surjective map all of whose restrictions to residues of rank at most m are isomorphisms (cf. [50, 233]). On the level of chamber systems, to be dealt with in Chap. 3, this notion appeared in [287]; see also [184, 260]. The existence of universal coverings is very relevant when it comes to classifications of geometries; often the best result that can be expected in such cases is that all residually connected geometries satisfying certain local properties are quotients of a given geometry (a universal cover) by groups of automorphisms in the vein of Exercise 1.9.16. An example can be found in [72, Main Theorem (c)].

Chapter 2
Diagrams

A diagram is a structure defined on a set of types I. This structure generally is close to a labelled graph and provides information on the isomorphism class of residues of rank two of geometries over I. This way diagrams lead naturally to classification questions like all residually connected geometries pertaining to a given diagram.

In Sect. 2.1, we start with one of the most elementary kinds of diagrams, the digon diagram. In Sect. 2.2, we explore some parameters of bipartite graphs that help distinguish relevant isomorphism classes of rank two geometries. Projective and affine planes can be described in terms of these parameters, but we also discuss some other remarkable examples, such as generalized m-gons; for $m = 3$, these are projective planes. The full abstract definition of a diagram appears in Sect. 2.3. The core interest is in the case where all rank 2 geometries are generalized m-gons, in which case the diagrams involved are called Coxeter diagrams, the topic of Sect. 2.4.

The significance of the axioms for geometries introduced via these diagrams becomes visible when we return to elements of a single kind, or, more generally to flags of a single type. The structure inherited from the geometry becomes visible through so-called shadows, studied in Sect. 2.5. Here, the key notion is that of a line space, where the lines are particular kinds of shadows. In order to construct flag-transitive geometries from groups with a given diagram, we need a special approach to diagrams for groups. This is carried out in Sect. 2.6. Finally in this chapter, a series of examples of a flag-transitive geometry belonging to a non-linear Coxeter diagram is given, all of whose proper residues are projective geometries.

2.1 The Digon Diagram of a Geometry

The digon diagram is a slightly simpler structure than a diagram and is useful in view of the main result, Theorem 2.1.6, which allows us to conclude that certain elements belonging to a given residue are incident. Let I be a set of types.

Definition 2.1.1 Suppose that $i, j \in I$ are distinct. A geometry over $\{i, j\}$ is called a *generalized digon* if each element of type i is incident with each element of type j.

F. Buekenhout, A.M. Cohen, *Diagram Geometry*,
Ergebnisse der Mathematik und ihrer Grenzgebiete. 3. Folge / A Series of
Modern Surveys in Mathematics 57, DOI 10.1007/978-3-642-34453-4_2,
© Springer-Verlag Berlin Heidelberg 2013

Let Γ be a geometry over I. The *digon diagram $\mathcal{I}(\Gamma)$* of Γ is the graph whose vertex set is I and whose edges are the pairs $\{i, j\}$ from I for which there is a residue of type $\{i, j\}$ that is not a generalized digon.

In other words, a generalized digon has a complete bipartite incidence graph. The choice of the name digon will become clear in Sect. 2.2, where the notion of generalized polygon is introduced.

Example 2.1.2 In the examples of Sect. 1.1, the cube, the icosahedron, a polyhedron, a tessellation of $\mathbb{E}^2$, and the Euclidean space $\mathbb{E}^3$ all have digon diagram

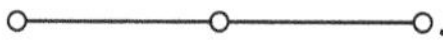

The digon diagram of Example 1.1.5 (tessellation of $\mathbb{E}^3$ by polyhedra) is

The digon diagram of a geometry Γ is a concise way of capturing some of the structure of Γ. This information is called *local* as it involves rank two residues only.

The usefulness of the digon diagram can be illustrated as follows. Assume that we are looking for all auto-correlations of Γ. It is obvious that each auto-correlation α of Γ permutes the types of Γ, inducing a permutation on I that is an automorphism of $\mathcal{I}(\Gamma)$. If $\mathcal{I}(\Gamma)$ is *linear*, that is, has the shape of single path, we see at once that $\mathcal{I}(\Gamma)$ has exactly two automorphisms: the identity and an involution permuting the endpoints of $\mathcal{I}(\Gamma)$. This implies that Γ has at most two families of auto-correlations: automorphisms and dualities (interchanging elements whose types are at the extreme ends of $\mathcal{I}(\Gamma)$, such as points and hyperplanes in projective geometries). The latter need not exist, as can be seen from the cube geometry of rank three; its dual geometry is the octahedron, which is not isomorphic to the original cube.

The digon diagram of a geometry is not necessarily linear. The digon diagram of Example 1.3.10, for instance, is a triangle. Actually every graph is the digon diagram of some geometry. Nevertheless, most of the geometries studied in this book have a digon diagram that is close to being linear. We now give some examples with non-linear digon diagrams.

Example 2.1.3 Consider the geometry constructed from a tessellation of $\mathbb{E}^3$ by cubes discussed in Example 1.1.5. Up to Euclidean isometry, there is a unique way to color the vertices with two colors b (for black) and w (for white), and the cubes (the cells) with two other colors g (for green) and r (for red) in such a way that two adjacent vertices (respectively, cubes) do not bear the same color. Let Γ be the geometry over $\{b, w, g, r\}$ on the bicolored vertices and cubes; incidence is symmetrized inclusion for vertices and cubes, and adjacency for two vertices, as well as for two cubes. The digon diagram of Γ is a quadrangle in which b and w represent opposite vertices and g and r likewise. Here, and elsewhere, a *quadrangle* is a circuit of length four without further adjacencies; in other words a complete bipartite graph with two parts of size two each.

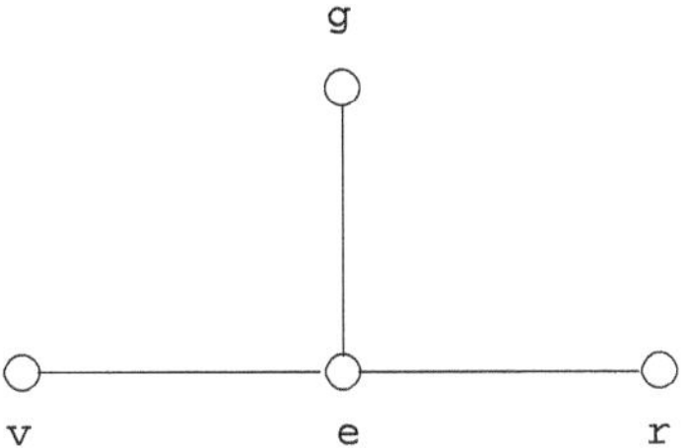

Fig. 2.1 The digon diagram
of Δ from Example 2.1.3

For a variation, consider the geometry Δ over $\{v, e, g, r\}$ whose elements of type g and r are as above, and whose elements of type v and e are the vertices and edges of the tessellation, respectively. Define incidence by the Principle of Maximal Intersection (cf. Exercise 1.9.20). Then the digon diagram of Δ is as indicated in Fig. 2.1.

There is a relation between the digon diagram of Γ and those of its residues. We will use the notion partial subgraph introduced in Definition 1.2.1.

Proposition 2.1.4 *Let Γ be a geometry over I and let F be a flag of Γ. The digon diagram $\mathcal{I}(\Gamma_F)$ is a partial subgraph of the digon diagram $\mathcal{I}(\Gamma)$.*

Proof It suffices to apply Proposition 1.5.3 and to see that every residue of type $\{i, j\}$ in Γ_F is also a residue of type $\{i, j\}$ in Γ. $\square$

Remark 2.1.5 It may happen that two vertices of $\mathcal{I}(\Gamma_F)$ are not joined while they are joined in $\mathcal{I}(\Gamma)$. Of course, the digon diagram $\mathcal{I}(\Gamma_F)$ is a subgraph of $\mathcal{I}(\Gamma)$ for all flags F if and only if, for any two distinct types i, j, either all or none of the residues in Γ of type $\{i, j\}$ are generalized digons. This property holds for flag-transitive geometries and for most of the geometries we study later. It gives rise to a useful heuristic trick. Given such a 'pure' geometry Γ, its digon diagram $\mathcal{I}(\Gamma)$, and a flag F, it suffices to remove the vertices of $\tau(F)$ from $\mathcal{I}(\Gamma)$ as well as the edges having a vertex in $\tau(F)$ in order to obtain the digon diagram of Γ_F.

Theorem 2.1.6 (Direct Sum) *Let Γ be a residually connected geometry of finite rank and let i and j be types in distinct connected components of the digon diagram $\mathcal{I}(\Gamma)$. Then every element of type i in Γ is incident with every element of type j in Γ.*

Proof Write $\Gamma = (X, *, \tau)$ and $r = \mathrm{rk}(\Gamma)$. We proceed by induction on r. For $r = 2$, the graph $\mathcal{I}(\Gamma)$ consists of the non-adjacent vertices i and j, so Γ is a generalized digon, for which the theorem obviously holds. Let $r \geq 3$, and let k be an element of $I = \tau(X)$ distinct from i, j. As i and j are in distinct connected components of $\mathcal{I}(\Gamma)$, we may assume that k and i are not in the same connected component of $\mathcal{I}(\Gamma)$. Let x_i (respectively, x_j) be an element of type i (respectively, j). By Lemma 1.6.3, there is an $\{i, j\}$-chain from x_i to x_j in Γ. We must show that x_i, x_j is such a path.

Suppose that we have $x_i * y_j * y_i * x_j$ with $y_j \in X_j$, $y_i \in X_i$. By Corollary 1.6.6 applied to the flag $\{y_i\}$, there is a $\{j, k\}$-chain from y_j to x_j in Γ_{y_i}, say $y_j = y_j^1, z_k^1, y_j^2, z_k^2, \ldots, z_k^s, x_j$, with $z_k^m \in X_k$ and $y_j^m \in X_j$ for all $m \in [s]$. By Proposition 2.1.4, the types i and k are in distinct connected components of $\mathcal{I}(\Gamma_{y_j})$ and by the induction hypothesis we obtain $x_i * z_k^1$. Similarly, in $\Gamma_{z_k^1}$, we find $x_i * y_j^2$. Repeated use of this argument eventually leads to $x_i * x_j$.

The preceding paragraph shows that an $\{i, j\}$-chain of length $m \geq 3$ between an element of X_i and an element of X_j can be shortened to an $\{i, j\}$-chain between the same endpoints of length $m - 2$. Hence the minimal length of such a path, being odd, must be 1. $\qquad\square$

In view of Exercise 2.8.20, the condition that I has finite cardinality is needed in Theorem 2.1.6.

Remark 2.1.7 Interesting geometries tend to have a connected digon diagram. However such a geometry may have residues whose digon diagrams are no longer connected. Take for instance a residually connected geometry Γ over [3] whose digon diagram is

If x is an element of type 2, then any element of type 1 and any element of type 3, both incident with x, are necessarily incident with each other. This is the kind of use we will make of the Direct Sum Theorem 2.1.6.

A consequence of the theorem is that any residually connected geometry of finite rank with a disconnected digon diagram can be seen as a direct sum of geometries whose digon diagrams are connected.

Definition 2.1.8 Let J be a set of indices and $(I_j)_{j \in J}$ a system of pairwise disjoint sets. Let $\Gamma_j = (X_j, *_j, \tau_j)$ be a geometry over I_j where $(X_j)_{j \in J}$ is a system of pairwise disjoint sets. The *direct sum* of the geometries Γ_j ($j \in J$) is the triple $\Gamma = (X, *, \tau)$ where $X = \bigcup_{j \in J} X_j$, $*|_{X_j} = *_j$, $x * y$ for $x \in X_j$, $y \in X_k$, $j \neq k$, and $\tau|_{X_j} = \tau_j$.

Example 2.1.9 A direct sum of two rank one geometries is a generalized digon. Any generalized digon is obtained in this way.

For the direct sum Γ as in Definition 2.1.8, it is clear that Γ is a geometry over $I = \bigcup_{j \in J} I_j$ whose digon diagram $\mathcal{I}(\Gamma)$ is the disjoint union of the digon diagrams $\mathcal{I}(\Gamma_j)$ for $j \in J$. Here is a converse of this statement.

Corollary 2.1.10 *Let Γ be a residually connected geometry of finite rank. Let $(I_j)_{j \in J}$ be the collection of all connected components of the digon diagram $\mathcal{I}(\Gamma)$. Then Γ is isomorphic to the direct sum of the I_j-truncations $_{I_j}\Gamma$ ($j \in J$) of Γ.*

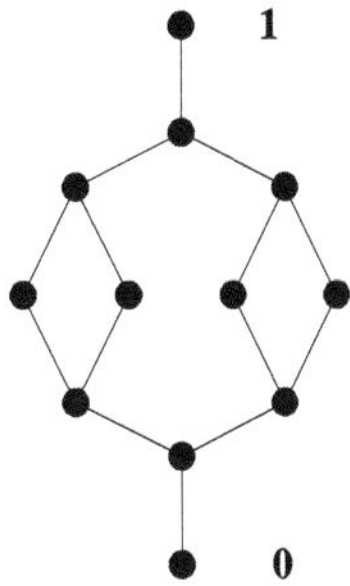

Fig. 2.2 A Jordan-Dedekind
poset that is neither firm nor
residually connected

Proof Set $\Gamma = (X, *, \tau)$ and $_{I_j}\Gamma = (X_j, *_j, \tau_j)$. The sets X_j $(j \in J)$ are pairwise disjoint and $X = \bigcup_{j \in J} X_j$. Also, $*|_{X_j} = *_j$ and $\tau|_{X_j} = \tau_j$. Finally, Theorem 2.1.6 forces $x * y$ for any $x \in X_j$ and $y \in X_k$ with $j \neq k$. $\square$

Example 2.1.11 Suppose that $(P, \leq)$ is a partially ordered set (sometimes abbreviated to *poset*) with a maximal element $\mathbf{1}$ and a minimal element $\mathbf{0}$. It is said to have the *Jordan-Dedekind property* if, for every pair $x, y \in P$ with $x \leq y$, all maximal totally ordered subsets of P with x and y as their minimal, respectively, maximal element, have the same finite cardinality. If $(P, \leq)$ satisfies the Jordan-Dedekind property, we denote by $\Gamma(P, \leq)$ the triple $(P \setminus \{\mathbf{0}, \mathbf{1}\}, *, \tau)$, where $*$ is symmetrized $\leq$, and, for any $x \in P$, the value $\tau(x) + 1$ is the cardinality of any maximal totally ordered subset of P with minimal element $\mathbf{0}$ and maximal element x. If $n = \tau(\mathbf{1}) - 1$, this is a geometry over $[n]$ with a digon diagram that is a partial subgraph of the graph on $[n]$ consisting of the single path $1, 2, \ldots, n$.

The geometry $\Gamma(P, \leq)$ need neither be firm nor residually connected; see Fig. 2.2, where the poset on the vertex set drawn is obtained from the figure by letting $a \leq b$ if and only if a appears at the end of a path downward from b.

Now, $\Gamma(P, <)$ is firm if and only if for all a, b, c in P with $a < b < c$ there exists $x \neq b$ in P such that $a < x < c$. We can characterize residual connectedness similarly. Define an interval (a, b) with $a < b$ in P, as the set of all x in P such that $a < x < b$. The elements of (a, b) constitute the vertices of a graph whose edges are the pairs $\{x, y\}$ of distinct vertices x, y such that either $x < y$ or $y < x$. Then, $\Gamma(P, \leq)$ is residually connected if and only if the graph of each interval in P containing at least two elements x, y with $x < y$, is connected. This follows readily from the Direct Sum Theorem 2.1.6.

Conversely, suppose that $\Gamma = (X, *, \tau)$ is a geometry over $[n]$ with a linear digon diagram. For $x, y \in X$, put $x \leq y$ if $\tau(x) \leq \tau(y)$ and $x * y$. Then $\leq$ is a transitive relation by the Direct Sum Theorem 2.1.6. Also, by the definition of incidence system, $x \leq y$ and $y \leq x$ imply $x = y$. Hence $(X \cup \{\mathbf{0}, \mathbf{1}\}, \leq)$, where $\leq$ is extended by the rule $\mathbf{0} \leq x \leq \mathbf{1}$ for all $x \in X$, is a partially ordered set with the Jordan-Dedekind property. Observe that $\Gamma(X \cup \{\mathbf{0}, \mathbf{1}\}, \leq)$ coincides with Γ. These examples show that the local structure (read off from the digon diagram) may be equivalent to the global structure (the ordering on X).

2.2 Some Parameters for Rank Two Geometries

In Sect. 2.1 we began with the idea that a diagram for a geometry Γ over a set of types I would assign to any pair $\{i, j\}$ of elements of I information on the residues of type $\{i, j\}$ of Γ, and went on to distinguish rank two residues that are generalized digons from arbitrary rank two geometries. We now introduce some refinements of this information, capturing more characteristics of rank two geometries. In the remainder of this section, I will be a finite index set and Γ a geometry over I. Often, I will be the set $\{\mathrm{p}, \mathrm{l}\}$ of cardinality two.

Definition 2.2.1 Let Δ be a graph. In Definition 1.6.1, we introduced the distance function d (or d_Δ). For a vertex x of Δ, the *diameter of Δ at x* is the largest distance from x to any other vertex of Δ. The *diameter δ_Δ* of Δ is the largest distance between two vertices of Δ. If Δ does not have finite diameter, then we put $\delta_\Delta = \infty$.

Often, for two vertices x, y of Δ, we write $x \perp y$ (or $x \perp_\Delta y$ if confusion is imminent) to denote $d(x, y) \leq 1$, that is, x and y are equal ($x = y$) or adjacent ($x \sim y$). For a vertex x and a set Y of vertices of Δ, we write $x^\perp$ for the set of all vertices y with $y \perp x$ and $Y^\perp$ for $\bigcap_{y \in Y} y^\perp$.

Let $\Gamma = (X, *, \tau)$ be an incidence system over I. If Δ is the incidence graph of Γ, then $x * y$ is equivalent to $x \perp y$, and $x^\perp = x^*$, etc. *Distance* in Γ is usually understood to be distance in Δ, and similarly for the diameter. For $j \in I$, the *j-diameter d_j* of Γ is the largest number occurring as a diameter of the incidence graph of Γ at some element of type j.

Now take $I = \{\mathrm{p}, \mathrm{l}\}$. The *collinearity graph* of Γ on $X_\mathrm{p} = \tau^{-1}(\mathrm{p})$ is the graph $(X_\mathrm{p}, \sim)$ with vertex set X_p in which x and y are adjacent (equivalently, $x \sim y$ holds) whenever they have distance two in Γ; equivalently, whenever they are distinct and there is a line $L \in X_\mathrm{l}$ such that $x * L * y$.

The *shadow* of $L \in X_\mathrm{l}$ on $\{\mathrm{p}\}$ is the set $L^* \cap X_\mathrm{p}$. It is a clique in the collinearity graph of Γ on X_p.

By the symmetry of the roles of p and l, we also have the notion of the collinearity graph on X_l. In this graph, two lines are adjacent whenever they are distinct and there is a point to which both are incident.

If Γ as above is connected (cf. Definition 1.6.1), then there are exactly two connected components in the graph on X in which adjacency is defined as having mutual distance two: the parts X_p and X_l. The subgraphs induced on these parts are the two collinearity graphs of Γ.

If $I = \{\mathrm{p}, \mathrm{l}\}$, the difference between d_p and d_l is at most one and the larger one is equal to the diameter δ of Γ. If δ is odd, then $d_\mathrm{p} = d_\mathrm{l} = \delta$, since both a point and a line are involved in a pair of elements at maximal distance.

Example 2.2.2 If Γ is a polygon with n vertices, considered as a geometry over $\{\mathrm{p}, \mathrm{l}\}$, then $d_\mathrm{p} = d_\mathrm{l} = n$. In a generalized digon, we have $d_\mathrm{p} = d_\mathrm{l} = 2$. In the real affine plane $\mathbb{E}^2$ (cf. Example 1.1.2), we have $d_\mathrm{p} = 3$, $d_\mathrm{l} = 4$, and in the real projective plane $\mathrm{PG}(\mathbb{R}^3)$ (cf. Example 1.4.9), we have $d_\mathrm{p} = d_\mathrm{l} = 3$. This means that the collinearity graph of the projective plane on the point set (and on the line set) is a clique. The collinearity graph of an affine plane on the line set has diameter two, while the collinearity graph on the point set is again a clique.

We introduce yet another parameter.

Definition 2.2.3 A *circuit* in the geometry Γ over $I = \{p, 1\}$ is a chain $x = x_0, x_1, x_2, \ldots, x_{2n} = x$ from x to x, with $x_i \neq x_{i+1}, x_{i+2}$ for $i = 0, \ldots, 2n$ (all indices taken modulo $2n$ and $n > 0$). Its length $2n$ is necessarily even. The minimal number $g > 0$ such that Γ has a circuit of length $2g$ is called the *girth* of Γ. If Γ has no circuits, we put $g = \infty$.

Lemma 2.2.4 *The girth g of a firm geometry Γ over $I = \{p, 1\}$ satisfies*

$$\text{either} \quad 2 \leq g \leq d_p \leq d_1 \leq d_p + 1$$
$$\text{or} \quad 2 \leq g \leq d_1 \leq d_p \leq d_1 + 1.$$

Proof This is a direct consequence of the observations preceding Example 2.2.2. The assertions also hold for $g = \infty$. $\square$

Definition 2.2.5 The *dual geometry* $\Gamma^{\vee}$ of a geometry Γ over $I = \{p, 1\}$ is the triple $(X, *, \tau^{\vee})$ where $\tau^{\vee}(x) = 1$ if and only if $\tau(x) = p$.

The girths of Γ and $\Gamma^{\vee}$ are equal while $d_p^{\vee} = d_1$ and $d_1^{\vee} = d_p$ (with the obvious interpretations of $d_p^{\vee}$ and $d_1^{\vee}$).

Definition 2.2.6 If Γ is a $\{p, 1\}$-geometry with finite diameter d and with girth g having the same diameter d_i at all elements of type i (for $i = p, 1$), then Γ is called a (g, d_p, d_1)-*gon* over $(p, 1)$. If, in addition, $g = d_p = d_1$, then Γ is called a *generalized g-gon*.

Allowing $g = \infty$ in the above definition of a generalized g-gon, we see that generalized ∞-gons have no circuits.

If Γ is a (g, d_p, d_1)-gon over $(p, 1)$, then it is a (g, d_1, d_p)-gon over $(1, p)$.

The definition of a generalized 2-gon coincides with that of a generalized digon in Definition 2.1.1. In view of the terminology below, the name digon fits a 2-gon.

Definition 2.2.7 Generalized 3-gons are also called *projective planes*, generalized 4-gons are called *generalized quadrangles*. Likewise, generalized 6-gons are called *generalized hexagons* and generalized 8-gons are called *generalized octagons*. *Generalized polygons* is the name used for all generalized g-gons ($g \geq 2$).

For $g \in \mathbb{N} \cup \{\infty\}$, the *(ordinary) g-gon* is defined as the thin generalized g-gon.

It is easy to see that the ordinary g-gon is indeed unique: it is isomorphic to the geometry consisting of the set $\mathbb{Z}/2g\mathbb{Z}$ with types even and odd and incidence

$x * y \iff x - y \in \{0, 1, -1\}$ for $x, y \in \mathbb{Z}/2g\mathbb{Z}$. Here, in case $g = \infty$, we interpret $\mathbb{Z}/2\infty\mathbb{Z}$ as $\mathbb{Z}$.

Generalized polygons are the building blocks of the classical geometries that we encounter in later chapters. Although there is no hope of describing all graphs that are (g, d_{p}, d_1)-gons for arbitrary g, d_{p}, d_1, a lot of structure can be pinned down if $g = d_{\mathrm{p}} = d_1$. A classification of all finite generalized g-gons with $g = 3$ or $g = 4$ can hardly be expected in the presence of so many wild examples. The examples in the remainder of this section provide some evidence for this. The occurrence of projective planes and generalized quadrangles as residues in geometries of higher rank usually imposes conditions which enable us to classify them.

Example 2.2.8 The Petersen graph (cf. Example 1.3.3) is a $(5, 5, 6)$-gon over $(\mathtt{vertex}, \mathtt{edge})$ and the cube is a $(4, 6, 6)$-gon over $(\mathtt{vertex}, \mathtt{edge})$. The Fano plane (cf. Exercise 1.9.7) is a generalized 3-gon.

Theorem 2.2.9 *Let Γ be a $\{\mathrm{p}, 1\}$-geometry. Then Γ is a projective plane if and only if*

(i) *every pair of points is incident with a unique line;*
(ii) *every pair of lines is incident with a unique point;*
(iii) *there exists a non-incident point-line pair.*

Proof Set $\Gamma = (X, *, \tau)$ and write $P = X_{\mathrm{p}}$ and $L = X_1$. First, suppose that Γ is a projective plane. Let $x, y \in P$ be distinct. As the point-diameter of Γ is 3, there is a path of length at most 3 from x to y. But this length must be even and strictly greater than 0, hence equal to 2. In other words, there is $h \in L$ incident with both x and y. If $m \in L$ is a line distinct from h also incident with x and y, then x, h, y, m, x is a 4-circuit in Γ, which contradicts that the girth is 6. Hence (i). Statement (ii) follows likewise. Finally (iii) is a consequence of the existence of a path of length three in the incidence graph of Γ.

Next suppose that Γ satisfies (i), (ii), and (iii). Then (i) and (ii) force $d_{\mathrm{p}} \leq 3$ and $d_1 \leq 3$, respectively, while (iii) implies that equality holds for both. If $g = 2$, then there are two points incident with two distinct lines, contradicting both (i) and (ii). Hence $g \geq 3$. But Γ is firm (it is a $\{\mathrm{p}, 1\}$-geometry) so, by the inequalities in Lemma 2.2.4, $g = 3$. $\square$

The geometries $\mathrm{PG}(\mathbb{D}^3)$ of Example 1.4.9 for any division ring $\mathbb{D}$ are examples. But there are more, as will be clear from Example 2.3.4.

Example 2.2.10 Let X_{p} be the set of all pairs from $[6]$ and let X_1 be the set of all partitions of $[6]$ into three pairs. The geometry $\Gamma = (X_{\mathrm{p}}, X_1, *)$, where $*$ is symmetrized containment, is a generalized quadrangle with 15 points and 15 lines. It is drawn in Fig. 2.3, where lines are represented by arcs and line segments.

Theorem 2.2.11 *Every generalized quadrangle with three points on each line and three lines through each point is isomorphic to the one of Example 2.2.10.*

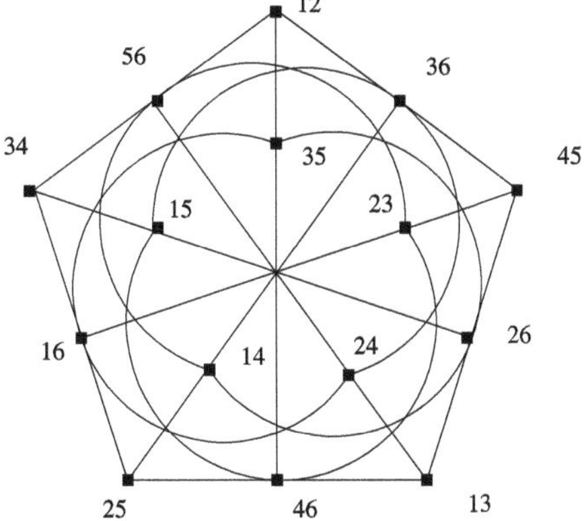

Fig. 2.3 A generalized quadrangle with 15 points and 15 lines

Proof Let $(X_p, X_1, *)$ be such a generalized quadrangle. It must have 15 points: if we fix one, say ∞, then there are $3 \cdot 2 = 6$ neighbors (two on each line through ∞) in the collinearity graph on the points; as each neighbor of ∞ is collinear with only one point that is also a neighbor of ∞, a count of edges from points at distance one to points at distance two, using $d_p = d_1 = 4$, shows that there are $6 \cdot 4/3 = 8$ points at distance two from ∞; finally, as $d_p = 4$, there are no points at distance greater than two from ∞.

The generalized quadrangle is determined by the collinearity graph on the point set X_p, as the lines correspond bijectively to the maximal cliques (of size three).

If a and b are distinct non-collinear points in X_p, then $\{a, b\}^\perp$ has exactly three points (for b is collinear with one point on each line incident with a). We claim that $(\{a, b\}^\perp)^\perp$ also has exactly three points. To see this, let x, y, z be the three points of $\{a, b\}^\perp$, and let c be the third point of $\{x, y\}^\perp$ distinct from a and b. It suffices to show $c \sim z$. If not, there would be four distinct lines on z: beside the lines incident with a and b, there would be lines incident with the third points on the line incident with c and y and on the line incident with c and x. Indeed, coincidence of any two of these four lines on z would lead to a contradiction with the girth being 4. But there are precisely three lines on z, so we must have $c \sim z$, as required for the claim.

We finish by identifying the collinearity graph on P with the graph on the pairs from [6] (see Example 2.2.10) in which two vertices are adjacent whenever they are disjoint. To this end, start with the configuration on the six points of the previous paragraph and label the points as follows: $a = 12$, $b = 23$, $c = 13$, $x = 45$, $y = 56$, $z = 46$. The subgraph of the collinearity graph of P induced on these points is as it should be. Each of the nine lines on two collinear points from the sextet has a third point as yet unaccounted for. Label these nine points by the pair that complements the two pairs from the points already assigned on that line. For instance, the third point of the line on a and x receives label 36. This way we have labelled all 15 points and the remaining lines are forced as indicated by Example 2.2.10. $\square$

Remark 2.2.12 Putting together the above uniqueness result and Example 2.2.10, we see that the symmetric group Sym_6 acts as a group of automorphisms on the generalized quadrangle Γ of Example 2.2.10. The uniqueness proof can also be used to show that $\mathrm{Aut}(\Gamma)$ is isomorphic to Sym_6. Besides, the flag transitivity of this group on Γ shows that the geometry can alternatively be described as $\Gamma(G, (C_G((1,2)), C_G((1,2)(3,4)(5,6))))$, where $G = \mathrm{Sym}_6$ and $C_G(x)$ denotes the centralizer in G of the element x of G.

Adding a formal point 0 to X_p, we can define addition on X_p as follows: for distinct $u, v \in X_\mathrm{p}$ we set $0 + u = u + 0 = u, u + u = 0$ and $u + v = w$, where w is the unique point other than u, v in $\{u, v\}^{\perp\perp} = (\{u, v\}^\perp)^\perp$. In view of the abundance of automorphisms, we only need check the existence and uniqueness of w for $u = 12$ and $v = 34$ or $v = 13$; in these respective cases, w is 56 or 23. The addition leads to an $\mathbb{F}_2$-vector space structure on $V := \{0\} \cup X_\mathrm{p}$, turning it into $\mathbb{F}_2^4$ (there being $1 + 15 = 2^4$ points). So the generalized quadrangle is a subgeometry of the $\{1, 2\}$-truncation of $\mathrm{PG}(\mathbb{F}_2^4)$ (see Examples 1.4.9 and 1.5.6).

There is a bilinear form $f : V \times V \to \mathbb{F}_2$ given by $f(x, y) = 1$ if $x, y \in X_\mathrm{p}$ are non-collinear points, and 0 otherwise. This form is nondegenerate and antisymmetric. The generalized quadrangle can now be described completely in terms of the vector space and the bilinear form: its points and its lines are the one, respectively, two dimensional singular subspaces of V (as defined in Example 1.4.13). It follows that the generalized quadrangle is an absolute geometry of the projective geometry $\mathrm{PG}(V)$.

Example 2.2.13 The uniqueness of the generalized quadrangle Γ of Example 2.2.10 implies the existence of a duality: the dual geometry $\Gamma^\vee$ also satisfies the properties of Theorem 2.2.11 and so must be isomorphic to Γ. This means that $\mathrm{Aut}(\Gamma)$ has index two in $\mathrm{Cor}(\Gamma)$. The bigger group contains 'outer' (as opposed to inner) involutions of Sym_6. Here, we will construct polarities of Γ in a geometric way, using ovoids and spreads.

An *ovoid* of Γ is a set O of points with the property that each line in X_1 is incident with exactly one point in O. A simple count shows that an ovoid is a set of five pairwise non-collinear points (and conversely). Dually, a *spread* of Γ is an ovoid of $\Gamma^\vee$, that is, a subset S of X_1 with the property that every point is on exactly one line in S. Denote by $\mathcal{O}$ and $\mathcal{S}$ the collection of ovoids and spreads, respectively, of Γ. It is easy to see that the ovoids are of the form $O_i = \{\{i, j\} \mid j \in [6] \setminus \{i\}\}$ and that $i \mapsto O_i$ is a bijection $[6] \to \mathcal{O}$. Any two members O_i and O_j meet in exactly one point, *viz.* $\{i, j\}$. Using the duality, corresponding statements for spreads can be derived.

The diagonal action of Sym_6 on $\mathcal{O} \times \mathcal{S}$ is transitive. For, in view of duality, every spread is left invariant by a subgroup of Sym_6 isomorphic to Sym_5, and, by taking a specific spread, it is easily seen that the stabilizer acts transitively on $[6]$ whence on $\mathcal{O}$.

Now, let (O, S) be a pair in $\mathcal{O} \times \mathcal{S}$. We claim that there is a unique polarity π of Γ mapping $x \in O$ to the unique line $\pi(x) \in S$ to which it belongs (thus $x \in \pi(x)$). By transitivity of Sym_6 on $\mathcal{O} \times \mathcal{S}$, it suffices to check this for a single pair (O, S). To determine the image of $y \in X_\mathrm{p} \setminus O$, consider the line $h \in S$ containing y, and the lines $m, n \in S$ distinct from h containing a point of O collinear with y. Then $\pi(y)$

Fig. 2.4 Distribution diagram of the double six

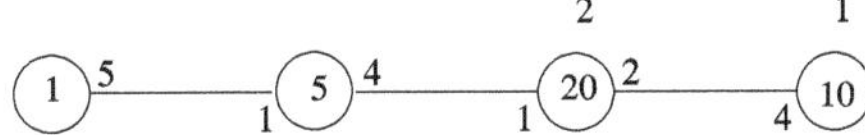

must be the line through the point $\pi(h) \in O$ (on h) and meeting both m and n. This, and the fact that π has order two, uniquely determines the polarity π on Γ.

As $\mathrm{Aut}(\Gamma)$ is isomorphic to Sym_6, we find that $\mathrm{Cor}(\Gamma)$ is a group isomorphic to $\mathrm{Sym}_6 \rtimes C_2$. The transposition $(1, 2)$ acts naturally on $[6]$, and, as $i \mapsto O_i$ is an equivalence of Sym_6-representations between $[6]$ and $\mathcal{O}$, also on $\mathcal{O}$. But this element has no fixed points on $\mathcal{S}$ (for, if $S \in \mathcal{S}$ would be fixed by $(1, 2)$, then the line in S on $\{1, 3\}$ will have a point $\{k, l\}$ with $k, l \neq 1, 2$, that is, a fixed point, so the line must be fixed by $(1, 2)$, contradicting that its point $\{1, 3\}$ is mapped to $\{2, 3\}$). Therefore, its image under the action on $\mathcal{S}$ is a product of three transpositions. This shows that the map $g \mapsto \pi g \pi$ $(g \in G)$ is an outer automorphism of G.

Example 2.2.14 On $\mathcal{O} \times \mathcal{S}$ of Example 2.2.13 as a vertex set, an interesting graph Δ arises by letting (O, S) and (O', S') be adjacent whenever $O \neq O'$ and $S \neq S'$ but $O \cup S$ and $O' \cup S'$ have a flag in common. This graph is known as the *double six*. By the above, the intersection of $O \cup S$ and $O' \cup S'$ always contains a point and a line. A fixed pair (O, S) has 5 neighbors (given the choice of a flag in $O \cup S$, the adjacent vertex meeting in that flag is uniquely determined). It requires a little elaboration to see that a vertex (O', S') for which the intersection consists of a non-incident point-line pair, is adjacent to a unique neighbor of (O, S), and that there are 20 of them. Finally, the 10 vertices (O, S') and (O', S) with $S' \neq S$ and $O' \neq O$ are at distance three from (O, S). Schematically, this information is conveyed in the distribution diagram (cf. Example 1.7.16) depicted in Fig. 2.4.

Another generalized quadrangle arises from the double six Δ. Consider the graph obtained from Δ on the same vertex set by letting two vertices be adjacent whenever they are at distance three in Δ. This graph is the 6×6-*grid*, that is, the Cartesian product of two cliques of size six. Its vertex set and its set of maximal cliques form a generalized quadrangle with six points per line but two lines per point.

Example 2.2.15 We construct a generalized hexagon with three points on each line and three lines on each point. Let $\mathbb{F}$ be $\mathbb{F}_9$, the field of 9 elements. Equip the vector space $\mathbb{F}^3$ with the standard unitary inner product

$$f(x, y) = \sum_{i=1}^{3} x_i^3 y_i \quad (x, y \in \mathbb{F}^3).$$

Let P be the set of nonsingular points of the underlying projective space (cf. Example 1.4.9), that is, $P = \{x\mathbb{F} \mid f(x, x) \neq 0\}$, and write $x\mathbb{F} \sim y\mathbb{F}$ whenever $f(x, y) = 0$. Maximal cliques in the graph $(P, \sim)$ have size three, and correspond to orthonormal bases of $\mathbb{F}^3$ (up to scalar multiples for the basis vectors). Let L be the collection of all these maximal cliques. We claim that $(P, L, *)$, where $*$ is symmetrized containment, is a generalized hexagon. It has 63 points and 63 lines.

Fig. 2.5 The distribution diagram of the generalized hexagon of Example 2.2.15

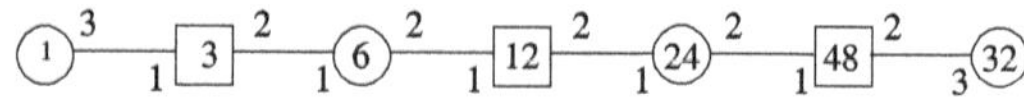

To verify this, observe that any pair $a\mathbb{F}$, $b\mathbb{F}$ of points with $f(a,b) = 0$ lies on a line. Suppose that $\{a\mathbb{F}, b\mathbb{F}, c\mathbb{F}\}$ is a line. Without loss of generality, we normalize $f(a,a) = f(b,b) = f(c,c) = 1$. Now $f(b+c, b+c) = f(b-c, b-c) = 2 \neq 0$, and $f(b \pm c, a) = f(b+c, b-c) = 0$, so $\{a\mathbb{F}, (b+c)\mathbb{F}, (b-c)\mathbb{F}\}$ is a line on a. It is readily seen that there is only one more line on a, viz. $\{a\mathbb{F}, (b+ic)\mathbb{F}, (b-ic)\mathbb{F}\}$, where i is a square root of -1 in $\mathbb{F}$. Thus, each point is on exactly 3 lines. Interchanging the roles of a, b, and c, we obtain the lines on $b\mathbb{F}$ and $c\mathbb{F}$ as well, and see that no point collinear with a neighbor of $a\mathbb{F}$ is collinear with a neighbor of $b\mathbb{F}$. In particular, the girth of $(P, L, *)$ is at least six.

Suppose $d\mathbb{F} \in P$ is not collinear with any neighbor of $a\mathbb{F}$, $b\mathbb{F}$, or $c\mathbb{F}$. Write $d = \alpha a + \beta b + \gamma c$, so $\alpha\beta\gamma \neq 0$. There is a common neighbor in P of $c\mathbb{F}$ and $d\mathbb{F}$ if and only if $\alpha \in \{\pm 1, \pm i\}\beta$, that is, $\alpha^4 = \beta^4$. Changing the roles of a, b, c again, and using that $\alpha^4 + \beta^4 + \gamma^4 \neq 0$, we see that there exists a unique point in $\{a, b, c\}$ that is at distance two to $d\mathbb{F}$. The conclusion is that the point diameter and line diameter are 6 and that the girth is at least 6. But then the girth must be precisely 6 by Lemma 2.2.4. This establishes the claim that $(P, L, *)$ is a generalized hexagon.

Figure 2.5 gives a pictorial description of this generalized hexagon, from which it is immediate that there are precisely 63 points and just as many lines.

There is a nice way to visualize automorphisms. To each $a\mathbb{F} \in P$, we associate the following linear transformation r_a of $\mathbb{F}^3$.

$$r_a(x) = x + af(a,a)^{-1}f(a,x) \quad (x \in \mathbb{F}^3).$$

Such a linear transformation is the special case of the unitary reflection $r_{a,\phi}$ with $\phi(x) = 2f(a,a)^{-1}f(a,x)$, presented in Exercise 1.9.31. Observe that r_a does not depend on the choice of vector in $\mathbb{F}a$. Being a member of the unitary group $\mathrm{U}(\mathbb{F}^3, f)$, it preserves the unitary inner product, so induces an automorphism of $(P, L, *)$. The reflection r_a fixes every vector in the hyperplane $\{x \in \mathbb{F}^3 \mid f(x,a) = 0\}$, whence all points of $(P, L, *)$ collinear with $a\mathbb{F}$. It moves each point at distance two from $a\mathbb{F}$ in the collinearity graph to its unique neighbor at distance two from $a\mathbb{F}$.

By use of these reflections, the transitivity of $G = \mathrm{U}(\mathbb{F}_9^3, f)$ on each of the sets $\{(a,b) \in P \times P \mid d(a,b) = i\}$ for $i = 2, 4, 6$ is readily established. The map $P \to G$ given by $a\mathbb{F} \mapsto r_a$ sends P to a conjugacy class of reflections in G. Moreover, three straightforward checks show that $r_a r_b$ is a transformation of order 2, 4, or 3 according to whether $a\mathbb{F}$ and $b\mathbb{F}$ have mutual distance 2, 4, or 6 in $(P \cup L, *)$. This gives an alternative description of $(P \cup L, *)$, entirely in terms of the group G: a line can be seen as the set of three nontrivial reflections in a subgroup S isomorphic to C_2^3 generated by reflections. As a consequence, the generalized hexagon can be described as $\Gamma(G, (C_G(r), N_G(S)))$, where $C_G(r)$ is the centralizer of a reflection r in S and $N_G(S) = \{g \in G \mid gSg^{-1} = S\}$, the *normalizer* of S in G.

For geometries with finite rank 1 residues, some more parameters are useful.

Definition 2.2.16 Let Γ be a geometry over I and fix $i \in I$. If each residue of Γ of type $\{i\}$ has the same finite size $s_i + 1$, then s_i is called the *i-order* of Γ. If Γ is a geometry over I having i-order s_i for each $i \in J$ for some subset J of I, then $(s_i)_{i \in J}$ is called the *J-order* of Γ. In case $J = I$ we just speak of the *order* of Γ. In particular, for a $\{\mathrm{p}, 1\}$-geometry in which both line and point order exist, we speak of the order (s_p, s_1) over $(\mathrm{p}, 1)$.

Thus, Example 2.2.10 gives a generalized quadrangle of order $(2, 2)$ and Example 2.2.15 a generalized hexagon of order $(2, 2)$.

Remark 2.2.17 The parameters g, d_p, d_1, s_p, s_1 can be used to enrich digon diagrams. We summarize the information they provide by a diagram such as

$$\overset{\mathrm{p} \quad d_\mathrm{p} \quad g \quad d_1 \quad 1}{\underset{s_\mathrm{p} \qquad\qquad\qquad\qquad s_1}{\circ\!\!-\!\!\!-\!\!\!-\!\!\!-\!\!\!-\!\!\!-\!\!\!-\!\!\!-\!\!\!-\!\!\!-\!\!\circ}}.$$

The diagram stands for the class of all (g, d_p, d_1)-gons of order (s_p, s_1) over $(\mathrm{p}, 1)$. A member of this class is said to *belong to the diagram*. If the subscripts s_p, s_1 are dropped, a geometry belongs to the diagram if and only if it is a (g, d_p, d_1)-gon over $(\mathrm{p}, 1)$. For example, the Petersen graph belongs to $\overset{\mathrm{p}\ 5 \quad 5 \quad 6\ 1}{\underset{1 \qquad\qquad 2}{\circ\!\!-\!\!\!-\!\!\!-\!\!\circ}}$ and the real affine plane to $\overset{\mathrm{p}\ 3 \quad 3 \quad 4\ 1}{\circ\!\!-\!\!\!-\!\!\!-\!\!\circ}$.

The Petersen graph is easily seen to be the unique geometry with the given diagram (including the specified orders). Replacing the order 2 by 6, we obtain a more complicated example.

Example 2.2.18 We construct a graph with a lot of symmetry known as the Hoffman-Singleton graph. Start with the set $X_\mathrm{p} = \binom{[7]}{3}$ of all triples from $[7]$. It has cardinality 35. There are 30 collections of 7 elements of X_p that form the lines of a Fano plane with point set $[7]$. The group Sym_7 acts transitively on this set of 30 Fano plane structures, but the alternating group Alt_7 has two orbits, of cardinality 15 each. Select one and call it X_1. Now consider the following graph HoSi constructed from the incidence graph $(X_\mathrm{p} \cup X_1, *)$ by additionally joining two elements of X_p whenever they have empty intersections. The resulting graph HoSi has 50 vertices, is regular of valency 7, and has a group of automorphisms isomorphic to Alt_7. But there are additional automorphisms, fusing the orbits of Alt_7 on the vertex set of sizes 35 and 15, as we will see in Theorem 2.2.19 below. This graph HoSi is the *Hoffman-Singleton graph*.

We give an alternative description of HoSi. At first sight it is not clear at all that we are dealing with the same graph (up to isomorphism), so we will call this graph HoSi$'$. Recall the double six Δ of Example 2.2.14 and the related collections $\mathcal{O}$ and $\mathcal{S}$ of ovoids and spreads. Select two types, o and s and build the tree T on 14 vertices with a central edge $\{\mathrm{o}, \mathrm{s}\}$ both of whose vertices are adjacent to 6 end nodes (that is, vertices of valency 1). Label the end nodes of T adjacent to o by the members of $\mathcal{O}$, and those adjacent to s by the members of $\mathcal{S}$. Now let HoSi$'$ be the graph whose vertex set is the disjoint union of the vertex sets of T and Δ and in which T and Δ are subgraphs; join the vertex $S \in \mathcal{S}$ of T to every (O', S) in Δ for $O' \in \mathcal{O}$, and, likewise, join the vertex $O \in \mathcal{O}$ of T to (O, S') for every $S' \in \mathcal{S}$. From this description, we see that $\mathrm{Aut}(\mathrm{Sym}_6)$ acts on HoSi'.

It is readily derived that the geometries of vertices and edges of both HoSi and HoSi' are $\{\mathrm{p}, 1\}$-geometries with parameters

$$\overset{\mathrm{p}}{\underset{1}{\circ}} \; \overset{5}{\rule{2em}{0.5pt}} \; \overset{5}{\rule{2em}{0.5pt}} \; \overset{6}{\rule{2em}{0.5pt}} \; \overset{1}{\underset{6}{\circ}}.$$

Here is a characterization of the Hoffman-Singleton graph.

Theorem 2.2.19 *There is a unique* $(5, 5, 6)$-*gon of order* $(1, 6)$ *up to isomorphism. It has a flag-transitive group of automorphisms.*

Proof Let $\Gamma = (X_{\mathrm{p}}, X_1, *)$ be a $(5, 5, 6)$-gon of order $(1, 6)$. As the girth is greater than two and $s_{\mathrm{p}} = 1$, we may view Γ as the geometry of points and edges of the collinearity graph $X = (X_{\mathrm{p}}, \perp)$. The line order is 6, so this graph has valency 7. Moreover, any point $x \in X_{\mathrm{p}}$ has 7 neighbors and each neighbor is adjacent to exactly 6 points at distance two from x in X, whereas no point at distance two has more than one neighbor in common with x. Hence, X has $1 + 7 + 7 \times 6 = 50$ vertices.

If x and y are nonadjacent vertices of X, then there is a unique path in the collinearity graph of length two from x to y, and each of the 6 edges on y not on this path of length two is on a unique path of length three from x to y. Thus every path of length two lies on exactly 6 pentagons. Let P be a pentagon in Γ and write $X_1(P)$ for the set of points outside P collinear with a point of P. The set $X_1(P)$ has $5 \times 5 = 25$ vertices (no vertex outside P can be adjacent to two members of P), and each of these lies on exactly two paths of length three between two nonadjacent members of P. Consequently, $X_1(P)$ is a regular subgraph of Γ of valency two. Set $X_2(P) = X_{\mathrm{p}} \setminus (P \cup X_1(P))$. Each vertex in this set has distance two to every member of P, so lies on precisely five edges having a vertex in $X_1(P)$. Consequently, $X_2(P)$ is also regular of valency two. Suppose that a, b, c is a path in $X_2(P)$. There are at least four paths from a to c of length three having both non-end points in $X_1(P)$; let a, d, e, c be one of them. Denote by f, g the unique neighbor of e, d, respectively, in $X_1(P) \setminus \{e, d\}$. Now f and b cannot be adjacent, so there must be a common neighbor x say. As x cannot be one of a or c, we must have $x \in X_1(P)$, so x is the unique neighbor of f in $X_1(P) \setminus \{e\}$. Tracing adjacencies with vertices of P, we see that x is also adjacent to the unique point z of P at distance two to both e and d, so that $\{x\} = \{z, b\}^{\perp}$. But then, arguing with g instead of f, we find that x is also adjacent to g, leading to the 5-circuit d, e, f, x, g in $X_1(P)$. Varying over the 4 edges in $X_1(P)$ on paths of length 3 from a to c, we see that $X_1(P)$ has at least four 5-circuits, whence consists entirely of five disjoint pentagons. Repeating the argument with P' a pentagon in $X_1(P)$, we find that $X_1(P') = P \cup X_2(P)$ also consists of five disjoint pentagons. Thus, there exists a partition Π of X_{p} into 10 pentagons.

Clearly, we get much more than that partition Π. Every pentagon A in Π determines the partition uniquely. Moreover, A is 'adjacent' to five other members of

Π that are pairwise non-adjacent. Here, adjacency of two pentagons from Π means that each vertex of the first is adjacent to one and only one from the second. On each pair of adjacent pentagons, the induced subgraph of X is a Petersen graph.

This forces a unique partition of Π in two sets of five pentagons say, Π_1, Π_2, with the property that each member of Π_i is adjacent to each member of Π_j for $j \neq i$. Recalling that the full graph has no circuits of length smaller than five and making a picture of the five pentagons in Π_1 next to those of Π_2, we find that, up to symmetry, there is indeed at most one graph with the required properties.

The statement on flag transitivity follows by comparing the two constructions given above. From HoSi, we derive the existence of a group of automorphisms isomorphic to Alt_7 with vertex orbits of lengths 15 and 35, and from HoSi$'$ a group of automorphisms isomorphic to $\mathrm{Aut}(\mathrm{Sym}_6)$ with orbits of lengths 2, 12 (in T), and 36 (in Δ). Thus, the automorphism group must be transitive on the vertex set. Let x be a vertex of X. By the first description, the stabilizer of x contains an element of order 7 permuting the 7 neighbors transitively. By the second description, the stabilizer of x and a neighbor y induces Sym_6 on the six remaining neighbors of x. Thus, the automorphism group is flag transitive on the corresponding geometry Γ. $\square$

Remark 2.2.20 The automorphism group G of the Hoffman-Singleton graph can be determined further. If p is a vertex of HoSi, the stabilizer in $\mathrm{Aut}(\mathrm{HoSi})$ of the set of 7 vertices adjacent to p contains a copy of Sym_7 and cannot be larger. Hence, G has order $50 \cdot 7! = 252000$. It can be shown that G has a simple subgroup H of index 2 which is isomorphic to $\mathrm{PSU}(\mathbb{F}_{25}^3, f)$, the quotient by the center of the subgroup $\mathrm{SU}(\mathbb{F}_{25}^3, f)$ of the unitary group $\mathrm{U}(\mathbb{F}_{25}^3, f)$ with respect to the standard unitary form f on $\mathbb{F}_{25}^3$ consisting of all transformations of determinant 1. This group is simple and G is obtained from it by adjoining the field automorphism (sending each matrix entry to its fifth power).

There are exactly 100 *cocliques* (that is, subgraphs without edges) in HoSi of size 15. A typical example is the Alt_7-orbit of length 15 in the vertex set of HoSi. The group $\mathrm{Aut}(\mathrm{HoSi})$ is transitive on this set of cocliques. Its index two subgroup H has two orbits, of size 50 each.

Notation 2.2.21 As we have mentioned before, the extreme cases where $g = d_{\mathrm{p}} = d_1$ will be studied most intensively. To economize on notation, the abbreviations indicated in Table 2.1 will be frequently used (with $g \in \mathbb{N} \cup \{\infty\}$).

2.3 Diagrams for Higher Rank Geometries

Theorem 2.2.9 shows how to capture the classical definition of a projective plane in terms of the parameters recorded in diagrams for rank two geometries. There is an analogue of this result for affine planes. We will discuss it as well as their connection with projective planes and show that not all affine planes are of the form $\mathrm{AG}(\mathbb{D}^2)$ for a division ring $\mathbb{D}$. The class of these rank two geometries serves as an example

Table 2.1 Pictorial abbreviations

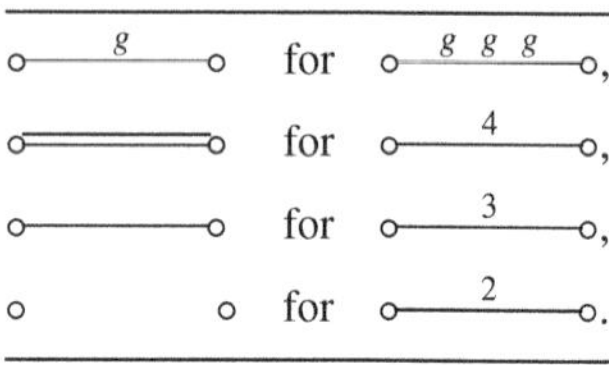

for describing geometries of higher rank by means of diagrams. The crux of the diagram information for a geometry Γ is that it gives us an isomorphism class to which specified rank two residues of Γ belong.

After the introduction of a diagram for geometries of arbitrary rank, we discuss variations on the notion of a graph that is locally isomorphic to a given graph. These variations are governed by a diagram for geometries of rank three.

Recall from Definition 1.2.3 that $x^* = \{x\}^*$ is the set of all elements of Γ incident with x. Throughout the section, we let I stand for a set of types.

Definition 2.3.1 A $\{p, 1\}$-geometry $\Gamma = (X_p, X_1, *)$ is called an *affine plane* if it satisfies the following three axioms.

(1) Any pair of distinct points is on a unique line.
(2) If $x \in X_p$ and $l \in X_1$ are non-incident, then there is a unique line in $X_1 \cap x^*$ at distance strictly greater than two to l. It is called the line through x *parallel* to l.
(3) There exists a non-incident pair in $X_p \times X_1$.

Example 2.3.2 The affine geometries $\mathrm{AG}(V)$, for V a 2-dimensional vector space, are examples. The *real affine plane* (of points and affine lines in $\mathbb{R}^2$) is an example, and so is the analogue over any field other than $\mathbb{R}$.

For q a prime power, the affine plane $\mathrm{AG}(\mathbb{F}_q^2)$ has q^2 points, $q^2 + q$ lines, and $q + 1$ classes of parallel lines. See Fig. 2.6 for $q = 4$. This affine plane has order $(q - 1, q)$. We often abbreviate this statement to saying that its order is q.

Exercise 2.8.9(a) shows a more general construction of an affine plane from a projective plane Π and a line h of it: the subgeometry induced on the set of elements of Π not incident with h. Conversely, Exercise 2.8.9(b) shows that the relation $\parallel$ on the line set of an affine plane $(X_p, X_1, *)$, defined by $l \parallel m$ if and only if $l = m$ or $d(l, m) = 4$, is an equivalence relation on X_1 and that the addition of a line (at 'infinity') built up from $\parallel$-equivalence classes leads to a projective plane.

Remark 2.3.3 Every affine plane is a $(3, 3, 4)$-gon, but not every $(3, 3, 4)$-gon is an affine plane. An example of a non-affine plane that is a $(3, 3, 4)$-gon is obtained by removing a point from the Fano plane. A geometry results having six points and seven lines, three of which have only two points.

Example 2.3.4 The most general construction of an affine plane, in the sense that each affine plane can be constructed in this way, makes use of a *ternary ring*. This is a set R with two distinguished elements 0 and 1 and a ternary operation $T : R^3 \to R$ satisfying the following properties for all $a, b, c, d \in R$.

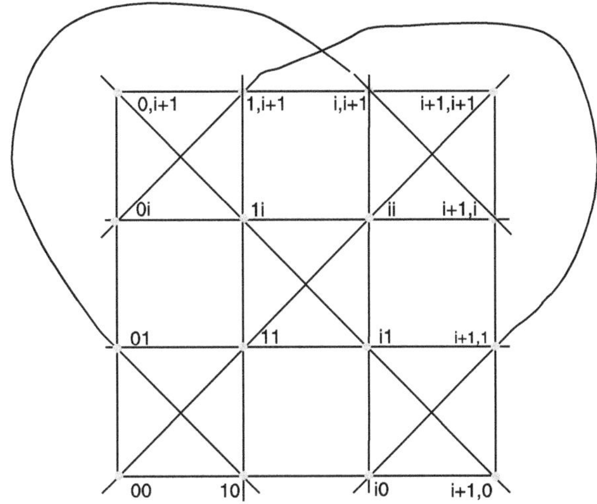

Fig. 2.6 The affine plane of order four with three of the five classes of parallel lines drawn

(1) $T(a, 0, c) = T(0, b, c) = c$.

(2) $T(a, 1, 0) = T(1, a, 0) = a$.

(3) If $a \neq c$, then there is a unique solution $x \in R$ to the equation $T(x, a, b) = T(x, c, d)$.

(4) There is a unique solution $x \in R$ to the equation $T(a, b, x) = c$.

(5) If $a \neq c$, then there is a unique solution $(x, y) \in R^2$ to the two equations $T(a, x, y) = b$ and $T(c, x, y) = d$.

Given a ternary ring (R, T), we construct an affine plane by taking as point set $P = R^2$ and letting the line set L consist of all sets of the forms $\{(x, y) \in P \mid y = T(x, a, b)\}$ and $\{(a, y) \in P \mid y \in R\}$ for $a, b \in R$. The triple $(P, L, *)$, where $*$ denotes symmetrized inclusion, is an affine plane. Any affine plane is isomorphic to one constructed in this way, and any two affine planes are isomorphic if and only if their ternary rings are isomorphic.

We will discuss three examples, where

$$T(a, b, c) = ab + c \qquad (2.1)$$

for a multiplication $(a, b) \mapsto ab$ and addition $(a, b) \mapsto a + b$ on R. Such examples are called *linear*.

First, if $R = \mathbb{D}$ is a division ring, then the construction of an affine plane using (2.1) copies the usual one for $\mathrm{AG}(\mathbb{D}^2)$.

For the second example, we let C be the 8-dimensional vector space over $\mathbb{R}$, with basis $e_0, e_1, \ldots, e_7$ and define multiplication as the bilinear operation on C determined by $e_0 = 1$, the identity element, $e_i^2 = -1$ for $i \in [7]$, and $e_i e_j = -e_j e_i = e_k$ whenever the 3-cycle (i, j, k) (in Sym_7) can be transformed to the 3-cycle $(1, 2, 4)$ under a power of the permutation $(1, 2, 3, 4, 5, 6, 7)$. For instance, $e_5 e_7 = e_4$ as $(5, 7, 4) = (4, 5, 7) = (i, 1 + i, 3 + i)$ for $i = 4$. The unordered triples arising in this way are the lines of the Fano plane as pictured in Fig. 1.21. The multiplication on C is not associative; cf. Exercise 2.8.11. The algebra C is known as the *Cayley division ring.* The ternary operation defined on it by means of (2.1) gives a ternary ring and hence an affine plane.

The third example is a finite one: the set $J = \{0, \pm 1, \pm i, \pm j, \pm k\}$ of size 9 with multiplication as in the quaternion group of order 8, with 0 added; so $ij = -ji = k$, and $i^2 = j^2 = k^2 = -1$ whereas $0x = x0 = 0$ for all $x \in J$. An Abelian group structure on J is determined by $1 + 1 = -1$, $1 + i = j$, $1 - i = k$, and $x + x + x = 0$ for all $x \in J$. The ternary ring on J determined by (2.1) leads to an affine plane on 81 points. It is not isomorphic to $\mathrm{AG}(\mathbb{F}_9^2)$.

If we specify point and line orders, we can characterize finite affine planes by means of a diagram.

Proposition 2.3.5 *Let $q \in \mathbb{N}$, $q \geq 2$. Every $\{\mathrm{p}, \mathrm{l}\}$-geometry belonging to the diagram* $\overset{\mathrm{p}}{\underset{q-1}{\circ}} \!\!\!\overset{3\ \ 3\ \ 4}{\rule{2.5cm}{0.4pt}}\!\!\! \overset{\mathrm{l}}{\underset{q}{\circ}}$ *is a finite affine plane. Conversely, every finite affine plane belongs to such a diagram.*

Proof Axioms (1) and (3) of Definition 2.3.1 are obviously satisfied. For Axiom (2) let x be a point, and m a line such that x and m are not incident. There are $q + 1$ lines on x, each containing $q - 1$ points distinct from x. Since any two points are collinear and no two points can be on more than one line (the girth is 3), the total number of points amounts to $1 + (q + 1)(q - 1) = q^2$. Moreover, each point of m is incident with a unique line on x. This accounts for q of the $q + 1$ lines on x, leaving a single line on x parallel to m. Hence Axiom (2).

The second assertion is immediate from the definition of affine plane, except for the order information. Let Π be a finite affine plane and suppose that m is a line of Π of size q. We need to show that each line of Π has size q and each point is on exactly $q + 1$ lines. If p is a point of Π not on m, there are exactly q lines on p meeting m in a point. Adding to this number the line n on p parallel to m, we find that there are exactly $q + 1$ lines on p. By a similar argument, each line disjoint from p has exactly q points. Now, by firmness, n must have a second point. Arguing for this point as for p, we see that each line, except possibly n has exactly q points. Hence, each point on m is on exactly $q + 1$ lines, and so, taking a point on m and arguing as before, we obtain that n has exactly q points. Hence the second assertion. $\qquad\square$

Notation 2.3.6 Since the class of affine planes is important, it is denoted by the separate diagram $\overset{\mathrm{p}}{\circ}\!\!\overset{\mathrm{Af}}{\rule{2.5cm}{0.4pt}}\!\!\overset{\mathrm{l}}{\circ}$.

We have come to the point where any class $\mathcal{K}$ of geometries over $\{\mathrm{p}, \mathrm{l}\}$ can be depicted by a labelled edge between vertices p and l, as in $\overset{\mathrm{p}}{\circ}\!\!\overset{\mathcal{K}}{\rule{2.5cm}{0.4pt}}\!\!\overset{\mathrm{l}}{\circ}$. This will be a useful convention for specifying rank two residues when working with geometries of rank greater than 2. A specific example of which we will make use later is $\overset{\mathrm{p}}{\circ}\!\!\overset{\mathrm{C}}{\rule{2.5cm}{0.4pt}}\!\!\overset{\mathrm{l}}{\circ}$ for the class C of all geometries over $\{\mathrm{p}, \mathrm{l}\}$ whose point collinearity graphs are complete and whose line order is one (so any two points are on a line of size two). A subtlety to note here is that the label depends on the directed edge $(p, 1)$, whereas the class of rank two geometries is defined over the unordered set $\{p, 1\}$. In particular, $\overset{1}{\circ}\!\!\overset{C}{\rule{2cm}{0.4pt}}\!\!\overset{p}{\circ}$ represents a (different) class of geometries over $\{p, 1\}$, namely those that are dual to members of C.

Here is the general notion of a diagram.

Definition 2.3.7 A *diagram* over I is a map D defined on $\binom{I}{2}$, the set of unordered pairs from I, that assigns to every pair $\{i, j\}$ some class $D(i, j) = D(j, i)$ of rank two geometries over $\{i, j\}$.

A geometry Γ *belongs to the diagram* D over I if, for all distinct types $i, j \in I$ and every flag F of Γ such that Γ_F is of type $\{i, j\}$, the residue Γ_F is isomorphic to a geometry in $D(i, j)$. In this case, Γ is also said to be of *type D*.

If Γ belongs to a diagram D over I and has order $(s_j)_{j \in J}$ for some subset J of I, then the parameters s_j are often given as subscripts in a pictorial description of D as in Remark 2.2.17.

Remark 2.3.8 We can view the diagram D over I as a complete labelled graph on the vertex set I, in which the label depends on the directed edge. The label of the directed edge (j, i) is determined by that of (i, j) by the prescription that the latter is assigned the dual geometry of the former (cf. Definition 2.2.5). Thus, given $D(j, i) = D(i, j)$, we need only draw one of them. For instance, if, for some i, $j \in I$, the class $D(i, j)$ consists of all $(5, 5, 6)$-gons over (i, j) (or, equivalently, all $(5, 6, 5)$-gons over (j, i)), we adorn the edge with the numbers 5, 5, and 6 in such a way that the node j is closest to 6 (cf. Remark 2.2.17). The dual geometry of a member of $D(i, j)$ is then a $(5, 6, 5)$-gon over (i, j) and does not belong to $D(j, i)$.

Example 2.3.9 Let Δ be a graph. We say that a graph Σ is *locally* Δ if, for every $x \in \Sigma$, the subgraph induced on the neighbors of x is isomorphic to Δ. For instance, the complete graph on n vertices is locally the complete graph on $n - 1$ vertices (and is the unique connected graph with that property). A graph that is locally Petersen (cf. Example 1.3.3) can be obtained on the vertex set of all transpositions (i.e., conjugates of $(1, 2)$) of Sym_7 by demanding that two vertices are adjacent whenever they commute. The fact that Σ is locally Δ can be expressed in diagram language. Consider the following diagram

$$D_\Delta \; := \; \overset{p}{\underset{1}{\circ}}\!\!\rule{2cm}{0.4pt}\!\!\overset{1}{\underset{1}{\circ}}\!\!\overset{\Delta}{\rule{1.5cm}{0.4pt}}\!\!\overset{c}{\circ}.$$

Here, we abbreviated $\{\Delta\}$ to Δ. Suppose that Σ is a connected graph such that each vertex lies in at least two cliques of size three. Take X_p, X_1, and X_c to be the sets of all vertices, all edges, and all cliques of size 3 of Σ, respectively, and define incidence $*$ by symmetrized containment to obtain a $\{p, 1, c\}$-geometry $(X_p, X_1, X_c, *)$. This geometry belongs to the above diagram of rank three if and only if Σ is locally Δ.

Example 2.3.10 Suppose that Γ is a $\{p, 1, c\}$-geometry belonging to the diagram D_Δ of Example 2.3.9. Consider the graph Σ whose vertex set is X_p and in which two vertices are adjacent if and only if there is a member of X_1 incident with both. The map $\phi : X_p \cup X_1 \to X_p \cup \binom{X_p}{2}$ which is the identity on X_p and maps any line in X_1 to the pair of points to which it is incident, is a homomorphism from the $\{p, 1\}$-truncation of Γ to Σ (viewed as a geometry whose vertices have type p and whose edge have 1). This homomorphism is surjective as a map between point sets.

However, Σ need not be locally Δ. An easy example that is not firm shows what may go wrong: take Γ to be the rank three geometry of the octahedron from which half the faces are removed; those that are white in a black and white coloring (such that no two adjacent faces have the same color). Then Σ is the octahedron graph, which is locally a quadrangle, but the graph Δ of the diagram D_Δ for Γ is the disjoint union of two cliques of size two rather than a quadrangle.

Example 2.3.11 We give a flag-transitive geometry Γ with diagram D_{Pet}, whose associated graph Σ is locally the complete graph on 10 points (instead of Pet). We construct it by use of a transitive extension of the alternating group Alt_5 in its action on Pet.

Suppose that G_0 is a group of automorphisms on a graph Δ with vertex set Ω_0. A *transitive extension* of (G_0, Ω_0) is a transitive permutation group G on the set $\Omega = \Omega_0 \cup \{o\}$ (the extension of Ω_0 by the single point o outside Ω_0) such that the stabilizer of o in G coincides with G_0 (in accordance with what the notation suggests). In general, such an extension need not exist. Suppose that (G, Ω) is a transitive extension of (G_0, Ω_0). Denote by $\mathcal{B}$ the G-orbit of the triple $\{o, a, b\}$, where $\{a, b\}$ is an edge of Δ. Suppose that Δ is connected. If G_0 is edge transitive on Δ, the geometry $\Gamma = (\Omega, \binom{\Omega}{2}, \mathcal{B}, *)$, where $*$ is symmetrized inclusion, is residually connected and belongs to D_Δ. Furthermore, the collinearity graph Σ of Ω is the complete graph on Ω, and so it is impossible to reconstruct $\mathcal{B}$ or Γ from Σ.

Now take $G_0 = \mathrm{Alt}_5$ acting on the Petersen graph $\Delta = \mathrm{Pet}$ (cf. Example 1.3.3) with the vertex set $\Omega_0 = \binom{[5]}{2}$. We know that G_0 is edge transitive on Pet. In order to construct a transitive extension, we pin down some necessary conditions. Consider the triple $B = \{o, 12, 34\} \in \mathcal{B}$. Its point-wise stabilizer in G lies in G_0, where it is readily seen to be $\{\mathrm{id}, (1, 2)(3, 4)\}$, of order two.

The set stabilizer in G_0 of the triple B contains the involution $(1, 3)(2, 4)$ inducing a transposition on B. As G_0 is transitive on the edge set of Pet, it is transitive on the set of triples in $\mathcal{B}$ containing o, so G is transitive on the collection of incident pairs from $\Omega \times \mathcal{B}$. The (set-wise) stabilizer G_B of B in G induces $\mathrm{Sym}(B)$ on B. Thus $G_B \cong C_2.\mathrm{Sym}_3$ (that means, there is a normal subgroup of order 2 in G_B with quotient group Sym_3). By Lagrange's Theorem, $|\mathcal{B}| = |G|/|G_B| = 11|G_0|/|2.\mathrm{Sym}_3| = 660/12 = 55$. The triples in $\mathcal{B}$ containing o correspond to the edges of Pet; there are precisely 15 of them. So there are 40 triples in $\mathcal{B}$ lying entirely in Ω_0; they form a union of G_0-orbits. On the other hand, the $\binom{10}{3} = 120$ triples in Ω_0 fall into G_0-orbits of lengths 10 (the neighbors of a vertex in Pet), 20 (cocliques of size 3), 30, 30 (for two orbits on triples carrying a single edge), and 30 (paths of length 3). The only way to make 40 is via $10 + 30$;

this indicates that the triple $\{34, 45, 35\}$ (consisting of the neighbors of 12 in Pet) must belong to $\mathcal{B}$.

Now switch the viewpoint to the residue of 12. The triples in $\mathcal{B}$ containing 12 determine the edges of a Petersen graph on $\Omega \setminus \{12\}$, say Pet_{12}. The neighbors of o are known from the triples in $\mathcal{B}$ containing o and 12; these are $\{o, 12, a\}$ for $a \in \{34, 45, 35\}$. The neighbors of 45 in Pet form the triple $\{12, 13, 23\}$, so 13 and 23 are adjacent in Pet_{12}. If the paths of length 3 in Pet were to be triples in $\mathcal{B}$, then both $\{45, 13\}$ and $\{45, 23\}$ are edges of Pet_{12}, leading to a clique on $\{45, 13, 23\}$, a contradiction as there are no cliques of size 3 in a Petersen graph. Hence the missing triples of $\mathcal{B}$ must come from one of the G_o-orbits of triples in Pet inducing a subgraph with a single edge. Since they are interchanged by Sym_5, it does not matter which one we take. Pick the one containing $\{12, 15, 35\}$. Inspection shows that the new edges coming from the triples containing 12 in this G_o-orbit indeed make Pet_{12} into a Petersen graph. Identification of the two Petersen graphs Pet and Pet_{12} can now be achieved via the permutation

$$g = (\mathrm{o}, 12)(13, 25)(14, 23)(15, 24) \in \mathrm{Sym}(\Omega).$$

As g preserves $\mathcal{B}$, it is readily checked that the group $G = \langle g, G_o \rangle$ generated by g and G_o is a transitive extension of G_o on Ω_o.

In fact, $G \cong \mathrm{PSL}(\mathbb{F}_{11}^2)$ (as introduced at the end of Example 1.8.16). As G acts flag transitively on the geometry, a direct existence proof of Γ is the following description

$$\Gamma\big(G, \big(G_o, \langle d, f, g \rangle, \langle d, e, g \rangle\big)\big)$$

in terms of subgroups of Sym_{11}, where $G = \langle d, e, f, g \rangle$, the permutation g is as above, and the permutations d, e, f correspond to the following elements of Alt_5 and $\mathrm{Sym}(\Omega)$.

Element	In Alt_5	Action on Ω
d	$(1, 2)(3, 4)$	$(13, 24)(14, 23)(15, 25)(35, 45)$
e	$(1,3)(2,4)$	$(12, 34)(14, 23)(15, 35)(25, 45)$
f	$(3,4,5)$	$(12, 13, 14)(25, 26, 27)(34, 45, 35)$

Flag transitivity also implies the existence of a c-order of Γ; it is 2.

2.4 Coxeter Diagrams

Geometries belonging to a so-called Coxeter diagram are a central theme in this book. Fix a set I.

Definition 2.4.1 A diagram D over I will be called a *Coxeter diagram* (over I) if, for each pair i, j of types, there is a number m_{ij} such that $D(i, j)$ is the class of all generalized m_{ij}-gons.

Table 2.2 Coxeter diagrams of some geometries

Geometry name	Coxeter diagram	Diagram name
Cube	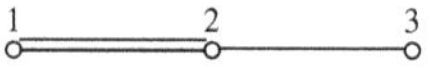	
Octahedron	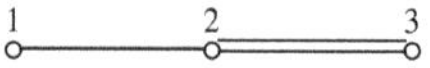	B_3
Icosahedron	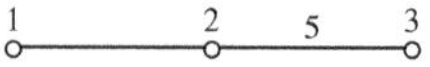	H_3
Dodecahedron	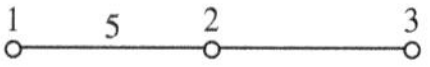	
$\mathbb{E}^2$ tiling by quadrangles	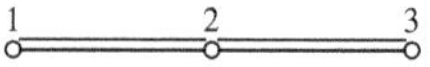	$\widetilde{B}_2$
$\mathbb{E}^2$ tiling by hexagons	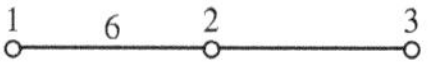	$\widetilde{G}_2$
$\mathbb{E}^2$ tiling by triangles	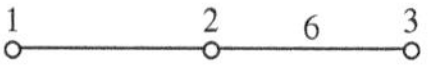	

In view of Table 2.1 it is no surprise that the Coxeter diagram is fully determined by a matrix satisfying the following definition.

Definition 2.4.2 A *Coxeter matrix* over I is a matrix $M = (m_{i,j})_{i,j \in I}$ where $m_{i,j} \in \mathbb{N} \cup \{\infty\}$ with $m_{i,i} = 1$ for $i \in I$, and $m_{i,j} = m_{j,i} > 1$ for distinct $i, j \in I$.

The Coxeter matrix $M = (m_{ij})_{i,j \in I}$ and the Coxeter diagram D for which $D(i, j)$ is the class of $m_{i,j}$-gons, determine each other. If Γ is a geometry of type D, we also say that Γ is of *Coxeter type* M.

Example 2.4.3 Table 2.2 contains a list of some geometries of Coxeter type that we have met before and a Coxeter diagram to which they belong. All orders are equal to 1. The labels near the nodes indicate types of the elements. The Coxeter matrix of the dodecahedron, for instance, is

$$\begin{pmatrix} 1 & 5 & 2 \\ 5 & 1 & 3 \\ 2 & 3 & 1 \end{pmatrix}.$$

Notice that dual (weakly isomorphic) geometries, like the icosahedron and the dodecahedron, have 'dual' diagrams. The hemi-dodecahedron of Example 1.3.4 also belongs to the dodecahedron diagram.

Example 2.4.4 The Coxeter diagram on the tricolored vertices of the $\mathbb{E}^2$ tiling by triangles of Example 1.3.10 is a triangle. The Coxeter diagram of the $\mathbb{E}^3$ tiling by bicolored cubes and bicolored vertices is a quadrangle, in accordance with its digon diagram discussed in Example 2.1.3. The Coxeter diagrams of the $\mathbb{E}^3$ tilings by cubes and bicolored cubes, respectively, are given in Fig. 2.7.

The 12 vertices, 30 edges, and 12 pentagons of an icosahedron, with incidence being symmetrized containment, lead to the thin geometry of the *great dodecahedron* with diagram

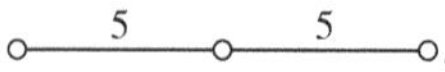

$$\circ \overset{5}{\rule{1.5cm}{0.4pt}} \circ \overset{5}{\rule{1.5cm}{0.4pt}} \circ.$$

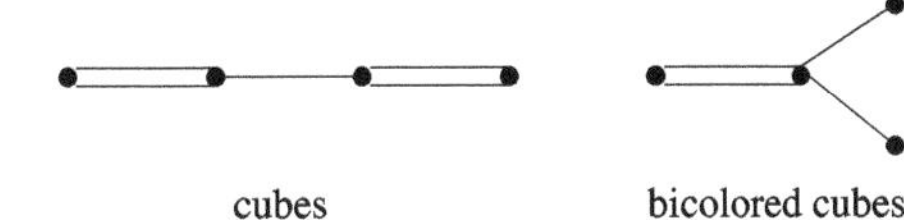

Fig. 2.7 $\mathbb{E}^3$ tilings and their Coxeter diagrams

cubes   bicolored cubes

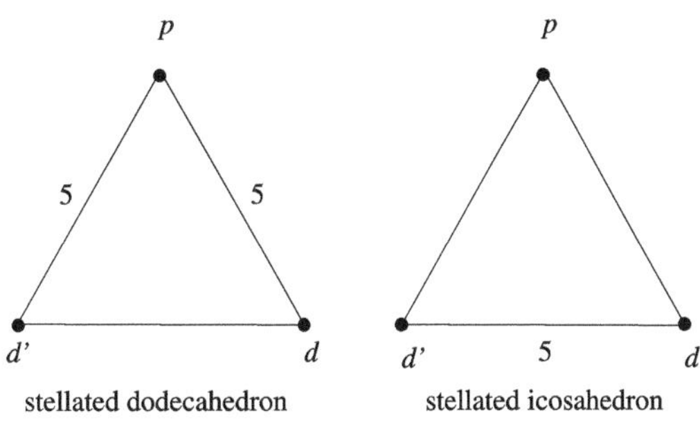

Fig. 2.8 Diagrams of the stellated dodecahedron and icosahedron

stellated dodecahedron   stellated icosahedron

It has a flag-transitive group of automorphisms, and a quotient geometry by a group of order two having the same diagram (see Exercise 2.8.13).

Example 2.4.5 Let D be the dodecahedron and D' the associated great stellated dodecahedron (see Example 1.3.4 and Fig. 1.12). Let X_p be the vertex set of D, and X_d, $X_{\mathrm{d}'}$ the set of all pentagons of D and D', respectively. Then the incidence system $\Gamma(X_\mathrm{p}, X_\mathrm{d} \cup X_{\mathrm{d}'})$ over $\{\mathrm{p}, \mathrm{d}, \mathrm{d}'\}$ built by use of the Principle of Maximal Intersection (cf. Exercise 1.9.20) is a geometry. It belongs to the diagram at the left hand side of Fig. 2.8.

The icosahedron can also be stellated. Taking X_p, X_d, and $X_{\mathrm{d}'}$ as for the dodecahedron, we find a geometry $\Gamma(X_\mathrm{p}, X_\mathrm{d} \cup X_{\mathrm{d}'})$ over $\{\mathrm{p}, \mathrm{d}, \mathrm{d}'\}$ of type the diagram at the right hand side of Fig. 2.8.

Example 2.4.6 Consider the tiling of $\mathbb{E}^2$ in Fig. 1.3. We derive another tiling T' from it. The elements of T' are the hexagons of T (type h), the rectangles obtained as the union of two adjacent squares of T (type r) and the triangles obtained as the union of four triangles of T (type t). Define incidence on the elements of T' by the Principle of Maximal Intersection (Exercise 1.9.20). We find a geometry of the following Coxeter type.

$\overset{\displaystyle \mathrm{t}}{\underset{\displaystyle 1}{\circ}} \overset{\displaystyle 6}{\rule{2cm}{0.4pt}} \overset{\displaystyle \mathrm{r}}{\underset{\displaystyle 1}{\circ}} \rule{2cm}{0.4pt} \overset{\displaystyle \mathrm{h}}{\underset{\displaystyle 1}{\circ}}$

Another geometry of the same Coxeter type and same orders can be constructed as follows. Let Δ be the graph on six vertices obtained from the complete graph by deletion of all edges of a single hexagon. There are nine edges and three hexagons in Δ. The vertices, edges, and hexagons give a rank three geometry whose automorphism group has order 12. This geometry is not flag transitive.

Recall from Example 1.4.9 the definition of the *projective geometry* PG(V).

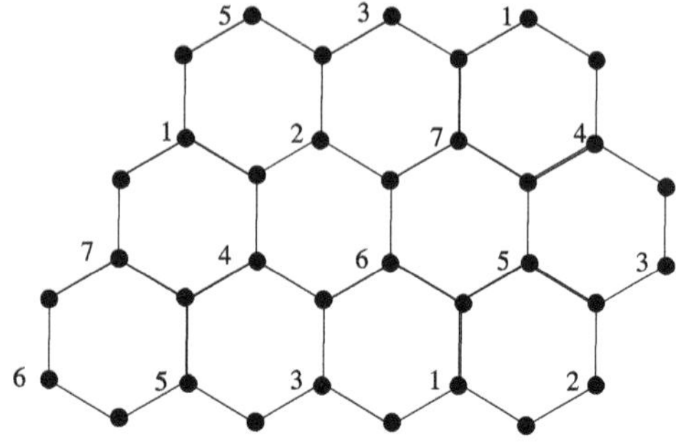

Fig. 2.9 The incidence graph of the Fano plane. Vertices that are not labelled represent lines. Labelled vertices represent points. Vertices with the same label need to be identified

Proposition 2.4.7 *Let $n > 0$. If V is a vector space over a division ring of finite dimension $n + 1$, then $\mathrm{PG}(V)$ is a thick $[n]$-geometry with Coxeter diagram*

$$\mathrm{A}_n = \overset{1}{\circ}\!-\overset{2}{\circ}\!\!\!\!-\overset{3}{\circ}\cdots\cdots\overset{n-1}{\circ}\!\!\!\!-\overset{n}{\circ}.$$

If the division ring has size q, then all i-orders are equal to q.

Proof Let V be defined over the division ring $\mathbb{D}$, so $V \cong \mathbb{D}^{n+1}$. Residual connectedness of $\mathrm{PG}(V)$ follows from the fact that every residue is a direct sum of geometries of the form $\mathrm{PG}(W)$ for W a vector space over $\mathbb{D}$. A projective plane $\mathrm{PG}(\mathbb{D}^3)$ is a generalized 3-gon as defined in Definition 2.2.6 and so belongs to the Coxeter diagram A_2. In $\mathrm{PG}(V)$, each rank two residue of type $\{i, i + 1\}$, for some $i \in [n]$, is also a projective plane isomorphic to $\mathrm{PG}(\mathbb{D}^3)$. The points on a line are in bijective correspondence with the disjoint union of $\mathbb{D}$ and a point 'at infinity'. Therefore, the 1-order, and, by dualization and induction, all other i-orders of $\mathrm{PG}(V)$ are one less than $|\mathbb{D}|$. This implies that the geometry is thick. $\square$

Example 2.4.8 The smallest thick projective geometry is $\mathrm{PG}(\mathbb{F}_2^3)$. Figure 1.21 provides a picture of it as a classical plane with points and lines, but Fig. 2.9 depicts its incidence graph.

Remark 2.4.9 The diagram A_n of $\mathrm{PG}(V)$ has only one nontrivial symmetry and so the order of the quotient of the correlation group of $\mathrm{PG}(V)$ by its automorphism group $\mathrm{Aut}(\mathrm{PG}(V))$ is at most two. If such a duality exists, then $\mathbb{D}$ must be isomorphic to $\mathbb{D}^{\mathrm{op}}$; cf. Exercise 1.9.13.

Affine geometries $\mathrm{AG}(V)$ as defined in Example 1.4.10 do not belong to a Coxeter diagram because affine subplanes are not generalized polygons. The rank two diagram Af for affine planes appearing in the diagram of Proposition 2.4.10 below was introduced in Notation 2.3.6.

Proposition 2.4.10 *The geometry $\mathrm{AG}(\mathbb{D}^n)$ is firm and residually connected and belongs to the diagram*

$$\mathrm{Af}_n: \quad \overset{1}{\circ}\!\!\overset{\mathrm{Af}}{-\!\!-}\overset{2}{\circ}\!\!-\overset{3}{\circ}\cdots\cdots\overset{n-1}{\circ}\!\!-\overset{n}{\circ}.$$

Proof The point residues of $AG(\mathbb{D}^n)$ are projective geometries isomorphic to $PG(\mathbb{D}^n)$. The i-orders for $i > 1$ are $|\mathbb{D}|$ as for $PG(\mathbb{D}^n)$, discussed in Proposition 2.4.7. The number of points on a line equals $\mathbb{D}$, and so $AG(\mathbb{D}^n)$ is firm with 1-order $|\mathbb{D}| - 1$. $\qquad\square$

The converses of Propositions 2.4.7 and 2.4.10, which recognize $PG(V)$ and $AG(V)$ as geometries of the indicated types, will appear in Chap. 6.

Example 2.4.11 We construct a remarkable geometry of Coxeter type B_3 (see Table 2.2). Let X_1 be the set [7], and let X_2 be the collection of all triples from X_1. Each Fano plane with X_1 as point set can be viewed as a collection of 7 triples from X_1, and hence as a set of 7 elements of X_2. As discussed in Example 2.2.18, the alternating group on X_1 has two orbits on the collection of all Fano planes, interchanged by an odd permutation of X_1; each orbit has 15 members. Let X_3 be one of these orbits. Then $\Gamma := (X_1, X_2, X_3, *)$, where $*$ is symmetrized containment, is a geometry belonging to the diagram

$$
\overset{\displaystyle 1}{\underset{\displaystyle 2}{\circ}}\!\!\!\!\rule[0.5ex]{4cm}{0.4pt}\!\!\!\!\overset{\displaystyle 2}{\underset{\displaystyle 2}{\circ}}\!\!\!\!=\!\!\!\!\overset{\displaystyle 3}{\underset{\displaystyle 2}{\circ}}
$$

on which $G := \mathrm{Alt}_7$ acts flag transitively. This accounts for the full group $\mathrm{Aut}(\Gamma)$ as restriction to X_1 is a faithful homomorphism, and $\mathrm{Sym}(X_1)$ is the only permutation group on X_1 properly containing G and does not preserve the selected class of 15 Fano planes. This rank three geometry is known as the *Neumaier geometry*. Observe that each 1-element is incident with each 3-element. By Theorem 2.2.11 or, by identification of the description in terms of pairs and partitions of [6], the residue of a 1-element is the generalized quadrangle of Example 2.2.10. In terms of subgroups, Γ is the geometry $\Gamma(G, (\mathrm{Alt}(\{2, \ldots, 6\}), G_l, L))$, where $l = \{1, 2, 4\}$, a line of the Fano plane described in Exercise 1.9.7, and L is the automorphism group of that plane. Corollary 1.8.13 applied to this description of Γ readily gives that the Neumaier geometry is residually connected.

There is an extraordinary relation with the projective space of $\mathbb{F}_2^4$ (cf. Examples 1.4.9 and 1.5.6). Consider the truncated geometry $_{\{2,3\}}\Gamma$. Any two distinct members of X_3 meet in a unique element of X_2 (to see this, use the fact that Alt_7 acts transitively on the collection of ordered pairs of distinct elements from X_3). Put $V = \{0\} \cup X_3$, and define addition of $u, v \in V$ on V by $v + v = 0$, $u + v = v + u = v$ if $u = 0$, and $u + v = w$ if $u, v \in X_3$ are incident with $l \in X_2$ and $\{u, v, w\} = X_3 \cap l^*$. This turns V into an additive group isomorphic to $\mathbb{F}_2^4$, so it can be viewed as a vector space. Moreover, X_3 and X_2 can be identified with the point and line set of this space. Since $\mathrm{Aut}(\Gamma)$ acts faithfully on V, we have obtained an embedding of Alt_7 in the general linear group $GL(V) = GL(\mathbb{F}_2^4)$. From this it is easy to derive the sporadic isomorphism $GL(\mathbb{F}_2^4) \cong \mathrm{Alt}_8$. For, $GL(V)$ has order $|\mathrm{Alt}_8|$, so G maps to a subgroup, H say, of $GL(V)$ isomorphic to Alt_7 and of index eight in $GL(V)$. By Theorem 1.7.5, $GL(V)$ has a transitive representation on $GL(V)/H$ of degree eight. Since $GL(V)$ is a simple group (a fact that is not hard to prove, but assumed for now), the representation is faithful, and it is an embedding in Alt_8. By comparison of orders, the embedding must be an isomorphism. The stabilizer in G of a triple

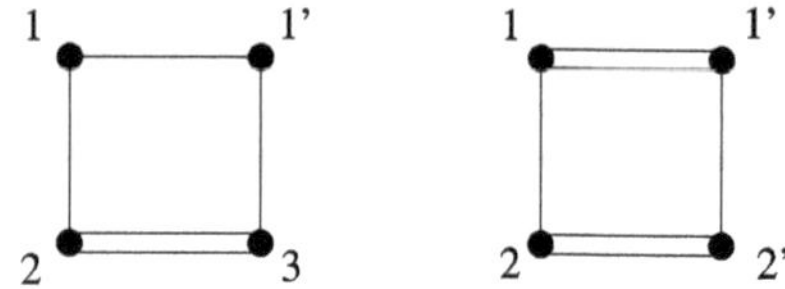

Fig. 2.10 Coxeter diagrams of geometries related to HoSi

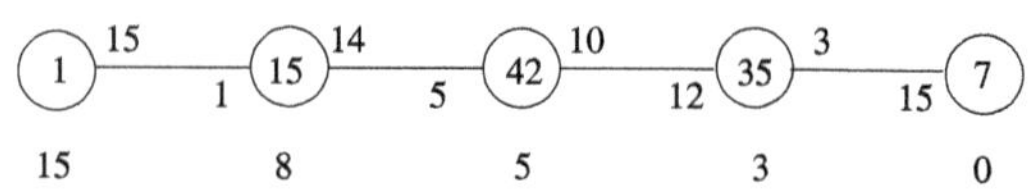

Fig. 2.11 Distribution diagram of the hundred 15-cocliques of HoSi

l of [7] (that is, an element of X_2, or a projective line of V) has orbits of lengths 1, 12, 18, 4 on X_2. There are precisely $3 \times 6 = 18$ projective lines meeting l in a point. Apparently, in terms of projective lines, the G_l-orbits of lengths 12 and 4 fuse to the single $GL(V)_l$-orbit of all lines disjoint from l.

Example 2.4.12 Consider once more the Hoffman-Singleton graph HoSi of Example 2.2.18. There are two geometries related to this graph with diagrams as depicted in Fig. 2.10.

Their constructions are based on a partition of the maximal cocliques (i.e., complements of cliques) of HoSi that are of size 15. Consider the graph on these cocliques in which two cocliques are adjacent if their intersection has size eight. This graph has the following distribution diagram. For each $i \in \{0, \ldots, 4\}$, the size of the intersection of two 15-cocliques at mutual distance i is given underneath the circle at distance i from the left hand node (which represents a fixed 15-coclique of HoSi).

In order to construct the two rank four geometries, take X_1 and $X_{1'}$ to be two copies of the vertex set of HoSi, take X_2 and X_3 to be the H-orbits of cocliques of size 15 in HoSi, where H is the index two subgroup of Aut(HoSi) described in Remark 2.2.20. It can be read off from the distribution diagram of Fig. 2.11 that the graph has no odd cycles, so it is bipartite. The sets X_2 and X_3 are the two classes of the unique partition. Finally, take $X_{2'}$ to be a copy of X_2. Define incidence $*$ for $a \in X_1, a' \in X_{1'}, b \in X_2, b' \in X_{2'}$, and $c \in X_3$, by

$$
\begin{aligned}
a * a' &\iff a \sim a' \quad \text{in HoSi}, & a * b &\iff a \in b, \\
a * b' &\iff a \notin b', & a * c &\iff a \notin c, \\
a' * b &\iff a' \notin b, & a' * b' &\iff a' \in b', \\
a' * c &\iff a' \in c, & b * b' &\iff b \cap b' = \emptyset, \\
b * c &\iff |b \cap c| = 8. &
\end{aligned}
$$

The *Wester HoSi geometry* is $(X_1, X_{1'}, X_2, X_3, *)$, and the *Neumaier HoSi geometry* is $(X_1, X_{1'}, X_2, X_{2'}, *)$. The residues of type B_3 are isomorphic to the Neumaier geometry of Example 2.4.11.

Remark 2.4.13 The graph obtained from the graph, say Δ, on the 15-cocliques of HoSi discussed in Example 2.4.12 can be used to construct another graph on the same vertex set with an interesting automorphism group. In the graph Δ' meant here, two vertices are adjacent whenever they are at distance one or four in Δ. This graph Δ' has 100 vertices and is regular of valency 22. Moreover, two adjacent vertices have no common neighbors and two non-adjacent vertices have six common neighbors. The automorphism group of Δ' has a unique index two subgroup; it is isomorphic to the sporadic simple group HS, named after Higman and Sims.

2.5 Shadows

We started Chap. 1 by saying that we want to abandon the usual physical viewpoint according to which each element of a geometry is a point or a set of points, and we have described a more abstract viewpoint. However, now that the latter has been developed, we want to point out how to recover the physical viewpoint, because the latter has an important role for instance in the construction of examples and in characterizations. We fix a finite set of types I, let $\Gamma = (X, *, \tau)$ be an I-geometry, and let J be a non-empty subset of I. Often, but not always, J will just be a singleton.

Definition 2.5.1 For every flag F of Γ, the J-*shadow* (of F) or *shadow* of F on J is the set $\mathrm{Sh}_J(F)$ of all flags of type J that are incident with F. If we need to emphasize the dependence on Γ of the shadow on J, we write $\mathrm{Sh}_J(F, \Gamma)$ instead of $\mathrm{Sh}_J(F)$. If $J = \{j\}$, we will often write $\mathrm{Sh}_j(F)$ instead of $\mathrm{Sh}_J(F)$.

For $j \in J$, the shadow of a flag of type $I \setminus \{j\}$ on J is said to be a j-*line* or a *line of type* j. Let T be a subset of J. The *shadow space* $\mathrm{ShSp}(\Gamma, J, T)$ of type (J, T) is the pair $(\mathrm{Sh}_J(\emptyset), L)$ where L is the collection of all j-lines for $j \in T$. The members of $\mathrm{Sh}_J(\emptyset)$ are called the *points* and the members of L are called the *lines* of $\mathrm{ShSp}(\Gamma, J, T)$. We write $\mathrm{ShSp}(\Gamma, J) = \mathrm{ShSp}(\Gamma, J, J)$ and call it the *shadow space* of Γ on J. If $J = \{j\}$, we also write $\mathrm{ShSp}(\Gamma, j)$ instead of $\mathrm{ShSp}(\Gamma, J)$, and speak of the shadow space on j instead of $\{j\}$.

In order to distinguish between shadow spaces viewed as line spaces and shadow spaces equipped with all shadows, we will refer to the latter as *full shadow spaces*.

If Γ is a firm geometry, then every shadow on J containing more than one point and not containing any other shadow satisfying this requirement, is a line of $\mathrm{ShSp}(\Gamma, J)$.

If Γ belongs to a diagram D over I, then we depict $\mathrm{ShSp}(\Gamma, J)$ in D by drawing a circuit around the set of vertices of I belonging to J.

Example 2.5.2 The geometry Γ of a cube belongs to the Coxeter diagram
$\overset{1}{\circ}\!=\!=\!=\!\overset{2}{\circ}\!-\!-\!-\!\overset{3}{\circ}$ with eight elements of type 1, twelve of type 2, and six of type 3. There are seven non-empty subsets J in $I = [3]$. Each of these gives rise to a J-space which can be represented by one of the seven semi-regular convex polyhedra having the same group of isometries as the cube in $\mathbb{E}^3$. The edges of the polyhedron representing $\mathrm{ShSp}(\Gamma, J)$ are precisely the lines while its faces are the other non-trivial shadows.

Fig. 2.12 The truncations of the cube

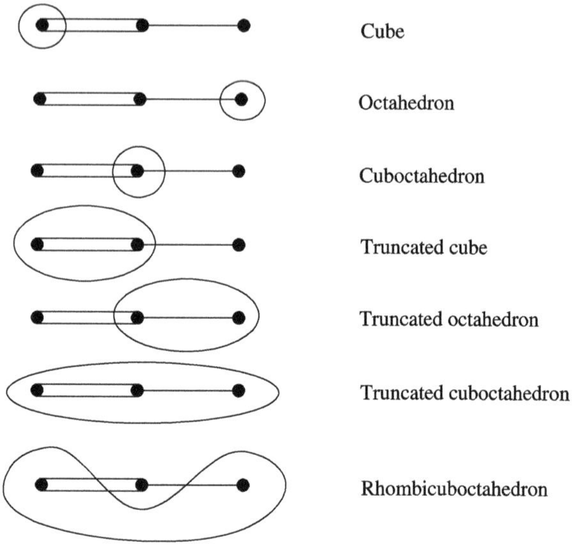

Cube

Octahedron

Cuboctahedron

Truncated cube

Truncated octahedron

Truncated cuboctahedron

Rhombicuboctahedron

Fig. 2.13 The cuboctahedron and the truncated cube

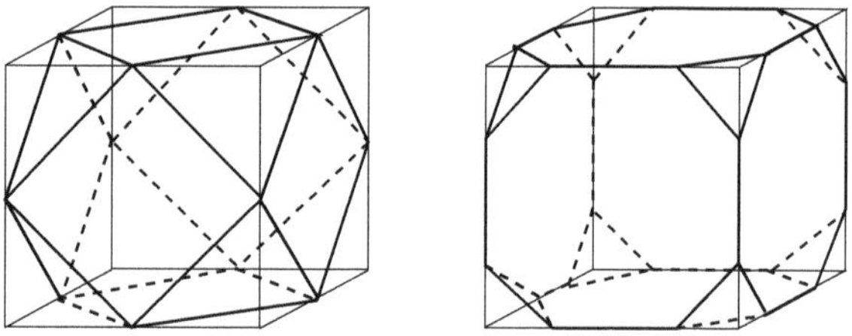

Fig. 2.14 The rhombicuboctahedron

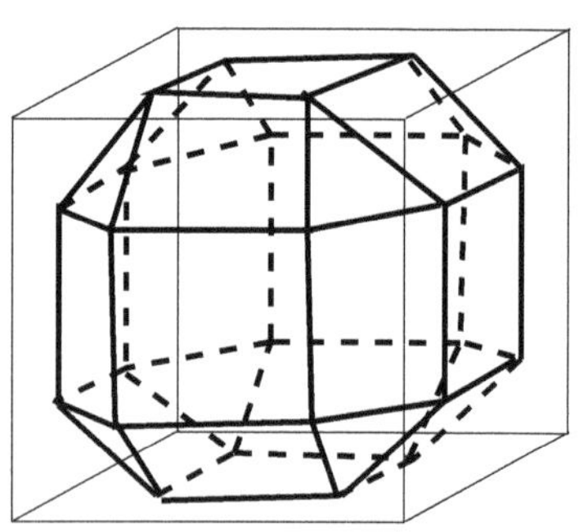

Figure 2.12 contains a list of these spaces with diagrams and names. Two of these truncations are drawn in Fig. 2.13, and another in Fig. 2.14.

The truncated cuboctahedron appeared in Fig. 1.1 and is redrawn in Fig. 3.1. The lines of various shadow spaces in the geometry of the cube are the edges drawn in Figs. 2.13 and 2.14. In the real affine geometry $AG(\mathbb{R}^3)$, with $J = \{1\}$, the physical

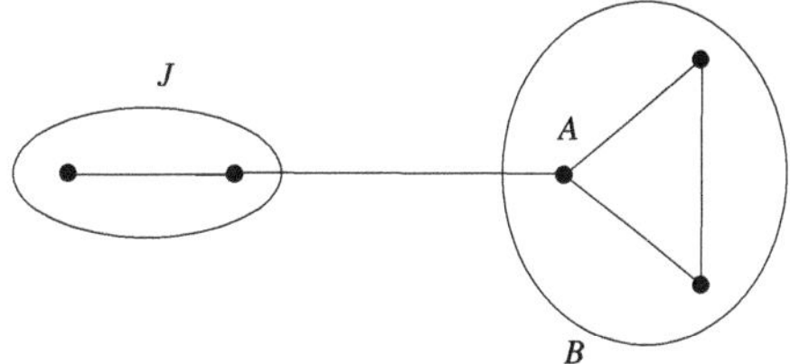

Fig. 2.15 The vertex set A separates J from B

viewpoint emerges from $\mathrm{ShSp}(\mathbb{E}^3, J)$, where points and lines are the 'usual' objects in $\mathbb{E}^3$.

Example 2.5.3 Shadow spaces $\mathrm{ShSp}(\Gamma, j)$ for a single type j are studied most intensively. Nevertheless, in projective geometry we need a case where $|J| = 2$, which appears in the projective geometry $\mathrm{PG}(V)$ of the vector space V of dimension $n + 1$ over a division ring (cf. Example 1.4.9). We take $J = \{1, n\}$, so the points of $\mathrm{ShSp}(\mathrm{PG}(V), J)$ correspond to incident point-hyperplane pairs. Two pairs (x, H) and (y, K), where x, y are 1-dimensional and H, K are n-dimensional, are 1-collinear (i.e., both incident with a 1-element) if and only if $H = K$ and n-collinear if and only if $x = y$. Verify that, in the former case, a 1-line on (x, H) and (y, H) is represented by all pairs (z, H) with z contained in the linear subspace spanned by x and y. Similarly, in the latter case, an n-line on (x, H) and (x, K) consists of all pairs (x, L) with L an n-dimensional subspace of V containing $H \cap K$.

In Definition 2.5.1, we did not allow $J = \emptyset$ because the resulting space is trivial and useless. In the examples where J is a singleton, i.e., consists of a single element of I, we see that the lines of $\mathrm{ShSp}(\Gamma, J)$ are shadows of flags of type $\bar{J}$ where $\bar{J}$ is the *neighborhood* of J in the digon diagram of Γ, i.e., the set of all $i \in I \setminus J$ such that i is on some edge of $\mathcal{I}(\Gamma)$ having a vertex in J. These considerations lead to the following development.

Definition 2.5.4 Let $\mathcal{I} = (I, \sim)$ be a graph. If J, A, B are subsets of I, then A *separates* J from B if no connected component of the subgraph of $\mathcal{I}$ induced on $I \setminus A$ meets both J and B. The J-*reduction* of B is the smallest subset of B separating J from B; it is readily seen to exist for all J and B.

An example is given in Fig. 2.15.

In the result below, the notion of separation will be applied to the digon diagram of a geometry.

Lemma 2.5.5 *If $F_1 \cup F_2$ is a flag of Γ such that $\tau(F_1)$ separates J from $\tau(F_2)$ in $\mathcal{I}(\Gamma)$, then $\mathrm{Sh}_J(F_1) \subseteq \mathrm{Sh}_J(F_2)$, so $\mathrm{Sh}_J(F_1 \cup F_2) = \mathrm{Sh}_J(F_1)$.*

Proof By the Direct Sum Theorem 2.1.6, each flag of type J contained in F_1^* is also contained in F_2^*. $\qquad\square$

Theorem 2.5.6 *Let Γ be an I-geometry of finite rank and let J be a non-empty subset of I. For $j \in J$, every j-line is the shadow on J of some flag of Γ of type J_j, where J_j is the J-reduction of $I \setminus \{j\}$. Conversely, the J-shadow of every flag of Γ of type J_j is a j-line.*

Proof Let $j \in J$ and suppose that l is a j-line. Then $l = \mathrm{Sh}_J(F)$ for some flag F of cotype $\{j\}$. If $H \subseteq F$ is the subflag of F of cotype J_j, then J_j separates J from the subset $I \setminus (\{j\} \cup J_j)$ of $J \setminus \{j\}$, so $\mathrm{Sh}_J(H) \subseteq \mathrm{Sh}_J(F \setminus H)$ by Lemma 2.5.5. This implies $\mathrm{Sh}_J(H) = \mathrm{Sh}_J(F) = l$, so H is a flag of type J_j such that $l = \mathrm{Sh}_j(H)$, as required.

In order to prove the converse, suppose that Y is a flag of type J_j. We show that its J-shadow is a j-line. Extend Y to a flag F of Γ of cotype $\{j\}$. By definition of J_j, the type $\tau(Y)$ separates J from $I \setminus \{j\}$ and hence also from $\tau(F \setminus Y)$, so Lemma 2.5.5 gives $\mathrm{Sh}_J(Y) = \mathrm{Sh}_J(F)$, which is a j-line. $\square$

Remark 2.5.7 Consideration of the (ordinary) digon shows that in an I-geometry with $j \in J \subseteq I$, the map $F \mapsto \mathrm{Sh}_J(F)$ from the set of flags of type J_j onto the set of j-lines need not be bijective.

In conclusion, from an abstract geometry, we have indicated a method to obtain various 'physical interpretations' in terms of points and lines. We refer to these interpretations as spaces. We would like to stress that our spaces are quite combinatorial in that they refer to collinearity, but not to a metric, topological, or differential structure. Starting from a space of points and lines, we can now construct diagram geometries, and go back to other spaces. We will do so in later chapters. The rest of this section is devoted to the basics of line spaces.

Definition 2.5.8 A *line space* is a pair (P, L) consisting of a set P, whose members are called *points*, and a collection L of subsets of P of size at least two, whose members are called *lines*. A line in L is called *thin* if it has exactly two points, and *thick* otherwise.

Let X be a subset of P. Then $L(X)$ denotes the collection of subsets $l \cap X$ of X of size at least 2 where $l \in L$. The resulting line space $(X, L(X))$ will be called the *restriction* of the line space (P, L) to X, or the line space *induced* on X.

A *subspace* X of (P, L) is a subset of P such that every line of L containing two distinct points of X is entirely contained in X (in other words, $L(X) \subseteq L$). Thus, $X = \emptyset$ and $X = P$ are trivial specimens of subspaces.

A *homomorphism* $\alpha : (P, L) \to (P', L')$ of line spaces is a map $\alpha : P \to P'$ such that the image under α of every line in L is contained in a line of L'. If it is injective and the image of every line in L is a line in L', then the homomorphism is also called an *embedding*. The notions *isomorphism* and *automorphism* are defined in the obvious way.

Often we will denote the lines space (P, L) by a single symbol, such as Z. We will then also abuse Z to indicate its point set as well, for instance, when writing

$p \in Z$ and $Z \setminus \{p\}$ to indicate that p is a point of Z and the set of points of Z distinct from p, respectively.

Finally, the *dual line space* of a line space (P, L) is the point shadow space of the dual geometry $\Gamma^\vee$ (cf. Definition 2.2.5) of the geometry $\Gamma = (P, L, *)$, where $*$ is symmetrized containment. It is a line space only if each point in P is on at least two members of L.

Example 2.5.9 Shadow spaces (cf. Definition 2.5.1) are examples of line spaces. In Chap. 5 we will study the line spaces that are shadow spaces on 1 of the geometries $AG(V)$ and $PG(V)$ for a vector space V.

Graphs are examples of line spaces all of whose lines have size two. Each subset of the vertex set is a subspace. The collinearity graph of the dual line space of a graph Δ is known as the line graph of Δ.

Lemma 2.5.10 *In a line space, the intersection of any set of subspaces is again a subspace.*

Proof Straightforward. $\square$

Definition 2.5.11 If X is a set of points in a line space Z, the subspace $\langle X \rangle$ of Z *generated* by X is the intersection of all subspaces of Z containing X.

It is possible to construct $\langle X \rangle$ from X by 'linear combination', see Exercise 2.8.21.

Definition 2.5.12 A line space (P, L) can be viewed as a rank two geometry $(P, L, *)$ by letting $*$ be symmetrized containment. This means that $*$ is determined by the rule that for $x \in P$, $y \in L$ we have $x * y$ (and $y * x$) if and only if $x \in y$. We will refer to this geometry as the *geometry of the space* (P, L). The line space (P, L) is called *connected* (*firm, thin, thick*) whenever the corresponding geometry $(P, L, *)$ is connected (firm, thin, thick, respectively).

The *collinearity graph* of a line space is the collinearity graph on the point set of the geometry of the space (introduced in Definition 2.2.1).

The geometry of the line space induced on a subset X of points of a line space (P, L) is a subgeometry of $(P, L, *)$ in the sense of Definition 1.4.1.

Definition 2.5.13 A subspace of a line space is said to be *singular* if any two of its points are on a line and *linear* if any two of its points are on a unique line. It is called *partial linear* if any two of its points are on at most one line. The unique line containing two collinear points p and q of a partial linear space is usually denoted by pq.

The definition of singularity is in accordance with Example 1.4.13, where the subspaces on which the form f defined there vanishes completely are indeed singular in the current sense.

Remark 2.5.14 The shadow spaces of affine and projective geometries are linear, but there are many other examples.

All the subspaces of a linear space are linear. All the subspaces of a partial linear space are partial linear. All singular subspaces of a partial linear space are linear. If X is a subset of a linear space Z and L is the set of lines of Z entirely contained in X, then (X, L) is a partial linear space.

The notions of point diameter and girth from Definitions 2.2.1 and 2.2.3 will be used to introduce a diagram for firm linear spaces on the basis of Definition 2.5.12.

Theorem 2.5.15 *Linear spaces relate to rank two geometries as follows.*

(i) *If (P, L) is a firm linear space, then the geometry $(P, L, *)$ of this space is a [2]-geometry with girth 3 and 1-diameter 3.*

(ii) *If Γ is a $\{\mathrm{p}, \mathrm{l}\}$-geometry with girth 3 and p-diameter 3, then the shadow space of Γ on p is a firm linear space.*

Proof (i) Suppose that (P, L) is a firm linear space. By definition of firmness for linear spaces, $\Gamma = (P, L, *)$ is firm. Since two distinct points cannot be incident with two distinct lines in (P, L), there are no circuits of length four in the incidence graph of Γ. There are circuits of length six as any two points are collinear and three non-collinear points can be found. Thus, the girth of Γ is three.

The point diameter of Γ is also three: given a point a, an argument similar to the above gives a line l not containing a, so $d(a, l) \geq 3$. As any point x distinct from a is on the line ax, it follows that $d(a, x) \leq 2$, so $d(a, l) = 3$. Therefore, the point diameter δ_{p} is equal to three.

(ii) Suppose that $\Gamma = (X_{\mathrm{p}}, X_1, *)$ is a firm geometry over $\{\mathrm{p}, \mathrm{l}\}$ with girth and point diameter both equal to three. Here, as usual, p stands for points and l for lines. An element (a line) in X_1 may be identified with its shadow on X_{p}: lines have at least two points and if two distinct lines l, m would both be incident with two distinct points $a, b \in X_{\mathrm{p}}$, then $a * l * b * m * a$ would be a 4-circuit, a contradiction with the girth being three. This argument also shows that two distinct points are on at most one line. Again, let a, b be distinct points. Then $d(a, b)$ (distance in Γ) is an even number; it is at most three as $\delta_{\mathrm{p}} = 3$. Hence $d(a, b) = 2$. Therefore there is a line containing a and b, which proves that the shadow space of Γ on p is linear. As it is clearly firm, this ends the proof of the theorem. $\qquad\square$

Notation 2.5.16 We denote by $\underset{\mathrm{p}\qquad\;\;\mathrm{l}}{\circ\!\!\overset{\mathrm{L}}{-\!\!\!-\!\!\!-}\!\!\circ}$ the collection of firm rank two geometries of the theorem. The subclass of all geometries $(P, L, *)$ where (P, L) is the complete graph will be denoted by $\underset{\mathrm{p}\qquad\;\;\mathrm{l}}{\circ\!\!\overset{\mathrm{C}}{-\!\!\!-\!\!\!-}\!\!\circ}$ (for clique or complete graph).

In Chap. 5, we will apply Theorem 2.5.15 to affine and projective geometries. In order to separate these nice examples from the other linear spaces, we look for conditions enabling us to build higher rank geometries by use of subspaces.

2.6 Group Diagrams

When a geometry is being constructed from a system of subgroups of a given group, the isomorphism types of the residues are not easy to describe beforehand. To remedy this, we adhere a diagram to a system of groups in such a way that the resulting coset geometry (if any) has the corresponding diagram in the sense of Definition 2.3.7. Fix a set I.

Definition 2.6.1 Let G be a group with a system of subgroups $(G_i)_{i \in I}$, and let D be a diagram over I. We say that G has *diagram* D over $(G_i)_{i \in I}$ if, for each pair $\{i, j\} \subseteq I$, the coset geometry $\Gamma(G_{I \setminus \{i,j\}}, (G_{I \setminus \{j\}}, G_{I \setminus \{i\}}))$, with $G_{I \setminus \{j\}}$ of type i and $G_{I \setminus \{i\}}$ of type j, belongs to $D(i, j)$.

Theorem 1.8.10 gives that $\Gamma(G_{I \setminus \{i,j\}}, (G_{I \setminus \{j\}}, G_{I \setminus \{i\}}))$ is indeed a geometry, for $G_{I \setminus \{i,j\}}$ acts flag transitively on it by Lemma 1.8.6.

If G has diagram D over $(G_i)_{i \in I}$, then G_J has diagram $D|_{(I \setminus J)}$ over $(G_{J \cup \{i\}})_{i \in I \setminus J}$. Here, $D|_{(I \setminus J)}$ stands for the restriction of D to the collection of pairs from $I \setminus J$. In the flag-transitive case, the above definition coincides with the ordinary diagram.

Proposition 2.6.2 *Suppose that I is finite and that G is a group having diagram D over a system of subgroups $(G_i)_{i \in I}$. If G is flag transitive on the coset incidence system $\Gamma = \Gamma(G, (G_i)_{i \in I})$, then Γ is a geometry belonging to D.*

Proof This is a direct consequence of Theorem 1.8.10(ii). $\qquad\square$

We analyze a case similar to the Direct Sum Theorem 2.1.6, in which generalized digons (cf. Definition 2.1.1) play a role.

Lemma 2.6.3 *Let I be finite and let G be a group with diagram D over a system of subgroups $(G_i)_{\in I}$. Suppose that the following two conditions hold.*

(i) *$G_J = \langle G_{J \cup \{i\}} \mid i \in I \setminus J \rangle$ for each $J \subseteq I$ with $|I \setminus J| \geq 2$.*
(ii) *I is partitioned into R and L in such a way that $D(r, l)$ consists of generalized digons for every $r \in R$ and $l \in L$.*

Then $G = G_L G_R$. In particular $G = G_l G_r$ for all $l \in L, r \in R$.

Proof If $|I| = 2$, then $R = \{r\}$ and $L = \{l\}$, so $\Gamma(G, (G_r, G_l))$ is a generalized digon. This means $gG_r \cap hG_l \neq \emptyset$ for each $g, h \in G$. In particular, if $x \in G$, then $xG_r \cap G_l \neq \emptyset$, so $x \in G_l G_r$. This settles $G = G_l G_r$ and establishes the rank two case. As G is a group, we can derive $G = G_r G_l$ by taking inverses.

In the case of arbitrary rank, we have

$$G = \langle G_i \mid i \in I \rangle = \langle G_J \mid |J| = 2, J \subseteq I \rangle = \cdots = \langle G_{I \setminus \{i\}} \mid i \in I \rangle.$$

But, for $r \in R$ and $l \in L$, by Definition 2.6.1 and what we have seen in the rank two case, $G_{I \setminus \{r,l\}} = G_{I \setminus \{l\}} G_{I \setminus \{r\}} = G_{I \setminus \{r\}} G_{I \setminus \{l\}}$, so

$$G = \langle G_{I \setminus \{i\}} \mid i \in I \rangle = \langle G_{I \setminus \{r\}} \mid r \in R \rangle \langle G_{I \setminus \{l\}} \mid l \in L \rangle \subseteq G_L G_R.$$

The last statement of the lemma follows as $G_L \subseteq G_l$ and $G_R \subseteq G_r$. $\square$

The case of a *linear diagram* D, which means that, as a graph, D is a path, lends itself to an easy criterion for flag transitivity.

Theorem 2.6.4 *Let I be finite and let G be a group with a system of subgroups $(G_i)_{i \in I}$ such that*

(i) $G_J = \langle G_{J \cup \{i\}} \mid i \in I \setminus J \rangle$ *for each $J \subseteq I$ with $|I \setminus J| \geq 2$;*
(ii) *G has a linear diagram over $(G_i)_{i \in I}$.*

Then $\Gamma(G, (G_i)_{i \in I})$ is a residually connected geometry on which G acts flag transitively.

Proof In view of Corollary 1.8.13, the incidence system $\Gamma(G, (G_i)_{i \in I})$ is a residually connected geometry if G acts flag transitively on it. The latter is immediate if $|I| \leq 2$. Suppose, therefore, $|I| \geq 3$.

In view of Theorem 1.8.10(iii) and induction on $|I|$, it suffices to show that, for every subset J of I of size three, G is transitive on the set of all flags of type J. Denote the linear diagram by D and write $J = \{i, j, k\}$ where i, j, k are chosen so that $D(j, k)$ consists of generalized digons (note that this is always possible as D is linear). Now, by Lemma 2.6.3 applied to G_i, we have $G_i = G_{\{i,j\}} G_{\{i,k\}}$, so $G_i G_k \cap G_j G_k = G_{\{i,j\}} G_k \cap G_j G_k$. But this is equal to $G_{\{i,j\}} G_k$ as $G_{\{i,j\}} \subseteq G_j$, so $G_i G_k \cap G_j G_k = (G_i \cap G_j) G_k$. By Lemma 1.8.9(ii), this implies the required transitivity. $\square$

Example 2.6.5 Let G be the group given by the following presentation.

$$G = \langle a, d, e, z, t \mid a^2 = d^3 = e^3 = z^2 = t^2 = 1,$$
$$[e, z] = [a, z] = [z, t] = [d, e] = [a, t]z = 1,$$
$$(dz)^2 = (et)^2 = (da)^5 = (dza)^5 = (ate)^3 = 1 \rangle.$$

A coset enumeration with respect to the subgroup $N = \langle a, d, z, eae^{-1} \rangle$ of G shows that we can also view G as the subgroup of Sym_{12} with generators

$$a = (3, 4)(5, 6)(7, 10)(8, 11),$$
$$d = (4, 5, 7)(6, 8, 9)(10, 12, 11),$$
$$e = (1, 3, 2)(6, 9, 8)(10, 12, 11),$$
$$z = (5, 7)(6, 10)(8, 11)(9, 12),$$
$$t = (1, 2)(6, 10)(8, 12)(9, 11).$$

Fig. 2.16 The diagram of the geometry associated with M_{11}

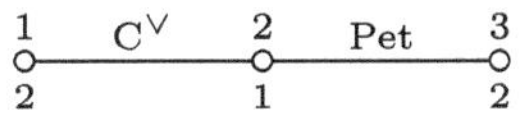

$$\overset{1}{\underset{2}{\circ}} \xrightarrow{\quad C^{\vee}\quad} \overset{2}{\underset{1}{\circ}} \xrightarrow{\quad \text{Pet}\quad} \overset{3}{\underset{2}{\circ}}$$

In fact, further computations show that G is 3-transitive on $[12]$ (cf. Exercise 2.8.12) and has order 7920. From this it is not hard to derive that G is isomorphic to the Mathieu group M_{11} (see Definition 5.6.4). But we will not need this. We distinguish the following subgroups.

$$G_1 = \langle a, d, z, t \rangle, \qquad G_2 = \langle d, e, z, t \rangle, \qquad G_3 = \langle a, e, z, t \rangle.$$

Observe that G_1 stabilizes the set $[2]$, that G_2 stabilizes $[3]$, and that G_3 stabilizes the set $[4]$. In fact, they are the full stabilizers; G_1 has index 66 in G, while G_2 has index 220, and G_3 has index 165.

Now $G_{12} = \langle d, z, t \rangle$ is a group of order 12, and $G_{13} = \langle a, z, t \rangle$ is a group of order 8, and $G_{23} = \langle e, z, t \rangle$ is a group of order 12. Consequently, $G_1 = \langle a, d, z \rangle = \langle G_{12}, G_{13} \rangle$, $G_2 = \langle d, e, z \rangle = G_{12} G_{23}$, $G_3 = \langle a, e, z \rangle = \langle G_{13}, G_{23} \rangle$. Since G_2 is factored into G_{12} and G_{23}, the group G has a linear diagram over (G_1, G_2, G_3). By Theorem 2.6.4, G is flag transitive on $\Gamma := \Gamma(G, (G_1, G_2, G_3))$ and Γ is a residually connected geometry. It is firm, as G_{123} has order four, whence index 3, 2, 3, in G_{12}, G_{13}, G_{23}, respectively.

The geometry belongs to the diagram depicted in Fig. 2.16. We already know that the diagram is linear. A residue of type $\{1, 2\}$ can be viewed as the geometry whose 1-elements are 2-sets and whose 2-elements are 3-sets in a given 4-set of $[12]$, whence the dual C diagram $C^{\vee}$. A residue of type $\{2, 3\}$ can be viewed as a graph on $\{3, \ldots, 12\}$, with the 1-elements being its vertices and the 2-elements being edges $\{x, y\}$ such that $\{1, 2, x, y\}$ belongs to the G_1-orbit of $\{1, 2, 3, 4\}$. There are precisely three such edges on $\{1, 2, 3\}$ and the group acting on this residue is $G_1 \cong \text{Sym}_5$; this implies that the graph must be the Petersen graph.

2.7 A Geometry of Type $\widetilde{A}_{n-1}$

So far, we have dealt with various constructions of geometries with linear diagrams from groups. In this section we provide an example having the non-linear Coxeter diagram $\widetilde{A}_{n-1}$ depicted in Fig. 2.17, where $n \in \mathbb{N}$, $n > 2$. For $n = 2$, the construction is also valid; the resulting diagram appears in Remark 2.7.15. Example 2.4.4 showed thin geometries of this type, the triangle ($n = 3$) and the quadrangle ($n = 4$). The geometries constructed in this section are thick and all of their rank $n - 1$ residues are isomorphic to $\text{PG}(V)$ for some vector space V of dimension n.

We first review some local ring theory from commutative algebra.

Definition 2.7.1 A commutative ring R is called a *discrete valuation ring* if it is a *principal ideal domain* (i.e., an associative ring with 1 and without zero divisors in which each ideal is generated by a single element), with a unique nonzero max-

Fig. 2.17 Coxeter type $\widetilde{A}_{n-1}$

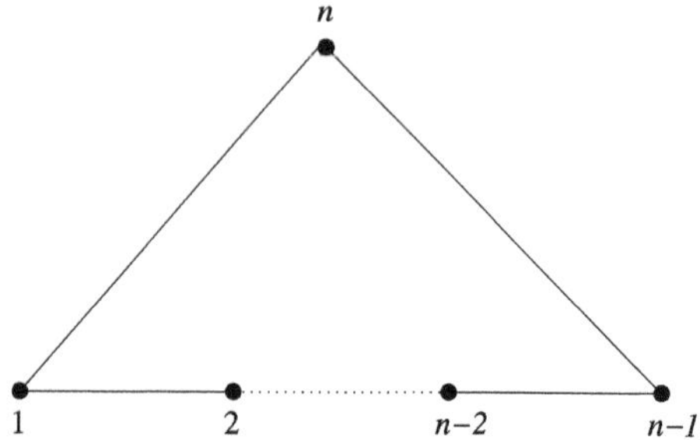

imal ideal. A generator of the maximal ideal is called a *local parameter*. The field obtained from R by modding out its maximal ideal is called its *residue field*.

Example 2.7.2 Let p be a prime.

(i) The ring of *p-local integers* consists of all rational numbers a/b with $\gcd(b, p) = 1$. The subset $\{pa/b \mid a, b \in \mathbb{Z}, \gcd(b, p) = 1\}$ is its unique maximal ideal and p is a local parameter (and so are $-p$ and $p(1 + p)$). The residue field of this ring is isomorphic to $\mathbb{F}_p$.

(ii) Let $\mathbb{F}$ be a field and set $R = \{f/g \mid f, g \in \mathbb{F}[X], \gcd(g, X) = 1\}$. Then $\{Xf/g \mid f, g \in \mathbb{F}[X], \gcd(g, X) = 1\}$ is the unique maximal ideal in R and X is a local parameter. The residue field of R is isomorphic to $\mathbb{F}$.

Lemma 2.7.3 *Let R be a discrete valuation ring with local parameter π. Then, for each nonzero element x of R there is a unique integer $i \in \mathbb{N}$ such that $x = \pi^i y$ with y an invertible element of R.*

Proof If $x \notin \pi R$, then the ideal generated by x coincides with R, so there is $z \in R$ with $xz = 1$; in other words, x is invertible in R. Since, obviously, invertible elements do not belong to πR, we find that $R \setminus \pi R$ is the set of invertible elements of R. In particular, for $x \notin \pi R$, the lemma holds with $i = 0$.

Observe that $(\pi^i R)_{i \in \mathbb{N}}$ is a strictly descending chain of ideals. For if $\pi^i R = \pi^{i+1} R$ for some $i \in \mathbb{N}$, then $\pi^i = \pi^{i+1} r$ for some $r \in R$ and so, as R is a domain, $1 = \pi r$, contradicting that π is not invertible.

We derive $\bigcap_{i \in \mathbb{N}} \pi^i R = \{0\}$. The left hand side is an ideal of R. As R is a principal ideal domain, there is $a \in R$ such that $\bigcap_{i \in \mathbb{N}} \pi^i R = aR$. We claim that $\pi aR = aR$. Let $x \in aR$. As $aR \subseteq \pi R$, there is an element y in R with $x = \pi y$. Let $j \in \mathbb{N}$ be arbitrary. As $x \in aR$, there is $z_j \in R$ such that $x = \pi^{j+1} z_j$. Now $\pi y = \pi^{j+1} z_j$, so $y = \pi^j z_j$. We find $y \in \bigcap_{i \in \mathbb{N}} \pi^i R = aR$ and $x \in \pi aR$. In particular, $\pi aR = aR$. This means that there is $b \in R$ with $b\pi a = a$. But $(b\pi - 1)$ is invertible as it is not in πR, so $a = 0$. Therefore, indeed, $\bigcap_{i \in \mathbb{N}} \pi^i R = aR = \{0\}$.

So, if $x \in R$, then $x \in \pi^i R \setminus \pi^{i+1} R$ for some $i \in \mathbb{N}$. Then $x = \pi^i y$ for some $y \in R \setminus \pi R$, and, by the above, y is invertible. $\square$

As a consequence, an element of a discrete valuation ring is invertible if and only if it does not belong to the maximal ideal. Being an integral domain, a discrete valuation ring has a field of fractions into which it embeds.

Corollary 2.7.4 *The elements of the field of fractions of a discrete valuation ring R are all of the form $\pi^i y$ with $y \in R \setminus \pi R$ and $i \in \mathbb{Z}$.*

Proof Suppose $v, w \in R$ with $w \neq 0$. By Lemma 2.7.3, there are $i, j \in \mathbb{N}$ and invertible elements $x, y \in R$ with $v = \pi^i x$ and $w = \pi^j y$, so $vw^{-1} = \pi^{i-j}xy^{-1}$, with $i - j \in \mathbb{Z}$ and xy^{-1} an invertible element of R, as required. $\qquad\qquad\square$

Definition 2.7.5 Let R be a discrete valuation ring with local parameter π and field of fractions $\mathbb{F}$. Consider $G = \mathrm{SL}(\mathbb{F}^n)$, with $n \geq 3$. For $i \in [n]$, take G_i to be the set of matrices in G, of the form

$$\begin{pmatrix} A_{i,i} & \pi^{-1} B_{i,n-i} \\ \pi C_{n-i,i} & D_{n-i,n-i} \end{pmatrix},$$

where $X_{k,l}$, for X one of A, B, C, D, denotes a $k \times l$-matrix with entries in R. In particular, $G_n = \mathrm{SL}(R^n)$. It is easily checked that each G_i is a subgroup of G. The corresponding coset incidence system $\Gamma = \Gamma(G, (G_i)_{i \in [n]})$ over $[n]$ is called the *π-adic geometry* of $\mathbb{F}^n$.

The word geometry in the name will be justified by Theorem 2.7.14 below. This group-theoretic definition of the incidence system is quite succinct. But, in order to derive properties of it, a geometric description will be used as well. The necessary objects stem from a generalization of lattices over the ring of integers $\mathbb{Z}$ to discrete valuation rings. For the introduction of lattices, we use the notion of a module over a ring, which is the well-known generalization of a vector space over a field.

Definition 2.7.6 Let V be a vector space over the field $\mathbb{F}$. For a subring R of $\mathbb{F}$, an *R-lattice* in V is an R-submodule of V generated by a vector space basis of V.

The notion of a lattice in the context of a poset, as used in Exercise 3.7.7, is an entirely different one.

Remark 2.7.7 A basis $b_1, \ldots, b_n$ of $\mathbb{F}^n$ determines the R-lattice

$$L = Rb_1 \oplus Rb_2 \oplus \cdots \oplus Rb_n$$

in V. We collect the vectors b_i as columns in a matrix B. These can be used to describe L as the image $B(R^n)$ of the map $R^n \to \mathbb{F}^n$, $x \mapsto Bx$. The matrix B belongs to $\mathrm{GL}(\mathbb{F}^n)$ as its columns constitute a basis of $\mathbb{F}^n$.

Lemma 2.7.8 *Suppose that the R-lattice L is generated by the columns of the matrix $B \in \mathrm{GL}(\mathbb{F}^n)$ and that $L' = B'(R^n)$ is the sublattice of L generated by the columns of the matrix $B' \in \mathrm{GL}(\mathbb{F}^n)$. The determinant $\det(B)$ is a divisor of $\det(B')$*

in R. Moreover, $L = L'$ if and only if the quotient $\det(B')/\det(B)$ is invertible in R. In particular, $\det(B)$ is determined by L up to multiplication by an invertible element of R.

Proof Denote the i-th column of B by b_i, and the i-th column of B' by b_i'. Since $\{B'x \mid x \in R^n\} \subseteq \{Bx \mid x \in R^n\}$, there are elements $x_{ij} \in R$ $(1 \leq i, j \leq n)$ such that $b_i' = \sum_{j=1}^n x_{ji} b_j$. This means that $X = (x_{ij})_{i,j \in [n]}$ is a matrix such that $B' = BX$, and so $\det(B') = \det(B)\det(X)$. This proves that $\det(B)$ divides $\det(B')$.

If $L = L'$, then there is also a matrix Y with entries in R such that $B = B'Y$. But then YX is the identity matrix and so X and Y are invertible, whence the last two statements of the lemma. $\qquad\square$

Definition 2.7.9 Two R-lattices L, M of the same vector space are called *homothetic* if there is a scalar $\lambda \in \mathbb{F}$ such that $L = \lambda M$.

Being homothetic is an equivalence relation on the set of all R-lattices in $\mathbb{F}^n$.

Notation 2.7.10 For an R-lattice in $\mathbb{F}^n$, we denote by $[L]$ its *homothety class*, that is, its equivalence class with respect to the relation of being homothetic. Furthermore, we write $[L] * [M]$ to denote that, for some $L' \in [L]$ and $M' \in [M]$, we have $\pi L' \subseteq M' \subseteq L'$.

We use $*$ for a geometric definition of the π-adic incidence system. Let $\mathcal{L}$ be the collection of homothety classes of R-lattices in $\mathbb{F}^n$ and, for $[L] \in \mathcal{L}$, denote by $\tau([L])$ the integer $i \in [n]$ for which there is a matrix B with entries in $\mathbb{F}$ such that $B(R^n) \in [L]$ and $\det(B) = \pi^i$.

Lemma 2.7.11 *For each $n \in \mathbb{N}$, $n \geq 2$, and discrete valuation ring R, the triple $(\mathcal{L}, *, \tau)$ is an incidence system over $[n]$.*

Proof We first verify that τ is well defined. Suppose, to this end, that $B(R^n) = \lambda B'(R^n)$ for two matrices B, B' in $\mathrm{GL}(\mathbb{F}^n)$ and $\lambda \in \mathbb{F}$. It follows that $\det(B) = \lambda^n \det(B')$ and $\lambda \neq 0$. Write $j = \tau([B(R^n)])$ and $k = \tau([B'(R^n)])$. If D is a diagonal matrix with all diagonal entries but one equal to 1 and the remaining entry equal to an invertible element z of R, then $B(R^n) = BD(R^n)$ and $\det(BD) = \pi^j z$. So, replacing B and B' by suitable matrices, we may assume $\det(B) = \pi^{j+pn}$ and $\det(B') = \pi^{k+qn}$ for certain $p, q \in \mathbb{Z}$. As $\lambda \neq 0$, by Lemma 2.7.3, there are $i \in \mathbb{Z}$ and $y \in R \setminus \pi R$ such that $\lambda = \pi^i y$. Consequently, $\pi^{j+pn} = \det(B) = \lambda^n \det(B') = \pi^{ni+k+qn} y^n$. By uniqueness of the exponent of π (see Lemma 2.7.3), we find $j - k = n(i + q - p)$. As both $j, k \in [n]$, it follows that $j = k$, which proves that τ does not depend on the choice of B.

Next, we show that $*$ is a symmetric relation. If $[L] * [M]$, then, by definition, there are $L' \in [L]$ and $M' \in [M]$ with $\pi L' \subseteq M' \subseteq L'$. Now $\pi(\pi^{-1}M') \subseteq L' \subseteq \pi^{-1}M'$, so $[\pi^{-1}M'] * [L]$. Since $[\pi^{-1}M'] = [M'] = [M]$, this proves symmetry of $*$.

Suppose, once more, $[L] * [M]$ but, in addition $\tau([L]) = \tau([M]) = i$. Choose L' and M' as before. Multiplying both lattices L and M by the same suitable power of π, we may assume $M = M'$. Let A and B be $n \times n$ matrices over $\mathbb{F}$ with $L' = A(R^n)$ and $M = B(R^n)$. Without loss of generality, we take $\det(B) = \pi^i$. By Lemma 2.7.8, there are $a, b \in R$ such that $\det(\pi A) = a \det(B)$ and $\det(B) = b \det(A)$. This gives $\pi^n \det(A) = \det(\pi A) = a \det(B) = ab \det(A)$, so $\pi^n = ab$. Now $\tau([L]) = i$ means that $\det(A) = \pi^{i+kn} z$ for some $k \in \mathbb{Z}$ and z is an invertible element of R. Furthermore, $\det(B) = b \det(A)$ leads to $\pi^i = b\pi^{i+kn} z$, and so $b = \pi^{-kn} z^{-1}$. Now $b \in R$ implies $k \leq 0$ and the fact that b divides π^n means that only $k = 0$ and $k = -1$ are possible. If $k = 0$, then $b \in R \setminus \pi R$ and Lemma 2.7.8 gives $M = L'$; if $k = -1$, then $a = z \in R \setminus \pi R$, so $\pi L' = M$. This proves $[L] = [M]$. We conclude that the restriction of $*$ to $\tau^{-1}(i)$ is the identity, so all conditions of Definition 1.2.2 are satisfied for $(\mathcal{L}, *, \tau)$. $\qquad\square$

A special chamber of $(\mathcal{L}, *, \tau)$ is the set $\{[L_i] \mid i \in [n]\}$, where

$$L_i = R\varepsilon_1 + \cdots + R\varepsilon_{n-i} + \pi R\varepsilon_{n-i+1} + \cdots + \pi R\varepsilon_n. \tag{2.2}$$

Here, $\varepsilon_1, \ldots, \varepsilon_n$ denotes the standard basis of $\mathbb{F}^n$. So $L_i = D_{n-i}(R^n)$, where D_i is the diagonal matrix with ones in the first i diagonal entries and π in the remaining $n - i$ diagonal entries. As $\det(D_{n-i}) = \pi^i$, we have $\tau([L_i]) = i$.

The group $\mathrm{GL}(\mathbb{F}^n)$ acts on the set of R-lattices via left multiplication: $A(R^n) \mapsto BA(R^n)$ for $A, B \in \mathrm{GL}(\mathbb{F}^n)$.

Lemma 2.7.12 *Let $\Gamma = (\mathcal{L}, *, \tau)$. The group $\mathrm{Cor}(\Gamma)$ of auto-correlations of Γ is transitive on $\mathcal{L}$. More specifically, the following statements hold.*

(i) *The group $\mathrm{SL}(\mathbb{F}^n)$ acts on Γ via $B([L]) = [BL]$ $(B \in \mathrm{SL}(\mathbb{F}^n))$.*

(ii) *The matrix*

$$\alpha = \begin{pmatrix} 0 & 1 & 0 & \cdot & & \cdot & 0 \\ 0 & 0 & 1 & \cdot & & \cdot & 0 \\ 0 & 0 & 0 & \ddots & & \cdot & \vdots \\ 0 & 0 & 0 & \ddots & & \ddots & 0 \\ 0 & 0 & 0 & \cdots & & 0 & 1 \\ \pi & 0 & 0 & \cdots & & \cdots & 0 \end{pmatrix}$$

belongs to $\mathrm{GL}(\mathbb{F}^n)$ and induces a correlation of Γ permuting the type set $[n]$ cyclically, according to $(1, 2, \ldots, n)$.

Proof The group $\mathrm{GL}(\mathbb{F}^n)$ preserves homothety classes, so its action on $\mathcal{L}$ is well defined. Also, the action is easily seen to preserve incidence.

If $B \in \mathrm{GL}(\mathbb{F}^n)$ and $\det(B) = \pi^i x$ for some $i \in \mathbb{Z}$ and some invertible $x \in R$, then $\det(BA) = \det(B)\det(A)$, so $\tau(B[A(R^n)]) \equiv i + \tau([A(R^n)])$ $(\mathrm{mod}\ n)$ for each $A \in \mathrm{GL}(\mathbb{F}^n)$. This shows that B maps elements of $\mathcal{L}$ of type j to elements of type $i + j$ modulo n.

(i) Suppose now $B \in \mathrm{SL}(\mathbb{F}^n)$. Then $\det(B) = 1$, so B preserves types.

An arbitrary representative of a homothety class in $\mathcal{L}$ is given by a matrix A whose columns are a basis of $\mathbb{F}^n$. Write $\det(A) = \pi^k x$ with $k \in \mathbb{Z}$ and invertible $x \in R$. After applying a suitable homothety, we may assume $k \in [n]$ without changing $[A(R^n)]$. But then $B = A D_{n-k}^{-1} \in \mathrm{SL}(\mathbb{F}^n)$ and $A(R^n) = B D_{n-k}(R^n) = B(L_k)$ in view of (2.2), so $\mathrm{SL}(\mathbb{F}^n)$ is transitive on $\tau^{-1}(k)$.

(ii) As $\det(\alpha) = \pm\pi$, we have $\alpha \in \mathrm{GL}(\mathbb{F}^n)$. Moreover, $\alpha L_i = L_{i+1}$ (indices taken mod n and in $[n]$), so the induced permutation of α on the type set $[n]$ is as stated.

Together with (i), it follows that the subgroup of $\mathrm{GL}(\mathbb{F}^n)$ generated by α and $\mathrm{SL}(\mathbb{F}^n)$ is transitive on $\mathcal{L}$. $\square$

The natural quotient map $R \to R/\pi R$ is a ring homomorphism onto the residue field $\mathbb{K} = R/\pi R$ of R. It leads to a group homomorphism $\phi : \mathrm{SL}(R^n) \to \mathrm{SL}(\mathbb{K}^n)$ given by reduction modulo πR in each entry. In turn, this homomorphism provides an action of $\mathrm{SL}(R^n)$ on $\mathbb{K}^n$.

Lemma 2.7.13 *The residue of each element of $\Gamma = (\mathcal{L}, *, \tau)$ is isomorphic to* $\mathrm{PG}(\mathbb{K}^n)$. *Let Γ_n denote the residue of $[R^n]$ in Γ. The map $\beta : \Gamma_n \to \mathrm{PG}(\mathbb{K}^n)$ given by $\beta([M]) = M/\pi R^n$, where M is chosen in such a way that $\pi R^n \subseteq M \subseteq R^n$, is an isomorphism establishing an equivalence of the natural action of the group $\mathrm{SL}(R^n)$ on Γ_n to the action via ϕ on $\mathrm{PG}(\mathbb{K}^n)$.*

Proof By Lemma 2.7.12 it suffices to prove the first statement for the element $[R^n]$. Suppose that $[M] \in \mathcal{L}$, of type $i \in [n-1]$, is in the residue Γ_n. We can choose the R-lattice M in such a way that $\pi R^n \subseteq M \subseteq R^n$. Now $M/\pi R^n$ is a subspace of $R^n/\pi R^n = \mathbb{K}^n$ of dimension $n - i$ (see Exercise 2.8.31).

We claim that the map β given by $\beta([M]) = M/\pi R^n$ is an isomorphism $\Gamma_n \to \mathrm{PG}(\mathbb{K}^n)$. First of all, β is injective because each homothety class has at most one representative M such that $\pi R^n \subseteq M \subseteq R^n$. Moreover, β is surjective because each subspace of $\mathbb{K}^n$ of dimension i is the image of R^n under an $n \times n$-matrix U over $\mathbb{K}$ of rank i and so each $n \times n$-matrix B over R whose entries map onto those of U defines a class $[M]$, where $M = B(R^n)$, that is, an element of Γ_n with $\beta([M]) = U(\mathbb{K}^n)$.

If $[M] * [M']$ for some element $[M']$ of Γ_n with $\pi R^n \subseteq M' \subseteq R^n$, then there is an integer j such that $M \subseteq \pi^j M' \subseteq \pi^{-1} M$. Now $\pi R^n \subseteq M, M' \subseteq R^n$ forces $j = 0, -1$. Interchanging the roles of M and M' if needed, we may take $j = 0$, which means $M \subseteq M'$ and implies $\beta(M) \subseteq \beta(M')$, proving $\beta([M]) * \beta([M'])$ in $\mathrm{PG}(\mathbb{K}^n)$. The reverse implication is proved similarly.

As for the group actions, let $B \in \mathrm{SL}(R^n)$ and let M be an R-lattice in $\mathbb{F}^n$ with $\pi R^n \subseteq M \subseteq R^n$. Then $\beta(B([M])) = \beta([BM]) = BM/\pi R^n = \phi(B)\beta(M)$. This establishes that β is an equivalence between the actions of $\mathrm{SL}(R^n)$ on Γ_n and $\mathrm{PG}(\mathbb{K}^n)$. $\square$

We now relate the incidence system $(\mathcal{L}, *, \tau)$ of Lemma 2.7.11 to the coset incidence system $\Gamma(G, (G_i)_{i \in [n]})$ of Definition 2.7.5. The type $\widetilde{\mathrm{A}}_{n-1}$ appearing in the theorem below is depicted in Fig. 2.17.

Theorem 2.7.14 *Let $n \in \mathbb{N}$, $n \geq 2$, and let R be a discrete valuation ring with local parameter π and field of fractions $\mathbb{F}$. The incidence system $(\mathcal{L}, *, \tau)$ is an $[n]$-geometry of type $\widetilde{A}_{n-1}$. Moreover, the action of $G = \mathrm{SL}(\mathbb{F}^n)$ on $(\mathcal{L}, *, \tau)$ is flag transitive with stabilizers G_i, so it is equivalent to the geometric representation on the π-adic geometry $\Gamma(G, (G_i)_{i \in [n]})$ over $\mathbb{F}^n$.*

Proof We show that G acts flag transitively on $\Gamma = (\mathcal{L}, *, \tau)$. Let J be a non-empty subset of $[n]$ and let $(M_i)_{i \in J}$ be R-lattices such that $([M_i])_{i \in J}$ is a flag of type J in Γ. We need to show that this flag is in the same G-orbit as $([L_i])_{i \in J}$ of (2.2). In view of Lemma 2.7.12, we may assume $n \in J$. But then $([M_i])_{i \in J \setminus \{n\}}$ is a flag of type $J \setminus \{n\}$ in Γ_n and so $(\beta([M_i]))_{i \in J \setminus \{n\}}$, where β is as in Lemma 2.7.13, is a flag of $\mathrm{PG}(\mathbb{K}^n)$. By flag transitivity of $\mathrm{SL}(\mathbb{K}^n)$ on this geometry (cf. Example 1.8.16), there is $A \in \mathrm{SL}(\mathbb{K}^n)$ such that $A(\beta([M_i]))_{i \in J \setminus \{n\}} = (\beta([L_i]))_{i \in J \setminus \{n\}}$. The equivalence of the two $\mathrm{SL}(R^n)$-actions of Lemma 2.7.13 gives that $([M_i])_{i \in J \setminus \{n\}}$ and $([L_i])_{i \in J \setminus \{n\}}$ are in the same $\mathrm{SL}(R^n)$-orbit. It follows that G acts flag transitively on Γ.

Fix $i \in [n]$. We verify that G_i is the stabilizer in G of the element $[L_i] \in \mathcal{L}$. Clearly, G_i stabilizes L_i and hence $[L_i]$. Let H_i be the stabilizer of $[L_i]$ in $\mathrm{SL}(\mathbb{F}^n)$ and take $A \in H_i$. Then H_i must stabilize L_i itself. If $j \leq n - i$, then $A\varepsilon_j \in L_i$ means that the j-th column of A has the first $n - i$ entries in R and the last i entries in πR. If $j > n - i$, then $A\varepsilon_j \in L_i$ means that the j-th column of A has the first $n - i$ entries in $\pi^{-1} R$ and the last i entries in R. Therefore, $H_i \subseteq G_i$ and so $H_i = G_i$.

Proposition 1.8.7 gives that the geometric representation of G over $(G_i)_{i \in [n]}$ is equivalent to the action on Γ. It follows from Theorem 1.8.10 that Γ is a geometry. By now it is easy to see that Γ has Coxeter diagram $\widetilde{A}_{n-1}$. For, the residue of each element is isomorphic to $\mathrm{PG}(\mathbb{K}^n)$ and so belongs to the diagram A_{n-1}.

It remains to show that Γ is residually connected. As all residues of non-empty flags are residues inside a geometry isomorphic to $\mathrm{PG}(\mathbb{K}^n)$, Lemma 1.8.9 shows that we only need to verify that G is generated by all G_i ($i \in [n]$). But this readily follows from the relation

$$\begin{pmatrix} \pi & 0 & 0 \\ 0 & \pi^{-1} & 0 \\ 0 & 0 & I_{n-2} \end{pmatrix} = \begin{pmatrix} 0 & -1 & 0 \\ 1 & 0 & 0 \\ 0 & 0 & I_{n-2} \end{pmatrix} \begin{pmatrix} 0 & \pi^{-1} & 0 \\ -\pi & 0 & 0 \\ 0 & 0 & I_{n-2} \end{pmatrix} \in G_n G_1. \qquad \square$$

Remark 2.7.15 (i) For $n = 2$, the above construction of $(\mathcal{L}, *, \tau)$ is still valid and leads to a flag-transitive $[2]$-geometry of type $\overset{1}{\circ}\!\!\overset{\infty}{\rule{2em}{0.4pt}}\!\!\overset{2}{\circ}$. This geometry is also known as a generalized ∞-gon (cf. Definition 2.2.6). The incidence graph of this geometry is a connected graph without circuits, also known as a *tree*.

(ii) If R is as in Example 2.7.2(i), then, due to Proposition 2.4.7, all i-orders of $(\mathcal{L}, *, \tau)$ are equal to p. Similarly, if $R = \mathbb{F}_q[[X]]$ (see Exercise 2.8.27), then all i-orders are equal to q.

2.8 Exercises

Section 2.1

Exercise 2.8.1 For the geometry Γ of Example 2.1.3 construct an auto-correlation that induces a permutation of order 4 on the set of types. Does the geometry Δ of Example 2.1.3 possess trialities?

Exercise 2.8.2 Let J be a finite index set. Prove that a direct sum of geometries Γ_j $(j \in J)$ is residually connected if and only if each Γ_j is residually connected.

Section 2.2

Exercise 2.8.3 Show that the (ordinary) m-gon, that is, with vertices as points and edges as lines, is the only generalized m-gon of order $(1, 1)$. Verify that there are no (m, d_p, d_1)-gons of order $(1, 1)$ unless $m = d_p = d_1$.

Exercise 2.8.4 Let Γ be a generalized m-gon. Take Δ to be the rank two geometry over $\{c, o\}$ whose o-elements are the elements (points and lines) of Γ and whose c-elements are the chambers of Γ; incidence is symmetrized containment. Show that Δ is a generalized $2m$-gon. Is it thick?

Exercise 2.8.5 Let Γ be a geometry of rank two satisfying the following conditions for some $m \in \mathbb{N}$ with $m \geq 2$.

(a) Every pair of elements is joined by a chain of length at most m and by at most one chain of length smaller than m.
(b) There exists an m-gon in Γ (that is, a subgeometry isomorphic to the ordinary m-gon).

Prove that Γ is a generalized m-gon.
(*Hint:* Let M be an m-gon in Γ and x an element of Γ. Then there is an element $y \in M$ such that x is at distance m from y in the incidence graph. Construct a bijection from Γ_x to Γ_y.)

Exercise 2.8.6 Show that the generalized hexagon of Example 2.2.15 admits no dualities.
(*Hint:* The part of the collinearity graph of the dual line space induced on the set of points at distance three from a given point differs from the original.)

Exercise 2.8.7 Consider the vector space $V = \mathbb{F}_9^3$ with nondegenerate bilinear form f given by $f(x, y) = x_1 y_1 + x_2 y_2 + x_3 y_3$ for $x = \varepsilon_1 x_1 + \varepsilon_2 x_2 + \varepsilon_3 x_3$ and $y = \varepsilon_1 y_1 + \varepsilon_2 y_2 + \varepsilon_3 y_3$ in $\mathbb{F}_9^3$. Put $G = O(V, f)$, the orthogonal group introduced in Exercise 1.9.31. Write $\mathbb{F} = \mathbb{F}_9$ and prove the following statements.

(a) The set P of projective points $a\mathbb{F} \in \mathbb{P}(V)$ for which $f(a, a)$ is a nonzero square in $\mathbb{F}$, is a single G-orbit of size 45.

(b) The set Q of orthogonal frames in P, that is, subsets $\{a\mathbb{F}, b\mathbb{F}, c\mathbb{F}\}$ of P with a, b, c mutually orthogonal, is a single G-orbit of size 30. The kernel of the action of G on Q is $Z(G)$, the center of G, which is a group of size two.

(c) Turn Q into a graph by letting $x \sim y$ for $x, y \in Q$ if and only $x \neq y$ and $x \cap y \neq \emptyset$. This graph has diameter four with 3, 6, 12, and 8 vertices at distance 1, 2, 3, and 4 from a given vertex, respectively. The group G acts *distance transitively* on Q (this means that G acts transitively on the sets of pairs of points at mutual distance i for each $i \in \mathbb{N}$). The distance distribution diagram coincides with the incidence graph of the rank two geometry of a generalized quadrangle of order $(2, 2)$. So the graph on Q is bipartite with two parts of size 15.

(d) The group G is isomorphic to $C_2 \times (\mathrm{Alt}_6 \rtimes \langle \sigma \rangle)$, where σ acts on Alt_6 as one of the outer automorphisms of Sym_6 given in Example 2.2.13.

(e) For $a\mathbb{F} \in P$, the orthogonal reflection $r_{\alpha,\phi} : V \to V$ from Exercise 1.9.31, with ϕ given by $\phi(x) = 2f(a, a)^{-1} f(a, x)$, has the form $r_{a,\phi}(x) = x + af(a, a)^{-1} f(a, x)$ $(x \in V)$ and belongs to G. The subgroup H of G generated by all such orthogonal reflections has order 720 and preserves the partition of Q into two cocliques. In particular, $Z(H) = Z(G)$ and $H/Z(H) \cong \mathrm{Alt}_6$.

Exercise 2.8.8 Let $G = \mathrm{Alt}_6$. This group has two conjugacy classes, say C_1 and C_2, of subgroups isomorphic to Alt_5. Show that the following graph is isomorphic to HoSi.

(1) Its vertex set consists of C_1, C_2, the twelve subgroups of G isomorphic to Alt_5 and the 36 Sylow 5-subgroups of G.

(2) The unordered pair $\{C_1, C_2\}$ is an edge. For $i \in [2]$, a subgroup of G isomorphic to Alt_5 is adjacent to C_i if it is a member of C_i. A subgroup of G isomorphic to Alt_5 and a Sylow 5-subgroup of G are adjacent if the latter is contained in the former. Two Sylow 5-subgroups of G are adjacent if together they generate G and there is an involution of G normalizing each. There are no further adjacencies.

Section 2.3

Exercise 2.8.9 (Cited in Example 2.3.2) Projective planes and affine planes are closely related.

(a) Let $\Pi = (P, L, *)$ be a thick projective plane. Fix $h \in L$. Show that the subgeometry of Π induced on $(P \setminus h^*) \cup (L \setminus \{h\})$ is an affine plane.

(b) Suppose that $A = (Q, M, *)$ is an affine plane. Prove that being parallel is an equivalence relation on M. For $m \in M$ we denote by $\overline{m}$ the parallel class (i.e., equivalence class) of m. Let P be the disjoint union of Q and the set h of equivalence classes of the parallelism relation on M. Let L be the disjoint union of M and the singleton $\{h\}$. Extend the relation $*$ to a symmetric and reflexive relation on all of $P \cup L$ by demanding that, for $l, m \in M$, we have

(1) $\overline{m} * h$ if and only if $\overline{m} \in h$;
(2) $l * \overline{m}$ if and only if $l \in \overline{m}$;
(3) $x * h$ for no $x \in Q$.

Prove that $(P, L, *)$ is a projective plane, whose subgeometry induced on $Q \cup M$ coincides with A.

Exercise 2.8.10 Show that every projective plane of order four is isomorphic to $\mathbb{P}(\mathbb{F}_4^2)$.

(*Hint:* Use Exercise 2.8.9 and prove that there is a unique affine plane of order four.)

Exercise 2.8.11 Let C be the Cayley division ring of Example 2.3.4.

(a) Prove that C is not associative, but satisfies the laws $x^2 y = x(xy)$ and $xy^2 = (xy)y$ for all $x, y \in C$ (here, of course, x^2 stands for xx). Algebras satisfying this law are called *alternative*.
(b) Define $\overline{x} = 2x_0 e_0 - x$ for $x = \sum_i x_i e_i \in C$ and prove that $N(x) = x\overline{x}$ is multiplicative (i.e., $N(xy) = N(x)N(y)$) and positive definite (i.e., $N(x) > 0$ whenever $x \neq 0$). (Algebras with such a map N are called composition algebras.)
(c) Derive from (b) that each nonzero element x of C has an inverse x^{-1}.
(d) Show that, if $x, y \in C$ with x nonzero, then $(yx)x^{-1} = y$. (This property helps to verify (3) and (5) of the definition of a ternary ring by means of (2.1).)

Exercise 2.8.12 (This exercise is used in Examples 2.6.5 and 6.2.7) Let $G \to \mathrm{Sym}(X)$ be a permutation representation of a group G on a set X. Let $t \in \mathbb{N}$ with $0 < t \le |X|$. The action of G on X is said to be *t-transitive* on X if G acts transitively on X and, whenever $t > 1$, the stabilizer G_x in G of a point $x \in X$ acts $(t-1)$-transitively on $X \setminus \{x\}$. Prove the following assertions.

(a) The group G acts 2-transitively on X if and only if there is $g \in G$ with $G = H \cup HgH$, where $H = G_x$.
(b) The group G (isomorphic to $\mathrm{PSL}(\mathbb{F}_{11}^2)$) of Example 2.3.11 acts 2-transitively on the collection of cosets of its subgroup G_0 isomorphic to Alt_5.
(c) For every division ring $\mathbb{D}$, the group $\mathrm{PGL}(\mathbb{D}^2)$ is 3-transitive on the set of points of $\mathbb{P}(\mathbb{D}^2)$.
(d) Let $\mathcal{B}$ be a G-orbit of subsets of X. Assume that G acts t-transitively on X. There is a positive integer λ such that each set of t elements of X lies in exactly λ members of $\mathcal{B}$.

Section 2.4

Exercise 2.8.13 Consider the Petersen graph Pet and its two Alt_5-orbits, B and R, say, of pentagons discussed in Example 1.3.4. Define a rank three geometry $\Gamma =$

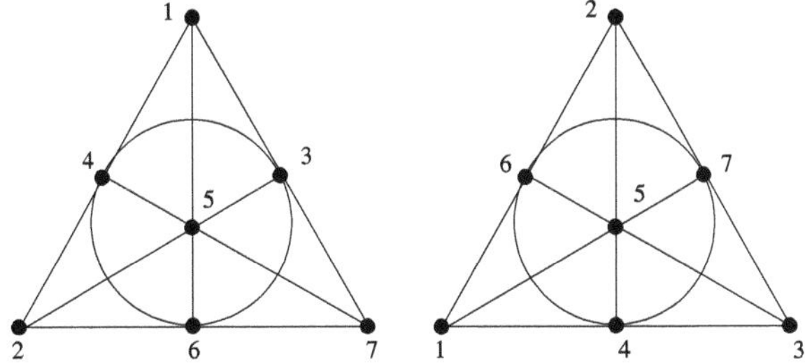

Fig. 2.18 Two disjoint Fano planes

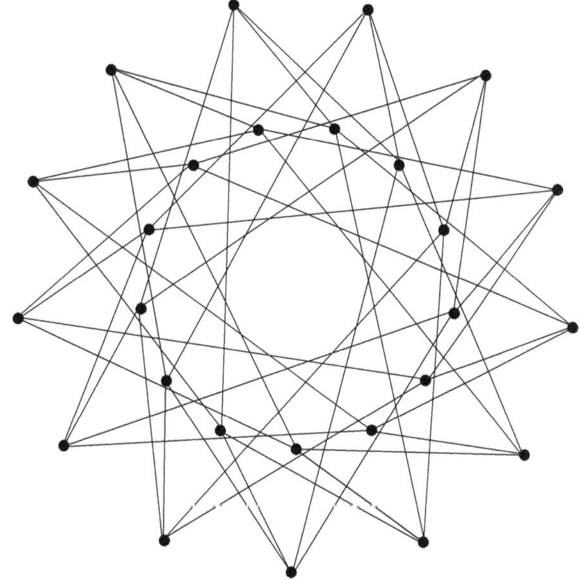

Fig. 2.19 The incidence graph of the projective plane $\mathbb{P}(\mathbb{F}_3^3)$. Rotations give a cyclic group of collineations of order 13

$(E, B, R, *)$ where E is the edge set of Pet, in such a way that it is isomorphic to the quotient of the great dodecahedron of Example 2.4.4 by a group of order two.

Exercise 2.8.14 (Cited in Remark 4.1.7) Let $\Omega = [7]$ and let G be the permutation group of order 42 consisting of all maps $x \mapsto ax + b$ $(x \in \Omega)$ for $a \not\equiv 0 \bmod 7$, where addition and multiplication are taken modulo 7. Consider the incidence system Γ over [3] whose 1-elements are all points of Ω, whose 2-elements are all unordered pairs of points, and whose 3-elements are all orbits of size three of subgroups of G of order three. Incidence is symmetrized inclusion. Prove the following three statements.

(a) The set of 3-elements of Γ consists of the 14 subsets of Ω of size three depicted as lines in the two disjoint Fano planes of Fig. 2.18.

(b) Γ is a thin [3]-geometry with diagram $\overset{1}{\underset{1}{\circ}} \!\!\rule[0.5ex]{2em}{0.4pt}\!\! \overset{2}{\underset{1}{\circ}} \overset{6}{\rule[0.5ex]{2em}{0.4pt}} \overset{3}{\underset{1}{\circ}}.$

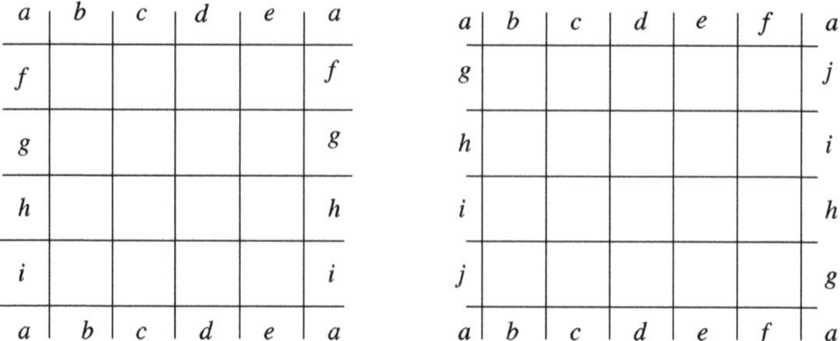

Fig. 2.20 *Left*: A toroidal geometry. *Right*: A Klein bottle geometry

(c) The group of automorphisms of Γ coincides with G. It is incidence transitive on Γ and has exactly two orbits on the set of chambers.

Exercise 2.8.15 Prove that the incidence graph of a projective plane of order 3 is necessarily as given in Fig. 2.19. Conclude that $\mathbb{P}(\mathbb{F}_3^3)$ is the unique projective plane of order three.

Exercise 2.8.16 Prove that $\mathbb{P}(\mathbb{F}_4^3)$ is the unique projective plane of order four.

Exercise 2.8.17 Consider a tiling T of the Euclidean plane by squares of unit side length. Let $u, v \in \mathbb{N}$, $u, v \geq 2$.

(a) Cut out a $u \times v$ rectangle from T that is covered by precisely uv tiles, and identify the border patches in the usual way to obtain a torus, as visualized in the left-hand side of Fig. 2.20 for $u = v = 5$. The result is $\mathbb{R}^2/(u\mathbb{Z} \times v\mathbb{Z})$, the quotient of $\mathbb{R}^2$ by the equivalence relation $\sim$ given by $(x, y) \sim (x', y')$ if and only if $x - x' \in u\mathbb{Z}$ and $y - y' \in v\mathbb{Z}$. Show that this quotient with vertices, edges, and tiles, leads to a geometry with diagram $\underset{1}{\circ}\!=\!=\!=\!\underset{1}{\circ}\!=\!=\!=\!\underset{1}{\circ}$ and that Γ is flag transitive if and only if $u = v$.

(b) Next, assume $v > 2$, start again with the $u \times v$ rectangle cut out from T and proceed to identify border patches as suggested by the right-hand side of Fig. 2.20 for $u = 6$ and $v = 5$. The result is the quotient of $\mathbb{R}^2$ by the equivalence relation $\sim$ given by $(x, y) \sim (x', y')$ if and only if either $x - x' \in u\mathbb{Z} \setminus 2u\mathbb{Z}$ and $y' + y \in v\mathbb{Z}$ or $x - x' \in 2u\mathbb{Z}$ and $y' - y \in v\mathbb{Z}$. It is the so-called *Klein bottle*. Show that the Klein bottle with vertices, edges, and tiles, leads to a geometry over $\underset{1}{\circ}\!=\!=\!=\!\underset{1}{\circ}\!=\!=\!=\!\underset{1}{\circ}$ whose automorphism group of order $\gcd(2, v)u$ is not transitive on the set of points.

Exercise 2.8.18 Consider the Petersen graph Pet. Its complement occurs in Example 1.7.16 for $\varepsilon = 1$ and $n = 5$. It is also the collinearity graph of the classical *Desargues configuration*; cf. Fig. 2.21. The depicted geometry Des has 10 lines of three points each. Observe that not all cliques of size three of the complement of Pet are represented by lines.

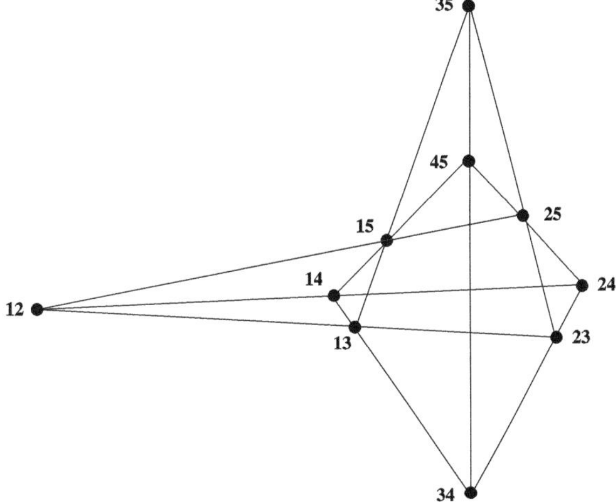

Fig. 2.21 Desargues
configuration Des

(a) Show that the geometry Des belongs to $\overset{\text{p}5}{\underset{2}{\circ}}\!\!\overset{3}{\rule{1.2cm}{0.4pt}}\!\!\overset{5}{\underset{2}{\overset{1}{\circ}}}$.

(b) Let Δ be the geometry over [3] whose 1-elements are the 10 points, whose 2-elements are the 30 pairs of collinear points, and whose 3-elements are the 15 induced quadrangles in the collinearity graph of Des (here, a quadrangle is a set of four points such that the collinearity induced on them is a 4-circuit). Incidence is symmetrized containment. Show that Δ belongs to $\underset{1}{\overset{1}{\circ}}\!\!\overset{2}{\rule{1.2cm}{0.4pt}}\!\!\underset{1}{\overset{6}{\circ}}\!\!\overset{3}{\rule{1.2cm}{0.4pt}}\!\!\underset{1}{\circ}$.

(c) Prove that $\mathrm{Aut}(\Delta)$ is isomorphic to Sym_5 and that this group acts flag transitively on Δ.

Section 2.5

Exercise 2.8.19 For $n, j \in \mathbb{N}$ with $1 \le j < n$, the *Johnson graph* with parameters (n, j) has vertex set $\binom{[n]}{j}$ and adjacency $x \sim y$ given by $|x \cap y| = j - 1$. Show that the Johnson graph is the shadow space on j of the geometry of rank $n - 1$ of Example 1.2.6.

Exercise 2.8.20 Consider the geometry Γ of Exercise 1.9.18, which has infinite rank.

(a) Show that Γ does not satisfy the conclusion of Theorem 2.1.6.
(b) Verify that, for each $n \in \mathbb{N}$ with $n > 1$, the space $\mathrm{ShSp}(\Gamma, \{1/n\})$ is the space whose point set is $X_{1/n}$ (the set of elements of type $1/n$), and whose lines are of the form L_b for $b \in \mathbb{R}$, where L_b is the set of all members of $X_{1/n}$ containing b.

Exercise 2.8.21 (This exercise is used in Lemma 9.2.4) Let Z be a linear space and $X \subseteq P$. Define the derived set $X^{(1)}$ of X as the union of X and of all lines of Z

having at least two points in X. Put $X^{(0)} = X$, and for $n \in \mathbb{N}$ with $n > 0$, define the n-th *derived set* $X^{(n)}$ as $(X^{(n-1)})^{(1)}$. Show that $\langle X \rangle = \bigcup_{n \in \mathbb{N}} X^{(n)}$.

Exercise 2.8.22 Give an example to demonstrate that a homomorphism $\alpha : (P, L) \to (P', L')$ of line spaces does not necessarily correspond to a homomorphism $(P, L, *) \to (P', L', *')$ of geometries. Show that if (P', L') is linear and $\alpha(l)$ has at least two points for every $l \in L$, there is a homomorphism $(P, L, *) \to (P', L', *')$ inducing α on the point shadow space.

Exercise 2.8.23 Let (P, L) and (P', L') be linear spaces having at least three points each. Suppose that $\alpha : P \to P'$ is a bijection. Show that α is an isomorphism if and only if both α and α^{-1} map each set of three collinear points to a set of three collinear points.

Exercise 2.8.24 Let $Z = (P, L)$ be a line space and $\equiv$ an equivalence relation on P. By $d(x, y)$ we denote the distance between x and y in the collinearity graph of Z and by $\overline{x}$ the equivalence class of x in P. The *quotient space* $Z/\equiv$ of Z by $\equiv$ is the pair (P', L') where P' consists of the equivalence classes in P and L' consists of the sets $\{\overline{x} \in P' \mid x \in l\}$ for some $l \in L$. We say that $\equiv$ is a *standard equivalence* if, whenever $x \equiv x'$ and $x \perp y$ for $x, x', y \in P$, there is $y' \in P$ such that $x' \perp y'$ and $y' \equiv y$. Prove the following two assertions for a standard equivalence $\equiv$.

(a) If $d(x, x') \geq 2$ for any two points x and x' in P with $x \equiv x'$, then $Z/\equiv$ is a line space.
(b) Let A be a group of automorphisms of Z. Show that the equivalence relation of being in the same A-orbit is standard.
(c) The *quotient space* Z/A of Z by A is defined as the quotient space $Z/\equiv$, where $\equiv$ is as in (b). As in Definition 1.6.1, we denote by $d(x, y)$ the distance between x and y in the collinearity graph of Z. Prove that, if Z is a partial linear space and $d(x, \sigma(x)) \geq 3$ for each $x \in P$ and $\sigma \in A \setminus \{1\}$, then Z/A is a partial linear space.
(d) Give a connected line space with a standard equivalence satisfying $d(x, y) \geq 3$ for all distinct $x, y \in P$ with $x \equiv y$, for which (c) is not valid.
  (*Hint:* Take $P = \{a_i, b_i, c_i, d_i \mid i \in [3]\}$; the line set L having $\{a_1, b_1, c_1\}$ and $\{b_2, c_2, d_2\}$ of size three and the following eight lines of size two: $\{b_1, d_3\}$, $\{c_1, d_1\}$, $\{b_3, d_1\}$, $\{a_2, c_2\}$, $\{a_2, b_3\}$, $\{a_3, b_2\}$, $\{a_3, c_3\}$, $\{c_3, d_3\}$; and take the equivalence of having the same letter in the name of the point.)

♯ **Exercise 2.8.25** (The tilde geometry) Put $\tau := \cos(\pi/5) = \frac{1}{4}(1 + \sqrt{5})$, $\rho := \cos(3\pi/5) = \frac{1}{4}(1 - \sqrt{5})$, and $\omega = e^{2\pi i/3}$, so $\omega^2 = -\omega - 1$. In unitary space $\mathbb{C}^3$ with standard hermitian inner product f and standard orthonormal basis $\varepsilon_1, \varepsilon_2, \varepsilon_3$, consider the vectors $\alpha_1 = \varepsilon_1$, $\alpha_2 = \omega^2 \tau \varepsilon_1 - \rho \varepsilon_2 + \omega/2 \varepsilon_3$, and $\alpha_3 = 1/2\varepsilon_1 + \rho \varepsilon_2 + \tau \varepsilon_3$. For $\alpha \in \{\alpha_1, \alpha_2, \alpha_3\}$, the unitary reflection $r_{\alpha, \phi} : \mathbb{C}^3 \to \mathbb{C}^3$ from Exercise 1.9.31, with ϕ given by $\phi(x) = 2f(\alpha, x)$, belongs to $U(\mathbb{C}^3, f)$. Denote by G the group generated by these three unitary reflections of order two.

(a) Prove that G leaves invariant the set

$$\Phi = \mu_6 \left\{ \varepsilon_i,\ \frac{1}{2}\varepsilon_i \pm \rho\varepsilon_j \pm \tau\varepsilon_k,\ \omega^2\tau\varepsilon_i \pm \rho\varepsilon_j \pm \frac{1}{2}\omega\varepsilon_k, \right.$$

$$\frac{1}{2}\varepsilon_i \pm \frac{1}{2}\omega^2\varepsilon_j \pm \left(\frac{1}{2}+\omega\tau\right)\varepsilon_k,\ \left(\frac{1}{2}+\omega^2\tau\right)\left(\varepsilon_i \pm \omega^2\varepsilon_j\right)$$

$$\left. \mid (i,j,k) \in \left\{(1,2,3),(3,1,2),(2,3,1)\right\} \right\}$$

containing α_1, α_2, α_3. Here, $\mu_6 = \{\pm 1, \pm\omega, \pm\omega^2\}$.
(b) Establish that Φ is a single G-orbit of size 6×45.
(c) Observe that $p = \{\pm\varepsilon_1, \pm\omega^2\varepsilon_2, \pm\omega\varepsilon_3\}$ is the only orthonormal basis up to signs of $\mathbb{C}^3$ inside $M = \mu_6\varepsilon_1 \cup \mu_6\varepsilon_2 \cup \mu_6\varepsilon_3$ containing ε_1 and invariant under the normalizer in G of M. Let P be the G-orbit of p and show it has size 45.
(d) Set

$$d = \varepsilon_2\left(-\omega^2\tau - 1/2\right) + \varepsilon_3\left(\omega\tau + 1/2\omega^2\right),$$

$$e = \varepsilon_2\left(\omega^2\tau + 1/2\right) + \varepsilon_3\left(\omega\tau + 1/2\omega^2\right).$$

Prove that every orthonormal basis of $\mathbb{C}^3$ containing ε_1, contained in Φ, and not contained in M lies in $N = \{\mu_6\varepsilon_1, \mu_6 d, \mu_6 e\}$ and that $l = \{\pm\varepsilon_1, \pm d, \pm e\}$ is the only orthonormal triple up to signs from N invariant under the normalizer in G of N. Let L be the G-orbit of l and show it has size 45.
(e) Show that the graph $(P \cup L, \sim)$, where $x \sim y$ if and only if $|x \cap y| = 1$, is a bipartite graph with parts P and L and with 3, 6, 12, 24, 24, 12, 6, 2 vertices at distance 1, 2, 3, 4, 5, 6, 7, 8, respectively, from a given vertex.
(f) Let Til be the incidence system $(P, L, *)$ over $\{\mathrm{p}, \mathrm{l}\}$, where $x * y$ if and only if $x \sim y$ or $x = y$. It is called the *tilde geometry*. Prove that Til is a connected and flag transitive geometry with parameters $d_\mathrm{p} = d_\mathrm{l} = 8$ and $g = 5$.
(g) Let A be the subgroup of G consisting of homotheties by a scalar from $\{1, \omega, \omega^2\}$. Verify that the elements of Til at distance eight from an element x are of the form ax for $a \in A \setminus \{1\}$. Conclude that being at distance eight is an equivalence relation on $P \cup L$.
(h) Consider the subring $\mathbb{Z}[\omega, \sqrt{5}, \frac{1}{2}]$ of algebraic numbers in $\mathbb{C}$ and the finite field $\mathbb{F}_9$ of order 9 with square root i of -1. Verify that there is a unique homomorphism $\sigma : \mathbb{Z}[\omega, \sqrt{5}, \frac{1}{2}] \to \mathbb{F}_9$ of rings determined by $\sigma(\frac{1}{2}) = -1$, $\sigma(\sqrt{5}) = i$, and $\sigma(\omega) = 1$.
(i) Observe that all components of vectors in Φ are actually in the ring $\mathbb{Z}[\omega, \sqrt{5}, \frac{1}{2}]$. By applying σ coordinatewise we obtain a map $\mathbb{Z}[\omega, \sqrt{5}, \frac{1}{2}]^3 \to \mathbb{F}_9^3$. The image of Φ under this map is a set $\overline{\Phi}$ of 45 vertices up to sign changes. Use this map to derive that the quotient incidence system Til$/A$ (cf. Definition 1.3.5) of Til by the group A is the generalized quadrangle of order $(2, 2)$ identified in Exercise 2.8.7.

Section 2.6

Exercise 2.8.26 Fix a finite-dimensional vector space V over the field $\mathbb{F}$, a natural number $k \leq \dim(V)/2$, and consider the graph Γ whose vertices are the k-dimensional vector spaces of V and in which two vertices X and Y are adjacent whenever $\dim(X \cap Y) = k - 1$.

(a) Prove that two vertices X, Y of Γ are at distance h in Γ (i.e., $d_\Gamma(X, Y) = h$) if and only if $\dim(X \cap Y) = k - h$.
(b) Let $h \in [k]$. Show that $GL(V)$ is transitive on the set of ordered pairs of vertices $\{(X, Y) \mid d_\Gamma(X, Y) = h\}$.
(c) Describe the distribution diagram of Γ for $\mathbb{F} = \mathbb{F}_q$, the finite field of order q.

A graph with a group of automorphisms satisfying (b) is called distance transitive.

Section 2.7

Exercise 2.8.27 Let $\mathbb{F}$ be a field and $\mathbb{F}[[X]]$ be the ring of formal power series in X with coefficients in $\mathbb{F}$. Prove that $\mathbb{F}[[X]]$ is a discrete valuation ring with local parameter X and residue field isomorphic to $\mathbb{F}$. Compare the result with Example 2.7.2(ii) and conclude that the residue field does not uniquely determine the discrete valuation ring.

Exercise 2.8.28 Let V and W be right vector spaces over a division ring $\mathbb{D}$. A map $g : V \to W$ is called *semi-linear* if, for each $x \in V$, there is an automorphism σ_x of $\mathbb{D}$ such that

$$g(u\lambda + v\mu) = (gu)\sigma_u(\lambda) + (gv)\sigma_v(\mu) \quad (\lambda, \mu \in \mathbb{D}; v, w \in V).$$

(a) Prove that σ_v is the same for all $v \in V$ with $gv \neq 0$; it is called the automorphism of $\mathbb{D}$ *induced* by g. Such a map g is called σ-*linear*, where $\sigma = \sigma_v$ whenever $gv \neq 0$.
(b) Show that the set of all invertible semi-linear maps $V \to V$, denoted $\Gamma L(V)$, is a subgroup of $\mathrm{Sym}(V)$ containing $GL(V)$ as a normal subgroup.
(c) For $\lambda \in \mathbb{D} \setminus \{0\}$, the *homothety* with respect to λ on V is the map $h_\lambda : V \to V$ given by $h_\lambda(v) = v\lambda$. Prove that the set of homotheties is a subgroup H of $\Gamma L(V)$, and that $H \subseteq GL(V)$ if and only if $\mathbb{D}$ is a field.
(d) Verify that the homothety class of an R-lattice as in Definition 2.7.9 is the same as its H-orbit.

Exercise 2.8.29 Suppose that R is a discrete valuation ring with local parameter π. Let $\mathrm{ord} : R \to \mathbb{N}$ be the map such that $x \in \pi^{\mathrm{ord}(x)} R \setminus \pi^{\mathrm{ord}(x)+1} R$ for all $x \in R$. Prove that ord satisfies the following rules for $x, y \in R$.

(1) $\mathrm{ord}(xy) = \mathrm{ord}(x)\mathrm{ord}(y)$,
(2) $\mathrm{ord}(x + y) \geq \min(\mathrm{ord}(x), \mathrm{ord}(y))$,
(3) $\mathrm{ord}(x) = 0$ if and only if x is invertible.

Exercise 2.8.30 (Cited in proof of Theorem 2.7.14) Prove that $[L] = \{\pi^j L \mid j \in \mathbb{Z}\}$. Derive from this that the stabilizers in $\mathrm{SL}(\mathbb{F}^n)$ of $[L]$ and of L coincide.

Exercise 2.8.31 (This exercise is used in Lemma 2.7.13) Let R be a discrete valuation ring, $n \in \mathbb{N}$, $n > 1$, and take $i \in [n - 1]$. Show that the map $M \mapsto M/\pi R^n$ is a bijective correspondence between the R-lattices M such that $\pi R^n \subseteq M \subseteq R^n$ and the $(n - i)$-dimensional subspaces of the vector space $(R/\pi R)^n$.

2.9 Notes

The concept of a diagram can be traced back to Schläfli in 1853 who exploited it in order to classify the regular convex polytopes. The discovery of the generalized polygons and of important classes of flag-transitive geometries over Coxeter diagrams is due to Tits; see [280, 281].

Section 2.1

The partially ordered set with the Jordan-Dedekind property, described in Example 2.1.11, are known as ranked posets in [267].

Section 2.2

The general concept of a diagram was coined by Buekenhout [44, 45] in order to encompass examples related to sporadic groups. For (g, d_p, d_1)-gons, see [46].

A wealth of information on generalized polygons can be found in [236, 294] (the terminology differs slightly in that lines in their generalized polygons are not allowed to have exactly two points (and dually); such generalized polygons are called weak).

In Definition 2.2.7, generalized g-gons were given a special name only for $g \in \{2, 3, 4, 6, 8\}$. The reason is that these are the only values for which thick finite generalized g-gons exist; see [36] for an overview. Besides, these are also the only values for which thick generalized g-gons exist that are Moufang. In [291] all generalized Moufang polygons are classified.

Good introductory texts to projective planes are [167, 240]. The usual definition of a projective plane in these books does not allow for lines to be thin.

Free constructions show that thick infinite generalized g-gons for all values of g occur; see for instance [275].

The construction of the generalized hexagon of order $(2, 2)$ in Example 2.2.15 is from [281]; see also [71]. In [83] it is proved that it is not isomorphic to its dual and that, up to isomorphism and duality, it is the unique one of its order (a fact already announced in [281]). A characterization of the known generalized octagons is given in [295].

The Hoffman-Singleton graph HoSi first appeared with uniqueness proof in [163]. More papers with details on HoSi are [17] and [92]. In the proof of Theorem 2.2.19, we follow [174]. The graph is a Moore graph (referring to a different person from his namesake mentioned in Sect. 1.10), whose definition and theory is summarized in [36].

Section 2.3

Proofs of most of the statements regarding affine planes can be found in [167]. The idea of a ternary ring goes back to M. Hall [146]. As a consequence of the strict correspondence, various geometric properties are reflected by properties of the ternary ring. The ternary ring is a division ring if and only if the corresponding projective plane is Desarguesian. The ternary ring is an alternative division ring if and only if the projective plane is Moufang, etc. Good introductions to alternative rings are [246, 265].

All locally Petersen graphs, of which only one has been mentioned in Example 2.3.10, have been determined in [148]. The notion of graphs being locally Δ was brought forward in 1980 [24] but appeared as early as 1963 in a question by Zykov. Several studies have been conducted since for special classes of graphs; see [25, 52, 81, 85, 121, 140, 150] for results related to geometry.

If the girth of the local graph is larger than five, free examples exist, whereas full characterizations seem often feasible in cases of smaller girth. See [300, 301], which show that the case of a local hexagon, the tiling of the plane by honeycombs, and the locally Petersen graphs, characterized in [148], are at the division line. Group-theoretical results in the same vein can be found in [176, Theorem D]. Some characterizations using extra hypotheses approach the verge of what is possible; examples are found in [78, 140].

Section 2.4

The Neumaier geometry of Example 2.4.11 can be found in [230]; for characterizations as a flag-transitive geometry with diagram B_3, see [6, 315]. The geometries in Example 2.4.12 are from [309] and [230], respectively; see also [214].

Heiss [158] established their uniqueness assuming that all residues with diagram
o══════o────────o are Neumaier geometries.

The graph Δ' of Remark 2.4.13 was constructed in [160] on the basis of a known
Steiner system, which is described in Exercise 5.7.34. The sporadic simple group
HS appears in Table 5.2.

More details on most of the sporadic geometries constructed in this chapter, as
well as additional references, can be found in [36].

Section 2.5

The concepts of linear space and linear subspace have been used in many places.
Around 1960 they received explicit recognition, thanks to the efforts of Libois
[206]. Theorem 2.5.15 is a slight generalization of Tits' characterization of projective planes as generalized 3-gons [281].

In [286], the original definition of a shadow space can be found.

Section 2.6

The special case of linear diagrams for group geometries, as appearing in Theorem 2.6.4 is treated in [217].

The coset enumeration mentioned in Example 2.6.5 is a systematic method of
producing the representation of a group G on a subgroup H when G is given by
finite sets of generators and relations, H by a finite set of generators expressed as
words in the generators of G, provided G/H is finite. See [84] for details. Example 2.6.5 is taken from [215]. It is only one example of a vast number of flag-
transitive geometries in which the Petersen graph occurs as a rank two residue; see
for example [171, 173], where the geometries are not only constructed but also classified as those with a given diagram and a flag-transitive automorphism group. For
instance, there are exactly eight flag-transitive Pet_n-geometries; these are geometries of type

$$\mathrm{Pet}_n = \overset{1}{\underset{2}{\circ}}\!\!\rule{1.5cm}{0.4pt}\!\!\overset{2}{\underset{2}{\circ}} \cdots\!\!\rule{1.5cm}{0.4pt}\!\!\overset{n-1}{\underset{2}{\circ}}\overset{\mathrm{Pet}}{\rule{1.5cm}{0.4pt}}\overset{n}{\underset{1}{\circ}}$$

where n is the rank. As in Example 2.3.9, the label Pet indicates a rank two geometry
isomorphic to the geometry of vertices and edges of the Petersen graph treated in
Example 2.2.8. The automorphism group of the unique example up to isomorphism
for $n = 2$ is isomorphic to Sym_5. For $n = 3$, they correspond to the Mathieu group
M_{22}, and its central 3-cover $3 \cdot M_{22}$, for $n = 4$ to the Mathieu group M_{23}, the second
Conway group Co_2, an extension of this group $3^{23}\mathrm{Co}_2$, and the fourth Janko group
J4 and for $n = 5$ to the Baby Monster group B and an extension of type 3^{4371}B.
Also, the McLaughlin group McL acts flag transitively on a geometry of type

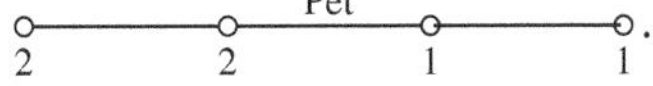

This accounts for six sporadic groups.

Section 2.7

The geometries of type $\widetilde{A}_{n-1}$ are the easiest nontrivial examples of buildings of affine type. Their analysis and classification (for rank at least four) is a major piece of work by Bruhat and Tits [39]. An excellent treatment of this material, together with new details on the classification, is given by Weiss [306]. Some examples of finite geometries of type $\widetilde{A}_2$ are given in [196, 242]. For other affine Coxeter diagrams, some finite examples of this kind are known as well; see for instance [183] for type $\widetilde{D}_4$. A survey of such constructions is given in [184], and a classification of finite flag-transitive quotient geometries of affine buildings, obtained by means of a discrete flag-transitive subgroup of the full automorphism group, is discussed in [185].

These geometries have 2-coverings of the above type, that is, which are buildings. Some examples of finite geometries of rank three having affine type $\widetilde{C}_2$ that are not covered by buildings are given in [182, 186].

Section 2.8

Part (a) of the definition of a generalized polygon in Exercise 2.8.5 refers to Tits' first definition of the notion in [281]. Part (b) was added later to avoid anomalies.

Exercise 2.8.8 was suggested to us by Ernest Shult.

The notion of standard equivalence in Exercise 2.8.24 is a slight variation of the one introduced in [107]. We owe the counterexample at the end of this exercise to Pasini.

The rank two geometry Til of Exercise 2.8.25 is discussed under the name Foster graph in [36, Sect. 13.2A]. See also [171, Sect. 6.2] and [234]. The automorphism group of Til is isomorphic to $3 \cdot \mathrm{Sym}_6 .2$. The group G of Exercise 2.8.25 is isomorphic to $C_2 \times (3 \cdot \mathrm{Alt}_6)$ and is identified in [69]. There are several interesting flag-transitive Til_n-geometries (see [171, 173]); these are geometries of type

$$
\mathrm{Til}_n = \overset{1}{\underset{2}{\circ}} \!\!\!\!\!\!\!\!\!\! \overset{2}{\underset{2}{\circ}} \cdots\cdots \overset{n-1}{\underset{2}{\circ}} \overset{\mathrm{Til}}{} \overset{n}{\underset{2}{\circ}}
$$

where n is the rank. Among the groups admitting flag-transitive Til_n-geometries there are an infinite series with group $3^{s(n)}\mathrm{Sp}(2n, 2)$, where $s(n) = (2^n - 1) \times (2^n - 2)/6$; furthermore, for rank $n = 3$, the Mathieu group M_{24} and the Held group He, for rank 4, the first Conway group Co_1, and for rank 5, the biggest sporadic simple group, called the Monster. Together with those from the diagram related to the Petersen graphs, this accounts for ten sporadic groups.

Chapter 3
Chamber Systems

The study of geometries can be developed starting from a different viewpoint than the diagram geometric one of the previous chapter. It corresponds to the structure induced on the set of maximal flags, also called chambers (cf. Definition 1.2.5), of a geometry. This slightly more abstract viewpoint has advantages for the study of thin geometries as well as group-related geometries.

We begin by exploring the above mentioned structure: the chamber system of a geometry. Then we study chamber systems in their own right, coming across several notions that have already been introduced for geometries, like residues, residual connectedness, and diagrams. These observations lead to the idea that geometries could be derived from chamber systems. Indeed, the main result of the present chapter is Theorem 3.4.6, which gives a correspondence between residually connected chamber systems and residually connected geometries.

Throughout this chapter, I is a set of types.

3.1 From a Geometry to a Chamber System

We introduce the notion of chamber system over I and show that the set of chambers of a geometry over I has such a structure. The correspondence is not bijective: Example 3.1.4 shows that not all chamber systems come from geometries and Example 3.1.8 gives two non-isomorphic geometries with isomorphic chamber systems. For $|I| = 2$, a criterion for a chamber system over I to be the chamber system of a geometry is given in Theorem 3.1.14. Furthermore, we introduce notions resembling those for geometries and graphs, such as chamber subsystems, (weak) homomorphisms, and quotients.

Let Γ be an incidence system over I.

Definition 3.1.1 A *chamber system over I* is a pair $\mathcal{C} = (C, \{\sim_i | i \in I\})$ consisting of a set C, whose members are called *chambers*, and a collection of equivalence relations $\sim_i$ on C indexed by $i \in I$. These relations are interpreted as graph structures

F. Buekenhout, A.M. Cohen, *Diagram Geometry*,
Ergebnisse der Mathematik und ihrer Grenzgebiete. 3. Folge / A Series of
Modern Surveys in Mathematics 57, DOI 10.1007/978-3-642-34453-4_3,
© Springer-Verlag Berlin Heidelberg 2013

Fig. 3.1 The chamber system
of the cube drawn on the cube

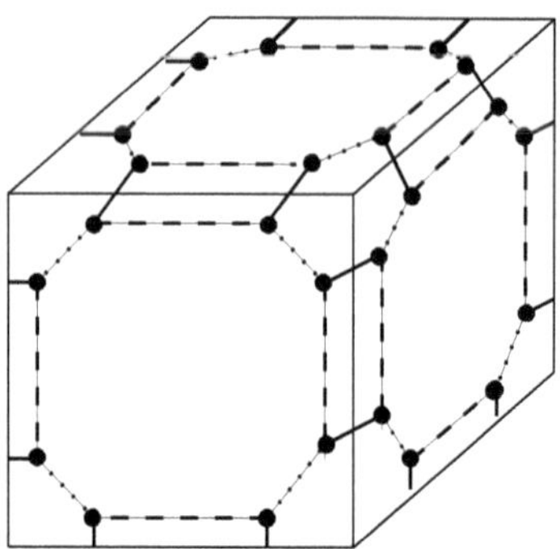

on C. For each $i \in I$, the graph $(C, \sim_i)$ is a disjoint union of cliques since $\sim_i$ is an equivalence relation. Two chambers c, d are called *i-adjacent* if $c \sim_i d$. For $i \in I$, each $\sim_i$-equivalence class is called an *i-panel*.

The *rank* of C is $|I|$. The chamber system C over I is called *firm*, *thick*, or *thin*, if, for each $i \in I$, every i-panel is of size at least two, at least three, or exactly two in the respective cases.

Incidence systems give rise to chamber systems in the following fashion.

Lemma 3.1.2 *Let C be the set of chambers of the incidence system Γ over I and, for $i \in I$, define the relation $\sim_i$ on C by $c \sim_i d$ if and only if, for each $j \in I \setminus \{i\}$, they have exactly the same j-element. The resulting pair $C(\Gamma) = (C, \{\sim_i \mid i \in I\})$ is a chamber system over I. If Γ is a firm, thick, or thin geometry, then $C(\Gamma)$ is firm, thick, or thin, respectively.*

Proof For each $i \in I$, the relation $\sim_i$ is an equivalence relation, so C is furnished with a collection of partitions indexed by I. This proves the first assertion. The second assertion follows from the fact that each $\sim_i$-equivalence class is in bijective correspondence with a residue of Γ of type i. $\square$

Example 3.1.3 Figure 1.1 shows the 48 chambers of the cube geometry as triangular faces. In Fig. 3.1, a chamber is visualized as a point on the face and close to the edge and vertex that it contains. It is joined by a dashed line to the unique chamber in the same face with the same nearest edge. It is joined by a dotted line to the unique chamber in the same face with the same nearest vertex, and with a full line to the unique chamber belonging to the same vertex and edge. Not all chambers are drawn: only those on the three visible faces of the cube.

Detaching the chamber system from the cube, we represent each chamber by a point close to the vertex, edge, and face belonging to it as in Fig. 3.2, and draw the edges representing the three distinct kinds of adjacency in distinct ways. We observe that each element x of the cube is represented by a 'face' of the chamber system and this face is the chamber system of the residue of x. Thus, vertices, edges, and faces of the cube are represented in the picture by hexagons, squares, and octagons.

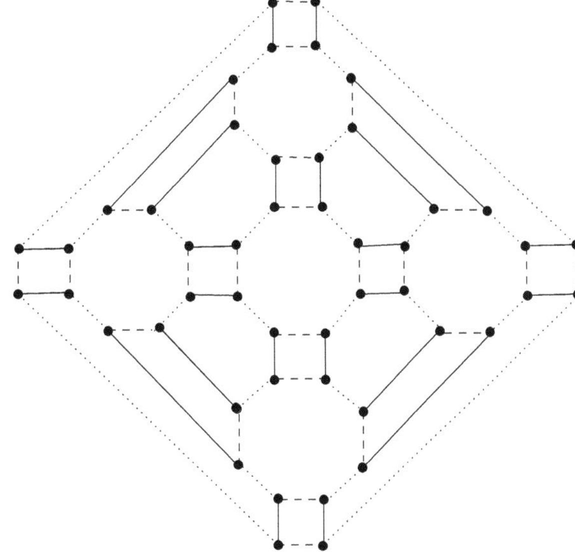

Fig. 3.2 The abstract chamber system of the cube

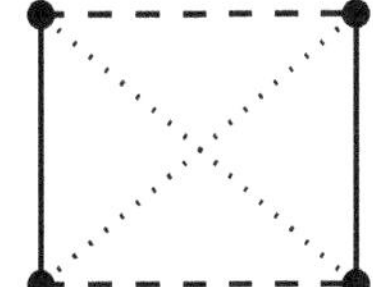

Fig. 3.3 A chamber system of rank three which is not the chamber system of a geometry

Example 3.1.4 Consider the rank three chamber system $\mathcal{C}$ with four chambers depicted in Fig. 3.3. There is no geometry Γ such that $\mathcal{C}$ is of shape $\mathcal{C}(\Gamma)$, not even an incidence system. Assume the contrary. Start constructing an incidence system Γ which is necessarily of rank three and draw an element of each type incident with one of the four given chambers. Then each of the three resulting rank two flags extends to one more chamber giving one further element of each type in Γ. This provides six elements of Γ, two of each type. There is no room for more chambers hence no further incidences among the six elements of Γ and we reach a contradiction as some adjacencies of the chamber system are still missing in Γ.

The preceding example shows that chamber systems do not correspond bijectively to either geometries or incidence systems.

Example 3.1.5 Consider the chamber system $\mathcal{C}$ of rank three described by Fig. 3.4. It has 8 chambers. Let us denote by 1, 2, 3, the types associated with full lines, speckled lines, and dotted lines, respectively, in Fig. 3.4. Is $\mathcal{C}$ of the form $\mathcal{C}(\Gamma)$ for some geometry Γ?

Again let us try to construct Γ. The chambers are labelled from 1 to 8 in such a way that 1, 2, 3, 4 are respectively 'opposite' to 5, 8, 7, 6. For a given 1-element

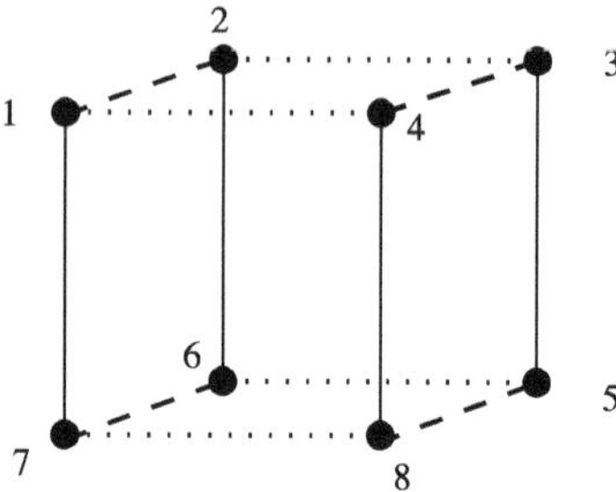

Fig. 3.4 A chamber system of rank three

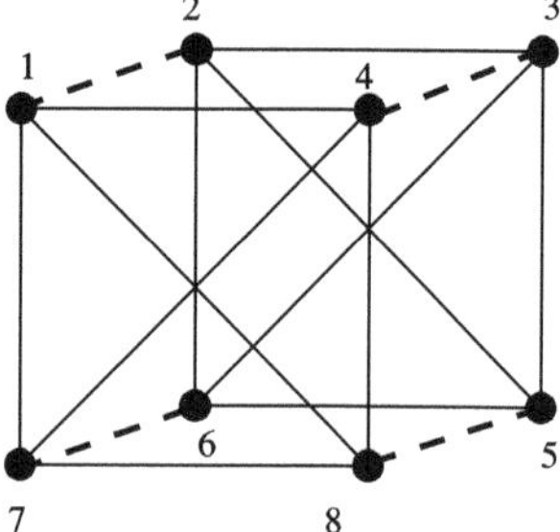

Fig. 3.5 A chamber system of rank two

a of Γ belonging to the chamber 1, the chambers that are connected to 1 by 2-adjacencies and 3-adjacencies, also contain a. This means that the chambers 1, 2, 3, 4 all contain a. Since there is no indication why the other chambers should contain a, we try and build up a from the set of chambers $\{1, 2, 3, 4\}$. The only other element of type 1 that can be built up in this way is the set of remaining chambers: $b = \{5, 6, 7, 8\}$. Continuing this way for the other two types, we find four more elements $c = \{1, 4, 7, 8\}$, $d = \{2, 3, 5, 6\}$, $e = \{1, 2, 6, 7\}$, $f = \{3, 4, 5, 8\}$ of Γ with c, d of type 2, and e, f of type 3. Indeed, now we can 'explain' the chambers of $\mathcal{C}$ as the maximal flags $1 = \{a, c, e\}$, $2 = \{a, d, e\}$, $3 = \{a, d, f\}$, $4 = \{a, c, f\}$, $5 = \{b, d, f\}$, $6 = \{b, d, e\}$, $7 = \{b, c, e\}$, and $8 = \{b, c, f\}$. This determines Γ completely: it is the geometry with two elements of each type in which each element is incident with all four elements of a different type.

Example 3.1.6 A chamber system of rank two is drawn in Fig. 3.5. It has 8 chambers. The elements whose types are indicated by speckled line segments are connected components with respect to the adjacency relation given by full line segments, and similarly with 'full' and 'speckled' interchanged. So, the chamber system is a generalized digon with four elements of type `full line` and two of type `speckled line`.

We will establish a general result about rank two chamber systems in Theorem 3.1.14. To express it, we need to define homomorphisms. As for geometries, we distinguish between (type preserving) homomorphisms and weak homomorphisms.

Definition 3.1.7 A *weak homomorphism* $\alpha : (C, \{\sim_i \mid i \in I\}) \to (C', \{\sim_{i'} \mid i' \in I\})$ of chamber systems over I is a map $\alpha : C \to C'$ for which a permutation π of I can be found such that, for all c, $d \in C$, the relation $c \sim_i d$ implies $\alpha(c) \sim_{\pi(i)} \alpha(d)$. If $\pi = \mathrm{id}$, the weak homomorphism is said to be a *homomorphism*. As usual, a bijective homomorphism whose inverse is also a homomorphism is called an *isomorphism* and an isomorphism from $\mathcal{C}$ to $\mathcal{C}$ is called an *automorphism* of $\mathcal{C}$. We denote by $\mathrm{Aut}(\mathcal{C})$ the group of all automorphisms of $\mathcal{C}$ and write $\mathcal{C} \cong \mathcal{C}'$ for two chamber systems $\mathcal{C}$ and $\mathcal{C}'$ to indicate that they are isomorphic.

Moreover, a bijective weak homomorphism whose inverse is also a weak homomorphism is called a *correlation* and a correlation from $\mathcal{C}$ to $\mathcal{C}$ is called an *auto-correlation* of $\mathcal{C}$. We denote by $\mathrm{Cor}(\mathcal{C})$ the group of all auto-correlations of $\mathcal{C}$.

Example 3.1.8 Let Γ_1 be the disjoint union of two tetrahedra. So Γ_1 has 8 vertices, 12 edges, and 8 faces. Let Γ_2 be the geometry obtained from Γ_1 by identifying a vertex from each tetrahedron. Thus, Γ_2 has 7 vertices, 12 edges, and 8 faces. Then $\mathcal{C}(\Gamma_1) \cong \mathcal{C}(\Gamma_2)$. We conclude that non-isomorphic geometries over I can have isomorphic chamber systems over I.

Lemma 3.1.9 *Let Γ be a geometry over I. If F is a flag of Γ of type $J \subseteq I$, and if $\mathcal{C}(\Gamma)_F$ denotes the set of all chambers of Γ containing F, endowed with the i-adjacencies induced by all $\sim_i$ with $i \in I \setminus J$, then $\mathcal{C}(\Gamma_F) \cong \mathcal{C}(\Gamma)_F$.*

Proof A chamber of Γ_F is a flag of type $I \setminus J$, so $G \mapsto F \cup G$ establishes a map $\mathcal{C}(\Gamma)_F \to \mathcal{C}(\Gamma)_F$. It is bijective, with inverse $c \mapsto c \setminus F$, and preserves i-adjacency for each $i \in I \setminus J$, as $G \sim_i H$ holds if and only if $(F \cup G) \sim_i (F \cup H)$. $\square$

A homomorphism $\mu : \Gamma' \to \Gamma''$ of geometries over I leads to a homomorphism $\mathcal{C}(\mu) : \mathcal{C}(\Gamma) \to \mathcal{C}(\Gamma')$ determined by $\mathcal{C}(\mu)(c) = (\mu(c_i))_{i \in I}$ for each chamber c of $\mathcal{C}$.

Proposition 3.1.10 *Suppose that A is a group of automorphisms of the chamber system $\mathcal{C} = (C, \{\sim_i \mid i \in I\})$ over I. Then the quotient C/A, that is, the set C/A of A-orbits in C, with the relations $\sim_i$ defined by*

$$Ax \sim_i Ay \iff \exists a \in A : ax \sim_i y \quad \text{for } x, y \in C$$

is again a chamber system over I.

Proof Let $i \in I$. Clearly, $\sim_i$ is a symmetric and reflexive relation on C/A. Suppose that $Ax \sim_i Ay \sim_i Az$. Then there are $a, c \in A$ with $ax \sim_i y \sim_i cz$. As $\sim_i$ is transitive on C, it follows that $ax \sim_i cz$, so $Ax \sim_i Az$. Therefore, $\sim_i$ is transitive on C/A, and an equivalence relation for each $i \in I$. $\square$

Clearly, there is a homomorphism $\mathcal{C} \to \mathcal{C}/A$. This justifies the following name.

Definition 3.1.11 The chamber system $\mathcal{C}/A$ of Proposition 3.1.10 is called the *quotient* of $\mathcal{C}$ by A.

Example 3.1.12 The chamber system $\mathcal{C}$ drawn in Fig. 3.4 has an automorphism a of order two interchanging opposite chambers (so, as a permutation, it is $(1, 5)(2, 8)(3, 7)(4, 6)$). The quotient $\mathcal{C}/\langle a \rangle$ is the chamber system drawn in Fig. 3.3.

Given a group of automorphisms, one might also consider the structure induced on fixed chambers. This is one reason for defining the notion of (partial) chamber subsystem.

Definition 3.1.13 If $C' = (C', \{\sim'_{i'} \mid i' \in I'\})$ is a chamber system over I' such that $I' \subseteq I$, $C' \subseteq C$, and $\sim'_i \subseteq \sim_i$ for each $i \in I'$, then C' is called a *partial chamber subsystem* of $\mathcal{C}$. If, moreover, for each $i \in I$, the restriction of $\sim_i$ to $C' \times C'$ coincides with $\sim'_i$ if $i \in I'$ and with the identity relation if $i \in I \setminus I'$, then C' is called a *chamber subsystem* of $\mathcal{C}$ (induced on C').

The relations 'is a chamber subsystem of' and 'is a partial chamber subsystem of' are transitive.

Let Γ be a geometry and F a flag of Γ. The set $\mathcal{C}(\Gamma)_F$ of all chambers of Γ containing F is a chamber subsystem of $\mathcal{C}(F)$ and, by Lemma 3.1.9, can be identified with the chamber system of Γ_F. For $i \in \tau(F)$, the restriction of $\sim_i$ to $\mathcal{C}(\Gamma)_F$ is the identity relation, so $\mathcal{C}(\Gamma)_F$ can indeed be seen as a chamber system over $I \setminus \tau(F)$.

Theorem 3.1.14 *Suppose that $\mathcal{C}$ is a rank two chamber system over $\{i, j\}$ such that the intersection of $\sim_i$ and $\sim_j$ is the identity relation. Then, up to isomorphism, there exists a unique geometry Γ over $\{i, j\}$ such that $\mathcal{C}(\Gamma)$ is isomorphic to $\mathcal{C}$.*

Proof Let $\mathcal{C} = (C, \sim_i, \sim_j)$. We take $X_i = (C/\sim_j) \times \{i\}$, the set of equivalence classes of $\sim_j$ 'marked' by i, and similarly $X_j = (C/\sim_i) \times \{j\}$. The type map τ takes value i on X_i and j on X_j. For $(x, a), (y, b) \in X_i \cup X_j$, define $(x, k) * (y, l)$ by $x \cap y \neq \emptyset$. If $k = l$, then $x \cap y \neq \emptyset$ implies $x = y$, so $\Gamma := (X_i \cup X_j, *, \tau)$ is an incidence system. Given an i-element (x, i) of Γ, the $\sim_i$-class y of a chamber in x gives a j-element (y, j) incident with (x, i), and similarly for i and j interchanged, so Γ is a geometry over $\{i, j\}$.

Suppose that $\{(x, k), (y, l)\}$ is a chamber of Γ. It follows that $\{k, l\} = \{i, j\}$. By the assumptions on $\mathcal{C}$, the intersection $x \cap y$ is a singleton, and so contains a unique element $\phi(\{(x, k), (y, l)\})$. This defines a map ϕ from the chamber set of $\mathcal{C}(\Gamma)$ to C. It is straightforward to check that ϕ is an isomorphism of chamber systems over $\{i, j\}$, and that the above construction of a geometry Γ from $\mathcal{C}$ such that $\mathcal{C}(\Gamma) \cong \mathcal{C}$ is the only one possible up to isomorphism. $\qquad \square$

Remark 3.1.15 The use of markers in the guise of second coordinates in the construction of the sets X_i and X_j in the proof of Theorem 3.1.14 is to prevent X_i and X_j, when defined as mere sets of chambers, from coinciding. For instance, for the chamber system $\mathcal{C} = (\{1, 2, 3\}, \sim_1, \sim_2)$ with $\sim_1 = \mathrm{id}$ and $\sim_2 = \{(2, 3), (3, 2)\} \cup \mathrm{id}$, the singleton $\{1\}$ would be an equivalence class of both $\sim_1$ and $\sim_2$, so that X_1 and X_2 would not be disjoint in the unmarked version.

Example 3.1.16 Let $m \in \mathbb{N}$, $m \geq 2$. If Δ is an m-gon, then its chamber system $\mathcal{C}(\Delta)$ satisfies the conditions of Theorem 3.1.14. As a consequence, Δ is isomorphic to the geometry Γ of the conclusion of the theorem.

3.2 Residues

In view of the fundamental role played by residues in incidence systems, we will also look for this concept in chamber systems. We start with the analysis of the chamber system of a geometry.

Remark 3.2.1 Let Γ be a geometry over I and F a flag of Γ. In order to capture F in the chamber system $\mathcal{C}(\Gamma)$ it seems appropriate to consider the chamber subsystem $\mathcal{C}(\Gamma)_F$ induced on the set of all chambers of Γ containing F, with the obvious i-adjacencies for $i \in I \setminus \tau(F)$. By Lemma 3.1.9, $\mathcal{C}(\Gamma)_F$ can be identified with $\mathcal{C}(\Gamma_F)$. This deserves two comments. First of all, the distinction between F and its residue Γ_F seems to vanish in $\mathcal{C}(\Gamma)$. Second, $\mathcal{C}(\Gamma)_F$ is non-empty (note that this requires Γ to be a geometry rather than an incidence system) and *closed* under i-adjacency for any $i \in I \setminus \tau(F)$ (i.e., if c is a chamber of $\mathcal{C}(\Gamma)_F$ and $d \in \mathcal{C}$ satisfies $c \sim_i d$, then $d \in \mathcal{C}(\Gamma)_F$).

The latter property does not suffice for a characterization of chamber systems coming from geometries. If F and F' are flags of Γ having the same type J, then $\mathcal{C}(\Gamma)_F \cup \mathcal{C}(\Gamma)_{F'}$ is again closed under j-adjacency for any $j \in I \setminus J$. So the counterpart of $\mathcal{C}(\Gamma)_F$ within the chamber system should be closed under i-adjacency for all $i \in I \setminus \tau(F)$. But even this property does not work out nicely in general: see Example 3.1.8 with F the singleton of the common point of the two tetrahedra. For residually connected geometries, however, it does suffice.

Definition 3.2.2 Let $\mathcal{C} = (C, \{\sim_i \mid i \in I\})$ be any chamber system over I. If $J \subseteq I$, then $\sim_J$ denotes the union of all $\sim_j$ with $j \in J$. Thus $c \sim_J d$ (in words: c and d are J-adjacent) if and only if c and d are j-adjacent for some $j \in J$. We also write $\sim$ instead of $\sim_I$. The pair $(C, \sim_I)$ is called the *graph of* $\mathcal{C}$. A path in this graph is called a *gallery*. A path of $(C, \sim_J)$ is called a *J-gallery*.

The chamber system $\mathcal{C}$ is called *connected* if its graph is connected. A connected component of $(C, \sim_J)$ is called a *J-cell* of $\mathcal{C}$. For $J \subseteq I$ and c a chamber, we denote by cJ^* the J-cell containing c.

For $i \in I$, the $(I \setminus \{i\})$-cells are called *objects* of $\mathcal{C}$ of type i, or *i-objects*.

Notice that $\{i\}$-cells, also referred to as i-cells, coincide with i-panels. By definition, a connected component is non-empty. A J-cell is nothing but a minimal connected chamber subsystem of $\mathcal{C}$, closed under i-adjacency for each $i \in J$. For $c \in C$, the J-cell cJ^* consists of all chambers of $\mathcal{C}$ that are the endpoint of a gallery starting at c whose adjacencies belong to $\sim_J$.

Lemma 3.2.3 *Let Γ be a residually connected geometry over the finite set I and let $\mathcal{C} = \mathcal{C}(\Gamma)$ be the corresponding chamber system. For any subset J of I, the following assertions hold.*

(i) *If F is a flag of Γ of type J, then $\mathcal{C}(\Gamma)_F$ is an $(I \setminus J)$-cell. In particular, $\mathcal{C}(\Gamma)$ is connected.*

(ii) *If $\mathcal{D}$ is an $(I \setminus J)$-cell of $\mathcal{C}$, then there exists a flag F of Γ of type J such that $\mathcal{C}(\Gamma)_F = \mathcal{D}$.*

Proof (i) It suffices to show that $\mathcal{C}(\Gamma)$ is connected. For then, this also holds for $\mathcal{C}(\Gamma_F)$ in view of the residual connectedness of Γ (use Theorem 1.6.5) and so $\mathcal{C}(\Gamma)_F$ is an $(I \setminus J)$-cell in view of Lemma 3.1.9. Clearly, the statement holds if $|I| = 1$, so assume that i and j are distinct elements of I.

If c and c' are chambers in $\mathcal{C}(\Gamma)$ with a common element x in Γ, then induction on $|I|$ and study of Γ_x give us what we need. For arbitrary chambers c, c' of Γ, we apply induction on the minimal length of an $\{i, j\}$-chain from an element x of c to an element x' of c'. By Lemma 1.6.3, which we may apply as I is finite, such a chain exists. The case of zero length has been dealt with. Fix an $\{i, j\}$-chain of minimal length from x to x'. Without loss of generality, we may assume that the neighbor y of x on this chain is a j-element. Take d to be a chamber of Γ on x and y. By the induction hypothesis on $|I|$, the chambers c and d are in the same connected component of $\mathcal{C}(\Gamma)$, and by the induction hypothesis on the chain length, the chambers d and c' are connected by a gallery in $\mathcal{C}(\Gamma)$. We conclude that c and c' are connected.

(ii) As $\mathcal{D}$ is non-empty, there is a chamber c in $\mathcal{D}$, whence a flag F of Γ, for instance $F = \{c\}$, such that $\mathcal{C}(\Gamma)_F \subseteq \mathcal{D}$ (here we interpret the chamber subsystems as subsets of chambers). Take F to be minimal with respect to this property. Since Γ is residually connected, (i) gives that any two chambers of $\mathcal{C}(\Gamma)_F$ can be connected by a gallery in $\mathcal{C}(\Gamma)_F$ with adjacencies $\sim_i$ for $i \in I \setminus \tau(F)$. In other words, $\mathcal{C}(\Gamma)_F$ is a connected chamber system over $I \setminus \tau(F)$. Hence, by minimality of F and the closure property of $\mathcal{D}$, we have $I \setminus J \subseteq I \setminus \tau(F)$, that is, $\tau(F) \subseteq J$. This implies that $\mathcal{C}(\Gamma)_F$ is closed under $\sim_i$ for all $i \in I \setminus J$. By minimality of $\mathcal{D}$, we find $\mathcal{D} = \mathcal{C}(\Gamma)_F$.

It remains to show that J and $\tau(F)$ coincide. Take G to be a maximal flag of Γ such that $\mathcal{D} = \mathcal{C}(\Gamma)_G$. If $j \in J \setminus \tau(G)$, then, by maximality of G, there are chambers c and d containing G whose elements of type j differ. By (i) there are distinct j-adjacent chambers a and b on a gallery from c to d in $\mathcal{C}(\Gamma)_G$. On the other hand, the closure property of $\mathcal{D}$ tells us that there is an $(I \setminus J)$-gallery from a to b. In particular, at no chamber in the latter gallery does the j-element change, so the j-elements of a and b must coincide, which is a contradiction. This proves $J \subseteq \tau(G)$. Now take F to be the subflag of G of type J. The chain $\mathcal{D} \subseteq \mathcal{C}(\Gamma)_G \subseteq \mathcal{C}(\Gamma)_F \subseteq \mathcal{D}$ shows that $\mathcal{C}(\Gamma)_F = \mathcal{D}$, as required. $\square$

Remark 3.2.4 The above lemma is not valid without the finiteness assumption on I, as the counterexample Γ of Exercise 1.9.18 shows. For, take $J = \{1\}$ and let F be the singleton consisting of a closed interval of length 1, so F is of type J; then

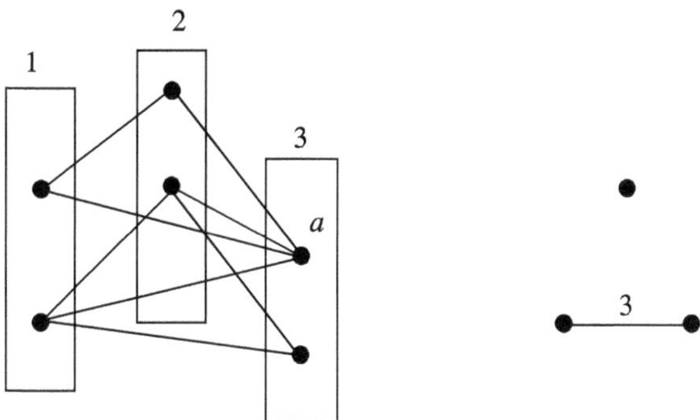

Fig. 3.6 A 'bad' geometry and its chamber system

$\mathcal{C}(\Gamma)_F$ consists of all chambers of Γ whose elements of type $0'$ are in F, and there is no gallery in $\mathcal{C}(\Gamma)_F$ between chambers in there with distinct elements of type $0'$.

Being interested in residually connected geometries, we use Lemma 3.2.3 as a source of inspiration for defining the counterpart of a residue in an arbitrary chamber system. Apparently, J-cells are the substitutes for flags of type $I \setminus J$ and residues of type J we were looking for, although they have to be big enough for the correspondence with geometries to work. For $i \in I$, objects of type i in chamber systems are the substitutes for elements of type i in geometries.

Example 3.2.5 If $\mathcal{C}$ is as in Example 3.1.4, then, for $|J| \geq 2$, the J-cells are all equal to $\mathcal{C}$. So here we have a rather poor residual structure.

Example 3.2.6 The left hand side of Fig. 3.6 shows a geometry Γ over [3] in which two pathological phenomena occur. Its chamber system $\mathcal{C}(\Gamma)$ is drawn at the right hand side. First, there is an element a of type 3 for which the set of all chambers containing it is not a [2]-cell (nor a J-cell for any other $J \subseteq [3]$). Second, there is a [2]-cell of $\mathcal{C}(\Gamma)$ (for instance, the singleton of the top chamber in the picture) that is not the set of chambers containing a flag (of type 3) of Γ.

3.3 From Chamber Systems to Geometries

Having pinned down the residue structure inside a chamber system $\mathcal{C}$, we proceed to derive an incidence system $\Gamma(\mathcal{C})$ from it. The end result of this section, Theorem 3.3.8, gives necessary and sufficient conditions for this incidence system to have a chamber system isomorphic to $\mathcal{C}$.

Definition 3.3.1 If $\mathcal{C}$ is a chamber system over I, the *incidence system* of $\mathcal{C}$, denoted $\Gamma(\mathcal{C})$, is the incidence system over I determined as follows. Its i-elements, for $i \in I$, are the pairs (x, i) with x an i-object of $\mathcal{C}$; two elements (x, k), (y, l) of $\Gamma(\mathcal{C})$ are incident if and only if $x \cap y \neq \emptyset$ in $\mathcal{C}$, i.e., x and y have a chamber in common.

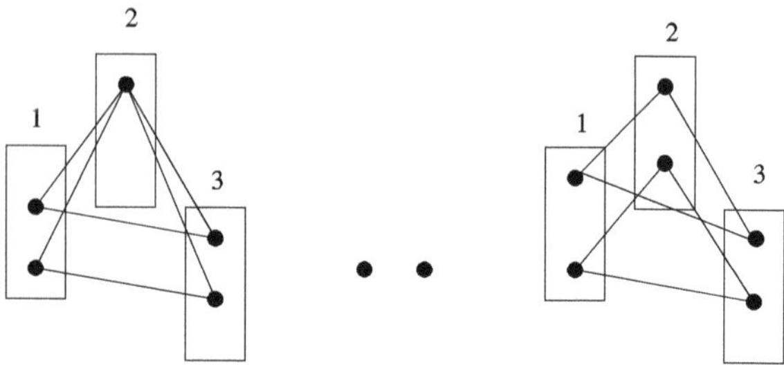

Fig. 3.7 From left to right: a geometry Γ over [3], its chamber system $\mathcal{C}(\Gamma)$ over [3], and the incidence system $\Gamma(\mathcal{C}(\Gamma))$ over [3]

The marking by type, established in the second component of each element of $\Gamma(\mathcal{C})$, helps to distinguish elements of different type, even if cells of different type of $\mathcal{C}$ coincide. Observe that $\Gamma(\mathcal{C})$ is indeed an incidence system because distinct i-objects of $\mathcal{C}$ are disjoint. In the cases of interest to us, $\Gamma(\mathcal{C})$ will be a geometry, but this is not true in general (see Exercise 3.7.5). For $|I| = 2$, the above construction occurs in Theorem 3.1.14.

Example 3.3.2 Let $\mathcal{C}$ be the chamber system of the cube as in Example 3.1.3. The elements of $\Gamma(\mathcal{C})$ are the circuits of length 4, 6, and 8 in the graph of $\mathcal{C}$. Their incidence obeys the Principle of Maximal Intersection 1.9.20 and $\Gamma(\mathcal{C})$ is the cube geometry.

Example 3.3.3 Let $\mathcal{C}$ be the chamber system given in Example 3.1.4. The incidence system $\Gamma(\mathcal{C})$ is a connected rank three geometry with a unique element of each type. Hence $\mathcal{C}(\Gamma(\mathcal{C}))$ consists of a unique chamber and so is not isomorphic to $\mathcal{C}$. This is no surprise, as we already saw that there is not such geometry.

Example 3.3.4 Figure 3.7 describes a rank three geometry Γ, the corresponding chamber system $\mathcal{C}(\Gamma)$, and the geometry $\Gamma(\mathcal{C}(\Gamma))$. Clearly, Γ and $\Gamma(\mathcal{C}(\Gamma))$ are not isomorphic.

Remark 3.3.5 To a given incidence system Γ, we assigned the chamber system $\mathcal{C}(\Gamma)$. To a given chamber system $\mathcal{C}$, we assigned the incidence system $\Gamma(\mathcal{C})$. We explore the connections. The maps $\Gamma \mapsto \mathcal{C}(\Gamma)$ and $\mathcal{C} \mapsto \Gamma(\mathcal{C})$ preserve homomorphisms. Indeed, let $\alpha : \mathcal{C} \to \mathcal{C}'$ be a homomorphism of a chamber system over I and define $\Gamma(\alpha) : \Gamma(\mathcal{C}) \to \Gamma(\mathcal{C}')$ by $\Gamma(\alpha)(c(I \setminus \{i\})^*, i) = (\alpha(c)(I \setminus \{i\})^*, i)$ for any chamber c of $\mathcal{C}$ and $i \in I$, where, we recall, $c(I \setminus \{i\})^*$ is the i-object of $\mathcal{C}$ containing c. Notice that $\Gamma(\alpha)$ is well defined since $\alpha(c)(I \setminus \{i\})^*$ is $(I \setminus \{i\})$-connected for each c. It is easy to check that $\Gamma(\alpha)$ is a homomorphism. Conversely, let $\beta : \Gamma \to \Gamma'$ be a homomorphism of incidence systems over I. The map $\mathcal{C}(\beta) : \mathcal{C}(\Gamma) \to \mathcal{C}(\Gamma')$ given by $\mathcal{C}(\beta)c = \{\beta x_i \mid x_i \in c, \, i \in I\}$ where c is a chamber of Γ, is clearly a homomorphism of chamber systems. (The image $\beta(c)$ of a chamber c of Γ is the chamber $\mathcal{C}(\beta)c$ of Γ'.)

We put these results, and slightly more, in the following lemma.

Lemma 3.3.6 *Let C, C' be chamber systems over I, and Γ, Γ' incidence systems.*

(i) *Suppose that $\alpha : C \to C'$ and $\alpha' : C' \to C''$ are homomorphisms of chamber systems. The maps $\Gamma(\alpha)$ and $\Gamma(\alpha')$ are homomorphisms of incidence systems and $\Gamma(\alpha'\alpha) = \Gamma(\alpha')\Gamma(\alpha)$.*

(ii) *Suppose that $\beta : \Gamma \to \Gamma'$ and $\beta' : \Gamma' \to \Gamma''$ are homomorphisms of incidence systems. The maps $C(\beta)$ and $C(\beta')$ are homomorphisms of chamber systems and $C(\beta'\beta) = C(\beta')C(\beta)$.*

Proof The proof, being straightforward, is left to the reader. $\square$

The preceding paragraphs explain that C and $C(\Gamma(C))$ are not exactly the same but closely related, and similarly for Γ and $\Gamma(C(\Gamma))$. We continue to explore this relationship.

Proposition 3.3.7 *Let C be a chamber system over I. For $c \in C$, let $\psi_C(c)$ be the set of all pairs (D, i) consisting of an i-object D containing c and a type $i \in I$. The map $c \mapsto \psi_C(c)$ is a homomorphism $C \to C(\Gamma(C))$.*

Proof It is clear that $\psi_C(c)$ is a flag of $\Gamma(C)$ (since incidence of elements (D, i) and (E, j) of $\Gamma(C)$ is defined by $D \cap E \ne \emptyset$) and that it has an element of each type $i \in I$ (since c is contained in an $(I \setminus \{i\})$-cell for each i), so it is a chamber of $\Gamma(C)$ and hence of $C(\Gamma(C))$.

Suppose that $c \sim_i d$ holds for two chambers c, d of C. For $j \in I$, the cells $c(I \setminus \{j\})^*$ and $d(I \setminus \{j\})^*$ coincide when $j \ne i$, so the chambers $\psi_C(c) = \{(c(I \setminus \{j\})^*, j) \mid j \in I\}$ and $\psi_C(d) = \{(d(I \setminus \{j\})^*, j) \mid j \in I\}$ of $C(\Gamma(C))$ are i-adjacent. This establishes that ψ_C preserves i-adjacency for each $i \in I$. $\square$

It is not difficult to characterize the situation where ψ_C as defined in Proposition 3.3.7 is an isomorphism.

Theorem 3.3.8 *If C is a chamber system over I, then canonical homomorphism $\psi_C : C \to C(\Gamma(C))$ is a bijection if and only if the following conditions hold.*

(i) *For any set $\{Z_i \mid i \in I\}$, with Z_i an i-object of C such that $Z_i \cap Z_j \ne \emptyset$ for all $i, j \in I$, we have $\bigcap_{i \in I} Z_i \ne \emptyset$.*

(ii) *For any two chambers c, d of C there is an object of C containing c but not d.*

If, moreover, for each $i \in I$ and each collection of j-objects Z_j ($j \in I \setminus \{i\}$) with $Z_j \cap Z_k \ne \emptyset$ for all $j, k \in J \setminus \{i\}$, the intersection $\bigcap_{j \ne i} Z_j$ is an i-panel, then ψ_C is an isomorphism.

Proof Condition (ii) is obviously equivalent to ψ_C being injective.

Let X be a chamber of $\Gamma(C)$. It consists of pairs (Z_i, i), where Z_i is an i-object for each $i \in I$, such that any two have a non-empty intersection. If the intersection of all members of X is non-empty, there is a chamber c in that intersection, and

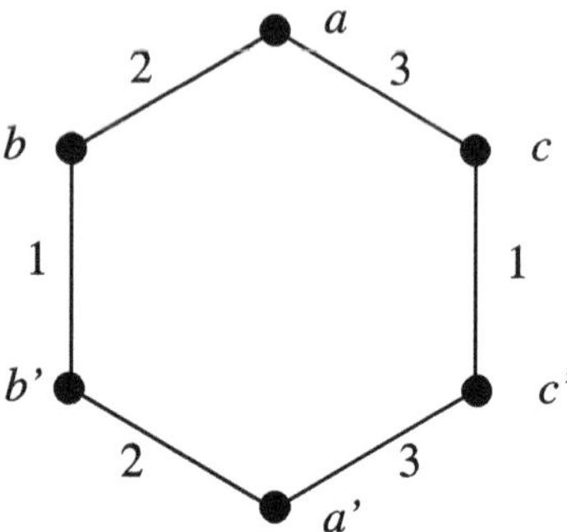

Fig. 3.8 A chamber system over [3]

$\psi_{\mathcal{C}}(c) = \{Z_i \mid i \in I\}$. Conversely, if there is such a chamber c, then obviously, the intersection over all members of X is non-empty. Thus, Assertion (i) is equivalent to $\psi_{\mathcal{C}}$ being surjective.

In order to prove the final statement, suppose that $Y := \{(Y_j, j) \mid j \in I\}$ and $Z := \{(Z_j, j) \mid j \in I\}$ are i-adjacent chambers of $\Gamma(\mathcal{C})$ for some $i \in I$. For each $j \in I \setminus \{i\}$, we have $Y_j = Z_j$, so $\bigcap_{j \in I \setminus \{i\}} Y_j = \bigcap_{j \in I \setminus \{i\}} Z_j$. By the hypothesis, this is an $\{i\}$-cell of $\mathcal{C}$. The inverse images c and d under $\psi_{\mathcal{C}}$ of Y and Z lie in $\bigcap_{j \in I} Y_j$ and $\bigcap_{j \in I} Z_j$, respectively, so they both belong to the $\{i\}$-cell mentioned above. This means $c \sim_i d$, and proves that $\psi_{\mathcal{C}}^{-1}$ is a homomorphism. We saw in Proposition 3.3.7 that $\psi_{\mathcal{C}}$ is a homomorphism and in the previous part of the theorem that it is a bijection, so it is an isomorphism indeed. $\square$

Example 3.3.9 (i) The chamber system $\mathcal{C}(\Gamma)$ in the middle of Fig. 3.7 is disconnected, and so is its geometry. It satisfies the conditions of Theorem 3.3.8, in accordance with the fact that $\mathcal{C}(\Gamma(\mathcal{C}(\Gamma)))$ is isomorphic to $\mathcal{C}(\Gamma)$.

(ii) The condition at the end of Theorem 3.3.8 for $\psi_{\mathcal{C}}^{-1}$ to be a homomorphism is necessary as will be clear from inspection of the chamber system $\mathcal{C}$ over [3] drawn in Fig. 3.8. Here, $\psi_{\mathcal{C}}(a)$ shares $(\{a, a', c, c'\}, 2)$ and $(\{a, a', b, b'\}, 3)$ with $\psi_{\mathcal{C}}(a')$, and so they are in the same 1-panel, while a and a' are not.

3.4 Connectedness

Comparing the geometries Γ and $\Gamma(\mathcal{C}(\Gamma))$, we have no counterpart of Theorem 3.3.8 providing a homomorphism from Γ into $\Gamma(\mathcal{C}(\Gamma))$. As we saw earlier, in Example 3.2.6, an element x of Γ does not necessarily provide a unique element of $\Gamma(\mathcal{C}(\Gamma))$. This ailment came from the fact that an i-object could well be a proper subset of the set of all chambers of Γ containing x. It can be remedied by assuming that Γ is residually connected, and so we now concentrate on translating these notions into chamber system language.

If $\mathcal{C}$ is a chamber system, the incidence system $\Gamma(\mathcal{C})$ need not be connected as was shown in Example 3.3.4. The next lemma shows that connectedness of $\mathcal{C}$ corresponds to connectedness of $\Gamma(\mathcal{C})$.

Lemma 3.4.1 *Suppose that C is a chamber system over an index set I with $|I| \geq 2$. It is connected if and only if $\Gamma(C)$ is connected.*

Proof Suppose that C is connected. Let x and y be elements of $\Gamma(C)$. Being objects of C, they contain chambers of C, say $X \in x$ and $Y \in y$. Since $(C, \sim_I)$ is connected, there is a gallery $X = X_0, X_1, \ldots, X_n = Y$ from X to Y. Then $X_0 \sim_i X_1$ for some $i \in I$ and for $j \neq i$, the j-object containing X_0 also contains X_1, so there is an element x_1 of $\Gamma(C)$ with $x * x_1$ and $X_1 \in x_1$. By iteration of this construction, we obtain a chain $x, x_1, \ldots, x_n, y$ with $X_{i-1}, X_i \in x_i$. In particular, x and y are connected by a chain in $\Gamma(C)$.

We now turn to the converse. Assume that $\Gamma(C)$ is connected and let c, c' be two chambers belonging to C. The chamber c gives rise to a set of objects $\psi_C(c) = \{c_i \mid i \in I\}$ containing c, one for each type $i \in I$. As we saw in Proposition 3.3.7, this is a chamber of $\Gamma(C)$. Similarly, $\psi_C(c')$ is a chamber $\{c_i' \mid i \in I\}$ of $\Gamma(C)$. Fix $i \in I$. As $\Gamma(C)$ is connected, there is a chain $c_i \sim_{i_1} c^1 \sim_{i_2} \cdots \sim_{i_s} c^s = c_i'$ of objects. Now, incidence of objects means having non-empty intersection, so we find a gallery from c to a member of $c_i \cap c^1$, via a chamber of C in $c^1 \cap c^2$, and so on towards a chamber in $c^{s-1} \cap c_i'$, which is i_s-adjacent to c'. $\qquad\square$

The geometry Γ of Example 3.2.6 (see Fig. 3.7) is connected while $C(\Gamma)$ is not. Hence there is no hope for a similar result involving Γ and $C(\Gamma)$ (in this generality).

Given a chamber system C over I, the corresponding incidence system $\Gamma(C)$ need not be residually connected. Thus the need arises to control residual connectedness from the chamber system point of view. The least we may ask for is that $(C, \sim)$ be connected. Since each J-cell, for $J \subseteq I$, is connected by definition, the usual notion of connectedness for the counterparts of residues in chamber systems does not properly translate residual connectedness of the geometry.

Lemma 3.4.2 *Suppose that I is a finite index set and J a subset of I. Let Γ be a residually connected geometry over I. For each $j \in J$, let Z_j be a j-object of $C(\Gamma)$ such that $Z_i \cap Z_j \neq \emptyset$ for $i, j \in J$. The intersection $\bigcap_{j \in J} Z_j$ is an $(I \setminus J)$-cell of $C(\Gamma)$.*

Proof By Lemma 3.2.3(ii), there are elements x_j of type $j \in J$ in Γ such that each chamber of Z_j contains x_j and such that Z_j is the set of all chambers containing x_j. Set $F = \{x_j \mid j \in J\}$. If $i, j \in J$, then, as $Z_i \cap Z_j \neq \emptyset$, there is a chamber whose elements of type i, j are x_i, x_j, respectively. Therefore, $x_i * x_j$. In particular, F is a flag and $Z = \bigcap_{j \in J} Z_j$ consists of all chambers containing F (observe that F is contained in a chamber because Γ is a geometry). By Lemma 3.2.3(i), Z is an $(I \setminus J)$-cell. $\qquad\square$

We use the lemma to define residual connectedness of a chamber system.

Definition 3.4.3 A chamber system over I is called *residually connected* if, for every subset J of I and every system of j-objects Z_j, one for each $j \in J$, with the property that any two have a non-empty intersection, it follows that $\bigcap_{j \in J} Z_j$ is an $(I \setminus J)$-cell.

Remark 3.4.4 Let us consider some consequences of the definition for a chamber system $\mathcal{C} = (C, (\sim_i)_{i \in I})$ over I.

(i) For $J = \emptyset$, the above condition states that $\mathcal{C}$ is non-empty and that $(C, \sim)$ is connected (as the empty intersection is by definition the whole set C).
(ii) For $J = I$, with $Z_j = c(I \setminus \{j\})^*$ for a given chamber c, we see that chambers which are in the same j-objects for all $j \in I$, are equal. In view of Theorem 3.3.8, this means that a residually connected chamber system $\mathcal{C}$ can be reconstructed from $\Gamma(\mathcal{C})$.
(iii) If $\mathcal{C}$ is residually connected and $J \subseteq I$, then each J-cell, considered as a chamber system over J, is also residually connected.
(iv) If the size of $I = \{i, j\}$ is two, then $\mathcal{C}$ is residually connected if and only if it is connected (whence non-empty) and $\sim_i \cap \sim_j = \mathrm{id}$. Compare this result with Theorem 3.1.14.

Definition 3.4.3 allows us to rephrase Lemma 3.4.2. Moreover, there is a converse.

Proposition 3.4.5 *Let I be a finite index set.*

(i) *If Γ is a residually connected geometry over I, then $\mathcal{C}(\Gamma)$ is a residually connected chamber system over I.*
(ii) *If $\mathcal{C}$ is a residually connected chamber system over I, then $\Gamma(\mathcal{C})$ is a residually connected geometry over I.*

Proof (i) was shown in Lemma 3.4.2.

(ii) If I is a singleton, there is nothing to prove. So assume that I has cardinality at least 2. Let $F = \{(Z_j, j) \mid j \in J\}$ be a flag of $\Gamma = \Gamma(\mathcal{C})$ of type J for some $J \subseteq I$. Then $Z = \bigcap_{j \in J} Z_j$ is non-empty, so taking $c \in Z$ and defining Z_i for $i \in I \setminus J$ to be the i-object of $\mathcal{C}$ containing c, we find a chamber $\{(Z_i, i) \mid i \in I\}$ of Γ containing F. This proves that Γ is a geometry.

Now, assume that the rank of Γ_F is at least 2. Suppose that (Y_i, i) and (Y_j, j) are elements of Γ_F of type i and j in $I \setminus J$, respectively. Then $Z \cap Y_i \neq \emptyset$ and $Z \cap Y_j \neq \emptyset$. Take $c \in Z \cap Y_i$ and $d \in Z \cap Y_j$. Since Z, being an $(I \setminus J)$-cell, is $(I \setminus J)$-connected, there is a gallery $c = c_0, c_1, \ldots, c_m = d$, entirely contained in Z, whose edges $\{c_r, c_{r+1}\}$ for $0 \leq r \leq m - 1$ have type $i_r \in I \setminus J$. For each r, let U_r be a k_r-object containing c_r, c_{r+1} for some $k_r \in I \setminus (\{i_r\} \cup J)$. Then (U_r, k_r) is an element of Γ_F incident with $(U_{r+1}, k + r + 1)$ as $c_{r+1} \in U_r \cap U_{r+1} \cap Z$, so $(Y_i, i), (U_0, k_0), \ldots, (U_{m-1}, k_{m-1}), (Y_j, j)$ is a chain in Γ_F from (Y_i, i) to (Y_j, j). This proves that Γ_F is connected. The conclusion is that Γ is residually connected. $\qquad\square$

We are now ready to formulate the main result of this chapter. Let $\mathcal{G}(I)$ denote the collection of residually connected geometries over I, and write $\mathcal{C}(I)$ for the collection of residually connected chamber systems over I.

Theorem 3.4.6 (Chamber System Correspondence) *Let I be a finite set of types. For each $\mathcal{C} \in \mathcal{C}(I)$ there is an isomorphism $\psi_{\mathcal{C}} : \mathcal{C} \to \mathcal{C}(\Gamma(\mathcal{C}))$ given by $\psi_{\mathcal{C}}(c) = \{(c(I \setminus \{i\})^*, i) \mid i \in I\}$. Each homomorphism $\alpha : \mathcal{C} \to \mathcal{C}'$ between members $\mathcal{C}, \mathcal{C}'$ of $\mathcal{C}(I)$ satisfies the relation $\mathcal{C}(\Gamma(\alpha)) = \psi_{\mathcal{C}'}\alpha\psi_{\mathcal{C}}^{-1}$.*

Similarly, for each $\Gamma \in \mathcal{G}(I)$ there is an isomorphism $\phi_{\Gamma} : \Gamma \to \Gamma(\mathcal{C}(\Gamma))$ given by $\phi_{\Gamma}(x) = c(I \setminus \{i\})^*$ whenever c is a chamber of Γ containing x and i is the type of x. Each homomorphism $\beta : \Gamma \to \Gamma'$ between members Γ, Γ' of $\mathcal{G}(I)$ satisfies the relation $\Gamma(\mathcal{C}(\beta)) = \phi_{\Gamma'}\beta\phi_{\Gamma}^{-1}$.

The theorem says that there is a bijective homomorphism preserving correspondence between the collections $\mathcal{G}(I)$ and $\mathcal{C}(I)$. Thus, each residually connected geometry over I corresponds to a unique residually connected chamber system over I (up to isomorphism), and has the same automorphism group as Γ, and vice versa. Before we prove this theorem, we first give a corollary.

Corollary 3.4.7 *Let Γ be a residually connected geometry over a finite set I, and let $\mathcal{C}$ be a residually connected chamber system over I. If either $\mathcal{C} = \mathcal{C}(\Gamma)$ or $\Gamma = \Gamma(\mathcal{C})$, then $\mathrm{Aut}(\mathcal{C}) \cong \mathrm{Aut}(\Gamma)$.*

Proof This is an immediate consequence of the discussion above. $\square$

Proof of Theorem 3.4.6 The map ϕ_{Γ} is well defined: if d is another chamber of Γ containing x, then c and d are in the same i-object of $\mathcal{C}(\Gamma)$ by Proposition 3.4.5(i) applied to Γ_x. It is readily verified that ϕ_{Γ} and $\psi_{\mathcal{C}}$ are isomorphisms. (For the latter, use Theorem 3.3.8.)

Next, if $\alpha : \mathcal{C} \to \mathcal{C}'$ is a homomorphism between members $\mathcal{C}, \mathcal{C}'$ of $\mathcal{C}(I)$, then, for every chamber c of $\mathcal{C}$, both $\mathcal{C}(\Gamma(\alpha))\psi_{\mathcal{C}}(c)$ and $\psi_{\mathcal{C}'}(\alpha(c))$ are equal to the set $\{\alpha(c)(I \setminus \{i\})^* \mid i \in I\}$, so that $\mathcal{C}(\Gamma(\alpha))\psi_{\mathcal{C}} = \psi_{\mathcal{C}'}\alpha$, proving the required expression for $\mathcal{C}(\Gamma(\alpha))$.

Similarly, if $\beta : \Gamma \to \Gamma'$ is a homomorphism between members Γ, Γ' of $\mathcal{G}(I)$, then, for each i-element x of Γ, both $\Gamma(\mathcal{C}(\beta))\phi_{\Gamma}(x)$ and $\phi_{\Gamma'}(\beta(x))$ are equal to the i-object of $\mathcal{C}'$ containing $\beta(c)$, where c is any chamber of Γ containing x, and the required expression for $\Gamma(\mathcal{C}(\beta))$ follows. This finishes the proof of the theorem. $\square$

To end this section, we formulate some useful criteria for residual connectedness of chamber systems. We begin by extending the notation cJ^* of Definition 3.2.2.

Notation 3.4.8 If Z is a set of chambers of the chamber system $\mathcal{C}$ over I and J is a subset of I, write $ZJ^* = \bigcup_{c \in Z} cJ^*$.

Thus, for instance, for $J, K \subseteq I$ and c a chamber, cJ^*K^* denotes the set of all chambers obtained as endpoints of galleries starting in c passing through the J-cell cJ^* and next through a K-cell.

Lemma 3.4.9 *Let C be a chamber system over a finite index set I which is non-empty and connected. Then the following four statements concerning C are equivalent.*

(i) *C is residually connected.*

(ii) *If J, K, L are subsets of I and if Z_J, Z_K, Z_L are J-, K-, L-cells which have pairwise non-empty intersections, then $Z_J \cap Z_K \cap Z_L$ is a $(J \cap K \cap L)$-cell.*

(iii) *If J, K, L are subsets of I and c is a chamber, then $cL^* \cap cJ^*K^* = c(L \cap J)^*(L \cap K)^*$.*

(iv) *If J, K, L are subsets of I and c is a chamber, then $cJ^*L^* \cap cK^*L^* = c(J \cap K)^*L^*$.*

Proof We first establish the equivalence of (i) and (ii). Then the equivalence of (ii), (iii), (iv) will be shown by means of the scheme (ii) $\Rightarrow$ (iii) $\Rightarrow$ (iv) $\Rightarrow$ (ii). But first, we observe that each of the three statements involved (applied with $J = K$ in (ii), (iii) and with $L = \emptyset$ in (iv)) yields that the intersection of any J_1-cell and any J_2-cell is either empty or a $(J_1 \cap J_2)$-cell.

(i) $\Rightarrow$ (ii) This is straightforward (write the cells as intersections of objects).

(ii) $\Rightarrow$ (i) Let $\{Z_j \mid j \in J\}$ be a system of j-objects indexed by a subset J of I having pairwise non-empty intersections. Then, for fixed $j_1, j_2 \in J$, the intersections $Z_{j_1} \cap Z_{j_2} \cap Z_j$, $j \in J \setminus \{j_1, j_2\}$, are cells with pairwise non-empty intersections. Since I is finite, we can invoke induction on $|J|$ to conclude that the whole intersection is a cell (and, in particular, non-empty).

(ii) $\Rightarrow$ (iii) Suppose that e belongs to $cL^* \cap dK^*$ for some $d \in cJ^*$. Then $dJ^* = cJ^*$, $eK^* = dK^*$ and $cL^* = eL^*$. Now $Z_L = cL^*$, $Z_J = dJ^*$ and $Z_K = eK^*$ satisfy the hypothesis of (ii), so there is a chamber f with $f(J \cap K \cap L)^* = cL^* \cap dJ^* \cap eK^*$. Now $f \in cL^* \cap eK^* = e(L \cap K)^*$ and $c \in cL^* \cap dJ^* = fL^* \cap fJ^* = f(L \cap J)^*$ by the above observation, so $e \in f(L \cap K)^* \subseteq e(L \cap J)^*(L \cap K)^*$. This shows $cL^* \cap cJ^*K^* \subseteq c(L \cap J)^*(L \cap K)^*$. The other inclusion is obvious.

(iii) $\Rightarrow$ (iv) Suppose $d \in cJ^*L^* \cap cK^*L^*$. Then there are chambers e_1, e_2 with $c \sim_J e_1 \sim_L d$ and $c \sim_K e_2 \sim_L d$, so by (iii) there is a chamber f in $e_1J^* \cap e_2K^* \cap dL^* = c(J \cap K)^* \cap dL^*$, whence $d \in fL^* \subseteq c(J \cap K)^*L^*$. This proves one inclusion. The other is obvious.

(iv) $\Rightarrow$ (ii) Let Z_J, Z_K, Z_L be as in the hypothesis of (ii). There are chambers $c \in Z_J \cap Z_K, d \in Z_K \cap Z_L$, and $e \in Z_L \cap Z_J$. Now $e \in cJ^* \cap cK^*L^* \subseteq c(J \cap K)^*L^*$ by (iv), so there is $f \in eL^* \cap c(J \cap K)^* \subseteq Z_L \cap Z_J \cap Z_K$. In view of the above observation it follows that $Z_L \cap Z_J \cap Z_K$ is non-empty and hence an $(L \cap J \cap K)$-cell. $\qquad\square$

The lemma is useful as, for abstract chamber systems, (iii) and (iv) may be easier to check than the original definition, especially when many automorphisms are available.

3.5 The Diagram of a Chamber System

Since the notion of chamber system is very general, it is very easy to construct objects of this kind. In order to decide whether the chamber system at hand is residually connected, and thus belongs to a residually connected geometry, we use the notion of diagram of a chamber system without recourse to the corresponding geometry.

Definition 3.5.1 Let C be a chamber system over I, and let D be a diagram over I as introduced in Definition 2.3.7. We say that D is a *diagram* for C, that C *belongs to the diagram* D, or that C *is of type* D if, for each subset $\{i, j\}$ of I of size two, every $\{i, j\}$-cell of C is the chamber system of a residually connected geometry over $\{i, j\}$ belonging to $D(i, j)$.

If C is of type D, then, for $K \subseteq I$, every K-cell is of type D_K. Some obvious connections with geometries of type D are formulated in Exercise 3.7.10.

Lemma 3.5.2 *Suppose that I is partitioned into subsets J and K in such a way that $D(j, k)$ consists of generalized digons for all $j \in J, k \in K$. If $C = (C, \{\sim_i \mid i \in I\})$ is a connected chamber system belonging to D, then $C = cJ^*K^*$ for any $c \in C$.*

Proof Suppose $c \sim_k d \sim_j e$ for $c, d, e \in C$. By the generalized digon properties of the $\{j, k\}$-cell containing all three, there also is a chamber d' with $c \sim_j d' \sim_k e$. Consequently, for $c \in C$, we have $cK^*J^* = cJ^*K^*$, whence $cJ^*K^* = cJ^*K^*J^* = \cdots$. Since C is connected, this leads to $C = cI^* = cJ^*K^*$. $\square$

The lemma gives rise to the following notion.

Definition 3.5.3 Let J and K be disjoint index sets. Suppose that C is a chamber system over J and D is a chamber system over K. The *direct sum* of C and D, denoted by $C \oplus D$, is the chamber system whose chambers are the pairs (c, d) with c a chamber of C and d a chamber of D in which $(c, d) \sim_i (c', d')$ if and only if either $i \in J, d = d'$, and $c \sim_i c'$ or $i \in K, c = c'$, and $d \sim_i d'$.

It is readily verified that $C \oplus D$ is indeed a chamber system over $J \cup K$. Clearly $C \oplus D \cong D \oplus C$, and direct sums of more than two chamber systems are easy to define. We connect the direct sum of chamber systems to the direct sum of geometries Γ and Δ introduced in Definition 2.1.8, which we denote $\Gamma \oplus \Delta$ here. The following result is to be compared with Exercise 2.8.2.

Proposition 3.5.4 *Let J and K be disjoint index sets.*

(i) *Let C be a chamber system over J, and D be a chamber system over K. The direct sum $C \oplus D$ is residually connected if and only if both C and D are residually connected.*

(ii) *If Γ and Δ are geometries over J and K, respectively, then $C(\Gamma \oplus \Delta)$ is isomorphic to $C(\Gamma) \oplus C(\Delta)$.*

Proof Straightforward. $\square$

Theorem 3.5.5 *Suppose that C is a chamber system belonging to a finite linear diagram. If, for all subsets J, K of I, every J-cell meets every K-cell either emptily or in a $(J \cap K)$-cell, then C is residually connected.*

Proof Let $J \subseteq I$, and suppose, for each $j \in J$, we are given a j-object Z_j. Let $Z_i \cap Z_j \neq \emptyset$ for all $i, j \in I$. We want to show that $\bigcap_{j \in J} Z_j$ is a non-empty $(I \setminus J)$-cell. This suffices for the proof of the theorem in view of Definition 3.4.3.

If $|J| \leq 2$, there is nothing to show. Proceeding by induction with respect to $|J|$, we may assume $|J| \geq 3$. Let D denote the linear diagram for C. Take $j \in J$ such that $I \setminus \{j\}$ is partitioned into two non-empty sets L and R with the property that $D(l, r)$ consists of generalized digons for every $l \in L, r \in R$ and $L \cap J \neq \emptyset \neq R \cap J$. Then, by the induction assumption applied to $J \setminus R$ and to $J \setminus L$, we can find chambers $c \in \bigcap_{i \in J \setminus R} Z_i$ and $d \in \bigcap_{i \in J \setminus L} Z_i$. According to the hypotheses, $\bigcap_{i \in J \setminus R} Z_i$ is the $(R \cup (L \setminus J))$-cell on c, and $\bigcap_{i \in J \setminus L} Z_i$ is the $(L \cup (R \setminus J))$-cell on d, so they contain the cell cR^* and dL^*, respectively. Applying Lemma 3.5.2 to the latter two cells, viewed as cells of Z_j, we obtain $c \in dL^* R^*$, whence $cR^* \cap dL^* \neq \emptyset$, so $\bigcap_{j \in J} Z_j$, which contains $cR^* \cap dL^*$, is non-empty. It readily follows that it is an $(I \setminus J)$-cell. $\square$

Example 3.5.6 Let $I = [n]$ and take $G = \mathrm{Sym}_{n+1}$, the symmetric group on $n + 1$ letters. It is generated by involutions $r_1, \ldots, r_n$ where r_i is the transposition $(i, i + 1)$. Let $C = (G, \{\sim_i \mid i \in [n]\})$ be the chamber system in which i-adjacency is given by $\sigma \sim_i \tau$ if and only if $\sigma \in \tau \langle (i, i + 1) \rangle$, for $\sigma, \tau \in G$. Then C belongs to the Coxeter diagram A_n, as $r_i r_j = r_j r_i$ for non-consecutive indices i, j in $[n]$ and $r_i r_j r_i = r_i r_j r_i$ whenever $j = i + 1 \in [n]$. For $J \subseteq [n]$, put $G^{(J)} = \langle (i, i + 1) \mid i \in J \rangle$. By Theorem 3.5.5, C is residually connected if and only if, for any $J, K \subseteq [n]$,

$$G^{(J)} \cap G^{(K)} = G^{(J \cap K)}. \tag{3.1}$$

For, if $cG^{(J)} \cap dG^{(K)}$ is a non-empty intersection of a J-cell and a K-cell, then it can be written as $e(G^{(J)} \cap G^{(K)})$ for a suitable element e of G.

The structure of $G^{(J)}$ depends on the maximal intervals (that is, the connected components of the subgraph of D induced on J) contained in J. If the interval $[k, l]$ of integers in $[n]$ is contained in J, then $G^{(J)}$ contains $\mathrm{Sym}([k, l + 1])$. Since this group commutes with all r_i for $i \in J \setminus [k - 1, l + 1]$, the group $G^{(J)}$ is the direct product of the symmetric groups on the maximal intervals in J. This observation makes it easy to see that $G^{(J)} \cap G^{(K)} = G^{(J \cap K)}$. We conclude that (3.1) is satisfied, and that C is indeed residually connected.

The geometry $\Gamma(C)$ is isomorphic to the geometry of rank n, say Γ, described in Example 1.2.6. The map $\alpha : C \to C(\Gamma)$ determined by

$$\alpha(\pi) = \left\{ \{\pi(1)\}, \{\pi(1), \pi(2)\}, \ldots, \{\pi(1), \pi(2), \ldots, \pi(n)\} \right\} \quad (\pi \in G)$$

is an isomorphism of chamber systems. It follows from Theorem 3.4.6 that $\Gamma(C)$ is isomorphic to Γ.

3.6 Groups and Chamber Systems

In this section, we consider chamber systems with a highly transitive group. Pursuing the translations of the geometric structures into group-theoretical ones begun in Sect. 1.8, we describe chamber systems by groups in Proposition 3.6.4 and residual connectedness in Theorem 3.6.9.

Definition 3.6.1 Let $\alpha : G \to \mathrm{Aut}(\mathcal{C})$ be a representation of a group G on a chamber system $\mathcal{C}$. When G is transitive on the set of chambers of $\mathcal{C}$, we say that G is *chamber transitive* on $\mathcal{C}$. We also say that $\mathcal{C}$ is *chamber transitive* if $\mathrm{Aut}(\mathcal{C})$ is chamber transitive.

If $\mathcal{C}$ is chamber transitive, an easy description of $\mathcal{C}$ can be given in terms of G and of some of its subgroups. It is somewhat less technical and more natural than the description for geometries with incidence-transitive groups of automorphisms via Proposition 1.8.7. The easy group-theoretic construction of such a chamber system is an important advantage of chamber systems over geometries.

Remark 3.6.2 Assume that Γ is an incidence system over I and that G is a group acting flag transitively on Γ. Let c be a chamber of Γ and consider the subgroup $B = G_c$ stabilizing c in G. By Theorem 1.7.5, we can identify the chambers with the left cosets gB ($g \in G$). For each $i \in I$, let c_i be the element of type i in c, and consider the subgroup $G^{(i)} = G_{c \setminus \{c_i\}}$ stabilizing the complement of $\{c_i\}$ in c. These groups contain B. If $d = gB$ and $e = hB$, for certain $g, h \in G$, are chambers of $\mathcal{C}(\Gamma)$, then $d \sim_i e$ if and only if $gG^{(i)} = hG^{(i)}$. For, $d \sim_i e$ is equivalent to $g^{-1}d \sim_i g^{-1}e$, whence to $g^{-1}h \in G^{(i)}$. So we can translate the structure of $\mathcal{C}(\Gamma)$ in terms of G, B, and the subgroup $G^{(i)}$ ($i \in I$). We are ready for a treatment of $\mathcal{C}$ without recourse to Γ.

Definition 3.6.3 Let G be a group, B a subgroup, $(G^{(i)})_{i \in I}$ a system of subgroups of G with $B \subseteq G^{(i)}$. The *coset chamber system of G on B with respect to* $(G^{(i)})_{i \in I}$, denoted $\mathcal{C}(G, B, (G^{(i)})_{i \in I})$, has the chamber set consisting of all cosets gB ($g \in G$), and i-adjacency determined by $gB \sim_i hB$ if and only if $gG^{(i)} = hG^{(i)}$.

For $i \in I$, the group $G^{(i)}$ is called the *standard parabolic subgroup* of type $I \setminus \{i\}$. The subgroup B of G is called the *Borel subgroup* of G on Γ.

Observe that $\sim_i$ as in the above definition is obviously an equivalence relation, so that $\mathcal{C}$ is indeed a chamber system. In the above setting, G has a canonical chamber-transitive representation on the chamber system $\mathcal{C}(G, B, (G^{(i)})_{i \in I})$ by means of left multiplication. Again, taking for granted the definition of *equivalence of chamber system actions*, we have the following converse. Here G_c is the stabilizer of a chamber c and G_X, for a set of chambers X, is the set-wise stabilizer in G of X.

Proposition 3.6.4 *If $\alpha : G \to \mathrm{Aut}(\mathcal{C})$ is a chamber-transitive representation of G on a chamber system $\mathcal{C}$ over I, then, for every chamber c of $\mathcal{C}$, the canonical representation of G on $\mathcal{C}(G, G_c, (G_{ci*})_{i \in I})$ is equivalent to α.*

Proof Denote by C the chamber set of $\mathcal{C}$. By Theorem 1.7.5, the map $G/G_c \to C$ given by $gG_c \mapsto \alpha(g)c$ is an equivalence of G-sets. We need to verify that this map and its inverse preserve i-adjacency. For $g, h \in G$, we have $\alpha(g)c \sim_i \alpha(h)c$ if and only if $c \sim_i \alpha(g^{-1}h)c$, which is equivalent to $g^{-1}h \in G_{ci^*}$, that is, i-adjacency of gG_c and hG_c in $\mathcal{C}(G, G_c, (G_{ci^*})_{i \in I})$. $\qquad\square$

Definition 3.6.3 gives us a huge family of chamber systems with a chamber-transitive group of automorphisms. Actually, it shows that this kind of structure is rather loose.

Examples 3.1.5 and 3.1.6 show chamber systems having chamber-transitive groups of automorphisms. Of course there are chamber systems whose automorphism groups are not chamber transitive. Figure 3.6 shows that Example 3.2.6 gives one.

Example 3.6.5 In Example 2.4.11, we saw that the group $G = \mathrm{Alt}_7$ acts on the Neumaier geometry Γ. Here we build the chamber system of this geometry. There are 15 planes, each of which has 7 lines of 3 points, so there are $15 \cdot 7 \cdot 3 = 315$ chambers. Recall that the point set can be described by $[7]$, the lines are all triples from $[7]$, and the planes are the Alt_7 images of a single plane Π. We will take Π to be the plane with lines $\{1, 2, 4\}$ and its images under the permutation $(1, 2, 3, 4, 5, 6, 7)$. This is the Fano plane depicted in Fig. 1.21.

Now consider the chamber $c = \{1, l, \Pi\}$, where $l = \{1, 2, 4\}$, as in Example 2.4.11. Since the automorphism group of the Fano plane has order 168 (cf. Exercise 1.9.7) and the stabilizer of the flag $\{1, \{1, 2, 4\}\}$ in there must have index $7 \cdot 3 = 21$, the stabilizer B in G of c has order eight. It is readily checked that $(3, 5)(6, 7)$, $(3, 6)(5, 7)$, and $(2, 4)(5, 6)$ generate B. The G-orbit of c is of size $|G|/|B| = 2520/8 = 315$, so G is transitive on the set of chambers of Γ. This reconfirms our findings in Example 2.4.11.

We can describe $\mathcal{C}(\Gamma)$ by starting with B as a Borel subgroup of G, and prescribing the standard parabolic subgroups $G^{(i)}$ ($i \in I$) for the determination of adjacencies. To find $G^{(1)}$, the standard parabolic fixing the line l and the plane Π, observe that the third generator of B interchanges 2 and 4. A similar element interchanges 1 and 2 and stabilizes the same line and plane: $(1, 2)(3, 6)$. Together with B, it generates the group $G^{(1)}$. Similar observations show that $G^{(2)} = \langle B, (2, 3)(4, 7)\rangle$, the group generated by B and $(2, 3)(4, 7)$. Moreover $(3, 5, 7) \in G$ changes Π and preserves both point and line of c, so $G^{(3)} = \langle B, (3, 5, 7)\rangle$. By Proposition 3.6.4, the chamber system $\mathcal{C}(\Gamma)$ is isomorphic to $\mathcal{D} = \mathcal{C}(G, B, (G^{(1)}, G^{(2)}, G^{(3)}))$.

In fact, Γ can easily be recovered from $\mathcal{D}$: its i-elements correspond to the objects of $\mathcal{D}$ of type i, that is, the cosets of $G^{(\{k, j\})} = \langle G^{(k)}, G^{(j)}\rangle$ in G, where $\{i, j, k\} = [3]$. For type 1, this means that the alternating group on $\{2, \ldots, 6\}$ (the stabilizer in G of 1) is generated by B, $(2, 3)(4, 7)$, and $(3, 5, 7)$. For type 2, it means that the stabilizer in G of l is generated by B, $(1, 2)(3, 6)$, and $(3, 5, 7)$. Finally, for type 3, it means that the automorphism group of Π is generated by B, $(2, 3)(4, 7)$, and $(1, 2)(3, 6)$.

Example 3.6.6 Let V be an n-dimensional vector space over a division ring $\mathbb{D}$. It determines the projective geometry $\mathrm{PG}(V)$ over $[n-1]$ introduced in Example 1.4.9. Put $\mathcal{C} = \mathcal{C}(\mathrm{PG}(V))$. If $c = \{V_i \mid i \in [n-1]\}$ is a chamber of this geometry, with V_i a subspace of V of dimension i, then we can find a basis $(v_i)_{1 \le i \le n}$ of V such that $v_1, \ldots, v_i$ span V_i. Since $\mathrm{GL}(V)$ is transitive on the set of all bases of V, a basis corresponding to one chamber can be mapped onto a basis corresponding to another. Hence $\mathrm{GL}(V)$ is chamber transitive on $\mathcal{C}$, a fact we already knew from Example 1.8.16.

We determine the Borel group and the standard parabolic subgroups of $\mathrm{GL}(V)$ for this geometry. Fix a chamber c and a basis $(v_i)_i$ of V determining it. With respect to this basis, the group B is the set of matrices that fix c:

$$B = \left\{ \begin{pmatrix} b_{1,1} & b_{1,2} & \cdot & \cdots & \cdots & b_{1,n} \\ 0 & b_{2,2} & b_{2,3} & \cdots & \cdots & b_{2,n} \\ \cdot & \ddots & \ddots & \ddots & \cdots & \cdot \\ \cdot & \cdot & \ddots & \ddots & \ddots & \cdot \\ 0 & \cdot & \cdots & 0 & b_{n-1,n-1} & b_{n-1,n} \\ 0 & 0 & \cdot & \cdots & 0 & b_{n,n} \end{pmatrix} \;\middle|\; \prod_{i=1}^{n} b_{i,i} \neq 0 \right\}$$

Now fix $j \in [n-1]$. The matrices stabilizing all V_i for $i \in [n-1] \setminus \{j\}$ are of the form

$$\begin{pmatrix} * & * & \cdot & \cdot & \cdot & \cdots & \cdots & * \\ 0 & * & * & \cdot & \cdot & \cdots & \cdots & * \\ 0 & \ddots & \ddots & \cdot & \cdot & \cdots & \cdots & \\ \cdot & \cdot & 0 & * & \cdot & \cdots & \cdots & \cdot \\ \cdot & \cdot & \cdot & * & * & \ddots & \cdots & \cdot \\ \cdot & \cdot & \cdot & \cdot & 0 & \ddots & \ddots & \cdot \\ 0 & \ddots & \cdot & \cdot & \cdot & \ddots & * & * \\ 0 & 0 & \cdot & \cdot & \cdot & \cdots & 0 & * \end{pmatrix}$$

Here, $*$ stands for an arbitrary value of the entry, with the understanding that the resulting matrix should be invertible. Thus, $G^{(j)}$ contains B and this containment is proper as the $(j+1, j)$ entry is allowed to be nonzero.

Lemma 3.6.7 *Let G be a group, B a subgroup, $(G^{(i)})_{i \in I}$ a system of subgroups of G containing B. The chamber system $\mathcal{C}(G, B, (G^{(i)})_{i \in I})$ is connected if and only if G is generated by the subgroups $G^{(i)}$ $(i \in I)$.*

Proof Suppose that c and d are chambers of the chamber system connected by gallery $c = c_0, c_1, \ldots, c_m = d$. By applying an appropriate automorphism (i.e., left multiplication by a suitable element of G), we may assume that $c = B$. Now $c \sim_i c_1$ means $c_1 \in ci^* = G^{(i)}/B$ (the set of left B-cosets in $G^{(i)}$), and

$c_2 \sim_j c_1$ then implies $c_2 \in ci^*j^* = G^{(i)}G^{(j)}/B$. Continuing along the gallery, we find $d \in G^{(i_1)} \cdots G^{(i_2)} \cdots G^{(i_m)}/B$. Now writing $d = gB$ for $g \in G$, we find that $g \in G^{(i_1)} \cdots G^{(i_2)} \cdots G^{(i_m)}$. The conclusion is that a chamber gB is connected to c in $\mathcal{C}(G, B, (G^{(i)})_{i \in I})$ if and only if g can be expressed as a product of elements of the $G^{(i)}$. This proves the lemma. $\qquad\qquad\qquad\qquad\qquad\qquad\square$

Notation 3.6.8 For G a group with a system of subgroups $(G^{(i)})_{i \in I}$ we write $G^{(J)} = \langle G^{(j)} \mid j \in J \rangle$.

For $J \subseteq I$, the chamber-transitive action of a group G on a chamber system $\mathcal{C}$ of Lemma 3.6.7 gives a nice description of the J-cell determined by the chamber B. By an argument as in the above proof, the J-cell BJ^* is the set of all left cosets of the form $g_1 g_2 \cdots g_m B$ with $g_i \in G^{(r_i)}$ for certain $r_i \in J$ ($i \in [m]$). This means that BJ^* is the $G^{(J)}$-orbit on B. In other words, it is the set of all gB with $g \in G^{(J)}$; in symbols, $BJ^* = G^{(J)}/B$. So the J-cells of $\mathcal{C}$ are the sets $xG^{(J)}/B$ with $x \in G$.

In view of these observations, a good strategy for checking residual connectedness may be to apply (iii) or (iv) in Lemma 3.4.9. It suffices to deal with the case $c = B$.

Theorem 3.6.9 *Let I be a finite index set. Suppose that G is a group with a system of subgroups $(G^{(i)})_{i \in I}$, and that B is a subgroup of $\bigcap_{i \in I} G^{(i)}$. The following statements are equivalent.*

(i) *The chamber system $\mathcal{C}(G, B, (G^{(i)})_{i \in I})$ is residually connected.*
(ii) *For all $J, K, L \subseteq I$, we have $G^{(L)} \cap G^{(J)}G^{(K)} = G^{(L \cap J)}G^{(L \cap K)}$.*
(iii) *For all $J, K, L \subseteq I$, we have $G^{(J)}G^{(L)} \cap G^{(K)}G^{(L)} = G^{(J \cap K)}G^{(L)}$.*

Proof This follows immediately from the discussion above and Lemma 3.4.9. $\quad\square$

Example 3.6.10 Consider the group-theoretic chamber system description of the Neumaier geometry in Example 3.6.5. Merely by computation, one can verify that $G^{(i)} \cap G^{(j)}G^{(k)} = B$ for each triple $\{i, j, k\} = [3]$. But the geometric interpretation is easier: it says that if one starts with a chamber c (which may be fixed, thanks to chamber transitivity), then changes the element of type j in c to obtain another chamber, d say, and next an element of type k of d to find a resulting chamber e, and if e can also be found by changing the element of type i of c, then $e = d$. Thus, although the group-theoretic conditions are readily stated, their geometric interpretation is easier to verify.

Example 3.6.11 Let us explore the construction of projective planes from groups. Let G be a group with subgroups $G^{(1)}$ and $G^{(2)}$. We wish to interpret the intersection $B = G^{(1)} \cap G^{(2)}$ as the stabilizer of a chamber of the projective plane, $G^{(1)}$ as the stabilizer of the line in this chamber, and $G^{(2)}$ as the stabilizer of the point in this chamber. In other words, we consider the chamber system $\mathcal{C} = \mathcal{C}(G, B, (G^{(1)}, G^{(2)}))$ and ask for sufficient conditions for $\mathcal{C}$ to be the chamber system of a projective plane.

Translating the generalized 3-gon properties into conditions on the subgroups involved, we find

(i) $G^{(1)}G^{(2)}G^{(1)} = G^{(2)}G^{(1)}G^{(2)} = G$;

(ii) $G^{(1)}G^{(2)} \cap G^{(2)}G^{(1)} = B$.

Definition 2.6.1 shows how to attach a diagram to a system of subgroups of a group G. Here is a similar approach, but now for a system of subgroups providing a chamber system rather than a geometry.

Definition 3.6.12 Let G be a group with a system of subgroups $(G^{(i)})_{i \in I}$ indexed by I, and let D be a diagram over I. We say that G has *chamber diagram* D over $(G^{(i)})_{i \in I}$ if, for any unordered pair $\{i, j\} \subseteq I$, the coset geometry $\Gamma(G^{(\{i,j\})}, (G^{(j)}, G^{(i)}))$, with $G^{(j)}$ of type i and $G^{(i)}$ of type j, belongs to $D(i, j)$.

Proposition 3.6.13 *Suppose that G is a group with a system of subgroups $(G^{(i)})_{i \in I}$ indexed by I. If G has chamber diagram D over $(G^{(i)})_{i \in I}$ and the coset chamber system $\mathcal{C} := \mathcal{C}(G, B, (G^{(i)})_{i \in I})$ over $B := \bigcap_{i \in I} G^{(I \setminus \{i\})}$ is residually connected, then the residually connected geometry $\Gamma(\mathcal{C})$ coincides with $\Gamma(G, (G_i)_{i \in I})$, where $G_i = G^{(I \setminus \{i\})}$, and belongs to the diagram D.*

Proof Because $\mathcal{C}$ is residually connected, we have $G^{(J)} \cap G^{(K)} = G^{(J \cap K)}$ for any two subsets J, K over I by Theorem 3.6.9. It follows that, for $J \subseteq I$, the subgroup G_J coincides with $\bigcap_{j \in J} G^{(I \setminus \{j\})} = G^{(I \setminus J)}$. An i-element of $\Gamma(\mathcal{C})$ is the collection of B-cosets in $gG^{(I \setminus \{i\})}$ for some $g \in G$, which corresponds bijectively with the i-element gG_i of $\Gamma(G, (G_i)_{i \in I})$. It is readily verified that this identification leads to the coincidence of $\Gamma(\mathcal{C})$ and $\Gamma(G, (G_i)_{i \in I})$. By the Chamber System Correspondence 3.4.6, these geometries are residually connected.

Since G has chamber diagram D over $(G^{(i)})_{i \in I}$, for $i, j \in I$, the geometry $\Gamma(G^{(\{i,j\})}, (i \mapsto G^{(j)}, j \mapsto G^{(i)}))$ belongs to $D(i, j)$. But this geometry is the same as $\Gamma(G_{I \setminus \{i, j\}}, (i \mapsto G_{I \setminus \{j\}}, j \mapsto G_{I \setminus \{i\}}))$ so, by Definition 2.6.1 and Proposition 2.6.2, $\Gamma(G, (G_i)_{i \in I})$ belongs to D. $\square$

Theorem 3.6.14 *Let $\mathcal{C}$ be a coset chamber system of the group G over B with respect to $(G^{(i)})_{i \in I}$, such that, for all $J \subseteq I$, the group $G^{(J)}$ is*

(i) *a proper subgroup of $G^{(J \cup \{i\})}$ for all $i \in I \setminus J$, and*

(ii) *a maximal subgroup of $G^{(J \cup \{i\})}$ for all but at most one $i \in I \setminus J$.*

If $\mathcal{C}$ has a linear diagram, then $\mathcal{C}$ is residually connected.

Proof We assert that, for $J, K \subseteq I$, each J-cell meets each K-cell either in a $(J \cap K)$-cell or not at all. By transitivity, we need only show that $G^{(J)} \cap G^{(K)} = G^{(J \cap K)}$. We prove this assertion by induction on $|I \setminus (J \cap K)|$. The cases where $J \subseteq K$ or $J \supseteq K$ are easily dealt with. Therefore, we assume $k \in K \setminus J$ and

$j \in J \setminus K$. Without harming generality, we may assume also that $G^{(J \cap K)}$ is a maximal subgroup of $G^{((J \cap K) \cup \{k\})}$ (for, otherwise, interchange the roles of J and K). Now by induction, $G^{(J \cup \{k\})} \cap G^{(K)} = G^{((J \cap K) \cup \{k\})}$, so

$$G^{(J \cap K)} \subseteq G^{(J)} \cap G^{(K)} \subseteq G^{((J \cap K) \cup \{k\})}.$$

If $G^{(J)} \cap G^{(K)} = G^{((J \cap K) \cup \{k\})}$, then $G^{(k)} \subseteq G^{(J)}$, and so $G^{(J)} = G^{(J \cup \{k\})}$, contradicting hypothesis (i). Thus, $G^{(J)} \cap G^{(K)}$ is a proper subgroup of $G^{(J \cap K) \cup \{k\}}$ containing the maximal subgroup $G^{(J \cap K)}$, and hence coincides with $G^{(J \cap K)}$. This establishes the assertion.

We next verify that G satisfies the conditions of Theorem 2.6.4 with respect to its subgroups $G_i = G^{(I \setminus \{i\})}$ ($i \in I$). As for (i), for $J \subseteq I$ with $|I \setminus J| \geq 2$, we have

$$G_J = G^{(I \setminus J)} = \langle G^{(I \setminus (J \cup \{i\}))} \mid i \in I \setminus J \rangle = \langle G_{J \cup \{i\}} \mid i \in I \setminus J \rangle,$$

as required. As for (ii), let i, j be distinct elements of I. The geometry $\Gamma(G_{I \setminus \{i,j\}},$ $(G_{I \setminus \{j\}}, G_{I \setminus \{i\}})) = \Gamma(G^{(\{i,j\})}, (G^{(j)}, G^{(i)}))$ is of type $D(i, j)$. Hence Theorem 2.6.4 applies, showing that $\Gamma(G, (G_i)_{i \in I})$ is residually connected. Consequently, $\mathcal{C}(\Gamma(G,$ $(G_i)_{i \in I})) = \mathcal{C}(G, \bigcap_{i \in I} G^{(i)}, (G^{(i)})_{i \in I})$ is residually connected. $\square$

Example 3.6.15 Let G be the group with presentation

$$G = \langle c, d, e, x, y, z \mid c^3, d^3, e^3, x^7, y^7, z^7,$$
$$x^{-1} c^{-1} d, y^{-1} d^{-1} e, z^{-1} e^{-1} c,$$
$$cxc^{-1} x^{-2}, dyd^{-1} y^{-2}, eze^{-1} z^{-2} \rangle,$$

and consider the chamber system $\mathcal{C} = \mathcal{C}(G, \{1\}, (\langle c \rangle, \langle d \rangle, \langle e \rangle))$ over [3]. The symmetry in the defining relations shows that there is an automorphism of G given by $c \mapsto d \mapsto e \mapsto c$ and $x \mapsto y \mapsto z \mapsto x$, which leads to a correlation of $\mathcal{C}$ (cf. Definition 3.1.7). We claim that $\mathcal{C}$ is a chamber system of type $\widetilde{A}_2$. To see this and keeping into account the automorphism just described, we only need to verify that $\mathcal{C}(\langle c, d \rangle, \{1\}, \langle c \rangle, \langle d \rangle)$ is the chamber system of a projective plane. On the one hand, the relations involving c and d in the above presentation of G show that the subgroup $G^{(\{1,2\})} = \langle c, d \rangle$ of G is a quotient of the Frobenius group of order 21. On the other hand, a coset enumeration of a quotient of G by the normal subgroup generated by $(cde)^2$ show that there is a homomorphism $G \to \mathrm{Alt}_8$ given by

$$c \mapsto (2, 4, 6)(3, 8, 5), \qquad d \mapsto (2, 3, 7)(4, 5, 8), \qquad e \mapsto (1, 2, 5)(3, 4, 8).$$

The image of the subgroup $\langle c, d \rangle$ of G under this homomorphism is readily seen to have size 21, and so its order must be precisely 21. But then Exercise 1.9.29 shows that $\mathcal{C}(\langle c, d \rangle, \{1\}, (\langle c \rangle, \langle d \rangle))$ is indeed the chamber system of the projective plane of order two.

Similarly, it can be verified that $G^{(\{1,2\})} G^{(\{2,3\})} \cap G^{(\{1,3\})} = \langle c \rangle \langle e \rangle = G^{(1)} G^{(3)}$, from which we derive by use of Lemma 3.6.9 that $\mathcal{C}$ is residually connected.

Remark 3.6.16 In Definition 3.6.3, a chamber system is constructed from a group and a system of subgroups. For a given diagram D over I, there is a more general notion that may lead to the construction of chamber systems of type D. The idea is to specify groups B, $G^{(i)}$ $(i \in I)$, and $G^{(\{i,j\})}$ $(i, j \in I; i \neq j)$ with embeddings $B \subseteq G^{(i)} \subseteq G^{(\{i,j\})}$ such that the geometry $\Gamma(G^{(\{i,j\})}, G^{(i)}, G^{(i)})$ over $\{i, j\}$ belongs to $D(i, j)$ for each choice of distinct $i, j \in I$, and to construct a universal group G in which these embeddings are realized. The system of groups and embeddings is a particular example of an amalgam. There is a unique universal construction for this group G, but it does not always provide for the required embeddings—these may collapse. This group is called the universal completion of the amalgam. Example 3.6.15, for instance, can be regarded as the universal completion of the system $B = 1$, $G^{(1)} = \langle c \mid c^3 \rangle$, $G^{(2)} = \langle d \mid d^3 \rangle$, $G^{(3)} = \langle e \mid e^3 \rangle$, $G^{(\{1,2\})} = \langle c, d \mid c^3, d^3, (c^{-1}d)^7, dc^{-1}(d^{-1}c)^2 \rangle$, $G^{(\{2,3\})} = \langle d, e \mid d^3, e^3, (d^{-1}e)^7, ed^{-1}(e^{-1}d)^2 \rangle$, $G^{(\{1,3\})} = \langle e, c \mid e^3, c^3, (e^{-1}c)^7, ce^{-1}(c^{-1}e)^2 \rangle$, with canonical embeddings as suggested by the notation.

3.7 Exercises

Section 3.1

Exercise 3.7.1 Consider the chamber system of the cube viewed as a rank three geometry (see Fig. 3.2).

(a) Prove that point reflection at the center of the cube in Euclidean space describes an automorphism α of the corresponding geometry, and that it induces an automorphism of its chamber system that interchanges 'opposite' chambers (i.e., chambers that are on the same line through the center of the cube).

(b) Verify that the quotient of the chamber system obtained by identifying opposite chambers (that is, chambers and their images under α) is residually connected and that it is locally isomorphic to the cube (in the sense that an element of the cube geometry has a residue isomorphic to that residue of an element of the quotient).

(c) Consider the quotient geometry of the cube with respect to $\langle \alpha \rangle$. What is its chamber system? Does it coincide with the quotient chamber system with respect to $\langle \alpha \rangle$?

Exercise 3.7.2 Let $\mathcal{C} = (C, \{\sim_i \mid i \in I\})$ be a chamber system. Proposition 3.1.10 can be generalized as follows. Let E be an equivalence relation on C such that $E \circ \sim_i = \sim_i \circ E$ for each $i \in I$. Here, $\circ$ is the usual composition of relations (so $A \circ B$ for relations A and B on C is the relation given by $x(A \circ B)y \iff \exists c \in C : x A c$ and $c B y$). Such a relation is called a *congruence* on $\mathcal{C}$.

(a) Verify that $\mathcal{C}/E := (C/E, \{(\sim_i \circ E)/E \mid i \in I\})$, where C/E is the set of E-equivalence classes on C and $(\sim_i \circ E)/E$ is the relation on C/E induced by $\sim_i$, is a chamber system over I. It is called the *quotient chamber system* of C by the congruence E.

(b) Prove that there exists a natural quotient homomorphism $C \to C/E$.

(c) Let $\mu : C \to C'$ be a surjective homomorphism of chamber systems over I such that for all $i \in I$, $x \in C$ and $v \in C'$ with $\mu(x) \sim_i' v$ there exists $y \in \mu^{-1}(v)$ with $y \sim_i x$. Show that C' is the quotient of C by a suitable congruence.

(d) For A a group of automorphisms of C, show that the chamber system C/A defined in Proposition 3.1.10 coincides with the quotient chamber system C/E, where E is the relation 'lying in the same A-orbit' on C.

Section 3.2

Exercise 3.7.3 Consider the vector space $U := \bigoplus_{i \in \mathbb{N}} \varepsilon_i \mathbb{R}$ with basis $(\varepsilon_i)_{i \in \mathbb{N}}$. We define the incidence system $\mathrm{PG}(U)$ similarly to Example 1.4.9: Its elements are all subspaces of U of finite positive dimension. Subspaces X and Y are incident if and only if either $X \subseteq Y$ or $Y \subseteq X$. The type of an element is its affine dimension, so the set of types is $\mathbb{N}$.

(a) Verify that $\mathrm{PG}(U)$ is a residually connected geometry over $\mathbb{N}$.

(b) Show that the conclusion of Lemma 1.6.3 is satisfied for $\mathrm{PG}(U)$.

(c) Prove that $C(\mathrm{PG}(U))$ is not connected.

Section 3.3

Exercise 3.7.4 (Cited in Corollary 11.3.11) Let $m \in \mathbb{N} \cup \{\infty\}$, $m \geq 2$. Show that a chamber system $C(\Gamma)$ over $I := [2]$ of a generalized m-gon Γ is characterized by the following four axioms. Here, a *simple gallery* is a gallery $x_1, x_2, \ldots, x_q$ without repetitions, so $x_i \neq x_{i+1}$ for $i \in [q-1]$.

(a) Each panel has at least two chambers.

(b) Its graph is connected of diameter m.

(c) If $n \in \mathbb{N}$ satisfies $0 < n < 2m$, then there is no simple closed gallery of length n and of type i, j, i, j, i, j, $\ldots$, where $I = \{i, j\}$.

(d) If $m < \infty$ and if there is a simple gallery of length m and of type i, j, i, j, $\ldots$ from x to y, then there is a simple gallery of length m and of type j, i, j, i, $\ldots$ with the same starting chamber x and end chamber y, where $\{i, j\} = I$.

Exercise 3.7.5 Show that Fig. 3.9 depicts an example of a chamber system C of rank four such that $\Gamma(C)$ is the unit geometry.

Exercise 3.7.6 Show that Fig. 3.10 depicts an example of a chamber system C of rank four such that $\Gamma(C)$ is not a geometry.

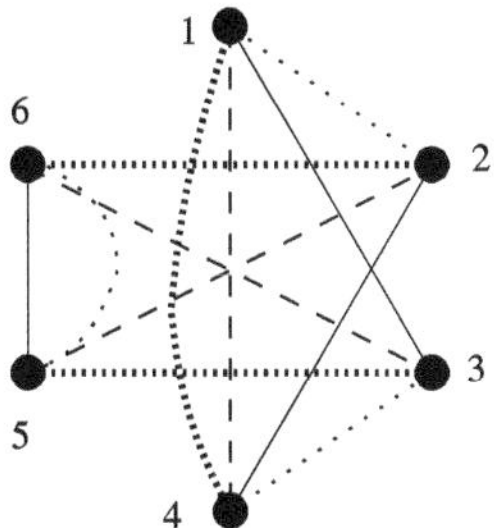

Fig. 3.9 A chamber system of rank four corresponding to the unit geometry

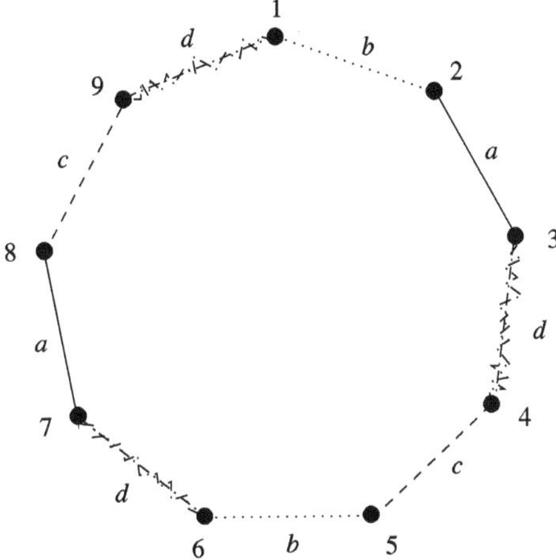

Fig. 3.10 A chamber system of rank four whose incidence system is not a geometry

(*Hint*: Show that the flag with elements $\{1, 2, 8, 9\}$ of type a, $\{2, 3, 4, 5\}$ of type b, and $\{5, 6, 7, 8\}$ of type c, cannot be extended to a chamber with any of the three elements of type d.)

Exercise 3.7.7 Let $\mathcal{C}$ be a chamber system over I, with chamber set C, satisfying the following condition.

> If $J \subseteq I$ and $(Z_j)_{j \in J}$ are j-objects of $\mathcal{C}$ with $Z_i \cap Z_j \neq \emptyset$ for all $i,\, j \in J$, then $\bigcap_{j \in J} Z_j \neq \emptyset$.

Let ϕ be the map $J \mapsto \{cJ^* \mid c \in C\}$ from the *power set* 2^I of I (that is, the collection of all subsets of I) to the collection $\mathcal{H}$ of partitions of C into cells of $\mathcal{C}$. Both 2^I and $\mathcal{H}$ are *lattices*, that is, they are partially ordered sets (2^I by inclusion, $\mathcal{H}$ by refinement) in which for every pair of elements x, y there is a least upper bound $x \cup y$ and a greatest lower bound $x \cap y$. A *lattice homomorphism* $2^I \to \mathcal{H}$ is a map preserving the partial orders and least upper bounds and greatest lower bounds of two elements.

(a) Prove that ϕ is a lattice homomorphism from 2^I to $\mathcal{H}$ if $\mathcal{C}$ is residually connected.

(b) Give a counterexample to show that the converse implication of (a) does not
    hold.

Section 3.4

Exercise 3.7.8 Show that the condition $|I| \geq 2$ is necessary in Lemma 3.4.1. Which
implication fails?

Exercise 3.7.9 Consider the three subgroups $G^{(1)} = \langle (1,3), (2,5) \rangle$, $G^{(2)} = \langle (1,2,3) \rangle$, $G^{(3)} = \langle (2,3), (4,5) \rangle$ of $G = \mathrm{Sym}_5$. As usual, we let G act on $[5]$ from
the left, so $(1,3)(2,3) = (1,3,2)$. Let $\mathcal{C}$ be the chamber system $\mathcal{C}(G, 1, (G^{(1)}, G^{(2)}, G^{(3)}))$ over $[3]$.

(a) Verify that $\mathcal{C}$ is connected.
(b) Verify that $G^{(i)} \cap G^{(j)} = 1$ for $1 \leq i < j \leq 3$.
(c) Verify that $(1,3,2) \in G^{(2)} \cap G^{(1)}G^{(3)}$.
(d) Conclude that $\mathcal{C}$ is not residually connected.

Section 3.5

Exercise 3.7.10 (Cited in Remark 11.2.13) Let D be a diagram over I such that all
geometries belonging to $D(i, j)$ for distinct $i, j \in I$ are residually connected. Prove
the following statements.

(a) If Γ is a geometry of type D, then $\mathcal{C}(\Gamma)$ is a chamber system of type D.
(b) If $\mathcal{C}$ is a residually connected chamber system of type D, then $\Gamma(\mathcal{C})$ is a resid-
    ually connected geometry of type D.

Exercise 3.7.11 (This exercise is used in Proposition 11.1.6) Let J and K be disjoint index sets. For diagrams D over J and E over K, let $D \oplus E$ be their *direct sum*,
that is, the disjoint union of the two labeled graphs, viewed as a diagram over $J \cup K$
(so $(D \oplus E)(j, k)$ consists of generalized digons whenever $j \in J$ and $k \in K$). Prove
that the direct sum of a chamber system of type D over J and a chamber system of
type E over K is a chamber system of type $D \oplus E$ over $J \cup K$.

Section 3.6

Exercise 3.7.12 Suppose that G is a group with a system of subgroups $(G^{(i)})_{i \in I}$
indexed by I and B is a subgroup of G contained in each $G^{(i)}$. Consider the coset
chamber system $\mathcal{C} = \mathcal{C}(G, B, (G^{(i)})_{i \in I})$ over B. Prove that $gG^{(J)}g^{-1}$ is the stabilizer in G of the J-cell $gG^{(J)}/B$ of $\mathcal{C}$.

Exercise 3.7.13 What can you say about the correspondence between $\mathcal{C}(G, B, (G^{(i)})_{i\in I})$ and $\Gamma(G, (G_i)_{i\in I})$, where $G_i = G^{(I\setminus\{i\})}$, in Proposition 3.6.13 if B is a subgroup of (but not necessarily equal to) $\bigcap_{i\in I} G^{(I\setminus\{i\})}$?

3.8 Notes

The study and development of chamber systems, especially in view of a local characterization of buildings was initiated by Tits in [287]. Other accounts on the subject are to be found in [51, 233].

Section 3.3

Example 3.3.9(ii) is due to Pasini.

Section 3.4

The chamber system correspondence can be phrased as an equivalence of categories between residually connected geometries over I and residually connected chamber systems over I.

Theorem 3.4.6, and various characterizations of residual connectedness in terms of chamber systems, have been worked out by various authors (see [217, 317] for some instances), including those of this book.

Section 3.6

Amalgams, as mentioned in Remark 3.6.16, are well described in [30]. The paper [139] generalizes results by Tits [288, 289] which show that the non-existence of certain proper coverings of flag-transitive geometries leads to presentations of groups of automorphisms as amalgams (cf. Remark 3.6.16) of flag stabilizers.

Chapter 4
Thin Geometries

As stated in Definition 1.2.7, a geometry is thin if each residue of rank one has exactly two elements. These geometries are firm in the most economical way. Residually connected examples are provided by the convex polyhedra in $\mathbb{E}^3$ and the tessellations of $\mathbb{E}^2$ and $\mathbb{E}^3$; all (residues of) faces are polygons, all (residues of) edges are digons, and all vertex residues are again polygons (see Examples 1.1.3, 1.1.4, 1.1.5). In Sect. 4.1, we study these examples in the Euclidean space. Next we look at them intrinsically, i.e., without reference to an embedding in $\mathbb{E}^n$. We are most interested in the regular examples, which are geometries of Coxeter type, the subject of Sect. 4.2. It turns out that these are all quotients of geometries corresponding to Coxeter groups, one for each Coxeter type. This leads to the study of Coxeter groups, which takes up Sects. 4.3 (where many groups generated by reflections are shown to be Coxeter groups), 4.4 (where Coxeter groups are shown to have a faithful linear representation as a group generated by reflections), 4.5 (where the centers of the perspectivities induced by these reflections are studied), and 4.6 and 4.7 (where the finite Coxeter groups are determined). Having gathered enough knowledge about Coxeter groups, we return to the regular polytopes in Sect. 4.8 and classify them by Coxeter types.

4.1 Being Thin

In this section, we primarily look at convex polytopes in $\mathbb{E}^3$ for examples of thin geometries of Coxeter types.

If a geometry is thin, then so are all of its residues. This leads to the following characterization of thin geometries.

Lemma 4.1.1 *A thin rank one geometry has exactly two elements. A geometry of rank at least two is thin if and only if all of its rank two residues are thin.*

Proof In the rank one case, the empty set is a flag of corank one, so the element set itself must be of size two. In the case of rank at least two, Proposition 1.5.3(ii)

F. Buekenhout, A.M. Cohen, *Diagram Geometry*,
Ergebnisse der Mathematik und ihrer Grenzgebiete. 3. Folge / A Series of
Modern Surveys in Mathematics 57, DOI 10.1007/978-3-642-34453-4_4,

shows that each residue of rank one can be obtained as the residue of an element inside a residue of rank two. $\square$

Thus, first order of business is to determine thin rank two geometries more closely. Of course, m-gons are thin for all $m \in \mathbb{N} \cup \{\infty\}$, $m \geq 2$. The converse is also true for connected geometries.

Theorem 4.1.2 *Every thin connected rank two geometry is an m-gon for some $m \in \mathbb{N} \cup \{\infty\}$, $m \geq 2$.*

Proof Let $\Gamma = (X, *, \tau)$ be a thin connected rank 2 geometry. If the incidence graph of Γ contains a circuit Y, it must be a circuit of even length, say, $2m$. This forces $X = Y$, as the incidence graph is connected, and Y, having valency two, is a connected component of X. Hence Γ is an m-gon.

We are left with the case where the incidence graph of Γ contains no circuits, i.e., is a tree of valency two. Take an element x_0 of Γ. This element is incident with exactly two other elements, x_{-1}, and x_1, say. Define a map $\delta : X \to \mathbb{Z}$ by

$$
\delta(x) = \begin{cases} d(x, x_0) & \text{if } d(x, x_{-1}) \geq d(x, x_1), \\ -d(x, x_0) & \text{otherwise,} \end{cases}
$$

where d stands for distance in the collinearity graph of Γ. Now δ establishes an isomorphism from Γ to the ordinary ∞-gon defined in Definition 2.2.6. $\square$

The restriction to connected geometries in the above theorem is not essential since every rank two geometry is the 'disjoint union' of connected geometries, namely the connected components of its incidence graph. In terms of Coxeter diagrams (cf. Sect. 2.4), we have the following consequence of the theorem.

Corollary 4.1.3 *Let Γ be a thin residually connected geometry over a set of types I. The geometry Γ is of Coxeter type if and only if, for each $i, j \in I$, all the residues of type $\{i, j\}$ are isomorphic to each other.*

Proof The 'only if' part follows from the fact that, for fixed distinct $i, j \in I$, the residues of type $\{i, j\}$ are thin generalized $m_{i,j}$-gons for some $m_{i,j} \geq 2$, and hence $m_{i,j}$-gons.

The 'if' part follows from Theorem 4.1.2 as each rank two residue of Γ is thin (see Lemma 4.1.1) and connected (cf. Corollary 1.6.6), so an m-gon for some $m \in \mathbb{N} \cup \{\infty\}$. As all the residues of type $\{i, j\}$ are isomorphic, the value m only depends on $\{i, j\}$. $\square$

The following construction shows that thin residually connected geometries of rank ≥ 3 can be rather wild objects and that there is little hope for their classification.

Fig. 4.1 A planar map whose
geometry is a thin, residually
connected geometry which is
not of Coxeter type

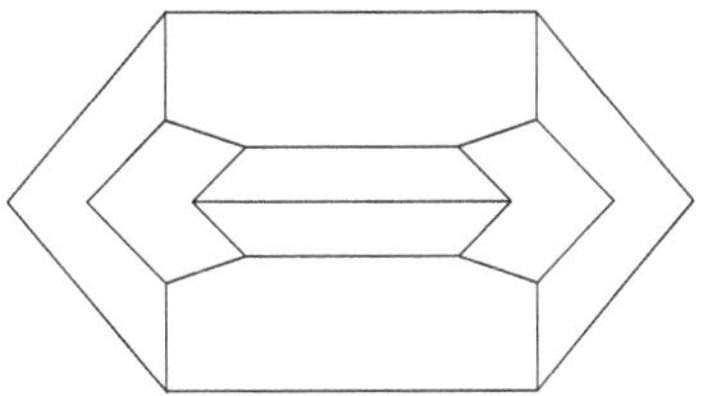

Definition 4.1.4 An *arc* in $\mathbb{E}^2$ is the image of an injective continuous piece-wise linear map from the unit segment $[0, 1]$ into $\mathbb{E}^2$. Here, *piece-wise linear* means that the image consists of a finite number of line segments. This assumption helps to avoid anomalies like space-filling curves and fractals.

By projecting the 2-dimensional sphere S^2, hanging above the Euclidean plane $\mathbb{E}^2$ in Euclidean space $\mathbb{E}^3$, from the north pole onto the plane, we can interpret $\mathbb{E}^2$ as the set of points of S^2 distinct from the north pole. A *planar map* is a triple (X_1, X_2, X_3) consisting of

(1) a discrete non-empty subset X_1 of $\mathbb{E}^2$ each point of which belongs to a finite number, at least two, of members of X_2;
(2) a collection X_2 of arcs in $\mathbb{E}^2$ whose intersections with X_1 consist of two distinct end points, with the property that no two arcs meet outside X_1;
(3) a collection X_3 of closed regions of $\mathbb{E}^2$ (each homeomorphic to a disk, possibly after adding the north pole) whose union is equal to $\mathbb{E}^2$; the interior of each region in X_3 is disjoint from all members of X_2; the border of each region in X_3 is the union of a finite number of members of X_2 and is homeomorphic to a circle; every member of X_2 is in the border of exactly two members of X_3.

If $P = (X_1, X_2, X_3)$ is a planar map, then the *geometry* associated with P is $\Gamma(P) := (X_1, X_2, X_3, *)$, where $*$ stands for symmetrized inclusion. This is a thin [3]-geometry whose digon diagram is linear.

See Figs. 4.1 and 4.2 for examples of planar maps.

Example 4.1.5 There are several analogues of the above construction. In the following examples, thin residually connected geometries arise, whose digon diagrams are non-linear.

Let $P = (X_1, X_2, X_3)$ be a planar map with face set X_3 admitting a *3-coloring* of the faces, that is, a coloring by three colors such that no two neighboring faces have a common color. Let $\Delta = (X_3, *, \tau)$ be the geometry of rank three where $\tau(x)$ is the color of $x \in X_3$ and $x * y \iff x \cap y \neq \emptyset$ for all $x, y \in X_3$. It is easy to determine conditions on P that are necessary and sufficient for Δ to be a thin residually connected geometry. A thin geometry of Coxeter type $\widetilde{A}_2$ (cf. Fig. 2.17) results if we take P to be the tiling of $\mathbb{E}^2$ by regular hexagons with the coloring depicted in Fig. 4.2.

Take the *tiling* of $\mathbb{E}^2$ by equilateral triangles and color its faces b (for black) and w (for white) in such a way that two adjacent faces have different colors. See

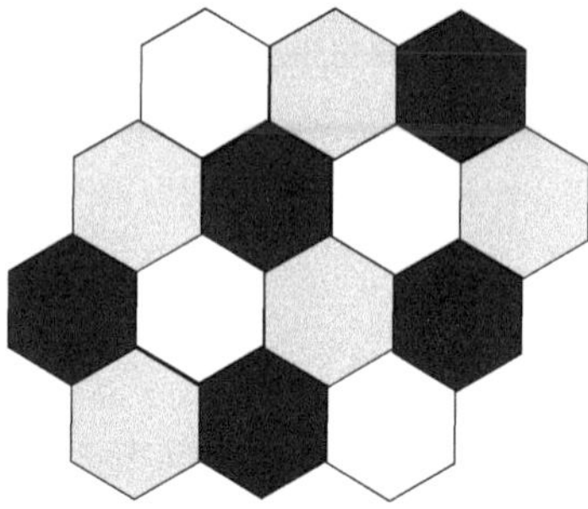

Fig. 4.2 A 3-colored tiling of $\mathbb{E}^2$ by regular hexagons

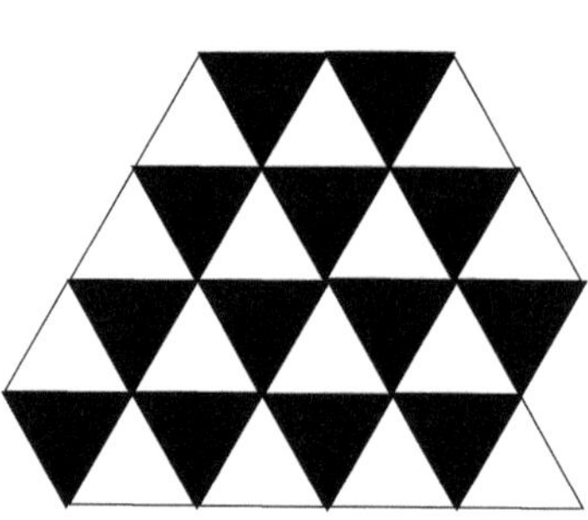

Fig. 4.3 A colored tiling of the plane by triangles

Fig. 4.3. Let Γ be the geometry over $\{\mathrm{o}, \mathrm{b}, \mathrm{w}\}$, where o-elements are the vertices, b-elements the black faces and w-elements the white faces, and where incidence for o-elements and faces is based on inclusion, while two faces of different colors are incident if they have an edge in common. Then Γ is a thin geometry all of whose rank two residues are triangles. It is again a thin geometry of Coxeter type $\widetilde{\mathrm{A}}_2$.

Remark 4.1.6 The finite planar maps can easily be seen as embedded in the sphere S^2. By suitably rotating the sphere, we can always arrange for the arc to miss a given point, and so the restriction that vertices or arcs of the map, as defined above, miss the north pole is not essential.

Analogues of finite planar maps in S^2 are obtained by replacing the sphere by other surfaces. The possible variations of residues of given type pertaining to such a geometry exhibits how capricious these objects are. In particular, the consequence of Theorem 4.1.2 according to which thin [2]-geometries have flag-transitive groups of automorphisms, does not carry over to thin [3]-geometries. We need to impose some regularity in order to isolate the thin geometries of our interest.

Remark 4.1.7 Finite planar maps occur as the boundaries of polyhedra. Regular polyhedra have been studied for about 2500 years, so it is no surprise that they have given rise to quite different treatments and definitions. Let us consider two opposing points of view on regularity, a weaker and a stronger one. In the 'weak sense', a thin [3]-geometry is said to be regular if, for all distinct $i, j \in [3]$, the residues of type $\{i, j\}$ are isomorphic; in other words (thanks to Corollary 4.1.3), if it is a thin [3]-geometry of Coxeter type.

Table 4.1 Parameters of Platonic solids

v_1	v_2	v_3	p	q	Name
2	m	m	2	m	Dual dihedron
m	m	2	m	2	Dihedron
4	6	4	3	3	Tetrahedron
6	12	8	3	4	Octahedron
8	12	6	4	3	Cube
12	30	20	3	5	Icosahedron
20	30	12	5	3	Dodecahedron

In the strong sense, a thin [3]-geometry will have a flag-transitive automorphism group. Such a geometry is obviously of Coxeter type. Both viewpoints apply to polyhedra and tilings. The geometry Γ of Exercise 2.8.14 is a thin [3]-geometry that is regular in the weak but not in the strong sense.

The five Platonic solids (the tetrahedron, cube, octahedron, dodecahedron and icosahedron, cf. Fig. 1.2) are examples of regular thin geometries in the strong sense. We now characterize these geometries as weakly regular geometries.

Theorem 4.1.8 *Let P be a finite planar map whose associated geometry $\Gamma(P)$ is of Coxeter type*

$$\overset{1}{\circ} \xrightarrow{\quad p \quad} \overset{2}{\circ} \xrightarrow{\quad q \quad} \overset{3}{\circ}.$$

For $i \in [3]$, let v_i be the number of elements of type i of $\Gamma(P)$.

(i) *The numbers p, q, v_1, v_2, v_3 are as given in one of the rows of the Table 4.1.*
(ii) *Each row of Table 4.1 corresponds to a unique thin [3]-geometry $\Gamma(P)$ up to isomorphism.*
(iii) *The geometry $\Gamma(P)$ is residually connected and has a flag-transitive automorphism group.*

Proof We freely use Euler's formula for P:

$$v_1 - v_2 + v_3 = 2. \tag{4.1}$$

It can be proved by induction on $N = v_1 + v_2$, starting with the case $N = 4$. Counting the number of chambers of $\Gamma(P)$ in three distinct ways we find

$$v_1 \cdot q \cdot 2 = v_2 \cdot 2 \cdot 2 = v_3 \cdot p \cdot 2$$

so that v_1 and v_3 can be expressed in terms of v_2, p, and q. Combining these equalities with Euler's formula (4.1), we find

$$\frac{1}{v_2} = \frac{1}{p} + \frac{1}{q} - \frac{1}{2}. \tag{4.2}$$

This gives $\frac{1}{p} + \frac{1}{q} > \frac{1}{2}$, and so $(p-2)(q-2) < 4$. It follows that the only possible values for p and q are those of Table 4.1. The values of v_1, v_2, v_3 follow from the above formulae. This gives (i).

The proofs of (ii) and (iii) are left to the reader. They are also consequences of later developments on Coxeter geometries. $\square$

The reader will have no difficulty drawing the maps P corresponding to the geometries $\Gamma(P)$ of Theorem 4.1.8 and relating them to the Platonic solids.

Remark 4.1.9 (i) The numbers appearing in Table 4.1 suggest a kind of duality interchanging the roles of points and faces. Given a planar map P as in the theorem, there is a *dual planar map* $P^\vee = (X_1^\vee, X_2^\vee, X_3^\vee)$ obtained as follows.

(1) $X_1^\vee$ consists of a point in each face in X_3.
(2) $X_2^\vee$ consists of arcs joining two points p_1, p_2 in adjacent members of X_3, one for each $a \in X_2$ in the intersection of the faces containing p_1 or p_2, in such a way that a is the unique member of X_2 meeting the arc.
(3) $X_3^\vee$ consists of the regions around points in X_1 bounded by the arcs in $X_2^\vee$.

Now $\Gamma(P^\vee)$ is correlated (but usually not isomorphic) to $\Gamma(P)$.

(ii) The three regular tilings of the plane $\mathbb{E}^2$, namely by regular triangles (cf. Fig. 4.3), by regular hexagons (cf. Fig. 4.2), and by squares, can be obtained by a proof analogous to the above, the difference being that the left hand side of (4.2) should be replaced by 0 (which can be interpreted as $1/\infty$ to connect with the polytope case). The Coxeter types are as in Theorem 4.1.8, with $(p,q) = (3,6)$, $(6,3)$, $(4,4)$, in the respective cases.

(iii) We recall that the hemidodecahedron was obtained in Example 1.3.4 by identifying opposite elements of the dodecahedron. The resulting geometry is of the same Coxeter type as the dodecahedron. But the tetrahedron has no proper quotients that are also maps.

Example 4.1.10 We list more examples of regular thin [3]-geometries of Coxeter type occurring in $\mathbb{E}^3$.

(i) The tiling (also called tessellation) of $\mathbb{E}^3$ by cubes provides a flag-transitive geometry over the diagram

$$\widetilde{C}_3 := \circ\!\!=\!\!=\!\!=\!\!=\!\!\circ\!\!-\!\!-\!\!-\!\!\circ\!\!=\!\!=\!\!=\!\!=\!\!\circ .$$

This diagram will recur in Table 4.3.

(ii) Similarly to the colored tiling of $\mathbb{E}^2$ by regular triangles (described in Example 4.1.2 and Fig. 4.3), the bicolored tiling of $\mathbb{E}^3$ by cubes, in which cubes of distinct colors are incident if and only if they intersect in a face, gives a flag-transitive geometry. It is of the Coxeter type depicted at the right hand side of Fig. 2.7 of Example 2.4.4.

(iii) *Stellation* is a process of constructing a new polyhedron from a given one by extending the faces of the given polyhedron; the result of the process is called the *stellated polyhedron*. The pentagram (a regular star shaped 5-gon), for instance,

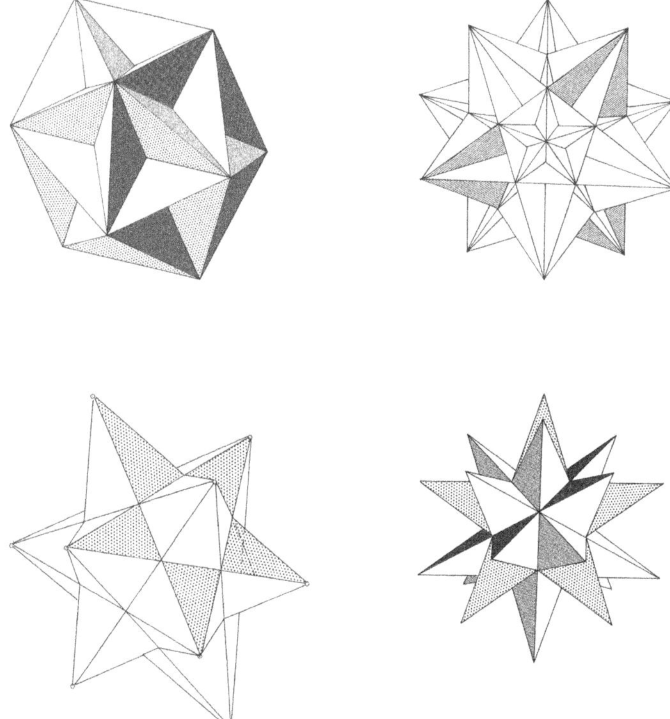

Fig. 4.4 Four regular stellated solids

is a stellation of the regular 5-gon in $\mathbb{E}^2$, and the star of David (built up of two regular triangles), is a stellation of the regular 6-gon in $\mathbb{E}^2$. Stellations of the dodecahedron and the icosahedron in $\mathbb{E}^3$ give rise to four regular stellated solids depicted in Fig. 4.4. The upper left one is known as the great dodecahedron. Continuing clockwise, we encounter the small stellated dodecahedron, the great stellated dodecahedron (which is also depicted in Fig. 1.12), and the great icosahedron. Each of these four leads to a thin regular rank three geometry. There are many more stellated dodecahedra and icosahedra, which are not regular.

We next exhibit higher-dimensional analogues of the Platonic solids.

Definition 4.1.11 Let $n > 0$. A *convex polytope* Π in the Euclidean affine space $\mathbb{E}^n$ is the convex hull of a finite set of points. The *polytope geometry* $\Gamma(\Pi)$ of Π is the following geometry over $[n]$. Its n-elements are the $(n-1)$-dimensional affine subspaces y of $\mathbb{E}^n$ spanned by n points on the boundary $\partial\Pi$ of Π and meeting Π only in $\partial\Pi$. The intersection of an n-element with Π is called a *face* of Π. It is a convex polytope in $\mathbb{E}^{n-1}$. For $i = n-1, \ldots, 1$, we define an i-element of $\Gamma(\Pi)$ recursively as an $(i-1)$-dimensional affine subspace y of $\mathbb{E}^n$ spanned by i points on the boundary of a convex polytope in $\mathbb{E}^i$ of the form $\Sigma := x \cap \Pi$, where x is

Table 4.2 Coxeter diagrams of finite regular polytopes

Name	Diagram	Example
A_n	 o——————o ·········· o——————o	n-simplex
B_n, C_n	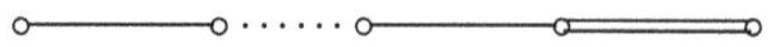 o——————o ······ o——————o═════o	n-cube, n-octahedron
F_4	 o——————o═════o——————o	24-cell
H_3	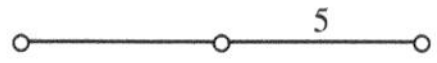 o——————o——5——o	Icosahedron, dodecahedron
H_4	 o——————o——————o——5——o	600-cell, 120-cell
$I_2^{(m)}$	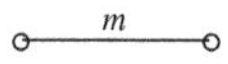 o——m——o	m-gon

an $i + 1$-element of $\Gamma(\Pi)$, such that y meets Σ only in $\partial \Sigma$. The intersection of an i-element with Π is called an $(i - 1)$-*face* of Π. Incidence of $\Gamma(\Pi)$ is symmetrized inclusion. The *vertices*, *edges*, and *faces* of Π are the 0-faces, 1-faces, and $(n - 1)$-faces, respectively.

If this geometry is of Coxeter type M (cf. Definition 2.4.2) and the rank of M equals n, we also say that Π itself is of *Coxeter type M*.

Proposition 4.1.12 *Let $n \geq 2$. The geometry of a convex polytope in $\mathbb{E}^n$ is thin and residually connected and has a linear digon diagram.*

Proof Let Π be a convex polytope satisfying the hypotheses. For $n = 2$, the proposition is trivial. Let $n > 2$. Fix $i \in [n]$ and let x be an i-element of $\Gamma(\Pi)$. It is not hard to find an $(n - i)$-dimensional affine subspace H of $\mathbb{E}^n$ disjoint from x such that the affine subspace $\langle x, v \rangle$ spanned by x and v meets H nontrivially for every vertex v of Π such that $\langle x, v \rangle$ is an $(i + 1)$-element of $\Gamma(\Pi)$. The convex polytope spanned by the points $\langle x, v \rangle \cap H$, for v as above (does not depend essentially on the choice of H and) is called the *star* of x. The residue of x in $\Gamma(\Pi)$ is the direct sum of the geometry of the convex polytope of rank $i - 1$ contained inside x and the geometry of the star of x. Therefore, the statement follows by induction on n. $\square$

In particular, if M is the Coxeter type of a convex polytope, then M is a linear diagram.

Example 4.1.13 We describe convex polytopes whose geometries have Coxeter types given in Table 4.2 (part of it appeared in Table 2.2). The last row lists the Coxeter types $I_2^{(m)}$ which belong to the regular m-gons in $\mathbb{E}^2$ and will not be discussed here further. For each $n \in \mathbb{N}$ with $n \geq 1$, let $\varepsilon_1, \ldots, \varepsilon_n$ be the standard basis of $\mathbb{R}^n$.

(i) The *hypertetrahedron* or *n-simplex* on $n + 1$ points is the convex polytope spanned by the $n + 1$ vectors

$$(n + 1)\varepsilon_j - \sum_{l=1}^{n+1} \varepsilon_l \qquad \left(j \in [n + 1] \right)$$

Fig. 4.5 The 4-dimensional hypercube

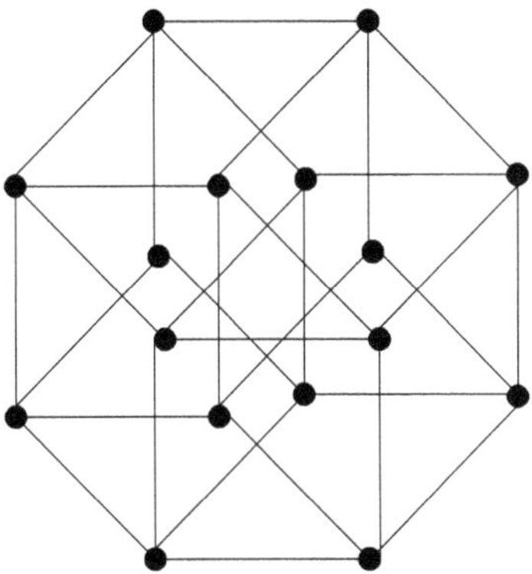

which belong to the hyperplane given by the equation $\sum_{i=1}^{n+1} x_i = 0$ in $\mathbb{E}^{n+1}$. These $n+1$ vectors are the 1-elements of the polytope geometry. Each set of $i+1$ elements of type 1 is the 1-shadow of an i-element of the geometry. The geometry has Coxeter type A_n and is isomorphic to the geometry of rank n of Example 1.2.6.

(ii) The *hypercube* or *n-dimensional measure polytope*. Its 1-elements are the 2^n vectors $\pm\varepsilon_1 \pm \varepsilon_2 \pm \cdots \pm \varepsilon_n$ in $\mathbb{R}^n$. The n-elements are the hyperplanes $x_j = 1$ or $x_j = -1$ for $j \in [n]$, so they are $2n$ in number. For $n = 2$ and $n = 3$, we obtain the 4-gon and the cube, respectively. Figure 4.5 shows a picture for $n = 4$. The geometry has Coxeter type B_n (also referred to as C_n; in terms of Coxeter diagrams, no distinction is made between the two). It is isomorphic to the geometry Γ over $[n]$ of Example 1.3.6. The node 1 occurs at the extreme left of the diagram Table 4.2, the other nodes are labeled in increasing order till n at the extreme right.

(iii) The *hyperoctahedron* or *n-dimensional cross polytope* is the dual of the hypercube and also has Coxeter type B_n. Whereas, in (ii), the nodes of B_n are labeled $1, \ldots, n$ from left to right, here, they are labeled $1, \ldots, n$ from right to left. Its 1-elements are the $2n$ vectors $\pm\varepsilon_i$ ($i \in [n]$). The 1-shadows of i-elements are the sets of 1-elements of size $i + 1$ that are pairwise orthogonal in $\mathbb{R}^n$.

(iv) In $\mathbb{E}^4$, take the 24 vectors $\pm\varepsilon_i$ ($i \in [4]$) and $\frac{1}{2}(\pm\varepsilon_1 \pm \varepsilon_2 \pm \varepsilon_3 \pm \varepsilon_4)$ which, as a set of points, is the disjoint union of a hypercube and a hyperoctahedron. The resulting convex polytope is the so-called *24-cell*. The polytope geometry has Coxeter type F_4.

As the pictures of the hypertetrahedron, hypercube, 24-cell (Fig. 4.6) and so on, gradually increase in complexity, we need another way to describe their main features. This is done in a distribution diagram (cf. Example 1.7.16) of the collinearity graph; see Fig. 4.7. We start at a vertex, say $p = \frac{1}{2}(\varepsilon_1 + \varepsilon_2 + \varepsilon_3 + \varepsilon_4) = +++ +$. There are eight vertices adjacent to it, whose angles with p are $60°$. Inside this set of eight (actually graphwise, a cube), each element is adjacent to three other points. The points at distance two (in the collinearity graph) from p split up into two sets of six and eight each; the first consists of all vertices orthogonal to p (precisely two coordinates are $+$); the latter is the 'cube' of points adjacent to the unique vertex at distance three from p, and consists of all vertices at 120 degrees to p. Each element

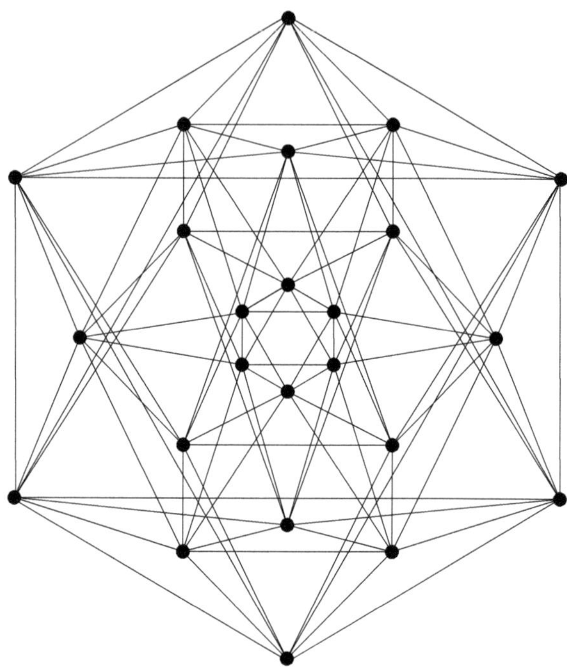

Fig. 4.6 The 24-cell: vertices and edges

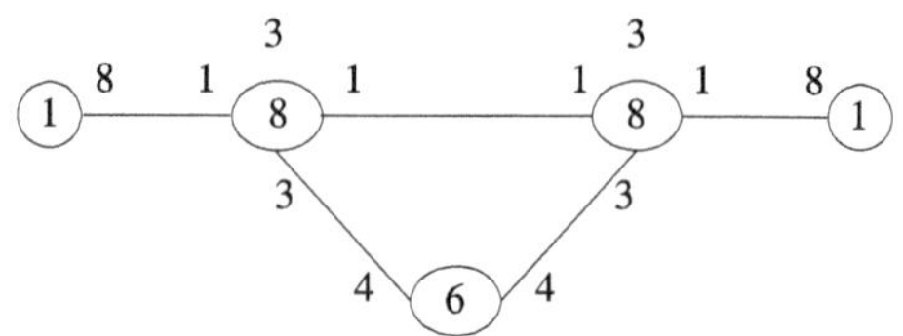

Fig. 4.7 A distribution diagram of the vertices with edges of the 24-cell

in the set of six (actually a coclique) is adjacent to exactly four of the eight vertices adjacent to p, and so on.

(v) The *600-cell* has 120 vertices which are those of the 24-cell of (iv) together with all points $\frac{1}{2}(\pm\tau, \pm 1, \pm\tau^{-1}, 0)$ where $\tau = \frac{1}{2}(\sqrt{5}+1)$ is the golden ratio, and all the vectors resulting from this one by even permutations of the coordinates. The distribution diagram of the 600-cell is given in Fig. 4.8. The polytope geometry has Coxeter type H_4. The residue of a vertex is the polytope geometry of a 3-dimensional polytope; it has Coxeter type H_3.

Definition 4.1.14 Let $n \in \mathbb{N}$, $n > 0$. A *tessellation* T of the Euclidean space $\mathbb{E}^n$ is a collection of convex polytopes in $\mathbb{E}^n$ whose union is $\mathbb{E}^n$ such that the non-empty intersection of any two members is a face of each of them. For $n = 2$, it is another name for a tiling. The *tessellation geometry* $\Gamma(T)$ of T is the following geometry over $[n+1]$. For $i \in [n+1]$, its elements of type i are the $(i-1)$-faces of the polytopes in T. The *vertices*, *edges*, and *polytopes* of T are the elements of type 1, 2, and $n+1$, respectively.

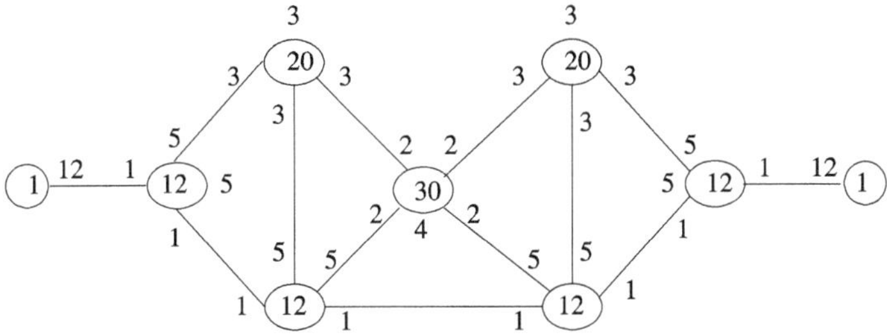

Fig. 4.8 The distribution diagram of the 600-cell (vertices and edges)

Table 4.3 Coxeter diagrams of tessellations by finite regular polytopes

Name	Diagram	Tiles
$\widetilde{A}_1$		1-simplices
$\widetilde{C}_n$		Hypercubes
$\widetilde{F}_4$		24-cells
$\widetilde{G}_2$		Hexagons, triangles

If this geometry is of Coxeter type M (cf. Definition 2.4.2) and the rank of M equals $n + 1$, we also say that T itself is of *Coxeter type M*.

Proposition 4.1.15 *Let $n \geq 2$. The geometry of a tessellation of $\mathbb{E}^n$ is thin and residually connected and has a linear digon diagram.*

Proof This is omitted, as it is very similar to the proof of Proposition 4.1.12.   □

Example 4.1.16 A *lattice* of a real vector space is understood to be an additive subgroup of the vector space generated by a basis of the vector space. In terms of Definition 2.7.6, this is the same as a $\mathbb{Z}$-lattice in V. We describe tessellations whose geometries have Coxeter types given in Table 4.3 (part of which appeared in Table 2.2). For each $n \in \mathbb{N}$, let $\varepsilon_1, \ldots, \varepsilon_n$ be the standard basis of $\mathbb{R}^n$. The first row corresponds to the tessellation of the Euclidean affine line $\mathbb{E}^1$ by segments, whose tessellation geometry is isomorphic to the ∞-gon.

(i) Consider the lattice

$$L = \mathbb{Z}2\varepsilon_1 \oplus \cdots \oplus \mathbb{Z}2\varepsilon_n$$

in $\mathbb{R}^n$ and the hypercube Π of Example 4.1.13(ii). Taking translates of Π with respect to the L, we obtain a tessellation of $\mathbb{E}^n$ by hypercubes. The corresponding tessellation geometry has Coxeter type $\widetilde{C}_n$.

(ii) Consider the lattice

$$L = \left\{ \frac{1}{2} \sum_{i=1}^{4} a_i \varepsilon_i \,\middle|\, a_i \in \mathbb{Z}, \sum_{i=1}^{4} a_i \in 2\mathbb{Z} \right\}$$

in $\mathbb{R}^4$. Taking translates of the 24-cell with respect to L, we obtain a tessellation of $\mathbb{E}^4$. Its geometry has Coxeter type $\widetilde{F}_4$.

(iii) The last row of Table 4.3 lists the Coxeter type $\widetilde{G}_2$, which belongs to the regular tiling of $\mathbb{E}^2$ with 3-gons as indicated in Example 4.1.9(ii). Its dual can be obtained as the translates of the 6-gon with vertices $\pm(1, 0)$, $\frac{1}{2}(\pm 1, \pm\sqrt{3})$ with respect to the lattice $\mathbb{Z}(0, \sqrt{3}) + \mathbb{Z}\frac{1}{2}(3, \sqrt{3})$.

In Theorems 4.8.2 and 4.8.4, we will see that the types of the polytope and tessellation geometries presented here are the only ones possible.

4.2 Thin Geometries of Coxeter Type

Let I be a finite set of types. Often we will assume I to be finite of size n and identify it with the set $[n]$. For the duration of this section, $M = (m_{i,j})_{i,j \in I}$ denotes a Coxeter matrix (cf. Definition 2.4.2) over I. We have interpreted a regular polytope as a thin I-geometry of Coxeter type. By Proposition 4.1.12 this implies that the digon diagram is linear. Here we drop the requirement that the digon diagram be linear and look for all thin I-geometries of Coxeter type. In Theorem 4.2.8, it turns out that, for each Coxeter type, there is a universal chamber system covering all others of the same type.

Remark 4.2.1 Let Γ be a thin I-geometry of Coxeter type M. Its chamber system $\mathcal{C}(\Gamma) = (C, \{\sim_i \mid i \in I\})$ is connected and each panel consists of exactly two elements. For $c \in C$ and $i \in I$, denote by ci the unique element i-adjacent to c.

Being thin implies $cii = (ci)i = c$, which shows that, for each $i \in I$, the map $c \mapsto ci$ is a permutation of C of order two. It need not be an automorphism of $\mathcal{C}(\Gamma)$. Rather, it tells us how to get from one chamber to another and as such it will be a useful ingredient to describe $\mathcal{C}(\Gamma)$ in terms of groups.

If $i, j \in I$, the word ij corresponds to the permutation sending c to the end chamber $c(ij) = (ci)j$ of the gallery $c, ci, (ci)j$. Obviously, for all $i, j \in I$ and $c \in C$,

$$\underbrace{c\,ij\cdots ij}_{2m_{i,j}} = c, \tag{4.3}$$

where the length of the word in i, j is $2m_{i,j}$. For $i = j$, we recover the equality $cii = c$.

The relations (4.3) do not tell us precisely which permutations of C obtained as compositions of the maps $c \mapsto ci$ ($c \in C$) for i varying over I, are the identity. They just imply that the $m_{i,j}$-th power of the composite of $c \mapsto ci$ and $c \mapsto cj$ is trivial.

With these permutations of C in mind, we introduce Coxeter groups.

Definition 4.2.2 The *Coxeter group* $W(M)$, or just W, of type M, is the group with presentation

$$\langle \{s_i \mid i \in I\} \mid (s_i s_j)^{m_{i,j}} = 1\rangle.$$

This means that W is freely generated by the set $S = \{s_i \mid i \in I\}$ subject to the relations $(s_i s_j)^{m_{i,j}}$ if $m_{i,j} \in \mathbb{N}$ and no such relation if $m_{i,j} = \infty$. The pair (W, S) is called a *Coxeter system* of type M.

The elements of S have order at most two (as $m_{i,i} = 1$). At this stage, it is not clear that s_i and s_j are distinct whenever $i, j \in I$ are distinct. It will be shown to hold later, in Theorem 4.4.6.

Example 4.2.3 Exercise 4.9.5 shows that $W(A_n)$ is isomorphic to Sym_{n+1}. For $|I| \leq 2$, the classification of Coxeter groups W is immediate: if $I = [1]$, then W is the unique group of order two; if $I = [2]$, then W is the unique dihedral group of order $2m_{12}$ (possibly infinite) with cyclic subgroup $\langle s_1 s_2 \rangle$ of index two.

Lemma 4.2.4 *Let Γ be a thin I-geometry of Coxeter type M, and let (W, S) be a Coxeter system of type M. The map $S \to \mathrm{Sym}(C)$ given by*

$$s_i \mapsto (c \mapsto ci) \quad (i \in I, \ c \in C)$$

determines a unique transitive action of W from the right on the set C of chambers of $\mathcal{C}(\Gamma)$.

Proof According to Remark 4.2.1, $s_i \in S$ acts on C as a permutation of order two (in accordance with $m_{i,i} = 1$). In view of (4.3) and the definition of W, these permutations determine a group homomorphism $W \to \mathrm{Sym}(C)$ assigning to $w = s_{i_1} \cdots s_{i_q}$ the bijection $c \mapsto cw = c i_1 \cdots i_q$.

Since $\mathcal{C}(\Gamma)$ is connected, any pair $c, d \in C$ can be connected by a gallery c, $ci_1, \ldots, ci_1 \cdots i_q = d$. Putting $w = s_{i_1} \cdots s_{i_q}$, we see that $d = cw$. This establishes that W acts transitively on C. $\qquad\square$

Theorem 1.7.5 will be used to describe C in terms of W. Fix $c \in C$ and write $A = W_c = \{w \in W \mid cw = c\}$. In words, A is the stabilizer of c in W. Each chamber cw can be identified with the right coset Aw of W consisting of all elements mapping c to cw. In other words, there is a bijective correspondence between the set $A \backslash\backslash W$ (cf. Exercise 1.9.22 for notation) and C given by the map $Aw \mapsto cw$.

For $i \in I$, we describe i-adjacency in $\mathcal{C}(\Gamma)$ in terms of W and A. Suppose that $w, w' \in W$ represent two (distinct) i-adjacent chambers and fix a chamber c. The equalities $cw' = (cw)i = cws_i$ imply $Aw' = Aws_i$. This argument can be reversed, showing that cw and cw' are i-adjacent if and only if $w \in Aw'\langle s_i \rangle$. Here, as usual, $\langle s_i \rangle$ denotes the subgroup of W generated by s_i. The conclusion is that $\mathcal{C}(\Gamma)$ is isomorphic to the chamber system $\mathcal{A}(W, S, A)$ over I defined as follows.

Definition 4.2.5 Let (W, S) be a Coxeter system of type M over I. By $\mathcal{A}(W, S, A)$ we denote the following chamber system over I.

(1) Its set of chambers is $A \backslash\backslash W$.
(2) The chambers Aw and Aw' are i-adjacent if and only if $w \in Aw'\langle s_i \rangle$. This is denoted by $Aw \sim_i Aw'$.

The chamber system $\mathcal{A}(W, S, 1)$ will be denoted $\mathcal{C}(M)$.

For an arbitrary Coxeter system (W, S) of type M and a subgroup A of W, the system $\mathcal{A}(W, S, A)$ is indeed a chamber system over I as $Au \sim_i Av$ with $u, v \in W$ is equivalent to $Au\langle s_i \rangle = Av\langle s_i \rangle$, which clearly defines an equivalence relation.

Lemma 4.2.6 *Let (W, S) be a Coxeter system of type M. If A is a subgroup of W, then $\mathcal{A}(W, S, A)$ coincides with the quotient chamber system $\mathcal{C}(M)/A$.*

Proof According to Proposition 3.1.10, the chambers of the quotient chamber system $\mathcal{A}(W, S, 1)/A$ are the A-orbits Aw ($w \in W$), which coincide with the elements of $A \backslash\backslash W$, that is, the chambers of $\mathcal{A}(W, S, A)$. By Definition 3.1.11, for u, v in W and $i \in I$, the chambers Au and Av are i-adjacent in $\mathcal{A}(W, S, 1)/A$ if and only if there exists an element $a \in A$ such that $au \sim_i v$, that is, $au\langle s_i \rangle = v\langle s_i \rangle$, where $S = \{s_i \mid i \in I\}$. As this is equivalent to $Au\langle s_i \rangle = Av\langle s_i \rangle$, which stands for i-adjacency in $\mathcal{A}(W, S, A)$, the two chamber systems coincide. $\square$

Realizing that all work has been done in the chamber system $\mathcal{C}(\Gamma)$ rather than in the geometry Γ we started from, we formulate our findings in terms of chamber systems in the theorem below.

Definition 4.2.7 A chamber system over I is said to be of *Coxeter type M* if it is connected and non-empty and if, for all distinct i and j in I, every $\{i, j\}$-cell is (the chamber system of) a generalized $m_{i,j}$-gon.

Thus, in a thin chamber of Coxeter type M all $\{i, j\}$-cells are $m_{i,j}$-gons. The chamber system $\mathcal{A}(W, S, A)$ above is not of type M for every choice of A; for instance $A = W$ leads to counterexamples.

Theorem 4.2.8 *Let (W, S) with $S = \{s_i \mid i \in I\}$ be a Coxeter system of type M.*

(i) *If Γ is an I-geometry of type M, then $\mathcal{C}(\Gamma)$ is a residually connected thin chamber system of type M.*
(ii) *Suppose that $\mathcal{C} = (C, \{\sim_i \mid i \in I\})$ is a thin chamber system of Coxeter type M. Fixing $c \in C$ and taking A to be the stabilizer W_c of c in the right action of W on C of Lemma 4.2.4, we obtain an isomorphism $\mathcal{C}(M)/A \to \mathcal{C}$ given by $Aw \mapsto cw$ for $w \in W$.*

Proof Assertion (i) is immediate from the Chamber System Correspondence 3.4.6 and the discussion above Definition 4.2.5. Assertion (ii) follows from the same arguments as used above for the chamber system of a thin geometry and Lemma 4.2.6. $\square$

So far, we have reduced the study of thin chamber systems of Coxeter type M and geometries of type M to a study of quotients $A \backslash\backslash W$ of the Coxeter group W of type M (a uniquely determined object!) by certain subgroups A. At this stage, it is not clear which choices of A suffice for the chamber system $\mathcal{A}(W, S, A)$ to be of type M, or for the geometry $\Gamma(\mathcal{A}(W, S, A))$ to be of type M. In Corollary 4.5.16 we will see that $\mathcal{C}(M)$ is residually connected, so that Theorem 4.2.8 can be phrased in terms of geometries thanks to Theorem 3.4.6 and the result below. A full answer for $M = A_n$ will be given in Lemma 6.5.1.

We next determine the automorphism group of $\mathcal{C} = \mathcal{C}(W, S, A)$. In view of Corollary 3.4.7, this group coincides with the automorphism group of the corresponding geometry provided $\mathcal{C}$ is residually connected.

Notation 4.2.9 By $N_W(A)$ we denote the *normalizer* of A in W, that is, $N_W(A) = \{g \in W \mid gAg^{-1} = A\}$. By α we denote the map $N_W(A) \to \mathrm{Sym}(A \backslash\backslash W)$ given by

$$\alpha_w = (Ax \mapsto Awx)$$

for $w \in N_W(A)$.

For $x, y \in W$ with $Ax = Ay$, we have $Awx = wAx = wAy = Awy$, so α_w is well defined for $w \in N_W(A)$.

Proposition 4.2.10 *Let (W, S) be a Coxeter system of type M. If A is a subgroup of W such that $\mathcal{C} = \mathcal{A}(W, S, A)$ is a thin chamber system over I, then $\alpha : N_W(A)/A \to \mathrm{Aut}(\mathcal{C})$ is an isomorphism.*

Proof To begin, observe that $A \backslash\backslash N_W(A) = N_W(A)/A$ as $xA = Ax$ for each $x \in N_W(A)$. Let $g \in \mathrm{Aut}(\mathcal{C})$. Then $g(Aws_i) = g(Aw)s_i$ for all $w \in W$ and $i \in I$, whence $g(Aw) = g(A)w$ for all $w \in W$. Take $w_0 \in W$ in such a way that $g(A) = Aw_0$. Then for all $a \in A$, we have $Aw_0a = g(A)a = g(Aa) = g(A) = Aw_0$, so $w_0 \in N_W(A)$, and $g(Aw) = g(A)w = Aw_0w$, so $g = \alpha_{w_0}$.

Conversely, suppose $w_0 \in N_W(A)$ and consider the map $\alpha_{w_0} : C \to C$. Then α_{w_0} belongs to $\mathrm{Aut}(\mathcal{C})$ because it maps a pair Aw, Aws_i of i-adjacent chambers onto the pair Aw_0w, Aw_0ws_i which is again i-adjacent, and because it is bijective with inverse $\alpha_{w_0}^{-1} = \alpha_{w_0^{-1}}$ of the same form.

Thus, α is a surjective map from $N_W(A)$ onto $\mathrm{Aut}(\mathcal{C})$. It is a homomorphism as $\alpha_{uv}(Aw) = Auvw = \alpha_u\alpha_v(Aw)$ for all $u, v \in N_W(A)$. Its kernel consists of all $u \in N_W(A)$ such that $Auw = Aw$ for every $w \in A$, and hence coincides with A. Therefore, $A \backslash\backslash N_W(A) \cong \mathrm{Aut}(\mathcal{C})$, proving the proposition. $\square$

The notation $\mathcal{C}(W, A, (\langle A \cup \{s_i\}\rangle)_{i \in I})$ in the corollary below was introduced in Definition 3.6.3.

Corollary 4.2.11 *For every thin chamber system $\mathcal{C}$ of Coxeter type M and every chamber c of $\mathcal{C}$, the following two statements hold.*

(i) *For each chamber d of C, there is at most one automorphism of C mapping c to d.*

(ii) *The chamber system C has a transitive (and hence regular) group of automorphisms if and only if the stabilizer A of c in $W(M)$ in the right action on the set of chambers of C is a normal subgroup of $W(M)$. In this case, $C \cong C(W, A, (\langle A \cup \{s_i\}\rangle)_{i \in I})$.*

Proof By Theorem 4.2.8 we may assume that $C = \mathcal{A}(W, S, A)$ for a subgroup A of W and that $c = A \in A \backslash\backslash W$.

(i) Let $d = Ax$ for some $x \in W$. By Proposition 4.2.10, there is a unique automorphism sending c to d if $x \in N_W(A)$ and none otherwise. Hence (i).

(ii) The transitivity is immediate from the observation that A is normal in W if and only if $A \backslash\backslash N_W(A) = A \backslash\backslash W$. The last part of (ii) is immediate from Proposition 3.6.4. $\qquad\square$

Once $C(M)$ is shown to be a thin chamber system, it will follow that α is a transitive action on it. This action should not be confused with the representation $c \mapsto cw$ of W on the chamber set C of Theorem 4.2.8.

Definition 4.2.12 Let (W, S) be a pair consisting of a group W and a generating subset S of W. A *minimal expression* for an element w of W is a product $r_1 r_2 \cdots r_q$ with $r_i \in S$ for each $i \in [q]$ that equals w in W and has minimal length q among all expressions with this property. The length of a minimal expression for w is called the *length* of w. It is denoted $l(w)$.

Remark 4.2.13 The distance $d_{C(M)}(x, y)$ between two chambers x, y of $C(M)$ (that is, the length of a minimal gallery from x to y) is equal to $l(x^{-1}y)$.

Example 4.2.14 If $M = B_2 = I_2^{(4)}$ (see Table 4.2), then $W(M)$ is the dihedral group of order eight, as discussed in Example 4.2.3. The set $S = \{s_1, s_2\}$ has cardinality two. Minimal expressions for the elements of $W(B_2)$ are $1, s_1, s_2, s_1 s_2, s_2 s_1, s_1 s_2 s_1$, $s_2 s_1 s_2$, and $s_1 s_2 s_1 s_2$. The lengths of the corresponding elements of W are 0, 1, 1, 2, 2, 3, 3, and 4. As a consequence, the diameter of $C(M)$ equals four.

Remark 4.2.15 Let (W, S) be a pair consisting of a group W and a subset S generating W. The definition of minimal expression for an element of W can be made more formal by use of the free monoid S^* on the alphabet S. Let $\zeta : S^* \to W$ be the homomorphism of monoids determined by $s_i \mapsto s_i$ $(i \in I)$. As elements of S^* are words in the alphabet S, the length of an element in S^* is easily defined. Now $l(w)$, for $w \in W$, is the minimum length of an element in $\zeta^{-1}(w)$. The corresponding function $l : W \to \mathbb{N}$ is called the *length function* on W with respect to S.

We collect some basic and useful observations on the length function l of a Coxeter system.

Lemma 4.2.16 *Let (W, S) be a Coxeter system over I. For $s \in S$ and $w \in W$, we have $l(sw) = l(w) \pm 1$.*

Proof Clearly, $l(sw) \leq 1 + l(w)$ and $l(w) = l(s(sw)) \leq 1 + l(sw)$, so $l(w) - 1 \leq l(sw) \leq l(w) + 1$. It remains to show that the parities of $l(w)$ and $l(sw)$ differ.

Write $S = \{s_i \mid i \in I\}$. By Exercise 4.9.8(a) there is a homomorphism $\mathrm{sg} : W \to \{\pm 1\}$ of groups (the latter being the multiplicative subgroup of the rationals of order two) determined by $\mathrm{sg}(r) = -1$ for each $r \in S$. If $w = s_1 \cdots s_q$ is an expression for w, then $\sigma(w) = \sigma(s_1) \cdots \sigma(s_q) = (-1)^q$. In particular, $(-1)^q$ does not depend on the expression chosen and equals $(-1)^{l(w)}$. This implies $(-1)^{l(sw)} = \mathrm{sg}(sw) = \mathrm{sg}(s)\sigma(w) = (-1)^{l(w)+1}$. Hence $l(sw) \equiv l(w) + 1 \pmod 2$, so the parities of $l(sw)$ and $l(w)$ differ. $\qquad\square$

Notation 4.2.17 Let (W, S) be a Coxeter system over a Coxeter diagram M over I. For $J \subseteq I$, we denote by $\langle J \rangle$ the subgroup of W generated by $\{s_j \mid j \in J\}$.

Definition 4.2.18 Suppose that (W, S) is a Coxeter system over I and let $J \subseteq I$. Write $S = \{s_i \mid i \in I\}$. If $w \in W$ satisfies $l(s_j w) > l(w)$ for all $j \in J$, then w is called *left J-reduced*. The set of all left J-reduced elements of W is denoted by $^J W$. Similarly, for $K \subseteq I$, an element $w \in W$ is called *right K-reduced* if $l(w s_k) > l(w)$ for all $k \in K$, and W^K denotes the set of all right K-reduced elements of W. As a consequence, $W = \langle J \rangle\, ^J W = W^J \langle J \rangle$.

Notation 4.2.19 For $J \subseteq I$, we denote by l_J the length function (cf. Definition 4.2.12) on the subgroup $\langle J \rangle$ of W with respect to the generating set $\{s_j \mid j \in J\}$.

Lemma 4.2.20 *Let (W, S) be a Coxeter system. For each $w \in W$ and $J \subseteq I$, the following properties hold.*

(i) *There are $u \in \langle J \rangle$ and $v \in\, ^J W$ such that $w = uv$ and $l(w) = l(u) + l(v)$.*
(ii) *If $w \in \langle J \rangle$, then $l(w) = l_J(w)$.*
(iii) $W^J = (^J W)^{-1}$.

Proof (i) Consider the subset D of $\langle J \rangle \times W$ consisting of all pairs (u, v) with $w = uv$ and $l(w) = l(u) + l(v)$ such that $l(u) = l_J(u)$. This set is nonempty, as it contains $(1, w)$. Let (u, v) be an element of D with $l(u)$ maximal. Suppose that $j \in J$ is such that $l(s_j v) < l(v)$. Then $v = s_j v'$ for some $v' \in W$ with $l(v) = l(v') + 1$ and so $w = (u s_j)v'$ is a decomposition of w with $l(w) \leq l(u s_j) + l(v') \leq (l(u) + 1) + (l(v) - 1) = l(w)$, whence $l(u s_j) = l(u) + 1$. Now $l(u s_j) \leq l_J(u s_j) \leq l_J(u) + 1 = l(u) + 1 = l(s_j u)$. Therefore, $(u s_j, v') \in D$ with $l(u s_j) > l(u)$, a contradiction. We conclude that $v \in\, ^J W$. This proves (i).

(ii) If $w \in \langle J \rangle$, then the proof of (i) gives a decomposition $w = uv$ with $u \in \langle J \rangle$ and $v \in\, ^J W$ such that $l(w) = l(u) + l(v) = l_J(u) + l(v)$. But now $v \in \langle J \rangle \cap\, ^J W = \{1\}$, so $w = u$ and $l(w) = l_J(w)$, as required for (ii).

(iii) As $s_i^{-1} = s_i$ for each $i \in I$, we have $l(w^{-1}) = l(w)$ for each $w \in W$. Therefore, $l(s_j w) < l(w)$ is equivalent to $l(w^{-1} s_j) < l(w^{-1})$. $\qquad\square$

The study of Coxeter groups can be reduced to irreducible types, as follows.

Definition 4.2.21 Let M be a Coxeter diagram over I. When referring to a connected component of M, we view M as a labelled graph. In other words, a *connected component* of M is a minimal non-empty subset J of I such that $m_{j,k} = 2$ for each $j \in J$ and $k \in I \setminus J$. If M has a single connected component, it is called *connected* or *irreducible*; otherwise, it is called *reducible*. A Coxeter group of type M is called *irreducible* (*reducible*) if M is irreducible (*reducible*, respectively).

If J is a set of nodes of M, then we also let J stand for the labelled graph induced on J by M, that is, the diagram $M|_J$ introduced in Definition 2.6.1. In particular, $W(J)$ is a Coxeter group of type J.

In Corollary 4.4.17 we will see that $\langle J \rangle$ (cf. Notation 4.2.17) is isomorphic to $W(J)$. Here, we are concerned with the special case in which M has more than one connected component.

Proposition 4.2.22 *Let* (W, S) *be a Coxeter system of type* M *over* I *and let* $I_1, \ldots, I_t$ *be a partition of* I *such that* $i \in I_r$, $j \in I_s$, $r \neq s$, *implies* $m_{i,j} = 2$. *Then* $W = \langle I_1 \rangle \times \cdots \times \langle I_t \rangle$, *and, for each* $k \in [t]$, *the pair* $(\langle I_k \rangle, \{s_i \mid i \in I_k\})$ *is a Coxeter system of type* I_k.

Proof Assume that I is partitioned in two subsets J, K such that M has no edges $\{j, k\}$ with $j \in J$, $k \in K$ (that is, $m_{j,k} = 2$). It follows that $s_j s_k = s_k s_j$ for all $j \in J$, $k \in K$, so $U := \langle J \rangle$ and $V := \langle K \rangle$ are normal subgroups of W centralizing each other. Therefore, $W = UV$. Moreover, the map $S \to W(J)$ sending s_i to 1 if $i \in K$ and to $s_i \in W(J)$ if $i \in J$ extends to a surjective homomorphism $\phi_J : W \to W(J)$ with kernel V. Indeed, as any of the defining relations for W is also satisfied by the images of the generators under ϕ_J, the map extends to a well-defined homomorphism $W \to W(J)$. It is clearly surjective. The restriction of ϕ_J to U is an isomorphism as the above method applied to the map $\{s_j \in W(J) \mid j \in J\} \to W$ gives the existence of a homomorphism $\psi_J : W(J) \to W$ with image U such that $\psi_J \circ \phi_J$ is the identity on U.

Each element of W can be written as a product uv with $u \in U$ and $v \in V$; it maps to $1 \in W(J)$ if and only if $\phi_J(u) = 1$, which is equivalent to $u = 1$ by the previous paragraph, and so implies $w \in V$. This proves $V = \mathrm{Ker}(\phi_J)$. Similarly, U is the kernel of the corresponding homomorphism $\phi_K : W \to V$ and there is a canonical isomorphism $\psi_K : W(K) \to V$. In particular, $u = \psi_J(\phi_J(w))$ and $v = \psi_K(\phi_K(w))$.

Suppose now $w \in U \cap V$. Then $w = \psi_J(\phi_J(w))\psi_K(\phi_K(w)) = \psi_J(1)\psi_K(1) = 1$, proving that the intersection of U and V is trivial. We conclude that $W = \langle J \rangle \times \langle K \rangle$ and that $(\langle s_j \mid j \in J \rangle, \{s_i \mid i \in J\})$ is a Coxeter system of type J. An easy induction extends these facts to the case of any finite number of connected components of M. $\square$

So, indeed, for a connected component J of M, the subgroup $\langle s_j \mid j \in J \rangle$ of $W(M)$ is isomorphic to $W(J)$. In view of Proposition 4.2.22, the analysis of Coxeter groups can often be reduced to the case where M is connected.

Example 4.2.23 Direct products of Coxeter groups appear in elementary geometry. For instance, let Π be a double pyramid in the Euclidean space $\mathbb{E}^3$: the convex closure of a regular m-gon for some $m \in \mathbb{N}$, $m \geq 3$, in a horizontal plane π, and two more vertices at the same distance to the center of gravity o of the m-gon on the vertical axis through o, but at different sides of π. Thus, π is the mirror of a reflection leaving the double prism invariant. The Coxeter group of type $A_1 \,\dot\cup\, I_2^{(m)}$ is a group of isometries of Π, with the reflection mentioned above corresponding to the first component of the diagram. If $m = 4$, then this Coxeter group need not be the full group of isometries of Π, but for $m > 4$, it is.

4.3 Groups Generated by Affine Reflections

Every group G generated by a set $\{\rho_i \mid i \in I\}$ of involutions is a homomorphic image of a Coxeter group W: simply take the Coxeter matrix of type $M = (m_{i,j})_{i,j \in I}$, where $m_{i,j}$ is the order of $\rho_i \rho_j$. Let (W, S), with $S = \{s_i \mid i \in I\}$, be the corresponding Coxeter system. The map $s_i \mapsto \rho_i$ $(i \in I)$ determines a surjective homomorphism $W \to G$ of groups. Theorem 4.3.6 shows that, when G is a subgroup of $\mathrm{AGL}(V)$ for a real vector space V and when the ρ_i $(i \in I)$ are suitable affine reflections on V, this surjective homomorphism is actually an isomorphism.

Definition 4.3.1 Let $\mathbb{D}$ be a division ring. An *affine reflection* on $\mathbb{D}^n$ is an element of $\mathrm{AGL}(\mathbb{D}^n)$ of order two fixing point-wise an *affine hyperplane* (that is, the image of a hyperplane of $\mathbb{D}^n$ under a translation), and fixing a parallel class of lines (a class being the collection of all cosets of 1-dimensional subspace of the additive group $\mathbb{D}^n$) having no members in the fixed-point hyperplane. The fixed-point affine hyperplane is also called its *mirror*.

By use of Exercise 1.9.30, we can give an explicit description of affine reflections. Recall the notation t_a for a translation over a from Example 1.8.17.

Lemma 4.3.2 *Let $n \geq 1$ and let $\mathbb{F}$ be a field of characteristic distinct from two. Each affine reflection on $V := \mathbb{F}^n$ is of the form $t_{\lambda a} r_{a,\phi}$ for certain $a \in V$, $\phi \in V^\vee$, and $\lambda \in \mathbb{F}$ with $\phi(a) = 2$. Its mirror is $\{x \in V \mid \phi(x) = \lambda\}$.*

Proof According to Exercise 1.9.30(e), a linear reflection has the form $r_{a,\phi}$ with $a \in V$ and $\phi \in V^\vee$ such that $\phi(a) = 2$. It follows that an affine reflection must have the form $t_v r_{a,\phi}$. Using that the square must be the identity, we find that $v = a\lambda$ for some $\lambda \in \mathbb{F}$. The mirror is easily computed to be as stated. $\qquad\square$

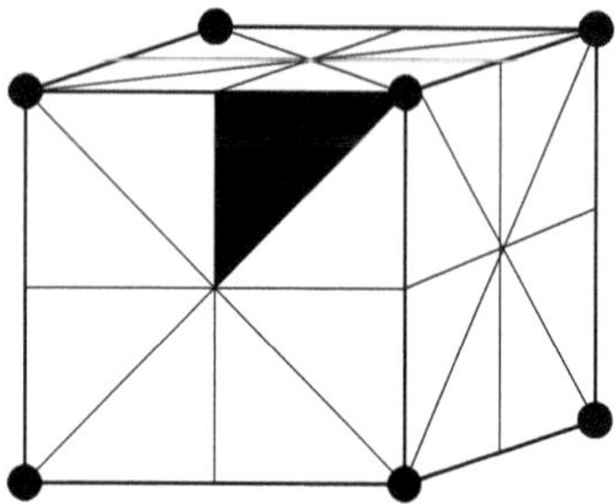

Fig. 4.9 The chambers of the
Coxeter group of type B_3
drawn on the cube, with one
chamber colored *black*

Example 4.3.3 Let Γ be the cube geometry of Example 1.1.1, whose vertices are the
vectors in $\mathbb{R}^3$ all of whose coordinates are ± 1. It is described in Example 4.1.13(ii),
where it is indicated to be of Coxeter type B_3. We fix the chamber c of Γ consisting
of the vertex $v_1 = \varepsilon_1 + \varepsilon_2 + \varepsilon_3$, the edge $v_2 = \{\varepsilon_1 \pm \varepsilon_2 + \varepsilon_3\}$, and the face $v_3 =
\{\varepsilon_1 \pm \varepsilon_2 \pm \varepsilon_3\}$. The latter two elements are described in terms of their 1-shadows.

In Fig. 4.9, we have drawn the cube. Choose the axes so that the positive x-axis
and y-axis are horizontal, with the former pointing toward us and the latter to the
right; the positive z-axis goes up. Now c corresponds to the black chamber at the
right hand top side of the front face. For $i = 1, 2, 3$, take ρ_i to be the reflection
stabilizing v_{i+1} and v_{i+2} (indices mod 3). The three reflections ρ_1, ρ_2, ρ_3 are automorphisms of Γ and have a mirror bounding the chamber c. They are the reflections
$r_{\alpha_i, \phi}$, where $\phi(x) = 2 f(\alpha_i, \alpha_i)^{-1} f(\alpha_i, x)$ $(x \in \mathbb{R}^3)$ and

$$\alpha_1 = \varepsilon_2, \qquad \alpha_2 = \varepsilon_2 - \varepsilon_3, \qquad \alpha_3 = \varepsilon_1 - \varepsilon_3.$$

Here, f denotes the standard Euclidean inner product. The 48 transforms of the
chamber c represent the 48 chambers of Γ and form a $W(B_3)$-orbit. We will see
below that this orbit is regular.

For arbitrary groups generated by affine reflections, we will be looking for a set
of domains similar to the above 48 chambers on which the group acts regularly.

Definition 4.3.4 If a group G acts on a set E, then a subset P of E is called a
prefundamental domain for G if $P \neq \emptyset$ and $P \cap gP = \emptyset$ for all $g \in G \setminus \{1\}$.

The existence of a prefundamental domain for G in E implies that the action of
G on E is faithful. Observe that a prefundamental domain need not quite be what
is classically called a fundamental domain as it is not required that the domain be
connected or contain a member of each G-orbit in E.

Example 4.3.5 Let $m \geq 2$ and let (W, S) be a Coxeter system of type $I_2^{(m)}$. The
latter denotes the Coxeter matrix over [2] whose off-diagonal entries are m. We will
realize W as a group generated by reflections.

Consider the two vectors $\alpha_1 = \sqrt{2}\varepsilon_1$ and $\alpha_2 = \sqrt{2}(-\cos(\pi/m)\varepsilon_1 + \sin(\pi/m)\varepsilon_2)$
in the Euclidean vector space $\mathbb{R}^2$ with standard inner product f. These two vectors

Fig. 4.10 A prefundamental domain for $W(\mathrm{I}_2^{(8)})$

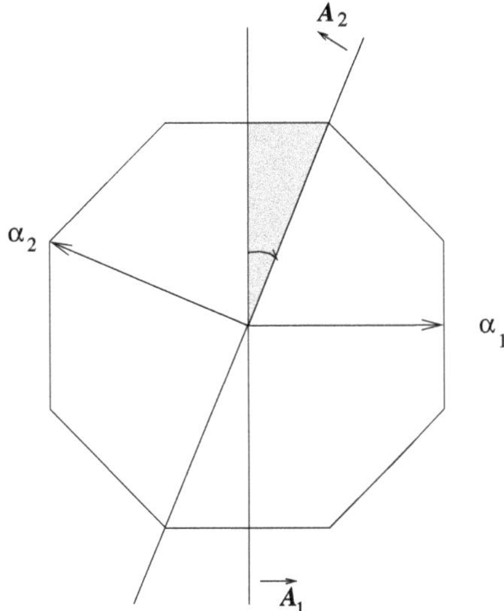

have squared norm 2 and make an angle of $\pi(1 - 1/m)$. The orthogonal reflections ρ_1 and ρ_2 (cf. Exercise 1.9.31) with mirrors $\alpha_1^{\perp} = \{x \in V \mid f(\alpha_1, x) = 0\}$ and $\alpha_2^{\perp}$, respectively, generate the dihedral group G of order $2m$. For, the product $\rho_1\rho_2$ is a rotation with angle $2\pi/m$ and is inverted by both reflections ρ_i. So there is a surjective group homomorphism $\gamma : W \to G$ determined by $\gamma(s_i) = \rho_i$ for $i = 1, 2$, where $S = \{s_1, s_2\}$. In fact, cardinality considerations show that this homomorphism is an isomorphism.

The two open half-planes A_1 and A_2, where $A_i = \{x \in \mathbb{R}^2 \mid f(\alpha_i, x) > 0\}$ meet in the cone (that is, the intersection of a finite number of half-spaces) A_{12} bounded to the left by $\mathbb{R}_{\geq 0}\varepsilon_2$ and to the right by $\mathbb{R}_{\geq 0}(\sin(\pi/m)\varepsilon_1 + \cos(\pi/m)\varepsilon_2)$. These half-lines make an angle of π/m and A_{12} is a prefundamental domain for G.

Here is a characteristic property of the length $l(w)$ of an element of w of W with respect to S that will be of use later.

$$l(s_i w) < l(w) \quad \Leftrightarrow \quad \gamma(w)A_{12} \subseteq \gamma(s_i)A_i. \tag{4.4}$$

It is readily verified by means of a picture like Fig. 4.10 that $l(w)$ is equal to the number of mirrors separating a vector in the interior of wA_{12} from a vector in the interior of A_{12}.

Theorem 4.3.6 *Let V be a real vector space and let $\{H_i \mid i \in I\}$ be a family of affine hyperplanes of $\mathrm{AG}(V)$ of V. For each $i \in I$, let A_i denote one of the two open half-spaces determined by H_i and write $A = \bigcap_{i \in I} A_i$. Assume that $A \neq \emptyset$. Furthermore, for each $i \in I$, let ρ_i be the affine reflection whose mirror in $\mathrm{AG}(V)$ is H_i. Assume that for distinct i, j in I, the intersection $A_{ij} = A_i \cap A_j$ is a prefundamental domain for the subgroup G_{ij} of $\mathrm{AGL}(V)$ generated by ρ_i and ρ_j.*

(i) *The set A is a prefundamental domain for the subgroup G of $\mathrm{AGL}(V)$ generated by $\{\rho_i \mid i \in I\}$.*

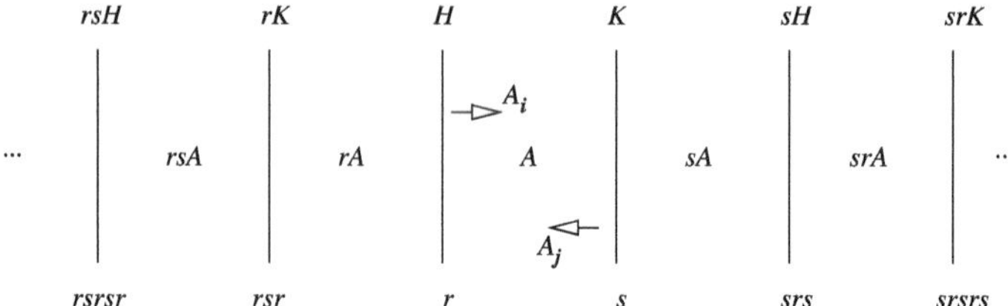

Fig. 4.11 The domains corresponding to the group generated by two affine reflections on the Euclidean plane with parallel mirrors. Observe that $\langle r, s \rangle = \mathrm{Dih}_\infty$

(ii) *The pair $(G, \{\rho_i \mid i \in I\})$ is a Coxeter system of type $M = (m_{i,j})_{i,j \in I}$, where $m_{i,j}$ is the order of $\rho_i \rho_j$. In particular, $G \cong W(M)$.*

Proof Let $M = (m_{i,j})_{i,j \in I}$ be as in (ii). Then M is a Coxeter matrix. Denote by (W, S) the corresponding Coxeter system, where $S = \{s_i \mid i \in I\}$, and by $\gamma : W \to G$ the natural homomorphism mapping $s_i \in S$ onto ρ_i. It is well defined as all defining relations for W are satisfied in G when the s_i are replaced by the ρ_i. The map γ establishes an action of W on $\mathrm{AG}(V)$; see Example 1.8.17. We will often write wX rather than $\gamma(w)X$ if $w \in W$ and $X \subseteq \mathrm{AG}(V)$.

Denote by l the length function on W with respect to S (cf. Remark 4.2.15). Here is a claim, called (P_q), for each $q \in \mathbb{N}$.

For all $i \in I$ and $w \in W$ with $l(w) \leq q$, the domain wA is contained in either A_i or $s_i A_i$; in the latter case, $l(s_i w) = l(w) - 1$.

We show that the truth of (P_q) for all q implies (i) and (ii). As for (i), assume that $w \in W$ satisfies $A \cap wA \neq \emptyset$. Then, for all i, we have $A_i \cap wA \neq \emptyset$ and so, by (P_q) with $q = l(w)$, as $A_i \cap s_i A_i = \emptyset$, even $wA \subseteq A_i$. Hence $wA \subseteq A$. But the assumption on w also implies $w^{-1}A \cap A \neq \emptyset$, so that, similarly, $w^{-1}A \subseteq A$, and hence $A \subseteq wA$. Consequently $A = wA$. Moreover, $s_i wA = s_i A \subseteq s_i A_i$. By (P_q) with $s_i w$ instead of w, so with $q = l(s_i w)$, this gives $l(s_i^2 w) = l(s_i w) - 1$, i.e., $l(s_i w) = l(w) + 1$ for all $i \in I$. This means that w is the identity. Thus (i) holds for W (instead of G) and hence also for G. The argument also shows that the kernel of the homomorphism γ is trivial, so γ is an isomorphism $W \to G$ and $\gamma(S)$ coincides with the set $\{\rho_i \mid i \in I\}$. This establishes (ii).

We next prove (P_q) in four steps, using induction on q.

STEP 1. (P_q) *holds if* $I = \{i, j\}$ *with* $i \neq j$.

We write r instead of ρ_i, s instead of ρ_j, H instead of H_i, and K instead of H_j. Distinguish the cases $H \parallel K$ and $H \nparallel K$. In case $H \parallel K$, the assumption that $A = A_{ij}$ is a prefundamental domain for G_{ij} implies that A is the set of points strictly between H and K. This determines the choice of the half-spaces A_i and A_j uniquely; for instance, in Fig. 4.11 A_i is to the right of H and A_j to the left of K. Now, clearly, $sA \subseteq A_i$, $rA \subseteq A_j$, $srA \subseteq sA_j$, etc., which finishes the proof of (P_q).

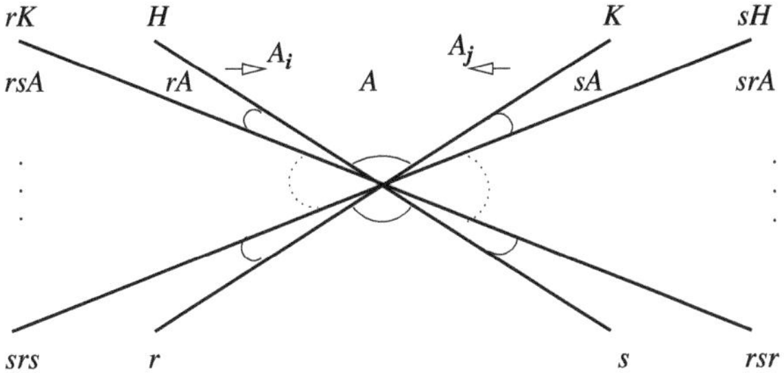

Fig. 4.12 The domains corresponding to the group generated by two affine reflections on the plane with intersecting mirrors. Observe that $\langle r, s \rangle = \mathrm{Dih}_{2m}$, where m is the order of rs

In the case where $H \nparallel K$, the intersection $H \cap K$ is a point (as $\dim(V) = 2$), and the assumption that A is a prefundamental domain forces that rs has finite order. Figure 4.12 shows how to establish (P_q).

STEP 2. *For each pair $i, j \in I$, the group G_{ij} satisfies the hypotheses of the theorem.*

This is obvious as the role of A is taken over by A_{ij}.

In view of Step 1, we have (P_q) for every choice of G_{ij} instead of G. We proceed by induction on q. For $q \leq 1$, the claim (P_q) is trivial.

STEP 3. *Suppose that (P_q) holds for some $q \geq 1$. For each $w \in W$ with $l(w) \leq q$ and $i \in I$, we have $l(s_i w) < l(w)$ if and only if $wA \subseteq s_i A_i$; and also $l(s_i w) > l(w)$ if and only if $wA \subseteq A_i$.*

By Lemma 4.2.16, $l(s_i w) = l(w) \pm 1$. If $l(s_i w) = l(w) - 1$, then $l(s_i(s_i w)) = l(s_i w) + 1$, so (P_{q-1}) gives $s_i w A \subseteq A_i$ and hence $wA \subseteq s_i A_i$. The assertion now follows from (P_q).

STEP 4. (P_q) *implies* (P_{q+1}).

Let $w \in W$ with $l(w) = q + 1$ and take $i \in I$. Choose $j \in I$ and $w' \in W$ such that $w = s_j w'$ and $l(w') = q$. By (P_q) and Step 3, $w'A \subseteq A_j$ and $l(s_j w') = l(w') + 1$. In particular, $wA = s_j w' A \subseteq s_j A_j$. If $j = i$, then we are done.

Suppose $j \neq i$. By Lemma 4.2.20, there exist $u \in \langle s_i, s_j \rangle$ and $v \in {}^{\{i,j\}}W$ such that $w' = uv$ and $l(w') = l_{\{i,j\}}(u) + l(v) = l(u) + l(v)$. Now $v \in {}^{\{i,j\}}W$ implies $l(s_i v) > l(v)$ and $l(s_j v) > l(v)$. Step 3 gives $vA \subseteq A_{ij}$, so $w'A \subseteq uA_{ij}$. Therefore, $wA = s_j w' A \subseteq s_j u A_{ij}$. Notice that $l_{\{i,j\}}(s_j u) \leq l_{\{i,j\}}(u) + 1 \leq l(w') + 1 = q + 1$. By Step 2, (P_{q+1}) holds for $s_j u \in \langle s_i, s_j \rangle$, so $s_j u A_{ij}$ is contained in either A_i or $s_i A_i$, and in the latter case, $l_{\{i,j\}}(s_i s_j u) = l_{\{i,j\}}(s_j u) - 1$. In particular, $wA \subseteq A_i$ or $wA \subseteq s_i A_i$, which is the first statement of (P_{q+1}). If $wA \subseteq s_i A_i$, then $wA \subseteq s_j u A_{ij} \subseteq s_i A_i$, and

$$l(s_i w) = l\big((s_i s_j u)(u^{-1} w')\big) \leq l(s_i s_j u) + l(v)$$

$$\leq l_{\{i,j\}}(s_i s_j u) + l(w') - l_{\{i,j\}}(u) \quad \text{(see above)}$$

$$\leq l_{\{i,j\}}(s_j u) - 1 + q - l_{\{i,j\}}(u) \leq q.$$

As $l(w) = q + 1$, we find $l(s_i w) = q = l(w) - 1$, and so (P_{q+1}) holds. $\square$

Table 4.4 Orders of some finite reflection groups related to polytopes

M	$|W(M)|$	Restriction
A_n	$(n+1)!$	$n \geq 1$
B_n	$2^n n!$	$n \geq 2$
F_4	$24 \cdot 48 = 1152$	
H_3	120	
H_4	$120 \cdot 120 = 14400$	
$I_2^{(m)}$	$2m$	$2 \leq m < \infty$

Corollary 4.3.7 *Retain the conditions of Theorem 4.3.6. Let $\gamma : W(M) \to G$ be the isomorphism found in the theorem. For all $i \in I$ and $w \in W$,*

$$\text{either} \quad \gamma(w)A \subseteq A_i \quad \text{and} \quad l(s_i w) = l(w) + 1,$$

$$\text{or} \quad \gamma(w)A \subseteq \gamma(s_i)A_i \quad \text{and} \quad l(s_i w) = l(w) - 1.$$

Proof This follows from the observation made at the beginning of the proof of the theorem. $\square$

Example 4.3.8 (i) With the notation of Example 4.3.3, the group of 48 isometries of the cube is isomorphic to the Coxeter group of type B_3. Here, A is a cone whose apex is the origin and whose rays are the half-lines starting at the apex and running through the chamber c visualized as a small black triangle in Fig. 4.9.

(ii) Each of the convex regular polytopes of the Euclidean space $\mathbb{E}^n$ gives rise to a group of isometries that is a Coxeter group. As a result, the Coxeter groups of the types represented in Table 4.4 are finite; their orders can be computed by a count of chambers in the same way as in Example 4.3.3(i). The geometries of the series with parameter n appear in Examples 1.2.6 and 1.3.6.

(iii) Each of the tessellations of a Euclidean space by regular convex polytopes obtained in Example 4.1.16 gives rise to an infinite group of isometries which is a Coxeter group. The diagrams of these infinite Coxeter groups are as in Table 4.3. The type $\widetilde{A}_1$ also emerges from the segments into which a horizontal line would be cut by the mirrors in Fig. 4.11.

(iv) Colorings of cells and/or vertices of some regular convex polytopes or tessellations of Euclidean spaces lead to additional finite and infinite Coxeter groups. From the hyperoctahedron or polytope of type B_n, $n \geq 3$, introduced in Example 4.1.13, we derive the finite Coxeter group of type D_n, whose order is $2^{n-1}n!$ (cf. Exercise 4.9.23). In Fig. 4.13 this phenomenon is shown for $n = 3$. The subgroup of the full group $W(B_3)$ of isometries of the octahedron preserving the bicoloring coincides with $W(D_3)$, which is isomorphic to $W(A_3)$ as the two diagrams are the same up to their labellings. From the tessellation of the Euclidean plane by regular triangles we find the infinite Coxeter group of type $\widetilde{A}_2$ (cf. Fig. 2.17). The tiling of the Euclidean space $\mathbb{E}^n$ for $n \geq 2$ by hypercubes and the related type $\widetilde{C}_n$ were discussed in Example 4.1.16. From the tiling of this affine space by bicolored

Fig. 4.13 A coloring of the faces of the octahedron by two colors; the *black* colored faces look *gray* from the inside and the *white* faces are transparent

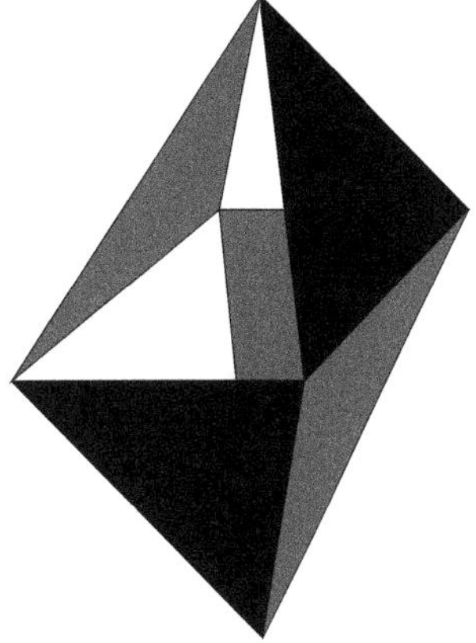

hypercubes, we find the Coxeter group of type

$$\widetilde{B}_n = \;\circ\!\!=\!\!=\!\!=\!\!\circ\!\!-\!\!-\!\!-\!\!\circ\cdots\cdots\circ\!\!-\!\!-\!\!-\!\!\circ\!\!-\!\!-\!\!-\!\!\circ\,. \qquad\qquad (4.5)$$

Compare these results with Examples 2.4.4 and 4.1.10. From the tiling of $\mathbb{E}^n$ with bicolored hypercubes and bicolored vertices, we find the Coxeter groups of type $\widetilde{D}_n$ (if $n \geq 4$) and $\widetilde{A}_3$.

4.4 Linear Reflection Representations

The main result of the previous section, Theorem 4.3.6, will be used to construct a faithful linear representation of each Coxeter group. Its distinguished generators are mapped to reflections.

Definition 4.4.1 Recall that a reflection on a vector space V is defined in Exercise 1.9.30 and that the fixed-point hyperplane of such a reflection (its axis) is called its mirror. A nontrivial eigenvector in the complement of the mirror is called a *root* of the reflection.

This means that the root spans the center of the corresponding perspectivity. The terminology is in accordance with affine reflections in Definition 4.3.1, which become linear reflections once a vector space structure on the affine space is chosen in such a way that its origin lies in the mirror of the affine reflection.

We will make use of quadratic forms. As these will be needed later for arbitrary vector spaces over fields, we give the general definition here.

Definition 4.4.2 Let V be a vector space over a field $\mathbb{F}$ and consider a homogeneous polynomial map κ of degree two on V. That is, there are linear functionals X_i $(i \in I)$ on V and elements $\kappa_{ij} \in \mathbb{F}$ such that $\kappa = \sum_{i,j \in I} \kappa_{ij} X_i X_j$. This means that, for $v \in V$, we have

$$\kappa(v) = \sum_{i,j \in I} \kappa_{ij} X_i(v) X_j(v).$$

Such a map is called a *quadratic form* on V. In particular,

$$\kappa(x\lambda) = \lambda^2 \kappa(x) \quad \text{for all } \lambda \in \mathbb{F}, \ x \in V,$$

$$\kappa(x+y) = \kappa(x) + \kappa(y) + f(x,y) \quad \text{for all } x, y \in V,$$

for a uniquely determined symmetric bilinear form f on V. We call f the *bilinear form* of κ.

Definition 4.4.3 Fix $n \geq 1$ and a Coxeter matrix $M = (m_{i,j})_{i,j \in [n]}$ over $[n]$. Let V be the real vector space with basis $(e_i)_{i \in [n]}$. Denote by f_M, or just f if M is clear from the context, the symmetric bilinear form on V determined by

$$f(e_i, e_j) = -2\cos(\pi / m_{i,j}) \tag{4.6}$$

for $i, j \in [n]$, with the understanding that $f(e_i, e_j) = -2$ if $m_{i,j} = \infty$. The form is indeed symmetric as $m_{i,j} = m_{j,i}$ for $i, j \in [n]$. We call f_M the *bilinear form* associated with M.

Recall from Example 1.4.13 that $\mathrm{Rad}(f)$, the radical of f, is the linear subspace of V consisting of all $x \in V$ such that $f(x, y) = 0$ for all $y \in V$.

Let κ_M, or just κ, be the quadratic form determined by f, i.e.,

$$\kappa(x) = \frac{1}{2} f(x, x)$$

for all $x \in V$. We call κ_M the *quadratic form* associated with M.

For the remainder of this section, we let n, M, V, e_i $(i \in [n])$, f, and κ be as in Definition 4.4.3.

For $x = \sum_i e_i x_i$ we have $\kappa(x) = -\sum_{i,j \in [n]} x_i x_j \cos(\pi / m_{i,j})$. The bilinear form f is linked to κ via

$$\kappa(x+y) = \kappa(x) + \kappa(y) + f(x, y).$$

We use the form f to define orthogonal reflections in $\mathrm{GL}(V)$ with respect to f as in Exercise 1.9.31. Here are some general properties of f and these reflections.

Proposition 4.4.4 *For the symmetric bilinear form f associated with the Coxeter matrix M, and for the linear transformations ρ_i $(i \in [n])$ given by*

$$\rho_i(x) = x - f(x, e_i)e_i \quad (x \in V), \tag{4.7}$$

the following assertions hold.

(i) $f(e_i, e_i) = 2$ *for all $i \in [n]$.*

(ii) $f(e_i, e_j) \leq 0$ *for all $i, j \in [n]$ with $i \neq j$, with equality if and only if $m_{i,j} = 2$.*

(iii) *For each $i \in [n]$, the transformation ρ_i is the orthogonal reflection on V with mirror $e_i^{\perp} := \{x \in V \mid f(x, e_i) = 0\}$ and root e_i.*

(iv) *For each $i \in [n]$, the transformation ρ_i is orthogonal with respect to f, that is, $\rho_i \in O(V, f)$.*

(v) *The order of $\rho_i \rho_j$ equals $m_{i,j}$.*

Proof (i) and (ii) are obvious from the definition of f.

(iii) This follows from Exercise 1.9.30.

(iv) By straightforward computation,

$$
\begin{aligned}
f(\rho_i x, \rho_i y) &= f\big(x - f(x, e_i)e_i, \, y - f(y, e_i)e_i\big) \\
&= f(x, y) - f(x, e_i)f(e_i, y) - f(x, e_i)f(y, e_i) \\
&\quad + f(x, e_i)f(y, e_i)f(e_i, e_i) \\
&= f(x, y).
\end{aligned}
$$

(v) The linear subspace $\mathbb{R}e_i + \mathbb{R}e_j$ of V is invariant under ρ_i and ρ_j. Writing $b = f(e_j, e_i)$ we can express the matrices of these linear transformations on the basis e_i, e_j as

$$
\rho_i : \begin{pmatrix} -1 & -b \\ 0 & 1 \end{pmatrix} \quad \text{and} \quad \rho_j : \begin{pmatrix} 1 & 0 \\ -b & -1 \end{pmatrix},
$$

so $\rho_i \rho_j$ has matrix

$$
\begin{pmatrix} -1 + b^2 & b \\ -b & -1 \end{pmatrix}.
$$

The characteristic polynomial with indeterminate λ of this matrix is $\lambda^2 - (b^2 - 2)\lambda + 1$, which, in view of (4.6), factors as $(\lambda - e^{2\pi i/m_{i,j}})(\lambda - e^{-2\pi i/m_{i,j}})$.

Suppose $m_{i,j} = \infty$. Then the above matrix is not the identity, so $(\lambda - 1)^2$ is the minimal polynomial of the matrix, and so the matrix must be of infinite order. Hence $\rho_i \rho_j$ has infinite order.

Suppose $m_{i,j} < \infty$. The restriction of κ to $\mathbb{R}e_i + \mathbb{R}e_j$ is

$$
\begin{aligned}
\kappa(x_i e_i + x_j e_j) &= x_i^2 - 2x_i x_j \cos(\pi/m_{i,j}) + x_j^2 \\
&= \big(x_i - x_j \cos(\pi/m_{i,j})\big)^2 + x_j^2 \sin^2(\pi/m_{i,j}).
\end{aligned}
$$

This computation shows that κ is positive definite on $\mathbb{R}e_i + \mathbb{R}e_j$, and so $V = (\mathbb{R}e_i + \mathbb{R}e_j) + (e_i^{\perp} \cap e_j^{\perp})$. In view of (iii), the order of $\rho_i \rho_j$ is the order of its restriction to $\mathbb{R}e_i + \mathbb{R}e_j$. The above formula for the characteristic polynomial of this restriction of $\rho_i \rho_j$ shows that its eigenvalues on that subspace are $e^{2\pi i/m_{i,j}}$ and $e^{-2\pi i/m_{i,j}}$, which are primitive $m_{i,j}$-th roots of unity. Therefore, the order of $\rho_i \rho_j$ is equal to $m_{i,j}$. $\square$

Example 4.4.5 For $M = B_3$, connected with the cube, we have

$$f(x, y) = 2x_1 y_1 + 2x_2 y_2 + 2x_3 y_3 - \sqrt{2}x_1 y_2 - \sqrt{2}x_2 y_1 - x_2 y_3 - x_3 y_2$$

with respect to the basis e_1, e_2, e_3. Consequently, f is positive definite (cf. Definition 4.6.1). After a coordinate transformation to an orthonormal basis, ρ_1, ρ_2, and ρ_3 can be seen to be the reflection symmetries of a regular cube. The vectors α_1, α_2, α_3 of Example 4.3.3 are roots of ρ_3, ρ_2, ρ_1, respectively.

Theorem 4.4.6 *Let* (W, S) *be a Coxeter system of type* M.

  (i) *The map* $w \mapsto \rho_w$ *given by* $\rho_w = \rho_1 \cdots \rho_q$ *if* $w = r_1 \cdots r_q$ *with* $r_j \in S$ $(j \in [q])$ *defines a linear representation of* W *on* V *preserving* f.
 (ii) *The map* $[n] \to \{\rho_i \mid i \in [n]\}$ *sending* i *to* ρ_i *is a bijection.*
(iii) *The restriction of* ρ *to the subgroup* $\langle s_i, s_j \rangle$ *of* W *is faithful for all* $i, j \in [n]$.

Proof (i) By Proposition 4.4.4(iii), (v), the subgroup of $\mathrm{GL}(V)$ generated by the ρ_i $(i \in [n])$, satisfies the defining relations of W. It follows that the map $s_i \mapsto \rho_i$ on $S = \{s_1, \ldots, s_n\}$ determines a unique group homomorphism $\rho : W \to \mathrm{GL}(V)$ obeying the equations stated. Finally, ρ preserves f thanks to Proposition 4.4.4(iv).

Assertions (ii) and (iii) follow directly from Proposition 4.4.4(v). $\square$

In principle, the presentation of a Coxeter group W on S might collapse in the sense that $s_i = 1$ or $s_i = s_j$ for certain $i, j \in I$. By Theorem 4.4.6(ii), however, this cannot happen. Consequently, the subset S of W is in bijective correspondence with i. The identification of s_i with $i \in I$ is consistent with the notation $\langle J \rangle$ for $\langle \{s_j \mid j \in J\} \rangle$, where $J \subseteq I$, introduced in Notation 4.2.17.

Definition 4.4.7 If (W, S) is a Coxeter system of type M, the corresponding linear representation $\rho : W \to \mathrm{O}(V, f)$ of Theorem 4.4.6 is called the *reflection representation* of W.

Proposition 4.4.8 *If* M *is an irreducible Coxeter diagram and if* E *is a proper invariant subspace of* V *with respect to the reflection representation* ρ *of* $W(M)$ *on* V, *then* E *is contained in the radical of* f_M.

Proof We claim that $e_i \notin E$ for $i \in [n]$. To see this, set $J = \{i \in [n] \mid e_i \in E\}$. We need to show that $J = \emptyset$. As E is a proper subspace of V, we have $J \neq I$. If $J \neq \emptyset$, then, since $W(M)$ is irreducible, we may assume that there exist $s \in J$, $t \in [n] \setminus J$ with $f(e_s, e_t) \neq 0$. Then $\rho_t e_s = e_s - f(e_s, e_t)e_t$ and so $f(e_s, e_t)e_t = e_s - \rho_t e_s$ is in E. Therefore $e_t \in E$, whence $t \in J$, a contradiction. Hence, $J = \emptyset$, so the claim holds.

Next, let $x \in E$. If $i \in [n]$, then $f(x, e_i)e_i = x - \rho_i x \in E$. But $e_i \notin E$, so $f(x, e_i) = 0$ for all $i \in [n]$. Thus, $x \in V^{\perp}$, which establishes $E \subseteq V^{\perp}$. $\square$

Exercise 4.9.9 shows that an abstract Coxeter group does not uniquely determine its Coxeter diagram.

Definition 4.4.9 In Definition 4.2.21 we introduced the notion of irreducibility for Coxeter groups. In the theory of linear representations, a linear representation of a group G on a vector space is called *irreducible* if there is no proper nontrivial linear subspace of the vector space invariant under G. A linear representation of G on the real vector space V is called *absolutely irreducible* if the representation remains irreducible after extension of the scalars of V to the complex numbers. Often, if there is no imminent confusion between representations, these terms are also applied to the underlying vector space.

Example 4.4.10 Let $M = \widetilde{A}_1 = I_2^{(\infty)}$, so $n = 2$ and $m_{1,2} = \infty$. The form f and the matrices ρ_1 and ρ_2 on the basis e_1, e_2 of the reflection representation ρ of $W(M)$ appear in the proof of Proposition 4.4.4(v) with $i = 1$ and $j = 2$. The vector $e_1 + e_2$ is orthogonal to all of V and spans the linear subspace $\mathrm{Rad}(f)$. In particular, ρ leaves invariant a nontrivial proper linear subspace of V, and so is a *reducible representation*. As $\mathbb{R}e_1 + \mathbb{R}e_2$ is the only 1-dimensional ρ-invariant subspace of V, the space V cannot be decomposed into a direct sum of irreducible ρ-invariant subspaces.

Proposition 4.4.4(ii) tells us that, if M is reducible, the reflection representation ρ of $W(M)$ is reducible. The converse does not hold as we saw in Example 4.4.10. There is however, the following partial converse.

Corollary 4.4.11 *For each Coxeter group W of type M and associated bilinear form f, the following three statements are equivalent.*

 (i) *M is irreducible and $\mathrm{Rad}(f) = 0$.*
 (ii) *The reflection representation of W is irreducible.*
 (iii) *The reflection representation of W is absolutely irreducible.*

Proof As usual, we denote by $\rho : W \to \mathrm{O}(V, f)$ the reflection representation.

(i) $\Rightarrow$ (ii) Suppose (i) and assume that E is a proper ρ-invariant subspace of V. By Proposition 4.4.8, we have $E \subseteq \mathrm{Rad}(f)$. As $\mathrm{Rad}(f) = 0$, we find $E = \{0\}$, as required.

(ii) $\Rightarrow$ (i) It is easily seen that $\mathrm{Rad}(f)$ is a ρ-invariant subspace of V. Clearly, $f \neq 0$, so irreducibility of W on V implies $\mathrm{Rad}(f) = 0$. Let K be a connected component of M. The linear span of $\{e_i \mid i \in K\}$ is clearly a ρ-invariant subspace of V. If ρ is irreducible, then this span must coincide with V, which forces $K = \int\!\!\int n$. Therefore, M is irreducible.

(iii) $\Rightarrow$ (ii) is direct from the definitions.

(ii) $\Rightarrow$ (iii) Suppose that ρ is irreducible. The argument of the proof of Proposition 4.4.8 applies equally well to the vector space V after extension of the field of scalars to a field containing $\mathbb{R}$, showing again that an invariant subspace of the vector space over the extended field lies in the radical of f. Since the radical of f over the extension field has the same dimension as it has over $\mathbb{R}$. The implication (ii) $\Rightarrow$ (i) (already proven) gives us $\mathrm{Rad}(f) = \{0\}$. Hence, also over the extension field, there are no invariant subspaces, so the representation is absolutely irreducible. $\square$

The prefundamental domain needed to apply Theorem 4.3.6 to Coxeter groups will be used for the contragredient representation rather than the reflection representation.

Definition 4.4.12 If $\rho : G \to \mathrm{GL}(V)$ is a linear representation on a vector space V, then the *contragredient* representation $\rho^\vee$ is defined by

$$\rho_g^\vee h = \big(v \mapsto h(\rho_{g^{-1}}v)\big)$$

for all $h \in V^\vee$, $g \in G$.

It is readily checked that $\rho^\vee$ is also a linear representation of G.

For an arbitrary Coxeter system (W, S) of type M and the corresponding reflection representation $\rho : W \to \mathrm{GL}(V)$, we will consider the contragredient representation $\rho^\vee$ (see Definition 4.4.12) to prove that ρ is faithful. This is justified by the following result.

Lemma 4.4.13 *If* $\dim(V) < \infty$, *then* $\mathrm{Ker}(\rho) = \mathrm{Ker}(\rho^\vee)$. *In particular, ρ is faithful if and only if $\rho^\vee$ is faithful.*

Proof If $x \in W$ satisfies $\rho_x = \mathrm{id}$, then $\rho_{x^{-1}} = \mathrm{id}$, so $(\rho_g^\vee h)v = h(\rho_{g^{-1}}v) = h(v)$ for all $v \in V$ and $h \in V^\vee$, proving $\rho_x^\vee = \mathrm{id}$. As $\dim(V) < \infty$, the representations $(\rho^\vee)^\vee$ and ρ are equivalent, so the converse follows immediately. $\square$

Example 4.4.14 We continue with Example 4.4.10, letting (W, S) be the Coxeter system of type $\widetilde{A}_1$, so $m_{1,2} = \infty$. On the basis e_1, e_2 of $\mathbb{R}^2$,

$$f(x, y) = 2x_1 y_1 + 2x_2 y_2 - 2x_1 y_2 - 2x_2 y_1 = 2(x_1 - x_2)(y_1 - y_2),$$

whence $\mathrm{Rad}(f) = \{x \in V \mid x_1 = x_2\}$. Moreover,

$$\rho_1 = \begin{pmatrix} -1 & 2 \\ 0 & 1 \end{pmatrix} \quad \text{and} \quad \rho_2 = \begin{pmatrix} 1 & 0 \\ 2 & -1 \end{pmatrix}$$

fix all points of $\mathrm{Rad}(f)$. There is no convenient choice for A as in Theorem 4.3.6. However, ρ has a contragredient representation $\rho^\vee$ on the dual vector space $V^\vee$ which behaves much better. In matrix form, with respect to a dual basis $(d_i)_i$ of $(e_i)_i$ (that is, $d_i(e_j) = 1$ if $i = j$ and 0 otherwise), we have $\rho_g^\vee = (\rho_g)^{-\top}$ and so

$$\rho_1^\vee = \begin{pmatrix} -1 & 0 \\ 2 & 1 \end{pmatrix} \quad \text{and} \quad \rho_2^\vee = \begin{pmatrix} 1 & 2 \\ 0 & -1 \end{pmatrix}.$$

Consider the set A of elements of $V^\vee$ which take strictly positive values on e_1 and e_2, i.e., the set of all $ad_1 + bd_2 \in \mathbb{R}^2$ with $a > 0$ and $b > 0$. The transforms $\rho_g^\vee(\overline{A})$ of the closure $\overline{A}$ of A, cover an open half-plane of $V^\vee$ bounded by the line $x_1 + x_2 = 0$ and the group acts regularly on the set of all those transforms, i.e., A is a prefundamental domain for this group; see Fig. 4.14.

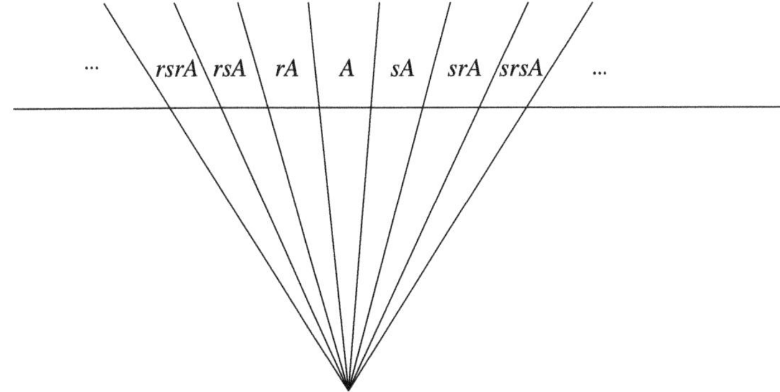

Fig. 4.14 The domain A and its transforms under $\rho^\vee(W)$

Example 4.4.15 We consider once more the case $M = \mathrm{I}_2^{(m)}$ with $m = m_{1,2} < \infty$. The form f is positive definite and the matrices of ρ_1 and ρ_2 in the reflection representation with respect to e_1, e_2 are

$$\text{for } \rho_1 : \begin{pmatrix} -1 & 2\cos(\pi/m) \\ 0 & 1 \end{pmatrix}, \qquad \text{for } \rho_2 : \begin{pmatrix} 1 & 0 \\ 2\cos(\pi/m) & -1 \end{pmatrix}.$$

On the other hand, in the Euclidean vector space $\mathbb{R}^2$ with standard inner product $(\cdot, \cdot)$ and standard orthonormal basis ε_1, ε_2, we can take the vectors $a = \varepsilon_1$ and $b = -\varepsilon_1 \cos(\pi/m) + \varepsilon_2 \sin(\pi/m)$ of unit length which make an angle $-\cos(\pi/m)$ and consider the corresponding orthogonal reflections r_a, r_b. Observe that the two pairs of reflections are actually the same up to a coordinate transformation sending the basis e_1, e_2 to $\sqrt{2}a$, $\sqrt{2}b$.

For $v \in V$, set $A_v = \{x \in V^\vee \mid x(v) > 0\}$. This is an open half-space in $V^\vee$.

Theorem 4.4.16 *Let (W, S) be a Coxeter system of type M and let $\rho : W \to \mathrm{GL}(V)$ be the reflection representation of W. Then, in $V^\vee$, the half-spaces $A_i = \{x \in V^\vee \mid x(e_i) > 0\}$ and the linear transformations $\rho_i^\vee \in \mathrm{GL}(V^\vee)$ satisfy the following conditions.*

(i) *For each $i \in [n]$, the transformation $\rho_i^\vee$ is a reflection on $V^\vee$ and A_i and $\rho_i^\vee(A_i)$ are the half-spaces separated by its mirror.*

(ii) *For distinct i, j in $[n]$, the intersection $A_{ij} = A_i \cap A_j$ is a prefundamental domain for the subgroup $G_{ij}^\vee := \langle \rho_i^\vee, \rho_j^\vee \rangle$ of $\mathrm{GL}(V^\vee)$.*

(iii) *The intersection $\bigcap_{i \in [n]} A_i$ is a prefundamental domain for the subgroup W acting on $V^\vee$.*

(iv) *The representation ρ is faithful.*

Proof (i) Let f be the bilinear form associated with M. For $i \in [n]$, put $a = f(\cdot, e_i)$ and $\phi = (h \mapsto h(e_i))$. According to Definition 4.4.12 and (4.7), we have, for

$h \in V^\vee$,

$$\rho_i^\vee h = h - h(e_i) f(\cdot, e_i) = r_{a,\phi} h,$$

with $\phi(a) = f(e_i, e_i) = 2$, so the assertion follows from Exercise 1.9.30.

(ii) Let $i, j \in [n]$ with $i \neq j$. Observe that $G_{ij}^\vee$ is the image under $\rho^\vee$ of $\langle s_i, s_j \rangle$. Consider $w \in \langle s_i, s_j \rangle$ with $A_{ij} \cap \rho_w^\vee A_{ij} \neq \emptyset$. We will show $w = 1$. Set $U = \mathbb{R} e_i + \mathbb{R} e_j$ in V. There is a canonical surjective linear map $\pi : V^\vee \to U^\vee$ obtained by restriction to U of each linear form on V. The subgroup $G_{ij} = \langle \rho_i, \rho_j \rangle$ of GL(V) leaves U invariant, since each ρ_k leaves invariant every linear subspace of V containing e_k, and $e_i, e_j \in U$. Let σ_w be the restriction of ρ_w to U, for $w \in \langle s_i, s_j \rangle$. Write $K_l = \{x \in U^\vee \mid x(e_l) > 0\}$ for $l \in \{i, j\}$. Then $K_l = \pi(A_l)$. Inspection of Examples 4.3.5 and 4.4.15 gives that the assertion holds if $n = 2$. Therefore, writing $K_{ij} = K_i \cap K_j$, we find that $K_{ij} \cap \sigma_w^\vee K_{ij} \neq \emptyset$ implies $w = 1$.

Now $h \in A_{ij} \cap \rho_w^\vee(A_{ij}) \neq \emptyset$ implies $h|_U \in \pi(A_{ij}) = K_{ij}$ and $h|_U \in \pi \rho_w^\vee(A_{ij}) = \sigma_w^\vee(K_{ij})$. Thus $h|_U \in K_{ij} \cap \sigma_w^\vee(K_{ij})$, so, as we saw in the previous paragraph, $w = 1$.

(iii) The intersection $\bigcap_{i \in [n]} A_i \neq \emptyset$ contains the linear form taking the value 1 on each e_i, and so is non-empty. As we have seen before, the restriction of $\rho^\vee$ to a subgroup $\langle s_i, s_j \rangle$ of W is faithful. In particular, $\rho_i^\vee \rho_j^\vee$ has order $m_{i,j}$. In view of (ii), we can apply Theorem 4.3.6 to conclude that $(\rho^\vee W, \{\rho_i^\vee \mid i \in [n]\})$ is a Coxeter system of type M and that A is a prefundamental domain.

(iv) The proof of (iii) implies that $\rho^\vee$ is faithful. The assertion now follows from Lemma 4.4.13. $\square$

Corollary 4.4.17 *If (W, S) is a Coxeter system and J a subset of S, then the subgroup $\langle J \rangle$ of W is a Coxeter group with Coxeter system $(\langle J \rangle, J)$.*

Proof The restriction of ρ to the subgroup $\langle s_j \mid j \in J \rangle$ of W is a linear representation of ρ on V with invariant subspace $U = \sum_{j \in J} e_j \mathbb{R}$. The reflection representation of the Coxeter system of type $M|_J$ (where M is the Coxeter type of (W, S)) is easily seen to factor through the linear representation of $\langle s_j \mid j \in J \rangle$ induced on U by ρ. Theorem 4.4.16 implies that it is faithful. This shows that the Coxeter group of type $M|_J$ is isomorphic to the subgroup $\langle s_j \mid j \in J \rangle$ of W. The rest of the theorem is an immediate consequence of these observations. $\square$

4.5 Root Systems

In this section, we study Coxeter groups as permutation groups and derive some consequences regarding special subgroups and the defining presentation by generators and relations. To this end, we will single out a union Φ of orbits of roots of reflections from W occurring in the reflection representation space V on which W acts faithfully. This set Φ, called a root system, gives information about the length of an element from W in terms of its action on Φ (see Corollary 4.5.5) and leads to a remarkable property of Coxeter systems, called the exchange condition, which is shown to characterize Coxeter groups in Theorem 4.5.10. In the remainder of the section, a host of useful properties of Coxeter groups are derived. In Corollary 4.5.16, we describe thin residually connected geometries of any Coxeter type M as quotients of a universal geometry of the same Coxeter type M.

Throughout this section, we let M be a Coxeter matrix over $[n]$ and (W, S) a pair consisting of a group W and a generating set $S = \{s_1, \ldots, s_n\}$. The pair need not always (cf. Theorem 4.5.10) be a Coxeter system of type M, although it will often be (explicitly) assumed. We think of S as a totally ordered set (by $s_i < s_j$ if and only if $i < j$) and frequently identify it with $[n]$; in particular, $e_{s_i} = e_i$ for $i \in [n]$.

Definition 4.5.1 Let (W, S) be a Coxeter system and let $\rho : W \to \mathrm{GL}(V)$ be its reflection representation. The basis $e_1, \ldots, e_n$ of the real vector space V consists of roots of the generating reflections $\rho(s_i) = \rho_i$ described in (4.7). Moreover the image of ρ lies in $\mathrm{O}(V, f)$, the stabilizer in $\mathrm{GL}(V)$ of the symmetric bilinear form f associated with M (cf. Definition 4.4.3). In this setting, the subset $\Phi = \bigcup_{s \in S} \rho(W) e_s$ of V is called the *root system* of W. The subsets

$$\Phi^+ = \Phi \cap (\mathbb{R}_{\geq 0} e_1 + \cdots + \mathbb{R}_{\geq 0} e_n) \quad \text{and} \quad \Phi^- = \Phi \cap (\mathbb{R}_{\leq 0} e_1 + \cdots + \mathbb{R}_{\leq 0} e_n)$$

of Φ are called the set of *positive roots* and the set of *negative roots* of W, respectively.

Figure 4.15 shows the root system of the Coxeter group of type $\widetilde{A}_2$. For the action of $w \in W$ on $v \in V$, we often write wv rather than $\rho(w)v$. Part (ii) of the following lemma justifies the name root system; each member of it is a root (cf. Definition 4.4.1).

Proposition 4.5.2 *Let (W, S) be a Coxeter system with root system Φ.*

(i) *W acts faithfully on Φ.*
(ii) *For $w \in W$ and $s \in S$, the vector $we_s \in \Phi$ is a root of the orthogonal reflection $\rho(wsw^{-1})$ with respect to the form f associated with M.*
(iii) *$\Phi = \Phi^+ \dot\cup \Phi^-$ and $\Phi^- = -\Phi^+$.*
(iv) *For $w \in W$ and $s \in S$ we have $we_s \in \Phi^-$ if and only if $l(ws) < l(w)$.*
(v) *If $\lambda \in \mathbb{R}$ and $\alpha \in \Phi$ satisfy $\lambda\alpha \in \Phi$, then $\lambda = \pm 1$.*

Proof (i) As Φ contains the basis $(e_s)_{s \in S}$ of V, this is immediate from the fact (see Theorem 4.4.16) that ρ is faithful.

(ii) Since $\rho(wsw^{-1})$ is a conjugate of $\rho(s)$, it is an orthogonal reflection with respect to f. Moreover $wsw^{-1}(we_s) = wse_s = -we_s$, so we_s is a root of wsw^{-1}.

(iii) Let A_i and A be as in Theorem 4.4.16. In terms of A, we have $\Phi^+ = \{x \in \Phi \mid \forall_{f \in A} f(x) \geq 0\}$ and $\Phi^- = \{x \in \Phi \mid \forall_{f \in A} f(x) \leq 0\}$.

Let $w \in W$ and $s \in S$, so $we_s \in \Phi$. Since $f(we_s) = (w^{-1}f)e_s$ for $f \in V^\vee$, we have $we_s \in \Phi^+$ if and only if $f \in A_s$ for each $f \in w^{-1}A$, which is of course equivalent to $w^{-1}A \subseteq A_s$. Similarly, $we_s \in \Phi^-$ if and only if $w^{-1}A \subseteq sA_s$. By

Fig. 4.15 The root system of
the Coxeter group of type A_2

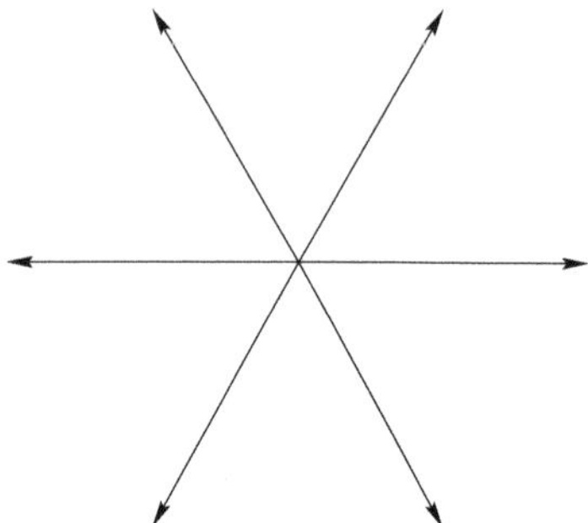

Theorem 4.4.16 and Corollary 4.3.7, either $w^{-1}A \subseteq A_s$ and $l(sw^{-1}) = l(w^{-1}) + 1$ or $w^{-1}A \subseteq sA_s$ and $l(sw^{-1}) = l(w^{-1}) - 1$. Therefore Φ is the union of Φ^+ and Φ^-. As $0 \notin \Phi$ this union is disjoint.

Finally, if $we_s \in \Phi^+$, then $-we_s = wse_s \in \Phi^-$, so $-\Phi^+ \subseteq \Phi^-$. Similarly $-\Phi^- \subseteq \Phi^+$, so $\Phi^- \subseteq -\Phi^+$. Therefore $\Phi^- = -\Phi^+$. This establishes (iii).

(iv) By what we have seen in the proof of (iii), $we_s \in \Phi^-$ if and only if $l(sw^{-1}) = l(w^{-1}) - 1$, which is equivalent to $l(ws) = l(w) - 1$; see Exercise 4.9.14. This proves (iv).

(v) Writing $\alpha = we_s$, we find $2 = f(e_s, e_s) = f(we_s, we_s)$, so $f(\alpha, \alpha) = 2$ for each $\alpha \in \Phi$. Consequently, $\alpha, \lambda\alpha \in \Phi$ gives $\lambda^2 2 = f(\lambda\alpha, \lambda\alpha) = 2$, and so $\lambda = \pm 1$. $\square$

Definition 4.5.3 In view of Proposition 4.5.2(ii), the members of the set $R = \{wsw^{-1} \mid w \in W, s \in S\}$ are called *reflections* of the Coxeter group W or, to be more precise, of the Coxeter system (W, S).

Remark 4.5.4 By Proposition 4.5.2(iii), (v), each reflection of a Coxeter group W has a unique positive root. Conversely, by Exercise 1.9.31, each member of Φ^+ is the positive root of a unique orthogonal reflection with respect to f. In other words, if $r, s \in S$ and $v, w \in W$ satisfy $we_s = ve_r$, then $wsw^{-1} = vrv^{-1}$.

Here is a more general assertion than the length law of Proposition 4.5.2(iv). For $w \in W$ and Φ the root system of W, we set

$$\Phi_w := \left\{ \alpha \in \Phi^+ \mid w\alpha \in \Phi^- \right\}. \tag{4.8}$$

Corollary 4.5.5 *Suppose that (W, S) is a Coxeter system with root system Φ. Let $w \in W$ with $q = l(w)$. If $r_1 \cdots r_q$ is a minimal expression for w, then*

$$\Phi_w = \left\{ r_q \cdots r_{i+1}\alpha_i \mid i \in [q] \right\},$$

where α_i is the root of r_i in Φ^+. In particular, $l(w) = |\Phi_w|$.

Proof Observe that the α_i are in $\{e_1, \ldots, e_n\}$, as each $r_i \in S$. Fix a minimal expression $r_1 \cdots r_q$ of w. As $l((r_q \cdots r_{i+1})r_i) > l(r_q \cdots r_{i+1})$, Proposition 4.5.2(iv) gives $r_q \cdots r_{i+1}\alpha_i \in \Phi^+$. Also, as $l((r_1 \cdots r_{i-1}r_i)r_i) < l(r_1 \cdots r_{i-1}r_i)$, the same result gives $w(r_q \cdots r_{i+1}\alpha_i) = r_1 \cdots r_{i-1}r_i\alpha_i \in \Phi^-$. This establishes $\{r_q \cdots r_{i+1}\alpha_i \mid i \in [q]\} \subseteq \Phi_w$.

Let $\gamma \in \Phi_w$. It remains to show that γ belongs to $\{r_q \cdots r_{i+1}\alpha_i \mid i \in [q]\}$. If $q = 0$, then $\Phi_w = \emptyset$ and the corollary is trivially true.

Suppose $q = 1$, so $w = r_1 = s \in S$. Write $\gamma = \sum_{t \in S} \lambda_t e_t$ with $\lambda_t \geq 0$. If $\gamma \neq \alpha_1$, there must be $u \in S$ distinct from s with $\lambda_u > 0$; cf. Proposition 4.5.2(v). But then $s\gamma = \sum_{t \in S} \lambda_t s e_t = \mu_s e_s + \sum_{t \in S \setminus \{s\}} \lambda_t e_t$ for some $\mu_s \in \mathbb{R}$. The coefficient of e_u in $s\gamma$ is again $\lambda_u > 0$, so $s\gamma \in \Phi^+$ by Proposition 4.5.2(iii). This means $\gamma \notin \Phi_{r_1}$ and so $\Phi_{r_1} = \{\alpha_1\}$, as required.

Assume $q > 1$ and $\gamma \in \Phi_w$. Since $w\gamma \in \Phi^-$ and $\gamma \in \Phi^+$, there is a maximal $i \in [q]$ such that $\gamma, r_q\gamma, \ldots, r_{i+1} \cdots r_q\gamma \in \Phi^+$ and $r_i \cdots r_q\gamma \in \Phi^-$. But then $r_{i+1} \cdots r_q\gamma$ lies in Φ_{r_i}, which we have seen coincides with $\{\alpha_i\}$ in the previous paragraph. Thus, $r_{i+1} \cdots r_q\gamma = \alpha_i$ and $\gamma = r_q \cdots r_{i+1}\alpha_i$. This proves that the two sets Φ_w and $\{r_q \cdots r_{i+1}\alpha_i \mid i \in [q]\}$ coincide.

As a consequence $|\Phi_w| \leq q$. To prove equality, assume, that for certain i, j with $i < j$ we have $r_q \cdots r_{i+1}\alpha_i = r_q \cdots r_{j+1}\alpha_j$. Then $\alpha_i = r_{i+1} \cdots r_j\alpha_j$. As $r_{i+1} \cdots r_j$ is a minimal expression, Proposition 4.5.2(iv) shows that the right hand side is in Φ^-, a contradiction with $\alpha_i \in \Phi^+$. Therefore, all elements of Φ_w are distinct, so $|\Phi_w| = q$. Hence the corollary. $\square$

Example 4.5.6 Let $M = A_{n-1}$. The associated form f satisfies $f(e_i, e_j) = -1$ if $|i - j| = 1$ and $f(e_i, e_j) = 0$ if $|i - j| > 1$. Let $\varepsilon_1, \ldots, \varepsilon_n$ be the standard basis of $\mathbb{R}^n$ and $(\cdot, \cdot)$ the standard inner product that makes the standard basis orthonormal. A convenient realization of the reflection representation is furnished by the vectors $e_i - \varepsilon_i - \varepsilon_{i+1}$ ($i \in [n - 1]$). These span the subspace $V - \mathbb{R}^n \cap \{\varepsilon_1 + \cdots + \varepsilon_n\}^\perp$ of $\mathbb{R}^n$ and f can be identified with the restriction of $(\cdot, \cdot)$ to V. The root system Φ for $W = W(A_{n-1})$ can be partitioned into

$$\Phi^+ = \{\varepsilon_i - \varepsilon_j \mid 1 \leq i < j \leq n\} \quad \text{and}$$

$$\Phi^- = \{\varepsilon_i - \varepsilon_j \mid 1 \leq j < i \leq n\}.$$

Each element in Φ^+ is of the form $\varepsilon_i - \varepsilon_j = e_i + \cdots + e_j$ for certain $1 \leq i < j \leq n$. Corollary 4.5.5 gives that the length of $w \in \mathrm{Sym}_n$ is equal to the number of pairs (i, j) in $[n] \times [n]$ with $i < j$ and $wi > wj$.

We continue by deriving an abstract property of Coxeter groups that is very powerful. Next we explain its power by showing that it is a characterizing property for Coxeter groups.

Definition 4.5.7 Let (W, S) be a pair consisting of a group W and a generating set S for W. The *exchange condition* for (W, S) is the following property.

If $s, r_1, \ldots, r_q \in S$ satisfy $w = r_1 \cdots r_q$ and $q = l(w) \geq l(sw)$, then there is $j \in [q]$ such that $sr_1 \cdots r_{j-1} = r_1 \cdots r_j$.

Theorem 4.5.8 (Exchange Condition) *If (W, S) is a Coxeter system, then it satisfies the exchange condition.*

Proof As the parities of $l(sw)$ and $l(w)$ differ (cf. Lemma 4.2.16), $l(sw) \leq l(w)$ implies $l(sw) = l(w) - 1$. By Corollary 4.5.5 and $|\Phi_{sw}| = l(sw) = q - 1 < q = |\Phi_w|$, there is $\beta \in \Phi_w$ such that $sw\beta \in \Phi^+$. Moreover, if $r_1 \cdots r_q$ is a minimal expression for w, then $\Phi_w = \{r_q \cdots r_{i+1}\alpha_i \mid i \in [q]\}$, where α_i is the root of r_i in Φ^+. Thus, for β, there is $j \in [q]$ with $\beta = r_q \cdots r_{j+1}\alpha_j$. Now $-w\beta \in \Phi^+$ and $s(-w\beta) \in \Phi^-$, so $-w\beta \in \Phi_s = \{\alpha_s\}$, so

$$\alpha_s = -w\beta = -(r_1 \cdots r_q)(r_q \cdots r_{j+1})\alpha_j = -r_1 \cdots r_j \alpha_j = r_1 \cdots r_{j-1}\alpha_j,$$

which implies $s = (r_1 \cdots r_{j-1})r_j(r_{j-1} \cdots r_1)$ by Remark 4.5.4, and establishes $sr_1 \cdots r_{j-1} = r_1 \cdots r_{j-1}r_j$, as required. $\qquad\square$

In order to prove the converse, we need the following useful lemma.

Lemma 4.5.9 *Suppose that (W, S) is a pair consisting of a group W and a set S of involutions generating W that satisfies the exchange condition. Let M be the matrix over S (assuming some total ordering on S) whose r, s-entry $m_{r,s}$ is the order of rs. Suppose that F is a monoid affording a map $\sigma : S \to F$ such that for any two distinct $r, s \in S$ we have*

$$\underbrace{\sigma(r)\sigma(s)\sigma(r) \cdots}_{m_{r,s}} = \underbrace{\sigma(s)\sigma(r)\sigma(s) \cdots}_{m_{r,s}} \quad \text{if } m_{r,s} < \infty.$$

Then σ can be extended to a uniquely determined map $W \to F$, also called σ, such that $\sigma(w) = \sigma(r_1) \cdots \sigma(r_q)$ whenever $r_1 \cdots r_q$ is a minimal expression for w.

Proof Let S^* denote the free monoid on S. Clearly, σ can be extended to a homomorphism of monoids $S^* \to F$. Thus, for $r = r_1 \cdots r_q \in S^*$, we have $\sigma(r) = \sigma(r_1) \cdots \sigma(r_q)$.

For $w \in W$, let D_w be the set of all minimal expressions for w in S^*. We want to show that $\sigma(r) = \sigma(r')$ for all $r, r' \in D_w$. We proceed by induction on $l(w)$. The case $l(w) \leq 1$ is trivial (as then $|D_w| = 1$), so assume $q = l(w) > 1$. Let $r = r_1 \cdots r_q$ and $r' = r_1' \cdots r_q'$ be two minimal expressions in S^* for w. Put $s = r_1'$. We have $l(sw) < q$, so the exchange condition gives $sr_1 \cdots r_{j-1} = r_1 \cdots r_j$ for some $j \leq q$. We obtain $r'' := sr_1 \cdots r_{j-1}r_{j+1} \cdots r_q \in D_w$. Comparing the first terms of r' and r'' and applying the induction hypothesis to sw, we find $\sigma(r') = \sigma(r'')$. If $j < q$, then, comparing the last terms of r'' and r and applying the induction hypothesis to wr_q, we find $\sigma(r) = \sigma(r'')$, and we are done.

It remains to consider the case where $j = q$. Then, replacing the pair r, r' by r'', r, and using the same arguments, we obtain $r''' = r_1 s r_1 \cdots r_{q-2} \in D_w$ with $\sigma(r''') = \sigma(r)$. Repeating this process, we arrive at $u = r_1 s r_1 \cdots$ and $v = s r_1 s \cdots \in D_w$, each word involving only s, r_1 alternately, with $\sigma(u) = \sigma(r)$ and $\sigma(v) = \sigma(r')$, while $w = s r_1 r \cdots = r_1 s r_1 \cdots$ (q terms). Now, by hypothesis, $\sigma(u) = \sigma(v)$, and so

$\sigma(r) = \sigma(r')$. In particular, the map σ is constant on D_w for each $w \in W$, and hence well defined on W. $\qquad\square$

Theorem 4.5.10 *Suppose that W is a group generated by a subset S of involutions.*

(i) *The pair (W, S) is a Coxeter system if and only if it satisfies the exchange condition.*

(ii) *If (W, S) is a Coxeter system, then for each $w \in W$, there is a unique subset S_w of S such that $S_w = \{r_1, \ldots, r_q\}$ for every minimal expression $r_1 \cdots r_q$ for w.*

Proof (i) By Theorem 4.5.8, a Coxeter system satisfies the exchange condition. Suppose that (W, S) satisfies the exchange condition. Let M be the matrix over S as given in Lemma 4.5.9. Denote by $(\overline{W}, \overline{S})$ the Coxeter system of type M. We will apply the lemma to the canonical map $r \mapsto \overline{r}$ from S to $\overline{S}$, taking F to be the monoid underlying the group $\overline{W}$. By definition of $(\overline{W}, \overline{S})$, this map satisfies the conditions of the lemma. Hence we obtain a map $w \mapsto \overline{w}$ from W to $\overline{W}$ such that $\overline{w} = \overline{r}_1 \cdots \overline{r}_q$ whenever $w = r_1 \cdots r_q$ and $q = l(w)$. We claim that $w \mapsto \overline{w}$ is a homomorphism of groups.

First, we show that $\overline{sw} = \overline{s}\,\overline{w}$ for all $s \in S$, $w \in W$. If $l(sw) = q + 1$, we have $\overline{sw} = \overline{s}\,\overline{r}_1 \cdots \overline{r}_q = \overline{s}\,\overline{w}$. If $l(sw) \le q$, Theorem 4.5.8 gives some $j \in [q]$ with $sr_1 \cdots r_{j-1} = r_1 \cdots r_j$, so $sw = r_1 \cdots r_{j-1} r_{j+1} \cdots r_q$, and $l(sw) = q - 1$. As $\overline{r}_j^2 = 1$, we find

$$\overline{sw} = \overline{r}_1 \cdots \overline{r}_{j-1} \overline{r}_{j+1} \cdots \overline{r}_q = \overline{r}_1 \cdots \overline{r}_{j-1} \overline{r}_j^2 \overline{r}_{j+1} \cdots \overline{r}_q$$

$$= \overline{r_1 \cdots r_j}\,\overline{r}_j \overline{r}_{j+1} \cdots \overline{r}_q = \overline{sr_1 \cdots r_{j-1}}\,\overline{r}_j \overline{r}_{j+1} \cdots \overline{r}_q$$

$$= \overline{s}\,\overline{r}_1 \cdots \overline{r}_{j-1} \overline{r}_j \overline{r}_{j+1} \cdots \overline{r}_q = \overline{s}\,\overline{r_1 \cdots r_q} = \overline{s}\,\overline{w}.$$

Next, we derive $\overline{u\,v} = \overline{u}\,\overline{v}$ for all $u, v \in W$ by induction on $l(u)$. The case $l(u) = 1$ has just been treated. Assume $l(u) > 1$. Then $u = s\,u'$ for some $s \in S$, $u' \in W$ with $l(u') < l(u)$, so

$$\overline{u\,v} = \overline{s(u'v)} = \overline{s}\,\overline{u'v} = \overline{s}(\overline{u'}\,\overline{v}) = (\overline{s}\,\overline{u'})\overline{v} = \overline{su'}\,\overline{v} = \overline{u}\,\overline{v},$$

proving that $w \mapsto \overline{w}$ is a homomorphism indeed. Finally the homomorphism is clearly surjective, and, since $\overline{W}$ is freely generated by the relations $(\overline{r}\,\overline{s})^{m_{r,s}} = 1$ $(r, s \in S)$, it must be an isomorphism.

(ii) We now apply the lemma to the map $r \mapsto \{r\}$ from S to the monoid 2^S of all subsets of S in which multiplication is given by set theoretic union (the empty set is the unit). As $\{r\} \cup \{s\} \cup \{r\} \cup \cdots = \{r, s\}$, the equality of Lemma 4.5.9 is satisfied. Therefore, the map can be extended to a uniquely determined map $w \mapsto S_w$ such that $S_w = \{r_1, \ldots, r_q\}$ for every minimal expression $r_1 \cdots r_q$ of w. $\qquad\square$

Recall the map $\zeta : S^* \to W$ of Remark 4.2.15.

Corollary 4.5.11 *Let (W, S) be a Coxeter system. Any sequence r in S^* can be reduced to a minimal element in S^* with the same image as $\zeta(r)$ in W by use of the rewriting rules*

(H) $rsr \cdots \rightsquigarrow srs \cdots$, *where both sides have length $m_{r,s} < \infty$,*
(R) $rr \rightsquigarrow 1$.

In $\zeta^{-1}(\zeta(r))$ such a minimal element is unique up to rewritings of the first kind.

Proof Rules (H) are homogeneous in the sense that the words on both sides have the same lengths. So if we take the quotient of S^* modulo the congruence generated by the corresponding relations on S^*, we obtain a monoid, denoted F, whose elements are equivalence classes $[r]$ of words $r \in S^*$ all of which have the same length. Applying Lemma 4.5.9 to the monoid F, we find a map $\phi : W \to F$ such that $\phi(w) = [r_1 \cdots r_q]$ whenever $r_1 \cdots r_q$ is a minimal expression of $w \in W$.

Suppose now that $r = r_1 \cdots r_q \in S^*$ is an expression for $w = \zeta(r)$ that is not minimal. Let $i \in [q - 1]$ be maximal with $l(r_i r_{i+1} \cdots r_q) < l(r_{i+1} \cdots r_q)$. Such an i exists as $l(w) < q$. Now $l(r_{i+1} \cdots r_q) = q - i$. By Theorem 4.5.8, there is $r_i r_{i+1} \cdots r_j = r_{i+1} \cdots r_{j+1}$ for some $j \in \{i + 1, \ldots, q - 1\}$. As $l(r_i r_{i+1} \cdots r_j) = j - i + 1$, the map ϕ satisfies $[r_i r_{i+1} \cdots r_j] = \phi(r_i r_{i+1} \cdots r_j) = \phi(r_{i+1} \cdots r_{j+1}) = [r_{i+1} \cdots r_{j+1}]$. This means that the two expressions involved can be rewritten to each other by Rule (H). In particular, $r = r_1 \cdots r_q \in S^*$ rewrites to $r_1 \cdots r_{i-1} r_{i+1} r_{i+1} r_{i+2} \cdots r_{j-1} r_{j+1} \cdots r_q$, so we have an occurrence of $r_{i+1} r_{i+1}$, and Rule (R) can be applied to shorten the word representing w. $\qquad\square$

Example 4.5.12 If (W, S) is the Coxeter system of type A_n, then W is doubly transitive on $W/\langle J \rangle$, where $J = \{s_2, \ldots, s_n\}$. For, the conclusion is equivalent to $W = \langle J \rangle \langle s_1 \rangle \langle J \rangle$, while the latter equality is readily derived as follows. For $n = 1$, it is trivial. Suppose $n \geq 2$ and use induction on n. Let $r \in S^*$ be a minimal expression such that $\zeta(r)$ is a minimal counterexample to the equality, i.e., $\zeta(r) \notin \langle J \rangle \langle s_1 \rangle \langle J \rangle$ and $l(r)$ is minimal with this property. Then $r = s_1 r_2 \cdots r_{q-1} s_1$ with $r_i \in J$ ($2 \leq i \leq q - 1$). Moreover $r_2 = r_{q-1} = s_2$. If $q = 3$, then $\zeta(r) = s_1 s_2 s_1 = s_2 s_1 s_2 \in \langle J \rangle s_1 \langle J \rangle$, a contradiction. Hence $q > 3$, and $r_2 \cdots r_{q-1}$ is a minimal expression for an element of $\langle J \rangle$. By induction on n and use of Corollary 4.4.17 we obtain $\zeta(r_2 \cdots r_{q-1}) \in \langle J \setminus \{s_2\} \rangle s_2 \langle J \setminus \{s_2\} \rangle$ so

$$\zeta(r) = \zeta(s_1 r_2 \cdots r_{q-1} s_1) \in s_1 \langle J \setminus \{s_2\} \rangle s_2 \langle J \setminus \{s_2\} \rangle s_1$$

$$= \langle J \setminus \{s_2\} \rangle s_1 s_2 s_1 \langle J \setminus \{s_2\} \rangle \subseteq \langle J \rangle s_1 \langle J \rangle,$$

leading to the final contradiction.

Corollary 4.5.13 *Let (W, S) be a Coxeter system. Suppose that J and K are subsets of S.*

(i) *If $w \in \langle J \rangle$, then $S_w \subseteq J$. In particular, $S \cap \langle J \rangle = J$.*
(ii) *$\langle J \rangle \cap \langle K \rangle = \langle J \cap K \rangle$.*

(iii) $J \subseteq K \iff \langle J \rangle \subseteq \langle K \rangle$.

Proof (i) If $r_1 \cdots r_q$ is a minimal expression for w, then $r_q \cdots r_1$ is a minimal expression for w^{-1}, so, by Theorem 4.5.10(ii), $S_w = S_{w^{-1}}$. Furthermore, by the exchange condition, $S_{rw} \subseteq \{r\} \cup S_w$ for any $r \in S$, so $S_{vw} \subseteq S_v \cup S_w$ for all $v, w \in W$. Hence $\{w \in W \mid S_w \subseteq J\}$ is a subgroup of $\langle J \rangle$ containing J. Consequently, $\langle J \rangle = \{w \in W \mid S_w \subseteq J\}$. Statement (i) follows directly.

   (ii) If $w \in \langle J \rangle \cap \langle K \rangle$, then by (i) the set S_w is contained in $J \cap K$. Hence $w \in \langle J \cap K \rangle$, and $\langle J \rangle \cap \langle K \rangle \subseteq \langle J \cap K \rangle$. The converse inclusion is obvious.

   (iii) This is a direct consequence of (ii). $\square$

Recall the notions left-reduced and right-reduced from Definition 4.2.18.

Definition 4.5.14 Let (W, S) be a Coxeter system and let $J, K \subseteq S$. Then ${}^J W^K :=$ ${}^J W \cap W^K$ is called the set of (J, K)-*reduced* elements of W.

The result below implies that ${}^J W^K$ is a set of double coset representatives with respect to the subgroups $\langle J \rangle$ and $\langle K \rangle$ of W.

Proposition 4.5.15 *Let (W, S) be a Coxeter system and let J, K, L be subsets of S.*

(i) *If $w \in \langle K \rangle \langle L \rangle$ then there are $x \in \langle K \rangle$ and $y \in \langle L \rangle$ such that $w = xy$ and $l(w) = l(x) + l(y)$.*
(ii) $\langle J \rangle \cap \langle K \rangle \langle L \rangle = (\langle J \rangle \cap \langle K \rangle)(\langle J \rangle \cap \langle L \rangle)$.
(iii) *For each $w \in W$ there is a unique $d \in {}^J W^K \cap \langle J \rangle w \langle K \rangle$ of minimal length. The set ${}^J W^K$ consists of all such d for $w \in W$.*
(iv) *For $w \in W$ and $d \in {}^J W^K \cap \langle J \rangle w \langle K \rangle$, each $u \in \langle J \rangle w \langle K \rangle$ can be written as $u = xdy$ with $x \in \langle J \rangle$, $y \in \langle K \rangle$ and $l(u) = l(x) + l(d) + l(y)$.*

Proof (i) Suppose $w = xy$ for certain $x \in \langle K \rangle$ and $y \in \langle L \rangle$. Obviously, $l(x) + l(y) \geq l(w)$. Let $r_1 \cdots r_t$ and $r_{t+1} \cdots r_q$ be minimal expressions for x and y, respectively, so $q = l(x) + l(y)$ and $l(x) = t$. If $q = l(w)$, we are done. Otherwise, take the largest $j \leq t$ such that $r_j \cdots r_q$ is not a minimal expression. Then, by Theorem 4.5.8 there is $k \in \{j+1, \ldots, q\}$ such that $r_j \cdots r_q = r_{j+1} \cdots r_{k-1} r_{k+1} \cdots r_q$. If $k \leq t$, then $x = r_1 \cdots r_{j-1} r_{j+1} \cdots r_{k-1} r_{k+1} \cdots r_t$, contradicting $l(x) = t$. Therefore $k > t$. Take $x_1 = r_1 \cdots r_{j-1} r_{j+1} \cdots r_t$ and $y_1 = r_{t+1} \cdots r_{k-1} r_{k+1} \cdots r_q$. Then $l(x_1) + l(y_1) = q - 2$, $x_1 \in \langle K \rangle$, $y_1 \in \langle L \rangle$ whereas $w = xy = x_1 y_1$, and we finish by induction on $l(x) + l(y)$.

   (ii) Clearly, $(\langle J \rangle \cap \langle K \rangle)(\langle J \rangle \cap \langle L \rangle) \subseteq \langle J \rangle \cap \langle K \rangle \langle L \rangle$. As for the converse, let $w \in \langle J \rangle \cap \langle K \rangle \langle L \rangle$. Then, by (i), there are $x \in \langle K \rangle$, $y \in \langle L \rangle$ with $w = xy$ and $l(x) + l(y) = l(w)$, so if $r_1 \cdots r_t$ and $r_{t+1} \cdots r_q$ are minimal expressions for x and y, respectively, then $r_1 \cdots r_q$ is a minimal expression for w. As $w \in \langle J \rangle$, Corollary 4.5.13(i) gives $r_i \in J$ for $1 \leq i \leq q$ so $x = r_1 \cdots r_t \in \langle J \rangle$ and similarly for y. This establishes $w \in (\langle J \rangle \cap \langle K \rangle)(\langle J \rangle \cap \langle L \rangle)$.

(iii) Let d be an element of minimal length in $\langle J \rangle w \langle K \rangle$. Then, clearly, $\langle J \rangle w \langle K \rangle = \langle J \rangle d \langle K \rangle$. Suppose $u \in \langle J \rangle d \langle K \rangle$. Then by the same kind of reasoning as in (i), we obtain that $u = xdy$ with $x \in \langle J \rangle$ and $y \in \langle K \rangle$ such that $l(u) = l(x) + l(d) + l(y)$. It follows that if $l(u) = l(d)$, then $u = d$. This settles uniqueness of d. Also, by minimality of $l(d)$, we have $d \in {}^J W^K$. Conversely, assume $v \in {}^J W^K$. Let d be the element of minimal length in $\langle J \rangle v \langle K \rangle$. Then $v = xdy$ for certain $w \in \langle J \rangle$, $y \in \langle K \rangle$ with $l(v) = l(x) + l(d) + l(y)$. If $l(x) > 0$, then there is $r \in J$ such that $l(rx) < l(x)$. But then $l(rv) < l(v)$ contradicting $v \in {}^J W^K$. Hence $l(x) = 0$. Similarly, $l(y) = 0$. Consequently $v = d$, proving that ${}^J W^K$ is the set of all elements of minimal length in the double cosets they represent.

(iv) This follows from the proof of (iii). $\qquad\square$

Corollary 4.5.16 *Let (W, S) be a Coxeter system of type M over I.*

(i) *The chamber system $\mathcal{C}(M)$ is thin and residually connected of type M with automorphism group W.*

(ii) *If Γ is a thin I-geometry of Coxeter type M, and A is the stabilizer in W of a chamber in the right action of W on the set of chambers of Γ, then $\Gamma \cong \Gamma(\mathcal{C}(M))/A$.*

Proof (i) By Corollary 4.4.17 and Lemma 4.1.1, $\mathcal{C}(M)$ is a thin chamber system of type M. To derive that it is residually connected, note that $\mathcal{C}(M) \cong \mathcal{A}(W, (\langle s \rangle)_{s \in S}, 1)$ by Corollary 4.2.11, and combine Proposition 4.5.15(ii) with Theorem 3.6.9. The part about the automorphism group follows from Proposition 4.2.10.

(ii) By Theorem 3.4.6 there is a thin residually connected chamber system $\mathcal{C}$ of type M such that $\Gamma = \Gamma(\mathcal{C})$. Theorem 4.2.8 yields that $\mathcal{C} \cong \mathcal{C}(M)/A$ for some subgroup A of W. According to Corollary 3.4.7, the subgroup A also acts on $\Gamma(\mathcal{C}(M))$ and $\Gamma(\mathcal{C}(M)/A) \cong \Gamma(\mathcal{C}(M))/A$. $\qquad\square$

Remark 4.5.17 As a consequence of Corollary 4.5.16, thin chamber systems of type M can be studied as quotients of $\mathcal{C}(M)$. For a quotient $\mathcal{C}(M)/A$, where A is a subgroup of the Coxeter group W, to be a chamber system of type M it is necessary and sufficient that $wAw^{-1} \cap \langle \{i, j\} \rangle = 1$ for all $w \in W$ and $i, j \in S$. Similarly, in view of Lemma 3.4.9, this quotient is residually connected if and only if $\langle J \rangle \langle K \rangle \cap wAw^{-1}\langle L \rangle \subseteq wAw^{-1}\langle J \cap L \rangle \langle K \cap L \rangle$ for all $w \in W$ and $J, K, L \subseteq S$.

The following lemma is recorded for later use in Chap. 11. Here, we write ${}^d X$ for dXd^{-1} when $d \in W$ and $X \subseteq W$. Moreover, S_d is as in Theorem 4.5.10. We will also use the following notation.

Notation 4.5.18 Let (W, S) be a Coxeter system. For $r \in S$ and $X \subseteq S$, we write $r^\perp := \{s \in S \mid sr = rs\} = \{s \in S \mid m_{r,s} \leq 2\}$ and $X^\perp := \bigcap_{s \in X} s^\perp$.

Lemma 4.5.19 *Let (W, S) be a Coxeter system. Suppose that K, L are subsets of S and $d \in {}^L W^K$. Let $r \in S \setminus L$ and write $L_i = \{s \in L \mid m_{r,s} = i\}$ for $i \geq 2$.*

(i) $\langle L \rangle \cap {}^d \langle K \rangle = \langle L \cap {}^d K \rangle$.
(ii) $\langle L \rangle \cap {}^d \langle K \rangle \subseteq \langle S_d \cup S_d^\perp \rangle = \langle S_d \rangle \langle S_d^\perp \rangle$.
(iii) $\langle L \rangle \cap {}^r \langle L \rangle = \langle L_2 \rangle = \langle L \cap r^\perp \rangle$.
(iv) $\langle L \rangle \cap r \langle L \rangle r \langle L \rangle r = \langle L_2 \rangle L_3 \langle L_2 \rangle$.

Proof (i) Clearly, $\langle L \cap {}^d K \rangle \subseteq \langle L \rangle \cap {}^d \langle K \rangle$. For the opposite inclusion, suppose that $w \in \langle L \rangle \cap {}^d \langle K \rangle$. If $l(w) = 0$, then $w = 1 \in \langle L \cap {}^d K \rangle$. We shall reason by induction on $l(w)$. Assume $l(w) \geq 1$. Then there is $s \in L$ with $l(sw) < l(w)$. Since $w \in {}^d \langle K \rangle$ there is $v \in \langle K \rangle$ with $dv = wd$. Now, since $d \in {}^L W^K$, we have

$$l(s\,d\,v) = l(s\,w\,d) = l(s\,w) + l(d) < l(w) + l(d) = l(w\,d) = l(d\,v),$$

so in view of the exchange condition we have $s\,d\,v \in d \langle K \rangle$. In particular, $d^{-1}s\,d \in \langle K \rangle v^{-1}$. Again since $d \in {}^L W^K$, this yields

$$1 + l(d) \geq l(sd) = l\big(d\big(d^{-1}sd\big)\big) = l(d) + l\big(d^{-1}sd\big).$$

Hence $l(d^{-1}sd) = 1$. Now $d^{-1}sd \in \langle K \rangle \cap S$, so it follows from Corollary 4.5.13(i) that $d^{-1}sd \in K$. Consequently $s \in \langle L \cap {}^d K \rangle$. Finally, $sw \in \langle L \rangle \cap {}^d \langle K \rangle$, and $l(sw) < l(w)$ so by induction, $sw \in \langle L \cap {}^d K \rangle$, whence $w = s(sw) \in \langle L \cap {}^d K \rangle$. This proves (i).

(ii) According to (i), the set $L \cap {}^d K$ generates the subgroup $\langle L \rangle \cap {}^d \langle K \rangle$, so, for the proof of the inclusion of (ii), it suffices to show $L \cap {}^d K \subseteq S_d \cup S_d^\perp$. Let $t \in L \cap {}^d K$. We will show $t \in S_d \cup S_d^\perp$. By assumption on t, there is $s \in K$ such that $td = ds$. As $t \in L$, $s \in K$ and $d \in {}^L W^K$, we have $l(td) = l(ds) = 1 + l(d)$, whence $\{t\} \cup S_d = S_{td} = S_{ds} = \{s\} \cup S_d$; see Theorem 4.5.10(ii). If $t \in S_d$ we are done, so assume $t \not\subseteq S_d$. Then $s = t$, so $d = tdt^{-1} \in \langle S_d \rangle \cap {}^t \langle S_d \rangle = \langle S_d \cap {}^t S_d \rangle$ in view of (i) and $t \in {}^{S_d} W^{S_d}$. But then $S_d \subseteq {}^t S_d$. Let $r \in S_d$. Then there is $s_1 \in S_d$ with $r = ts_1 t$. Now $r \neq t$ (for otherwise $t = s_1 \in S_d$, a contradiction), so $\{r\} = S_r = S_{ts_1 t} \setminus \{t\} = \{s_1\}$. It follows that $r = s_1$ and $r = trt$, whence $t \in r^\perp$. The conclusion is that t belongs to $\bigcap_{r \in S_d} r^\perp = S_d^\perp$. This proves the inclusion. Since the equality in (ii) is a trivial consequence of the fact that S_d and $S_d^\perp$ commute, this ends the proof of (ii).

(iii) By (i) and (ii) we have $\langle L \rangle \cap {}^r \langle L \rangle = \langle L \cap {}^r L \rangle \cap (\langle r \rangle \langle r^\perp \rangle)$, so, by Proposition 4.5.15(ii),

$$\langle L \rangle \cap {}^r \langle L \rangle = \big(L \cap {}^r L \cap \{r\} \big)\big(L \cap {}^r L \cap r^\perp \big) = \big(L \cap r^\perp \big).$$

(iv) Set $F = L_2$. Since $\langle F \rangle \subseteq \langle L \rangle$, it suffices to show $r \langle L \rangle r \langle L \rangle r \cap {}^F W^F = L_3$. Since, for $s \in L_3$, we have $s = rsrsr$, the right hand side is contained in the left hand side. As for the other inclusion, suppose $w \in r \langle L \rangle r \langle L \rangle r \cap {}^F W^F$. Then rwr is a minimal expression, for otherwise, by the Exchange Condition 4.5.8, $w = rwr \in \langle L \rangle r \langle L \rangle$, a contradiction. On the other hand, as $rwr \in \langle L \rangle r \langle L \rangle$, there are $w_1, w_2 \in \langle L \rangle$ such that $rwr = w_1 r w_2$ is a minimal expression. By commuting factors with r, we may assume, without loss of generality, that $w_1 \in W^F$. Consider $wr = rw_1 r w_2$. Now $l(rw_1 r w_2) = l(w_1 r w_2) - 1$, so by the exchange condition, there must be $v_1, v_2 \in \langle L \rangle$ and $s \in L$ with $v_1 s v_2$ a minimal expression for w_2

and $rw_1rw_2 = w_1rv_1v_2$. Hence $w_1^{-1}w = rv_1v_2r$ belongs to $\langle L\rangle \cap r\langle L\rangle r$, which by (i) coincides with $\langle F\rangle$. Now $w_1 \in w\langle F\rangle \cap W^F = \{w\}$, so $w_1 = w$. This implies $v_1v_2 = 1$, and, as v_1v_2 is a minimal expression, even $v_1 = v_2 = 1$. Consequently, $w_2 = s$ and we find $wr = rwrs$, which rewrites to $w^{-1}rw = rsr$. By the exchange condition, we find $w \in \langle F\rangle s \cap {}^F W^F = \{s\}$, proving $w = s$. The equation now reads $srs = rsr$, so $w = s \in L_3$, as required. $\qquad\square$

The lemma shows that the stabilizer of the two flags $x\langle J\rangle$, $y\langle K\rangle$ of $\Gamma(W)$ of type $S \setminus J$, $S \setminus K$, respectively, can be determined as follows. Let $d \in {}^J W^K$ be in the double coset of $x^{-1}y$. Then the stabilizer in $W = \mathrm{Aut}(\Gamma(W))$ of $x\langle J\rangle$ and $y\langle K\rangle$ is

$$^x\langle J\rangle \cap {}^y\langle K\rangle = {}^x\big(\langle J\rangle \cap {}^d\langle K\rangle\big) = {}^x\big\langle J \cap {}^d K\big\rangle.$$

Proposition 4.5.20 *Let (W, S) be a Coxeter system with set of reflections R. For $t \in R$ and $w \in W$, the following three statements are equivalent.*

(i) $l(tw) \leq l(w)$.

(ii) $\alpha_t \in \Phi_{w^{-1}}$.

(iii) *If $r_1 \cdots r_q$ is an expression for w (not necessarily minimal), then there is $i \in [q]$ such that $tr_1 \cdots r_i = r_1 \cdots r_{i-1}$.*

Proof (ii) $\Rightarrow$ (iii) Suppose that $w = r_1 \cdots r_q$ with $r_i \in S$. Let $\alpha_t \in \Phi_{w^{-1}}$. Then $\alpha_t \in \Phi^+$ and $w^{-1}\alpha_t \in \Phi^-$. There is $i \in [q]$ with $r_{i-1} \cdots r_1\alpha_t \in \Phi^+$ and $r_i \cdots r_1\alpha_t \in \Phi^-$. By Corollary 4.5.5, $r_{i-1} \cdots r_1\alpha_t \in \Phi_{r_i} = \{\alpha_i\}$, where $\alpha_i = \alpha_{r_i}$, so, by Remark 4.5.4, $r_{i-1} \cdots r_1tr_1 \cdots r_{i-1} = r_i$, whence $tr_1 \cdots r_{i-1} = r_1 \cdots r_{i-1}r_i$, proving (iii).

(iii) $\Rightarrow$ (i) Take $q = l(w)$ in (iii), so $r_1 \cdots r_{i-1}r_{i+1} \cdots r_q$ is an expression for tw for some $i \in [q]$. The length of this expression is $q - 1$.

(i) $\Rightarrow$ (ii) The two implications just proved show that each $w \in W$ and $t \in S$ with $w^{-1}\alpha_t \in \Phi^-$ satisfy $l(tw) \leq l(w)$. Suppose now that (ii) does not hold; then $w^{-1}\alpha_t \in \Phi^+$. As $t^2 = 1$, we then have $(tw)^{-1}\alpha_t \in \Phi^-$, so, replacing w by tw in the conclusion of the first sentence of this paragraph, we find $l(t(tw)) \leq l(tw)$, proving $l(tw) \geq l(w)$, which means that (i) does not hold (equality does not occur in view of the difference in parity between $l(w)$ and $l(tw)$). This establishes the required implication. $\qquad\square$

Remark 4.5.21 Thanks to Proposition 4.5.20, the roots of a Coxeter system (W, S) of type M can be fully described in terms of the thin chamber system $\mathcal{C}(M)$. The technical details are in Exercise 4.9.22. A reflection $t \in R$ determines the set of panels $\{v, vs\}$ with $s \in S$ and $v \in W$ such that $t = vsv^{-1}$, which is known as a wall. Such panels are on the boundary of each of the two roots in $\mathcal{C}(M)$ associated with t. The set of eight vertical panels, drawn parallel in the middle Fig. 3.2, is an example of a wall in $\mathcal{C}(B_3)$. Two chambers c, d of $\mathcal{C}(M)$ each lie in a different root corresponding to t if and only if every gallery from c to d in $\mathcal{C}(M)$ contains a panel from the wall of t.

Corollary 4.5.22 *Let (W, S) be a Coxeter system with set of reflections R. If $t \in R$ and $v, w \in W$ satisfy $l(tw) \leq l(w)$ and $l(tv) \leq l(v)$, then $l(v^{-1}w) < l(v^{-1}tw)$.*

Proof By Proposition 4.5.20, for t, v, and w as in the hypotheses, we have $\alpha_t \in \Phi_{w^{-1}} \cap \Phi_{v^{-1}}$. Consequently, $-v^{-1}\alpha_t$ is the positive root of the reflection $v^{-1}tv$, so $-v^{-1}\alpha_t = \alpha_{v^{-1}tv}$, and

$$\left(w^{-1}v\right)\alpha_{v^{-1}tv} = -\left(w^{-1}v\right)v^{-1}\alpha_t = -w^{-1}\alpha_t \in \Phi^+.$$

Proposition 4.5.20 gives $l((v^{-1}tv)(w^{-1}v)^{-1}) \geq l((w^{-1}v)^{-1})$, which is equivalent to $l(v^{-1}tw) \geq l(v^{-1}w)$. $\qquad\square$

Remark 4.5.23 Let (W, S) be a pair consisting of a group W and a generating set S for W. The *strong exchange condition* for (W, S) is the following property, where $R = S^W = \{w^{-1}sw \mid w \in W, s \in S\}$.

> If $t \in R$ and $r_1, \ldots, r_q \in S$ satisfy $l(tw) \leq l(w)$ and $w = r_1 \cdots r_q$, then there is $j \in [q]$ such that $tr_1 \cdots r_{j-1} = r_1 \cdots r_j$.

By Proposition 4.5.20 this condition holds in a Coxeter system. But the condition is stronger than the exchange condition of Definition 4.5.7 as q is no longer required to be the length of w and t varies over all reflections of (W, S) (cf. Definition 4.5.3) rather than S. Hence, in view of Theorem 4.5.10, the strong exchange condition also characterizes Coxeter groups.

4.6 Finiteness Criteria

Let V be a real vector space of finite dimension n. The purpose of this section is to establish that every finite subgroup G of $GL(V)$ generated by reflections is a Coxeter group. The approach is to find a set S of reflections in G such that (G, S) is a Coxeter system. Following this result, which is Theorem 4.6.4, we give some criteria (in Theorem 4.6.6) for a Coxeter group to be finite. Finally, the study of the longest element in a finite Coxeter group leads to a polarity on the thin geometry of some of these groups (see Remark 4.6.11).

We recall some elementary facts on finite subgroups of $GL(V)$.

Definition 4.6.1 A symmetric bilinear form f on a real vector space V is called *positive definite* if its corresponding quadratic form is positive definite, that is, $f(x, x) \geq 0$ for all $x \in V$ with equality only if $x = 0$.

Below we will use the notion of homothety as defined in Exercise 2.8.28.

Lemma 4.6.2 *Let $\rho : G \to GL(V)$ be a linear representation of a finite group G on a finite-dimensional real vector space V. There is a positive-definite symmetric bilinear form f on V that is invariant under G. If, moreover, ρ is absolutely irreducible, then*

(i) *each linear map $V \to V$ commuting with $\rho(G)$ is a homothety;*
(ii) *the form f is the unique G-invariant bilinear form on V up to positive scalar multiples.*

Proof As before, for the action of an element $g \in G$ on a vector $v \in V$, we will often suppress ρ and write gv rather than $\rho(g)v$. Take any positive-definite symmetric bilinear form f_1 on V and consider the sum f over all of its transforms by elements of G:

$$f(x, y) := \sum_{g \in G} f_1(gx, gy).$$

Then f is a positive definite symmetric bilinear form on V that is invariant under G.

(i) Suppose that A is a linear map $V \to V$ commuting with $\rho(G)$, and take an eigenvalue $\lambda \in \mathbb{C}$ of A. Then, after extension of the field of scalars to $\mathbb{C}$, we obtain the G-invariant subspace $\mathrm{Ker}(A - \lambda\,\mathrm{id}_V)$ of V. As A has an eigenvector with respect to λ, this subspace is nontrivial. Since G is absolutely irreducible, the subspace must be all of V, whence $A = \lambda\,\mathrm{id}_V$. This establishes (i).

(ii) Suppose that f_2 is yet another G-invariant bilinear form. Since V is finite dimensional and f is nondegenerate, any element of $V^\vee$ is of the form $y \mapsto f(z, y)$ for a unique $z \in V$. In particular, for each $x \in V$, there is a unique vector $u(x)$ in V such that $f_2(x, y) = f(u(x), y)$ for each $y \in V$. It is readily seen that $u : V \to V$ is a linear map. Since f_2 and f are invariant under G, the transformation u commutes with each member of G. Indeed,

$$f\big(u(gx), y\big) = f_2(gx, y) = f_2\big(x, g^{-1}y\big) = f\big(ux, g^{-1}y\big) = f\big(g(ux), y\big)$$

for all $x, y \in V$. Since f is nondegenerate, $ug(x) = gu(x)$ for all $x \in V$. By (i), u is a homothety, say $u = \alpha \cdot \mathrm{id}$, with $\alpha \in \mathbb{R}$. Then $f_2 = \alpha f$, as required. $\square$

The following lemma covers part of the 2-dimensional case of the general result stated in Theorem 4.6.4. It uses the notion of root introduced in Definition 4.4.1.

Lemma 4.6.3 *Let V be a real 2-dimensional vector space supplied with a positive-definite symmetric bilinear form f and let G be a finite subgroup of $\mathrm{O}(V, f)$ generated by two reflections with distinct mirrors. Write Φ for the set of roots $\alpha \in V$ of reflections in G with $f(\alpha, \alpha) = 2$. Suppose that $h \in V$ is a vector such that $f(h, \alpha) \neq 0$ for all roots $\alpha \in \Phi$. There is a unique pair $\alpha, \beta \in \Phi$ such that $f(h, \alpha) > 0$, $f(h, \beta) > 0$ and each root $\gamma \in \Phi$ with $f(h, \gamma) \geq 0$ is a linear combination with non-negative coefficients of α and β. Moreover $f(\alpha, \beta) = 2\cos(\pi(1 - 1/m)) = -2\cos(\pi/m)$ for some $m \in \mathbb{N}$.*

Proof Let $\alpha, \beta \in \Phi$ be such that the orthogonal reflections r_α and r_β with respect to f in the roots α and β, respectively, generate G. Then G is isomorphic to Dih_{2m}, where m is the order of $r_\alpha r_\beta$ and we recover the setting of Example 4.3.5. The result now follows from a comparison of $G\alpha \cup G\beta$ with the 2-dimensional root system of Dih_{2m}. $\square$

Theorem 4.6.4 *Let V be a real vector space of dimension $n < \infty$ and G a finite subgroup of $\mathrm{GL}(V)$ generated by reflections.*

(i) *There is a set Δ of linearly independent roots of reflections in G such that each reflection in G has a root that is a linear combination with nonnegative coefficients of roots from Δ.*
(ii) *There is a set S of $|\Delta|$ reflections in G, each having a root from Δ as in (i), such that (G, S) is a Coxeter system.*

Proof By Lemma 4.6.2 there is a positive definite symmetric bilinear form f such that G is a subgroup of $\mathrm{O}(V, f)$. In particular, each reflection of G is an orthogonal reflection with respect to f. Let Φ be the set of roots α of these reflections with $f(\alpha, \alpha) = 2$. Then Φ is finite as G is finite. Consequently, the union of all mirrors of reflections in G does not cover V; in other words, there is a vector h in V not contained in any mirror of G. Let Φ^+ be the intersection of Φ with the open half-space $\{x \in V \mid f(h, x) > 0\}$.

Take Δ to be the set of roots in Φ^+ that cannot be written as a linear combination with positive coefficients of at least two elements of Φ^+. Let $\alpha, \beta \in \Delta$. By Lemma 4.6.3 we find $f(\alpha, \beta) = -2\cos(\pi/m_{\alpha,\beta})$ for some $m_{\alpha,\beta} \in \mathbb{N}$. Consequently, the conditions of Theorem 4.3.6 are satisfied for the subgroup W of G generated by $S = \{r_\alpha \mid \alpha \in \Delta\}$, with $A_s = \{x \in V \mid f(e_s, x) \geq 0\}$ for each $s \in S$. This proves (ii) provided we show $W = G$.

(i) We prove that the roots in Δ are linearly independent. Suppose that $\sum_{\alpha \in \Delta} \lambda_\alpha \alpha = 0$ for certain $\lambda_\alpha \in \mathbb{R}$. Put $\Sigma = \{\alpha \in \Delta \mid \lambda_\alpha > 0\}$ and $\Pi = \{\beta \in \Delta \mid \lambda_\beta < 0\}$. Then $v := \sum_{\alpha \in \Sigma} \lambda_\alpha \alpha = \sum_{\beta \in \Pi}(-\lambda_\beta)\beta$ satisfies $0 \leq f(v, v) = -\sum_{\alpha \in \Sigma, \beta \in \Pi} \lambda_\alpha \lambda_\beta f(\alpha, \beta) \leq 0$, so $f(v, v) = 0$, and hence $v = 0$. Now $0 = f(h, v) = \sum_{\alpha \in \Sigma} \lambda_\alpha f(h, \alpha) \geq 0$, so $\Sigma = \emptyset$, and similarly $\Pi = \emptyset$. Therefore, $\lambda_\alpha = 0$ for $\alpha \in \Delta$, which establishes that the roots in Δ are linearly independent. This completes the proof of (i).

(ii) By construction, $W \subseteq G$. Let $\gamma \in \Phi^+$. In order to establish $G \subseteq W$, it suffices to show $\gamma = w\alpha$ for some $\alpha \in \Delta$ and $w \in W$, for then $r_\gamma = wr_\alpha w^{-1} \in W$ and we are done as G is generated by reflections r_γ for $\gamma \in \Phi^+$. Suppose the contrary, and let γ be such that $f(h, \gamma)$ is minimal for all choices of γ in $\Phi^+ \setminus \bigcup_{\alpha \in \Delta} W\alpha$. Write $\gamma = \sum_{\alpha \in \Delta} c_\alpha \alpha$ with $c_\alpha \geq 0$. As $\sum_{\alpha \in \Delta} c_\alpha f(\gamma, \alpha) = f(\gamma, \gamma) > 0$, there exists $\alpha \in \Delta$ with $f(\gamma, \alpha) > 0$ and $c_\alpha > 0$. Now $r_\alpha \gamma = \sum_{\beta \in \Delta \setminus \{\alpha\}} c_\beta \beta + (c_\alpha - f(\gamma, \alpha))\alpha$ with $c_\beta > 0$ for some $\beta \in \Delta \setminus \{\alpha\}$ (for otherwise $\gamma = \alpha$, contradicting the choice of γ), and so $r_\alpha \gamma \in \Phi^+$. Moreover, $f(r_\alpha \gamma, h) = f(\gamma, r_\alpha h) = f(\gamma, h) - f(\gamma, \alpha)f(h, \alpha) < f(\gamma, h)$, a contradiction with the minimality of $f(h, \gamma)$. This gives $\gamma \in \bigcup_{\alpha \in \Delta} W\alpha$, as required. $\qquad\square$

The intersection of all mirrors of reflections of G coincides with the intersection of all mirrors of reflections having roots in Δ. Hence, it is an $(n - |\Delta|)$-dimensional subspace of V each of whose vectors is fixed by G. Besides, it is orthogonal to the $|\Delta|$-dimensional subspace spanned by Δ. Therefore, we can restrict our attention to the latter subspace and, by the theorem, identify it with the reflection representation space of the corresponding Coxeter group.

In view of Theorem 4.6.4, a full determination of finite reflection groups hinges on the classification of finite Coxeter groups.

Example 4.6.5 Consider the cube as drawn in Fig. 4.9 of Example 4.3.3. There are 9 reflections leaving the cube invariant. Let G be the group they generate. Every choice of a vector h in the black part of the front face of the figure leads to the choice of S as the set of reflections whose mirrors bound the black part. More precisely, set $\alpha_1 = \varepsilon_1 - \varepsilon_3$, $\alpha_2 = \varepsilon_3 - \varepsilon_2$, and $\alpha_3 = \sqrt{2}\varepsilon_2$ (so the three reflections ρ_1, ρ_2, ρ_3 of Example 4.3.3 have respective roots α_3, α_2, α_1). The black area coincides with the part of the cube surface consisting of all vectors x with $(x, \alpha_i) \geq 0$ for $i \in [3]$. Theorem 4.6.4 gives that (G, S) is a Coxeter system, where S consists of the reflections with mirrors α_1, α_2, and α_3. Now $(\alpha_1, \alpha_3) = 0 = -2\cos(\pi/2)$, $(\alpha_1, \alpha_2) = -1 = -2\cos(\pi/3)$, and $(\alpha_2, \alpha_3) = -\sqrt{2} = -2\cos(\pi/4)$, so the diagram B_3 emerges.

Recall from Definition 4.5.3 that the set of reflections of a Coxeter system (W, S) is the set $R = \bigcup_{w \in W} wSw^{-1}$. For $t \in R$, denote by α_t the unique positive root of t in Φ; cf. Remark 4.5.4. The above details regarding the root system lead to several finiteness criteria for Coxeter groups.

Theorem 4.6.6 *The following statements regarding a Coxeter system (W, S) with S finite and R the set of all reflections of W are equivalent.*

  (i) *The group W is finite.*
 (ii) *The root system Φ of W is finite.*
(iii) *There is a longest element in W (with respect to l).*
(iv) *There is a unique element $w \in W$ with $l(tw) < l(w)$ for all $t \in R$.*
 (v) *There is a unique longest element in W (with respect to l).*

Moreover, the elements of (iii)–(v) coincide and have length $|\Phi^+|$.

Proof (i) $\Rightarrow$ (ii) A look at Definition 4.5.1 shows that the root system Φ is the union of at most n orbits of W, so, if W is finite, then so is Φ.

   (ii) $\Rightarrow$ (iii) If Φ is finite, then, by Corollary 4.5.5, there is an upper bound to the values of the length function on W, so there is an element in W of greatest length.

   (iii) $\Rightarrow$ (iv) According to (iii), there is a longest element w_0 of W. By definition, $l(tw_0) < l(w_0)$ for all $t \in R$. Suppose that w_1 is an element of W with $l(tw_1) < l(w_1)$ for all $t \in R$. By Corollary 4.5.22, for $t \in R$, we have $l(w_0^{-1}w_1) < l(w_0^{-1}tw_1)$. Applying this inequality with the reflection $w_0 t w_0^{-1} \in R$ instead of t, we find $l(w_0^{-1}w_1) < l(t(w_0^{-1}w_1))$ for all $t \in R$. This implies $w_0^{-1}w_1 = 1$, and so $w_0 = w_1$.

   (iv) $\Rightarrow$ (i) Let w be as in (iv). By Proposition 4.5.20 each positive root belongs to $\Phi_{w^{-1}}$, so, by Corollary 4.5.5, $|\Phi^+| = l(w^{-1})$ is finite. In view of Proposition 4.5.2(iii), $|\Phi| = 2|\Phi^+|$ is finite as well. According to Proposition 4.5.2(i), W acts faithfully on Φ and so W is finite.

(iv)$\Leftrightarrow$(v) follows directly from the fact (cf. Proposition 4.5.20) that any $w \in W$ has the maximal possible length $|\Phi^+|$ if and only if $l(tw) < l(w)$ for all $t \in R$.

The final statement is a consequence of the proofs of (iv) $\Rightarrow$ (i) and (iv)$\Leftrightarrow$(v). $\square$

Corollary 4.6.7 *Let (W, S) be a Coxeter system with $|W| < \infty$. The longest element of W is the unique element of W with $l(sw) < l(w)$ for all $s \in S$.*

Proof Let $w \in W$ satisfy $l(sw) < l(w)$ for each $s \in S$ and let $t \in R$. Then $\alpha_t \in \Phi^+$, so there are $\lambda_s \in \mathbb{R}_{\geq 0}$ such that $\alpha_t = \sum_{s \in S} e_s \lambda_s$. By Proposition 4.5.2(iv), $w^{-1}\alpha_t = \sum_{s \in S} w^{-1} e_s \lambda_s$ with $w^{-1} e_s \in \Phi^-$. Hence $w^{-1}\alpha_t \in \sum_{s \in S} e_s \mathbb{R}_{\leq 0}$, so $w^{-1}\alpha_t \in \Phi^-$ and, by Proposition 4.5.20, $l(tw) < l(w)$. In particular, w satisfies the condition of Theorem 4.6.6(iii), hence also (iv), which states that w is the unique longest element of W. $\square$

Definition 4.6.8 If $W(M)$ is finite, then we say that M is *spherical*. If $T \subseteq S$ is such that $M|_T$ is spherical, we also say that T is spherical. In this case, we denote by w_T the unique longest element of $\langle T \rangle$.

The following result gives more information on the coset decomposition than its predecessor Lemma 4.2.20.

Corollary 4.6.9 *If $w \in W$ and $T \subseteq S$ satisfy $l(tw) \leq l(w)$ for each $t \in T$, then the subgroup $\langle T \rangle$ of W is finite and there is $v \in {}^T W$ such that $w = w_T v$ with $l(w) = l(w_T) + l(v)$.*

Proof By Lemma 4.2.20 there are $u \in \langle T \rangle$ and $v \in {}^T W$ with $w = uv$ and $l(w) = l(u) + l(v)$. Let $s \in T$. By assumption, $l(suv) \leq l(uv)$, so by the exchange condition, which is valid due to Theorem 4.5.8, we either have $suv = \widehat{u}v$ for some $\widehat{u} \in \langle T \rangle$ with $l(\widehat{u}) < l(u)$ or $suv = u\widehat{v}$ for some $\widehat{v} \in W$ with $l(\widehat{v}) < l(v)$. In the latter case, we find $\widehat{v} = u^{-1}suv$ to be a shorter coset representative of $\langle T \rangle v$ than v, a contradiction with the choice $v \in {}^T W$, so the first case prevails. This means that $l(su) \leq l(u)$ holds for all $s \in T$. Now Corollary 4.6.7 implies $u = w_T$. $\square$

Corollary 4.6.10 *If W is finite and $S \neq \emptyset$, then its longest element w_S satisfies the following properties.*

(i) $l(w_S) = |\Phi^+|$.
(ii) w_S *is an involution.*
(iii) *For each $w \in W$, we have $l(ww_S) = l(w_S) - l(w) = l(w_S w)$.*
(iv) *The map $x \mapsto w_S x w_S$ $(x \in W)$ is an automorphism of W leaving invariant the subset S.*

Proof (i) is already stated in Theorem 4.6.6.

(ii) According to Exercise 4.9.14, $l(w_S^{-1}) = l(w_S)$, so it follows from Theorem 4.6.6 that $w_S^{-1} = w_S$, that is, $w_S^2 = 1$. Clearly, $w_S \neq 1$ as $S \neq \emptyset$.

Table 4.5 Longest elements of finite irreducible Coxeter groups. Here, $[ijk]^r$ stands for $(s_i s_j s_k)^r$

Y_n	Longest element w_{Y_n}
A_n	$w_{A_{n-1}}[n(n-1)\cdots 1] = 1[21][321][4321]\cdots[n(n-1)(n-2)\cdots 1]$
B_n	$[12\cdots n]^n = w_{A_n}[n(n-1)\cdots 2]\cdots[n(n-1)][n]$
D_n	$[12\cdots n]^{n-1}$
E_6	$[123456]^6$
E_7	$[1234567]^9$
E_8	$[12345678]^{15}$
F_4	$[1234]^6$
G_2	$[12]^3$
H_3	$[123]^5$
H_4	$[1234]^{15}$
$I_2^{(m)}$	$[12]^m$

(iii) Fix $w \in W$. We apply Exercise 4.9.15 with $v = w_S w^{-1}$. As $\Phi_w \subseteq \Phi^+ = \Phi_{w_S}$, we find $l(w_S) = l(w_S w^{-1}) + l(w) = l(w w_S) + l(w)$, proving the first equality. As w_S is an involution (see (ii)), the second equality follows by inverting the arguments and replacing w by its inverse.

(iv) By (ii), w_S is an involution, so the map is conjugation by w_S. Let $s \in S$. Then $l(w_S s w_S) = l(w_S) - l(s w_S) = l(w_S) - l(w_S) + l(s) = 1$ by a double application of (iii). This implies $w_S s w_S \in S$. $\qquad\square$

Remark 4.6.11 The fact that w_S preserves S under conjugation implies that it induces an automorphism on M. This automorphism is called the *opposition* on M. Opposition is easily seen to be an auto-correlation of $\mathcal{C}(M)$. It is the identity if and only if w_S is in the center of the group W. Otherwise, it is a polarity on $\Gamma(\mathcal{C}(M))$. Table 4.5 shows the longest elements of known finite Coxeter groups.

4.7 Finite Coxeter Groups

We have already given several examples of finite Coxeter groups. In this section, we will classify all of them. In view of Proposition 4.2.22, there is no harm in restricting ourselves to irreducible Coxeter groups; cf. Definition 4.2.21.

Lemma 4.7.1 *Let (W, S) be a Coxeter system of type M. Then W is finite if and only if S is finite and $\langle J \rangle$ is finite for each connected component $J \subseteq S$ of the labelled graph M.*

Proof This is immediate from Proposition 4.2.22. $\qquad\square$

The quadratic form κ_M associated with M was introduced in Definition 4.4.3. The following finiteness criteria for $W(M)$ are to be compared with Corollary 4.4.11.

Proposition 4.7.2 *For each Coxeter system (W, S) of irreducible type M with S finite, the following properties are equivalent.*

  (i) *M is spherical (so W is finite).*
 (ii) *The reflection representation of W is irreducible.*
(iii) *The quadratic form κ_M associated with M is positive definite.*

Proof As usual, we let $\rho : W \to O(V, f)$ be the reflection representation of W, where $V = \bigoplus_{s \in S} \mathbb{R}e_s$ and f is the bilinear form associated with M (cf. Definition 4.4.3).

(i) $\Rightarrow$ (ii) Since W is finite, S is finite and the vector space V is finite dimensional. By Lemma 4.6.2 there is a positive-definite symmetric bilinear form g on V invariant under $\rho(W)$. Suppose that E is a proper nontrivial invariant subspace of V. Then so is its perpendicular $D = \{x \in V \mid g(x, E) = 0\}$ with respect to g. By Proposition 4.4.8, both D and E are in the radical of f. As $V = D \oplus E$ (use the fact that g is positive definite), this leads to $f = 0$, a contradiction with $f(e_1, e_1) = 2$. We have shown that ρ is irreducible.

(ii) $\Rightarrow$ (iii) In view of Corollary 4.4.11 and Lemma 4.6.2(ii), the symmetric bilinear form f is a nonzero scalar multiple of a positive-definite form. As $f(e_1, e_1) > 0$, this scalar must be positive, and so f is positive definite as well.

(iii) $\Rightarrow$ (i) The linear map sending $y \in V$ to $Dy \in V^{\vee}$ defined by $(Dy)x = f(y, x)$ is an isomorphism of W-modules. For, it clearly is an isomorphism of vector spaces and $(D(wy))x = f(wy, x) = f(y, w^{-1}x) = (Dy)(w^{-1}x) = (w(Dy))x$ for all $x, y \in V$, so $Dw = wD$ for all $w \in W$. Recall from Theorem 4.4.16(iii) that $A = \bigcap_{s \in S} \overline{A_s}$, where $A_s = \{h \in V^{\vee} \mid h(e_s) > 0\}$, is a prefundamental domain for W in $V^{\vee}$. As D is an isomorphism of W-modules, $D^{-1}A$ is a prefundamental domain for W in V, so all wA, for $w \in W$, are distinct. Now $D^{-1}A$ is non-empty and coincides with the intersection of the half-spaces $\{x \in V \mid f(x, e_s) \geq 0\}$. In particular, its intersection with the unit ball $\mathbb{B} := \{x \in V \mid f(x, x) \leq 1\}$ has positive volume, say μ. Now $\bigcup_{w \in W} wD^{-1}A \cap \mathbb{B}$ has volume $\mu|W|$ and lies in $\mathbb{B}$, so $\mu|W|$ is bounded from above by the volume of the unit ball $\mathbb{B}$, which proves that $|W|$ is finite. $\quad\square$

Theorem 4.7.3 *An irreducible Coxeter group is finite if and only if its Coxeter diagram occurs in Tables 4.2 or 4.6.*

Proof As usual, we let $M = (m_{i,j})_{i,j \in S}$ be a Coxeter diagram of rank n and (W, S) a Coxeter system of type M. There is no harm in assuming $n \geq 2$. We proceed in thirteen steps.

STEP 1. *If M occurs in Tables 4.2 or 4.6, then W is finite.*

This can be derived directly from Proposition 4.7.2 by checking that κ_M is positive definite.

Table 4.6 Nonlinear Coxeter diagrams of irreducible finite Coxeter systems

Name	Diagram
D_n $(n \geq 3)$	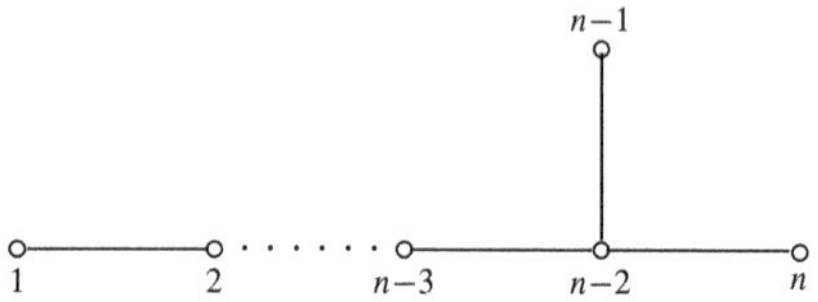
E_6	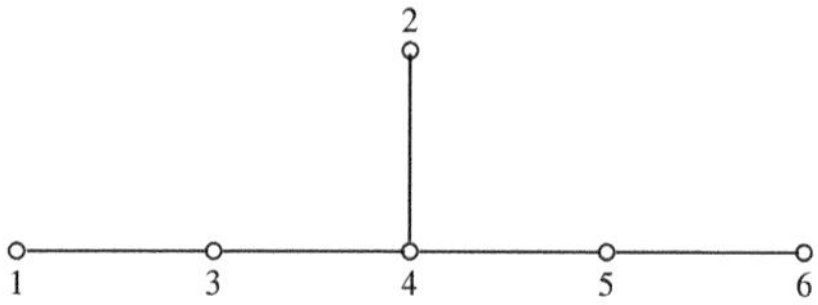
E_7	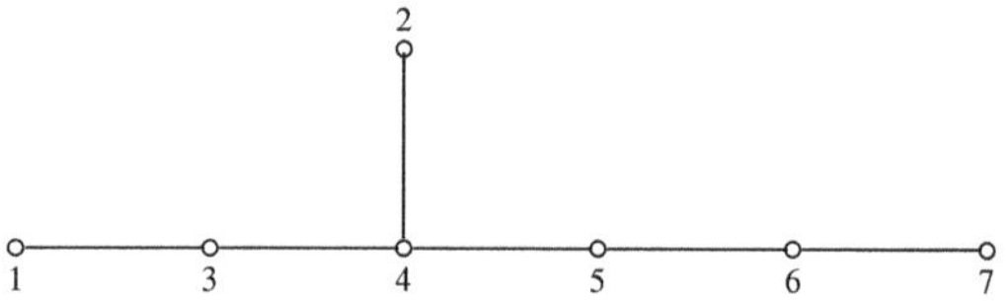
E_8	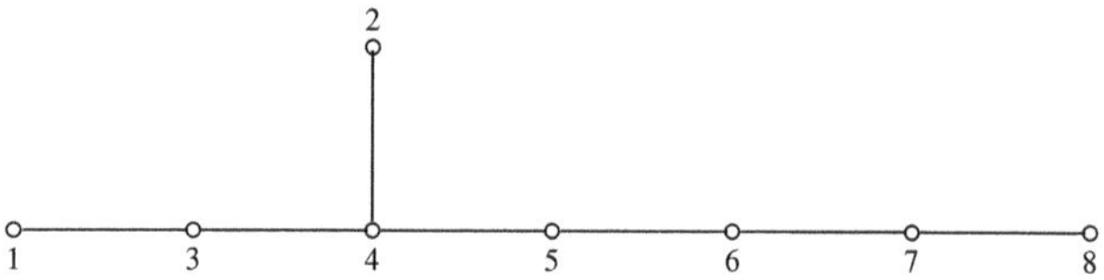

From now on, let W be a finite irreducible Coxeter group of type M.

STEP 2. *Each subdiagram of M is spherical.*

This follows from Corollary 4.4.17.

STEP 3. *There are no circuits in M.*

Suppose (after suitably relabeling the indices) that M has a circuit on $[k]$ for some $k \geq 3$. The vector $e = \sum_{i=1}^{k} e_i$ satisfies $\kappa_M(e) = k - 2\sum_{i<j}\cos(\pi/m_{i,j})$. Since there are at least k pairs $\{i, j\}$ with $\cos(\pi/m_{i,j}) \geq \frac{1}{2}$ while the other pairs provide a non-positive contribution to $\kappa_M(e)$, we find $\kappa_M(e) \leq 0$, which contradicts that κ_M is positive definite. An alternative proof is indicated in Exercise 4.9.29.

STEP 4. *The diagram M cannot have a subdiagram of the form $\widetilde{A}_1$, $\widetilde{B}_n$ $(n \geq 2)$, $\widetilde{C}_n$ $(n \geq 2)$, $\widetilde{D}_n$ $(n \geq 4)$, $\widetilde{F}_4$, or $\widetilde{G}_2$.*

By Examples 4.3.8(iii), (iv), these diagrams have infinite groups.

STEP 5. *For each $i \in [n]$, we have $\sum_{j \neq i} f(e_i, e_j)^2 < 4$.*

Indeed, let J be the set of $j \in [n]$ such that $m_{i,j} \geq 3$. By Step 3, $m_{j,k} = 2$ for all $j, k \in J$, and so $\{e_j \mid j \in J\}$ is an orthogonal set. Now

$$\kappa_M \left(e_i - \frac{1}{2} \sum_{k \in J} f(e_i, e_k) e_k \right) = 1 + \frac{1}{4} \sum_{k \in J} f(e_i, e_k)^2 - \frac{1}{2} \sum_{k \in J} f(e_i, e_k)^2$$

$$= 1 - \frac{1}{4} \sum_{k \neq i} f(e_i, e_k)^2,$$

so the statement states that this value must be positive.

STEP 6. *An element $i \in [n]$ cannot be on more than three edges of M.*

If $\{i, j\}$ is an edge of M, then $f(e_i, e_j)^2 = 4\cos^2(\pi/m_{i,j}) \geq 1$ and so it suffices to apply Step 5.

STEP 7. *If an element $i \in [n]$ is on three edges of M then all these edges have label 3.*

Obvious by Step 5.

STEP 8. *If $m_{i,j} \geq 6$, then $n = 2$.*

If $n \geq 3$, then, as M is connected, there is a node in $\{i, j\}$, say i, with another node, say k, adjacent to it. Now Step 5 gives $4 = 1 + 3 \leq f(e_i, e_k)^2 + f(e_i, e_j)^2 < 4$, a contradiction.

STEP 9. *If $i \in [n]$ is on three edges of M, then all edges of M are of multiplicity three.*

Otherwise, by Steps 8 and 7, there exists a subdiagram

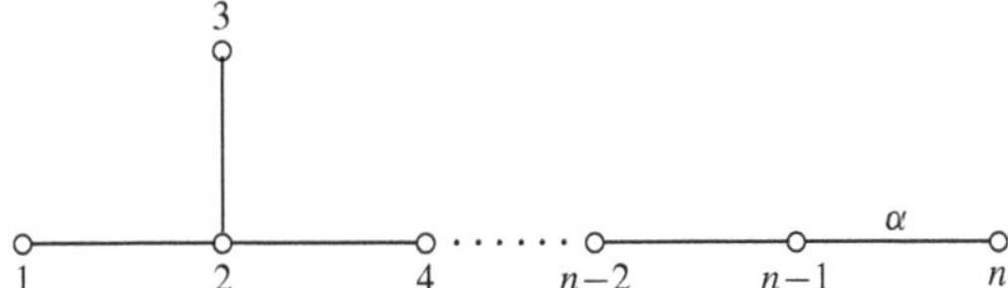

with $\alpha \in \{4, 5\}$, and so $\kappa_M(e_1 + e_2 + 2(e_3 + e_4 + \cdots + e_{n-1}) + \sqrt{2}e_n) \leq 0$.

STEP 10. *If $m_{i,j} = 5$, then i is on at most one more edge and, if so, this edge, say $\{i, k\}$, has label $m_{i,k} = 3$.*

This is due to Step 5 as $\cos(\pi/5)^2 = (6 + 2\sqrt{5})/16 \approx 0.65$.

STEP 11. *There is at most one $i \in [n]$ which is on three edges of M.*

For otherwise, by Steps 6, 7, and 9, there is a subdiagram

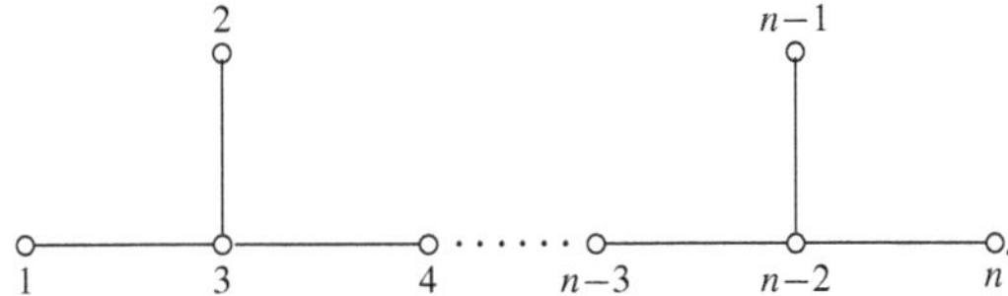

Now, κ_M vanishes on $e_1 + e_2 + 2(e_3 + e_4 + \cdots + e_{n-2}) + e_{n-1} + e_n$, which contradicts that it be positive definite.

STEP 12. *The diagram M has no subdiagram of the form*

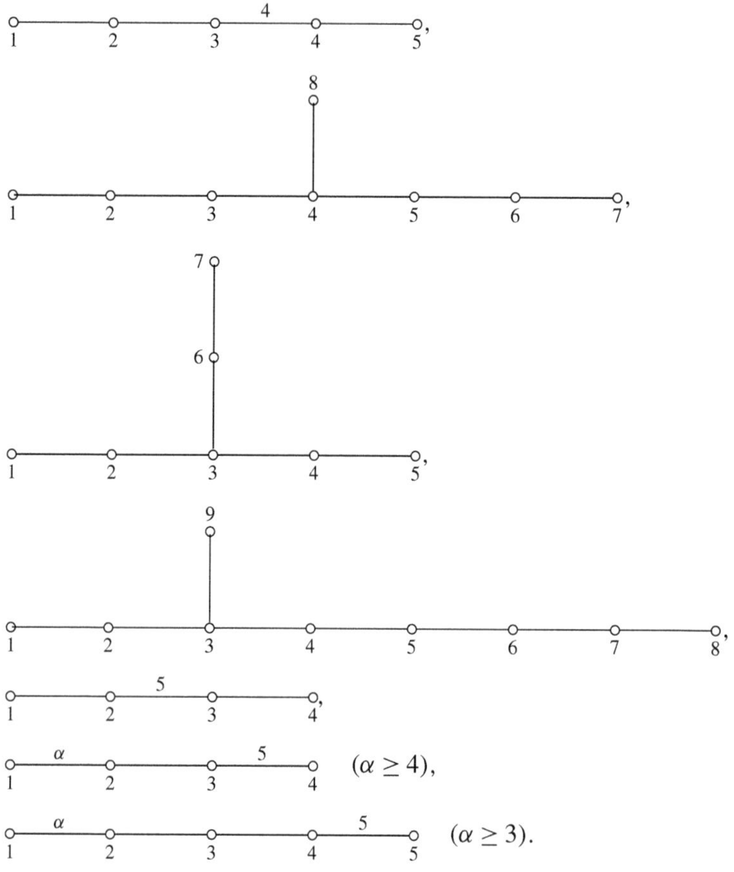

Consider the vector x given by, in the respective cases,

$$x = e_1 + 2e_2 + 3e_3 + 2\sqrt{2}e_4 + \sqrt{2}e_5,$$

$$x = e_1 + 2e_2 + 3e_3 + 4e_4 + 3e_5 + 2e_6 + e_7 + 2e_8,$$

$$x = e_1 + 2e_2 + 3e_3 + 2e_4 + e_5 + 2e_6 + e_7,$$

$$x = 2e_1 + 4e_2 + 6e_3 + 5e_4 + 4e_5 + 3e_6 + 2e_7 + e_8 + 3e_9,$$

$$x = e_1 + 2e_2 + 2e_3 + e_4,$$

$$x = e_1 + 2e_2 + 2e_3 + e_4,$$

$$x = e_1 + 2e_2 + 3e_3 + 4e_4 + \frac{5}{2}(\sqrt{5} - 1)e_5.$$

Table 4.7 Finite Coxeter groups of type E_n

Type	Group order	Root system size
E_6	51,840	72
E_7	2,903,040	126
E_8	696,729,600	240

A straightforward calculation gives that $\kappa_M(x) \leq 0$ in each of these cases.

STEP 13. *The only connected diagrams satisfying the conditions of Steps 3–12 are those in the list of the theorem.*

This is easily verified. $\square$

The cases $\widetilde{B}_n$, $\widetilde{D}_n$, $\widetilde{F}_4$, and $\widetilde{G}_2$ were eliminated in Step 4, and once more in Steps 9, 11, 12, and 8, respectively.

Remark 4.7.4 By analysis of the root system and the corresponding Coxeter group action it can be shown that the sizes of the Coxeter groups of type E_n $(n = 6, 7, 8)$ and of their root systems are as in Table 4.7.

The corresponding numbers for D_n are in Exercise 4.9.23 and the orders for most other irreducible spherical types are in Table 4.4.

4.8 Regular Polytopes Revisited

Thanks to the description of geometries of Coxeter type in Theorem 4.2.8 and Lemma 4.2.6 as quotients of universal examples coming from Coxeter systems, and a little topology, we can classify regular convex polytopes by their Coxeter types. The topology comes in at the point where we view the boundary of the polytope in $\mathbb{E}^n$ as an $(n-1)$-sphere. We state a similar result for regular tessellations.

The current section is not needed for what follows in this book. Rather, it is the closure of the questions on regular polytopes raised in the beginning of this chapter, which motivated us to study Coxeter groups. We assume familiarity with the basic concepts of topology.

Definition 4.8.1 A surjective continuous map $\phi : X \to Y$ of topological spaces is called a *covering map* if each point of Y has a neighborhood U such that $\phi^{-1}(U)$ is a disjoint union of open sets, each of which is mapped homeomorphically by ϕ onto U.

The polytope geometry $\Gamma(\Pi)$ of a convex polytope Π was introduced in Definition 4.1.11. Recall from Example 4.1.13(i) that an n-simplex is the convex hull of $n+1$ points of $\mathbb{E}^n$ in *general position* (that is, each point lies outside the affine space generated by the other n points). As a topological space, it is homeomorphic to the unit ball in $\mathbb{E}^n$ and so its boundary is homeomorphic to the $(n-1)$-sphere.

Theorem 4.8.2 *Let $n > 0$, and let M be a Coxeter matrix over $[n]$. If Π is a convex polytope in $\mathbb{E}^n$ of Coxeter type M, then $\Gamma(\Pi)$ is isomorphic to $\Gamma(\mathcal{C}(M))$.*

Proof Let Π be a convex polytope in $\mathbb{E}^n$ whose polytope geometry $\Gamma(\Pi)$ is of Coxeter type M. If $n = 1$, then Π is the convex hull of two points, and the assertions are readily verified. Also the case $n = 2$ is easy. By Theorem 4.1.8 the theorem holds for $n = 3$, but we use induction on n and assume $n \geq 3$, so as to cover the case $n = 3$ once more.

By Proposition 4.1.12, the polytope geometry $\Gamma(\Pi)$ is residually connected. According to Theorem 3.4.6 this implies that the chamber system $\mathcal{C} := \mathcal{C}(\Gamma(\Pi))$ is residually connected and fully determines the isomorphism type of $\Gamma(\Pi)$. Fix a Coxeter system (W, S) of type M. By Theorem 4.2.8, the group W has a subgroup A such that $\mathcal{C}$ is isomorphic to $\mathcal{C}(M)/A$. Denote by $\phi : \mathcal{C}(M) \to \mathcal{C}$ the corresponding map. As a consequence, two chambers x, y of $\mathcal{C}(M)$ satisfy $\phi(x) = \phi(y)$ if and only if $x \in Ay$. As $\mathcal{C}(M)$ is residually connected (see Corollary 4.5.16), there is a residually connected geometry $\widetilde{\Gamma}$ over $[n]$ such that $\mathcal{C}(\widetilde{\Gamma}) \cong \mathcal{C}(M)$. Besides, W acts on $\widetilde{\Gamma}$ as a group of automorphisms. By Theorem 3.4.6 again, $\widetilde{\Gamma}/A \cong \Gamma(\Pi)$. Hence, for the proof of the theorem, it suffices to show that A is trivial.

Let $i \in [n]$. Each $(i-1)$-face x of Π determines a convex polytope Π_x in $\mathbb{E}^{i-1}$ and has the topology of a closed ball in $\mathbb{E}^{i-1}$. But also the space $\mathbb{E}^n/\langle x \rangle$ obtained by taking the quotient of $\mathbb{E}^n$ with respect to the $(i-1)$-dimensional affine subspace $\langle x \rangle$, contains a convex polytope that can be identified with the star of x described in the proof of Proposition 4.1.12. It is the convex hull of the vertices in $\mathbb{E}^n/\langle x \rangle$ that are i-faces of Π containing x and we denote it by Π/x. By Proposition 4.1.12 the residue $\Gamma(\Pi)_x$ is the direct sum of the polytope geometries of Π_x and Π/x. The induction hypothesis gives that both components of the disjoint union are isomorphic to geometries of chamber systems of Coxeter groups. It follows from Proposition 3.5.4 that the chamber system of the residue $\Gamma(\Pi)_x$ is isomorphic to $\mathcal{C}(M|_{[n]\setminus\{i\}})$. On the other hand, by the description of $\mathcal{C}$ as $\mathcal{C}(M)/A$, the chamber system of the residue $\Gamma(\Pi)_x$ is isomorphic to the quotient $\mathcal{C}(M|_{[n]\setminus\{i\}})/(gAg^{-1} \cap \langle [n] \setminus \{i\} \rangle)$ for some $g \in W$ (cf. Exercise 4.9.31), so $gAg^{-1} \cap \langle [n] \setminus \{i\} \rangle = 1$. In fact, as every $g \in W$ gives rise to such an isomorphism, the latter equality holds for any $g \in W$. We will use this fact to construct a covering space of the boundary of Π.

We have already seen that the boundary Σ of Π is homeomorphic to an $(n-1)$-sphere in $\mathbb{E}^n$. We realize $\mathcal{C}$ by means of $(n-1)$-simplices in Σ as follows. We represent each face x of $\Gamma(\Pi)$ by a *barycentric* point p_x in Σ: for each 0-face x, we choose p_x to be the point x itself; for each i-face x of Π with $i \in [n-1]$, we choose a point p_x in the interior of x, and hence off each of the j-faces of Π contained in x for $j < i$. If c is a chamber of Σ, the convex hull s_c of all p_x for $\langle x \rangle \in c$ (where $\langle x \rangle$ is the i-element spanned by x) is an $(n-1)$-simplex in Σ. Two distinct chambers c and d are i-adjacent if and only if $s_c \cap s_d$ is the convex hull of all p_x for $\langle x \rangle \in c$ with type distinct from i (equivalently, x is not an $(i-1)$-face). This way, $\mathcal{C}$ is realized inside Σ by means of a set of $(n-1)$-simplices, which form a tiling of Σ. This construction is much like the one for the cube in Fig. 4.9 of Example 4.3.3, where a single 2-simplex representing a chamber is colored black.

We now construct a topological covering space $\widetilde{\Sigma}$ of Σ. To each chamber c of $\mathcal{C}(M)$, we assign a copy s_c of the $(n-1)$-simplex $s_{\phi(c)}$ by means of a homeomorphism $\phi_c : s_c \to s_{\phi(c)}$ of $(n-1)$-simplices. The space $\widetilde{\Sigma}$ is the union of all s_c over the chambers c of $\mathcal{C}(M)$, with the convention that, whenever two chambers c and d, say, are i-adjacent, we glue the simplices s_c and s_d just as $\phi_c(s_c) = s_{\phi(c)}$ and $\phi_d(s_d) = s_{\phi(d)}$ are glued, which is along their unique $(n-2)$-simplices corresponding to an i-panel. As a consequence, for distinct chambers c, d of $\mathcal{C}(M)$, two points $p \in s_c$ and $q \in s_d$ are identified if and only if there is $i \in [n]$ such that c and d lie in a common $([n] \setminus \{i\})$-cell of $\mathcal{C}(M)$ and $\phi_c(p) = \phi_d(q)$. The construction gives us a topological space $\widetilde{\Sigma}$ as well as a continuous surjective map $\widetilde{\phi} : \widetilde{\Sigma} \to \Sigma$ determined by $\widetilde{\phi}(p) = \phi_c(p)$ whenever $p \in s_c$.

By the equality $g A g^{-1} \cap \langle [n] \setminus \{i\} \rangle = 1$ derived above for each $g \in W$, the restriction of $\widetilde{\phi}$ to the union of all $(n-1)$-simplices s_c for c varying over an $([n] \setminus \{i\})$-cell of $\mathcal{C}(M)$ is a homeomorphism. For each point $p \in s_{\phi(c)}$, where c is a chamber of $\mathcal{C}(M)$, this enables us to find an open neighborhood U of p in Σ such that U is homeomorphic to an open neighborhood in $\widetilde{\Sigma}$ of a point in $s_c \cap \widetilde{\phi}^{-1}(p)$. We conclude that $\widetilde{\phi} : \widetilde{\Sigma} \to \Sigma$ is a topological covering. But Σ is simply connected and $\widetilde{\Sigma}$ is connected, so $\widetilde{\phi}$ is a homeomorphism. As $\widetilde{\phi}(s_c) = \widetilde{\phi}(s_{ac})$ for each $a \in A$, we conclude $A = 1$, as required. $\square$

Corollary 4.8.3 *The connected Coxeter diagrams M for which there is a convex polytope of type M are as indicated in Table* 4.2.

Proof Proposition 4.1.12 shows that the diagram is linear. Theorem 4.8.2 shows that the geometry is uniquely determined by M and has $|W|$ chambers. This number must be finite and so Theorem 4.7.3 applies. As a consequence, the only types M that can occur are as in Tables 4.2 and 4.6. But those of Table 4.6 are not linear. Example 4.1.13 shows that all types of Table 4.2 are actually realized. $\square$

Theorem 4.8.4 *Let $n > 0$ and let M be a Coxeter matrix over $[n+1]$. If T is a tessellation by convex polytopes in $\mathbb{E}^n$ of Coxeter type M, then $\Gamma(T)$ is isomorphic to $\Gamma(\mathcal{C}(M))$, where (W, S) is a Coxeter system of type M.*

The proof is similar to the proof of Theorem 4.8.2 and will be omitted. We just mention that the topological space Σ is $\mathbb{E}^n$ rather than the $(n-1)$-sphere.

Corollary 4.8.5 *The connected Coxeter diagrams M for which there is a tessellation of type M are as indicated in Table* 4.3.

We only give an indication of the proof. Proposition 4.1.15 shows that the diagram of the geometry of a tessellation in $\mathbb{E}^n$ is linear of rank $n + 1$. Theorem 4.8.4 shows that the diagram itself does not belong to Table 4.2 but that its subdiagrams on $[n]$ and on $[n+1] \setminus \{1\}$ do. This limits the possibilities for Coxeter diagrams M,

but does not exclude diagrams like

One approach uses the observation that each tessellation of Coxeter type M is isomorphic to a tessellation obtained from the contragredient $\rho^\vee$ of the reflection representation ρ of the Coxeter group of type M whose corresponding bilinear form f_M (see Definition 4.4.3) is semi-positive definite with a 1-dimensional radical. To be a little more precise, this representation $\rho^\vee$ is $(n+1)$-dimensional, and induces a group generated by affine reflections on an affine hyperplane. Figure 4.14 illustrates this for the case $M = \widetilde{A}_1$. Identifying this affine hyperplane with $\mathbb{E}^n$ and choosing an n-simplex as the intersection of this hyperplane with a prefundamental domain as in Theorem 4.4.16, we can rebuild the original tessellation up to isomorphism from suitable unions of these chambers. The condition that f_M be semi-positive definite with a 1-dimensional radical limits the possibilities for M as stated. For instance, the form f_M of the Coxeter diagram M depicted above has three positive eigenvalues and a negative one, and so it is excluded. Example 4.1.9(ii) gives a topological argument for the case $n = 2$.

Example 4.1.16 shows that all types of Table 4.3 are actually realized.

4.9 Exercises

Section 4.1

Exercise 4.9.1 Let m be a natural number.

(a) Show that the shadow space on 1 of a thin geometry of Coxeter type

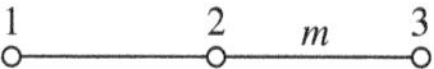

is a graph which is locally an m-gon.

(b) Show that there are infinitely many thin geometries of the Coxeter type of (a) if $m = 6$.

(*Hint*: They can be obtained as quotients of the regular hexagon tiling of $\mathbb{E}^2$.)

Exercise 4.9.2

(a) Show that a cube of $\mathbb{E}^3$ is a disjoint union of two regular tetrahedra (when the cube and tetrahedra are considered as sets of vertices).

(b) Show that a hypercube of $\mathbb{E}^4$ is a disjoint union of two regular hyperoctahedra (as sets of vertices).

(c) How about hypercubes of $\mathbb{E}^n$?

(d) Use (b) to derive a thin geometry over the diagram D_4 (see Table 4.6).

Exercise 4.9.3 Show that the group G of automorphisms of the 24-cell has two subgroups of the form $2.\mathrm{Alt}_4$ (a group with a normal subgroup of order 2 and a quotient isomorphic to Alt_4) which are sharply transitive on the vertices, whose intersection is the center $Z(G)$ and which generate a normal subgroup $2.(\mathrm{Alt}_4 \times \mathrm{Alt}_4)$ of index 4 in G.

Exercise 4.9.4 Show that the group G of automorphisms of the 600-cell has two subgroups of the form $2.\mathrm{Alt}_5$ which are sharply transitive on the vertices, whose intersection is $Z(G)$ and which generate a subgroup $2.(\mathrm{Alt}_5 \times \mathrm{Alt}_5)$ of index 2 in G.

Section 4.2

Exercise 4.9.5 (Symmetric groups are Coxeter groups) Let $n \geq 1$. Consider the Coxeter system (W, S) of type A_n. Verify that the map $S \to \mathrm{Sym}_{n+1}$ given by $s_i \mapsto (i, i+1)$ for $i \in [n]$ extends to a group homomorphism $\phi : W \to \mathrm{Sym}_{n+1}$. Show that ϕ is an isomorphism.

(*Hint*: First observe that the defining relations of W are satisfied, so that ϕ is a well defined homomorphism. A look at the image of S readily gives that it is surjective. Prove that every element of W can be written as a product of s_i ($i \in [n]$) with at most one occurrence of s_n, and next that if $w = r_1 \cdots r_n$ maps to 1 under ϕ, and all r_i are of the form s_j with exactly one of them equal to s_n, then $\phi(s_n) \in \langle \phi(s_1), \ldots, \phi(s_{n-1}) \rangle = \mathrm{Sym}_{n-1}$, which is impossible. Finish by induction on n.)

♯ **Exercise 4.9.6** Consider the group $\overline{W} = \mathrm{PSL}(\mathbb{F}_{19}^2)$. It is a simple group of order 3420, generated by the four elements

$$\bar{s}_1 = \begin{pmatrix} 1 & 9 \\ 4 & -1 \end{pmatrix}, \qquad \bar{s}_2 = \begin{pmatrix} 0 & 7 \\ 8 & 0 \end{pmatrix},$$

$$\bar{s}_3 = \begin{pmatrix} 0 & 8 \\ 7 & 0 \end{pmatrix}, \qquad \bar{s}_4 = \begin{pmatrix} -1 & 4 \\ 9 & 1 \end{pmatrix}.$$

(a) Verify that $\overline{W}$ is a quotient of the Coxeter group W of type

$$
\overset{1}{\circ} \overset{5}{\rule{1.2cm}{0.4pt}} \overset{2}{\circ} \overset{}{\rule{1.2cm}{0.4pt}} \overset{3}{\circ} \overset{5}{\rule{1.2cm}{0.4pt}} \overset{4}{\circ},
$$

via the group homomorphism sending s_i to $\bar{s}_i$ ($i \in [4]$).

(b) Check that $(s_1 s_2 s_3)^{10} = (s_2 s_3 s_4)^{10} = 1$, that $(\bar{s}_1 \bar{s}_2 \bar{s}_3)^5 = (\bar{s}_2 \bar{s}_3 \bar{s}_4)^5 = 1$, and that the latter equalities do not hold when the bars are removed.

(c) Determine the isomorphism type of the group $\langle \bar{s}_1, \bar{s}_2, \bar{s}_3 \rangle$.

(d) Show that $\bar{s}_1 \bar{s}_2 \bar{s}_3 \bar{s}_4$ has order 9.

(e) Is the chamber system $\mathcal{C}(M)/\mathrm{Ker}(\phi)$ residually connected?

(f) Determine the full group of auto-correlations of $\mathcal{C}(M)/\mathrm{Ker}(\phi)$.

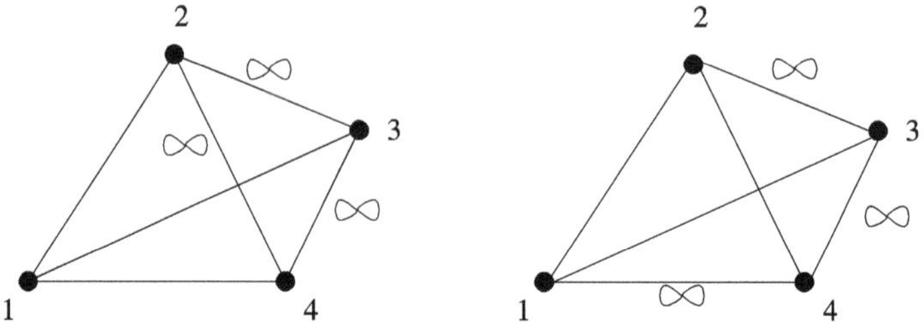

Fig. 4.16 Two Coxeter diagrams whose Coxeter groups are isomorphic

Exercise 4.9.7 Let W_n be the Coxeter group of type B_n (cf. Table 4.4). Use Corollary 4.4.17 to verify that W_{n-1} is a subgroup of W_n. Prove that the coset graph $\Delta(W_n, W_{n-1}, s_n)$ on W_n / W_{n-1} determined by $W_{n-1} s_n W_{n-1}$ (see Definition 1.7.13) is the graph of the n-dimensional hypercube (see Example 4.1.13).

Exercise 4.9.8 (Cited in Lemma 4.2.16) Let (W, S) be a Coxeter system of type M.

(a) Prove that there is a homomorphism of groups $\mathrm{sg} : W \to \{\pm 1\}$ determined by $\mathrm{sg}(s) = -1$ for each $s \in S$.
(b) Let K be a connected component of the graph M' obtained from M by omitting all edges with an even label. Prove that there is a homomorphism of groups $\mathrm{sg}_K : W \to \{\pm 1\}$ determined by $\mathrm{sg}_K(s) = -1$ for each $s \in K$ and $\mathrm{sg}_K(s) = 1$ for each $s \in S \setminus K$.
(c) Show that two members of S are conjugate in W if and only if they belong to the same connected component of the graph M' of (b).

Exercise 4.9.9 The Coxeter diagram of a Coxeter group is not uniquely determined by the abstract group. Here are two counterexamples.

(a) Show that the groups $W(A_1 \dot{\cup} I_2^{(3)})$ and $W(I_2^{(6)})$ are both isomorphic to the dihedral group of order 12.
(b) Consider the two diagrams of Fig. 4.16. Prove that the two Coxeter groups are isomorphic.
    (*Hint*: Show that replacement of the reflection s_4 by $s_1 s_2 s_4 s_2 s_1$ in the left hand Coxeter group leads to the same Coxeter group, but a presentation corresponding to the diagram at the right hand side, and that replacement of s_4 by $s_2 s_1 s_4 s_1 s_2$ in the right hand Coxeter group leads to the same Coxeter group, but a presentation corresponding to the diagram at the left hand side.)

Section 4.3

Exercise 4.9.10 Consider the Coxeter system (W, S) of type H_3, with distinguished generator set $S = \{s_1, s_2, s_3\}$.

(a) Show that $z = (s_1 s_2 s_3)^5$ is the central element of W.
(b) Derive that the automorphism group of $\mathcal{C}(H_3)/\langle z \rangle$ is isomorphic to Alt_5.
(c) Show that W is isomorphic to the direct product $\mathrm{Alt}_5 \times C_2$ of the alternating group on five letters and the cyclic group of order 2. Conclude that $W(H_3)$ is not isomorphic to $W(A_4)$, as the latter is isomorphic to Sym_5, a proper semidirect product of Alt_5 and C_2.

Exercise 4.9.11 By $\mathrm{PGL}(2, \mathbb{Z})$ we denote the quotient of the group $\mathrm{GL}(2, \mathbb{Z})$ of invertible 2×2 matrices with integer entries by the central subgroup consisting of the identity element and its negative. Take

$$\rho_1 = \pm \begin{pmatrix} 0 & 1 \\ 1 & 0 \end{pmatrix}, \qquad \rho_2 = \pm \begin{pmatrix} -1 & 1 \\ 0 & 1 \end{pmatrix}, \qquad \rho_3 = \pm \begin{pmatrix} 1 & 0 \\ 0 & -1 \end{pmatrix} \qquad \text{in } \mathrm{PGL}(2, \mathbb{Z}).$$

(a) Prove that $\{\rho_1, \rho_2, \rho_3\}$ is a generating set for $\mathrm{PGL}(2, \mathbb{Z})$.
(b) Use the generating set in (a) to verify that $\mathrm{PGL}(2, \mathbb{Z})$ is a quotient of the Coxeter group of type

$$\underset{1}{\circ} \rule{1.5cm}{0.4pt} \underset{2}{\circ} \overset{\infty}{\rule{1.5cm}{0.4pt}} \underset{3}{\circ}.$$

In fact, Exercise 4.9.12 establishes that $\mathrm{PGL}(2, \mathbb{Z})$ is isomorphic to this Coxeter group.
(c) Derive from the above that $W(A_3)$ is isomorphic to $\mathrm{PGL}(\mathbb{F}_3^2)$ and that $W(H_3)$ is isomorphic to $\mathrm{PGL}(\mathbb{F}_5^2)$.

Section 4.4

Exercise 4.9.12 We continue Exercise 4.9.11, retaining the notation ρ_i for $i = 1, 2, 3$ and writing M for the Coxeter diagram of Part (b) of that exercise. There it was shown that the map $\{s_1, s_2, s_3\} \to \{\rho_1, \rho_2, \rho_3\}$ extends to a surjective group homomorphism $\psi : W(M) \to \mathrm{PGL}(2, \mathbb{Z})$. Set $V = \mathbb{R}^2$.

(a) By $V \otimes V$ we denote the tensor product of V with itself. Prove that there is a unique linear transformation $\sigma \in \mathrm{GL}(V \otimes V)$ of order 2 with $\sigma(x \otimes y) = y \otimes x$ for all $x, y \in V$.
(b) Let V^{2+} be the subspace of $V \otimes V$ linearly spanned by $d_1 = e_1 \otimes e_1$, $d_2 = \frac{1}{2}(e_1 \otimes e_2 + e_2 \otimes e_1)$, and $d_3 = e_2 \otimes e_2$ for a fixed basis e_1, e_2 of V. Prove that V^{2+} coincides with the subspace of $V \otimes V$ consisting of the fixed vectors of σ.
(c) Prove that there is a unique group homomorphism

$$\phi : \mathrm{GL}(V) \to \mathrm{GL}(V \otimes V)$$

determined by $\phi(g)(x \otimes y) = (gx) \otimes (gy)$ for $x, y \in V$.

(d) Show that for $g \in \mathrm{GL}(V)$, the map $\phi(g)$ leaves invariant V^{2+} and satisfies

$$f\big(\phi(g)x, \phi(g)y\big) = \det(g)^2 f(x, y) \qquad (x, y \in V^{2+}),$$

where f is the symmetric bilinear form on V^{2+} given by the matrix

$$\begin{pmatrix} 0 & 0 & 2 \\ 0 & -1 & 0 \\ 2 & 0 & 0 \end{pmatrix}$$

on the basis d_1, d_2, d_3.

(e) Verify that the respective images of ρ_1, ρ_2, ρ_3 under ϕ are

$$\begin{pmatrix} 0 & 0 & 1 \\ 0 & 1 & 0 \\ 1 & 0 & 0 \end{pmatrix}, \qquad \begin{pmatrix} 1 & -1 & 1 \\ 0 & -1 & 2 \\ 0 & 0 & 1 \end{pmatrix}, \qquad \begin{pmatrix} 1 & 0 & 0 \\ 0 & -1 & 0 \\ 0 & 0 & 1 \end{pmatrix},$$

and that these transformations are orthogonal reflections with respect to f (see Exercise 1.9.31) on V^{2+} with respective roots $d_1 - d_3$, $d_1 + 2d_2$, and d_2.

(f) Prove that $\phi(\langle \rho_1, \rho_2, \rho_3 \rangle)$ is a Coxeter group of type M and deduce that $\phi \circ \psi$ is an isomorphism. Deduce that $\mathrm{PGL}(2, \mathbb{Z})$ is isomorphic to $W(M)$.

Exercise 4.9.13 Let $\sigma_1, \sigma_2, \sigma_3$ be the unitary reflections on $\mathbb{C}^3$ considered in Exercise 2.8.25 and let M be the Coxeter matrix

$$\begin{pmatrix} 1 & 5 & 3 \\ 5 & 1 & 3 \\ 3 & 3 & 1 \end{pmatrix}.$$

We consider the Coxeter system (W, S) of type M.

(a) Verify that $(\sigma_1\sigma_2)^5 = (\sigma_1\sigma_3)^3 = (\sigma_2\sigma_3)^3 = 1$. Conclude that the map $s_i \mapsto \sigma_i$ $(i \in [3])$ determines a linear representation $W \to \mathrm{U}(\mathbb{C}^3, h)$, where h is the standard hermitian form on $\mathbb{C}^3$.

(b) Verify that $(\sigma_1\sigma_2\sigma_1\sigma_3)^4 = 1$.

(c) Let $\rho : W \to \mathrm{O}(\mathbb{R}^3, f)$, where f is the bilinear form associated with M, be the reflection representation. Use it to verify that $(s_1 s_2 s_1 s_3)^4 \neq 1$.

(d) Conclude that σ is not faithful, that Theorem 4.3.6 does not hold for groups generated by complex reflections and that Theorem 4.4.16 does not extend to arbitrary complex representations of Coxeter groups whose generators are mapped to reflections.

Section 4.5

Exercise 4.9.14 (Cited in proof of Proposition 4.5.2) Show that $l(w^{-1}) = l(w)$ and $\Phi_{w^{-1}} = -w\Phi_w$ for each element w of a Coxeter group W.

Exercise 4.9.15 (Used in Corollary 4.6.10) Let v and w be elements of the Coxeter group W. Prove the following statements.

(a) The equality $l(vw) = l(v) + l(w)$ holds if and only if $\Phi_w \subseteq \Phi_{vw}$.
(b) If $l(vw) = l(v) + l(w)$, then $\Phi_{vw} = \Phi_w \cup w^{-1}\Phi_v$.

Exercise 4.9.16 (Cited in the proof of Theorem 11.2.11) Let (W, S) be a Coxeter system and let J, K, L be subsets of S. Show that ${}^J W^K \cap \langle J \rangle \langle L \rangle \langle K \rangle \subseteq \langle L \rangle$.

Exercise 4.9.17 Let (W, S) be a Coxeter system and $J, K \subseteq S$.

(a) Prove that there is a bijection from ${}^J W^K$ onto the set of orbits of $\langle J \rangle$ on $W/\langle K \rangle$.
(b) Prove that there is a bijection between the set of orbits of $\langle J \rangle$ on $W/\langle K \rangle$ and the set of orbits of $\langle K \rangle$ on $W/\langle J \rangle$.

Exercise 4.9.18 Let $n \geq 3$. Let (W, S) be the Coxeter system of type A_n and set $J = [n] \setminus \{1, n\}$. Prove that ${}^J W^J$ has size 7 and give minimal expressions for its elements.

Exercise 4.9.19 Let (W, S) be the Coxeter system of type A_n and set $J = [n] \setminus \{k\}$ for some $k \in [n]$ with $k \leq n/2$.

(a) Prove that ${}^J W^J$ has size $k + 1$ and give minimal expressions for its elements.
(b) Derive that $\langle J \rangle$ has $k + 1$ orbits on $W/\langle J \rangle$.

Exercise 4.9.20 (Cited in the proof of Proposition 11.1.9) Let (W, S) be the Coxeter system of type B_n ($n \geq 2$). Prove that, for each element $w \in W$, there is a minimal expression of w in S^* with at most two occurrences of s_1.

Exercise 4.9.21 Let $n \geq 2$ and let (W, S) be the Coxeter system of type B_n (defined in Example 4.3.8(i)). Set $J = S \setminus \{1\}$. Show that ${}^J W^J$ has size three and give minimal expressions for its elements.

Exercise 4.9.22 Let (W, S) be a Coxeter system of spherical type M and let R be the set of reflections in W. Then $R_t^+ := \{w \in W \mid l(tw) > l(w)\}$ is called a *positive root* and $R_t^- := \{w \in W \mid l(tw) < l(w)\}$ is called a *negative root* of the thin building $\mathcal{C}(M)$. Prove the following statements on these roots.

(a) For each $t \in R$, the set W is the disjoint union of the two subsets R_t^+ and R_t^-, each of which is connected in $\mathcal{C}(M)$. Moreover, these parts are interchanged by t.
(b) The map $\rho : \Phi \to 2^W$ given by $\rho(\alpha) = R_{r_\alpha}^\varepsilon$, where $\varepsilon = +$ if $\alpha \in \Phi^+$ and $\varepsilon = -$ otherwise, is injective.
(c) If $c \in R_t^+$ and $d \in R_t^-$ are the chambers of a panel of $\mathcal{C}(M)$, then $t = cd^{-1}$. (*Hint*: Apply Corollary 4.5.22 with $v = tc$ and $w = d$.)

(d) If $\{c, d\}$ is a panel of $\mathcal{C}(M)$, then $\{e \in \mathcal{C}(M) \mid d_{\mathcal{C}(M)}(c, e) < d_{\mathcal{C}(M)}(d, e)\}$
(cf. Remark 4.2.13) is a *root* (that is, a positive or a negative root) of $\mathcal{C}(M)$,
and every root is obtained in this way.
(*Hint*: Use Corollary 4.5.22.)

Section 4.6

Exercise 4.9.23 (Cited in Theorem 4.7.3) Let $n \geq 4$ and consider the following set
of $2n(n-1)$ vectors in $\mathbb{R}^n$ with standard basis $\varepsilon_1, \ldots, \varepsilon_n$.

$$\Phi = \{\pm\varepsilon_i \pm \varepsilon_j \mid 1 \leq i < j \leq n\}$$

Here, the two $\pm$ signs stand for independent variations, so one expression indicates
four distinct elements. Let $(\cdot, \cdot)$ be the standard inner product on $\mathbb{R}^n$.

(a) Prove that each orthogonal reflection r_α (cf. Exercise 1.9.31) with root $\alpha \in \Phi$
leaves Φ invariant. Denote by W the group generated by all r_α for $\alpha \in \Phi$.
(b) Prove that

$$\Delta = \{\varepsilon_1 - \varepsilon_2, \varepsilon_2 - \varepsilon_3, \ldots, \varepsilon_{n-1} - \varepsilon_n, \varepsilon_{n-1} + \varepsilon_n\}$$

is a subset of Φ with the property that each element of Φ is a linear combination
of members of Δ all of whose nonzero coefficients have the same sign.
(c) Derive that W is a finite Coxeter group of type D_n (see Table 4.6 for the diagram).
(d) Prove that $|W| = 2^{n-1} n!$
(*Hint*: Interpreting the roots of Φ modulo 2, one obtains a surjective homomorphism $W \to W(\mathrm{A}_{n-1})$; determine the order of the kernel and use the first
isomorphism theorem.)

Exercise 4.9.24 (Cited in Example 4.3.8(iv)) Let (W, S) be a Coxeter system of
type B_n. Put $t_n = s_n s_{n-1} s_n$ and set $T = (S \cup \{t_n\}) \setminus \{s_n\}$. Prove that $(\langle T \rangle, T)$ is a
Coxeter system of type D_n.

Exercise 4.9.25 Consider the real vector space $V = \mathbb{R}^8$, its standard basis $\varepsilon_1, \ldots,$
ε_8, and the standard inner product $(\cdot, \cdot)$. Let Φ be the subset of V consisting of

- all $\pm\varepsilon_i \pm \varepsilon_j$, $i < j$ (with the same convention for the two $\pm$ signs as in Exercise 4.9.23), and
- all $\frac{1}{2}\sum_i (-1)^{k_i} \varepsilon_i$ with $\sum_i k_i$ even.

(a) Verify that $|\Phi| = 240$.
(b) Prove that Φ is invariant under each orthogonal reflection r_α with respect to
$(\cdot, \cdot)$ having a root α in Φ; cf. Exercise 1.9.31.

(c) Consider the following elements of Φ.

$$\alpha_1 = \frac{1}{2}(\varepsilon_1 - \varepsilon_2 - \varepsilon_3 - \varepsilon_4 - \varepsilon_5 - \varepsilon_6 - \varepsilon_7 + \varepsilon_8),$$

$$\alpha_2 = \varepsilon_1 + \varepsilon_2, \quad \alpha_3 = \varepsilon_2 - \varepsilon_1, \quad \alpha_4 = \varepsilon_3 - \varepsilon_2,$$

$$\alpha_5 = \varepsilon_4 - \varepsilon_3, \quad \alpha_6 = \varepsilon_5 - \varepsilon_4, \quad \alpha_7 = \varepsilon_6 - \varepsilon_5,$$

$$\alpha_8 = \varepsilon_7 - \varepsilon_6.$$

Verify that the *Gram matrix* of $\alpha_1, \ldots, \alpha_8$, that is, the matrix of inner products (α_i, α_j) for $i, j \in [8]$, is equal to the Gram matrix of the basis $e_1, \ldots, e_8$ of the reflection representation of the Coxeter group of type E_8 with respect to the symmetric bilinear form f_{E_8} (see Table 4.6 for the diagram). Conclude that $(\langle S \rangle, S)$, where $S = \{r_{\alpha_i} \mid i \in [8]\}$, is a Coxeter system of type E_8.

(d) Prove that the subgroup $\langle S \rangle$ of (c) coincides with $\langle r_\alpha \mid \alpha \in \Phi \rangle$ and that $W(E_8)$ is finite.

Exercise 4.9.26 Let $n \in \mathbb{N}$, $n \geq 3$, and consider the Coxeter system (W, S) of type B_n with $S = \{s_1, \ldots, s_n\}$.

(a) Take $\varepsilon_1, \ldots, \varepsilon_n$ to be the standard orthonormal basis of $\mathbb{R}^n$ with respect to the standard inner product $(\cdot, \cdot)$. Prove that, up to a coordinate transformation,

$$\left\{ \varepsilon_i \pm \varepsilon_j, \sqrt{2}\varepsilon_k \mid i, j, k \in [n], i < j \right\}$$

is the set of positive roots of the root system for W in its reflection representation.

(b) Show that the subgroup D of W generated by $s_1, \ldots, s_{n-1}, s_n s_{n-1} s_n$ is a homomorphic image of the Coxeter group of type D_n.

(c) Derive from the results of Exercise 4.9.23 that D is isomorphic to $W(D_n)$.

(d) Prove that D has index two in W and conclude that $|W| = 2^n n!$.

Exercise 4.9.27 Let (W, S) be a Coxeter system of irreducible spherical type M. Prove that the following statements are equivalent.

(a) The action of the opposition on S is trivial.

(b) There is an element in W that maps to scalar multiplication by -1 under the reflection representation.

(c) The longest element of W maps to scalar multiplication by -1 under the reflection representation.

Exercise 4.9.28 Let (W, S) be a Coxeter system of irreducible spherical type M. Prove that the opposition on M is trivial if and only if M is one of A_1, B_n $(n \geq 3)$, D_n (n even), E_7, E_8, F_4, $I_2^{(m)}$ (m even), H_n $(n \in \{3, 4\})$.

Section 4.7

Exercise 4.9.29 (Cited in Theorem 4.7.3) Let (W, S) be a Coxeter system of type M and let $k \geq 3$. Suppose that $1, 2, \ldots, k$ is a circuit in M (so $m_{1,k} \geq 3$ and $m_{i,i+1} \geq 3$ for $i \in [k-1]$). Show that the word $(12 \cdots k)^i$ is a minimal expression in S^* for each $i \in \mathbb{N}$. Conclude that W is an infinite group.
(*Hint*: Use Corollary 4.5.11.)

♯ **Exercise 4.9.30** By means of an example it will become clear in this exercise that finite complex linear groups generated by reflections need not be Coxeter groups. Put $\sigma = (1 + \sqrt{-7})/2$ and consider the following set of vectors in $\mathbb{C}^3$ with standard basis $\varepsilon_1, \varepsilon_2, \varepsilon_3$.

$$\Phi = \pm\left\{\varepsilon_i, (\varepsilon_i \pm \varepsilon_j)\frac{\overline{\sigma}}{2}, (\varepsilon_i \pm \varepsilon_j \pm \sigma\varepsilon_k)\frac{1}{2} \,\middle|\, \{i, j, k\} = [3]\right\}$$

Supply $\mathbb{C}^3$ with the standard hermitian inner product $(x, y) = \sum_i \overline{x_i} y_i$. For $\alpha \in \Phi$, let r_α be the unitary reflection with respect to $(\cdot, \cdot)$ having root α, that is, $r_\alpha = r_{\alpha,\phi}$ in the notation of Exercise 1.9.30, where $\phi(x) = 2(\alpha, \alpha)^{-1}(\alpha, x)$ $(x \in \mathbb{C}^3)$. Prove the following assertions.

(a) The size of Φ is equal to 42.
(b) For each $\alpha \in \Phi$, the reflection r_α leaves Φ invariant.
(c) The group $W = \langle r_\alpha \mid \alpha \in \Phi \rangle$ is generated by $S := \{r_{\alpha_1}, r_{\alpha_2}, r_{\alpha_3}\}$, where

$$\alpha_1 = \varepsilon_2, \qquad \alpha_2 = (\varepsilon_2 + \varepsilon_3)\frac{\overline{\sigma}}{2}, \qquad \alpha_3 = (\varepsilon_1 + \varepsilon_2 - \sigma\varepsilon_3)\frac{1}{2}.$$

(d) W is a homomorphic image of the Coxeter group with Coxeter matrix

$$\begin{pmatrix} 1 & 4 & 3 \\ 4 & 1 & 3 \\ 3 & 3 & 1 \end{pmatrix}.$$

(e) The order of W is equal to 336.
(f) The pair (W, S) is not a Coxeter system.
(g) The center of W has order two.
(h) The group W is not isomorphic to any finite Coxeter group.

Section 4.8

Exercise 4.9.31 (This exercise is used in Theorem 4.8.2) Let M be a Coxeter diagram over $[n]$ and let A be a subgroup of the Coxeter group W of type M. Show that, for each $J \subseteq [n]$, the J-cell containing the chamber Ag of $C(M)/A$ is isomorphic to $C(J)/(g^{-1}Ag \cap \langle J \rangle)$.
(*Hint*: The map $(g^{-1}Ag \cap \langle J \rangle)x \mapsto Agx$ $(x \in \langle J \rangle)$ is an isomorphism.)

4.10 Notes

Many basic ideas about geometries are rooted in the theory of polytopes of $\mathbb{E}^n$. As for $\mathbb{E}^3$, these go back to Euclid and Archimedes among others. In Schläfli's classification of the regular convex polytopes, as early as 1852, the study of residues plays a major role.

The presentation given here is due to DuVal [123]. This reference includes the fascinating description of the 24-cell and the 600-cell as multiplicative groups of 24 and 120 quaternions.

The concept of a thin geometry and the abstract approach to polytopes and regularity are due to Tits who used them as a central piece for his monumental building theory [284, 286].

Another leading figure in the history of this chapter is Coxeter, whose influence appears almost in every section. In particular, he originated the classification of finite reflection groups.

Section 4.1

Well-known books on regular polytopes are [99] and, for unitary space, [100]. More recently, a theory of abstract regular polytopes and its realizations in Euclidean space has been developed, most of which is covered well in [213]. All sufficiently small examples are covered in [153]. As recent as 2007, McMullen [212] constructed new realizations of regular polytopes in $\mathbb{E}^4$.

There are many more interpretations of regularity in geometries than the weak and strong versions discussed in Remark 4.1.7. Examples are equality of orders (see Definition 2.3.7) or regularity (in the sense that the valency of each vertex only depends on its type) of their incidence graphs. Most versions of regularity are essentially distinct. An interesting regularity study occurs in the work of Walsh [299]. Thin rank 3 geometries over a linear Coxeter diagram are, up to terminology, the subject of Errera's work [124]. Our regular thin rank 3 geometries in the maximal sense appear as 'reflexible maps' in [101].

A modern classic on polytopes that are not necessarily regular is [316].

Euler's formula, used in the proof of Theorem 4.1.8, can be found in [99, 1.61] and many other places.

The four regular stellated solids of Fig. 4.4 are the so-called *Kepler-Poinsot* solids. They were discovered long before, but were characterized by Cayley in 1859; see [12].

As mentioned in Example 4.1.13, no distinction is made between the Coxeter types B_n and C_n. This is due to the fact that the notation is adopted from Elie Cartan [61], who used the labels A_n ($n \geq 1$), B_n ($n \geq 2$), C_n ($n \geq 3$), D_n ($n \geq 4$), E_n ($n = 6, 7, 8$), F_4, and G_2 in the classification of complex simple Lie groups. The Dynkin types, B_n and C_n differ; this is related to the difference in lattices left invariant by the corresponding Coxeter group in the reflection representation and explains the distinction between $\widetilde{B}_n$ and $\widetilde{C}_n$ for $n \geq 3$.

Section 4.2

Theorem 4.2.8 is a special case of a theorem in [287].

The name Coxeter groups refers to the fact that Coxeter studied presentations for finite linear groups generated by reflections; see [96–99].

The general theory of abstract Coxeter groups including the terminology in honor of Coxeter, is due to Tits who wrote a paper entitled Groupes et Géométries de Coxeter (Notes polycopiées, I.H.E.S., Bures-sur-Yvette, juin 1961, 26 pages) intended for Bourbaki's book [29]. This became a very thorough and basic reference for Coxeter groups. Tits' paper was finally published by Hirzebruch and Manin in the Wolf Prize Volume (2001).

A very good textbook on the subject is [169]. The combinatorial aspects are the main focus of [22]. Most textbooks on buildings, such as [2, 244, 305], devote attention to Coxeter groups.

The basic theory on presentations of groups by generators and relations is assumed known. It is discussed in [101]. A very accessible introduction to this subject is [178].

The process of coset enumeration is not an algorithm as it will not always terminate. If the finitely presented group is known to be finite, then the process will terminate in a finite number of steps; see [84] for a gentle introduction.

In its greatest generality, the word problem for groups and monoids is undecidable in the sense that there is no algorithm (for a Turing machine—the most common theoretic model for computers) deciding whether two words represent the same element in the quotient group or monoid. The solution to the word problem for Coxeter groups, presented in Corollary 4.5.11, is due to Tits [285] and Matsumoto [210]. The method is relatively convenient when applied by hand, but algorithmically cumbersome: the number of minimal expressions for the longest word in $W(A_{n-1}) = \mathrm{Sym}_n$ is

$$\frac{\binom{n}{2}!}{1^{n-1}3^{n-2}5^{n-3}\cdots(2n-1)^0}.$$

according to [266]. Faster algorithms are known; cf. [63, 64].

Section 4.3

Theorem 4.3.6 is due to Coxeter, who was the first to state results of this kind, and Tits, who proved the theorem in its full generality.

Section 4.4

Although versions were undoubtedly known to Coxeter, the proof of Theorem 4.4.16 in its full generality should be credited to Tits. The current proof is adapted from [29] by use of Lemma 4.2.20, which is due to Casselman (presented at CRM lectures in 2002). An alternative proof of linearity is given in [116].

Proofs that the order of $s_i s_j$ is equal to $m_{i,j}$ avoiding the reflection representation have been explored in [227]. A proof using more algebraic facts is given in [220, 221].

Section 4.5

The usual definition of root system is given in [29]. It differs from the one used in this section in that it has an integrality and a finiteness condition. The approach using the root system to prove properties of minimal expressions was known and probably introduced by Deodhar and Howlett; see [115].

For Weyl groups (finite Coxeter groups leaving invariant an integral lattice in the reflection representation), Papi [232] has given a characterization of subsets of Φ^+ of the form Φ_w for some $w \in W$ as the subsets that are closed and whose complements in Φ^+ is also closed. He also shows that the element w is uniquely determined by Φ_w.

The exchange condition can be found in [29]. The strong exchange condition appears in [116].

Section 4.6

All finiteness characterizations are in [29]. There are interesting combinatorial properties of Φ that we will not go into in this book. For instance, for each finite Coxeter group of rank n, there exist positive integers $d_1, \ldots, d_n$ with $d_1 \leq d_2 \leq \cdots \leq d_n$ such that Φ has precisely $d_1 + \cdots + d_n - n$ elements, $d_1 d_2 \cdots d_n = |W|$, and the sum $d_i + d_{n+1-i}$ is constant for $i \leq \lfloor n/2 \rfloor$. If $M - A_n$, then these integers are $2, 3, \ldots, n+1$.

If (W, S) is a Coxeter system in which W is finite, then the product of all elements of S (in any order) is called a Coxeter element (Coxeter transformation in [29]). It acts regularly on the corresponding root system. If its order h (known as the Coxeter number of (W, S)) is even, then its $h/2$-th power is the longest element of W. This explains the form of many longest elements in Table 4.5.

Section 4.7

Most of this material is dealt with in [29]. There, and in [169], classifications of Coxeter diagrams M for which κ_M is hyperbolic are also given.

The proof of Proposition 4.7.2 did not need the precise value of the volume of the n-dimensional unit ball, which is well known to be $\pi^{\frac{n}{2}} \Gamma(\frac{n}{2} + 1)$.

Section 4.8

The results on polytopes and tessellations of Coxeter type, notably Theorems 4.8.2, 4.8.4, and Corollary 4.8.5, are treated in [213]. The proof of Theorem 4.8.2 given here uses similar topological arguments but stays closer to the chamber systems of Coxeter type developed in this chapter. The elementary results on topological coverings used in the proof can be found in many textbooks, like [129]. The fact that $\widetilde{\Sigma}$ and Σ are homeomorphic, established in the proof of Theorem 4.8.2, can also be derived by using [129, Lemma 11.17, Corollary 13.15] and showing that the fundamental group of Σ coincides with A.

Several series of groups have been considered as suitable quotients of Coxeter groups for providing abstract regular polytopes. See, for instance, [201, 202] for the groups $\mathrm{PSL}(\mathbb{F}_q^2)$ and $\mathrm{PGL}(\mathbb{F}_q^2)$, where, for rank at least four, only a very limited number of values for q turns out to work; the example in Exercise 4.9.6 of rank four with $q = 19$ is among these.

Section 4.9

Exercise 4.9.9(b) is due to Mühlherr [226]. Exercise 4.9.30 contains an example of a finite complex reflection group, just like Exercises 2.8.25 and 4.9.13. The classification of these groups is due to Shephard and Todd [254]; it was simplified in [69], and is treated in the textbook [203]. The group of Exercise 4.9.30 is isomorphic to $C_2 \times \mathrm{PSL}(\mathbb{F}_7^2)$, as will become clear in Exercise 10.7.30.

Chapter 5
Linear Geometries

In Example 1.4.9 we introduced the projective geometry PG(V) and in Example 1.4.10 the affine geometry AG(V) associated with a vector space V of finite dimension n. In Proposition 2.4.7 the geometry PG(V) was shown to have a linear Coxeter diagram A_{n-1}, and in Proposition 2.4.10 the geometry AG(V) was shown to belong to the linear diagram Af_n. We now turn our attention to the more general class of all geometries with a linear diagram. The shadow spaces on 1 of our motivating examples PG(V) and AG(V) are linear line spaces (in the sense that any two points are on a unique line; cf. Definition 2.5.13), and we will restrict ourselves mostly to geometries with this property. Within this class there are combinatorial structures such as matroids and Steiner systems.

In Sect. 5.1, we introduce affine space as an abstraction of a vector space, and in Sect. 5.2 we do the same for projective space. The point shadows of the geometries AG(V) are examples of affine spaces and those of PG(V) are examples of projective spaces. In Sect. 5.3 we make the connection between geometries with a linear diagram and matroids, whereas in Sect. 5.4 we show how to build a geometry with a linear diagram from a matroid. In Sect. 5.5 we devote attention to Steiner systems and in particular the unique Steiner system $S(5, 8, 24)$ related to the Golay code. Finally, in Sect. 5.6, we study its automorphism group, which is a sporadic simple group, the Mathieu group on 24 letters. Although the last two sections are not needed for the remainder of the book, the treatment of this Mathieu group (as well as the other four Mathieu groups, which appear as groups of automorphisms of substructures in the Steiner system $S(5, 8, 24)$) shows how they appear naturally in diagram geometry. In Sect. 5.8, the notes to this chapter, we give a linear diagram for a geometry of each of the 26 sporadic finite simple groups, with references to the literature.

5.1 The Affine Space of a Vector Space

Affine spaces, the subject of this section, are abstractions of vector spaces. We formulate the abstract definition of an affine space and show that the well-known affine

F. Buekenhout, A.M. Cohen, *Diagram Geometry*,
Ergebnisse der Mathematik und ihrer Grenzgebiete. 3. Folge / A Series of
Modern Surveys in Mathematics 57, DOI 10.1007/978-3-642-34453-4_5,
© Springer-Verlag Berlin Heidelberg 2013

space associated with a vector space is an example. In Chap. 6, the abstract definition will be used for an axiomatic characterization of the affine space of a vector space.

We state some facts without proofs because they do not go beyond elementary linear algebra. We take the scalars of a vector space to come from a division ring rather than a field. The main reason for this greater generality is that the class of affine spaces over a division ring will be characterized by a simple set of axioms and that the restriction to fields would not imply an essential simplification.

For the duration of the section, let $\mathbb{D}$ be a division ring and let V be a right vector space of dimension n (possibly $n = \infty$) over $\mathbb{D}$.

Recall the notion of affine plane from Definition 2.3.1 and bear in mind that, by Theorem 2.5.15, its point shadow space is a linear space fully determining the affine plane itself.

Definition 5.1.1 A *space with parallelism* $(P, L, \|)$ is a line space (P, L) provided with an equivalence relation $\|$, called *parallelism*, on L such that each equivalence class forms a partition of P. A *parallel-closed* subspace of such a space $(P, L, \|)$ is a subspace X of (P, L) such that, for each $l \in L$ with $l \subseteq X$ and each $x \in X$, the parallel to l containing x lies in X.

For an arbitrary space with parallelism $(P, L, \|)$, the collection of all parallel-closed subspaces is closed under intersection. Therefore, we can speak of the parallel-closed subspace *generated* by a subset X of P as the intersection of all parallel-closed subspaces containing X.

An *affine space* is a space with parallelism in which the parallel-closed subspace generated by any three non-collinear points is an affine plane.

Notation 5.1.2 For $p, q \in V$, we write $p + (q - p)\mathbb{D}$ to denote $\{p + (q - p)\lambda \mid \lambda \in \mathbb{D}\}$, and similarly for sums with more terms. By $\mathbb{A}(V)$ we denote the following space with parallelism.

(1) A *point* of $\mathbb{A}(V)$ is a vector of V.
(2) A *line* of $\mathbb{A}(V)$ is an affine line of V, that is, a coset of a 1-dimensional subspace of V. So, if p and q are distinct points, then $pq := p + (q - p)\mathbb{D}$ is the unique line containing p and q.
(3) Two lines are *parallel* if they are cosets of the same 1-dimensional subspace of V.

Proposition 5.1.3 *The parallel-closed subspace generated by any three non-collinear points of* $\mathbb{A}(V)$ *is isomorphic to* $\mathbb{A}(\mathbb{D}^2)$. *In particular,* $\mathbb{A}(V)$ *is an affine space.*

Proof Clearly, the line space $\mathbb{A}(V)$ is linear. Parallelism is an equivalence relation and all parallels to a given line partition the set of points. The parallel-closed subspace of $\mathbb{A}(V)$ generated by three non-collinear points p, q, and r is the affine plane $p + (q - p)\mathbb{D} + (r - p)\mathbb{D}$, isomorphic to $\mathbb{A}(\mathbb{D}^2)$. $\square$

Fig. 5.1 The affine space
$\mathbb{A}(\mathbb{F}_3^3)$ with 3 of its 13 parallel
classes of lines

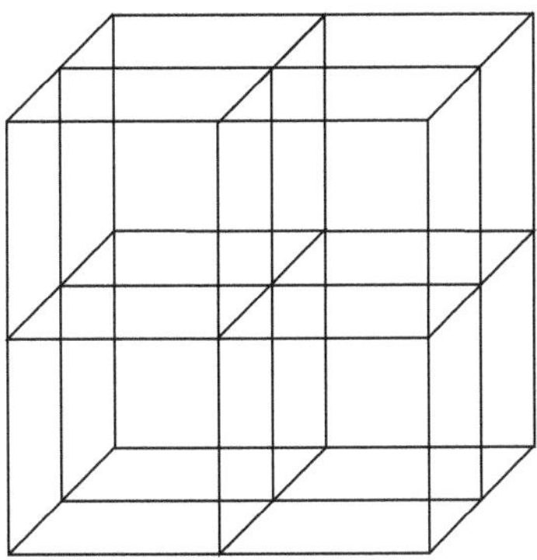

In earlier chapters (for instance, in Example 1.1.2), we have been using the term
affine space in a way that is consistent with $\mathbb{A}(V)$ being the affine space on V.
Theorem 2.5.15 shows that, if $n < \infty$, the affine space $\mathbb{A}(V)$ without the parallelism
can be identified with the [2]-truncation of $AG(V)$. (The restriction to finite n is due
to the limitations on the construction of the geometry $AG(V)$; see Remark 1.4.11
for a construction using finite-dimensional affine subspaces of V only for which the
identification is still correct.) For $i \in [n]$, an affine subspace of V of dimension $i - 1$
is an i-element of $AG(V)$ and a parallel-closed subspace of $\mathbb{A}(V)$. Each parallel-
closed subspace of $\mathbb{A}(V)$ is an affine subspace of V.

Example 5.1.4 The 3-dimensional affine space $\mathbb{A}(\mathbb{F}_3^3)$ of order three has 27 points.
Figure 5.1 shows three parallel classes of lines (from a total of 13 classes). Also,
three classes of planes (from a total of 13) are visible.

Remark 5.1.5 When $|\mathbb{D}| \geq 3$, each subspace of $Z := \mathbb{A}(V)$ is parallel closed, and
the parallelism of Z is uniquely determined by its lines. Given a line l and a point p
in $Z \setminus l$, there is a unique line in Z contained in the affine plane of Z generated by l
and p, containing p, and disjoint from l. This line is parallel to l.

For $\mathbb{D} = \mathbb{F}_2$, the requirement that subspaces are parallel closed is crucial. If, in
addition $n \geq 3$, then the parallelism is not uniquely determined by the lines: there is
another parallelism than the one of Proposition 5.1.3. For instance in $\mathbb{F}_2^3$, we can start
with the usual parallelism and decree that the two disjoint lines $\{(1, 0, 1), (0, 1, 1)\}$
and $\{(0, 0, 1), (1, 1, 1)\}$ in the affine hyperplane $x_3 = 1$ are assigned to the parallel class of $\{(0, 0, 0), (0, 1, 0)\}$, whereas the two disjoint lines $\{(1, 0, 1), (1, 1, 1)\}$
and $\{(0, 0, 1), (0, 1, 1)\}$ in the same hyperplane are assigned to the parallel class
of the line $\{(0, 0, 0), (1, 1, 0)\}$, and that nothing else changes. Each of the classes
of the resulting equivalence relation on lines is again a partition of $\mathbb{F}_2^3$. In fact,
the line space $\mathbb{A}(\mathbb{F}_2^n)$ without parallelism is nothing but the complete graph on 2^n
points.

The exploration of affine spaces will go hand in hand with projective spaces,
to be introduced in the next section. In the remainder of this section we study the

automorphism group of $\mathbb{A}(V)$. For this purpose, but also for the later treatment of matroids, the notion of bases is useful.

Definition 5.1.6 An *affine basis* of an affine space Z is a set B of points of Z such that the parallel-closed subspace generated by B is Z itself and such that no proper subset of B has the same property.

If B is an affine basis of $\mathbb{A}(V)$, then, for each $v \in B$, the set $\{b - v \mid b \in B \setminus \{v\}\}$ is a basis of V.

Definition 5.1.7 In the vein of Definition 1.7.9, the group $\mathrm{Aut}(\mathbb{A}(V))$ of all *automorphisms* of $\mathbb{A}(V)$ is the group of all permutations of the points mapping every line to a line and preserving parallelism.

Recall from Example 1.8.17 that a translation of V is a map of the form t_a for some $a \in V$, where $t_a(x) = x + a$ $(x \in V)$, and that $\mathrm{T}(V)$ is the subgroup of $\mathrm{Aut}(\mathbb{A}(V))$ consisting of all translations of V.

Lemma 5.1.8 *If* $\dim(V) \geq 2$, *the group* $\mathrm{T}(V)$ *is a normal subgroup of* $\mathrm{Aut}(\mathbb{A}(V))$.

Proof This follows directly from the fact that a non-trivial translation can be characterized as an automorphism of $\mathbb{A}(V)$ fixing each parallel class of lines and fixing no point.

An alternative way of seeing this runs as follows. Let p, q, r be points of $\mathbb{A}(V)$ and $g \in \mathrm{Aut}(\mathbb{A}(V))$. We first claim

$$g(p + q - r) = g(p) + g(q) - g(r). \tag{5.1}$$

To prove the claim, we first suppose that p, q, and r are non-collinear. The parallel-closed subspace spanned by p, q, and r is an affine plane in which $p + q - r$ is the unique point on both the line through q and parallel to pr and the line through p and parallel to qr. In particular, since automorphisms preserve lines and parallelism, we have $g(p + q - r) = g(p) + g(q) - g(r)$. Suppose next that p, q, and r are distinct and collinear. Take a point s off the line pq, so the three points $p + s$, q, and $r + s$ are non-collinear. By the above, $g(p + q - r) = g((p + s) + q - (r + s)) = g(p + s) + g(q) - g(r + s)$. As p, $q + s$, and $r + s$ are also non-collinear, we find $g(p + q - r) = g(p) + g(q + s) - g(r + s)$ as well. Combining these two equalities, we find $g(p + s) - g(q + s) = g(p) - g(q)$. Applying this result to p, r, and s instead of p, q, and s, we conclude $g(p + s) - g(r + s) = g(p) - g(r)$, and so $g(p + q - r) = g(p + s) + g(q) - g(r + s) = g(p) + g(q) - g(r)$, establishing (5.1) for distinct p, q, and r. Clearly, (5.1) also holds if $p = r$ or $q = r$. Assume, therefore, that $p = q$ is distinct from r. Pick a point u in $\mathbb{A}(V)$ distinct from p and r; now $g(p + p - r) = g(p + (p + u - r) - u) = g(p) + g(p + u - r) - g(u) = g(p) + g(p) - g(r)$, again. This proves the claim.

Now, suppose $t \in \mathrm{T}(V)$ and $g \in \mathrm{Aut}(\mathbb{A}(V))$. There is $a \in V$ such that $t = t_a$. For $x \in V$, (5.1) gives

$$gt_a g^{-1}(x) = g\left(g^{-1}x + a\right) = g\left(g^{-1}x\right) + g(a) - g(0) = t_{g(a)-g(0)}(x), \quad (5.2)$$

whence $gTg^{-1} \subseteq T$. $\qquad\square$

In the case where $\dim V = 1$, these arguments collapse (cf. Remark 5.1.14). See Exercise 5.7.4 for a remedy.

The definition of a semi-linear map and of $\Gamma\mathrm{L}(V)$ are given in Exercise 2.8.28.

Lemma 5.1.9 *Let V be a right vector space over $\mathbb{D}$.*

(i) *The group $\Gamma\mathrm{L}(V)$ is a subgroup of $\mathrm{Aut}(\mathbb{A}(V))$ and the map $\sigma : \Gamma\mathrm{L}(V) \to \mathrm{Aut}(\mathbb{D})$ assigning to g the automorphism σ_g of $\mathbb{D}$ induced by it is a homomorphism of groups.*

(ii) *The kernel of σ is the group $\mathrm{GL}(V)$ of all linear transformations of V. If B is a basis of V, then the vector-wise stabilizer K of B in $\Gamma\mathrm{L}(V)$ is a group isomorphic to $\mathrm{Aut}(\mathbb{D})$, and $\Gamma\mathrm{L}(V)_B = \mathrm{GL}(V)_B \rtimes K$. Moreover, $\Gamma\mathrm{L}(V) \cong \mathrm{GL}(V) \rtimes \mathrm{Aut}(\mathbb{D})$.*

(iii) *The group $\mathrm{GL}(V)$ is transitive on the set of bases of V.*

Proof (i) If $g_1 \in \Gamma\mathrm{L}(V)$ induces $\sigma_1 \in \mathrm{Aut}(\mathbb{D})$ and $g_2 \in \Gamma\mathrm{L}(V)$ induces $\sigma_2 \in \mathrm{Aut}(\mathbb{D})$, then $g_1 g_2 \in \Gamma\mathrm{L}(V)$ induces $\sigma_1 \sigma_2$ and $g_1^{-1} \in \Gamma\mathrm{L}(V)$ induces σ_1^{-1}. This shows that $\Gamma\mathrm{L}(V)$ is indeed a subgroup of $\mathrm{Aut}(\mathbb{A}(V))$ and that σ is a homomorphism.

(ii) The kernel of σ is clearly the group $\mathrm{GL}(V)$ of all linear transformations of V. The homomorphism σ is surjective: fixing a basis B of V, we can realize $\tau \in \mathrm{Aut}(\mathbb{D})$ as the automorphism σ_g induced by the element $g \in \Gamma\mathrm{L}(V)$ defined by

$$g\left(\sum_{b \in B} b\lambda_b\right) = \sum_{b \in B} b\tau(\lambda_b) \quad (\lambda_b \in \mathbb{D}).$$

For fixed B, the set of all such g forms a subgroup K of $\Gamma\mathrm{L}(V)$ fixing B vector-wise, which complements $\mathrm{GL}(V)$. Clearly, K is the full vector-wise stabilizer of B in $\Gamma\mathrm{L}(V)$, and is isomorphic to $\mathrm{Aut}(\mathbb{D})$. We conclude that $\Gamma\mathrm{L}(V)$ is the semi-direct product of the normal subgroup $\mathrm{GL}(V)$ by K.

(iii) This is standard linear algebra. If $B = \{b_1, \ldots, b_n\}$ and $B' = \{b'_1, \ldots, b'_n\}$ are bases of V, then there is a unique linear transformation $g \in \mathrm{GL}(V)$ such that $gb_i = b'_i$ ($i \in [n]$), namely the map $\sum_i b_i \lambda_i \mapsto \sum_i b'_i \lambda_i$. It sends B to B'. $\qquad\square$

Definition 5.1.10 The subgroup of $\mathrm{Aut}(\mathbb{A}(V))$ generated by $\mathrm{T}(V)$ and $\Gamma\mathrm{L}(V)$, usually denoted by $\mathrm{A}\Gamma\mathrm{L}(V)$, is called the *affine group* of V.

Remark 5.1.11 Since $\mathrm{T}(V)$ and $\mathrm{GL}(V)$ intersect trivially and $\mathrm{T}(V)$ is normalized by $\mathrm{GL}(V)$ according to Lemma 5.1.8, the subgroup $\mathrm{AGL}(V)$ of $\mathrm{A}\Gamma\mathrm{L}(V)$ generated by both is a semidirect product of the two.

For $\lambda \in \mathbb{D}$, the homothety h_λ (see Exercise 2.8.28) belongs to $\Gamma\mathrm{L}(V)$ and satisfies $\sigma_{h_\lambda}(x) = \lambda^{-1}x\lambda$. The homotheties form a subgroup H of $\Gamma\mathrm{L}(V)$, but, if $\mathbb{D}$ is not commutative, not of $\mathrm{GL}(V)$. The group H is isomorphic to the multiplicative group of $\mathbb{D}^{\mathrm{op}}$, the opposite division ring of $\mathbb{D}$.

Theorem 5.1.12 *Let V be a vector space of dimension n over $\mathbb{D}$.*

(i) *If $n = 1$, then* $\mathrm{Aut}(\mathbb{A}(V)) = \mathrm{Sym}(V)$;
(ii) *if $n \geq 2$, then* $\mathrm{Aut}(\mathbb{A}(V)) = \mathrm{A\Gamma L}(V)$.

If $n < \infty$, this group acts flag transitively on $\mathrm{AG}(V)$.

Proof Since $\mathbb{A}(V)$ and $\mathrm{AG}(V)$ determine each other uniquely, $\mathrm{Aut}(\mathbb{A}(V)) = \mathrm{Aut}(\mathrm{AG}(V))$. Clearly, $\mathrm{A\Gamma L}(V)$ is a subgroup of $\mathrm{Aut}(\mathbb{A}(V))$. The final statement was already proven in Example 1.8.17.

(i) is obvious as there is only one line.

(ii) Let α be an automorphism of $\mathbb{A}(V)$. We must show that $\alpha \in \mathrm{A\Gamma L}(V)$. As before, write t_p for the translation mapping 0 to $p \in V$. Then $\beta = t_{-\alpha(0)}\alpha$ is an automorphism of $\mathbb{A}(V)$ fixing 0. It suffices to show that $\beta \in \mathrm{\Gamma L}(V)$. As $n \geq 2$, we can apply (5.2), which gives $\beta(x + y) = \beta t_y \beta^{-1} \beta(x) = t_{\beta y - \beta 0}(\beta x) = \beta x + \beta y - \beta 0 = \beta x + \beta y$ for all $x, y \in V$. We conclude that β preserves addition.

The set H of all homotheties can be characterized as the set of all automorphisms of $\mathbb{A}(V)$ fixing the point 0 and every affine line through 0. Hence, it is a normal subgroup of the stabilizer $\mathrm{Aut}(\mathbb{A}(V))_0$. In particular, $\beta h_\lambda \beta^{-1} = h_{\rho\lambda}$, where $\rho : \mathbb{D} \to \mathbb{D}$ is a permutation of $\mathbb{D}$ fixing 0. Now, for $\lambda \in \mathbb{D}$ and $x \in V$,

$$\beta(x\lambda) = \beta h_\lambda x = \left(\beta h_\lambda \beta^{-1}\right)(\beta x) = h_{\rho\lambda}(\beta x) = (\beta x)(\rho\lambda),$$

so, with the notation of Lemma 5.1.9(i), we have $\sigma_\beta = \rho$. It remains to show that $\rho \in \mathrm{Aut}(\mathbb{D})$. On the one hand, for all $\lambda, \mu \in \mathbb{D}$,

$$h_{\rho(\lambda\mu)} = \beta h_{\mu\lambda}\beta^{-1} = \beta h_\mu h_\lambda \beta^{-1} = h_{\rho\mu} h_{\rho\lambda} = h_{(\rho\lambda)(\rho\mu)}$$

and so $\rho(\lambda\mu) = (\rho\lambda)(\rho\mu)$. On the other hand, for $\lambda, \mu \in \mathbb{D}$, we have $h_{\rho(\lambda+\mu)} = \beta h_{\lambda+\mu}\beta^{-1}$; this, applied to any $x \in V$, gives

$$\begin{aligned}
x\rho(\lambda + \mu) &= h_{\rho(\lambda+\mu)}x = \beta h_{\lambda+\mu}\beta^{-1}x \\
&= \beta\big((\beta^{-1}x)(\lambda + \mu)\big) = \beta\big((\beta^{-1}x)\lambda + (\beta^{-1}x)\mu\big) \\
&= \beta\big(h_\lambda(\beta^{-1}x) + h_\mu(\beta^{-1}x)\big) = \beta h_\lambda\beta^{-1}x + \beta h_\mu\beta^{-1}x \\
&= h_{\rho(\lambda)}x + h_{\rho(\mu)}x = x\big(\rho(\lambda) + \rho(\mu)\big),
\end{aligned}$$

and so $\rho(\lambda + \mu) = \rho\lambda + \rho\mu$. This suffices for the proof of $\rho \in \mathrm{Aut}(\mathbb{D})$. $\qquad\square$

The result that the automorphism group of $\mathbb{A}(V)$ acts flag transitively on $\mathrm{AG}(V)$ if $\dim(V) < \infty$ can be extended to the point shadow space of the geometry of denumerable rank described in Remark 1.4.11.

With the methods of proof of Theorem 5.1.12, it is also possible to show the following more general result.

Corollary 5.1.13 *Suppose that V is a vector space of dimension $n \geq 2$ over the division ring $\mathbb{D}$ and V' is a vector space of dimension $n' \geq 2$ over the division ring $\mathbb{D}'$. The affine spaces $\mathbb{A}(V)$ and $\mathbb{A}(V')$ are isomorphic (i.e., there is a bijection from the set of points of $\mathbb{A}(V)$ onto the set of points of $\mathbb{A}(V')$ preserving lines and* *parallelism) if and only if* $\mathbb{D}$ *is isomorphic to* $\mathbb{D}'$ *and* V *is isomorphic (by a semilinear map after identification of* $\mathbb{D}$ *with* $\mathbb{D}'$*) to* V'*. If* n *and* n' *are finite, then this is equivalent to* $\mathbb{D} \cong \mathbb{D}'$ *and* $n = n'$*.*

Remark 5.1.14 In the one-dimensional case, the translations do not form a normal subgroup of $\mathrm{Aut}(\mathbb{A}(V))$. For $5 \le |V| < \infty$, the only proper nontrivial normal subgroup of $\mathrm{Sym}(V)$ is $\mathrm{Alt}(V)$. The group $\mathrm{T}(V)$ of translations of V, however, has order $|V|$ and so is not a normal subgroup of $\mathrm{Sym}(V)$.

Example 5.1.15 Let $\mathbb{F}_q$ be the field of order $q = p^m$ where p is a prime number and m a non-negative integer. It is known that $\mathrm{Aut}(\mathbb{F}_q)$ is a cyclic group of order m, generated by the Frobenius automorphism $\mathbb{F}_q \to \mathbb{F}_q$, which maps x to x^p. If $V = \mathbb{F}_q^n$, then $\mathrm{T}(V)$ has order q^n and $\mathrm{GL}(V)$ has order

$$\left(q^n - 1\right)\left(q^n - q\right)\left(q^n - q^2\right) \cdots \left(q^n - q^{n-1}\right).$$

Consequently, $\mathrm{A\Gamma L}(V)$ has order

$$mq^n \prod_{i=0}^{n-1} \left(q^n - q^i\right).$$

In particular, the affine plane of order three (Example 2.3.2) has a group of automorphisms of order 432 while the affine plane of order four (Example 2.3.2) has a group of order 5760.

5.2 The Projective Space of a Vector Space

In this section we explore the shadow space on 1 of the projective geometry of a vector space. Throughout this section, $\mathbb{D}$ is a division ring and V is a right vector space of dimension $n + 1$ over $\mathbb{D}$. We allow for n to be ∞.

Projective planes were introduced in Definition 2.2.7. They are related to line spaces as indicated in Theorems 2.2.9 and 2.5.15.

Definition 5.2.1 A linear space is called a *projective space* if it contains three non-collinear points and if the subspace generated by any three non-collinear points is a projective plane.

The *projective space* on V, denoted $\mathbb{P}(V)$, is the space whose points are the 1-dimensional subspaces of V and whose lines are the sets of 1-dimensional subspaces contained in a 2-dimensional subspace of V.

Proposition 5.2.2 *The subspace of* $\mathbb{P}(V)$ *generated by any three non-collinear points of* $\mathbb{P}(V)$ *is isomorphic to the projective plane* $\mathbb{P}(\mathbb{D}^3)$*. In particular, if* $n \ge 2$*, then* $\mathbb{P}(V)$ *is a projective space. If* $2 \le n < \infty$*, the corresponding [2]-geometry is isomorphic to the [2]-truncation of* $\mathrm{PG}(V)$*.*

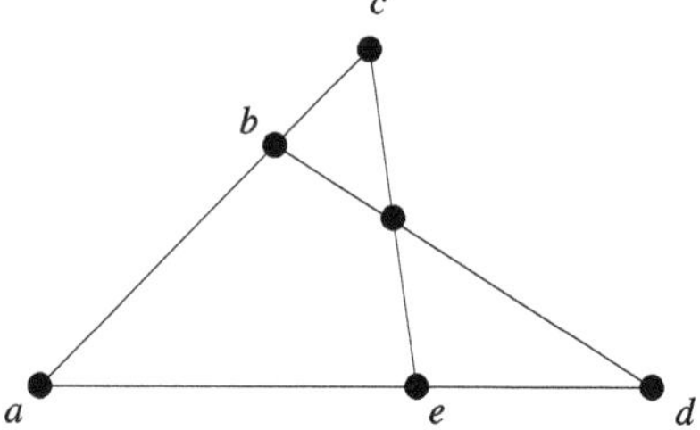

Fig. 5.2 Pasch's axiom: the lines ce and bd intersect

Proof If p and q are distinct points of $\mathbb{P}(V)$, then the set pq of points $(p\lambda + q\mu)\mathbb{D}$ with $\lambda, \mu \in \mathbb{D}$ (not both zero) is the unique line on p and q. So $\mathbb{P}(V)$ is linear indeed. The subspace of $\mathbb{P}(V)$ generated by three non-collinear points clearly consists of points and lines inside the 3-dimensional vector subspace W of V spanned by the three points. This readily implies that these three points generate $\mathbb{P}(W)$, which is isomorphic to $\mathbb{P}(\mathbb{D}^3)$, a projective plane.

The final statement follows directly from the definition of lines in $\mathbb{P}(V)$. $\qquad\square$

Notice that $\mathrm{Aut}(\mathrm{PG}(V)) = \mathrm{Aut}(\mathbb{P}(V))$, as $\mathrm{PG}(V)$ and $\mathbb{P}(V)$ determine each other uniquely.

A line space consisting of a single line is trivially a projective space. However, we do not refer to such a space as a projective line, but reserve that name for a richer structure, to be defined in Sect. 6.2.

Remark 5.2.3 Corollary 6.3.2 will show that any thick projective space satisfying a few additional conditions is isomorphic to $\mathbb{P}(V)$ for some vector space V. We will see that projective spaces are simple matroids and as such will give rise to geometries of type A_n. All of this shows how a concise set of geometric properties suffices for reconstructing the vector space and the underlying division ring. In comparison, the usual algebraic definition involves about 20 axioms.

Curiously enough, we can still improve on the definition of a projective space (Definition 5.2.1) when we use 'Pasch's axiom'. Recall from Definition 2.5.13 that a space is called linear if any two points are on a unique line.

Definition 5.2.4 Let Z be a linear space. A *Pasch configuration* in Z is a set of four distinct collinear triples of points whose union has cardinality six (see Fig. 5.2).

We say that Z satisfies *Pasch's axiom* if, for all points a, b, c, d, e of Z such that $\{a, b, c\}$ and $\{a, d, e\}$ are collinear triples on distinct lines, the lines bd and ce have a common point.

If, in the definition of Pasch's axiom, the common point is called f, then $\{b, d, f\}$ and $\{c, e, f\}$ are two more collinear triples besides the two already mentioned, and so the four form a Pasch configuration.

Clearly, a projective space satisfies Pasch's axiom, since every projective plane does and, by Definition 5.2.1, any two intersecting lines lie in a projective plane. We will prove the converse, for which we need the notion of a maximal subspace.

Definition 5.2.5 A *maximal subspace* of a line space Z is understood to be a maximal proper subspace of Z.

A proof of the existence of maximal subspaces in a projective space will be deferred till Proposition 5.2.10. For each hyperplane H of V, the projective space $\mathbb{P}(H)$ is a maximal subspace of $\mathbb{P}(V)$ and conversely, every maximal subspace of $\mathbb{P}(V)$ is equal to $\mathbb{P}(B)$ for some hyperplane B of V.

If x, y are points of a linear space, we denote by xy the (unique) line containing these.

Theorem 5.2.6 *Let Z be a linear space satisfying Pasch's axiom. Then Z is a projective space and the following properties hold.*

(i) *For each non-empty subspace X and every point $p \in Z \setminus X$, the subspace $\langle X, p \rangle$ of Z is the union of all lines px where $x \in X$.*

(ii) *For each maximal subspace H and each line l of Z, the subspace $H \cap l$ is non-empty.*

(iii) *Given two distinct maximal subspaces H, H', the intersection $H \cap H'$ is a maximal subspace of both H and H'.*

Proof Let l be a line and p a point of Z with $p \notin l$. Consider the union U of all lines px with $x \in l$. We will show that U is the plane (subspace) generated by p and l and that U is a projective plane. Let a, b be distinct points in U. By Pasch's axiom, ab intersects l in some point x. So, for $y \in ab$ with $y \neq p$, Pasch's axiom implies that py intersects l, hence $y \in U$ and so $ab \subseteq U$. Thus U is a (line) subspace. Moreover, every line in U intersects l. If $p \notin ab$, the preceding arguments also show that U is the union of all lines joining p to a point of ab, i.e., l may be replaced by ab. Therefore every line in U intersects ab and U is a projective plane. Consequently, Z is a projective space.

(i) Let U be the union of all lines xp, for x running in X. Clearly $X \cup \{p\} \subseteq U \subseteq \langle X, p \rangle$. Therefore it suffices to show that U is a subspace of Z. Let a, b be distinct points in U. We show that the line ab is in U. This is obvious if p, a, and b are collinear. Assume that p, a, b are non-collinear and let π be the projective plane which they generate. Let a', b' be points of X such that $a \in pa'$, $b \in pb'$. Then a', b' are distinct points in π, so $a'b'$ is contained in π. For $x \in ab$, the line px is in π and intersects $a'b'$ in some point x'. Then $x' \in X$ and so $x \in xp \subseteq U$, showing that x, and hence ab, is contained in U.

(ii) We (may) assume $l \not\subseteq H$. If $p \in l \setminus H$, then $\langle H, p \rangle = Z$, so, by (i), the point set of Z is the union of all lines joining p to some point in H. Thus l is a line joining p to a point in H.

(iii) By (ii), every line in H intersects $H \cap H'$ in some point. Hence, if $p \in H \setminus H'$, then $\langle H \cap H', p \rangle$ contains every line on p in H, whence, in view of (i), coincides with H. $\square$

Theorem 5.2.6(ii) above is important enough for it to be caught in a definition.

Definition 5.2.7 A *geometric hyperplane* of a line space Z is a proper subspace of Z such that each line meets it. If Z is a projective space, we also write *hyperplane* instead of geometric hyperplane.

Lemma 5.2.8 *Each geometric hyperplane of a linear space Z is a maximal subspace of Z.*

Proof If H is a geometric hyperplane of Z, and a is a point of Z outside H, then every point b distinct from a lies on the line ab meeting H in a point distinct from a and so is contained in $\langle H \cup \{a\} \rangle$. Consequently, $\langle H \cup \{a\} \rangle = Z$ and H is a maximal subspace of Z. $\square$

Proposition 5.2.9 *A subspace of a projective space is a hyperplane if and only if it is maximal.*

Proof By Theorem 5.2.6(ii), a maximal subspace of a projective space is a geometric hyperplane. The converse is proved in Lemma 5.2.8. $\square$

The terminology hyperplane is justified by the correspondence of geometric hyperplanes of $\mathbb{P}(V)$ with hyperplanes of V for any vector space V alluded to before.

In Exercise 5.7.21 it is shown that a geometric hyperplane of a line space is not necessarily a maximal subspace. The other implication is also not valid in general. For instance, a line of an affine plane all of whose lines are thick, is a maximal subspace but not a geometric hyperplane.

The following result establishes that projective spaces possess maximal subspaces.

Proposition 5.2.10 *In a projective space every subspace is the intersection of some family of hyperplanes.*

Proof Let Z be a subspace of the projective space (P, L). If $Z = P$, then Z is the intersection of the empty family of maximal subspaces. If $Z \neq P$, it suffices to show that for each $p \in P \setminus Z$, there exists a maximal subspace containing Z and not containing p.

Let $\mathcal{Z}$ be the family of all subspaces of P containing Z but not p. Then $\mathcal{Z}$ is non-empty as $Z \in \mathcal{Z}$. If $\mathcal{T}$ is a subset of $\mathcal{Z}$ which is totally ordered by inclusion, then $\bigcup_{T \in \mathcal{T}} T$ does not contain p and it is a subspace. Hence $\bigcup_{T \in \mathcal{T}} T \in \mathcal{Z}$ and, by Zorn's lemma, $\mathcal{Z}$ has a maximal element, say H. Clearly $p \notin H$. Let us show that H is a maximal subspace of (P, L). Suppose $x \in P \setminus H$. Then $\langle H, x \rangle$ must contain p, so, by Theorem 5.2.6(i), p lies on a line xu joining x with a point $u \in H$. But then x lies on the line pu, which is in $\langle H, p \rangle$. We have shown that $\langle H, p \rangle = \langle H, x \rangle = P$ for all x in $P \setminus H$. As H is a subspace, we conclude that it is a maximal subspace containing Z but not p. $\square$

Remark 5.2.11 There is a classical link between affine spaces and projective spaces, extending the planar case, which was dealt with in Exercise 2.8.9. Here we review how an affine space $\mathbb{A}(V)$ can be completed to a projective space $\mathbb{P}(\mathbb{A}(V))$. In Sect. 6.4 we will deal with this correspondence on the level of abstract affine and projective spaces.

(1) A *point* of $\mathbb{P}(\mathbb{A}(V))$ is either a point of $\mathbb{A}(V)$ or a *point at infinity*, that is, the set of all lines parallel to a given line.
(2) A *line* of $\mathbb{P}(\mathbb{A}(V))$ is either a line of $\mathbb{A}(V)$ completed by the point at infinity to which it belongs or a *line at infinity*, i.e., the set of all points at infinity in an affine plane of $\mathbb{A}(V)$.

The *hyperplane at infinity* of $\mathbb{A}(V)$ is the set of all points at infinity. This set is a hyperplane of $\mathbb{P}(\mathbb{A}(V))$ whose lines are the lines at infinity.

Identification of $\mathbb{A}(V)$ with the set of points and lines of $\mathbb{P}(V \oplus \mathbb{D})$ that are not contained in the projective subspace $\mathbb{P}(V)$, leads to the following result. Here, the term projective hyperplane of $\mathbb{P}(V)$ is used for a subspace of $\mathbb{P}(V)$ of the form $\mathbb{P}(U)$, where U is a hyperplane of V.

Proposition 5.2.12 *There is a natural isomorphism* $\mathbb{P}(V \oplus \mathbb{D}) \to \mathbb{P}(\mathbb{A}(V))$ *such that the projective hyperplane* $\mathbb{P}(V)$ *of* $\mathbb{P}(V \oplus \mathbb{D})$ *maps onto the hyperplane at infinity of* $\mathbb{A}(V)$.

Proof The isomorphism is established by means of the map $(v, \lambda)\mathbb{D} \mapsto v\lambda^{-1}$ for $\lambda \in \mathbb{D} \setminus \{0\}$ and $(v, 0)\mathbb{D} \mapsto \{a + v\mathbb{D} \mid a \in V\}$ (the class of lines parallel to $v\mathbb{D}$) for $v \in V$. Details are left to the reader. $\square$

The proposition shows that the two constructions $\mathbb{P}(V)$ and $\mathbb{P}(\mathbb{A}(V))$ of projective spaces from vector spaces lead to the same class of spaces.

Notation 5.2.13 For $x = \varepsilon_1 x_1 \cdots + \varepsilon_{n+1} x_{n+1} \in \mathbb{D}^{n+1}$, the projective point $x\mathbb{D}$ of $\mathbb{P}(\mathbb{D}^{n+1})$ will often be denoted by $(x_1 : \ldots : x_{n+1})$.

Example 5.2.14 For every integer $n \geq 1$, we have $\mathbb{P}(\mathbb{D}^{n+1}) \simeq \mathbb{P}(\mathbb{A}(\mathbb{D}^n))$, cf. Proposition 5.2.12. This corresponds to the classical passage from n Cartesian coordinates in $\mathbb{A}(\mathbb{D}^n)$ to $n + 1$ *homogeneous coordinates* in $\mathbb{P}(\mathbb{D}^{n+1})$. Now, on the complement in $\mathbb{P}(\mathbb{A}(\mathbb{D}^n))$ of the hyperplane at infinity of $\mathbb{A}(\mathbb{D}^n)$, the map $\mathbb{P}(\mathbb{A}(\mathbb{D}^n)) \to \mathbb{P}(\mathbb{D}^{n+1})$ establishing the isomorphism is determined by

$$\varepsilon_1 x_1 + \cdots + \varepsilon_n x_n \mapsto (x_1 : \ldots : x_n : 1) \quad \left(x \in \mathbb{D}^n \right),$$

while on the hyperplane at infinity, realized as the set of the lines through 0 in $\mathbb{A}(\mathbb{D}^n)$ and therefore by points of $\mathbb{P}(\mathbb{D}^n)$, we have

$$(x_1 : \ldots : x_n) \mapsto (x_1 : \ldots : x_n : 0) \quad \left(x \in \mathbb{D}^n \setminus \{0\} \right).$$

We next describe the group $\mathrm{Aut}(\mathbb{P}(V))$ of all automorphisms of $\mathbb{P}(V)$ (cf. Definition 2.5.8). Each element of $\Gamma\mathrm{L}(V)$ induces an automorphism of $\mathbb{P}(V)$, so there is a homomorphism of groups $\psi : \Gamma\mathrm{L}(V) \to \mathrm{Aut}(\mathbb{P}(V))$. Homotheties were discussed in Exercise 2.8.28.

Lemma 5.2.15 *If $n \geq 1$, then* $\mathrm{Ker}(\psi)$ *is the group of homotheties in* $\Gamma\mathrm{L}(V)$.

Proof Denote the group of homotheties by H. Suppose that $g \in \Gamma\mathrm{L}(V)$ belongs to the kernel of ψ. Then, for each $v \in V$, there is $\lambda_v \in \mathbb{D}^{\mathrm{op}}$ such that $gv = v\lambda_v$. Taking $v, w \in V$ linearly independent, we obtain

$$(v + w)\lambda_{v+w} = g(v + w) = gv + gw = v\lambda_v + w\lambda_w,$$

whence $\lambda_v = \lambda_{v+w} = \lambda_w$. This readily implies that $\lambda = \lambda_v$ is independent of the choice of $v \in V \setminus \{0\}$ (for linearly dependent vectors v and w, the choice of a vector linearly independent of v and recourse to the previous case will do the job). This forces $g = h_\lambda \in H$ and $\ker(\psi) \subseteq H$, the subgroup of all homotheties of V. It is easy to see that H actually coincides with this kernel. $\qquad\square$

Thanks to the lemma, the quotient group $\mathrm{P}\Gamma\mathrm{L}(V)$ of $\Gamma\mathrm{L}(V)$ by H can be viewed as a subgroup of $\mathrm{Aut}(\mathbb{P}(V))$.

Theorem 5.2.16 *Let V be a vector space of dimension $n + 1$ over $\mathbb{D}$.*

 (i) *If $n \in \{0, 1\}$, then* $\mathrm{Aut}(\mathbb{P}(V)) = \mathrm{Sym}(\mathbb{P}(V))$.
(ii) *If $n \geq 2$, then* $\mathrm{Aut}(\mathbb{P}(V)) = \mathrm{P}\Gamma\mathrm{L}(V)$.

Proof (i) is obvious since $\mathbb{P}(V)$ has only one line in this case.

(ii) Let H be a hyperplane in $\mathbb{P}(V)$ and 0 a point outside H. Take any automorphism α of $\mathbb{P}(V)$. By Lemma 5.1.9(iii), there exists $t \in \mathrm{PGL}(V)$ such that $t(H) = \alpha(H)$ and $t(0) = \alpha(0)$. Consider $\beta := t^{-1}\alpha$. It is an automorphism of $\mathbb{P}(V)$ stabilizing H and 0. Therefore, β induces an automorphism in the affine space $\mathbb{P}(V) \setminus H$ which fixes 0. Now Theorem 5.1.12 applies to the vector space W underlying $\mathbb{P}(V) \setminus H$ with zero vector 0, so the restriction of β to $\mathbb{A}(W)$ belongs to the stabilizer $\Gamma\mathrm{L}(W)$ of 0 in $\mathrm{A}\Gamma\mathrm{L}(W)$. By use of the isomorphism $V \cong W \oplus U$ of Proposition 5.2.12 for a 1-dimensional subspace U of V, we can embed $\Gamma\mathrm{L}(W)$ in $\Gamma\mathrm{L}(V)$ in such a way that $g(d) = d$ for each $g \in \Gamma\mathrm{L}(W)$ and $d \in U$. Thus, there is an element $g \in \mathrm{P}\Gamma\mathrm{L}(V)$ such that $\gamma := \beta g$ fixes $\mathbb{A}(W)$ point-wise. But then, γ also fixes $\mathbb{P}(V) = \mathbb{P}(A(W))$, so $\beta = g^{-1} \in \mathrm{P}\Gamma\mathrm{L}(V)$. We conclude $\alpha = t\beta \in \mathrm{P}\Gamma\mathrm{L}(V)$. $\square$

Notation 5.2.17 The image of $\mathrm{GL}(V)$ in $\mathrm{P}\Gamma\mathrm{L}(V)$ is denoted by $\mathrm{PGL}(V)$.

For $\lambda \in \mathbb{D}$, the homothety h_λ in $\mathrm{Aut}(\mathbb{P}(V))$ is linear if and only if $\lambda \in \mathrm{Z}(\mathbb{D})$, so $H \cap \mathrm{GL}(V) = \mathrm{Z}(\mathrm{GL}(V)) \cong \mathrm{Z}(\mathbb{D}) \setminus \{0\}$, where $\mathrm{Z}(\mathrm{GL}(V))$ is the center of $\mathrm{GL}(V)$. Consequently, $\mathrm{PGL}(V) \cong \mathrm{GL}(V)/\mathrm{Z}(\mathrm{GL}(V))$.

The reader may care to use the proof of the theorem to show that if V and V' are finite-dimensional vector spaces of dimension 3 over the division ring $\mathbb{D}$ and $\mathbb{D}'$, respectively, then the projective spaces $\mathbb{P}(V)$ and $\mathbb{P}(V')$ are isomorphic if and only if $\mathbb{D} \cong \mathbb{D}'$ and $V \cong V'$ (by a semilinear map).

Definition 5.2.18 A *projective basis* of $\mathbb{P}(V)$ is a set B of points such that, for each $b \in B$, the difference set $B \setminus \{b\}$ is of the form $\{x\mathbb{D} \mid x \in E\}$ for a basis E of V. If

$n = \dim(V) - 1 < \infty$, we relate a projective basis in $\mathbb{P}(V)$ to a basis of V as follows. If $B = \{p_1, \ldots, p_{n+2}\}$ is an ordered projective basis of $\mathbb{P}(V)$, then a *normal basis* of V corresponding to B is an ordered basis $e_1, \ldots, e_{n+1}$ of V such that $\langle e_i \rangle = p_i$ for $i \in [n+1]$ and $\langle \sum_{i=1}^{n+1} e_i \rangle = p_{n+2}$.

Lemma 5.2.19 *Suppose $n = \dim(V) - 1 < \infty$. Each ordered projective basis of $\mathbb{P}(V)$ has a corresponding normal basis E of V. If F is another such normal basis, then there is a scalar λ with $E = F\lambda$.*

Proof Let $B = \{p_1, \ldots, p_{n+2}\}$ be an ordered projective basis of $\mathbb{P}(V)$ and let b_i be vectors of V such that $p_i = \langle b_i \rangle$ for $i \in [n+2]$. Then $b_1, \ldots, b_{n+1}$ constitute a basis of V and so there are scalars μ_i such that $b_{n+2} = \sum_{i=1}^{n+1} b_i \mu_i$. As B is a projective basis, we must have $\mu_i \neq 0$ for $i \in [n+1]$. Putting $e_i = b_i \mu_i$, for $i \in [n+1]$, we obtain $b_{n+2} = \sum_{i=1}^{n+1} e_i$, and $E = (e_1, \ldots, e_{n+1})$ is a normal basis of V.

If $F = (f_1, \ldots, f_{n+1})$ is another normal basis of V for B, then $e_i = f_i \lambda_i$ for certain scalars λ_i and $b_{n+2} = (\sum_i f_i)\lambda$ for some λ. Now $b_{n+2} = \sum_i e_i = \sum_i f_i \lambda_i$ forces $\lambda = \lambda_1 = \cdots = \lambda_n$, so $e_i = f_i \lambda$ for all $i \in [n+2]$. $\qquad\square$

Recall the definition of opposite division ring from Exercise 1.9.11.

Proposition 5.2.20 *Suppose $3 \leq \dim(V) = n+1 < \infty$. The group $\mathrm{PGL}(V)$ is transitive on the set of ordered projective bases of $\mathbb{P}(V)$. The stabilizer in $\mathrm{PGL}(V)$ of an ordered projective basis is isomorphic to $\mathrm{Int}(\mathbb{D})$, the group of all inner automorphisms of $\mathbb{D}$.*

Proof Let $e_1, \ldots, e_{n+1}$, and $f_1, \ldots, f_{n+1}$ be two normal bases of V corresponding to ordered projective bases $p_1, \ldots, p_{n+2}$, and $q_1, \ldots, q_{n+2}$, respectively of $\mathbb{P}(V)$. There is an element $a \in \mathrm{GL}(V)$ such that $a(e_i) = f_i$ for $i \in [n+1]$ and so its image in $\mathrm{PGL}(V)$ maps p_i to q_i for $i \in [n+1]$. It follows from Lemma 5.2.19 that a maps p_{n+2} to q_{n+2}. This proves the first assertion.

Let B be a projective basis of $\mathbb{P}(V)$ corresponding to the normal basis $e_1, \ldots, e_{n+1}$ of V. Suppose that $a \in \mathrm{GL}(V)$ fixes each element of B. It maps a normal basis of V for B onto a corresponding normal basis for B, so Lemma 5.2.19 gives a unique element $\lambda \in \mathbb{D}$ such that $a(e_i) = e_i \lambda$ for each $i \in [n+1]$. Retaining the above notation, we see that if $x = \sum_i e_i x_i$, then $a(x) = \sum_i a(e_i) x_i = \sum_i (e_i \lambda) x_i = \sum_i e_i (\lambda x_i)$, and so $a(x)$ has coordinates λx_i. In particular, the composition $h_\lambda^{-1} a$ of a with the homothety h_λ^{-1} induces the inner automorphism $\mu \mapsto \lambda \mu \lambda^{-1}$ of $\mathbb{D}$ on each coordinate and fixes the normal basis $e_1, \ldots, e_{n+1}$ vector-wise. Denote by $\mathrm{PGL}(V)_{[B]}$ the elementwise stabilizer of B in $\mathrm{PGL}(V)$. By Lemma 5.2.15 and Theorem 5.2.16, the homothety h_λ belongs to the kernel of the homomorphism $\Gamma\mathrm{L}(V) \to \mathrm{P}\Gamma\mathrm{L}(V)$, so the map $\phi : \mathrm{PGL}(V)_{[B]} \to \mathrm{Int}(\mathbb{D})$ determined by $\phi(a) = (\mu \mapsto \lambda \mu \lambda^{-1})$ is well defined. It is clearly a surjective homomorphism. The kernel of ϕ consists of all elements $(H \cap \mathrm{GL}(V))a$ corresponding to some $\lambda \in Z(\mathbb{D}) \setminus \{0\}$. This shows that $\ker(\phi) = H \cap \mathrm{GL}(V)$, so ϕ is an isomorphism. $\qquad\square$

If the division ring $\mathbb{D}$ is commutative, then $\mathrm{Int}(\mathbb{D})$ consists of the identity only. Hence the following consequence.

Corollary 5.2.21 *If* $\mathbb{D}$ *is a field and* $n \geq 2$, *then the projective linear group* $\mathrm{PGL}(V)$ *is sharply transitive on the set of ordered projective bases of* $\mathbb{P}(V)$.

Later, in Sect. 6.2, we will impose more structure on the projective line (corresponding to the case $n = 1$) in such a way that its automorphism group is no longer the whole symmetric group.

5.3 Matroids

We present a straightforward definition of the dimension of a subspace of a linear space. It will help us to distinguish projective and affine spaces from arbitrary linear spaces. This is put into effect by matroids, which will supply the common ground for the well-behaved linear spaces. The section ends with Theorem 5.3.11, which assigns an $[n]$-geometry with a linear digon diagram to a matroid of dimension n.

Definition 5.3.1 If X is a subspace of the line space Z, the *dimension* of X, denoted $\dim(X)$, is the smallest integer $n \geq -1$ such that every chain of subspaces (ordered by inclusion) contained in X has at most $n + 2$ members. The *dimension* of Z is the dimension of the subspace of all points of Z.

The dimension of the empty space is -1, the dimension of a singleton is 0. The lines of a linear space are its subspaces of dimension one. A subspace of dimension two is called a *plane* of Z.

Example 5.3.2 For the projective space $\mathbb{P}(V)$ associated with a vector space V of finite dimension $n + 1$, these definitions fit rather nicely. The dimension n of the space $\mathbb{P}(V)$ coincides with the rank n of the geometry $\mathrm{PG}(V)$, whose elements of type i are the subspaces of $\mathbb{P}(V)$ of dimension $i - 1$ for $i \in [n]$.

If V is defined over a division ring of size at least 3 and $n = \dim(V) \geq 2$, then, according to Remark 5.1.5, the subspaces of $\mathbb{A}(V)$ of dimension $i - 1$ are parallel closed and coincide with the elements of $\mathrm{AG}(V)$ of type i for $i \in [n]$. Thus, it is tempting to define a plane of a linear space to be a subspace generated by a set of three non-collinear points. But if V is defined over $\mathbb{F}_2$, the subspaces of $\mathbb{A}(V)$ of dimension $i - 1$ are just the sets of points of size i.

It is not true in general that three non-collinear points of a linear space are contained in a unique plane. Moreover, there are quite interesting situations where we would like to consider planes which are not necessarily the linear subspaces generated by three non-collinear points. A typical example is provided by a sphere S in $\mathbb{E}^3$. The Euclidean planes in $\mathbb{E}^3$ induce circles on S, the lines induce pairs of points. So the linear space on S induced by lines is the complete graph (Example 1.2.1), while the linear space induced by planes is much more interesting: it consists of the circles on S. Hence we need a completely different approach to the concepts of plane and subspace. To this end, we introduce yet another notion which generalizes affine and projective spaces.

Definition 5.3.3 Consider a pair $(P, \mathcal{F})$ consisting of a set P of points and a collection $\mathcal{F}$ of subsets of P such that any intersection of members of $\mathcal{F}$ belongs to $\mathcal{F}$. The elements of $\mathcal{F}$ are called *flats*.

If $X \subseteq P$ is a set of points, then the intersection of all flats containing X, denoted $\langle X \rangle_{\mathcal{F}}$, is called the flat *generated* by X. It is the smallest flat containing X. A *basis* of a flat F is a subset B of F with $\langle B \rangle_{\mathcal{F}} = F$ such that no proper subset of B has the same property.

In an affine or projective space, the flats would be distinguished (or all) subspaces of the linear space.

In order to avoid the trouble of planes included in other planes and similar phenomena with higher-dimensional subspaces, we require that a flat generated by some flat f and an additional point p admits f as a maximal proper subflat. This leads to the following notion.

Definition 5.3.4 A *matroid* is a pair $(P, \mathcal{F})$ where P is a set of *points* and $\mathcal{F}$ is a collection of subsets of P, called *flats*, with the following properties.

(1) If $\mathcal{G} \subseteq \mathcal{F}$, then the intersection of all members of $\mathcal{G}$ belongs to $\mathcal{F}$. In particular (taking $\mathcal{G} = \emptyset$) we have $P \in \mathcal{F}$.
(2) If $v \in \mathcal{F}$ and $x \in P$ but $x \notin v$, then there is a unique flat containing v and x, denoted $\langle v, x \rangle_{\mathcal{F}}$, with the property that v is maximal among all flats that are properly contained in $\langle v, x \rangle_{\mathcal{F}}$.
(3) If $B \subseteq P$, then B is a basis of $\langle B \rangle_{\mathcal{F}}$ if and only if every finite subset X of B is a basis of $\langle X \rangle_{\mathcal{F}}$.

A matroid $(P, \mathcal{F})$ is called *simple* if $\{x\} \in \mathcal{F}$ for each $x \in P$ and if $\emptyset \in \mathcal{F}$.

The *dimension* of a flat v of a matroid, denoted $\dim(v)$, is the smallest cardinality $n \geq -1$ such that every chain of flats (ordered by inclusion) contained in v has at most $n + 2$ members. The *dimension* of $(P, \mathcal{F})$ is the dimension of the flat P.

For a matroid $(P, \mathcal{F})$ of dimension n, the collection of all elements of $\mathcal{F}$ of dimension i $(-1 \leq i \leq n)$ will be denoted by $\mathcal{F}_i$. The members of $\mathcal{F}_1, \mathcal{F}_2, \mathcal{F}_{n-1}$ are called *lines*, *planes*, *hyperplanes* of $(P, \mathcal{F})$, respectively.

Remark 5.3.5 The notation $\langle B \rangle_{\mathcal{F}}$ makes sense in view of the first axiom; it is the intersection of all flats containing B.

If $(P, \mathcal{F})$ is a matroid, then $\langle v, x \rangle_{\mathcal{F}} = \langle v \cup \{x\} \rangle_{\mathcal{F}}$. The intersection of all flats of $\mathcal{F}$, often the empty set, has dimension -1. In a simple matroid, singletons have dimension 0.

Example 5.3.6 A vector space of dimension $n \geq 1$ together with its subspaces as flats, is a non-simple matroid. Below are some simple ones.

(i) Full shadow spaces $\mathrm{ShSp}(\Gamma, 1)$ (cf. Definition 2.5.1), where $\Gamma = \mathrm{AG}(V)$ or $\mathrm{PG}(V)$ with V a finite-dimensional vector space, are simple matroids. But even if $\dim(V) = \infty$, taking $\mathcal{F}$ to be the collection of all subspaces of a projective space $\mathbb{P}(V)$, and the collection of all affine subspaces of an affine space $\mathbb{A}(V)$, we obtain simple matroids. Later, in Theorem 5.3.8, we will see that each projective space with its subspaces is a matroid. As affine spaces with lines of size at least four will turn out to be restrictions of projective spaces (see Proposition 6.4.2), they will also become matroids when supplied with their subspaces.

(ii) Also the sphere S in $\mathbb{E}^3$ of Example 5.3.2 fits into the present context. Take P to be the set of points of S, and set $\mathcal{F}_0 = \{\{x\} \mid x \in P\}$. Let $\mathcal{F}_1$ be the set of all pairs of distinct points of S. Finally, let $\mathcal{F}_2$ be the set of all circles contained in S. The resulting pair $(P, \mathcal{F})$, where $\mathcal{F} = \{\emptyset\} \cup \mathcal{F}_0 \cup \mathcal{F}_1 \cup \mathcal{F}_2 \cup \{P\}$, is a simple matroid of dimension 3.

(iii) Firm linear spaces are simple matroids of dimension two, and conversely.

Matroids provide the right context for a development of the theory of linear independence in linear algebra. We briefly mention some key properties of matroids and leave the proofs to the reader.

Theorem 5.3.7 *Each matroid $(P, \mathcal{F})$ satisfies the following properties.*

(i) *For each subset B of P that is a basis of $\langle B \rangle_{\mathcal{F}}$ and each point $x \in P \setminus \langle B \rangle_{\mathcal{F}}$, the subset $B \cup \{x\}$ is a basis of $\langle B, x \rangle_{\mathcal{F}}$.*

(ii) *If $X \subseteq P$ and A is a subset of X such that A is a basis of $\langle A \rangle_{\mathcal{F}}$, then there exists a basis B of $\langle X \rangle_{\mathcal{F}}$ with $A \subseteq B \subseteq X$.*

(iii) *If F is a flat, then any two bases of F have the same cardinality.*

(iv) *If B is a basis of $\langle B \rangle_{\mathcal{F}}$, then the dimension of $\langle B \rangle_{\mathcal{F}}$ is equal to $|B| - 1$.*

(v) *If F_1 and F_2 are flats of finite dimension, then*

$$\dim \langle F_1 \cup F_2 \rangle + \dim(F_1 \cap F_2) \leq \dim(F_1) + \dim(F_2).$$

Thanks to Example 5.3.6(i), the following result is already known in the case where the projective space is of the form $\mathbb{P}(V)$.

Theorem 5.3.8 *For each projective space, its point set supplied with all its subspaces is a simple matroid.*

Proof Let Z be a projective space with point set P. We check the properties of a simple matroid given in Definition 5.3.4. Obviously, every intersection of subspaces is a subspace. Next, if V is a subspace and p a point not on V, then V is a maximal proper subspace of $\langle V, p \rangle$ because, by Theorem 5.2.6, $\langle V, p \rangle$ is the union of all lines joining p to some point of V. Finally let F be a set of points. Then $\langle F \rangle$ is the union U of all $\langle X \rangle$ where X is a finite subset of F because $F \subseteq U \subseteq \langle F \rangle$ and U is a subspace of Z (observe that if $x \in \langle X \rangle$ and $y \in \langle Y \rangle$ for finite subsets X, Y of F, then xy belongs to $\langle X \cup Y \rangle$, while $X \cup Y$ is a finite subset of F). Hence, if every finite subset X of F is a basis of $\langle X \rangle$, we see that F is a basis of $\langle F \rangle$. Consequently,

P and its subspaces constitute a matroid. This matroid is simple as the empty set and every singleton in P are subspaces of Z. $\square$

Remark 5.3.9 Let $\mathbb{P}$ be a projective space. The dimension of $\mathbb{P}$ according to Definition 5.3.1 coincides with the dimension of the matroid of all subspaces of $\mathbb{P}$ appearing in Theorem 5.3.8 according to Definition 5.3.4. If $\dim(\mathbb{P}) = n$ is finite, then the maximal subspaces of $\mathbb{P}$ are the flats of the corresponding matroid of dimension $n - 1$.

Example 5.3.10 We give three ways of building new simple matroids from a given simple matroid $(P, \mathcal{F})$ of dimension n. It is no coincidence that the notions truncation of Definition 1.4.6 and quotient of Definition 1.3.5 reappear.

(i) Fix $m \in [n - 1]$ and set $\mathcal{F}_{\leq m} = \bigcup_{j \leq m} \mathcal{F}_j$. The pair $(P, \mathcal{F}_{\leq m} \cup \{P\})$ is a simple matroid of dimension $m + 1$, called *truncation*, of $(P, \mathcal{F})$.

(ii) Fix a subset X of P. A flat in $\mathrm{Restr}_\mathcal{F}(X)$ is any set $v \cap X$ such that $v \in \mathcal{F}$ and v is generated by $v \cap X$. The pair $(X, \mathrm{Restr}_\mathcal{F}(X))$ is a simple matroid of dimension $\dim\langle X\rangle_\mathcal{F}$, called *restriction* of $(P, \mathcal{F})$.

(iii) Fix $v \in \mathcal{F}_m$ $(-1 \leq m \leq n)$. Set $P/v = \{\langle v, x\rangle_\mathcal{F} \mid x \in P \setminus v\}$ and let $\mathcal{F}/v$ be the collection of all subsets of P/v of the form $w/v := \{\langle v, x\rangle_\mathcal{F} \mid x \in w \setminus v\}$ for w running over all $w \in \mathcal{F}$ with $v \subseteq w$. The pair $(P/v, \mathcal{F}/v)$ is a simple matroid of dimension $n - m - 1$, called *quotient* of $(P, \mathcal{F})$.

The next result shows that these restrictions and quotients are in fact residues.

Theorem 5.3.11 *Let $n \in \mathbb{N}$, $n \geq 2$, and let $(P, \mathcal{F})$ be a simple matroid of dimension n.*

(i) *The pair $(P, \mathcal{F}_1)$ is a firm linear space in which each member of $\mathcal{F}$ is a linear subspace.*

(ii) *The geometry $\Gamma(P, \mathcal{F}) = (\mathcal{F}_0, \dots, \mathcal{F}_{n-1}, *)$, where $*$ stands for symmetrized containment, is firm and residually connected over the diagram*

$$\mathrm{L}_n: \quad \overset{1}{\circ}\!\!\overset{\mathrm{L}}{\rule{1.5cm}{0.4pt}}\!\!\overset{2}{\circ}\!\!\overset{\mathrm{L}}{\rule{1.5cm}{0.4pt}}\!\!\overset{3}{\circ}\cdots\cdots\overset{n-1}{\circ}\!\!\overset{\mathrm{L}}{\rule{1.5cm}{0.4pt}}\!\!\overset{n}{\circ}.$$

Proof (i) is a direct consequence of the definition. For, if x and y are distinct members of P, then $\langle x, y\rangle_\mathcal{F}$ is the unique line containing these. Moreover, if $v \in \mathcal{F}$ contains both x and y, then $\langle x, y\rangle_\mathcal{F} \cap v$ is a member of $\mathcal{F}$ containing $\{x, y\}$ and contained in $\langle x, y\rangle_\mathcal{F}$, whence equal to $\langle x, y\rangle_\mathcal{F}$. This proves that v is a subspace of $(P, \mathcal{F}_1)$. Firmness follows from the assumption $n \geq 2$.

(ii) Set $\Gamma = \Gamma(P, \mathcal{F})$. If $n = 2$, the statement follows from (i) and Theorem 2.5.15. Therefore, we assume $n \geq 3$ and proceed by induction on n. Assume $B \in \mathcal{F}_{i-1}$, $C \in \mathcal{F}_i$, and $D \in \mathcal{F}_{i+1}$ satisfy $B \subset C \subset D$ for some $i \in [n - 2]$. Taking $x \in D \setminus C$, we find that $C' := \langle B, x\rangle_\mathcal{F}$ also satisfies $B \subset C' \subset D$ and is distinct from C, as $x \in C' \setminus C$. Therefore, Γ is firm.

Fix $x \in P$, and let $v \in \mathcal{F}_{i-1}$ for some $i \in [n]$. Take $y \in v$ such that $y \neq x$. The flat $\langle \{y\}, x\rangle_\mathcal{F} \in \mathcal{F}_1$ is incident with both $\{x\}$ and $\{y\}$, and $\{y\}$ is incident with v, so each element of Γ is connected to $\{x\}$. Hence Γ is connected. Retain $v \in \mathcal{F}$, so v is an element of Γ of type i. As the digon diagram of Γ is obviously linear, the residue Γ_v is a direct sum of two geometries Γ' and Γ'', say, whose elements are the members $u \in \mathcal{F}$ with $u \subseteq v$ and the members $w \in \mathcal{F}$ with $w \supseteq v$, respectively. It is easily seen that Γ' is the geometry associated with the simple matroid $\mathrm{Restr}_{\mathcal{F}}(v)$ (cf. Example 5.3.10(ii)) and that Γ'' is the geometry associated with $(P/v, \mathcal{F}/v)$ (cf. Example 5.3.10(iii)). The induction hypothesis applied to Γ' and Γ'', and Theorem 2.1.6 yields that Γ_v is a firm, residually connected geometry with diagram $\mathrm{L}_{i-1} \,\dot{\cup}\, \mathrm{L}_{n-i}$. The required properties of Γ follow directly. □

5.4 Matroids from Geometries

In Sect. 5.3 we went from a global structure (the simple matroid, involving no residues) to information about rank 2 residues, the local structure captured in the diagram of a geometry. Here we proceed conversely. Theorem 5.4.3 shows how to derive a matroid from a residually connected geometry with a suitable linear diagram. We fix a set I of types.

Notation 5.4.1 Let Γ be a geometry over I. Recall from Definition 2.5.1 that, for $J \subseteq I$, the J-shadow of a flag F in Γ is denoted by $\mathrm{Sh}_J(F)$. Here we extend the notation to a union G of flags of Γ by writing $\mathrm{Sh}_J(G)$ for $\bigcap_{g \in G} \mathrm{Sh}_J(\{g\})$. This notation is justified by the fact that it holds when G is a flag. Moreover, for $i \in I$, we also write $\mathrm{Sh}_i(G)$ rather than $\mathrm{Sh}_{\{i\}}(G)$.

Proposition 5.4.2 *Let $\Gamma = (X_1, X_2, \ldots, X_n, *)$ be an $[n]$-geometry belonging to the diagram L_n for some $n \in \mathbb{N}$, $n \geq 2$. Suppose $x \in X_i$ and $y \in X_j$, where $i, j \in [n]$ satisfy $i + j \leq n$.*

- (i) *There is an $(i + j)$-element of Γ in $\{x, y\}^*$.*
- (ii) *If $\mathrm{Sh}_1(\{x, y\}) = \emptyset$, then there is a unique $(i + j)$-element z in $\{x, y\}^*$ and $\{x, y\}^*$ consists of all elements of type $\geq i + j$ incident with z.*
- (iii) *The map $u \mapsto \mathrm{Sh}_1(u)$ is an injective map from $\bigcup_i X_i$ to the collection of subspaces of $\mathrm{ShSp}(\Gamma, 1)$.*
- (iv) *If U is a non-empty set of elements of Γ with $\mathrm{Sh}_1(U) \neq \emptyset$, then there is $u \in \mathrm{Sh}_i(U)$ for some i smaller than or equal to the type of any element from U such that $\mathrm{Sh}_1(U) = \mathrm{Sh}_1(u)$.*

Proof For $n = 2$, these statements are immediate from Theorem 2.5.15. We proceed by induction on n. Let $n > 2$.

(i) Assume $i + j = 2$, so $i = j = 1$. We need to show that $\mathrm{Sh}_2(\{x, y\})$ is non-empty. By Lemma 1.6.3, residual connectedness of Γ provides a $\{1, 2\}$-path from x to y. Induction on the length of such a path allows us to assume that there are a, $b \in X_2$ and $u \in X_1$ such that $x * a * u * b * y$. If $a = b$, then $a \in \mathrm{Sh}_2(\{x, y\})$, so we may restrict to the case where a and b are distinct. Now they are distinct elements

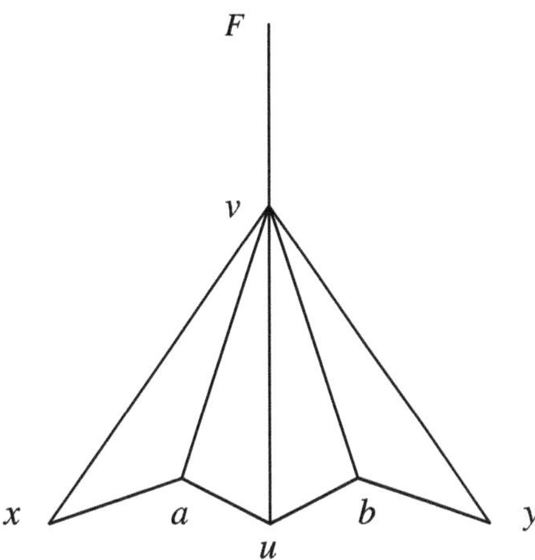

Fig. 5.3 Incidences between some elements of Γ

of Γ_u, so, by the induction hypothesis, we can find $v \in X_3$ incident with a, b, and u. Let F be a flag of type $\{3, \ldots, n\}$ containing v.

We know that a and b belong to F^*. Since $\{a, x\} \cup F$ and $\{y, b\} \cup F$ are chambers, we find $x, y \in F^*$ (cf. Fig. 5.3). Thus, Γ_F is a firm geometry of rank two with diagram of type L_2 (defined in Theorem 5.3.11) containing x and y. Hence, there is a 2-element in $\{x, y\}^* \cap F^*$, as required.

We proceed by induction on $i + j$ and assume $i + j > 2$. Without loss of generality we take $j \geq i$, so that $j > 1$. Pick $w \in \mathrm{Sh}_{j-1}(y)$. By induction on $i + j$ we can find $v \in \mathrm{Sh}_{i+j-1}(\{x, w\})$. Taking a flag G of type $[j - 1]$ containing w and applying the induction hypothesis (with respect to n) to Γ_G, we obtain an element $z \in \mathrm{Sh}_{i+j}(\{v, y\})$. By Theorem 2.1.6, $z \in x^*$, so $z \in \mathrm{Sh}_{i+j}(\{x, y\})$, proving (i).

(ii) By (i) there exists an $(i + j)$-element z in $\{x, y\}^*$. Let $z' \in \mathrm{Sh}_k(\{x, y\})$ with $k \geq i + j$. It suffices to establish $z' \in z^*$.

Assume $i = j = 1$ and $k = 2$. By (i) applied to Γ_x, there is a 3-element v in $\{z, z'\}^*$. Using Theorem 2.1.6 in Γ_z, we find $x, y \in v^*$. If F is a flag of type $\{3, \ldots, n\}$ containing v, then Γ_F is a rank two geometry containing x, y, z, and z'. By Theorem 2.5.15 (applied to Γ_F) and the hypothesis $X_1 \cap \{x, y\}^* = \emptyset$, this implies $z = z'$.

If $i = j = 1$ and $k > 2$, take a flag G of type $\{k, k + 1, \ldots, n\}$ containing z'. Then Γ_G contains x, y, so by the induction hypothesis (on n) there is a unique 2-element in $\{x, y\}^* \cap G^*$. But by the first paragraph, this 2-element must be z. Since $z' \in G \subseteq z^*$, it follows that $z' \in z^*$.

Having shown (ii) for $i + j = 2$, we continue by induction on n and on $i + j$ and assume $i + j > 2$. Without loss of generality we (may) take $j > 1$. Pick $p \in \mathrm{Sh}_1(y)$. According to (i) applied to the residue Γ_F of a flag F of type $\{k, \ldots, n\}$ containing z', there is $u \in \{x, p, z'\}^* \cap X_{i+1}$. As $\mathrm{Sh}_1(\{x, p\}) = x^* \cap \{p\} \subseteq \mathrm{Sh}_1(\{x, y\}) = \emptyset$ and $u \in \mathrm{Sh}_{i+1}(\{x, p\})$, we may apply the induction hypothesis (on $i + j$) to x and p, which gives that every element in $\{x, p\}^*$ is incident with u. In particular, z is incident with u. Take H to be a flag of Γ that is maximal with respect to the properties of having type in $[j - 1]$, containing p, and lying inside $\{u, y\}^*$. Then

$\{u, y\}^* \cap H^* \subseteq H$. We contend that $H = \{p\}$. For, suppose v is an element of H of type h, where h is chosen in such a way that H has type $[h]$. As $i + h < i + j$ and $\{x, v\}^* \cap X_1 \subseteq \{x, y\}^* \cap X_1 = \emptyset$, the induction hypothesis shows that there is a unique element $w \in \{x, v\}^*$ of type $i + h$ such that each element of $\{x, v\}^*$ is incident with w. In particular, $u \in \{x, v\}^* \subseteq w^*$, and $i + 1$, the type of u, is at least $i + h$, the type of w. Therefore, $h = 1$, so $H = \{v\} = \{p\}$ and the contention holds.

As z is of type $i + j - 1$ in Γ_p, which is the sum of the types i and $j - 1$ of u and y in Γ_p, the elements u, y, z of Γ_H satisfy the hypotheses of x, y, z in (ii), respectively. By the induction hypothesis (on n) applied to Γ_p, every element of Γ_p incident with u and y must be incident with z as well. Since $z' \in p^* \cap \{u, y\}^* \subseteq \{p, u, y\}^*$, this gives $z' * z$, which finishes the uniqueness part of (ii).

It remains to show that there is no element in $\{x, y\}^*$ of type smaller than $i + j$. Suppose that $a \in \{x, y\}^*$ is of type $t < i + j$. If $t < i$, then either $t < j$ and $a^* \cap X_1$ is a nonempty subset of $\mathrm{Sh}_1(\{x, y\})$, or $t \geq j$ and $y^* \cap X_1$ is a nonempty subset of $\mathrm{Sh}_1(\{x, y\})$. Both cases contradict the hypothesis $\mathrm{Sh}_1(\{x, y\}) = \emptyset$, and, similarly, $t < j$ is ruled out. Therefore, we may assume $t \geq i, j$, in which case $a^* \cap X_{i+j}$ is a subset of $\{x, y\}^* \cap X_{i+j} = \{z\}$ of size at least two (as Γ is firm), contradicting the uniqueness of z. We conclude $t \geq i + j$, which ends the proof of (ii).

(iii) Suppose that $u, v \in \bigcup_i X_i$ are distinct and satisfy $\mathrm{Sh}_1(u) = \mathrm{Sh}_1(v)$. The shadow $\mathrm{Sh}_1(u)$ contains at least one point, p say. If $l \in \mathrm{Sh}_2(\{u, p\})$, then, due to firmness, there is $q \in \mathrm{Sh}_1(l) \setminus \{p\}$. Now $q \in \mathrm{Sh}_1(u) = \mathrm{Sh}_1(v)$ and, by (ii), v is incident with l, whence $l \in \mathrm{Sh}_2(\{v, p\})$. Using symmetry in u and v, we conclude that the 2-shadows of u and v in Γ_p coincide. By the induction hypothesis, it follows that $u = v$. Hence (iii).

(iv) If $\mathrm{Sh}_1(U)$ is a singleton, say $\{x\}$, then $\mathrm{Sh}_1(U) = \mathrm{Sh}_1(x)$ and we are done. Otherwise, there are distinct elements x and y in $\mathrm{Sh}_1(U)$. Now (ii) applied to x and y gives an element $l \in \mathrm{Sh}_2(\{x, y\}) \cap \mathrm{Sh}_2(U)$, so $\mathrm{Sh}_2(U) \neq \emptyset$. By the induction hypothesis applied to Γ_x, there is an element u of Γ_x of type at most the minimum over all types of elements from $U \setminus \{x\}$ such that $\mathrm{Sh}_2(\{x\} \cup U) = \mathrm{Sh}_2(\{x, u\})$. Let $w \in X_1 \setminus \{x\}$. The unique element in $\mathrm{Sh}_2(\{x, w\})$ whose existence is guaranteed by (ii), is contained in $\{x, u\}^*$ if and only if it belongs to $x^* \cap U^*$. Consequently, again by (ii), $w \in u^*$ if and only if $w \in U^*$. Hence $\mathrm{Sh}_1(u) = \mathrm{Sh}_1(U)$. $\square$

Full shadow spaces were introduced in Definition 2.5.1.

Theorem 5.4.3 *Let $\Gamma = (X_1, \ldots, X_n, *)$ be an $[n]$-geometry belonging to the diagram L_n for some $n \in \mathbb{N}$ with $n \geq 2$. The full shadow space $\mathrm{ShSp}(\Gamma, 1)$ is a simple matroid of dimension n whose flats of dimension $i - 1$ are the 1-shadows of elements of X_i for $i \in [n]$. Up to isomorphism, it is the unique simple matroid $(P, \mathcal{F})$ of dimension n such that $\Gamma(P, \mathcal{F}) \cong \Gamma$.*

Proof For $n = 2$, the statements are trivial or trivial consequences of Theorem 2.5.15. Let $n > 2$. Put $P = X_1$ and, for $i \in [n]$, set $\mathcal{F}_{i-1} = \{\mathrm{Sh}_1(u) \mid u \in X_i\}$. In addition, write $\mathcal{F}_{-1} = \{\emptyset\}$ and $\mathcal{F}_n = \{P\}$. The pair $(P, \mathcal{F})$, where $\mathcal{F} = \bigcup_{-1 \leq i \leq n} \mathcal{F}_i$, is the full shadow space $\mathrm{ShSp}(\Gamma, 1)$. We show that it is a simple matroid.

Suppose $\mathcal{G} \subseteq \mathcal{F}$. By definition of $\mathcal{F}$, there is a set U of elements of Γ such that $\mathcal{G} = \{\mathrm{Sh}_1(v) \mid v \in U\}$. Denote by V the intersection of all members of $\mathcal{G}$. We want to show $V \in \mathcal{F}$. We may suppose $V \neq \emptyset$ and $V \neq P$. Now $V = \mathrm{Sh}_1(U)$, so, according to Proposition 5.4.2(iv), there is an element u of Γ with $V = \mathrm{Sh}_1(u) \in \mathcal{F}$. This establishes $V \in \mathcal{F}$, and hence Definition 5.3.4(1).

Next let $U \in \mathcal{F}$ and $x \in P$. By definition of $\mathcal{F}$, there is $u \in X_i$ for some $i \in [n]$ such that $U = \mathrm{Sh}_1(u)$. Now $x \in U$ if and only if $x * u$. In this case, $\langle x, U \rangle_{\mathcal{F}} = U \in \mathcal{F}$ (the left hand side makes sense thanks to Definition 5.3.3). Otherwise, if $i \leq n - 1$, there exists a unique element $w \in \mathrm{Sh}_{i+1}(\{x, u\})$ by Proposition 5.4.2(ii), and $\langle x, U \rangle_{\mathcal{F}} = \mathrm{Sh}_1(w)$. If $i = n$ and still $x \notin U$, we have $\langle x, U \rangle_{\mathcal{F}} = P$, whence Definition 5.3.4(2), the second axiom. The third axiom holds trivially. This establishes that $(P, \mathcal{F})$ is a matroid. Moreover, $\mathcal{F}_0 = \{\mathrm{Sh}_1(u) \mid u \in X_1\} = \{\{x\} \mid x \in X_1\}$, so the matroid is simple.

In view of Proposition 5.4.2(iii), the map $u \mapsto \mathrm{Sh}_1(u)$ is a bijection from $\bigcup_{i \in [n]} X_i$ onto $\bigcup_{i \in [n]} \mathcal{F}_{i-1}$. Uniqueness of $(P, \mathcal{F})$ follows. $\square$

The following corollary summarizes the axiomatic description of projective spaces obtained so far. The diagram A_n has been introduced in Table 4.2 (see also Proposition 2.4.7). In Definition 2.2.7, a projective plane is defined as a generalized 3-gon. Often, however, we will use Theorem 2.5.15 to identify it with its shadow space on points, in which case a projective plane is a line space.

Corollary 5.4.4 *Let Γ be an $[n]$-geometry belonging to the diagram A_n for some $n \in \mathbb{N}$, $n \geq 2$. There is a unique projective space (P, L) of dimension n (up to isomorphism) such that $(P, \mathcal{F})$, where $\mathcal{F}$ is the collection of all subspaces of (P, L), is a simple matroid satisfying $\Gamma(P, \mathcal{F}) \cong \Gamma$.*

Proof Let Γ be as stated and let $(P, \mathcal{F})$ be the simple matroid as in Theorem 5.4.3. This implies that P is the set of 1-elements of Γ. We take $L := \mathcal{F}_1$, the set of 1-shadows of 2-elements of Γ.

By Proposition 5.4.2(ii), the elements of $\mathcal{F}$ are subspaces of $(P, \mathcal{F}_1)$. We prove by induction on n that $\mathcal{F}$ coincides with the collection of all subspaces of (P, L). If $n = 2$, there is nothing to show.

Let U be a non-empty proper subspace of (P, L). First consider the case where U is a plane of (P, L). We claim that it is the 1-shadow of a 3-element of Γ. This follows by consideration of the case where U has dimension 2, so U is generated by a line l and a point $x \in U \setminus l$. By the definition of dimension of a space, $\langle l \cup \{x\} \rangle = U$. There is a 2-element h of Γ with $l = \mathrm{Sh}_1(h)$. By Proposition 5.4.2(i), (ii), there is a unique 3-element $u \in x^* \cap h^*$. As x and l are contained in the subspace $\mathrm{Sh}_1(u)$ of (P, L), we have $U \subseteq \mathrm{Sh}_1(u)$. The residue of u in Γ is a [2]-geometry of type A_2, that is, a projective plane. But l is a maximal subspace of this plane strictly contained in U, so $U = \mathrm{Sh}_1(u) \in \mathcal{F}_2$. It also follows that any three non-collinear points generate a projective plane, so (P, L) is a projective space.

For U of dimension > 2, take $x \in U$ and consider the residue Γ_x. This is an $[n-1]$-geometry of type A_{n-1}. The corollary applied to this residue leads to the quotient matroid $(P/x, \mathcal{F}/x)$ of Example 5.3.10(iii). By use of the previous paragraph, the corresponding projective space $(P/x, \mathcal{F}/x)$ can be identified with the lines on x and pencils of lines on x in a plane on x. The induction hypothesis gives that the subspace U/x of this space is a flat of $(P/x, \mathcal{F}/x)$. Hence there is a flat $U' \in \mathcal{F}$ with $U/x = U'/x$. This implies $U = U'$ and we are done. $\square$

Here is a converse of Corollary 5.4.4.

Corollary 5.4.5 *If (P, L) is a projective space of finite dimension n, then $\Gamma(P, \mathcal{F})$, where $\mathcal{F}$ is the collection of subspaces of (P, L), is an $[n]$-geometry with diagram A_n whose members of type i are the $(i - 1)$-dimensional subspaces of (P, L).*

Proof By Theorem 5.3.8, $\Gamma(P, \mathcal{F})$ is a matroid. It follows from Theorem 5.3.11 that $\Gamma(P, \mathcal{F})$ belongs to the diagram L_n. For $n = 2$, three non-collinear points generate the whole space, which is, according to the hypotheses, a projective plane, and so belongs to the diagram A_2.

In view of induction on n, it suffices to establish that $\Gamma(P, \mathcal{F})$ belongs to the diagram A_3 if $n = 3$. To this end, consider a point $x \in P$ and two distinct projective planes u, v containing x. Take $y \in u \setminus v$. In view of dimensions, v and y generate P. We claim that P consists of the union of v and all lines passing through y and meeting v in a point. For, this union is a subspace (due to the fact that any three points are contained in a projective plane), it contains both y and v, and it is obviously contained in P. Applying the claim to u, we see that $l = u \cap v$ is a subspace of u with the property that each line on y in u meets it nontrivially. This implies that l is a line. As l is incident with x, it follows that $d(u, v) = 2$ in Γ_x. This establishes that Γ_x is a projective plane. The structures of the other rank two subdiagrams are obvious, so we conclude that Γ is of type A_3, as required. $\square$

We continue with an analogue for affine geometry. Recall that Af_n denotes the diagram over $[n]$ described in Proposition 2.4.10.

Corollary 5.4.6 *Let $n \in \mathbb{N}, n \geq 2$.*

(i) *If Γ is an $[n]$-geometry belonging to the diagram Af_n, then there is a unique simple matroid $(P, \mathcal{F})$ of dimension n (up to isomorphism) such that $\Gamma(P, \mathcal{F}) \cong \Gamma$. Moreover, any three non-collinear points of P generate an affine plane in $(P, \mathcal{F})$.*

(ii) *Conversely, if $(Q, \mathcal{G})$ is a simple matroid of dimension n in which the subspace generated by any three non-collinear points is an affine plane, then $\Gamma(Q, \mathcal{G})$ is an $[n]$-geometry with diagram*

$$\mathrm{AfL}_n: \quad \overset{}{\underset{1}{\circ}} \overset{\mathrm{Af}}{\rule{1.5cm}{0.4pt}} \overset{}{\underset{2}{\circ}} \overset{\mathrm{L}}{\rule{1.5cm}{0.4pt}} \overset{}{\underset{3}{\circ}} \cdots\cdots \overset{}{\underset{n-2}{\circ}} \overset{\mathrm{L}}{\rule{1.5cm}{0.4pt}} \overset{}{\underset{n-1}{\circ}} \overset{\mathrm{L}}{\rule{1.5cm}{0.4pt}} \overset{}{\underset{n}{\circ}}.$$

Proof The proof of (i) is very similar to the proof of Corollary 5.4.4. Part (ii) follows from Theorem 5.3.11. $\square$

Truncations of the usual affine geometries show that Part (ii) no longer holds if AfL_n is replaced by Af_n.

From Corollary 5.4.4 we gained the insight that the geometries of types A_n can be fully reconstructed from the associated line spaces. As for the analogue for Af_n, we haven't yet discussed how to obtain information on the parallelism in order to conclude that the shadow space is an affine space (cf. Definition 5.1.1). In the case of thick lines, Exercise 5.7.16 allows us to conclude that any three collinear points generate an affine plane in the shadow space, but the parallelism is still absent. If lines have size at least four, Theorem 6.4.8 will produce the required parallelism from a geometry with diagram AfL_3.

5.5 Steiner Systems

This section is not needed for the main developments of the book, but continues in the next section. It focuses on combinatorial objects with high symmetry that arise from matroids whose geometries have residues isomorphic to affine or projective geometries. These objects are Steiner systems and the extended Golay code. Their automorphism groups are connected with the five sporadic simple groups known as the Mathieu groups (mentioned in Remark 5.6.4).

Definition 5.5.1 A *Steiner system* with parameters (t, k, v), an $S(t, k, v)$ for short, is a finite set S of v *points* together with a collection B of subsets of S of size k, called *blocks*, such that each subset of S of size t is contained in a unique member of B.

Example 5.5.2 Here are some constructions of Steiner systems.

(i) Trivial Steiner systems are those with $k = v$, having a unique block, and those with $t = k$, in which every set of t points is a block.

(ii) An $S(2, k, v)$ is a linear space with a point set of size v all of whose lines have k points. Thus, an $S(2, k, v)$ is a geometry with the following diagram, where $m = (v - k)/(k - 1)$

$$\overset{1}{\underset{k-1}{\circ}} \overset{\text{L}}{\rule{2cm}{0.4pt}} \overset{2}{\underset{m}{\circ}} \, .$$

If q is a prime power, then $\mathbb{A}(\mathbb{F}_q^n)$ gives an $S(2, q, q^n)$ and $\mathbb{P}(\mathbb{F}_q^{n+1})$ an $S(2, q + 1, (q^{n+1} - 1)/(q - 1))$.

(iii) Consider a matroid of dimension three whose geometry (as in Theorem 5.3.11) belongs to the diagram

$$\overset{1}{\underset{1}{\circ}} \overset{\text{C}}{\rule{1.5cm}{0.4pt}} \overset{2}{\underset{k-2}{\circ}} \overset{\text{L}}{\rule{1.5cm}{0.4pt}} \overset{3}{\underset{m}{\circ}}$$

for $k, m \in \mathbb{N}$ with $k \geq 3$ and $m \geq 1$. The point order 1 indicates that lines have two points. Call the elements of type 3 (the flats of dimension 2) blocks. Each block has exactly k points. Every set of three points is contained in a unique block, so we find an $S(3, k, v)$ with $v = 2 + (m + 1)(k - 2)$.

The last example generalizes to $t \geq 2$. Recall from Example 1.2.6 that the collection of all subsets of size i of a set S is denoted $\binom{S}{i}$.

Proposition 5.5.3 *For arbitrary $t \geq 2$, an $S(t, k, v)$ with point set S and block set B yields a $[t]$-geometry $\Gamma = \left(S, \binom{S}{2}, \ldots, \binom{S}{t-1}, B, *\right)$, where $*$ stands for symmetrized containment. It belongs to the following diagram, where $m = (v - k)/(k - t + 1)$*

$$\underset{1}{\overset{1}{\circ}} \xrightarrow{\quad C \quad} \underset{1}{\overset{2}{\circ}} \cdots\cdots\cdots \underset{1}{\overset{t-2}{\circ}} \xrightarrow{\quad C \quad} \underset{k-t+1}{\overset{t-1}{\circ}} \xrightarrow{\quad L \quad} \underset{m}{\overset{t}{\circ}}.$$

Proof The Steiner system (S, B) determines a simple matroid $M(S)$ of dimension t whose flats of dimension i for $i \in \{0, \ldots, t - 2\}$ are the subsets of S of size $i + 1$ in S and whose flats of dimension $t - 1$ are the blocks in B. The geometry $\Gamma(M(S))$ of Theorem 5.3.11(ii) coincides with the geometry Γ. The orders of Γ as displayed in the diagram follow from the Steiner system parameters (t, k, v).

The residues of type $\{1, 2\}$ are clearly of type C. The case $t = 2$ is covered by Example 5.5.2(ii). Let $t \geq 3$. If p is a point of S, then the quotient matroid S/p of Example 5.3.10(iii) consists of all points distinct from p, together with the collection of subsets $Y \setminus \{p\}$ for $Y \in B$. As S_p is an $S(t - 1, k - 1, v - 1)$ whose corresponding matroid satisfies $M(S_p) = (M(S))_p$, the diagram is as stated. $\square$

Definition 5.5.4 A Steiner system with parameters (t, k, v) is called *locally projective* if the corresponding geometry Γ of Proposition 5.5.3 has the following diagram for some $m \in \mathbb{N}$, $m \geq 1$

$$\underset{1}{\overset{1}{\circ}} \xrightarrow{\quad C \quad} \underset{1}{\overset{2}{\circ}} \cdots\cdots\cdots \underset{1}{\overset{t-2}{\circ}} \xrightarrow{\quad C \quad} \underset{m}{\overset{t-1}{\circ}} \xrightarrow{\quad\quad} \underset{m}{\overset{t}{\circ}}.$$

For such an $S(t, k, v)$, we have $(v - k)/(k - t + 1) = m = k - t + 1$, and so $v = m^2 + k$. We will use this strong numerical information to pin down the structure of such systems.

Theorem 5.5.5 *The geometry of a nontrivial locally projective $S(t, k, v)$ with $t \geq 3$ has one of the following parameter sets*

$$(t, k, v) = \left(3, m + 2, m^2 + m + 2\right) \quad \text{with } m \in \{2, 4, 10\}, \ (4, 7, 23), \ (5, 8, 24).$$

Proof Let $t = 3$, so $k = m + 2$ and $v = m^2 + m + 2$. The number v_2 of blocks of an $S(3, m + 2, m^2 + m + 2)$ equals $b_3 = (m^2 + m + 2)(m^2 + m + 1)(m^2 + m)/(m + 2)(m + 1)m$. This implies $(m + 2) \mid m^2(m^2 - 1)$, so $(m + 2) \mid 12$, forcing $m \in \{2, 4, 10\}$.

Consideration of quotients gives the same restrictions on m for $t \geq 4$. If $t = 4$, the number of blocks is $(m^2 + m + 3)b_3/(m + 3)$, which excludes the possibilities $m = 2$ and $m = 10$. Therefore, $m = 4$ if $t \geq 4$. A similar computation shows that the number of blocks is no longer an integer if $t \geq 6$, so the result follows. $\square$

Remark 5.5.6 In the case where $t = 3$ and $m = 2$, Exercise 5.7.30 shows the existence of an $S(3, 4, 8)$. Existence of an $S(3, 12, 112)$ would imply the existence of a projective plane of order 10, which is contradicted by an exhaustive computer search. The possibility $m = 4$ for $t = 3$ remains. The Steiner system, being an $S(3, 6, 22)$, occurs as the quotient of an $S(4, 7, 23)$, which in turn is the quotient of a unique $S(5, 8, 24)$. The Steiner system with parameters $(5, 8, 24)$ will be studied below.

We also consider Steiner systems that extend affine planes.

Definition 5.5.7 A Steiner system $S(t, k, v)$ is called *locally affine* if the corresponding matroid has the following diagram for some $m \geq 2$

$$\underset{1}{\overset{1}{\circ}} \xrightarrow{C} \underset{1}{\overset{2}{\circ}} \cdots\cdots \underset{1}{\overset{t-2}{\circ}} \xrightarrow{C} \underset{m-1}{\overset{t-1}{\circ}} \xrightarrow{\mathrm{Af}} \underset{m}{\overset{t}{\circ}} .$$

By Proposition 5.5.3, the t-order m of such an $S(t, k, v)$ satisfies $k - t + 2 = m = (v - k)/(k - t + 1)$, so $k = m + t - 2$ and $v = m^2 + t - 2$.

Theorem 5.5.8 *A nontrivial locally affine Steiner system* $S(t, k, v)$ *with* $t \geq 4$ *satisfies one of the following conditions with* $v = (k - t + 2)^2 + t - 2$.

(i) $t = 4$ *and* $k \in \{5, 6, 10, 12, 15, 20, 30, 60\}$.
(ii) $t = 5$ *and* $k \in \{6, 11\}$.
(iii) $t \in \{6, \ldots, 12\}$ *and* $k = 6 + t$.

Proof Suppose $t \geq 3$ and consider an affine plane A of order $m = k - t + 2$ (cf. Example 2.3.2 for terminology). The plane A has m^2 points. In order to have a nontrivial $S(t, k, v)$ whose quotient with respect to the set of $t - 2$ points is isomorphic to A, we need $m > 2$. The number of blocks of such an $S(t, k, v)$ equals

$$\frac{(m^2 + t - 2)(m^2 + t - 3) \cdots (m^2 - 1)}{(m + t - 2)(m + t - 3) \cdots (m - 1)} .$$

If $t = 3$, this number is $(m^2 + 1)m$. If $t = 4$, we find $(m^2 + 2)(m^2 + 1)m/(m + 2)$, which implies that 60 is a multiple of $m + 2$, from which the conditions of (i) follow. Proceeding this way with the count of blocks for $t \geq 5$, we reach divisibility conditions leading to (ii) and (iii). $\qquad\square$

Remark 5.5.9 Examples of thick finite projective planes are only known for orders equal to a power q of a prime. In view of Exercise 2.8.9, this means we can only expect examples of non-trivial locally affine $S(4, k, v)$ if $m = k - 2 \in \{3, 4, 8, 13\}$. In fact, it is known that these are the only values of m for which a Steiner system as indicated may occur. Nothing is known about $m = 13$. In the cases where $m \in \{4, 8\}$, Steiner systems $S(4, 6, 18)$ and $S(4, 10, 66)$ are known not to exist. As a consequence, locally affine $S(5, 11, v)$ fail to exist and Case (iii) of Theorem 5.5.8 does not occur. The case $m = 3$ is well behaved: there are unique Steiner systems $S(3, 4, 10)$, $S(4, 5, 11)$, and $S(5, 6, 12)$.

Example 5.5.10 An *ovoid* in a projective space $\mathbb{P}$ is a non-empty set of points O such that every line intersects O in at most two points and such that the union of all lines of $\mathbb{P}$ whose intersection with O consists of a given point p, is a projective hyperplane, denoted O_p, of $\mathbb{P}$.

Let $\mathbb{D}$ be a division ring and let O be an ovoid in $\mathbb{P}(\mathbb{D}^4)$. The restriction $\mathrm{Restr}(O)$ of the matroid on $\mathbb{P}(\mathbb{D}^4)$ to O (cf. Example 5.3.10(ii)) is a simple matroid of dimension three and determines a geometry belonging to

$$\overset{1}{\underset{}{\circ}} \xrightarrow{\quad \mathrm{C} \quad} \overset{2}{\underset{}{\circ}} \xrightarrow{\quad \mathrm{AG}(\mathbb{D}^2) \quad} \overset{3}{\underset{}{\circ}} .$$

In the finite case, where $\mathbb{D} = \mathbb{F}_q$ for some prime power q, the restriction $\mathrm{Restr}(O)$ of the ovoid O in $\mathbb{P}(\mathbb{F}_q^4)$, together with its elements of types 1 and 3, is an $S(3, q+1, q^2 + 1)$. It is a locally affine Steiner system.

Besides the trivial cases where $k = v$ or $k = t$, no Steiner system with $t \geq 6$ is known. Below, in Theorem 5.5.21, we construct a remarkable $S(5, 8, 24)$ giving in addition remarkable $S(4, 7, 23)$, $S(3, 6, 22)$, $S(5, 6, 12)$, $S(4, 5, 11)$. We begin with a representation of a set and its subsets over $\mathbb{F}_2$. The field $\mathbb{F}_2$ has a unique feature when we deal with representations. Consider the finite set $[v]$ of v points. We can identify the points of $[v]$ with the vectors ε_i $(i \in [v])$ of the standard basis of the vector space $\mathbb{F}_2^v$.

Definition 5.5.11 Let $v \in \mathbb{N}$. For any subset A of $[v]$, we represent A by the vector $\varepsilon_A := \Sigma_{i \in A} \varepsilon_i \in \mathbb{F}_2^v$. Thus, every vector $x = \varepsilon_1 x_1 + \cdots + \varepsilon_v x_v \in \mathbb{F}_2^v$ represents a unique subset of $[v]$, called the *support* $\mathrm{sup}(x)$ of x, consisting of all points $i \in [v]$ such that $x_i = 1$. The *weight* $\mathrm{wt}(x)$ of x is the cardinality of its support.

Remark 5.5.12 Addition in $\mathbb{F}_2^v$ can be expressed in terms of the set $[v]$. Indeed, if A, B are subsets of $[v]$, then $\varepsilon_A + \varepsilon_B$ represents the *symmetric difference* $(A \setminus B) \cup (B \setminus A)$ of A and B. In other words, $\varepsilon_A + \varepsilon_B = \varepsilon_{A \cup B} + \varepsilon_{A \cap B}$. If x, y are vectors in $\mathbb{F}_2^v$ we have $\mathrm{sup}(x + y) = (\mathrm{sup}(x) \cup \mathrm{sup}(y)) \setminus (\mathrm{sup}(x) \cap \mathrm{sup}(y))$.

We provide $\mathbb{F}_2^v$ with the standard inner product f: if

$$x = \sum_i x_i \varepsilon_i, \qquad y = \sum_i y_i \varepsilon_i$$

then

$$f(x, y) = \sum_i x_i y_i. \tag{5.3}$$

The form f is symmetric and nondegenerate. As in Example 1.4.13, we write $x \perp y$ if $f(x, y) = 0$. If $X \subseteq \mathbb{F}_2^v$, then $X^\perp$ denotes the set of vectors $y \in \mathbb{F}_2^v$ such that $y \perp x$ for all $x \in X$. This is always a subspace of $\mathbb{F}_2^v$. If $x \neq 0$ is a vector, then $x^\perp := \{x\}^\perp$ is a hyperplane of $\mathbb{F}_2^v$. Observe that the standard basis is orthogonal. The set-theoretic meaning of f is obvious: the cardinality of $\mathrm{sup}(x) \cap \mathrm{sup}(y)$ is even or odd according to whether $f(x, y) = 0$ or 1.

Definition 5.5.13 In coding theory, any subset of $\mathbb{F}_2^v$ is called a *code* of length v (or a *binary code* to indicate that the underlying field is $\mathbb{F}_2$). A vector of the code is called a *code word*. The *minimum distance* of a code C is the smallest nonzero weight of the difference of any two elements of the code. If C is *linear*, that is, a linear subspace of $\mathbb{F}_2^v$, then the minimum distance is the minimum weight of a nonzero vector in C.

The *dual code* of C is the code $C^\perp$. This is a linear code. A code C is called *selfdual* if $C = C^\perp$.

An *automorphism* of a code C is a linear transformation of $\mathbb{F}_2^v$ leaving both C and the standard basis invariant.

Let $n \in \mathbb{N}$, $n > 0$, and let N be an $n \times n$-matrix all of whose entries are in $\mathbb{F}_2$. The *code $C(N)$ of the matrix N* is the subset of $\mathbb{F}_2^{2n}$ consisting of all (x_1, x_2) with $x_1 \in \mathbb{F}_2^n$ and $x_2 = Nx_1 \in \mathbb{F}_2^n$.

Lemma 5.5.14 *The code $C(N)$ of an $n \times n$-matrix N over $\mathbb{F}_2$ is linear and of dimension n. It is selfdual if N is an orthogonal matrix.*

Proof The code $C(N)$ is a linear subspace of $\mathbb{F}_2^{2n}$ because $N(x_1 + y_1) = N(x_1) + N(y_1)$ whenever $x_1, y_1 \in \mathbb{F}_2^n$. Also, $\dim(C(N)) = n$ because the standard basis $\varepsilon_1, \ldots, \varepsilon_n$ of $\mathbb{F}_2^n$ provides n linearly independent vectors $(\varepsilon_i, N\varepsilon_i)$ in $C(N)$, and if $x_1 = \sum_{i=1}^n \varepsilon_i \lambda_i$, then $(x_1, Nx_1) = (\sum_{i=1}^n \varepsilon_i \lambda_i, \sum_{i=1}^n N(\varepsilon_i)\lambda_i)$ is a linear combination of the vectors $(\varepsilon_i, N\varepsilon_i)$ $(i \in [n])$.

Let N be as in Definition 5.5.13 and assume N is an orthogonal matrix. We denote the standard inner products in $\mathbb{F}_2^n$ and $V = \mathbb{F}_2^{2n}$ introduced in (5.3) by the same letter f. The orthogonality of N means $N \in O(\mathbb{F}_2^n, f)$ and we find $f((x_1, x_2), (y_1, y_2)) = f(x_1, y_1) + f(x_2, y_2) = f(x_1, y_1) + f(Nx_1, Ny_1) = 0$. Consequently, $C(N) \subseteq C(N)^\perp$. As f is nondegenerate, $\dim(C(N)^\perp) = 2n - \dim(C(N)) = n$, and so $C(N) = C(N)^\perp$. $\qquad\square$

Example 5.5.15 By J_n we denote the $n \times n$-matrix over $\mathbb{F}_2$ all of whose entries are 1. Put $N = J_4 - I_4$. Then N is orthogonal, so $C(N)$ is a selfdual code of dimension 4 and of length 8. Self-duality of $C(N)$ implies that the supports of any two code words meet in an even number. The sum of all four vectors $(\varepsilon_i, N\varepsilon_i)$ $(i \in [4])$ of the code equals $\sum_{i=1}^8 \varepsilon_i$, the all-one word, so if a word belongs to $C(N)$, then so does its complement. Beside 0 and the all-one word of weight eight, there are 14 words of weight 4. This accounts for all 16 words of $C(N)$, so the minimal distance of $C(N)$ is 4.

Now we make a particular choice for N that leads to a mathematical gem.

Definition 5.5.16 Let $\mathcal{I}$ be the icosahedron with vertex set $[12]$ as indicated in Fig. 5.4. Let A be the *adjacency matrix* of $\mathcal{I}$; this means that the rows and columns label the 12 points of $\mathcal{I}$, that $A_{ij} = 1$ if i, j are distinct and adjacent vertices of $\mathcal{I}$, and that $A_{ij} = 0$ otherwise. Put $N = J_{12} - A$, where J_{12} is as in Example 5.5.15. The resulting code $C(N)$ is called the *extended (binary) Golay code*.

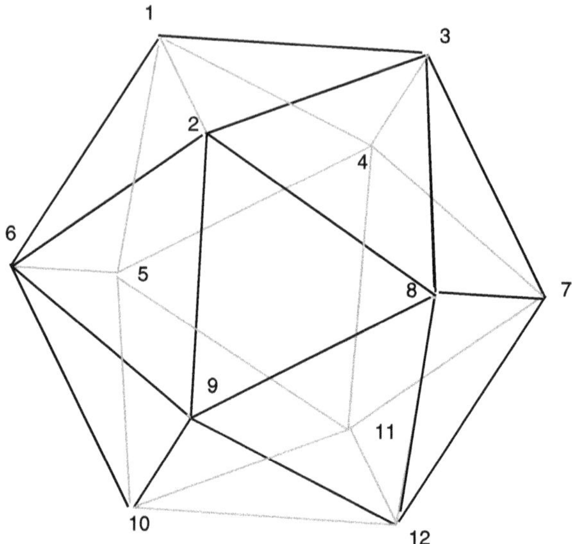

Fig. 5.4 Icosahedron with labels

The adjective extended is meant to distinguish the code from its truncation on the first 23 coordinates. Here, truncation means that we throw away the last coordinate. The resulting code is famous for being perfect: every vector in $\mathbb{F}_2^{23}$ is at distance at most three from a unique code word. See Exercise 5.7.33.

Lemma 5.5.17 *The extended Golay code is selfdual. In particular, code words intersect evenly.*

Proof Observe that N is a symmetric matrix. It has seven entries equal to 1 on each row. The support of a row of N consists of two opposite vertices a, a' and a pentagon all of whose vertices are adjacent to a. We call such a support a *co-pentagon* since its complement in $\mathcal{I}$ is a pentagon (namely, the one all of whose vertices are adjacent to a').

The automorphism group of $\mathcal{I}$ has three orbits on the set of pairs of distinct vertices, whence also on the set of pairs of distinct co-pentagons. Inspection of the three cases shows that two co-pentagons intersect in two or in four vertices. Hence two rows of N share an even number of entries equal to 1 and therefore they are orthogonal. So, N is orthogonal. By Lemma 5.5.14, the code C is selfdual. Hence any two vectors in C have supports intersecting in an even number of points in $[v]$. $\square$

Write $j_m := \varepsilon_1 + \cdots + \varepsilon_m$ for the all one vector in $\mathbb{F}_2^m$.

Lemma 5.5.18 *The extended Golay code C has the following properties.*

(i) *All weights of C are divisible by four.*
(ii) *The all one vector j_{24} belongs to C.*
(iii) *If $(x_1, x_2) \in C$, then $(x_2, x_1) \in C$.*

Proof (i) The weights of the elements from the basis $(\varepsilon_i, N\varepsilon_i)_{i\in[12]}$ of C are $1 + 7 = 8$, which are multiples of four. Suppose that $c_1, c_2 \in C$ have weights $4a_1$, $4a_2$, respectively, where a_1, a_2 are integers. By Lemma 5.5.17, the supports of c_1, c_2 have an even number of common points, say $2t$, so their symmetric difference has $4a_1 - 2t + 4a_2 - 2t$ points and the weight of $c_1 + c_2$ is $4(a_1 + a_2 - t)$. Thus, the weight of every linear combination of the basis vectors of C is divisible by four.

(ii) Each row of N has an odd number of nonzero entries, so $Nj_{12} = j_{12}$ (where $j_{12} = \varepsilon_1 + \cdots + \varepsilon_{12}$). This implies $j_{24} = (j_{12}, Nj_{12}) \in C$.

(iii) Since N is symmetric and orthogonal, $N^2 = I_{12}$ (the identity matrix). If $(x_1, x_2) \in C$, then by definition $x_2 = Nx_1$ and so $Nx_2 = x_1$, establishing $(x_2, x_1) \in C$. $\qquad\square$

Proposition 5.5.19 *The extended Golay code C has neither vectors of weight 4 nor of weight* 20.

Proof By Lemma 5.5.18 it suffices to show that there are no vectors of weight four. If $x = (x_1, x_2) \in C$ has weight four, then we are faced with one of the following possibilities.

Case	I	II	III	IV	V
$\text{wt}(x_1)$	0	1	2	3	4
$\text{wt}(x_2)$	4	3	2	1	0

In view of Lemma 5.5.18(iii), we need only rule out cases I, II, III. Since $x_1 = 0$ forces $Nx_1 = x_2 = 0$, case I does not occur.

In case II, x_1 is a standard basis vector with label say i, and so the support of $x_2 = Nx_1$ is the number of vertices of $\mathcal{I}$ that are not adjacent to i, that is $12 - 5 = 7$, a contradiction with $\text{wt}(x_2) = 3$.

In case III, with $\text{wt}(x_1) = 2$, the fact that two rows (respectively, columns) of N share either two or four entries equal to 1 (see the proof of Lemma 5.5.17), implies that Nx_1 has weight 10 or 6. So also case III does not occur. $\qquad\square$

By counting arguments we derive the main result of this section about the extended Golay code C.

Definition 5.5.20 The support sets of code words of $C(N)$ of weight eight are called *blocks* and those of weight twelve are called *dodecads*.

Theorem 5.5.21 *Let C be the extended Golay code. Denote by n_i the number of code words of weight i.*

(i) $n_0 = n_{24} = 1$, $n_8 = n_{16} = 759$, and $n_{12} = 2576$; *the other n_i are zero. In particular, C has minimal distance eight.*

(ii) *The set* [24] *of points together with the collection of blocks is a locally projective Steiner system* $S(5, 8, 24)$.

Proof Let sup(C) be the subset of [24] consisting of all the supports of vectors in C (see Definition 5.5.11). By Lemma 5.5.18, the full set [24] and the empty set are in sup(C). Now, by Proposition 5.5.19, each nontrivial member of sup(C) has size 8, 12, or 16.

By Lemma 5.5.18, $n_i = n_{24-i} = 1$. By Proposition 5.5.19, $n_0 + n_{24} + n_8 + n_{16} + n_{12} = 2^{12} = 4096$. Therefore, $2n_8 + n_{12} = 4094$. We proceed in eight steps.

STEP 1. *On five distinct points there is at most one block.*

Indeed, if B_1 and B_2 share five points, their symmetric difference has less than six points and it is in sup(C), so it is empty and $B_1 = B_2$.

STEP 2. $n_8 \leq 759$.

Indeed, by Step 1, we have $\binom{24}{5} \cdot 1 \geq \binom{8}{5} \cdot n_8$, so $n_8 \leq 759$.

Let $n_{8,i}$ be the number of vectors $x = (x_1, x_2) \in C$ such that $\text{wt}(x) = 8$ and $\text{wt}(x_1) = i$.

STEP 3. $n_{8,i} = n_{8,8-i}$, $n_{8,0} = 0$, $n_{8,1} = 12$, *and* $n_8 = 24 + 2n_{8,2} + 2n_{8,3} + n_{8,4}$.

The first equality follows from Lemma 5.5.18. As $x_1 = 0$ forces $x_2 = 0$, we have $n_{8,0} = 0$. The value of $n_{8,1}$ follows from the argument in the proof of Lemma 5.5.18(i). The last equality follows directly.

STEP 4. $n_{8,2} = 60$.

Indeed, let i, j be distinct vertices of $\mathcal{I}$ and let a_{ij} be the number of co-pentagons of $\mathcal{I}$ containing i and j. Then $a_{ij} = 2$ or 4 according to whether i and j are opposite or not. Therefore, $\text{wt}(\varepsilon_i + \varepsilon_j, N(\varepsilon_i + \varepsilon_j)) = 2 + (7 - 2) + (7 - 2) = 12$ if i, j are opposite and $2 + (7 - 4) + (7 - 4) = 8$ if i, j are not opposite. The latter situation occurs $12 \cdot 10/2 = 60$ times.

STEP 5. $n_{8,3} \geq 180$.

Indeed, if $x_1 = \varepsilon_i + \varepsilon_j + \varepsilon_k$ where i, j, k are distinct vertices of $\mathcal{I}$, then there are contributions to $n_{8,3}$ from three cases, up to a permutation of indices:

(1) $d(i, j) = 1, d(j, k) = 1$, and $d(i, k) = 2$,
(2) $d(i, j) = 1, d(j, k) = 2$, and $d(i, k) = 2$,
(3) $d(i, j) = 1, d(j, k) = 2$, and $d(i, k) = 3$.

Here, d denotes the graph-theoretic distance in $\mathcal{I}$, as in Definition 1.6.1. A simple count gives 60 triples for each case.

STEP 6. $n_{8,4} \geq 255$.

For $x_1 = \varepsilon_i + \varepsilon_j + \varepsilon_k + \varepsilon_l$ with i, j, k, l distinct, the following configurations contribute to $n_{8,4}$ to the amount indicated.

(1) $d(i, j) = d(j, k) = d(i, k) = d(l, k) = d(i, l) = 1$ and $d(j, l) = 2$ gives $12 \cdot 5/2 = 30$ vectors.
(2) $d(i, j) = d(j, k) = d(i, k) = 1$, $d(i, l) = 3$ and $d(j, l) = d(k, l) = 2$ gives $12 \cdot 5 = 60$ vectors.

(3) $d(i,j) = d(i,k) = d(k,l) = 1$ and $d(i,l) = d(j,l) = d(j,k) = 2$ gives $12 \cdot 5 = 60$ vectors.

(4) $d(i,j) = d(i,k) = 1$, $d(i,l) = 3$ and $d(j,k) = d(k,l) = d(j,l) = 2$ gives $12 \cdot 5 = 60$ vectors.

(5) $d(i,j) = 1$ and $d(i,k) = d(i,l) = d(j,k) = d(j,l) = d(k,l) = 2$ gives $12 \cdot 5/2 = 30$ vectors.

(6) $d(i,j) = d(l,k) = 1$, $d(i,k) = d(j,l) = 2$ and $d(i,l) = d(j,k) = 3$ gives $12 \cdot 5/4 = 15$ vectors.

STEP 7. *Assertion* (i) *holds.*

By Steps 3, 4, 5, 6, we have $n_8 = 24 + 2n_{8,2} + 2n_{8,3} + n_{8,4} \geq 24 + 120 + 180 + 255 \geq 759$, and so, by Step 2, $n_8 = 759$. According to Lemma 5.5.18, $n_{16} = 759$ and $n_{12} = 4096 - 2 - 2 \cdot 759 = 2576$.

STEP 8 *Assertion* (ii) *holds.*

Every set of five points is on a unique block. This is immediate from Steps 2 and 7.

Each residue of type $\{4,5\}$ of the geometry of the resulting Steiner system as in Proposition 5.5.3 consists of the points and blocks of an $S(2,5,21)$, which is necessarily a projective plane of order four. Therefore, the Steiner system is locally projective. $\qquad\square$

We now look for the geometry induced on a dodecad D.

Corollary 5.5.22 *Fix a dodecad D of the* $S(5,8,24)$. *Let a small block of D be any set of the form $B \cap D$ for B a block having six points in D. There are 132 small blocks. They form a locally affine Steiner system $S(5,6,12)$ on D.*

Proof If B is a block, Theorem 5.5.21 shows that $B \cap D$ can only have 6, 4, or 2 points. (For instance, if $B \subset D$, then the symmetric difference would be a code word of weight 4, a contradiction.) By Theorem 5.5.21, every subset of D of five points is in a unique small block. Thus, there are $\binom{12}{5}/\binom{6}{5} = 132$ small blocks. It is immediate that they form a Steiner system $S(5,6,12)$. Analyzing the residues of type $\{4,5\}$ in the geometry of this Steiner system, formed as in Proposition 5.5.3, we find they are $S(2,3,9)$, which must be affine planes of order 3. $\qquad\square$

5.6 Geometries Related to the Golay Code

Let S be the Steiner system $S(5,8,24)$ of Theorem 5.5.21(ii) and let Γ be the corresponding geometry as in Proposition 5.5.3. In this section, we determine its automorphism group and prove uniqueness of S as a Steiner system with the parameters $(5,8,24)$. As a consequence, the Golay code is uniquely determined by its construction in $\mathbb{F}_2^{24}$ as the linear span of all ε_B for B ranging over all blocks of S (cf. Definition 5.5.11). We use the information obtained on S to analyze some geometries related to it. The most important new geometry is the Shult-Yanushka near-hexagon of Corollary 5.6.7.

Theorem 5.6.1 *The automorphism group M of S acts 5-transitively on the point set* [24] *of S and flag transitively on Γ. It has order* $24 \cdot 23 \cdot 22 \cdot 21 \cdot 20 \cdot 48$.

Proof Consider three distinct points a, b, c of S. There are 21 blocks on a, b, c. These blocks are the blocks of the quotient matroid $S/\{a,b,c\}$, a Steiner system $S(2,5,21)$, that is, a projective plane π of order four on the set $[24] \setminus \{a,b,c\}$. We will use the fact that this plane is isomorphic to $\mathbb{P}(\mathbb{F}_4^3)$, which is the content of Exercise 2.8.16.

There are 56 blocks on a and b that do not contain c. They have six points in the projective plane π. A line of π intersects such a block in either 0 or 2 points (for the block determined by the line has exactly two points outside π in common with it and the two blocks meet in 0, 2, or 4 points). Therefore, the set of six points in π of such a block is a hyperoval as defined in Exercise 5.7.35. Two of the 56 hyperovals intersect in an even number of points since blocks do so. Since there are three pairs of points in $\{a,b,c\}$, we find three families, denoted by H_{ab}, H_{ac}, H_{bc}, of 56 hyperovals each. The indices refer to the pair of points from $\{a,b,c\}$ that they contain. Hyperovals from distinct families intersect in an odd number of points, so the sets H_{ab}, H_{ac}, H_{bc} are the classes of the equivalence relation 'meeting in an even number of points'.

We proceed in 7 steps.

STEP 1. *Every hyperoval of π belongs to $H_{ab} \cup H_{ac} \cup H_{bc}$.*

This follows from Exercise 5.7.35.

STEP 2. *The group $\mathrm{Aut}(\pi)$ acts on $\{H_{ab}, H_{ac}, H_{bc}\}$.*

Indeed, given $h, k \in H_{ab}$, then $|h \cap k|$ is even, and so $|g(h) \cap g(k)| = |g(h \cap k)|$ is even for every $g \in \mathrm{Aut}(\pi)$. As such g preserve the equivalence relation of meeting in an even number of points, they preserve the set of corresponding equivalence classes.

STEP 3. *There are 120 blocks containing a (respectively, b, c) only among the points a, b, c. Each of these has 7 points in π, and a line in π intersecting such a block in two points has exactly three points in common with it. This provides three families B_a, B_b, B_c of 120 so-called* Baer subplanes *of π.*

Indeed, the number of blocks containing a is $23 \cdot 22 \cdot 21 \cdot 20 / 7 \cdot 6 \cdot 5 \cdot 4 = 253$, and the number of blocks intersecting $\{a,b,c\}$ in a only is $253 - 56 - 56 - 21 = 120$. Since any two blocks intersect in an even number of points and four points at most (see Theorem 5.5.21), the other statement follows.

STEP 4. *In π there are exactly 360 Baer subplanes, and so each of those is in one of the three families B_a, B_b, B_c. Moreover, $\mathrm{Aut}(\pi)$ acts on $\{B_a, B_b, B_c\}$.*

To show this, we observe that the number of subplanes is at most $21 \cdot 20 \cdot 16 \cdot 9 / 7 \cdot 6 \cdot 4 = 360$, hence the first statement. Next, two members of B_a (respectively

B_b, B_c) intersect in an odd number of points by Step 3, while two members of different collections among those three intersect in an even number of points.

STEP 5. *There are 210 blocks of S lying entirely in π. Each of these is the symmetric difference of two lines in π. The group* Aut(π) *is transitive on this collection.*

By Steps 1, 3, and 4, there are $759 - 360 - 168 - 21 = 210$ blocks with no point in $\{a, b, c\}$. If B_1 and B_2 are two blocks on $\{a, b, c\}$, then $B_1 \cap B_2$ has size four and so these blocks have a unique point of π in common. Their symmetric difference $B_1 \oplus B_2$ is a block (this follows immediately from Definition 5.5.20) and equals $(B_1 \setminus \{a, b, c\}) \oplus (B_2 \setminus \{a, b, c\})$, the symmetric difference of two lines of π. As there are $21 \cdot 20/2 = 210$ pairs of distinct lines in π, each of the 210 blocks entirely contained in π is of this kind. The transitivity of Aut(π) is a direct consequence of the transitivity of Aut($\mathbb{P}(\mathbb{F}_4^3)$) on the set of non-collinear triples.

A substantial part of Aut(π) extends to automorphisms of S. Let l be a line in π and let A be the affine plane $\pi \setminus l$ (cf. Exercise 2.8.9(a)). The set $B = [24] \setminus A = \{a, b, c\} \cup l$ is a block. The group T(A) of translations of A (cf. Example 1.8.17) has order 16. For any translation $t \in$ T(A), let $\bar{t}$ be its extension to S fixing B point-wise. We call $\bar{t}$ a *translation* of S.

STEP 6. *Every translation of S belongs to M.*

Let $\bar{t}$ be a translation. Then $\bar{t}$ maps a line of π onto a line and a symmetric difference of two lines on a similar set. Since $\bar{t}\,|_\pi$ is in Aut(π), it leaves invariant $\{h \cap \pi \mid h \in H_{ab} \cup H_{bc} \cup H_{ac}\}$ and $\{h \cap \pi \mid h \in B_a \cup B_b \cup B_c\}$ by Steps 3 and 4. Let $h \in H_{ab}$ where h has two points on l and four in A. Since t is of order two and fixes no point of A, the sets $\bar{t}(h)$ and h share an even number of points in [24], so $\bar{t}(h) \in H_{ab}$. Similarly, for $x \in B_a$, we may assume that x has three points on l and four further points in A. Now $\bar{t}(x)$ and x share an odd number of points, so $\bar{t}(x) \in B_a$.

STEP 7. *The theorem holds.*

Using the fact that there is a unique plane of order four (Exercise 2.8.16), we see that each block is fixed point-wise by a group of 16 translations. Combining these for different blocks we instantly see that M is 5-transitive on the point set. Hence, $|M| = 24 \cdot 23 \cdot 22 \cdot 21 \cdot 20 \cdot \alpha$, where α is the order of the stabilizer of five points (and hence of the unique block containing these points) in M. Let H be the image in Sym(B) of the action of the stabilizer in M of a block B. By 5-transitivity of M and uniqueness of B among the blocks containing five given points, H is 5-transitive on B, and so its index in Sym(B) is at most six. However, Alt$_8$ is a simple subgroup of Sym$_8$ of index 2 and therefore Sym(B) $\cong$ Sym$_8$ cannot act transitively by left translation on a set of six or three cosets of H. Hence, the index of H in Sym(B) is at most two and Alt$_8 \subseteq H$. Thus $3|\alpha$. Moreover, the point-wise stabilizer $M_{[B]}$ of B in M contains the group T(A) of 16 translations, so $48 \mid \alpha$. If $\alpha > 48$, then there is $g \in M$ fixing six points, a, b, c and, say, d, e, f on B and a point $p \notin B$. The element g has order dividing four and fixes one further point on each of the three blocks containing $\{a, b, c, d, p\}$, $\{a, b, c, e, p\}$, and $\{a, b, c, f, p\}$, respectively. Clearly now, g fixes all points, a contradiction. We conclude that $\alpha = 48$.

Finally, M acts transitively on the set of blocks, and the stabilizer of a block induces Alt_8 on it, so it is 5-transitive, whence transitive, on the set of maximal flags of its residue. $\qquad\square$

Corollary 5.6.2 *Each* $S(5, 8, 24)$ *is isomorphic to the Steiner system S.*

Proof The proof of Theorem 5.6.1 only assumed about S that it is an $S(5, 8, 24)$ and pinned down the structure of blocks entirely in terms of the unique projective plane of order four and the unique substructures of hyperovals and Baer planes. $\quad\square$

We next revisit the geometry of a dodecad as in Corollary 5.5.22.

Corollary 5.6.3 *The automorphism group of a dodecad of S has order* $12 \cdot 11 \cdot 10 \cdot 9 \cdot 8$. *It acts 5-transitively on a dodecad and flag transitively on the rank 5 geometry underlying the Steiner system $S(5, 6, 12)$ induced on it.*

Proof If D is a dodecad and if $g \in M$ leaves D invariant and fixes five points of D, then g fixes a sixth point of D. Hence, by the fact that the stabilizer of a block in M induces Alt_8 on it, g fixes the eight points of a block B intersecting D in six points. Moreover, the symmetric difference of D and B is another block B', invariant under g. If a, b, c are points of D fixed by g, then, in the projective plane obtained by removing a, b, c, the block B' is the symmetric difference of two lines whose common point is also fixed by g. Therefore, g is a translation and, in fact, the identity.

Hence, D is invariant under at most $12 \cdot 11 \cdot 10 \cdot 9 \cdot 8$ elements of M and D has at least $24 \cdot 23 \cdot 22 \cdot 21 \cdot 20 \cdot 48 / 12 \cdot 11 \cdot 10 \cdot 9 \cdot 8 = 2576$ transforms under M. By Theorem 5.5.21 there are exactly 2576 dodecads. So, all statements follow at once. $\qquad\square$

Remark 5.6.4 The five sporadic simple *Mathieu groups* M_{24}, M_{23}, M_{22}, M_{12} and M_{11} occur as follows in the automorphism groups of the Steiner systems constructed.

(1) $M_{24} = M$, the automorphism group of S.
(2) M_{23} is the stabilizer of a point of S in M_{24}, and so acts on an $S(4, 7, 23)$. It is the full automorphism group of this Steiner system.
(3) M_{22} is the stabilizer of two points of S in M. It is a subgroup of index 2 in the full (set-wise) stabilizer in M of an unordered pair of points in S, which is the full automorphism group of the quotient Steiner system $S(3, 6, 22)$.
(4) $M_{12} = \mathrm{Aut}(D)$, the stabilizer of a dodecad D in M as discussed in Corollary 5.6.3. It is the full automorphism group of the Steiner system $S(5, 6, 12)$ of Corollary 5.5.22.
(5) M_{11} is the stabilizer of a point of $S(5, 6, 12)$ in M_{12}. It is the full automorphism group of the resulting $S(4, 5, 11)$.

Table 5.1 Mathieu groups and diagrams. The subscripts of nodes indicate, from top to bottom, the order, the number of elements, and the structure of the stabilizer of an element of the corresponding type

i	$|M_i|$	Diagram
24	$24 \cdot 23 \cdot 22 \cdot 21 \cdot 20 \cdot 48$	Nodes: $\begin{smallmatrix}1\\24\\M_{23}\end{smallmatrix}$ — $\begin{smallmatrix}1\\276\\M_{22}\cdot 2\end{smallmatrix}$ — $\begin{smallmatrix}1\\ \\ \end{smallmatrix}$ —^{C}— $\begin{smallmatrix}4\\ \\ \end{smallmatrix}$ — $\begin{smallmatrix}4\\759\\2^4\,\mathrm{Alt}_8\end{smallmatrix}$
23	$23 \cdot 22 \cdot 21 \cdot 20 \cdot 48$	Nodes: $\begin{smallmatrix}1\\23\\M_{22}\end{smallmatrix}$ — $\begin{smallmatrix}1\\253\\M_{21}\cdot 2\end{smallmatrix}$ —^{C}— $\begin{smallmatrix}4\\1771\\ \end{smallmatrix}$ — $\begin{smallmatrix}4\\253\\2^4\,\mathrm{Alt}_7\end{smallmatrix}$
22	$22 \cdot 21 \cdot 20 \cdot 48$	Nodes: $\begin{smallmatrix}1\\22\\M_{21}\end{smallmatrix}$ —^{C}— $\begin{smallmatrix}4\\231\\ \end{smallmatrix}$ — $\begin{smallmatrix}4\\77\\2^4\,\mathrm{Alt}_6\end{smallmatrix}$
12	$12 \cdot 11 \cdot 10 \cdot 9 \cdot 8$	Nodes: $\begin{smallmatrix}1\\12\\M_{11}\end{smallmatrix}$ — $\begin{smallmatrix}1\\66\\ \end{smallmatrix}$ — $\begin{smallmatrix}1\\220\\ \end{smallmatrix}$ —^{C}— $\begin{smallmatrix}2\\495\\ \end{smallmatrix}$ —^{Af}— $\begin{smallmatrix}3\\132\\\mathrm{Sym}_6\end{smallmatrix}$
11	$11 \cdot 10 \cdot 9 \cdot 8$	Nodes: $\begin{smallmatrix}1\\11\\M_{10}\end{smallmatrix}$ — $\begin{smallmatrix}1\\55\\ \end{smallmatrix}$ —^{C}— $\begin{smallmatrix}2\\165\\ \end{smallmatrix}$ —^{Af}— $\begin{smallmatrix}3\\55\\\mathrm{Sym}_5\end{smallmatrix}$

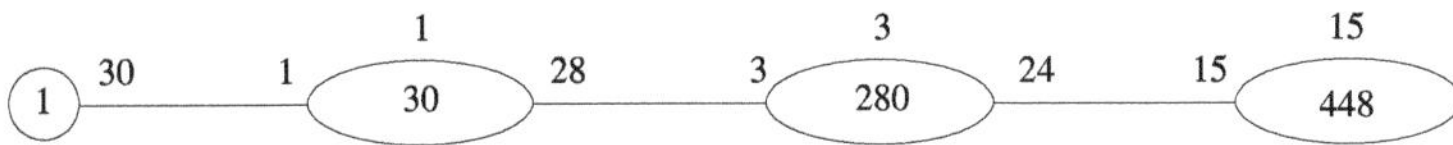

Fig. 5.5 The distribution diagram of the collinearity graph of (R, L). The intersection of two blocks has size 8, 0, 4, 2 according to whether they are at mutual distance 0, 1, 2, 3 in R

Table 5.1 lists some information on these groups, which can be used to reconstruct the geometries as group geometries by means of subgroups of the Mathieu groups.

The stabilizer of a point of $S(4, 5, 11)$ in M_{11} is often referred to as M_{10}. It is not a sporadic group, but isomorphic to $P\Gamma L(\mathbb{F}_9^2)$ (which has a subgroup of index two isomorphic to Alt_6, but is not isomorphic to Sym_6).

We next investigate how M acts on the set of ordered pairs of blocks of S and obtain an interesting geometry and line space, involving the projective space $\mathbb{P}(\mathbb{F}_2^4)$.

Definition 5.6.5 By Lemma 5.5.18, the extended Golay code C of Definition 5.5.16 contains the all one vector j_{24}. Let R denote the set of blocks of S. If B and B' are disjoint blocks, the vector $j_{24} + \varepsilon_B + \varepsilon_{B'}$ also belongs to C and has weight 8, so its support is another block, say B'' of S. The *Shult-Yanushka space* has point set R and line set L consisting of the unordered triples $\{B, B', B''\}$ of mutually disjoint members of R.

Proposition 5.6.6 *The distribution diagram of the collinearity graph of the Shult-Yanushka space is as in Fig. 5.5. Moreover, for each distance $i \in [3]$, the group $\mathrm{Aut}(S)$ acts transitively on the set of ordered pairs of vertices at mutual distance i.*

Proof Put $M = \text{Aut}(S)$. The 5-transitivity of M on the point set of S, proved in Theorem 5.6.1, gives that M is transitive on R. Let B be a block and let $G = M_B$ be its stabilizer in M; its structure is $C_2^4 \cdot \text{Alt}_8$. We proceed in seven steps.

STEP 1. *The group G acts transitively on the set of 280 blocks intersecting B exactly in four points.*

There are indeed $\binom{8}{4} \cdot (5 - 1) = 280$ such blocks because there are five blocks on each set of four points. If B', B'' are two such blocks, the fact that G induces Alt_8 on B, allows us to assume that $B \cap B' = B \cap B''$, and then there is a translation of S (cf. Step 6 of the proof of Theorem 5.6.1) fixing B point-wise and mapping B' onto B''.

STEP 2. *The group G acts transitively on the set of 448 blocks intersecting B exactly in two points.*

Given two points of B, there are $(6 \cdot 5/2) \cdot 4 = 60$ blocks meeting B in four points and containing the two given points. Hence, the number of blocks meeting B in exactly two points is $(8 \cdot 7/2)(77 - 1 - (6 \cdot 5/2) \cdot 4) = 448$. If B', B'' are such blocks, we may assume without loss of generality that $B \cap B' = B \cap B''$. Now, the symmetric difference $B \oplus B'$ is a dodecad D' and $B \oplus B''$ is a dodecad D''. Here, B intersects D' and D'' in six points. As M is transitive on the set of dodecads and $\text{Aut}(D)$ is transitive on the set of blocks intersecting a dodecad in six points (see Corollary 5.6.3), the statement follows.

STEP 3. *There are 30 blocks disjoint from B, providing the affine hyperplanes of an affine space $\mathbb{A}(\mathbb{F}_2^4)$ on the 16 points off B.*

There are indeed $759 - 1 - 280 - 448 = 30$ such blocks. Recall from the proof of Theorem 5.6.1 that we may view B as the subset $\{a, b, c\} \cup l$, where l is a line of the projective plane π induced on the set complement of $\{a, b, c\}$ (of size 21). By Step 5 of that proof, every block disjoint from B is the symmetric difference of two lines. The intersection of these two lines belongs to l, and so their union is contained in the affine plane A on $\pi \setminus l$. The group $T(A)$ of 16 translations fixing B point-wise determines the structure of an affine space $\mathbb{A}(\mathbb{F}_2^4)$ on A. Each symmetric difference of two lines of π meeting l in a point is an affine hyperplane of this space. An easy count shows that there are $5 \times \binom{4}{2} = 30$ such symmetric differences, thus accounting for all affine hyperplanes of $\mathbb{A}(\mathbb{F}_2^4)$.

STEP 4. *The group G acts transitively on the set of 30 blocks disjoint from B and Alt_8 is isomorphic to $\text{GL}(\mathbb{F}_2^4)$.*

If p is a point off B, then G_p is isomorphic to Alt_8. Therefore, the latter is a subgroup of the linear group $\text{GL}(\mathbb{F}_2^4) = \text{SL}(\mathbb{F}_2^4) = \text{PSL}(\mathbb{F}_2^4)$, namely the stabilizer of p in the affine group of the affine space $\mathbb{A}(\mathbb{F}_2^4)$ (cf. Theorem 5.1.12(ii)). Since Alt_8 and $\text{GL}(\mathbb{F}_2^4)$ have the same order, these groups are isomorphic. Finally, $\text{GL}(\mathbb{F}_2^4)$ acts transitively on the set of hyperplanes containing p, so G acts transitively on the set of 30 affine hyperplanes.

STEP 5. *The group M acts transitively on the set of pairs of vertices at mutual distance two in (R, L), and the vertices at distance two from B are the 280 blocks intersecting B in four points.*

The neighborhood N of the vertex B of R has 30 vertices and the graph induced on N has connected components made up of two vertices. Any two vertices in N not in the same component represent non-parallel hyperplanes of the affine space induced by $T(A)$ and so meet in four points. This shows that two vertices are at mutual distance two if and only if they meet in four points of S. As M is transitive on R, it follows from Step 1 that M is also transitive on the set of all pairs of vertices of R at mutual distance two.

STEP 6. *If B and B' are blocks intersecting in four points, then $B' \setminus B$ is an affine plane in the affine space $\mathbb{A}(\mathbb{F}_2^4)$ induced by $T(A)$ on $[24] \setminus B$.*

Indeed, every translation fixing B point-wise and mapping some point of $B' \setminus B$ to some point of $B' \setminus B$, must leave B' invariant and so leaves $B' \setminus B$ invariant. Hence $B' \setminus B$ is an affine subspace on four points, that is, an affine plane.

STEP 7. *If B and B' are blocks intersecting in four points, then there are exactly three blocks disjoint from both of these.*

It suffices to observe that in the affine space induced by $T(A)$, a plane is disjoint from three affine hyperplanes.

In conclusion, in Steps 3, 4, and 5, the G-orbits of blocks distinct from B have been found, and in Step 7, the number of blocks adjacent to the two blocks B and B' at mutual distance two has been determined as 3. As these are on distinct lines on B', there are at least three blocks (one more on each line on B' having a block collinear with B) at distance two from B and collinear with B'. It readily follows that this is the exact number, and so, there must be $30 - 3 - 3 = 24$ blocks collinear with B' and meeting B in exactly two points. The remaining parameters of the distribution diagram can be derived directly from those found. $\square$

Corollary 5.6.7 *The Shult-Yanushka space (R, L) belongs to the diagram*
$$\overset{\displaystyle p}{\underset{\displaystyle 2}{\circ}}\ \overset{\displaystyle 6\quad 4\quad 6}{\rule{2cm}{0.4pt}}\ \overset{\displaystyle 1}{\underset{\displaystyle 14}{\circ}}.$$
Moreover, for each line $l \in L$ and each point $p \in R$, there is a unique point q on l nearest p.

Proof All statements are straightforward consequences of the information given by the distribution diagram. $\square$

We will see that the geometry of the Shult-Yanushka space can be viewed as a truncation of an interesting rank three geometry.

Definition 5.6.8 A space (R, L) satisfying the 'nearest point on a line' property stated at the end of Corollary 5.6.7 is called a *near-hexagon*.

If B_1 and B_2 are blocks meeting in four points, then $B_1 \cup B_2$ has size 12, and so does its complement $[24] \setminus (B_1 \cup B_2)$. By Step 7 of the proof of Proposition 5.6.6, there are three blocks contained in $[24] \setminus (B_1 \cup B_2)$. Take two of these, say B_3, B_4. They meet in four points. The six sets $B_1 \cap B_2$, $B_1 \setminus B_2$, $B_2 \setminus B_1$, $B_3 \cap B_4$, $B_3 \setminus B_4$, $B_4 \setminus B_3$ of size four partition $[24]$. Moreover, the union of any two of these is a block of S. Such a partition of $[24]$ in six subsets E_i $(i \in [6])$ of size four each, such that the union of any two is a block, is called a *sextet* of S. We let Q be the collection of all sextets of S.

The *Shult-Yanushka geometry* is the incidence system $(R, L, Q, *)$ with the following incidence $*$, where $C \in R$, $B := \{B_i \mid i \in [3]\} \in L$, and $E := \{E_i \mid i \in [6]\} \in Q$.

(1) $E * C$ if there are $i, j \in [6]$ such that $C = E_i \cup E_j$.
(2) $E * B$ if $B_k * E$ for each $k \in [3]$.
(3) $B * C$ if $C = B_i$ for some $i \in [3]$.

Theorem 5.6.9 *The Shult-Yanushka geometry* $\Gamma = (R, L, Q, *)$ *is a* $[3]$*-geometry belonging to the diagram*

$$
\begin{array}{ccccc}
1 & & 2 & \text{L} & 3 \\
\circ & \!\!\!=\!\!\!=\!\!\!=\!\!\! & \circ & \rule{1.5em}{0.4pt} & \circ. \\
2 & & 2 & & 6 \\
759 & & 3795 & & 1771
\end{array}
$$

Its point residues are isomorphic to the $\{1, 2\}$*-truncation of* $\mathrm{PG}(\mathbb{F}_2^4)$ *with* 15 *points and* 35 *lines. The* $\{1, 2\}$*-truncation of* Γ *is the Shult-Yanushka near-hexagon. Moreover,* $\mathrm{Aut}(S)$ *acts flag transitively on* Γ.

Proof The statement on the $\{1, 2\}$-truncation is immediate from Corollary 5.6.7.

Let $E := \{E_i \mid i \in [6]\}$ be a sextet. The residue Γ_E of E in Γ has 15 blocks of the form $E_i \cup E_j$ for distinct $i, j \in [6]$ and 15 lines consisting of unordered triples of mutually disjoint blocks of Γ_E. By Proposition 5.6.6, the mutual distances between blocks in the collinearity graph of (R, L) are at most two. The incidence structure is readily identified with the generalized quadrangle of order $(2, 2)$ of Example 2.2.10. This explains the double bond in the diagram of the theorem.

In view of Proposition 5.6.6, the number of sextets is $759 \cdot 280/15 \cdot 8 = 1771$ and $\mathrm{Aut}(S)$ acts transitively on these. Since two blocks meeting in four points are incident with a unique sextet, and, by Proposition 5.6.6, $\mathrm{Aut}(S)$ is transitive on the set of such pairs of blocks, the stabilizer in $\mathrm{Aut}(S)$ of a sextet is transitive on the set of blocks. Moreover, the stabilizer of a point and a sextet incident with it, is easily seen to be transitive on the set of three lines through the block and incident with the sextet. Hence the transitivity results.

It remains to consider the residue Γ_B of a block B. From the distribution diagram in Fig. 5.5 we read off that there are 30 blocks collinear with B, and hence 15 lines on B. Suppose $l := \{B, B_1, B_2\}$ and $m := \{B, C_1, C_2\}$ are lines on B. As $|B_i \cap C_j| \leq 4$, we find $|B_i \cap C_j| = 4$ for all $i, j \in [2]$, so l and m contain blocks from the unique sextet E determined by B_1 and C_1 as in Definition 5.6.8. This proves that Γ_B is the geometry of a linear space with 15 points and 35 lines of size three.

In order to prove that Γ_B is isomorphic to the geometry of $\mathbb{P}(\mathbb{F}_2^4)$, the $[2]$-truncation of $\mathrm{PG}(\mathbb{F}_2^4)$, we use Proposition 1.8.7 to identify the residue Γ_B with the coset incidence system $\Gamma(G_B, (G_{B,l}, G_{B,E}))$, where $\{B, l, E\}$ is the chamber of Γ of the previous paragraph. The group G_B has order $|M/759| = 16 \cdot |\mathrm{Alt}_8|$. By Steps 3 and 4 of Proposition 5.6.6, it has a normal subgroup $T := \mathrm{T}(A)$, where $A = \pi \setminus B$, for a projective plane π on the complement in [24] of an unordered triple $\{a, b, c\}$ of points of B, is as in the proof of Theorem 5.6.1. As each block of S disjoint from B is the symmetric difference of two lines of π meeting in a point of B, the group T lies in the kernel of the action of G_B on the set of lines of Γ containing B. Therefore, the quotient of G_B by T, which is isomorphic to $H := \mathrm{PSL}(\mathbb{F}_2^4)$ by Step 4 of the proof of Proposition 5.6.6, acts on the set of lines of Γ containing B. The stabilizer of the line l of Γ_B in G_B/T coincides with the stabilizer of a point p of $\mathrm{PG}(\mathbb{F}_2^4)$ in H. Similarly, T acts trivially on sextets incident with B, and the stabilizer of the sextet E of Γ_B in G_B/T coincides with the stabilizer of a line n on p in H. We conclude $\Gamma_B \cong \Gamma(G_B, (G_{B,l}, G_{B,E})) \cong \Gamma(G_B/T, (G_{B,l}/T, G_{B,E}/T)) \cong \Gamma(H, (H_p, H_n)) \cong \mathrm{PG}(\mathbb{F}_2^4)$, where the second isomorphism is direct from the definition of coset incidence systems and the third is another consequence of Proposition 1.8.7. $\qquad\square$

Remark 5.6.10 The argument at the end of the proof of Theorem 5.6.9 is needed as $\mathbb{P}(\mathbb{F}_2^4)$ is not the only linear space on 15 points all of whose lines have size three. To see this, consider the linear space obtained from $\mathbb{P}(\mathbb{F}_2^4)$ by replacing the lines of a plane by the lines of another model of the Fano plane.

Exercise 5.7.37 shows the remarkable feature of this geometry that it is not the [3]-truncation of a rank 4 geometry whose 4-elements in residues of flags of type 1 correspond to planes of $\mathrm{PG}(\mathbb{F}_2^4)$.

5.7 Exercises

Section 5.1

Exercise 5.7.1 Show that the affine space of a 2-dimensional vector space (cf. Proposition 5.1.3) is an affine plane as defined in Definition 2.3.1.

Exercise 5.7.2 Let V be a vector space. In line with Definition 4.3.1, an affine hyperplane in $\mathbb{A}(V)$ is a coset $a + U$ of a hyperplane U in V (i.e., a linear subspace of V of codimension 1). Suppose that $L = b + W$ is a line of $\mathbb{A}(V)$ with $b \in V$ and W a 1-dimensional subspace of V. Show that $L \cap H$ is a unique point unless $W \subseteq U$, in which case L is either disjoint from H or contained in it.

Exercise 5.7.3 Let $\mathbb{D}$ be a division ring and let V be a right vector space over $\mathbb{D}$. Establish the assertions made in Remark 5.1.5:

(a) Suppose $|\mathbb{D}| \geq 3$. Prove that if Z is a subset of V with the property that, for every two distinct members p, q of Z, all points of the affine line pq also belong to Z, then Z is an affine subspace of $\mathbb{A}(V)$.

(*Hint:* If $x \in Z$ and l is a line in Z, then each point y on the line parallel to l and through x is on a line through the intersection of a line through x meeting l and a line through y meeting l.)

(b) Suppose $|\mathbb{D}| = 2$ and $\dim(V) = 2$. Show that the property in (a) fails for $Z = \{(0,0), (0,1), (1,0)\}$.

Exercise 5.7.4 Let V be a right vector space over the division ring $\mathbb{D}$. The construction of the affine space $\mathbb{A}(V)$ in Proposition 5.1.3 leads to a trivial structure if $\dim(V) = 1$, while V is not trivial. A way to improve on this is to define $\mathbb{A}(V)$ on the basis of points, lines, parallelism (as in Definition 5.1.1), and *dilatations*, which are maps $V \to V$, $x \mapsto xa + v$, where $a \in \mathbb{D} \setminus \{0\}$ and $v \in V$.

(a) Show that the dilatations constitute a group, which we denote by $\mathrm{Dil}(\mathbb{A}(V))$.
(b) Show that, when $\dim(V) \geq 2$, the group $\mathrm{Dil}(\mathbb{A}(V))$ consists precisely of those automorphisms of $\mathbb{A}(V)$ that map every line to a parallel.
(b) From now let $\dim(V) = 1$, and require of an automorphism of $\mathbb{A}(V)$ not only that it be as in Definition 5.1.7 but also that it maps every dilatation to a dilatation, i.e., that it normalizes $\mathrm{Dil}(\mathbb{A}(V))$. (In the case where $\dim(V) \geq 2$ this is immediate from Theorem 5.1.12.) Show that the group $\mathrm{T}(V)$ of all translations of V is a normal subgroup of $\mathrm{Aut}(\mathbb{A}(V))$. Show also that $\mathrm{Aut}(\mathbb{A}(V)) = \mathrm{T}(V) \rtimes \mathrm{\Gamma L}(V)$.
(c) Show that a dilatation fixing two points is the identity.
(d) Is a dilatation without fixed points necessarily a translation?
(e) Observe that there are translations and dilatations with a unique fixed point in Examples 2.3.2 and 5.1.4.

Exercise 5.7.5 If $\mathbb{A}(V)$ is an affine space, A its automorphism group, U an affine subspace of $\mathbb{A}(V)$ and A_U^U the group induced on U by the stabilizer A_U of U, then $A_U^U \cong \mathrm{Aut}(U)$. Prove this. (For the case where $\dim(U) = 1$, take Exercise 5.7.4 into account.)

Exercise 5.7.6 Let $n \in \mathbb{N}$. Show that the affine group $\mathrm{AGL}(\mathbb{F}^n)$ is isomorphic to a subgroup of $\mathrm{GL}(\mathbb{F}^{n+1})$.
(*Hint:* View the affine space $\mathbb{A}(\mathbb{F}^n)$ as the affine hyperplane $\{x \in \mathbb{F}^{n+1} \mid x_{n+1} = 1\}$ of $\mathbb{A}(\mathbb{F}^{n+1})$ and extend the elements $\mathrm{AGL}(\mathbb{F}^n)$ to elements of $\mathrm{AGL}(\mathbb{F}^{n+1})$ fixing $0 \in \mathbb{F}^{n+1}$.)

Exercise 5.7.7 Assume that we allow $\mathbb{D} = \mathbb{Z}$ (a ring but not a division ring) and $V = \mathbb{Z}^2$ in Proposition 5.1.3. What goes wrong and what does still work for $\mathbb{A}(\mathbb{Z}^2)$?

Exercise 5.7.8 In this exercise, we continue with the dilatation structure introduced in Exercise 5.7.4. Let V and W be right vector spaces over the division ring $\mathbb{D}$. A *dilatation homomorphism* from $\mathbb{A}(V)$ to $\mathbb{A}(W)$ is a map α from the point set of $\mathbb{A}(V)$ to the point set of $\mathbb{A}(W)$ such that

(a) if l is a line of $\mathbb{A}(V)$ and a, b are points of l with $\alpha(a) \neq \alpha(b)$, then $\alpha(l)$ is a line of $\mathbb{A}(W)$ and α restricted to the point set of l is an injection;

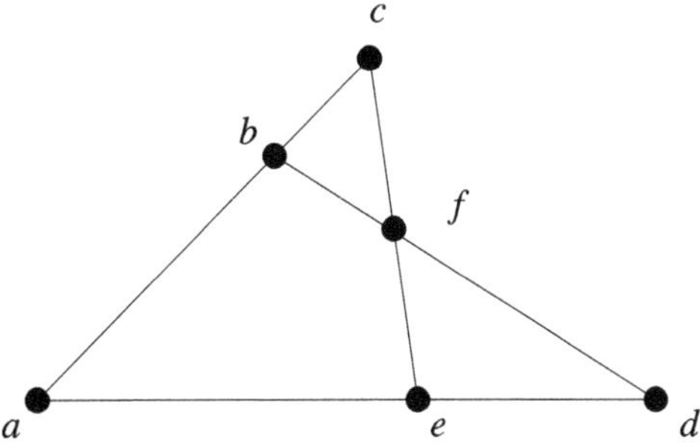

Fig. 5.6 The dual affine plane of order two

(b) if l, l' are parallel lines of $\mathbb{A}(V)$ and if $\alpha(l)$ is a line, then $\alpha(l')$ is a line parallel to $\alpha(l)$;
(c) if δ is a dilatation of $\mathbb{A}(V)$ fixing a point a, and X is a subspace of $\mathbb{A}(V)$ containing a such that $\alpha|_X$ is injective, then the bijection $\alpha|_X \delta \alpha|_X^{-1}$ is induced on X by a dilatation of $\mathbb{A}(W)$ fixing $\alpha(a)$.

Show that, for every semi-linear map $\beta : V \to W$ (cf. Exercise 2.8.28) and every $w \in W$, the map $\mathbb{A}(V) \to \mathbb{A}(W)$, $x \mapsto \beta(x) + w$ is a dilatation homomorphism, and that every dilatation homomorphism is obtained in this way.

Exercise 5.7.9 Verify that the line space on six points with four lines depicted in Fig. 5.6 is the dual line space of the affine plane of order two. Show that its automorphism group is not transitive on the set of triples generating the line space.

Section 5.2

Exercise 5.7.10 Let $\mathbb{P} = \mathbb{P}(\mathbb{F}^2)$ be the projective line over a field $\mathbb{F}$ and consider $G := \mathrm{PGL}(\mathbb{F}^2)$ in its action on $\mathbb{P}$.

(a) Show that G is transitive on the set of ordered triples of points from $\mathbb{P}$.
(b) Show that $G_{0,1,\infty}$, the stabilizer in G of the points $0 = (0 : 1)$, $1 = (1 : 1)$, and $\infty = (1 : 0)$, is the trivial group.
(c) Derive from the above that, if $\mathbb{F} = \mathbb{F}_q$, the order of G equals $q(q^2 - 1)$ (in accordance with Example 5.1.15).

Exercise 5.7.11 (This exercise is used in Theorem 9.5.7) Let Z be a linear space satisfying Pasch's axiom and let X be a set of points of Z such that $\langle p, q \rangle \cap \langle r, s \rangle$ is nonempty for all $p, q, r, s \in X$ with $p \neq q$ and $r \neq s$. Prove that the subspace of Z generated by X is a projective plane.

Exercise 5.7.12 (This exercise is used in Lemma 7.1.8) Let Z be a linear space. The *hyperplane dual* of Z, denoted $Z^\star$, is the line space whose point set consists of all geometric hyperplanes of Z and whose lines are *pencils*, a pencil being the set of all geometric hyperplanes containing the intersection $H \cap H'$ of two distinct geometric hyperplanes H and H' of Z.

(a) Prove that $Z^\star$ is a linear space.
(b) If all lines of Z have three points, prove that $Z^\star$ is a projective space.

Exercise 5.7.13 Consider a division ring $\mathbb{D}$ and its opposite $\mathbb{D}^{\mathrm{op}}$ (cf. Exercise 1.9.11) at the hyperplane dual $\mathbb{P}(V)^\star$ of $\mathbb{P}(V)$ introduced in Exercise 5.7.12.

(a) Show that $\mathbb{P}(V)^\star$ is isomorphic to $\mathbb{P}(V^\vee)$. (Recall from Exercise 1.9.13 that $V^\vee$ is the dual vector space of V.)
(b) Prove that if $\mathbb{D}$ is isomorphic to $\mathbb{D}^{\mathrm{op}}$ and V is finite dimensional, then $\mathbb{P}(V)$ is *selfdual*, i.e., $\mathbb{P}(V) \cong \mathbb{P}(V)^\star$.

Exercise 5.7.14 Let V be a vector space and P the projective space $\mathbb{P}(V)$.

(a) Show that a direct sum decomposition $V = A \oplus B$ provides *complementary subspaces* of $\mathbb{P}$, i.e., subspaces $\mathbb{P}(A)$, $\mathbb{P}(B)$ of $\mathbb{P}$ such that

$$\langle \mathbb{P}(A), \mathbb{P}(B) \rangle = \mathbb{P} \quad \text{and} \quad \mathbb{P}(A) \cap \mathbb{P}(B) = \emptyset.$$

(b) Verify that, conversely, every pair of complementary subspaces of $\mathbb{P}$ is derived from a direct sum decomposition in V.
(c) Let C, D be complementary subspaces of $\mathbb{P}$. The *projection* of $\mathbb{P}$ on D from C, denoted by ρ_{CD}, is defined on $\mathbb{P} \setminus C$, and maps a point x to $\langle C, x \rangle \cap D$, which is a point of C. Show that ρ_{CD} maps every subspace of $\mathbb{P}$ not contained in C onto a subspace of D. Show also that ρ_{CD} corresponds to the natural homomorphism of vector spaces $V \twoheadrightarrow V/C$, where C is identified with the subspace of V corresponding to it.

Exercise 5.7.15 Let $\mathbb{F}_q$ be the finite field of order q and let $n \in \mathbb{N}$.

(a) Show that the affine space $\mathbb{A}(\mathbb{F}_q^n)$ has

$$\frac{q^n \left(q^n - 1 \right) \cdots \left(q^n - q^{i-1} \right)}{q^i \left(q^i - 1 \right) \cdots \left(q^i - q^{i-1} \right)}$$

subspaces of dimension i for $1 \le i \le n$.
(b) Prove that its group of dilatations (cf. Exercise 5.7.4) has order $q^n(q-1)$.
(c) Show that the projective space $\mathbb{P}(\mathbb{F}_q^{n+1})$ has

$$\frac{q^{n+1} - 1}{q - 1} = q^n + q^{n-1} + \cdots + 1$$

points and establish similar formulae for its number of subspaces of dimension i.

Exercise 5.7.16 (Cited in Remark 6.4.6) Prove that, if Z is an affine plane and each line of Z has at least three points, then each line of Z is a maximal subspace.

Exercise 5.7.17 Show that, in a thick projective space, all lines have the same cardinality.

Exercise 5.7.18 Let X and Y be distinct non-empty subspaces of a projective space. Show that the subspace $\langle X, Y \rangle$ is the union of all lines joining some point of X to some point of Y.

♮ **Exercise 5.7.19** Let Z be a linear space and $\mathcal{H}$ a collection of geometric hyperplanes of Z satisfying the following two properties.

(1) For any two geometric hyperplanes H, H' in $\mathcal{H}$, every point of $Z \setminus (H \cap H')$ belongs to a member of $\mathcal{H}$ containing $H \cap H'$.
(2) For any point p of Z, there exists a geometric hyperplane in $\mathcal{H}$ not containing p.

Prove that Z is a projective space.

Exercise 5.7.20 Let (P, B) be a finite set P of points together with a family B of subsets of P called blocks such that P is not the union of two blocks and there is some integer $\lambda \geq 1$ with the property that every pair of points is contained in λ blocks. Let a line be the intersection of all blocks containing two given points. Assume that every block intersects every line. Prove that P is the set of points of a projective space whose hyperplanes are the blocks and whose lines are the lines of (P, B).

Exercise 5.7.21 Let H be the generalized hexagon of Example 2.2.15.

(a) Suppose $n \in \mathbb{N}$ is even and positive. Verify that, in a generalized n-gon, the set of points at distance at most $\frac{n}{2} - 1$ in the point collinearity graph from a given point is a geometric hyperplane.
(b) Show that, in H, the subgraph of the collinearity graph induced on the set of points at distance three from a given point x has two connected components.
(c) Prove that joining one of the components of (b) to the set of points at distance at most two to x in the collinearity graph of H, we obtain a geometric hyperplane of H strictly containing the one described in (a).
(d) Conclude that, in general, a geometric hyperplane of a generalized hexagon need not be a maximal subspace.

Section 5.3

Exercise 5.7.22 Let $\mathcal{B}$ be the set of bases of flats of a matroid $(P, \mathcal{F})$ of finite dimension.

(a) Derive the following properties of $\mathcal{B}$.
  (1) $\emptyset \in \mathcal{B}$.
  (2) If $B \in \mathcal{B}$ and $A \subseteq B$, then $A \in \mathcal{B}$.

(3) If A, $B \in \mathcal{B}$ with $|A| = |B| + 1 < \infty$, then there is $x \in A \setminus B$ such that $B \cup \{x\} \in \mathcal{B}$.

(b) As in Exercise 3.7.7, we write 2^P for the power set of P. Define the *dimension map* dim: $2^P \to \mathbb{Z}$ by

$$\dim(A) = -1 + \max\{|X| \mid X \subseteq A,\ X \in \mathcal{B}\}.$$

Prove that $\dim(P)$ is the dimension of $(P, \mathcal{F})$.

(c) Call $X \subseteq P$ *closed* with respect to dim if, for all $x \in P \setminus X$, we have $\dim(X \cup \{x\}) = \dim(X) + 1$. Show that $\mathcal{F}$ is the collection of closed subsets of P. Conclude that $(P, \mathcal{F})$ is determined by $(P, \mathcal{B})$.

Exercise 5.7.23 Fix a number $n \in \mathbb{N}$. Let P be a set and let $\mathcal{B}$ be a collection of subsets of P of size at most n satisfying (1), (2), (3) of Exercise 5.7.22(a). Define the dimension map dim: $2^P \to \mathbb{Z}$ by use of $\mathcal{B}$ as in Part (b) of that exercise and let $\mathcal{F}$ be the collection of all subsets of P that are closed with respect to dim. Prove that $(P, \mathcal{F})$ is a matroid. Relate $\mathcal{B}$ to the set of its bases.

Exercise 5.7.24 Let $G = (V, E)$ be a finite graph. Let $\mathcal{B}$ be the collection of subsets X of E such that (V, X) is a *forest* (that is, a graph without circuits). Verify that the pair $(E, \mathcal{B})$ satisfies (1), (2), (3) of Exercise 5.7.22(a). Describe the flats of the corresponding matroid as in Exercise 5.7.23.

Exercise 5.7.25 (This exercise is used in Proposition 9.1.7) In Example 5.3.10, we encountered matroid quotients. According to Theorem 5.3.8 these apply to projective spaces. Let K be a subspace of a projective space $\mathbb{P}$. Show that the matroid quotient with respect to K of the matroid obtained from $\mathbb{P}$ is not the quotient space described in Exercise 2.8.24, but the map $\mathbb{P} \setminus K \to \mathbb{P}/K$ corresponding to natural homomorphism $V \to V/K$ of vector spaces as in Exercise 5.7.14(c).

Section 5.4

Exercise 5.7.26 Let (P, L) be the 4×4-grid, which is a generalized quadrangle with point order three and line order one. Verify that the pair $(P, \mathcal{F})$, where $\mathcal{F}$ is the collection of subspaces of (P, L), is not a matroid.
(*Hint:* Consider a set v of three mutually non-collinear points, and a point $x \notin v$ with $\langle v, x \rangle = P$ in the notation of Definition 2.5.11.)

Exercise 5.7.27 Define a *Hilbert geometry* to be a firm and residually connected geometry Γ over the diagram

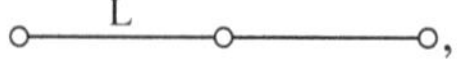

in which any two planes are incident with some line. These assumptions are equivalent to the incidence axioms of Hilbert in his Foundations of Geometry.

(a) Give examples of Hilbert geometries. Think of a rank three projective space and a set Z of points in it such that no line intersects Z in a single point.
(b) Call Γ *dihedral* if there is a line which is incident with exactly two planes. Show that each of these planes is projective (any two lines have a common point in it) and that at least one of these planes has a line of two points. Derive a complete classification of the dihedral Hilbert geometries.
(c) Suppose that Γ is not dihedral. Show that all point residues are thick projective planes of the same order n and that there are $n^3 + n^2 + n + 1$ planes in Γ.

Section 5.5

Exercise 5.7.28 Let X be a set of points in $\mathbb{P}(\mathbb{F}_q^4)$, where $q > 2$. Prove that X is an ovoid if and only if no three points of X are collinear and $|X| = q^2 + 1$.

Exercise 5.7.29 Let q be a prime power. Show that $\mathbb{P}(\mathbb{F}_q^n)$ has no ovoids if $n > 4$.
(*Hint:* Show that such an ovoid O would have $q^{n-2} + 1$ points and that each hyperplane H meeting it in more than one point meets it in an ovoid of H. Count the number of pairs (p, H) of a point $p \in O$ and a hyperplane H on p meeting O in more than one point to derive that the number of hyperplanes meeting O in an ovoid equals $(q^{n-2} + 1)(q^{n-1} - q)/((q - 1)(q^{n-3} + 1))$, which is not an integer if $n > 4$.)

Exercise 5.7.30 (This exercise is used in Remark 5.5.6) Let S be the set of eight vectors of $\mathbb{F}_2^3$ and let $\mathcal{B}$ be the set of 2-dimensional linear subspaces of $\mathbb{F}_2^3$ and their translates. Prove that $(S, \mathcal{B})$ is an $S(3, 4, 8)$.

Exercise 5.7.31 Let G be a t-transitive permutation group on a set X of size $v > t$ for some $t \geq 2$ and let B be a subset of X of size k. Show that $(X, \mathcal{B})$, where $\mathcal{B}$ is the G-orbit of B, is an $S(t, k, v)$ if and only if $|\mathcal{B}| = \binom{v}{t}/\binom{k}{t}$ (cf. Exercise 2.8.12(d)).

Exercise 5.7.32 Consider the point set X of $\mathbb{P}(\mathbb{F}_{11}^2)$ with the natural action of the group $G := \mathrm{PSL}(\mathbb{F}_{11}^2)$. Set $B = \{(1 : i) \mid i = 0, 1, 3, 4, 5, 9\}$ and let $\mathcal{B}$ be the G-orbit of B in X. Prove that $(X, \mathcal{B})$ is an $S(5, 6, 12)$.
(*Hint:* Notice that i runs through the squares of $\mathbb{F}_{11}$.)

Exercise 5.7.33 Consider the *Golay code*, that is, the code C' in $\mathbb{F}_2^{23}$ obtained from the extended Golay code by truncating the last coordinate.

(a) Prove that the Golay code C' is a linear code of minimal weight 7.
(b) Show that each vector in $\mathbb{F}_2^{23}$ is at Hamming distance (which is the weight of the symmetric difference) at most three from a unique code word. In coding terms, this says that the Golay code is a perfect 3-error correcting code.
(*Hint:* Count the number of words at distance at most three from a code word and verify that it equals 2^{23}.)

Exercise 5.7.34 Let Δ' be the graph on 100 vertices of Remark 2.4.13. Fix a vertex ∞ of Δ' and take X to be the set of vertices of Δ' adjacent to ∞. Let B be the collection of all subsets of X that are the common neighbor set of ∞ and a vertex of Δ' at distance two from ∞. Prove that (X, B) is an $S(3, 6, 22)$ Steiner system.

Section 5.6

Exercise 5.7.35 (This exercise is used in Theorem 5.6.1) Consider the projective plane $\mathbb{P}$ of order four. By Exercise 2.8.16, we may take $\mathbb{P} = \mathbb{P}(\mathbb{F}_4^3)$. A *hyperoval* in $\mathbb{P}$ is a set H of six points such that each line of $\mathbb{P}$ meets H in either 0 or 2 points.

(a) Let $\kappa : \mathbb{F}_4^3 \to \mathbb{F}_4$ be a quadratic form (cf. Definition 4.4.2) and let $\mathbb{P}_\kappa$ be the set of zeros $\langle x \rangle \in \mathbb{P}$ of the quadratic equation $\kappa(x) = 0$. Denote by f the bilinear form of κ. Show that, if $\mathbb{P}_\kappa$ has no lines and contains at least two points, then it has five points, and the union of it with $\mathrm{Rad}(f)$ (a singleton) is a hyperoval.
(b) Prove that $\mathrm{Aut}(\mathbb{P})$ is transitive on the set of triples of non-collinear points. Conclude that, if H is a hyperoval, then, up to a change of H by a member from its $\mathrm{Aut}(\mathbb{P})$-orbit, we may assume that $\langle \varepsilon_i \rangle$ belongs to H for each $i \in [3]$. Determine the quadratic forms κ such that $\langle \varepsilon_i \rangle \in \mathbb{P}_\kappa$ and show that H is as in (a) for one of these forms.
(c) Derive that $\mathrm{Aut}(\mathbb{P})$, which is isomorphic to $\mathrm{P\Gamma L}(\mathbb{F}_4^3)$, is transitive on the collection of hyperovals, whereas its subgroup isomorphic to $\mathrm{PSL}(\mathbb{F}_4^3)$ has three orbits of hyperovals, each of size 56.

Exercise 5.7.36 Let D be a dodecad of the $S(5, 8, 24)$ Steiner system S. Denote its complement by D' and its stabilizer in $\mathrm{Aut}(S)$ by M_{12}. The stabilizer G in $\mathrm{Aut}(S)$ of the pair $\{D, D'\}$ admits M_{12} as a normal subgroup of index two.

(a) Show that G is the automorphism group of M_{12}.
(b) If $p \in D$ and M_{11} is the stabilizer of p in M_{12}, show that M_{11} acts as a 3-transitive group on D'.
(c) Consider the 3-transitive action of M_{11} on the set D'. Let $p' \in D'$ and let $(\mathrm{M}_{11})_{p'}$ be its stabilizer. Show that $(\mathrm{M}_{11})_{p'}$ acting on $D' \setminus \langle p' \rangle$ is the group $\mathrm{PSL}(\mathbb{F}_{11}^2)$, acting 2-transitively on this set. Use this example to show that M_{11} acts flag transitively on a geometry of diagram

$$\underset{\substack{1 \\ 12}}{\overset{1}{\circ}} \text{———} \underset{\substack{1 \\ 66}}{\overset{2}{\circ}} \text{———} \underset{\substack{1 \\ 220}}{\overset{3}{\circ}} \overset{5 \ 5 \ 6}{\text{———}} \underset{\substack{2 \\ 165}}{\overset{4}{\circ}} .$$

Here, underneath each node of the diagram, we have listed its order on top of the total number of elements of the corresponding type.

Exercise 5.7.37 Show that the Shult-Yanushka geometry for M_{24} (Definition 5.6.8) does not extend to a geometry of type

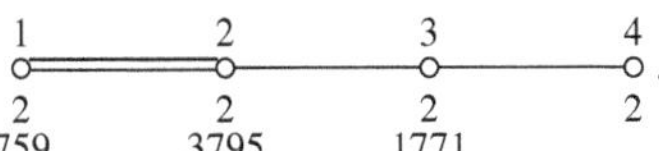

5.8 Notes

Around 300 B.C., Euclid of Alexandria, who lends his name to Euclidean geometry, published thirteen books, usually referred to as the Elements, on various topics, ranging from number theory to geometry. He was not the first to study axioms for geometry related to the real physical space, but probably the first to publish them. In 1899, Hilbert proposed 21 axioms for Euclidean space. He was inspired by (amongst others) Moritz Pasch, who had published an ancestor of Pasch's axiom (Definition 5.2.4) in terms of segments in 1882, as part of a proof that Euclid's axioms were incomplete. The axiomatic characterization of a general projective (3-dimensional) space was given soon after, notably by Veblen and Young, cf. [296].

Section 5.1

Non-Desarguesian planes have been the subject of an enormous amount of work. See for instance [113, 147, 167, 240]. In the literature, systems of axioms for affine spaces appeared much later than systems for projective spaces. The result mentioned in our proposition, which provides a convenient system of axioms for affine geometry, is essentially due to [205].

The facts stated in Example 5.1.15 on $\mathrm{Aut}(\mathbb{F}_q)$ and $\mathrm{GL}(\mathbb{F}_q^n)$ are proved in many textbooks, like [88], to name one.

Section 5.2

The material on geometries in Sect. 5.2 is based on a long and intricate history, an excellent account of which is provided by Baer [11].

Pasch's axiom (Definition 5.2.4) is also (and deservedly) called Veblen's axiom.

The diagrams of projective spaces and the first 'From local to global' Theorem are due to Tits [280]. The present extension to more general diagrams is due to Buekenhout [44].

Section 5.3

Matroids and the abstract theory of linear dependence were introduced by Whitney [310]. See [21] for a discussion of the relations to other subjects. An excellent introduction to matroid theory is [308]. Often, matroids are introduced via bases, as indicated in Exercises 5.7.22 and 5.7.23.

Section 5.4

Geometries over the diagram

with $n \geq 4$, embed in a projective space of dimension n by a result due to Kantor [181].

Section 5.5

Steiner systems were first coined by Steiner [270] in 1853. A Steiner system $S(t, k, v)$ coincides with what is known as a t-$(v, k, 1)$-design. Classical texts on this topic are [10, 20, 113, 207]; good introductions are [18, 89].

The nonexistence of a projective plane of order 10, indicated in Remark 5.5.6, is based on computer computations by Lam et al. [198]. This leaves open the intriguing question whether there are thick finite projective planes whose order is not a prime power.

The study of affinely planar spaces with lines of cardinality three was initiated by Marshall Hall [147], who also gave the first exotic examples. He was followed by many authors.

Steiner systems with diagram

are called *inversive planes* in view of the classical model over the reals, which underlies the study of inversions (a kind of transformation). Example 5.5.10 shows how to obtain inversive planes. No other inversive planes than those constructed in this way are known. Dembowski [113, Theorem 6.2.14] showed that every finite inversive plane with even 3-order q can be obtained from an ovoid in $\mathbb{P}(\mathbb{F}_q^4)$. The nonexistence, mentioned in Remark 5.5.9, of locally affine systems $S(4, k, v)$ with $k \in \{12, 20, 30, 60\}$, as appearing in the conclusion of Theorem 5.5.8, is an immediate consequence. It is proved by Kantor [181] that $k = 10$ does not lead to a locally affine $S(4, k, v)$.

The known ovoids in $\mathbb{P}(\mathbb{F}_q^4)$ comprise the class of quadrics and of Suzuki ovoids, constructed by Tits [283] following the discovery of the Suzuki groups; see [162, 293] for more background.

In this section, we have given constructions of all known locally projective and affine Steiner systems. All of these are uniquely determined by their parameters.

In the years 1861–1873 Mathieu found the first five sporadic simple groups. In 1931, Carmichael [60] found the $S(5, 8, 24)$ Steiner system, which is also known as the Witt design. The relationship between $S(5, 8, 24)$ and M_{24} was worked out by Witt in [314]. The construction of $S(5, 8, 24)$ from the extended Golay code follows [9]. The construction of the extended Golay code in Definition 5.5.16 follows [32], which gives a concise overview of block designs. See [209] for uniqueness proofs of these systems as Steiner systems with the given parameters.

For $t = 2$, there seem to be a lot of Steiner systems. Take $k = 3$. According to Kirkman (in 1847) and a long series of other authors, in particular Rick Wilson [312], an $S(2, 3, v)$ exists if and only if $v = 6n + 1$ or $6n + 3$ for some integer n, so the small values of interest for v are 7, 9, 13, 15, 19, 21, ... Let $N(v)$ be the number of pairwise nonisomorphic $S(2, 3, v)$. It is known that $N(7) = N(9) = 1$, $N(13) = 2$, $N(15) = 80$, and $N(19) = 11084874829$ (cf. [191]). An extensive treatment of this subject can be found in [90].

For $t = 3$ and $k = 4$, a lot of nontrivial Steiner systems $S(t, k, v)$ are known. The existence of an $S(3, 4, v)$ for all $v \equiv 2$ or 4 (mod 6) is due to [151]. See [122, 154, 192] for estimates of the numbers of isomorphism classes. For $t \geq 4$ only finitely many nontrivial $S(t, k, v)$ are known; for $t \geq 6$ none.

Section 5.6

The Mathieu groups are extensively treated in [142]. Hyperovals (cf. Exercise 5.7.35) in $\mathbb{P}(\mathbb{F}_4^3)$ are discussed in [162]. Up to automorphisms, these are all of the form $\{\varepsilon_1 + \lambda\varepsilon_2 + \lambda^2\varepsilon_3 \mid \lambda \in \mathbb{F}_4\} \cup \{\varepsilon_2, \varepsilon_3\}$. This means that the hyperoval is the union of a quadric and the radical of the corresponding symmetric bilinear form. For fields of even order higher than eight, this is not always the case; see, for instance, [66].

The Shult-Yanushka near-hexagon appears in [262]. Proposition 5.6.6 shows that the collinearity graph of the Shult-Yanushka space is distance transitive in the terminology explained in [36].

The second generation proof of the classification of finite simple groups is being rewritten in [133–137], and further forthcoming volumes. In individual papers, such as [15], the role of diagram geometry appears to be useful for certain uniqueness proofs of sporadic groups.

In Sect. 2.9 we mentioned sporadic groups that act flag transitively on geometries with a rank 2 residue isomorphic to the Petersen graph or the tilde geometry. Here we give a rough indication of the diagram score on the 26 sporadic groups (cf. [91]). For most of these groups, firm residually connected geometries of high rank have been found. Table 5.2 lists linear diagrams of high rank pertaining to geometries on which the sporadic group acts flag transitively. For instance, for J_1, a systematic search in [138] has shown that two firm residually connected geometries of rank four exist on which J_1 acts flag transitively; the example in Table 5.2 appeared in [172] as a residue of a rank five geometry for the sporadic group ON.

It will be clear that there is no universal picture here for the sporadic groups (which justifies the adjective sporadic). Of course, the diagrams do not represent all residually connected geometries that have been found for the sporadic groups. Several more can be found in [44, 47, 170, 172], to name a few references.

Table 5.2 The sporadic groups

Year	Discoverer	Name	Diagram	Ref.
1861	Mathieu	M_{11}, M_{12}	cf. Table 5.1	Remark 5.6.4
1873	Mathieu	M_{22}, M_{23}, M_{24}	Pet_3, Pet_4, Til_3	[171, 173]
1965	Janko	J_1	 o—5—o—o—P—o (1 1 1 2)	[172]
1968	Higman-Sims	HS	 o—C—o—o—C*—o (1 4 4 1)	[160, 199]
1968	Conway	Co_1, Co_2, Co_3	Til_4, Pet_4	[171, 173]
1969	Suzuki	Suz	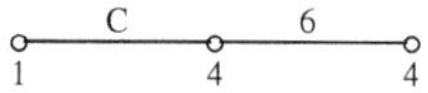 o—C—o—6—o (1 4 4)	[44]
1969	Janko	J_2	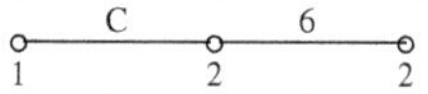 o—C—o—6—o (1 2 2)	[44]
1969	Janko	J_3	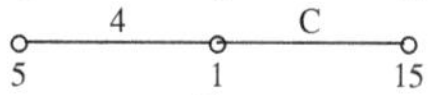 o—4—o—C—o (5 1 15)	[15]
1969	McLaughlin	McL	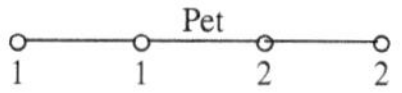 o—o—Pet—o—o (1 1 2 2)	[16]
1969	Held	He	Til_3	Section 2.9
1971	Fischer	$Fi_{22}, Fi_{23}, Fi_{24}$	Fischer space	Section 11.9
1972	Lyons	Ly	$\widetilde{G}_2$	[182]
1973	Rudvalis	Ru	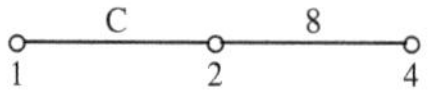 o—C—o—8—o (1 2 4)	[216, 304]
1973	Fischer	BM	Pet_5	[173]
1973	Griess	M	Til_5	[173]
1973	Thompson	Th	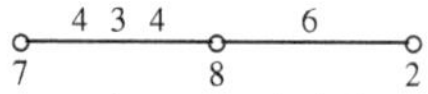 o—4 3 4—o—6—o (7 8 2)	[47]
1973	Norton	NH	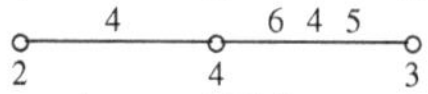 o—4—o—6 4 5—o (2 4 3)	[47]
1976	O'Nan	ON	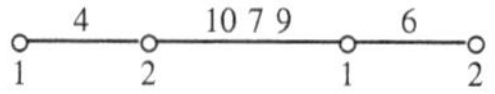 o—4—o—10 7 9—o—6—o (1 2 1 2)	[47]
1976	Janko	J_4	Pet_4	[173]

Thanks to the Feit-Thompson Theorem [133], each non-Abelian finite simple group has a conjugacy class of involutions. The permutation action of G on such a class often leads to an interesting graph on which the group acts. Such graphs have been studied extensively for the Fischer groups, as well as for the Monster M and several of its subgroups which are sporadic.

The Fischer groups play a special role in that they can be viewed as extreme cases of root filtration spaces, which will be introduced in Sect. 6.7; see Chap. 11. In Table 5.2 we have mentioned this fact rather than the following linear diagrams over which flag transitive geometries exist that can be found in [44]:

$$Fi_{22}: \quad \text{o—C—o—o—4—o},$$

$$Fi_{23}: \quad \text{o—C—o—C—o—o—4—o},$$

$$Fi_{24}: \quad \text{o—C—o—C—o—C—o—o—4—o}.$$

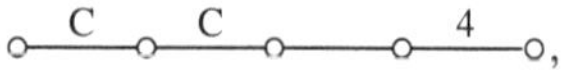

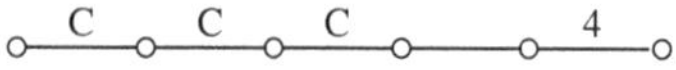

The overall conclusion is that each sporadic group acts flag transitively on a residually connected geometry of rank at least three with a linear diagram.

Section 5.7

Exercises 5.7.12 and 5.7.19 stem from work of Teirlinck [273]. Exercise 5.7.20 is due to Dembowski and Wagner [114].

Contrary to what Exercise 5.7.21 might suggest, it does not happen often that the complement of a geometric hyperplane in the collinearity graph of a generalized polygon is not connected; see [31].

The Hilbert geometry of Exercise 5.7.27 reflects the combination of axioms = namely, seven of the 20 axioms (actually 21, but one turned out redundant) used in Hilbert's Foundations of Geometry [161].

Chapter 6
Projective and Affine Spaces

In Definitions 5.2.1 and 5.1.1, projective and affine spaces were introduced by means of axioms, and in Propositions 5.2.2 and 5.1.3, the spaces $\mathbb{P}(V)$ and $\mathbb{A}(V)$, where V is a vector space, were shown to be examples. In this chapter we show that by and large there are no further examples.

For projective spaces, this result is stated in Theorem 6.3.1. Table 6.1 of Remark 6.3.5 gives an overview of the relations established between projective spaces and projective geometries. The result for affine spaces depends on it and occurs in Corollary 6.4.3. The case of a projective line is given special attention in Sect. 6.2. In Theorem 5.2.16(i), it became clear that extra structure is needed to characterize the projective line coming from a division ring $\mathbb{D}$. This is taken care of by perspectivities, which were introduced in Exercise 1.9.30 and will be further investigated in Sect. 6.1. They will be used in the characterization of abstract projective spaces as projective spaces of vector spaces.

Once the classification work is over, we look at different aspects of projective geometries. In Sect. 6.5, we prove the existence of many thin subgeometries of $\mathrm{PG}(V)$ of type A_n, where V is a vector space of dimension $n + 1$. These subgeometries are called apartments and play a role in the study of $\mathrm{Aut}(\mathrm{PG}(V))$ as well as in the theory of buildings, to be developed in Chap. 11. In Sect. 6.6 we briefly analyze other physical viewpoints of $\mathrm{PG}(V)$: the shadow spaces of these projective geometries on i for $i \in \{2, \ldots, n - 1\}$. Finally, Sect. 6.7 is devoted to shadow spaces of $\mathrm{PG}(V)$ on $\{1, n\}$; these satisfy certain universal axioms collected under the name root filtration space. We explore some of the basic properties of these root filtration spaces, which will be used later (in Sect. 7.9 for polar spaces, and in Sect. 11.6 for root shadow spaces of spherical buildings). In Theorem 6.7.26, these shadow spaces are characterized as special root filtration spaces.

6.1 Perspectivities

Recall from Definition 2.5.12 that a line space is *thick* if each line has at least three points and if every point is on at least three lines. In this section (Theorem 6.1.5) we

F. Buekenhout, A.M. Cohen, *Diagram Geometry*,
Ergebnisse der Mathematik und ihrer Grenzgebiete. 3. Folge / A Series of
Modern Surveys in Mathematics 57, DOI 10.1007/978-3-642-34453-4_6,
© Springer-Verlag Berlin Heidelberg 2013

show that thick projective spaces of dimension at least three (cf. Definition 5.3.1) admit many automorphisms.

We begin by extending the notion of perspectivity from $\mathbb{P}(V)$, introduced in Exercise 1.9.30, to arbitrary linear spaces.

Definition 6.1.1 Let H be a geometric hyperplane and let c be a point of a linear space Z. A *perspectivity* of Z with *center* c and *axis* H is an automorphism of Z fixing all points of H and mapping every line containing c onto itself.

A perspectivity of a thick linear space necessarily fixes its center.

Example 6.1.2 Let V be a right vector space over the division ring $\mathbb{D}$. Let ϕ be a nondegenerate linear form on V and H the maximal subspace of $\mathbb{P}(V)$ consisting of all points on which ϕ vanishes, i.e., $H = \mathbb{P}(\mathrm{Ker}(\phi))$. Let p, a, b be vectors in V such that $\langle p \rangle, \langle a \rangle, \langle b \rangle$ are distinct collinear points of $\mathbb{P}(V)$. Without loss of generality, we may assume that $p = a - b$. Assume furthermore that $\langle a \rangle$ and $\langle b \rangle$ are not in H. The map

$$\alpha : V \to V, \qquad x \mapsto x - (a - b)\phi(a)^{-1}\phi(x),$$

that is, $\alpha = r_{(a-b)\phi(a)^{-1},\phi}$ in the notation of Exercise 1.9.30, induces a perspectivity on $\mathbb{P}(V)$ with center $\langle p \rangle$ and axis H, mapping $\langle a \rangle$ to $\langle b \rangle$. Indeed, α is a linear map with $\alpha(a) = b$. If $x \in \mathrm{Ker}(\alpha)$, then $x = (a - b)\phi(a)^{-1}\phi(x)$, whence $\phi(x) = \phi(a - b)\phi(a)^{-1}\phi(x) = \phi(x) - \phi(b)\phi(a)^{-1}\phi(x)$; since $\phi(b) \neq 0$ this implies $\phi(x) = 0$, whence $x = 0$, showing that α is invertible. Furthermore, for $y \in \mathrm{Ker}(\phi)$, we have $\alpha(y) = y$, so H is an axis. Clearly, $\langle y \rangle, \langle \alpha(y) \rangle, \langle p \rangle$ are collinear, so $\langle p \rangle$ is a center.

Lemma 6.1.3 *Let H be a geometric hyperplane in a thick linear space Z.*

  (i) *Each automorphism of Z fixing each point of H as well as a point outside H is a perspectivity with axis H and center c.*

 (ii) *Each nontrivial perspectivity of Z has a unique center and a unique axis.*

(iii) *If o, a, b are collinear points with a, b not in $H \cup \{o\}$, then $\mathrm{Aut}(Z)$ has at most one perspectivity α with axis H and center o such that $\alpha(a) = b$.*

Proof Let α be a nontrivial automorphism of Z fixing point-wise a geometric hyperplane H.

(i) Assume that α fixes a point c outside H. As H is a geometric hyperplane, every line l on c intersects H in a point h, whence $\alpha(l) = \alpha(c)\alpha(h) = ch = l$. Thus, c is a center of α.

(ii) Suppose that α is a perspectivity with distinct centers c and c', and let d be a point of Z outside cc' (it exists because of thickness). Then $\alpha(d) = d$ as $\alpha(d)$ is on the distinct lines cd and $c'd$ meeting at d. Let $x \in cc'$. Then dx contains at least a third point d', which is also fixed by α in view of the first argument. So $\alpha(d\,d') = d\,d'$ and $\{\alpha(x)\} = d\,d' \cap c\,c' = \{x\}$, forcing α to be the identity. This contradiction shows that the center is unique.

Suppose that α is a perspectivity with distinct axes H and H'. Let x be a point of Z outside $H \cup H'$. If p is a point in $H \setminus H'$, then p is distinct from x and px meets H' in some point $p' \neq p$ (as H' is a geometric hyperplane). In view of $\alpha(pp') = \alpha(p)\alpha(p') = pp'$, we must have $\alpha(x) \in pp'$. Since Z is thick, there is a point $q \in H \setminus H'$ with $q \neq p$. In particular, x lies on the two distinct α-invariant lines pp', qq', so $\alpha(x) = x$. It follows that α is the identity, proving uniqueness of the axis.

(iii) If α and α' satisfy all assumptions, then $\beta = \alpha^{-1}\alpha'$ is a perspectivity with center o and axis H fixing a. By (i), a is a second center of β. Now (ii) implies that β is the identity, so $\alpha = \alpha'$. $\qquad\square$

In the case of projective spaces, we can characterize perspectivities as follows.

Proposition 6.1.4 *Let α be an automorphism of a thick projective space Z.*

(i) *If α fixes all points of some maximal subspace (i.e., has an axis), then α is a perspectivity.*
(ii) *If α fixes all lines through a point c (i.e., has a center), then α is a perspectivity.*

Proof (i) Let H be a maximal subspace of Z that is point-wise fixed by α. According to Proposition 5.2.9, the subspace H is a hyperplane. If α fixes some point of Z off H, then Lemma 6.1.3(i) shows that α is a perspectivity. Therefore, we assume that α fixes no point outside H.

For a point p of Z off H, the line $p\alpha(p)$ intersects H in some point c and $\alpha(pc) = \alpha(p)c$; hence $p\alpha(p)$ is invariant under α. We claim that c is a center of α. Let x be a point of Z off H. We show that α leaves cx invariant. If $x \in pc$, then $cx = p\alpha(p)$, and we are done. Suppose therefore that x is not on pc. Let Π be the plane generated by p, c, x. Then px is in Π and meets H in a point, d say, distinct from c. We have $\Pi = \langle p, c, x \rangle = \langle p, c, d \rangle$ and $\alpha(\Pi) = \langle \alpha(p), c, d \rangle = \Pi$. Hence $\alpha(x) \in \Pi$ and, by Pasch's Axiom, the lines $p\alpha(p)$ and $x\alpha(x)$ meet in a point e in Π. These two lines are α-invariant, so e is fixed by α, which forces $e = c$.

(ii) This is left to the reader; see Exercise 6.8.1. $\qquad\square$

Theorem 6.1.5 *Suppose that Z is a thick projective space of dimension at least three. If H is a maximal subspace and o, a, b are collinear points with a, b not in $H \cup \{o\}$, then there exists a perspectivity with center o, axis H, mapping a onto b.*

Proof We construct a perspectivity α with the required properties. We (may) assume $a \neq b$, for otherwise the identity is appropriate. For $x \in H \cup \{o\}$ we put $\alpha(x) = x$; moreover, $\alpha(a) = b$. For any point x of Z outside $H \cup ab$, we let $\alpha(x)$ be determined by $\{\alpha(x)\} = ox \cap yb$, where y is the unique point in $ax \cap H$. To complete the construction of α, we choose an auxiliary point e of Z outside $H \cup ab$. For $x \in ab \setminus (H \cup \{a, o\})$, we put $\alpha_e(x) = ab \cap h_e\alpha(e)$ where h_e is the unique point in $ex \cap H$. It is readily checked that α_e is a well-defined permutation of the point set of Z.

To show that α_e does not depend on the choice of $e \in Z \setminus (H \cup ab)$, retain $x \in ab \setminus (H \cup \{a, o\})$ and let $f \in Z \setminus (H \cup ab \cup \{e\})$. We claim $\alpha_e(x) = \alpha_f(x)$ if $ef \cap ab = \emptyset$. To see this, let m be the line $\langle a, e, f \rangle \cap H$ and put $m' = \langle e, f, x \rangle \cap H$. Notice that, due to the assumption $ef \cap ab = \emptyset$, we have $h_e \neq h_f$, so $m' = h_e h_f$ is a line as well. Moreover, the planes $\langle a, e, f \rangle$ and $\langle e, f, x \rangle$ do not coincide, so they meet in ef, which has a point, h say, in common with H. Thus, $\{h\} = m \cap m' = ef \cap H = ef \cap m$.

By construction, the point $ea \cap H$ lies in $b\alpha(e)$ and $fa \cap H$ lies in $b\alpha(f)$, so the line m lies in the plane $\langle b, \alpha(e), \alpha(f) \rangle$ and so coincides with $\langle b, \alpha(e), \alpha(f) \rangle \cap H$. Therefore, $m = \langle a, e, f \rangle \cap \langle b, \alpha(e), \alpha(f) \rangle$ and the lines ef and $\alpha(e)\alpha(f)$ meet in the point h of m. In particular, $\langle \alpha(e), m' \rangle = \langle \alpha(e)h, hh_e \rangle = \langle \alpha(f)h, hh_f \rangle = \langle \alpha(f), m' \rangle$, so $\langle \alpha(e), m' \rangle = \langle \alpha(f), m' \rangle$.

Now $\{\alpha_e(x)\} = ab \cap \alpha(e)h_e = ab \cap \langle \alpha(e), m' \rangle$ (for otherwise $ab \subseteq \langle \alpha(e), h_e, h_f \rangle$, so, as $e \in xh_e \subseteq \langle ab, h_e \rangle$ and $f \in \langle ab, h_f \rangle$, the line ef lies in the plane $\langle \alpha(e), h_e, h_f \rangle$ containing ab, a contradiction with $ab \cap ef = \emptyset$), and similarly, $\{\alpha_f(x)\} = ab \cap \langle \alpha(f), m' \rangle$. Therefore, $\{\alpha_e(x)\} = ab \cap \langle \alpha(e), m' \rangle = ab \cap \langle \alpha(f), m' \rangle = \{\alpha_f(x)\}$. We have established the claim that $\alpha_e(x) = \alpha_f(x)$ whenever $ef \cap ab = \emptyset$.

If $ef \cap ab \neq \emptyset$, we can select a point g of Z outside $H \cup ab$ such that both eg and fg are disjoint from ab, and a double application of the claim shows $\alpha_e(x) = \alpha_g(x) = \alpha_f(x)$. The conclusion is that, indeed, α_e does not depend on the choice of e. We write $\alpha(x) = \alpha_e(x)$, thus obtaining a permutation α of the point set of Z.

It remains to show that α maps every line onto a line. If $l \cap ab = \emptyset$, then $\langle a, l \rangle$ is a plane intersecting H in a line l' (cf. Theorem 5.2.6) and so $\alpha(l) \subseteq \langle \alpha(a), l' \rangle$ while $\alpha(l) \subseteq \langle o, l \rangle$. Since these two planes are distinct, $\alpha(l)$ is contained in a line. On the other hand, every point of $\langle \alpha(a), l' \rangle \cap \langle o, l \rangle$ is the image under α of a point in $\langle a, l' \rangle \cap \langle o, l \rangle = l$, so $\alpha(l)$ coincides with the line $\langle \alpha(a), l' \rangle \cap \langle o, l \rangle$.

If $l \cap ab$ is a point d, assume first that l is not in $\langle o, ae \rangle$, where e is the auxiliary point introduced earlier. Set $\Pi = \langle o, l \rangle$, $l' = \langle e, l \rangle \cap H$, and $l'' = \langle \alpha(e), l' \rangle \cap \Pi$. Then l', and l'' are necessarily lines. For each $x \in l \setminus d$, we have $\alpha(x) \in l''$. Also, $\alpha(d) \in l''$ by construction of $\alpha(d)$. Thus $\alpha(l) \subseteq l''$, and equality follows as in the previous case. Finally, assume $l \subseteq \langle o, a, e \rangle$. Let q be a point outside the plane $\langle o, a, e \rangle$. (Observe that q exists as Z has more than one plane.) Set $l' = \langle q, l \rangle \cap H$, $l'' = \langle \alpha(q), l' \rangle \cap \langle o, l \rangle$. Then l' and l'' are lines, again. For each $x \in l$, we have $\alpha(x) \in l''$, and so $\alpha(l) \subseteq l''$, allowing us to finish as earlier. $\square$

Remark 6.1.6 The perspectivity α found in the proof of the theorem seems to depend on the choice of the auxiliary point e. However, Lemma 6.1.3(iii) shows that this is not the case. This independence of choice is related to the Desargues configuration, which was studied before, in Exercise 2.8.18, as a geometry by itself. Figure 2.21 is redrawn in Fig. 6.1 with the points relabelled to $1, \ldots, 10$. The remarkable configuration is now used in connection with the existence of perspectivities on projective planes.

Fig. 6.1 Desargues
configuration

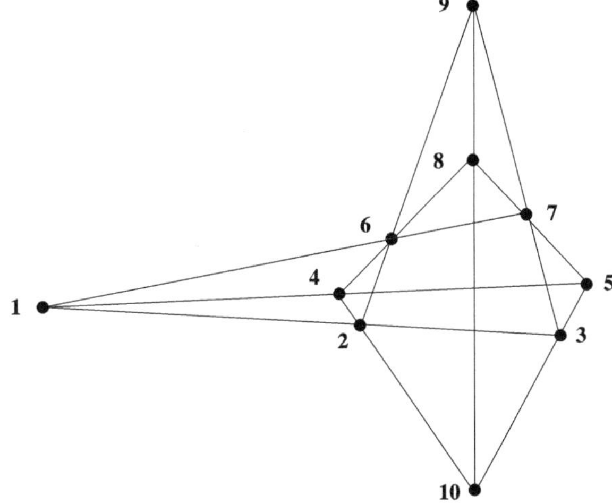

Definition 6.1.7 A thick projective plane is called *Desarguesian* if, for every set of ten distinct points $1, 2, \ldots, 10$ in which the following nine sets of triples are collinear,

$\{1, 2, 3\}$; $\{1, 4, 5\}$; $\{1, 6, 7\}$ on three distinct lines,
$\{2, 4, 10\}$; $\{4, 6, 8\}$; $\{2, 6, 9\}$ on three distinct lines,
$\{3, 5, 10\}$; $\{5, 7, 8\}$; $\{3, 7, 9\}$ on three distinct lines,

the (tenth) triple $\{8, 9, 10\}$ is collinear as well (cf. Fig. 6.1).

Theorem 6.1.8 *A thick projective plane is Desarguesian if and only if it has a perspectivity with center o and axis H mapping a to b, for all maximal subspaces H and distinct collinear points o, a, b with a, b not in H.*

Proof Since a perspectivity maps lines onto lines, we only need show that it can be well defined. But this is exactly what the Desargues configuration proves. Indeed, in Fig. 6.1, choose 1 to be the center o, let H contain the triple 8, 9, 10, and put $a = 4$ and $b = 5$. Now consider the point 6. The line 46 through it meets H in 8, so the image of 6 under a perspectivity α with center o and axis H mapping a to b would have to be on the line 58. On the other hand it must be on the line 16 and so it must be 7. Similarly α will map 2 to the unique point 3 on the lines 12 and 5(10). But suppose now, that we would have used the fact that α maps 6 to 7 rather than 4 to 5 in order to find the image of 2. The image under α would be the unique point on the lines 12 and 79. Thus, for α to be well defined it is necessary and sufficient that the Desargues configuration holds, which means that, building up nine of the ten points with the corresponding lines, we find that the tenth point (here, 9) is incident with each of the three remaining lines of the configuration (here, 26, 8(10), and 37) of which only two points are drawn. $\qquad\square$

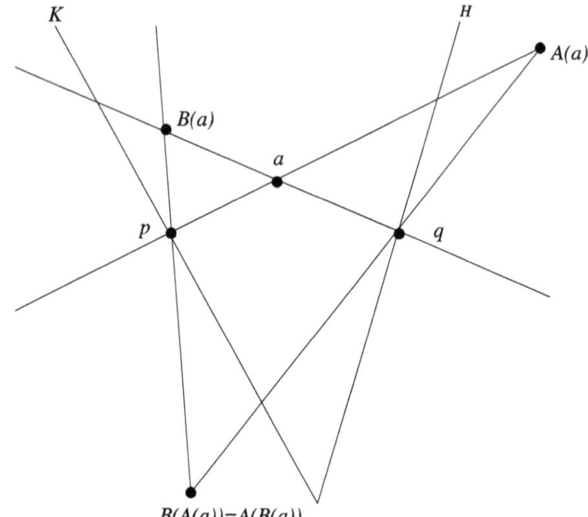

Fig. 6.2 Coincidence of the points $AB(a)$ and $BA(a)$, where A is a perspectivity with center p and axis H and B is a perspectivity with center q and axis K

Remark 6.1.9 In view of Theorem 6.1.5, every thick projective space containing at least two planes is Desarguesian in the sense that there are many perspectivities. Accordingly, all of its planes are Desarguesian. However, not all projective planes are Desarguesian; see Example 2.3.4 and Exercise 6.8.3 for counterexamples.

We finish this section with some properties of perspectivities which will help us later on (Sect. 6.2) in the study of the projective line. Besides, they are of interest in their own right.

Proposition 6.1.10 *Let Z be a thick projective space. For a hyperplane H and a point p of Z, denote by $\pi(p, H)$ the set of perspectivities with center p and axis H.*

(i) *$\pi(p, H)$ is a permutation group of the point set of Z fixing p and H and if a is a point of Z outside $\{p\} \cup H$, then the stabilizer $\pi(p, H)_a$ in $\pi(p, H)$ of a is trivial.*

(ii) *Every automorphism α of Z satisfies $\alpha\pi(p, H)\alpha^{-1} = \pi(\alpha(p), \alpha(H))$.*

(iii) *If $q \in H$ and K is a hyperplane containing p, then $\pi(p, H)$ centralizes $\pi(q, K)$.*

(iv) *The sets $\pi(p)$ of all perspectivities with center p and $\pi^{\vee}(H)$ of all perspectivities with axis H are subgroups of $\mathrm{Aut}(Z)$.*

Proof Parts (i), (ii) and (iv) are left to the reader—they are easy consequences of Lemma 6.1.3 and Proposition 6.1.4.

The proof of Part (iii) is illustrated in Fig. 6.2. Let $A \in \pi(p, H)$ and $B \in \pi(q, K)$. Clearly, A and B both fix p and q. Consider a point a of Z outside $H \cup K$, so $a \neq p, q$. Its image $A(a)$ lies in $ap \setminus (H \cup K)$ and the point $BA(a)$ lies in $qA(a) \setminus (H \cup K)$. Thanks to the two points $A(a)$ and $BA(a)$, the image $B(a)$ is determined by $\{B(a)\} = aq \cap pBA(a)$ (as $a \in aq \cap pA(a)$) and $AB(a)$ is the unique point of

$pB(a) \cap qA(a)$ (as $B(a) \in pB(a) \cap qa$), which forces $AB(a) = BA(a)$. The same equation is readily verified for $a \in H \cup K$, so we conclude $AB = BA$, as required. $\square$

6.2 Projective Lines

At first sight, a line of a projective space has no structure. But if it belongs to a bigger projective space, perspectivities induce a structure on each line that it leaves invariant. Using this structure, we are able to prove that every Desarguesian projective line (a thick projective line for which the structure guarantees many perspectivities) is of a specific form that is completely determined by a division ring. This is the content of Theorem 6.2.11.

The following definition is inspired by Proposition 6.1.10. It is employed only for thick lines. As the difference between maximal subspaces and points, both used in the definition of perspectivities, vanishes for projective lines, the definition involves pairs of points.

Definition 6.2.1 A *projective line* is a pair (P, π) consisting of a set P of points of cardinality at least three and a map $\pi : P \times P \to \mathrm{Sym}(P)$ assigning to each ordered pair (p, q) a set $\pi(p, q)$ of permutations of P, whose members are called *perspectivities* of P with *center* p and *axis* q, satisfying the following four properties.

(1) The set $\pi(p, q)$ is a permutation group of P fixing p and q, and if a is a point in $P \setminus \{p, q\}$, then $\pi(p, q)_a = 1$.
(2) Every perspectivity α is an automorphism in the sense that $\alpha\pi(p, q)\alpha^{-1} = \pi(\alpha(p), \alpha(q))$ for all $p, q \in P$.
(3) If $p \neq q$, then $\pi(p, q)$ centralizes $\pi(q, p)$.
(4) The sets $\pi(p) := \bigcup_{q \in P} \pi(p, q)$ of all perspectivities with center p and $\pi^{\vee}(q) = \bigcup_{p \in P} \pi(p, q)$ of all perspectivities with axis q are subgroups of $\mathrm{Sym}(P)$.

We refer to $\pi(p, q)$ as the *perspectivity set* with center p and axis q.

An *isomorphism* $\phi : (P, \pi) \to (R, \rho)$ between two projective lines (P, π) and (R, ρ) is a bijection $\phi : P \to R$ such that, for all $p, q \in P$, we have $\phi\pi(p, q)\phi^{-1} = \pi(\phi(p), \phi(q))$.

Often, π is clear from the context, in which case we also write P instead of (P, π) for the projective line.

Example 6.2.2 (Structure induced from higher-dimensional space) A trivial example of a projective line on a set P of cardinality at least three is the pair (P, π) where π assigns the singleton consisting of the identity map on P to each pair from P.

In order to exhibit the motivating example for Definition 6.2.1, we let Z be a thick projective space of dimension at least two (possibly a non-Desarguesian plane) and select a line P of Z. We will show that Z induces on P the structure of a projective line.

For $q \in P$, let $\mathcal{H}_q$ be the family of all maximal subspaces of Z on q which do not contain P. Now, for $(p, q) \in P^2$, the set $\pi(p, q)$ is defined as the set of restrictions to P of perspectivities of Z with center p and axis $H \in \mathcal{H}_q$. Then Proposition 6.1.10 shows that this defines indeed a projective line on P. We refer to this structure as the *projective line* on P *induced* from Z.

Remark 6.2.3 A *Moufang set* is a set X of cardinality at least three together with a family U_x ($x \in X$) of subgroups of $\mathrm{Sym}(X)$ such that

(1) for each $x \in X$, the group U_x fixes x and is regular on $X \setminus \{x\}$;
(2) for all $x, y \in X$ and $g \in U_y$, we have $g U_x g^{-1} = U_{gx}$.

The subgroup $\langle U_y \mid y \in X \rangle$ of $\mathrm{Sym}(X)$ is called the *projective group* of the Moufang set. If (P, π) is a Desarguesian projective line (see below), then the collection $\pi(p, p)$ ($p \in P$) is a Moufang set. The converse is false. Exercise 6.8.5 gives a sufficient (but not a necessary) criterion for a Moufang set to be a Desarguesian projective line.

Definition 6.2.4 If P is a projective line, with perspectivity sets $\pi(p, q)$, for all $p, q \in P$, then the *dual projective line* $P^\vee$ has the same points as P and perspectivity sets $\pi^\vee(p, q)$ given by $\pi^\vee(p, q) = \pi(q, p)$ for all $p, q \in P$.

It is readily verified that $P^\vee$ is indeed a projective line. Observe that the notation $\pi^\vee(q)$ of Definition 6.2.1 is consistent with the one of Definition 6.2.4. Obviously, $P^{\vee\vee} = P$.

So far, the structure of a line can still be entirely trivial in that $\pi(p, q)$ may consist of the identity only for each pair $(p, q) \in P^2$. We therefore focus on nontrivial cases.

Definition 6.2.5 A projective line (P, π) is *Desarguesian* if $\pi(p, q)$ acts transitively on $P \setminus \{p, q\}$ for all $p, q \in P$.

Taking $p = q$, we see that even if P has only three points, nontrivial perspectivities exist. Also, for $|P| \leq 4$, the group generated by all perspectivities coincides with $\mathrm{Sym}(P)$ (for it contains the transpositions). But, as we will see later, for $|P| \geq 5$ the group $\mathrm{Aut}(P)$ is a proper subgroup of $\mathrm{Sym}(P)$.

The result below comes back to Example 6.2.2.

Proposition 6.2.6 *If Z is a Desarguesian projective plane or a projective space of dimension at least three, and P is a line of Z, then the projective line on P induced from the perspectivities on Z is a Desarguesian projective line.*

Proof This is immediate from Theorems 6.1.8 and 6.1.5.   □

It will become clear later (Sect. 6.3) that the choice of the line P in a Desarguesian projective plane does not affect the structure of the projective line induced on P from Z.

Example 6.2.7 Let V be a 2-dimensional right vector space over the division ring $\mathbb{D}$. The projective space $\mathbb{P}(V)$ has the following structure of a Desarguesian projective line. For p, q in $\mathbb{P}(V)$, let $\pi(p, q)$ be

(1) the image in PGL(V) of the group of all linear transformations of V fixing the subspace q of V point-wise and the subspace p of V set-wise, if $p \neq q$;
(2) the image in PGL(V) of the group of all transvections of V fixing p point-wise, i.e., those linear transformations α for which $\alpha(p) = p$ and $\alpha(x) - x \in p$ for all $x \in V$, if $p = q$.

The check that this indeed defines a projective line is relegated to Exercise 6.8.4. This projective line is denoted $\mathbb{P}^1(\mathbb{D})$.

Let us describe the perspectivity sets $\pi(p, q)$ in terms of matrices. If $p = (1 : 0)$ and $q = (0 : 1)$ (cf. Notation 5.2.13), then, $\pi(p, q)$ is the image in PGL(V) of the set of matrices

$$\begin{pmatrix} \lambda & 0 \\ 0 & 1 \end{pmatrix} \quad \text{for } \lambda \in \mathbb{D} \setminus \{0\},$$

and $\pi(p, p)$ is the image in PGL(V) of the set of matrices

$$\begin{pmatrix} 1 & \lambda \\ 0 & 1 \end{pmatrix} \quad \text{for } \lambda \in \mathbb{D}.$$

For any other pair p', q' of points in $\mathbb{P}(V)$, there is a linear transformation $T \in \mathrm{GL}(V)$ mapping p to p' and q to q', and so we can compute $\pi(p', q')$ as $\pi(Tp, Tq) = T\pi(p, q)T^{-1}$. For instance, taking T to be the 2×2 matrix corresponding to the permutation $(1, 2)$, we find

$$\pi(q, p) = \begin{pmatrix} 0 & 1 \\ 1 & 0 \end{pmatrix} \begin{pmatrix} \lambda & 0 \\ 0 & 1 \end{pmatrix} \begin{pmatrix} 0 & 1 \\ 1 & 0 \end{pmatrix} = \begin{pmatrix} 1 & 0 \\ 0 & \lambda \end{pmatrix} \quad \text{for } \lambda \in \mathbb{D} \setminus \{0\}.$$

There is yet another way of describing this projective line, namely by means of the affine picture. Then, as in Example 5.2.14, we present $\mathbb{P}(V)$ by means of $\mathbb{D} \cup \{\infty\}$, thereby identifying λ with $(\lambda : 1)$ and ∞ with $(1 : 0)$. For instance, $p = \infty$ and $q = 0$. Then $\pi(\infty, \infty)$ is the set of *translations* of $\mathbb{D}$, that is, the maps $t_a : \mathbb{D} \cup \{\infty\} \to \mathbb{D} \cup \{\infty\}$ given by $t_a(x) = x + a$ for $a \in \mathbb{D}$. Moreover, for $a \in \mathbb{D}$, the perspectivity set $\pi(a, \infty)$ consists of the maps $x \mapsto x\lambda + a(1 - \lambda)$, where $\lambda \in \mathbb{D}$, $\lambda \neq 0$. For $a = 0$, we retrieve the above description of $\pi(q, p) = \pi(0, \infty)$. In order to describe the group of automorphisms of the projective line generated by all perspectivity sets, it suffices to take one more generator: the element $w \in \mathrm{PGL}(V)$ given by $w(x) = -x^{-1}$. The reason lies in the facts that w interchanges 0 and ∞, and that PGL(V) is 2-transitive on the projective line with stabilizer $B :=$ $\langle \pi(\infty, \infty), \pi(0, \infty), w\pi(0, \infty)w^{-1} \rangle = \langle \pi(\infty, \infty) \rangle \rtimes \langle \pi(0, \infty), \pi(\infty, 0) \rangle$ of ∞, so $\mathrm{PGL}(V) = B \cup BwB = B \cup \pi(\infty, \infty)wB$ (cf. Exercise 2.8.12, where PGL(V) is seen to act even 3-transitively).

Proposition 6.2.8 *For each division ring $\mathbb{D}$, the projective line $\mathbb{P}^1(\mathbb{D})$ is Desarguesian. The group generated by its perspectivities is* $\mathrm{PGL}(\mathbb{D}^2)$.

Proof Let $p = (1 : 0)$ and $q = (0 : 1)$. It is clear from the presentation of $\pi(p, q)$ in Example 6.2.7 that the $\pi(p, q)$-orbit of $(1 : 1)$ is the set $\{(\lambda : 1) \mid \lambda \in \mathbb{D} \setminus \{0\}\}$. But this coincides with $\mathbb{P}(\mathbb{D}^2) \setminus \{p, q\}$. Similarly, the $\pi(p, p)$-orbit of $(0 : 1)$ is the set $\{(\lambda : 1) \mid \lambda \in \mathbb{D}\}$, which coincides with $\mathbb{P}(\mathbb{D}^2) \setminus \{p\}$. For other points p', q' the transitivity of $\pi(p', q')$ and of $\pi(p', p')$ follows by conjugation with a coordinate transformation in $\mathrm{PGL}(\mathbb{D}^2)$. We conclude that $\mathbb{P}^1(\mathbb{D})$ is Desarguesian.

The element $w \in \mathrm{PGL}(\mathbb{D}^2)$ of Example 6.2.7 has a preimage in $\mathrm{GL}(\mathbb{D}^2)$ that can be expressed as

$$\begin{pmatrix} 1 & -1 \\ 0 & 1 \end{pmatrix} \begin{pmatrix} 1 & 0 \\ 1 & 1 \end{pmatrix} \begin{pmatrix} 1 & -1 \\ 0 & 1 \end{pmatrix}.$$

This implies that it is a product of three perspectivities. The subgroup B of $\mathrm{PGL}(\mathbb{D}^2)$ of Example 6.2.7 is already known to be generated by perspectivities, so the second statement of the proposition follows from the expression of $\mathrm{PGL}(\mathbb{D}^2)$ as the union of the two double cosets B and BwB. $\qquad\square$

Below we prove the converse: every Desarguesian projective line can be constructed in this way.

Definition 6.2.9 For $t \in \mathbb{N}$, $t > 0$, the action of a group on a set X is said to be *sharply t-transitive* if it is t-transitive on X and the point-wise stabilizer of a set of t points in X is trivial. Instead of sharply 1-transitive, we also say just *sharply transitive* or *regular*.

Property (1) of Definition 6.2.1 expresses the fact that $\pi(p, q)$ is sharply transitive if P is a Desarguesian projective line containing p and q.

Lemma 6.2.10 *If P is a Desarguesian projective line and p is a point, then $\pi(p)$ is a sharply 2-transitive permutation group on $P \setminus \{p\}$.*

Proof By Definition 6.2.5, the group $\pi(p, p)$ is transitive. If q is a point distinct from p, then $\pi(p)_q = \pi(p, q)$, and this group is sharply transitive on $P \setminus \{p, q\}$. $\square$

Theorem 6.2.11 *Let (P, π) be a Desarguesian projective line and let 0, 1, and ∞ be three distinct members of its point set P. There is a unique division ring $\mathbb{D}$ such that*

(i) *as a set, $\mathbb{D} = P \setminus \{\infty\}$;*
(ii) *0 is the zero element of $\mathbb{D}$;*
(iii) *1 is the unit element of $\mathbb{D}$;*
(iv) *for $a, b \in \mathbb{D}$, we have $a + b = t_b(a)$, where t_b is the unique member of $\pi(\infty, \infty)$ mapping 0 to b;*
(v) *for $a \in \mathbb{D} \setminus \{0\}$ and $b \in \mathbb{D}$, we have $ab = \lambda_a(b)$, where λ_a is the unique member of $\pi(0, \infty)$ mapping 1 to a.*

Moreover, $(P, \pi) \cong \mathbb{P}^1(\mathbb{D})$.

Proof Since the underlying set $\mathbb{D}$ and all essential operations (subtraction and division are uniquely determined by addition and multiplication) are described in the assertion, uniqueness is immediate. We proceed in fifteen steps.

STEP 1. *Addition on $\mathbb{D}$.*

For $a, b \in \mathbb{D}$, by Lemma 6.2.10, there is a unique member t_b of $\pi(\infty, \infty)$ mapping 0 to b; now $a + b$ is defined to be $t_b(a)$. It belongs to $\mathbb{D}$ as $t_b(a) \neq t_b(\infty) = \infty$.

STEP 2 *Multiplication on $\mathbb{D}$.*

For $a, b \in \mathbb{D}$, by Lemma 6.2.10, if $b \neq 0$, there is a unique member λ_b of $\pi(0, \infty)$ mapping 1 to b; now ba is defined to be $\lambda_b(a)$. It lies in $\mathbb{D}$ as $ba = \lambda_b(a) \neq \lambda_b(\infty) = \infty$. Furthermore, set $\lambda_0(a) = 0$, so $0a = 0$. Now $a0 = \lambda_a(0) = 0$ because $\lambda_a \in \pi(0, \infty)$.

STEP 3. *The element 1 is the identity element of $\mathbb{D}$.*

Indeed, λ_1 is the unique element of $\pi(0, \infty)$ mapping 1 to 1 and so is the identity. Hence $1x = x$ for all $x \in \mathbb{D}$. Moreover, $x1 = x$ as both sides are equal to $\lambda_x(1)$.

STEP 4. $\mathbb{D} \setminus \{0\}$ *is a group with respect to multiplication.*

First note that $\lambda_a \lambda_b$ and λ_{ab} are both perspectivities in $\pi(0, \infty)$ mapping 1 to ab, so they coincide. Therefore,

$$(ab)c = \lambda_{ab}(c) = \lambda_a \lambda_b(c) = a(bc),$$

proving that multiplication is associative. If $b \in \mathbb{D} \setminus \{0\}$, then λ_b^{-1} is a perspectivity in $\pi(0, \infty)$ mapping b to 1. Put $b^{-1} = \lambda_b^{-1}(1)$. Then $bb^{-1} = \lambda_b \lambda_b^{-1}(1) = 1$ and $b^{-1}b = \lambda_b^{-1}\lambda_b(1) = 1$. Thus each nonzero element of $\mathbb{D}$ has an inverse.

STEP 5. $\mathbb{D}$ *is a group with respect to addition.*

The argument is similar to the one for Step 4.

STEP 6. *The distributivity laws for multiplication and addition on $\mathbb{D}$ hold.*

By Condition (2) of Definition 6.2.1, the perspectivity λ_a is an automorphism satisfying $\lambda_a t_y \lambda_a^{-1} = t_{\lambda_a y}$. Evaluating both sides at $\lambda_a(x) \in \mathbb{D}$, we find $\lambda_a t_y(x) = t_{\lambda_a y}(\lambda_a x)$, that is, $a(x + y) = ax + ay$, whence left distributivity.

Right distributivity can be shown to hold as follows. Consider the set $\pi(\infty, 0)$ of perspectivities with center ∞ and axis 0. By (3) of Definition 6.2.1, it centralizes the maps λ_a for $a \in \mathbb{D}$. Let $\sigma \in \pi(\infty, 0)$ be the element mapping 1 onto b. Then $a\sigma(x) = \lambda_a \sigma(x) = \sigma(\lambda_a x) = \sigma(ax)$ and so, taking $x = 1$, we find $\sigma(a) = ab$ for each $a \in \mathbb{D}$. Moreover, $\sigma t_a \sigma^{-1} = t_{\sigma a}$ applied to $\sigma(x)$ gives $\sigma(x + a) = \sigma(x) + \sigma(a)$ for all $x \in \mathbb{D}$, whence $(x + a)b = xb + ab$, which proves right distributivity.

STEP 7. *The additive group on $\mathbb{D}$ is commutative.*

For $0 = 0(a + b) = (1 + (-1))(a + b) = (a + b) + (-1)(a + b) = a + b + (-1)a + (-1)b = a + b - a - b$. Adding first b and then a to each side, we find $b + a = a + b$.

STEP 8. *For $b \in \mathbb{D} \setminus \{0\}$, the unique element μ_b of $\pi(\infty, 0)$ mapping 1 to b satisfies $\mu_b(x) = xb$ for all $x \in \mathbb{D}$.*

As $\mu_b \in \pi(0, \infty)$, it leaves $\mathbb{D}$ invariant and fixes 0. In particular, $\mu_b(0) = 0b$. Let $x \in \mathbb{D} \setminus \{0\}$. By Condition (3) of Definition 6.2.1, the element μ_b commutes with every element of $\pi(0, \infty)$, so $\mu_b(x) = \mu_b \lambda_x(1) = \lambda_x \mu_b(1) = x\mu_b(1) = xb$, as required.

STEP 9. *For $a \in \mathbb{D} \setminus \{0\}$, the unique element u_a of $\pi(0, 0)$ mapping ∞ to a^{-1} satisfies $u_a(x) = a^{-1}u_1(ax)$ and $u_x(a) = u_{a^{-1}}(x^{-1})$ for all $x \in \mathbb{D} \setminus \{0\}$.*

The map λ_a normalizes $\pi(0, 0)$, so, there is $c \in \mathbb{D} \setminus \{0\}$ such that $\lambda_a u_b \lambda_a^{-1} = u_c$. Now $ab^{-1} = \lambda_a(b^{-1}) = \lambda_a u_b(\infty) = \lambda_a u_b \lambda_a^{-1}(\infty) = u_c(\infty) = c^{-1}$ gives $c = ba^{-1}$, so $\lambda_a u_b \lambda_a^{-1} = u_{ba^{-1}}$. Taking $b = 1$ and replacing a by its inverse, we find the first equality.

It follows from Step 7 that $\pi(0, 0)$, being a conjugate of $\pi(\infty, \infty)$ by Condition (2) of Definition 6.2.1, is commutative. Thus, for all $a, x \in \mathbb{D} \setminus \{0\}$, we have $u_{a^{-1}}(x^{-1}) = u_{a^{-1}}u_x(\infty) = u_x u_{a^{-1}}(\infty) = u_x(a)$, which settles the second equality.

STEP 10. *If $x \in \mathbb{D} \setminus \{0\}$, then $u_1(x) = xu_1(x^{-1})$, and so $u_1(-1) = \infty$.*

Using both formulas of Step 9, we find $u_1(x) = xu_x(1) = xu_1(x^{-1})$, as required for the first equality. For $x = -1$, this implies $u_1(-1) = -u_1(-1)$. As $u_1(-1) \neq 0$, this implies $u_1(-1) = \infty$ if $\mathrm{char}(\mathbb{D}) \neq 2$. So, assume that the characteristic of $\mathbb{D}$ is even. If $u_1(x) = \infty$, then we would have $\infty = u_1(x) = xu_1(x^{-1})$, and so $u_1(x^{-1}) = \infty$, which would imply $x = x^{-1}$ as u_1 is injective on $\mathbb{D} \setminus \{0\}$, and so $x = 1 = -1$. Therefore, the second equality holds for all characteristics.

STEP 11. *Let w be the element $t_{-1}u_1t_{-1}$ or $u_{-1}t_1u_{-1}$ of $\langle U_0, U_\infty \rangle$. This element satisfies $w(\infty) = 0$, $w(0) = \infty$, and $w(1) = -1$.*

Simple computations give these results:

$$u_{-1}t_1u_{-1}(\infty) = u_{-1}t_1(-1) = u_{-1}(0) = 0;$$

$$u_{-1}t_1u_{-1}(0) = u_{-1}(1) = u_1(-1) = \infty;$$

$$u_{-1}t_1u_{-1}(1) = u_{-1}t_1(\infty) = u_{-1}(\infty) = -1;$$

$$t_{-1}u_1t_{-1}(\infty) = t_{-1}u_1(\infty) = t_{-1}(1) = 0;$$

$$t_{-1}u_1t_{-1}(0) = t_{-1}u_1(-1) = t_{-1}(\infty) = \infty;$$

$$t_{-1}u_1t_{-1}(1) = t_{-1}u_1(0) = t_{-1}(0) = -1.$$

STEP 12. *Each $w \in G$ with $w(\infty) = 0$, $w(0) = \infty$, and $w(1) = -1$ satisfies the following equalities for all $a, b \in \mathbb{D}$.*

$$w\lambda_a w^{-1} = \mu_{-w(a)} \quad \text{if } a \neq 0$$

$$w(ab) = -w(b)w(a)$$

$$w(-1) = 1$$

$$w\big(a^{-1}\big) = w(a)^{-1}$$

$$w(-a) = -w(a)$$

$$w t_a w^{-1} = u_{w(a)^{-1}}$$

$$u_1(b) = w\big(w^{-1}(b) - 1\big)$$

The equalities trivially hold if $a = 0$ or $b = 0$, so we assume $a, b \neq 0$.

Condition (2) of Definition 6.2.1 implies $w\pi(0, \infty)w^{-1} = \pi(\infty, 0)$, so, for each $a \in \mathbb{D} \setminus \{0\}$, there is $z \in \mathbb{D} \setminus \{0\}$ such that $w\lambda_a w^{-1} = \mu_z$. Applying both sides to -1, we find $w(a) = -z$, so $w\lambda_a w^{-1} = \mu_{-w(a)}$. This shows that $w(aw^{-1}(b)) = -bw(a)$. Substitution of b by $w(b)$ gives $w(ab) = -w(b)w(a)$.

Substituting $b = a^{-1}$, $a = b = -1$, and $b = -1$ in the product formula just obtained, we find the three equalities $w(a^{-1}) = w(a)^{-1}$, $w(-1) = 1$, and $w(-a) = -w(a)$, respectively.

Condition (2) of Definition 6.2.1 also implies $w\pi(\infty, \infty)w^{-1} = \pi(0, 0)$, so, for each $a \in \mathbb{D} \setminus \{0\}$, there is $y \in \mathbb{D} \setminus \{0\}$ such that $w t_a w^{-1} = u_y$. Applying both sides to ∞, we find $w(a) = y^{-1}$, so $w t_a w^{-1} = u_{w(a)^{-1}}$. By Step 9, it follows that $w(w^{-1}(b) + a) = w(a)u_1(w(a)^{-1}b)$. Setting $a = -1$ gives $u_1(b) = w(w^{-1}(b) - 1)$. This settles Step 12.

STEP 13. *The elements $t_{-1}u_1t_{-1}$ and $u_{-1}t_1u_{-1}$ of G coincide and are involutions of G with the properties of w in Step 12.*

Fix $w_1 := t_{-1}u_1t_{-1}$ and $w_2 := u_{-1}t_1u_{-1}$. By Step 11, these elements of G are as w of Step 12. Notice that w interchanges 1 and -1 as well as 0 and ∞, so that w^{-1} also satisfies the conditions on w of Step 12. Therefore, $w_2 = wt_1(w^{-1}t_1w)t_1w^{-1} = wt_1u_{w^{-1}(1)}t_1w^{-1} = wt_1u_{-1}t_1w^{-1} = ww_1^{-1}w^{-1}$, where we used $u_{-1}^{-1} = u_1$ (which follows from $u_{-1} = wt_1w^{-1}$). We found $w_2 = ww_1^{-1}w^{-1}$. Substituting $w = w_1$, we see $w_2 = w_1^{-1}$.

By use of the formula for u_1 of Step 12, the equality $w_1 = t_{-1}u_1t_{-1}$ applied to $a \in \mathbb{D}$ gives $w_1(a) + 1 = w_1(w_1^{-1}(a - 1) - 1)$. Applying w_1^{-1} to both sides gives $w_1^{-1}(w_1(a) + 1) = w_1^{-1}(a - 1) - 1$. Substituting $-w_1(a)$ for a leads to $w_1^{-1}(-w_1^2(a) + 1) = w_1^{-1}(-w_1(a) - 1) - 1 = -w_1^{-1}(w_1(a) + 1) - 1 = -w_1^{-1}(a - 1)$, where the latter equality is obtained from the equality in the previous sentence and we used the equality $w(-x) = -w(x)$ of Step 12. We conclude $w_1^{-1}(-w_1^2(a) + 1) = w_1^{-1}(1 - a)$, so, $-w_1^2(a) + 1 = -a + 1$, proving $w_1^2 = \mathrm{id}$. It follows that $w_1 = w_1^{-1} = w_2$.

STEP 14. *The involution $w = t_{-1}u_1t_{-1}$ of Step 13 satisfies $w(a) = -a^{-1}$ for each $a \in \mathbb{D} \setminus \{0\}$.*

For $x \in \mathbb{D} \setminus \{0\}$, we have $w(x) = t_{-1}u_1t_{-1}(x) = w(w^{-1}(x - 1) - 1) - 1$. After addition of 1 to both ends and application of w, this gives

$$w\big(w(x) + 1\big) = w(x - 1) - 1. \tag{6.1}$$

Put $b = w(a)$ and $y = ab$. If $y = -1$, then $w(a) = b = a^{-1}y = -a^{-1}$, as required. So it suffices to establish $y = -1$. By Step 12, we have $w(y) = -w(b)w(a) = -y$. Substitution of this result in (6.1) gives $w(1 - y) = w(y - 1) - 1$. Suppose $y \neq 1$. Then $w(1 - y) = -w(y - 1) \in \mathbb{D} \setminus \{0\}$, so $2w(y - 1) = 1$. If $\mathbb{D}$ has characteristic two, this gives the contradiction $0 = 1$, so $y = -1$, and we are done. Otherwise, we find $y - 1 = w(1/2) = -2$ (for (6.1) gives $w(w(2) + 1) = w(1) - 1 = -2$, so $w(2) + 1 = -w(2)$, whence $w(2) = -1/2$ and $w(1/2) = -2$), so $y = -1$, indeed.

It remains to consider the case where $\mathbb{D}$ has characteristic distinct from two and $y = aw(a) = 1$ for some $a \in \mathbb{D} \setminus \{0\}$. Clearly, $a \neq 1, -1$. Observe that the above yields $w(x) = \pm x^{-1}$ for each $x \in \mathbb{D} \setminus \{0\}$. Equation (6.1) for $x = a$ and $x = a^{-1}$ gives $w(a^{-1} + 1) = w(a - 1) - 1$ and $w(a + 1) = w(a^{-1} - 1) - 1$. Extracting factors a from arguments of w, we find

$$
\begin{aligned}
w(a + 1) = w\big((a^{-1} + 1)a\big) &= -w(a)w\big(a^{-1} + 1\big) = -a^{-1}\big(w(a - 1) - 1\big) \\
&= -a^{-1}\big(w\big((1 - a^{-1})a\big) - 1\big) = -a^{-1}\big(-w(a)w\big(1 - a^{-1}\big) - 1\big) \\
&= -a^{-1}\big(a^{-1}w\big(a^{-1} - 1\big) - 1\big) = -a^{-1}\big(a^{-1}\big(w(a + 1) + 1\big) - 1\big),
\end{aligned}
$$

which leads to $(a + a^{-1})w(a + 1) = 1 - a^{-1}$. We know that there is a sign $\varepsilon \in \{\pm\}$, such that $w(a + 1) = \varepsilon(a + 1)^{-1}$. Substituting this in the equation just obtained after multiplying both sides by $w(a + 1)^{-1}$ from the right, we find $a + a^{-1} = \varepsilon(1 - a^{-1})(1 + a) = \varepsilon(a - a^{-1})$, a contradiction for both sign choices. Therefore, $w(a) = -a^{-1}$ for all $a \in \mathbb{D} \setminus \{0\}$.

STEP 15. $(P, \pi) \cong \mathbb{P}^1(\mathbb{D})$.

The isomorphism is determined by the identity map on $P = \mathbb{D} \cup \{\infty\}$. Step 14 shows that w acts on P as the element w of Example 6.2.7 and Proposition 6.2.8. By the theory of 2-transitive groups, this fixes the complete description of (P, π) in terms of $\pi(\infty, \infty)$, $\pi(0, \infty)$, and w. $\square$

Remark 6.2.12 The maps t_b and λ_a of Theorem 6.2.11 come from linear transformations on $\mathbb{D}^2$ with matrices as described in Example 6.2.7, so the group generated by all perspectivities of the Desarguesian projective line $\mathbb{P}^1(\mathbb{D})$ is contained in $\mathrm{PGL}(\mathbb{D}^2)$. By use of Theorem 6.2.11, one easily finds that the full automorphism group of the projective line coincides with $\mathrm{P\Gamma L}(\mathbb{D}^2)$. This shows that the goal set out in the beginning of this section has been achieved.

Definition 6.2.13 If $\mathbb{D}$ is a division ring, then the equation $x^2 = 1$ has one solution if $\mathbb{D}$ has characteristic two and two solutions otherwise. The group $\pi(\infty, 0)$ of automorphisms of the Desarguesian projective line $\mathbb{P}^1(\mathbb{D})$ consists of all maps $(y : 1) \mapsto (xy : 1)$ for some $x \in \mathbb{D}$, $x \neq 0$. In particular, there is a unique involution $\sigma_{\infty,0}$ in $\pi(\infty, 0)$ if the characteristic of $\mathbb{D}$ is not even. Thus, $\sigma_{\infty,0} = (y : 1) \mapsto (-y : 1)$. By double transitivity of the group generated by all projectivities, there is a unique involution $\sigma_{p,q}$ for any two points p and q of $\mathbb{P}^1(\mathbb{D})$. For any ordered triple (p, q, r) of points of $\mathbb{P}^1(\mathbb{D})$, the ordered quadruple $(p, q, r, \sigma_{p,q}(r))$ is called a *harmonic quadruple*.

Uniqueness of the involution makes that isomorphisms of projective lines preserve harmonic quadruples.

Corollary 6.2.14 *If (p,q,r,s) is a harmonic quadruple, then (p,q,s,r), and (q,p,s,r), and (r,s,p,q) are also harmonic. If $p = (a:1)$, $q = (b:1)$, and $r = (c:1)$, then $\sigma_{p,q}(r) = (d:1)$ where*

$$d = a + \left(2(b-a)^{-1} - (c-a)^{-1}\right)^{-1}.$$

Proof The matrix

$$t = \begin{pmatrix} a & 1 + a(b-a)^{-1} \\ 1 & (b-a)^{-1} \end{pmatrix}$$

is invertible and satisfies $t(\infty) = p$ and $t(0) = q$. Moreover, $\sigma_{\infty,0}$ is induced by the diagonal matrix with entries -1, 1, so $\sigma_{p,q}(r) = t\sigma_{\infty,0}t^{-1}(r) = (d:1)$. This proves the last assertion. The harmonicity of the three quadruples listed is readily checked by means of the equation for d. $\qquad\square$

6.3 Classification of Projective Spaces

We reach a major (and classical) result, namely the connection of projective spaces of dimension at least three with vector spaces over some division ring. The proof will be quite long since so many axioms need to be checked in verifying that a structure is a vector space over a division ring.

The reader may want to compare the proof below with Theorem 6.2.11 for projective lines, and may even look for a shorter proof using that result. In contrast to the one-dimensional case, we need to distinguish between the vector space V and the division ring $\mathbb{D}$.

Theorem 6.3.1 *Let Z be a thick projective space of dimension at least two with point set P. Let H be a maximal subspace of Z and o a point of $P \setminus H$. Set $V = P \setminus H$. Assume that the group $\pi(o, H)$ of perspectivities of Z with center o and axis H is transitive on $\ell \setminus \{o, h\}$, where ℓ is a line on o and $\{h\} = \ell \cap H$, and that, for each $a \in V$ there is a perspectivity t_a with axis H and center in H mapping o to a. There is a unique left vector space structure on V and a unique division ring $\mathbb{D}$ satisfying the following five properties.*

  (i) *As a set, $\mathbb{D} = \pi(o, H) \cup \{0\}$, where 0 is regarded as the function $V \to V$ mapping each point to o, and where $\pi(o, H)$ is identified with its restriction to V.*
 (ii) *The point o is the identity in the additive group of V.*
(iii) *Each $a, b \in V$ satisfy $a + b = t_a(b)$.*
 (iv) *For $\lambda \in \mathbb{D}$ and $b \in V$, we have $\lambda b = \lambda(b)$, i.e., the image of b under λ.*
  (v) *For $\lambda, \mu \in \mathbb{D}$, we have $\lambda\mu = \lambda \circ \mu$, i.e., the composition of the maps λ and μ.*

Moreover, Z is isomorphic to $\mathbb{P}(\mathbb{A}(V))$.

The conditions may seem somewhat technical, but the main point is that the required perspectivities exist if there are at least two projective planes in Z (cf. Theorem 6.1.5) or if Z is a Desarguesian plane (cf. Theorem 6.1.8).

For those who would expect us to find a right vector space, we refer to Exercise 5.7.13, where it is shown that a left vector space over $\mathbb{D}$ amounts to a right vector space over $\mathbb{D}^{\mathrm{op}}$.

As for $\pi(o, H)$, there is no harm in regarding its members as maps $V \to V$, because they fix every point of H.

Proof First, we verify that the left vector space structure on V is uniquely determined by the conditions. The only structure not specified is the addition on $\mathbb{D}$. But it is forced to be the map

$$f + g : x \mapsto f(x) + g(x) \quad (x \in V),$$

in view of distributivity of scalar multiplication on V. (At this point, there is no guarantee that $f + g$ is an element of $\mathbb{D}$.)

So far for the uniqueness of the structure on V of a left vector space over $\mathbb{D}$. In the following six steps, we verify that, with the specified operations, V is indeed a left vector space over $\mathbb{D}$. By T we denote the set of all perspectivities with center in H and axis H. By Proposition 6.1.4(i) it is a group, so that each element of this group is of the form t_a for some $a \in V$. For, if $a, b \in V$, then $t_a t_b$ is a perspectivity with axis H and so, if it does not belong to T, has a center p, say, in V. But then t_b and t_a^{-1} are perspectivities with center in H and coinciding on $H \cup \{p\}$, so must be equal, whence $t_a t_b = t_o \in T$.

STEP 1. *The set V together with addition is a commutative group with identity o.*

Let $a, b \in V$. Observe that $a + o = t_a o = a$. The perspectivity with axis H and center in H mapping a onto o, belongs to T and is the inverse of t_a. Hence it is of the form t_{-a} for some element $-a \in V$, and $(-a) + a = t_{-a} t_a(o) = o$.

It remains to show that $+$ is commutative. By Proposition 6.1.10(iii), T is commutative, so $a + b = t_a t_b(o) = t_b t_a(o) = b + a$, and Step 1 is proved.

STEP 2. *Multiplication in $\mathbb{D} \setminus \{0\}$ is a group structure with identity t_o.*

This comes down to the obvious statement that the set of all perspectivities with center o and axis H, form a group.

STEP 3. *We check the axioms relating $\mathbb{D}$ and V. Let $f, g \in \mathbb{D}$ and $x, y \in V$. The relations $f(g(x)) = fg(x)$ and $(f + g)x = f(x) + g(x)$ are clear from the construction of multiplication and addition on $\mathbb{D}$. Also $1 = t_o \in \mathbb{D}$ satisfies $t_o(x) = x$.*

STEP 4. *For $x, y \in V$ and $f \in \mathbb{D}$ we have $f(x) + f(y) = f(x + y)$.*

After a little rewriting, this equation reads $f t_x f^{-1} = t_{f(x)}$ for $x \in V$ and $f \in \mathbb{D} \setminus \{0\}$. But, in view of Lemma 6.1.3(iii), this is immediate from the observation that the corresponding automorphisms of Z at both sides of the equation are perspectivities with the same axis and center that map o to $f(x)$.

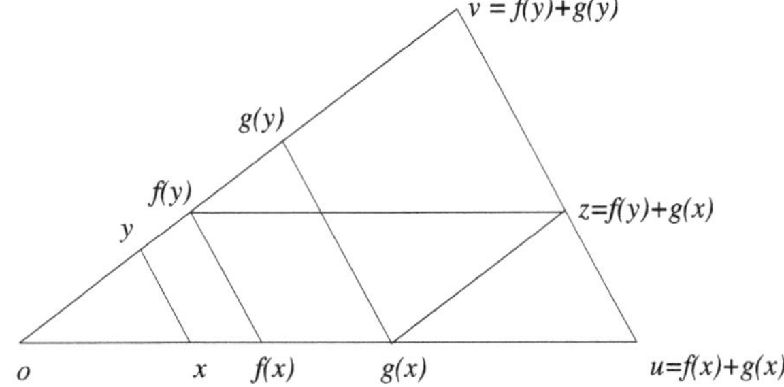

Fig. 6.3 The sum of two scalars f and g

STEP 5. *Addition in $\mathbb{D}$ is an Abelian group structure with identity* 0.

We need only show that the sum of two elements in $\mathbb{D}$ is again in $\mathbb{D}$ and that each element has an inverse, all other verifications being routine. The proofs of these two facts are developed along the same argument. Let f, $g \in \mathbb{D}$ with $f \neq o$, $g \neq o$, and let x, $y \in V$ be points such that o, x, y are non-collinear. The lines xy, $f(x)f(y)$, $g(x)g(y)$ of Z are 'parallel' in the sense that they meet in a point of H. Consider the points $z = f(y) + g(x)$, $u = f(x) + g(x)$ (on ox), and $v = f(y) + g(y)$ (on oy) (cf. Fig. 6.3).

Several times below, we will use the fact that, for s a perspectivity with axis H and m, n points of V, the lines mn and $s(m)s(n)$ are parallel. Applying this observation with $s = t_{g(x)}f$, and $m = o$, $n = y$, we see that $g(x)z$ is parallel to oy; similarly, with $s = t_{f(y)}g$, we find that $f(y)z$ is parallel to ox. The translation $t_{g(x)}$ transforms $f(x)f(y)$ in uz and so uz is parallel to xy. Similarly, the translation $t_{f(y)}$ transforms $g(x)g(y)$ to zv, and so zv is parallel to xy. Hence uz and zv are parallel, so they have a common point in H. Since they also meet in z, this forces $uz = zv$, whence uv is parallel to xy.

If $f(x) + g(x) \neq o$, then, $f(y) + g(y) \neq o$ for all $y \in V \setminus \{o\}$, for otherwise $f(x) + g(x)$ is on both ox and the line from o to $xy \cap H$, whence equal to o. Hence the perspectivity h of center o, axis H, that maps x onto $f(x) + g(x)$, transforms y to $f(y) + g(y)$. By repeating the argument for y as before and $x' \in ox$, we find that h also sends x' to $f(x') + g(x')$. This shows that $f + g = h$ belongs to $\mathbb{D}$.

Suppose, on the other hand, that $f(x) + g(x) = o$. Then $f(y) + g(y) = o$ for all $y \in V$ and g is an additive inverse of f. In particular, the inverse of each nonzero $f \in \mathbb{D}$ exists: if o, x, f are given, we take $-f$ to be an element of $\mathbb{D}$ that transforms x to $-(f(x))$ (the image of o under the translation mapping $f(x)$ on o).

STEP 6. *With the above structure, $\mathbb{D}$ is a division ring.*

Let $p \in V$ and $a, b, c \in \mathbb{D}$. Then, by Steps 4 and 5,

$$(a(b+c))p = a\big((b+c)(p)\big) = a\big(b(p) + c(p)\big) = at_{c(p)}b(p)$$

$$= at_{c(p)}a^{-1}ab(p) = \big(at_{c(p)}a^{-1}\big)\big(ab(p)\big)$$

$$= t_{ac(p)}\big(ab(p)\big) = ac(p) + ab(p)$$

$$= (ac + ab)(p),$$

whence $a(b + c) = ab + ac$. Also, $((a + b)c)p = (a + b)(cp) = a(cp) + b(cp) = (ac)p + (bc)p$ for all $p \in V$, and so $(a + b)c = ac + bc$.

We have established that V is a left vector space over $\mathbb{D}$. Finally, it is readily checked that the map sending a point of $V = P \setminus H$ onto itself and a point p of H onto the point at infinity (i.e., the parallel class) of lines containing p, determines an isomorphism $Z \to \mathbb{P}(\mathbb{A}(V))$. $\square$

Corollary 6.3.2 *Every thick projective space of dimension at least three and every thick Desarguesian projective plane is isomorphic to $\mathbb{P}(V)$ for some vector space V over a division ring. Every thick Desarguesian projective line is isomorphic to $\mathbb{P}^1(\mathbb{D})$ for some division ring $\mathbb{D}$.*

Proof Let Z be a thick projective space of dimension at least three or a thick Desarguesian projective plane. As a consequence of Proposition 5.2.10, there is a maximal subspace H of Z. If $\dim(Z) \geq 3$, then Theorem 6.1.5 shows that the conditions of Theorem 6.3.1 are satisfied. If Z is Desarguesian, then so does Theorem 6.1.8. Therefore, Theorem 6.3.1 applies in these cases, and shows that there is a vector space U over a division ring $\mathbb{D}$ such that Z is isomorphic to $\mathbb{P}(\mathbb{A}(U))$. Setting $V := U \oplus \mathbb{D}$, we conclude from Proposition 5.2.12 that Z is isomorphic to $\mathbb{P}(V)$.

The final statement is a direct consequence of Theorem 6.2.11. $\square$

Remark 6.3.3 If we drop the existence condition on perspectivities, the projective plane needs no longer be associated with a(n associative) division ring; see Example 2.3.4.

Theorem 6.3.1 also has an important consequence for homomorphisms as introduced in Definition 2.5.8 and isomorphisms of projective lines as introduced in Definition 6.2.1.

Corollary 6.3.4 *Suppose that $\phi : \mathbb{P}(V) \to \mathbb{P}(W)$ is a homomorphism of thick projective spaces of vector spaces V and W of dimension at least three over division rings $\mathbb{D}$ and $\mathbb{E}$, respectively, mapping lines onto lines. There is an isomorphism $\sigma : \mathbb{D} \to \mathbb{E}$ and a σ-linear map $V \to W$, whose projectivization is ϕ. The same conclusion holds for every isomorphism $\phi : \mathbb{P}^1(\mathbb{D}) \to \mathbb{P}^1(\mathbb{E})$ for division rings $\mathbb{D}$ and $\mathbb{E}$.*

Proof This follows directly from the uniqueness statements in Theorem 6.3.1. $\square$

Remark 6.3.5 We summarize the projective results by means of Table 6.1.

At the left hand side the algebraic constructions of the geometric objects $\mathbb{P}(\mathbb{D}^{n+1})$ and $\mathrm{PG}(\mathbb{D}^{n+1})$ appear; the right hand side contains the axiomatic notions of a projective space and a geometry of type A_n. The top contains line spaces; the bottom deals with geometries.

We can go around as follows. A thick projective space of finite dimension $n \geq 3$ is of the form $\mathbb{P}(\mathbb{D}^{n+1})$ by Corollary 6.3.2. The latter is a projective space (Definition 5.2.1) by Proposition 5.2.2. A projective space of dimension n corresponds

Table 6.1 Dependencies among projective geometry constructs

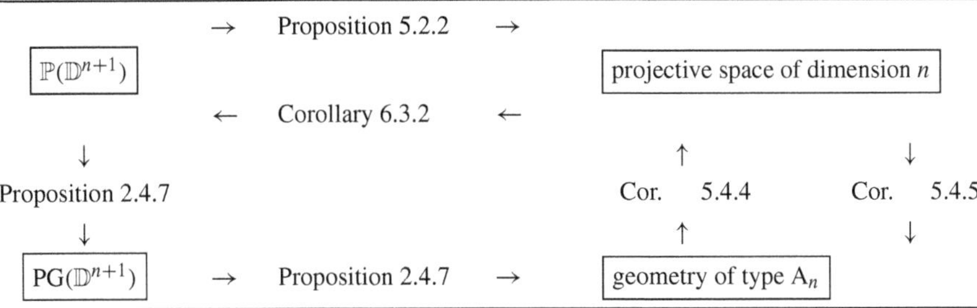

to a residually connected geometry of type A_n by Corollaries 5.4.4 and 5.4.5. In the particular case where the space is $\mathbb{P}(\mathbb{D}^{n+1})$, the geometry is $PG(\mathbb{D}^{n+1})$ due to Proposition 2.4.7. The geometry $PG(\mathbb{D}^{n+1})$ is of type A_n by Proposition 2.4.7.

The last subject of this section is the classification of non-thick projective spaces. Consider the following construction of new projective spaces out of given ones.

Definition 6.3.6 The *direct sum* of a given family $\{(P_j, L_j)\}_{j\in J}$ of non-empty line spaces is the space (P, L), where

(1) $P = \bigcup_{j\in J} P_j$ (the disjoint union over all P_j), and
(2) $L = \bigcup_j L_j \cup \{\{p, q\} \mid p \in P_j,\ q \in P_k,\ j \neq k;\ j, k \in J\}$.

A line space is called *irreducible* if it is not isomorphic to the direct sum of at least two non-empty line spaces.

If each P_j is a singleton, (the L_j are empty and) the space (P, L) is the complete graph on P, i.e., the thin projective space of dimension $|J| - 1$.

Proposition 6.3.7 *The direct sum of at least two non-empty projective spaces is a non-thick projective space.*

Proof Easy. $\square$

The following converse extends the classification of thick projective spaces of the previous section to arbitrary projective spaces.

Theorem 6.3.8 *Each projective space is the direct sum of a collection of non-empty projective spaces which are either a single point, a single thick line, or a thick projective space. This direct sum decomposition is unique.*

Proof Let Z be a projective space. Call two points of Z adjacent if they are distinct and on a thick line. This defines a graph on the point set of Z. Let $\{Z_j\}_{j\in J}$ be the connected components of this graph. They partition the point set of Z. Moreover, every line having a point in Z_i and one in Z_j, for distinct $i, j \in J$, has only two points. Clearly, for each $j \in J$, the subset Z_j is a subspace of P and so it is a projective space. Moreover, Z is isomorphic to the direct sum of the spaces Z_j for $j \in J$.

It remains to show that any two points of Z_j are adjacent. If not, there is a pair of non-adjacent points p, q which are adjacent to some point o. Let p', q' be points distinct from p, q on the lines op and oq, respectively. By Pasch's Axiom 5.2.4, the lines pq and $p'q'$ in the plane $\langle p, o, q \rangle$ must intersect in a point distinct from both p and q. But then pq has a third point so p and q are adjacent, a contradiction. Thus, each line in Z_j is thick, and so Z_j is either a singleton, a thick line, or a thick space (cf. Exercise 6.8.11). It follows that the partition $\{Z_j\}_{j \in J}$ of Z is unique. $\square$

6.4 Classification of Affine Spaces

Since affine spaces (cf. Definition 5.1.1) are closely related to projective spaces, it is not a big surprise that we make use of projective spaces for their classification. Nevertheless, there are remarkable differences. In Theorem 6.4.8 of this section, we will recognize an affine space from a diagram geometry.

According to Exercise 2.8.9, the restriction of a projective plane to the complement of a line is an affine plane. The example below generalizes this result to projective spaces, in the sense that its restrictions to complements of hyperplanes give affine spaces.

Example 6.4.1 Let Z be a thick projective space, and let H be a maximal subspace of Z. Consider the space with parallelism $\mathbb{A}(Z, H)$ consisting of the restriction of Z to the point set $Z \setminus H$ (in the sense of Definition 2.5.8) together with the *parallelism* on the set of lines, defined by $l \parallel l'$ if and only if the line of Z containing l and the line of Z containing l' have a common point that belongs to H. By use of Exercise 2.8.9 it is easy to see that $\mathbb{A}(Z, H)$ is an affine space. Proposition 6.4.2 below states the converse.

If Z is Desarguesian, then, by Theorem 6.3.1, there is a vector space V such that $Z \cong \mathbb{P}(\mathbb{A}(V))$, from which it follows that $\mathbb{A}(Z, H)$ is the affine space $\mathbb{A}(V)$ introduced in Notation 5.1.2.

Proposition 6.4.2 *If A is an affine space of dimension at least two, then, up to isomorphism, there is a unique pair of a thick projective space Z and a hyperplane H of Z such that $A \cong \mathbb{A}(Z, H)$.*

Proof Take H to be the set of equivalence classes of parallel lines of A. If l is a line of A, then its class $[l]$ of parallels will be called the point at infinity of l. The point set P is the disjoint union of H and the point set of A. A line at infinity is defined to be the set of all points at infinity of the lines of some affine plane in A. The line set L consists of lines of A extended by their points at infinity, and the lines at infinity. Clearly, the resulting line space $Z := (P, L)$ is thick and H is a hyperplane of Z.

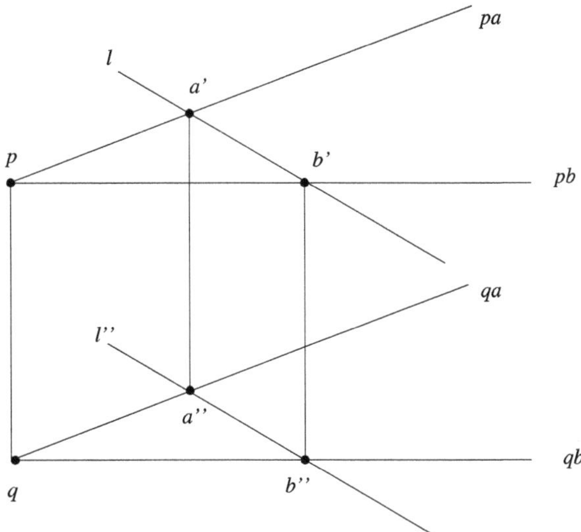

Fig. 6.4 The two affine planes $pa'b'$ and $qa''b''$, and some additional lines

Having given the construction of Z and H, we prove that Z is a linear space. The only difficulty is to show that two points at infinity, say a, b, are on one and only one line (at infinity). Given a point p of A, there are uniquely determined lines pa, pb of A on p admitting a and b, respectively, as points at infinity.

The lines pa and pb of A determine a unique affine plane pab on p; its line at infinity contains a and b. We must show that if q is a point of A not in pab, then pab and qab have the same line at infinity. In view of symmetry it suffices to show that each line l of pab admits some parallel in qab. We (may) assume that the line l of pab does not pass through p, a, or b. This implies that l intersects pa in some point $a' \neq p$ and pb in some point $b' \neq p$. In the affine plane qpa', the line qa and the parallel to pq on a' intersect in some point a''. Similarly there is a point $b'' \in qb$ such that $b'b'' \parallel pq$. As $\parallel$ is an equivalence relation, we have $a'a'' \parallel b'b''$, so these lines are contained in some affine plane π. The lines $l = a'b'$ and $l'' := a''b''$ are also contained in π, so if they are not parallel, then there is a point $o \in l \cap l''$. Now $o \in pa'b' \cap qa''b'' \setminus (qa \cup qb)$, so the parallels m, n to qa, qb, respectively, on o are contained in $qa''b''$. Since m, n must be parallel to pa, pb (by transitivity of $\parallel$), they are even in $pa'b' \cap qa''b''$. The assumption $a \neq b$ implies that m, n are distinct, leading to the absurdity that m, n generate an affine plane coinciding with both $pa'b'$ and $qa''b''$. Therefore l'' is a parallel to l contained in $qa''b''$, as required. This is illustrated in Fig. 6.4.

Next we show that all planes of Z are projective planes. This is easy for planes with some point in A since these appear as the union of an affine plane and of its line at infinity. Hence we focus on a plane π of Z (cf. Definition 5.3.1) all of whose points are at infinity. Let p be a point of A and let $U := \langle p, \pi \rangle$ be the subspace of Z generated by p and π. We claim that U is the union of all lines px with $x \in \pi$. Indeed, if x, y are distinct points in π and u, v are distinct points in $A \cap U \setminus \{p\}$ with $u \in px$, $v \in py$, then the line uv of A lies in the affine plane pxy and its point at infinity lies in the line xy at infinity of the affine plane pxy. Therefore, the line of Z containing u and v and the line xy of Z meet in a point, say z. We conclude that the line pz, as well as each line pw with $w \in uv$, is of the form pd with $d \in xy$, which proves the claim.

Take three non-collinear points a, b, c generating π as a subspace of Z. In U, let a', b', c' be points of A distinct from p that are on pa, pb, pc, respectively. Then a', b', c' generate an affine plane α in A because a, b, c are non-collinear. The union of α and of its line at infinity, is a projective plane $\overline{\alpha}$ as we have seen earlier. Now $\overline{\alpha}$ is contained in U since a', b', c' are in U and $\overline{\alpha}$ is the subspace of Z generated by a', b', c'. Thus the map $i : \overline{\alpha} \to \pi$ given by $x \mapsto px \cap \pi$ is well defined. We claim that i is an isomorphism. Indeed, i is injective (observe that $p \notin \overline{\alpha}$ since $p \notin \alpha$). It maps a line onto a line (the line uv of above maps onto xy), and $i(\overline{\alpha})$ is a subspace of π containing a, b, c, whence $i(\overline{\alpha}) = \pi$. It follows that i is surjective, and therefore bijective. The conclusion is that π is a projective plane.

Finally, uniqueness follows from an argument close to the proof of Proposition 5.2.12: the affine space structure on $Z \setminus H$ uniquely determines Z. $\square$

Corollary 6.4.3 *Every affine space of dimension at least three is isomorphic to the affine space of a vector space.*

Proof By Proposition 6.4.2, there is a thick projective space Z and a hyperplane H of Z such that $A \cong \mathbb{A}(Z, H)$. As the dimension hypothesis implies $\dim(Z) \geq 3$, Theorem 6.1.5 shows that Z satisfies the hypotheses of Theorem 6.3.1, so there is a vector space V such that $Z \cong \mathbb{P}(\mathbb{A}(V))$. It follows that $\mathbb{A}(Z, H) \cong \mathbb{A}(V)$. $\square$

A fortiori, every affine space is a matroid and so belongs to a linear diagram.

By Corollary 5.4.4, the 1-shadow of a [3]-geometry of type A_3 is a projective space. What about the analogue for affine spaces? Suppose that (P, L) is the 1-shadow of a [3]-geometry of type

$$\text{AfL}_3: \quad \circ \overset{\text{Af}}{\rule{2cm}{0.4pt}} \circ \overset{\text{L}}{\rule{2cm}{0.4pt}} \circ \; .$$

Does there exist an equivalence relation $\parallel$ on L such that $(P, L, \parallel)$ is an affine space? If the lines of (P, L) have size at least three, this question makes sense, since then affine planes are determined by their structure as line spaces (cf. Remark 5.1.5). In the remainder of this section we will see that the answer is yes if some line of L has cardinality at least four, and no otherwise.

Example 6.4.4 We consider the case of lines of size two. Let S be a Steiner system $S(3, 4, v)$ (cf. Definition 5.5.1). Take X_1, X_2, X_3 to be the set of points, pairs, and blocks of S, respectively, so $v = |X_1|$. The geometry $\Gamma = (X_1, X_2, X_3, *)$, where $*$ stands for symmetrized containment, is of type AfL_3. As mentioned in Sect. 5.8, such a Steiner system S exists if and only if $v = 2$ or $4 \pmod 6$. Thus the smallest possible values of v for non-trivial $S(3, 4, v)$ are 8, 10, 14, 16. In view of Corollary 6.4.3, every $S(3, 4, v)$ which is the [3]-truncation of the geometry of an affine space of dimension at least three requires v to be a power of 2. There is a unique $S(3, 4, 8)$, consisting of the points and planes of the affine space $\mathbb{A}(\mathbb{F}_2^3)$. For our purpose, which is to exhibit the existence of a geometry with diagram AfL_3 whose 1-shadow is not an affine space, it suffices to construct an $S(3, 4, 10)$. The inversive plane derived from an ovoid in $\mathbb{P}(\mathbb{F}_3^4)$, cf. Example 5.5.10, does the job.

Example 6.4.5 We consider the case of lines of size three. Let A be the set of points of a [3]-geometry of type AfL_3 all of whose lines have cardinality 3. Fix $0 \in A$, and mimick the construction of a vector space over the field $\mathbb{F}_3$ with underlying set A and origin 0. In this vein, define $a + 0 = 0 + a$ for all $a \in A$; for $a \neq 0$, put $a + a = -a$ if and only if $0, a, -a$ are distinct collinear points; for a, b non-collinear with 0, let $a + b$ be the unique point c such that $-c$ is collinear with a and b and distinct from them. In general $(A, +)$ need not be a commutative group, but we can show that it is a commutative Moufang loop with identity 0. Here, $(A, +)$ is called a *Moufang loop* if $+$ is a binary operation on A with identity element 0 and inverse $-a$ for each $a \in A$ such that every equation $a + x = b$ with $a, b \in A$ has a unique solution and the following weak form of associativity holds: $-a + (a + b) = b$ and $(a + b) + (c + a) = a + ((b + c) + a)$ for all $a, b, c \in A$.

Conversely, given a commutative Moufang loop $(M, +)$ all of whose elements are of order three (a so-called 3-Moufang loop), it can be turned into a geometry of type AfL_3 with lines of cardinality three, along the same lines as the construction of $AG(V)$ from a vector space V. If M is not a group, the space on M is not an affine space. The number of elements of every finite 3-Moufang loop is a power of 3. Exercise 6.8.12 shows that there exist (commutative) 3-Moufang loops that are not groups.

In view of these examples it is surprising that no further non-affine spaces exist all of whose planes are affine, as we will see in Theorem 6.4.8.

Remark 6.4.6 Let Γ be a [3]-geometry of type AfL_3 whose lines have at least three points. According to Theorem 5.4.3, Γ is the geometry of a matroid with point set X_1 whose non-empty flats are shadows on 1 of elements of Γ. As a consequence, up to isomorphism, $\Gamma = (P, L, \Pi, *)$, where Π is the collection of all subspaces of dimension two of the linear space (P, L) and $*$ is symmetrized containment. Each member of Π is indeed an affine plane in view of Exercise 5.7.16.

We have seen that the linear space (P, L) underlying $(P, L, \|)$ is uniquely determined and our remaining concern is to find a convenient parallelism $\|$ on L. But $l \parallel m$ for distinct $l, m \in L$ can be defined by l and m being disjoint and generating a member of Π. This observation indicates how to define $\|$.

In order to prove that the relation $\|$ as suggested in Remark 6.4.6 is transitive, we need the following lemma, which requires that lines have size at least four.

Lemma 6.4.7 *Let $\Gamma = (P, L, \Pi, *)$ be a [3]-geometry of type AfL_3 whose lines have at least four points. For each pair $(l, \pi) \in L \times \Pi$ such that $l \cap \pi$ is a singleton, the union of all planes containing l and a line of π is the subspace of (P, L)*

*generated by l and π. In particular, every plane on l contained in this subspace
meets π in a line.*

Proof Let a be the single point of $l \cap \pi$, and denote by U the union of all planes
containing l and a line of π. Clearly, we only need establish that U is a subspace.
We will use the relation $\|$ on L suggested by Remark 6.4.6, so two lines m and n
satisfy $m \| n$ if and only if they are parallel in a member of Π.

For $v \in U \setminus l$, we denote by $l(v)$ the point which is on l and on the parallel to
the line $\langle l, v \rangle \cap \pi$ through v. (Of course, $l(v)$ exists since $v \in U$ and $\langle v, l \rangle$ is the
only plane on v containing l.) If $p \in l \setminus \{l(v), a\}$, then the line pv meets $\langle l, v \rangle \cap \pi$
in a point denoted by $p(v)$. We have $p(v) \neq a$. Let m be a line having two distinct
points, x and y say, in U, so $m = xy$. We will show $m \subseteq U$ in three steps.

If l and m are coplanar, this is immediate. Therefore, we (may) assume that l
and m are not coplanar. Take a point $p \in l \setminus \{l(x), l(y), a\}$; such a point exists by
the assumption that lines have cardinality at least four. The points $p(x)$ and $p(y)$
are distinct (for otherwise the lines l and m would be coplanar), and generate a line
denoted by $p(m)$.

STEP 1. *If $m \| p(m)$, then $m \subseteq U$.*

Suppose $m \| p(m)$. If $z \in m$, then the line pz meets $p(m)$ in some point u. As
p, u, l are coplanar, the line pu is in U, so $z \in U$ and $m \subseteq U$. Hence Step 1.

From now on, we (may) assume that m is not parallel to $p(m)$. As $p(m)$ lies in
the plane $\langle p, m \rangle$, it meets the line m in some point b. Observe that $a \notin p(m)$ by the
assumption that l and m are not coplanar. Let c_p be the unique point common to m
and the parallel to $p(m)$ passing through p. Clearly, all points of $m \setminus \{c_p\}$ belong
to U.

Suppose $m \not\subseteq U$. By the above argument, there is a unique point $c := c_p \in m \setminus U$.

STEP 2. $l(x) \neq l(y)$.

Suppose $l(x) = l(y)$. The existence of c implies that m is not contained in π. The
intersection $\langle l(x), m \rangle \cap \pi$ is contained in a line. But if it were a line, one of the lines
$xl(x), yl(y)$ on $l(x) = l(y)$ in $\langle l(x), m \rangle$ would meet it nontrivially, which is absurd
in view of the definition of $l(x)$ and of $l(y)$. As b is a member of the intersection, it
follows that $\langle l(x), m \rangle \cap \pi$ is the singleton $\{b\}$. In particular, if $x' \in xl(x)$ is distinct
from x and $l(x)$, then the line $x'y$ is a line on y in $\langle l(x), m \rangle$ distinct from m, so
$x'y \cap \pi = \emptyset$, whence $x'y \subseteq U$ due to Step 1. Since there are at least two choices
for x', there are at least two lines on y in $\langle l(x), m \rangle \cap U$. Therefore every point of
$\langle l(x), m \rangle$ lies on a line joining $l(x)$ to a point of U, and hence belongs to U. It
follows that $\langle l(x), m \rangle$, and in particular m, is contained in U, contradicting $m \not\subseteq U$.
This proves Step 2.

Let $p \in l \setminus \{a\}$. As $m \cap U$ has at least three points, Step 2 gives a point $u \in
m \setminus \{b, c\}$ with $l(u) \neq p$, so $p(u)$ is a well-defined point of π. In particular, $p(m) =
bp(u) = \langle p, m \rangle \cap \pi$ is a line, and $c = c_p$. In view of Step 1, there is a unique point
p' on $p(m)$ with $pp' \| m$ inside $\langle p, m \rangle$. Let $n(p)$ denote the line on a in π parallel
to $p(m)$. Finally, write $\alpha(p) = \langle l, p' \rangle$ and $\beta(p) = \langle l, n(p) \rangle$. Both planes lie in U.

STEP 3. *The planes on l disjoint from m and meeting π in a line are precisely $\alpha(p)$ and $\beta(p)$.*

Let γ be a plane on l meeting π in a line n, say. If $\gamma \cap m = \emptyset$, then either $n \cap p(m) = \{p'\}$ or $n \parallel p(m)$, so γ is equal to $\alpha(p)$ or to $\beta(p)$.

Conversely, as $a \in \alpha(p) \setminus p(m)$, we have $\alpha(p) \cap m \subseteq \alpha(p) \cap \langle p, m \rangle = pp' \parallel m$, so $\alpha(p) \cap m = \emptyset$. Suppose $q \in \beta(p) \cap m$. If $q \neq c$, then pq meets $p(m)$ in a point. This point lies in $\beta(p) \cap p(m) = n(p)$. As $n(p) \parallel p(m)$, we obtain $n(p) = p(m)$, contradicting $a \notin p(m)$. Consequently, $q = c$, and so $c \in \beta(p) \subseteq U$, a contradiction. Hence Step 3.

We now derive the final contradiction. Let p_1, p_2, p_3 be three distinct points of $l \setminus \{a\}$. Take distinct i, j from [3]. If $\beta(p_i) = \beta(p_j)$, then $n(p_i) = n(p_j)$, so $p_i(m)$ and $p_j(m)$ (being parallel in π and containing b) coincide; but then $\langle p_i, m \rangle = \langle p_i(m), m \rangle = \langle p_j(m), m \rangle = \langle p_j, m \rangle$, so $p_i = p_j$, a contradiction. According to Step 3, this forces $\alpha(p_i) = \beta(p_j)$ for each pair i, j from [3], which is clearly impossible. This establishes $m \subseteq U$.

As for the last assertion, suppose that ρ is a plane on l inside U. Pick a point $w \in \rho \setminus l$. As $w \in U$, it lies on a plane in U containing l and a line of π. This plane contains w and l, so coincides with ρ, which shows that ρ meets π in a line. This ends the proof of the lemma. $\square$

Theorem 6.4.8 *Let Γ be a [3]-geometry of type AfL_3 whose lines have at least four points. Then Γ is isomorphic to the geometry $(P, L, \Pi, *)$ obtained from an affine space $(P, L, \parallel)$, where Π is the collection of all affine planes in $(P, L, \parallel)$ and $*$ is symmetrized containment. Moreover, $(P, L, \parallel)$ is unique up to isomorphism.*

Proof As indicated in Remark 6.4.6, we may view Γ as the geometry $(P, L, \Pi, *)$ where Π is the collection of all subspaces of dimension two of the linear space (P, L) and $*$ is symmetrized containment. For $l, m \in L$, set $l \parallel m$ if and only if either $l = m$ or l and m arc disjoint and generate a member $\langle l, m \rangle$ of Π. Clearly, these properties for $\parallel$ are necessary for $(P, L, \parallel)$ to be an affine space, so the last assertion will follow once the first is proved.

Given a line l and a point x off l, there is a unique line m on x with $m \parallel l$ (the line on x in $\langle l, x \rangle$ disjoint from l), so the parallels of l partition P, as required. In order to prove that $(P, L, \parallel)$ is indeed an affine space, it suffices to establish that $\parallel$ is a transitive relation, all other conditions being immediate. To this end, let l, m, n be distinct lines with $l \parallel m$ and $m \parallel n$. Consider the plane $\pi = \langle l, m \rangle$. If n lies in π, it follows immediately that $l \parallel n$, so assume $n \not\subseteq \pi$. Pick a line bc joining $b \in m$ and $c \in n$. Denote by U the subspace of (P, L) generated by π and c. Let l' be the parallel to l passing through c, and l'' the parallel to l' passing through b. Then $l' \subseteq \langle l, c \rangle \subseteq U$, so applying Lemma 6.4.7 with $l = bc$, we obtain that $\langle l', b \rangle \cap \pi$ is a line. If l' has a point of π, then it is entirely contained in π (as $l' \parallel l$), whence $c \in \pi$, and $n \subseteq \pi$ (because $c \in n \parallel m \subseteq \pi$), proving $l \parallel n$ by transitivity of the restriction of $\parallel$ to π. Therefore, we (may) assume $l' \cap \pi = \emptyset$. Consequently, $\langle l', b \rangle \cap \pi$, being a line in $\langle l', b \rangle$ on b disjoint from l', must coincide with l''.

Suppose $l'' \neq m$. Then l'' meets l in a point a, say. Observe that $a \notin l'$ since $l' \cap \pi = \emptyset$. We have $\langle l, c \rangle = \langle l', a \rangle = \langle l'', c \rangle$. Since $b \in l'' \setminus l$ (in view of the assumption that l and m are distinct), we obtain $\pi = \langle l, b \rangle \subseteq \langle l'', c \rangle$ (for l is contained in $\langle l', a \rangle$ as it contains a and is parallel to l'). Hence $\pi = \langle l'', c \rangle = \langle l', a \rangle$ and $l' \subseteq \pi$, contradicting $l' \cap \pi = \emptyset$. This shows that $l'' = m$. But then $l' = n$, so $l \parallel n$, as required. $\square$

6.5 Apartments

In this section, we consider geometries of type A_n and derive transitivity of their automorphism groups on sets involving thin subgeometries called apartments. The precise statement is given in Proposition 6.5.6. This property will reappear in greater generality in Chap. 11.

Let n be a positive integer and let (W, S) be a Coxeter system of type A_n. In Lemma 4.5.16 it became clear that the thin chamber system $\mathcal{C}(A_n)$ is residually connected. By the Chamber System Correspondence 3.4.6 it determines a thin $[n]$-geometry of type A_n. This geometry is also described in Example 4.1.13(i). By Theorem 4.2.8 any other thin $[n]$-geometry of type A_n is a proper quotient of $\Gamma(\mathcal{C}(A_n))$. The following lemma shows that no such quotients exist.

Lemma 6.5.1 *If $\Delta = (X_1, \ldots, X_n, *)$ is a thin $[n]$-geometry of type A_n, then $|X_1| = n + 1$ and $\Delta \cong (Y_1, Y_2, \ldots, Y_n, *)$ where $Y_i = \binom{X_1}{i}$ for each $i \in [n]$ and $*$ is symmetrized inclusion.*

Proof If $n = 1$, then $|X_1| = 2$, due to thinness and the lemma follows. We proceed by induction on n. Suppose $n > 1$. By Proposition 5.4.2, and thinness, lines (elements of X_2) can be viewed as unordered pairs from X_1. So, if $x \in X_1$, then, by the induction hypothesis for Γ_x, there are n lines l on x, each of which accounts for a single point in $\mathrm{Sh}_1(l) \setminus \{x\}$. By Theorem 5.4.3, the $n + 1$ points (including x) so obtained, constitute all of X_1. It readily follows that X_i can be identified with the collection of subsets of X_1 of size i. $\square$

We will reformulate the result in terms of chamber systems and make the connection with the Coxeter system (W, S) more explicit. Recall from Definition 4.2.5 that the chambers of $\mathcal{C}(M)$ are the members of W and that, for v, $w \in W$ and $s \in S$, we have $v \sim_s w$ if and only if $vs = w$. By Exercise 4.9.5 we may take the Coxeter system of type A_n to be (W, S) with $W = \mathrm{Sym}_{n+1}$ and $S = \{(i, i + 1) \mid i \in [n]\}$. There is no difference in the result if we start from the seemingly more general diagram L_n instead of A_n.

Recall from Remark 4.2.15 that l denotes the length function on W with respect to S.

Proposition 6.5.2 *Let $n \in \mathbb{N}$, $n \geq 2$, and let $\Delta = (X_1, \ldots, X_n, *)$ be a thin $[n]$-geometry of type L_n. The chamber system $\mathcal{C}(\Delta)$ over $[n]$ is isomorphic to $\mathcal{C}(A_n)$.*

Let (W, S) be the Coxeter system of type A_n consisting of the group $W = \mathrm{Sym}_{n+1}$ and the generating set $S = \{(i, i+1) \mid i \in [n]\}$. There is a right action of W on the chamber set of $\mathcal{C}(\Delta)$ satisfying the following assertions for all chambers c, d of $\mathcal{C}(\Delta)$.

(i) *The element $w \in W$ with $d = cw$ is determined by*

$$w(i) = \min\{j \in [n+1] \mid c_j \cap d_i \not\subseteq c_j \cap d_{i-1}\},$$

where c_i denotes the element of Δ of type i in c, with the understanding that $c_{-1} = \emptyset$ and $c_{n+1} = X_1$.

(ii) *The distance between c and d in $\mathcal{C}(\Delta)$ is given by*

$$l(w) = \left|\{(i, j) \in [n+1] \times [n+1] \mid i < j \text{ and } w(i) > w(j)\}\right|,$$

where w is as in (i).

Proof The shadow space $\mathrm{ShSp}(\Delta, 1)$ (cf. Definition 2.5.1) is a linear space whose lines are thin. In any linear space whose lines are thin, three distinct points span a plane isomorphic to a clique on three elements, which is a projective plane. So, Corollary 5.4.4 applies and Δ is a geometry of type A_n. By Lemma 6.5.1 and Exercise 4.9.5 this shows that $\mathcal{C}(\Delta)$ is isomorphic to $\mathcal{C}(A_n)$.

Explicitly, let $X = [n+1]$ be the point set of Δ. Then a chamber is of the form $c = \{w_c(1), \{w_c(1), w_c(2)\}, \ldots, \{w_c(1), \ldots, w_c(n)\}\}$, where $w_c \in \mathrm{Sym}_{n+1}$ and $\{w_c(1), , \ldots, w_c(i)\} \in X_i$. With this notation we can describe the isomorphism $\mathcal{C}(\Delta) \to \mathcal{C}(A_n)$ as the map $c \mapsto w_c$.

(i) Using the above isomorphism, we find $\{w_d(i)\} = d_i \setminus d_{i-1}$ and $w_c^{-1}(x) = \min\{j \in [n+1] \mid x \in c_j\}$, so

$$\begin{aligned}
\{w_c^{-1} w_d(i)\} &= \min\{j \in [n+1] \mid w_d(i) \in c_j\} \\
&= \min\{j \in [n+1] \mid c_j \cap (d_i \setminus d_{i-1}) \neq \emptyset\} \\
&= \min\{j \in [n+1] \mid c_j \cap d_i \not\subseteq c_j \cap d_{i-1}\},
\end{aligned}$$

whence Statement (i).

(ii) In view of Lemma 4.2.4, the distance between c and d is $l(w)$. By Corollary 4.5.5, this length equals the size of Φ_w where Φ is a root system of type A_n. Using the presentation of Φ^+ as the set of all $\varepsilon_i - \varepsilon_j \in \mathbb{R}^{n+1}$ with $1 \leq i < j \leq n+1$ as in Example 4.5.6, we derive that Φ_w is the number of such roots for which $w(\varepsilon_i - \varepsilon_j) = \varepsilon_{w(i)} - \varepsilon_{w(j)}$ satisfies $w(i) > w(j)$. This establishes (ii). $\square$

In particular, every thin $[n]$-geometry of type L_n is of Coxeter type. Let us now address the occurrence of thin $[n]$-geometries of Coxeter type in arbitrary geometries related to simple matroids.

Definition 6.5.3 Suppose that $\Gamma = (X_1, X_2, \ldots, X_n, *)$ is an $[n]$-geometry of type L_n, and Δ is a thin residually connected subgeometry of Γ of the same type. Then

$B := \mathrm{Sh}_1(\emptyset, \Delta)$, the shadow on 1 of the empty flag in the geometry Δ, is a subset of X_1 of size $n + 1$ such that the flat $\langle B \rangle_{\mathcal{F}}$ of the simple matroid $\mathcal{F} := \mathrm{ShSp}(\Gamma, 1)$ (cf. Theorem 5.4.3) coincides with X_1. The geometry Δ coincides with the subgeometry of Γ induced on $\{\langle Y \rangle_{\mathcal{F}} \mid Y \subseteq B\}$. Conversely, each subset B of X_1 of size $n + 1$ with $\langle B \rangle_{\mathcal{F}} = X_1$ determines a unique thin subgeometry with the same diagram as Γ. We call such a set B a *frame* of $\mathcal{F}$ and such a thin subgeometry Δ an *apartment* of Γ.

Example 6.5.4 If Γ is the geometry of a projective or affine space of a vector space V of finite dimension, then a frame of $\mathrm{ShSp}(\Gamma, 1)$ corresponds to a basis of the space interpreted as a matroid (introduced in Definition 5.3.3). For the projective case, this means that the frame is strictly smaller than the projective basis of Definition 5.2.18. Thus, in $\mathbb{P}(V)$, a frame is nothing but a set of projective points of a basis of V. In $\mathbb{A}(V)$, a frame is nothing but a collection of points corresponding to a basis of $\mathbb{A}(V)$. In both cases, the automorphism group is transitive on the set of frames. Also, every chamber belongs to an apartment. But $\mathbb{A}(V)$ and $\mathbb{P}(V)$ differ when it comes to pairs of chambers. In $\mathbb{A}(V)$, the existence of parallel lines prevents two arbitrary chambers from belonging to a common apartment. In fact, if Γ is a [2]-geometry of type L, then the statement that any two chambers can be embedded in an apartment is equivalent to the statement that Γ is a projective plane.

If Γ is a geometry of Coxeter type A_n, the advantage over type L_n, apart from what we have just seen, is that given two elements u, v, we can apply Theorem 5.4.3 in two ways: both the full shadow space $\mathrm{ShSp}(\Gamma, 1)$ and $\mathrm{ShSp}(\Gamma, n)$ are simple matroids of dimension n. This is due to the fact that a Coxeter diagram of type A_n is also of type L_n (and even of type A_n) when read 'backwards', i.e., when the types i and $n - i + 1$ are interchanged for all $i \in [n]$. This leads to elements $u \cap v$ and $\langle u, v \rangle$ that have 'dual properties'; for instance, their types add up to $n + 1$. We exploit these advantages to prove existence of apartments.

Theorem 6.5.5 *Let $\Gamma = (X_1, X_2, \ldots, X_n, *)$ be an $[n]$-geometry of type A_n. Suppose that c and d are chambers of Γ.*

(i) *There is an apartment containing both c and d.*
(ii) *The map $w_{c,d} : [n + 1] \to [n + 1]$ given by*

$$w_{c,d}(i) = \min\left\{ j \in [n + 1] \mid c_j \cap d_i \not\subseteq c_j \cap d_{i-1} \right\}$$

   is a permutation of $[n + 1]$.
(iii) *If e is a chamber on a minimal gallery from c to d, then e is a member of each apartment containing both c and d.*

Proof (i) The proof is by induction on the rank n. For $n = 2$, the statement is easily verified (and referred to above). So, let $n > 2$. By the induction hypothesis, we can find a frame $\{x_2, \ldots, x_{n+1}\}$ in Γ_{d_n} such that the flags $\{d_1, \ldots, d_{n-1}\}$ and

$\{c_2 \cap d_n, \ldots, c_n \cap d_n\}$ are contained in the apartment corresponding to it. (Here, the intersection refers to the unique element found in Proposition 5.4.2(iv).)

Suppose that c_1 and d_n are non-incident. Since $c_1 \notin \langle x_2, \ldots, x_{n+1} \rangle = d_n$ (for otherwise c_1 and d_n would be incident), the set $\{c_1, x_2, \ldots, x_{n+1}\}$ is a frame of Γ whose apartment Δ contains both c_1 and d. We will show that it also contains c. Moreover, given $j \in [n]$, we have $\langle c_1, c_j \cap d_n \rangle \subseteq c_j$. Also, since $c_1 \notin c_j \cap d_n$ and the type of $c_j \cap d_n$ is at most one less than the type of c_j, the elements $\langle c_1, c_j \cap d_n \rangle$ and c_j have the same (matroid) dimension, so they coincide. As c_1 and $c_j \cap d_n$ belong to Δ, so does $c_j = \langle c_1, c_j \cap d_n \rangle$. Hence c is contained in Δ.

Next, suppose $c_n = d_n$. For $x_1 \in X_1 \setminus d_n$, the apartment determined by the frame $\{x_1, \ldots, x_{n+1}\}$ is as required.

Finally, suppose that c_1 is incident with d_n and c_n is distinct from d_n. Take $i \in [n]$ to be minimal such that c_i is not incident with d_n. Now $i > 1$ and $c_{i-1} = c_i \cap d_n$. We can finish by adding any $x_1 \in c_i \setminus c_{i-1}$ to $x_2, \ldots, x_{n+1}$. Details are as before.

(ii) By (i), there is an apartment containing c and d. This is a thin residually connected subgeometry of type A_n and so Proposition 6.5.2 applies. This gives the result.

(iii) The proof is by induction on the distance $d_{\mathcal{C}}$ in $\mathcal{C}(\Gamma)$ between c and d. We take d and e distinct; otherwise there is nothing to show. Let Δ be an apartment containing c and d. It suffices to treat the case where $e \sim_k d$ for some $k \in [n]$. Thus, we have $d_i = e_i$ for all $i \neq k$, whence $w_{c,d}(i) = w_{c,e}(i)$ for all $i \neq k, k+1$. As d and e are distinct, this implies $w_{c,d} = w_{c,e}(k, k+1)$. In view of Proposition 6.5.2(ii),

$$\left| l(w_{c,d}) - l(w_{c,e}) \right| = 1. \tag{6.2}$$

Suppose now that $d_{\mathcal{C}}(c, e) = d_{\mathcal{C}}(c, d) - 1$. Then, by the induction applied to the pair c, e, we have $d_{\mathcal{C}}(c, e) = l(w_{c,e})$. Consequently, $l(w_{c,e}) \leq l(w_{c,d}) - 1$, and, by (6.2), equality follows. Thus, $w_{c,d}(k+1) = w_{c,e}(k) < w_{c,d}(k)$. Writing $j = w_{c,d}(k+1)$, we derive from (ii):

$$c_j \cap e_{k-1} \subseteq c_j \cap e_k \quad \text{and} \quad c_j \cap e_{k-1} = c_j \cap d_k.$$

Consequently, $c_j \cap d_k = c_j \cap e_{k-1}$ is contained in $c_j \cap e_k \subseteq c_j \cap d_{k+1}$. Comparing types (ranks), we see that $c_j \cap e_k = c_j \cap d_{k+1}$, whence $e_k = \langle e_{k-1}, c_j \cap e_k \rangle = \langle d_{k-1}, c_j \cap d_{k+1} \rangle$ (for $c_j \cap e_k \not\subseteq e_{k-1}$). Since d_{k-1} and $c_j \cap d_{k+1}$ belong to Δ, so does e_k. It follows that e is contained in Δ. $\square$

Proposition 6.5.6 *Let n be a positive integer and let V be a vector space of dimension $n + 1$ over a division ring. In its action on $\mathrm{PG}(V)$, the group $\mathrm{GL}(V)$ is transitive on the set of all pairs of a chamber and an apartment containing the chamber.*

Proof It is known from Proposition 5.2.20 that the group $G := \mathrm{GL}(V)$ is transitive on the set of bases of V. Since an apartment is uniquely determined by a frame, which in turn is uniquely determined by a basis of V (see Example 6.5.4), this implies that G acts transitively on the set of apartments of $\mathrm{PG}(V)$. Fix a standard basis $\varepsilon_1, \ldots, \varepsilon_{n+1}$ of V and let N be the stabilizer of the frame in $\mathbb{P}(V)$ determined by it.

The chambers of the apartment Δ determined by this frame correspond bijectively to the permutations of Sym_{n+1} in the manner described in Proposition 6.5.2. This means that $w \in \mathrm{Sym}_{n+1}$ corresponds to the chamber

$$c_w := \left\{ \langle \varepsilon_{w(1)} \rangle, \ldots, \langle \varepsilon_{w(1)}, \ldots, \varepsilon_{w(n)} \rangle \right\}$$

of Δ. As a consequence, the linear transformation corresponding to the permutation w on the standard basis represents an element of N sending the chamber c_1 of Δ to c_w. This shows that N acts transitively on the set of chambers incident with Δ. $\square$

6.6 Grassmannian Geometry

In this section we introduce other ways of looking at projective spaces, namely shadow spaces of the geometry of subspaces of a projective space. Recall the notions of shadow and J-space of a geometry from Definition 2.5.1.

Definition 6.6.1 Let Γ be a thick residually connected geometry of type A_n for some $n \in \mathbb{N}$, $n \geq 1$. For $j \in [n]$, the *Grassmannian* of Γ of type j is the shadow space $\mathrm{ShSp}(\Gamma, j)$ on j.

By Corollary 5.4.4 and Theorem 6.3.1, a geometry Γ as in Definition 6.6.1 with $n \geq 3$ is of the form $\mathrm{PG}(\mathbb{D}^{n+1})$ for some division ring $\mathbb{D}$. If $j = 1$ or $j = n$, the Grassmannian of type j is the usual projective space $\mathbb{P}(\mathbb{D}^{n+1})$ or its hyperplane dual $\mathbb{P}((\mathbb{D}^{\mathrm{op}})^{n+1})$. In general, its points are the j-dimensional subspaces of $\mathbb{D}^{n+1}$ and its lines are the non-empty collections of j-dimensional subspaces containing a $(j-1)$-dimensional and contained in a $(j+1)$-dimensional subspace of $\mathbb{D}^{n+1}$.

Proposition 6.6.2 *Let Γ be a thick residually connected geometry of type A_n for some $n \in \mathbb{N}$, $n \geq 3$. For each $j \in \{2, \ldots, n-1\}$, the Grassmannian $Z = \mathrm{ShSp}(\Gamma, j)$ satisfies the following properties.*

(i) *The distance between two points x, y of Z is $j - \tau(x \cap y)$, where τ is the type function of Γ.*
(ii) *The collinearity graph of Z is connected of diameter $\min(j, n - j + 1)$.*
(iii) *Suppose that x is a point and l is a line of Z such that no point of l is collinear with x. If there is a point collinear with x and with each point of l, then the set of all such points is a line of Z.*

Proof By Corollary 6.3.2, we may assume $\Gamma = \mathrm{PG}(V)$ for some vector space V of dimension $n + 1 \geq 4$. For each $i \in [n]$, elements of Γ of type i are subspaces of V of (affine) dimension i. When viewed as points of Z, two j-dimensional subspaces x and y of V are collinear, if and only if $\dim(x \cap y) = j - 1$. As Γ is residually connected, the collinearity graph of Z is connected (cf. Corollary 1.6.6).

(i) Put $k := d(x, y)$. For $k = 0$, this is obvious. Suppose $k > 0$. If x and y have distance k, there is a j-space z collinear with y such that the distance between x and

z is $k - 1$. By induction on k, the dimension of $x \cap z$ equals $j - k + 1$ and by the observation above, $\dim(y \cap z) = j - 1$. Hence $\dim(x \cap y) \geq \dim(x \cap z \cap y) \geq j - k$. But $\dim(x \cap y) > j - k$ would imply, by induction on k, that x and y are at mutual distance less than k, a contradiction. Hence $\dim(x \cap y) = j - k$. Conversely, if $\dim(x \cap y) = j - k$, pick a point $p \in x \setminus y$ and let z be the subspace of V spanned by p and a hyperplane of y containing $x \cap y$. The subspace z of V is j-dimensional whereas $\dim(x \cap z) \geq j - k + 1$ and $\dim(y \cap z) \geq j - 1$, so the distance between x and z is at most $k - 1$ and the distance between z and y is at most 1, which shows that the distance between x and y is at most k. Again, using the induction hypothesis we can show by way of contradiction that the distance is not smaller than k.

(ii) Let x be a point of Z, that is, a j-dimensional subspace of V. The number $\min(j, n + 1 - j)$ is the maximal possible codimension of the intersection of x with another j-dimensional subspace of V. By (i) it follows that it is the diameter of Z.

(iii) Let v_{j-1} and v_{j+1} be subspaces of V, incident in $\mathbb{P}(V)$, and of dimensions $j - 1$ and $j + 1$, respectively, such that l consists of all j-dimensional subspaces incident with the flag $\{v_{j-1}, v_{j+1}\}$. Suppose that u is a point of Z collinear with x and with each point of l. It is incident with either v_{j-1} or v_{j+1}. Going over to the dual geometry if necessary, we may (and will) assume that u contains v_{j-1}. As x and u are collinear and distinct, $x \cap u$ is a $(j - 1)$-dimensional subspace of V. Since $x \cap u$ and v_{j-1} are both hyperplanes of u, they meet in a $(j - 2)$-dimensional subspace $x \cap u \cap v_{j-1}$. But $x \cap v_{j-1}$ is also $(j - 2)$-dimensional as x is not collinear with a point of l, so $x \cap u \cap v_{j-1} = x \cap v_{j-1}$. Write w for the subspace of V spanned by x and v_{j-1}. As they meet in a $(j - 2)$-dimensional space, the dimension of w is $j + 1$. Denote by m the line of Z determined by the flag $\{v_{j-1}, w\}$.

We claim that each point collinear with x and with each point of l belongs to a line m consisting of all j-dimensional subspaces of V incident with $\{v_{j-1}, w\}$. Let p be such a point. If p is contained in v_{j+1}, then $x \cap p$, which is a $(j - 1)$-dimensional subspace of V, coincides with $x \cap v_{j+1}$ (by a reasoning analogous to the one that showed $x \cap u \cap v_{j-1} = x \cap v_{j-1}$ above, but for the dual geometry). But then $x \cap p$ and v_{j-1} span a j-dimensional subspace of v_{j+1} (for the dimension is obviously j or $j + 1$, while in the latter case, $w = v_{j+1}$, contradicting $u \in l$), which is a point on l collinear with x, a contradiction. Now, p does not lie in v_{j+1} and, being collinear with each point of l, must contain v_{j-1}. A fortiori, p contains $x \cap v_{j-1}$. Moreover, p is spanned by $x \cap p$ and v_{j-1} and hence contained in the space spanned by x and v_{j-1}, which is w. Thus, being incident with both v_{j-1} and w, the point p lies on m. $\qquad\square$

6.7 Root Filtration Spaces

In this section we study one more type of shadow space of a geometry of type A_n. These are shadow spaces on $\{1, n\}$, so their point sets are the incident pairs of a point and a hyperplane of a projective space. Although it may seem a surprise at first sight, the resulting line space fits best with other geometries related to groups of Lie type. The reason is that they all share a single axiomatic description: the root filtration space, introduced in Definition 6.7.2.

Notation 6.7.1 Let x be an element of a set E. For a relation X on E, we denote by $X(x)$ the set of all elements $y \in E$ with $(x, y) \in X$. If, in addition, $y, z \in E$ and $Y \subseteq E$, we write $X(x, y)$ for $X(x) \cap X(y)$, $X(x, y, z)$ for $X(x) \cap X(y) \cap X(z)$, and $X(Y)$ for $\bigcap_{y \in Y} X(y)$, etc.

We will be using a system $(E_i)_{i \in I}$ of relations on E. Using an ordering $<$ of the indices on the relations, we write $E_{\leq i}$ for $\bigcup_{j \leq i} E_j$.

Definition 6.7.2 Let (E, F) be a partial linear space. We call (E, F) a *root filtration space* if there is a quintuple $(E_i)_{-2 \leq i \leq 2}$ of disjoint symmetric relations partitioning $E \times E$ that satisfy the following six conditions.

(1) The relation E_{-2} is equality on E.
(2) The relation E_{-1} is collinearity (the adjacency relation in the collinearity graph; see Definition 2.5.12) of distinct points of E.
(3) There is a map $E_1 \to E$, denoted by $(u, v) \mapsto [u, v]$ such that, if $(u, v) \in E_1$, then $E_i(u) \cap E_j(v) \subseteq E_{\leq i+j}([u, v])$.
(4) For each $(x, y) \in E_2$, we have $E_{\leq 0}(x) \cap E_{\leq -1}(y) = \emptyset$.
(5) For each $x \in E$, the subsets $E_{\leq -1}(x)$ and $E_{\leq 0}(x)$ are subspaces of (E, F).
(6) For each $x \in E$, the subset $E_{\leq 1}(x)$ is either all of E or a geometric hyperplane of (E, F).

Given such a root filtration space (E, F), we call a pair $(x, z) \in E_i$ *hyperbolic* if $i = 2$, *special* if $i = 1$, and *polar* if $i = 0$. We also say that x and z *commute* if $i \leq 0$. In line with Definition 2.2.1, we often write $x \sim y$ instead of $(x, y) \in E_{-1}$.

If (E, F) is a root filtration space satisfying the following two conditions with respect to the quintuple $(E_i)_{-2 \leq i \leq 2}$, it is called *nondegenerate*.

(7) For each $x \in E$, the set $E_2(x)$ is not empty.
(8) The graph (E, E_{-1}) is connected.

Axiom (3) applied to $x \in E_i(u) \cap E_j(v)$, will be referred to as the *filtration around* x, Condition (4) applied to $(x, y) \in E_2$ and $(x, z) \in E_{\leq 0}$ with the conclusion $(y, z) \notin E_{\leq -1}$, as the *triangle condition* on x, y, z.

Remark 6.7.3 In general, the quintuple of relations $(E_i)_{-2 \leq i \leq 2}$ need not be uniquely determined by (E, F), but Remark 6.7.17 will show that this is the case for nondegenerate root filtration spaces.

Condition (3) gives a kind of filtration around each point. According to Lemma 6.7.8(ii) below, $[u, v]$ is the unique point in $E_{\leq -1}(u, v)$, so the map $[\cdot, \cdot]$ is uniquely determined by the relations $(E_i)_{-2 \leq i \leq 2}$.

Condition (4) can be replaced by the statement that, for each $(x, y) \in E_2$, we have $E_{\leq i}(x) \cap E_{\leq j}(y) = \emptyset$ whenever $i + j < 0$; see Exercise 6.8.17.

Condition (5) can be replaced by the statement that $E_{\leq -i}(x)$ is a subspace of (E, F) for each i; see Lemma 6.7.8(i) below. A space in which, for every line l and every point p, the set of points of l collinear with p has size 0, 1 or coincides with l, is called a *gamma space*. Condition (5) implies that (E, F) is a gamma space.

Example 6.7.4 Consider the collinearity graph (E, F) on the vertices of the 24-cell, the 4-dimensional polytope of Example 4.1.13(iv). It is a space with lines of size two. We claim it is a root filtration space and point out the relations E_i. Fix a vertex x and consult Fig. 4.6 and the distribution diagram in Fig. 4.7. There are 8 neighbors forming a cube; these form the set $E_{-1}(x)$. There are also 8 vertices at distance two to x sharing exactly one neighbor with x; these constitute $E_1(x)$. For $y \in E_1(x)$, the vertex $[x, y]$ is the unique common neighbor of x and y. There are 6 more vertices at distance two from x; the common neighbors of x and such a vertex form a quadrangle. These 6 vertices form $E_0(x)$. Finally, $E_2(x)$ is the singleton consisting of the unique vertex opposite (at distance three from) x, and, of course $E_{-2}(x) = \{x\}$. This determines the system $(E_i)_{-2 \le i \le 2}$ completely.

Definition 6.7.5 Two projective spaces $\mathbb{P}$ and $\mathbb{H}$ are said to be *in duality* if $\mathbb{H}$ is isomorphic to a subspace of the hyperplane dual $\mathbb{P}^\star$ of $\mathbb{P}$ in such a way that the intersection in $\mathbb{P}$ of all hyperplanes of $\mathbb{P}$ corresponding to members of $\mathbb{H}$ is empty.

Let $\mathbb{P}$ and $\mathbb{H}$ be projective spaces in duality. Take E to be the subset of $\mathbb{P} \times \mathbb{H}$ consisting of incident pairs. Let F be the collection of all subsets of E of the form $\{x\} \times L$ for x a point and L a line of $\mathbb{H}$ or of the form $l \times \{H\}$ for l a line of $\mathbb{P}$ and H a point of $\mathbb{H}$. The pair (E, F) is a line space, denoted $E(\mathbb{P}, \mathbb{H})$.

Remark 6.7.6 The definition suggests symmetry in the roles of $\mathbb{P}$ and $\mathbb{H}$. According to Exercise 6.8.16, this is indeed the case.

Suppose that, in the above definition, $\mathbb{P}$ has finite dimension n. Then $\mathbb{H}$ must be the hyperplane dual of $\mathbb{P}$, so, $E(\mathbb{P}, \mathbb{H}) \cong E(\mathbb{P}, \mathbb{P}^\star)$. Moreover, $E(\mathbb{P}, \mathbb{P}^\star)$ is the shadow space of type $\{1, n\}$ of the geometry of type A_n associated with $\mathbb{P}$ as in Corollary 5.4.5. This is the setting of Example 2.5.3.

Theorem 6.7.7 *If $\mathbb{P}$ and $\mathbb{H}$ are projective spaces of dimension at least two in duality, then $E(\mathbb{P}, \mathbb{H})$ is a nondegenerate root filtration space.*

Proof Write $(E, F) := E(\mathbb{P}, \mathbb{H})$. The space is obviously partial linear. We partition $E \times E$ into five symmetric relations E_i on E, indexed by $i \in \{-2, -1, 0, 1, 2\}$ as in Table 6.2, where (p, H) and $(q, K) \in E$, the point set of $E(\mathbb{P}, \mathbb{H})$.

We verify the conditions of a root filtration space. Conditions (1) and (2) are immediate from the definition.

As for (3), suppose $((p, H), (q, K)) \in E_1$. We define the map $[\cdot, \cdot]$ by decreeing $[(p, H), (q, K)] = (q, H)$ if $q \in H$, and $[(p, H), (q, K)] = (p, K)$ if $p \in K$.

For (4), suppose $((p, H), (q, K)) \in E_2$. Then $p \notin K$ and $q \notin H$. If $(z, L) \in E_{\le 0}((p, H))$, then $z \in H$ and $p \in L$. So, if in addition $(z, L) \in E_{\le -1}((q, K))$, then $z = q$ or $K = L$. The former is impossible as $z \in H$ and $q \notin H$, and the latter is impossible as $p \notin K$ and $p \in L$. Hence $E_{\le 0}((p, H)) \cap E_{\le -1}((q, K)) = \emptyset$, proving (4).

Table 6.2 The relations E_i for $E(\mathbb{P}, \mathbb{H})$

i	Definition of $((p, H), (q, K)) \in E_i$
-2	$p = q$ and $H = K$
-1	$p = q$ or $H = K$ but not both
0	$p \in K$ and $q \in H$ but $p \neq q$ and $H \neq K$
1	$p \in K$ or $q \in H$ but not both
2	$p \notin K$ and $q \notin H$

The proof of (5) is left to the reader. We continue by verifying (6) with $x = (p, H)$. Up to an interchange of the roles of points and hyperplanes for the pair $(\mathbb{P}, \mathbb{H})$ in duality, we may assume that the line is of the form $l \times \{L\}$ for some line l of $\mathbb{P}$ and a fixed $L \in \mathbb{H}$. As $H \cap L$ is a hyperplane of L and l is contained in L, there is a point $r \in l \cap H$. Now (r, L) is a point of the line $l \times \{L\}$ belonging to $E_{\leq 1}((p, H))$, as required.

Let $(p, H) \in E(\mathbb{P}, \mathbb{H})$. As $\mathbb{P}$ and $\mathbb{H}$ are in duality, there is a hyperplane $K \in \mathbb{H}$ that does not contain p. Consequently, it is distinct from H and so there is $q \in K \setminus H$. Now $(q, K) \in E_2((p, H))$, proving (7).

Again, let $(p, H) \in E(\mathbb{P}, \mathbb{H})$ and suppose that (q, K) is another point of this space. If $(q, K) \in E_2((p, H))$, then there is a point $r \in H \cap K$ due to the assumption on the dimension. Now $(r, K) \in E_1((p, H)) \cap E_{-1}((q, K))$. If $(q, K) \in E_1((p, H))$, then without loss of generality, we may assume $p \in K$, and then $(p, K) \in E_{-1}((p, H)) \cap E_{-1}((q, K))$. If $(q, K) \in E_0((p, H))$, then $p \in K$ and $q \in H$, so (q, H) is in $E_{-1}((p, H)) \cap E_{-1}((q, K))$. This proves that the graph (E, E_{-1}) is connected. Hence (8). $\qquad \square$

Recall from Definition 2.5.13 that a singular subspace of a space is a subspace in which any two points are collinear. So, in a partial linear space, singular subspaces are linear.

Lemma 6.7.8 *In a root filtration space (E, F) the following properties hold.*

(i) *For each $i \in \{-2, \ldots, 2\}$ and each $x \in E$, the subset $E_{\leq i}(x)$ is a subspace of (E, F).*

(ii) *If $(u, v) \in E_1$, then $[u, v]$ is the unique common neighbor of both u and v in the collinearity graph (E, E_{-1}) of (E, F).*

(iii) *If $(u, v) \in E_1$, then $E_0(u) \cap E_2(v) \subseteq E_1([u, v])$.*

(iv) *If $(x, y) \in E_0$ and $z \in E_{-1}(y)$, then either $z \in E_{\leq 0}(x)$, or $z \in E_1(x)$ and $E_{-1}(x, y, z) = \{[x, z]\}$.*

(v) *If (x, q) and (u, z) belong to E_1 whereas $u = [x, q]$ and $q = [u, z]$, then $(x, z) \in E_2$.*

(vi) *If P is a pentagon in the collinearity graph (E, E_{-1}) (that is, an induced subgraph isomorphic to a pentagon), then each pair of distinct non-collinear points of P is polar.*

(vii) *If $(u, v) \in E_1$, then $E_{-1}(u) \cap E_0([u, v]) \subseteq E_1(v)$.*

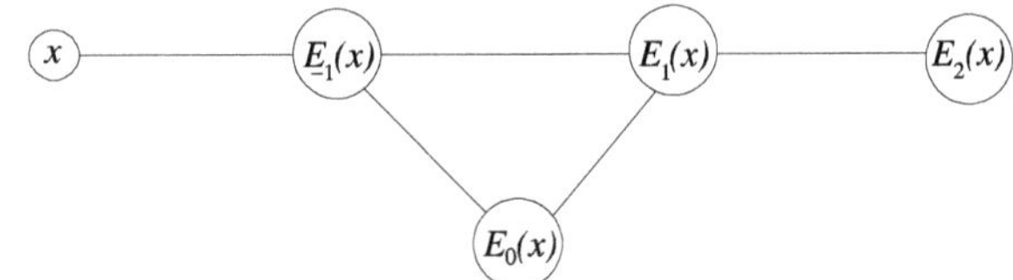

Fig. 6.5 The adjacencies of the distribution diagram of a root filtration space

Proof (i) This is stated in Definition 6.7.2(5) for $i = -1, 0$. It is trivial for $i = -2$ and $i = 2$ since singletons and the whole set E are subspaces. As a geometric hyperplane is a subspace, it follows from (6) for $i = 1$.

(ii) By filtration around u, we have $[u, v] \in E_{\leq -1}(u)$, and similarly $[u, v] \in E_{\leq -1}(v)$. By the disjointness assumption on E_1 and E_{-1}, the points u and v are not collinear, so $[u, v]$ does not coincide with u or v. Suppose now that y is a point collinear with both u and v. Then, by (3) we have $[u, v] \in E_{\leq -2}(y)$, which implies $y = [u, v]$ by (1).

(iii) By (ii), $[u, v] \in E_{-1}(u, v)$. Let $x \in E_0(u) \cap E_2(v)$. Applying (4) to x and $[u, v]$ we find $[u, v] \notin E_2(x)$ and applying (4) to x and v, we find $[u, v] \notin E_{\leq 0}(x)$. Therefore, $x \in E_1([u, v])$.

(iv) By (4), $z \in E_{\leq 1}(x)$. Suppose $z \in E_1(x)$, so $[x, z]$ is a point collinear with x and z. Then, by (3) applied to y, we find $[x, z] \in E_{-1}(y)$, so by (ii) the point $[x, z]$ is in $E_{-1}(x, y, z)$, and the only point of $E_{\leq -1}(z, x)$ is $[z, x]$.

(v) Observe that x, u, q, z is a path in the collinearity graph of (E, F). If $(x, z) \in E_{\leq 1}$, then by the filtration around x, we find $q = [u, z] \in E_{\leq 0}(x)$, contradicting the assumption $(q, x) \in E_1$. Hence, $(x, z) \in E_2$.

(vi) Let a, b, c, d, e be the points of P, chosen so that $\{e, a\}$ and successive pairs of points are collinear. By (v) and the fact that a pair of hyperbolic points has mutual distance at least three (possibly ∞) in the collinearity graph (a consequence of (4)), no two consecutive non-collinear pairs of the pentagon can be special. So at most two non-collinear pairs can be special. Therefore, after a suitable renaming of the points of P, the pairs (a, c) and (a, d) may be assumed not special, whence polar. Now $b, e \in E_{-1}(a)$ and $c, d \in E_0(a)$, so P is contained in $E_{\leq 0}(a)$. If $(b, d) \in E_1$, then, by (ii) and (iv), the point $c = [b, d]$ is collinear with a, contradicting that P is a pentagon. Therefore $(b, d) \in E_0$. Similarly, it can be shown that $(e, c) \in E_0$ and, finally, that $(b, e) \in E_0$.

(vii) Let $y \in E_{-1}(u) \cap E_0([u, v])$. In view of the triangle condition on $[u, v]$, y, v, we must have $y \in E_{\leq 1}(v)$. If y and v commute, then $(y, [u, v]) \in E_{\leq -1}$ by the filtration around y, a contradiction. Therefore, $y \in E_1(v)$. $\square$

As a consequence of Lemma 6.7.8, the point distribution diagram of a root filtration space has the form described in Fig. 6.5. It is no coincidence that the picture resembles Fig. 4.7 for the geometry of type F_4 of Example 4.1.13. The relation is explained in Example 6.7.4.

Example 6.7.9 Here are some degenerate examples of root filtration spaces. Every linear space is a trivial example of a root filtration space with $E_i = \emptyset$ for $i \geq 0$.

Every space without lines is a trivial example of a root filtration space with $E_i = \emptyset$ for $-2 < i < 2$ and E_2 the relation of being distinct. Even if we keep $E_1 = E_{-1} = \emptyset$ and allow for the relation of being distinct to be partitioned in E_0 and E_2, the result is a root filtration space. For example, if (E, L) is a generalized quadrangle, taking E_2 to be the non-collinearity relation, E_{-2} the identity relation, and E_0 the complement in $E \times E$ of $E_{-2} \cup E_2$, we obtain a root filtration space $(E, \emptyset)$ with $E_{-1} = E_1 = \emptyset$. Nondegeneracy of the generalized quadrangle (E, L) is equivalent to Condition (7) for $(E, \emptyset)$. This shows that nondegeneracy conditions like (7) and (8) are needed for root filtration spaces.

Example 6.7.10 Let (E, F) be a generalized hexagon and let E_i for $i = 1, 2$ be the set of pairs of points at mutual distance $i + 1$. Put $E_0 = \emptyset$. For $(x, y) \in E_1$ denote by $[x, y]$ the unique point collinear with both x and y. The space (E, F) is a nondegenerate root filtration space with respect to $(E_i)_{-2 \le i \le 2}$. Conversely, if (E, F) is a nondegenerate root filtration space with $E_0 = \emptyset$, then it is a generalized hexagon.

No generalized octagon is a nondegenerate root filtration space; cf. Exercise 6.8.18.

Example 6.7.11 Suppose that $(E^{(1)}, F^{(1)})$ and $(E^{(2)}, F^{(2)})$ are root filtration spaces. Let E be the disjoint union of $E^{(1)}$ and $E^{(2)}$ and let F be the disjoint union of $F^{(1)}$ and $F^{(2)}$. Then (E, F) is a root filtration space with filtration $E_0 = E_0^{(1)} \cup E_0^{(2)} \cup (E^{(1)} \times E^{(2)}) \cup (E^{(2)} \times E^{(1)})$ and $E_i = E_i^{(1)} \cup E_i^{(2)}$ for $i \ne 0$.

Lemma 6.7.12 *Suppose that (E, F) is a nondegenerate root filtration space. Let $(x, z) \in E_2$ and $(u, v) \in E_{-1} \cup E_0$ be such that $x \in E_{-1}(u, v)$ and $z \in E_1(u, v)$. For $u' = [u, z]$ and $v' = [v, z]$, we have $(u, v') \in E_0 \cup E_1$, $u' \ne v'$, and u' commutes with v'. Moreover, one of the following two cases prevails.*

(i) $(u, v), (u', v') \in E_{-1}$ *and* $(u, v'), (u', v) \in E_0$;
(ii) $(u, v), (u', v') \in E_0$ *and* $(u', v), (u, v') \in E_1$, *while* $[u, v'] = [u', v]$.

Proof By the triangle condition on u, v, v' and on u, v, u', we find $(u, v'), (v, u') \notin E_2$. As $(x, v') \in E_1$, the point $v = [x, v']$ is the unique element in $E_{-1}(x, v')$ by Lemma 6.7.8(ii). This proves $(u, v'), (v, u') \notin E_{-1}$. In particular, $(u, v') \in E_0 \cup E_1$, and similarly for (v, u').

As $[x, u'] = u \ne v = [x, v']$, we have $u' \ne v'$. The only alternative to $(u', v') \in E_{\le 0}$ would be $(u', v') \in E_1$ with $[u', v'] = z$, in which case the filtration around u would imply $(u, z) \in E_{\le 0}$ (notice that $u \in E_{-1}(u') \cap E_{\le 1}(v')$), a contradiction. Therefore, $(u', v') \in E_{\le 0}$. Similarly, $(u, v) \in E_{\le 0}$.

(i) Suppose now $u \sim v$. By Lemma 6.7.8(vi), the fact that z, v', v, u, u' cannot be a pentagon forces $u' \sim v'$. Since u and v' are distinct points collinear with both u' and v, we find $(u', v) \in E_0$, and similarly $(u, v') \in E_0$.

(ii) Suppose next $(u, v) \in E_0$. If $(u', v) \in E_0$, then Lemma 6.7.8(iv) gives $E_{-1}(u', v, x) = \{u\}$, so $u \sim v$, a contradiction. Hence $(u', v) \in E_1$. Similarly,

$(u, v') \in E_1$. If $u' \sim v'$, then as in (i), also $u \sim v$, a contradiction. Therefore, $(u', v') \in E_0$. Write $y = [u, v']$. By Lemma 6.7.8(iv), $E_{-1}(u, v, v') = \{y\}$, so $y \sim v$. Similarly, $y \sim u'$, so $y \in E_{-1}(u', v)$. It follows that $y = [u', v]$. □

Lemma 6.7.13 *The following three conditions regarding a root filtration space are equivalent.*

 (i) *For each $(x, u) \in E_{-1}$, there exists a point in $E_2(x) \cap E_1(u)$.*
 (ii) *For each $(x, u) \in E_{-1}$, there exists a point in $E_{-1}(u) \cap E_1(x)$.*
 (iii) *For each point x with $E_{-1}(x) \neq \emptyset$, there exists a point in $E_2(x)$.*

If the root filtration space is nondegenerate, then these conditions hold.

Proof Once the equivalences are established, the final statement follows as (iii) is immediate from (8) of Definition 6.7.2.

   (i) $\Rightarrow$ (ii) Let $(x, u) \in E_{-1}$. Now take $z \in E_2(x) \cap E_1(u)$ as in (i). By Lemma 6.7.8(ii), $v = [u, z]$ is collinear with u and by filtration around x it belongs to $E_{\leq 1}(x)$. But $v \in E_{\leq 0}(x)$ would contradict (4), the triangle condition on x, v, and z. Therefore, $v \in E_1(x)$.

   (ii) $\Rightarrow$ (iii) Assume $(x, u) \in E_{-1}$. By (ii), there are $q \in E_{-1}(u) \cap E_1(x)$ and $z \in E_{-1}(q) \cap E_1(u)$. By Lemma 6.7.8(v), $z \in E_2(x)$.

   (iii) $\Rightarrow$ (i) Let $(x, u) \in E_{-1}$. Take $a \in E_2(x)$. As $E_{\leq 1}(a)$ is a hyperplane there is a unique point y of $E_{\leq 1}(a)$ on the line xu. We are done if $y = u$. Therefore we assume that $u \in E_2(a)$. As in the proof of implication (i)$\Rightarrow$(ii) we see that $y \in E_1(a)$ and for $v' = [a, y]$ we have $v' \in E_1(u) \cap E_{-1}(y)$. Let $w \in E_2(y)$. There is a unique point v of $E_{\leq 1}(w)$ on the line $v'y$. Observe that $(v, u) \in E_1$. Again, $(v, w) \in E_1$ and for $z' = [v, w]$ we have $z' \in E_1(y)$. Together with the previous observation, by Lemma 6.7.8(v), this implies $(z', u) \in E_2$. We also have $z' \in E_{-1}(w)$ and hence there is a unique point $z \neq z'$ on the line $z'w$ which is in $E_{\leq 1}(u)$. As $w \in E_2(y)$ the point z' is the unique point of $z'w \cap E_{\leq 1}(y)$, so $z \in E_2(y)$. This implies that u is the unique point of $E_{\leq 1}(z)$ on the line $xy = xu$. In particular, as $x \neq u$, we have $x \in E_2(z)$. Thus, $z \in E_2(x) \cap E_1(u)$, as required. □

   The third condition means that $E_{\leq 1}(x)$ is a geometric hyperplane if x lies on a line.

Lemma 6.7.14 *Let (E, F) be a root filtration space with $(u, v) \in E_1$ and $y \in E_0([u, v], u, v)$. If $E_{-1}(v, y) \neq \emptyset$, then $E_{-1}([u, v], y) \neq \emptyset$.*

Proof Put $x = [u, v]$. Suppose $E_{-1}(x, y) = \emptyset$ and $w \in E_{-1}(v, y)$. Then $w \notin E_{\leq -1}(x)$. By the triangle condition for u, y, w, we must have $u \in E_{\leq 1}(w)$. Assume that $u \in E_{\leq 0}(w)$. Then the filtration around w gives $x = [u, v] \in E_{\leq -1}(w)$, contradicting $w \notin E_{\leq -1}(x)$. Consequently, $(u, w) \in E_1$, so, by Lemma 6.7.8(vi), the 5-circuit $x, u, [u, w], w, v$ cannot be a pentagon. This forces $[u, w] \in E_{\leq -1}(x)$ (observe that $[u, w] \notin E_{\leq -1}(v)$, for otherwise $[u, w] = [u, v] = x$ would be collinear with w) and, in view of the filtration around y, also $[u, w] \in E_{\leq -1}(y)$, a contradiction. □

Under mild conditions, an arbitrary root filtration space can be deconstructed in the vein of Example 6.7.11.

Lemma 6.7.15 *If (E, F) is a root filtration space satisfying (7), then (E, F) is the disjoint union of connected subspaces B_i such that $B_i \times B_j \subseteq E_0$ whenever $i \neq j$ unless $B_i \times B_j \subseteq E_2$, in which case B_i and B_j are singletons. Moreover, if $x, y \in B_i$ for some i and $(x, y) \in E_0$ then $E_{-1}(x, y) \neq \emptyset$.*

Proof We first show that $E_1(x) \cup E_2(x)$ is contained in the connected component of x in (E, E_{-1}), except possibly when there is no line on x. If $z \in E_1(x)$, then there is the path $x, [x, z], z$. If $z \in E_2(x)$, and l is a line on x then there is a path $x, v, [v, z], z$ where $\{v\} = l \cap E_{\leq 1}(z)$ by (6). Therefore, either z is connected to x by a path in (E, E_{-1}) or $\{x\}$ is a connected component in (E, E_{-1}) (in which case the same argument can be applied with the roles of z and x interchanged).

Suppose $(x, y) \in E_0$ and $E_{-1}(x, y) = \emptyset$. We show that x and y lie in different components of (E, E_{-1}). Let $v \in E_{-1}(x)$. Then $v \in E_{\leq 0}(y)$ by the triangle condition and Lemma 6.7.8(iv). By Lemma 6.7.13, there exists $u \in E_1(v)$ such that $x = [u, v]$. By the argument for v applied to u, we also have $(u, y) \in E_0$. Lemma 6.7.14 gives $E_{\leq -1}(y, v) = \emptyset$. So the pair (v, y) inherits the property of having no common collinear points from the pair (x, y). Since v was chosen to be an arbitrary point collinear with x, we find that $(v, y) \in E_0$ and $E_{-1}(v, y) = \emptyset$ for all points of the connected component of x in (E, E_{-1}). Varying y in the same way, we find the required assertion. $\qquad\square$

Lemma 6.7.16 *Suppose that (E, F) is a nondegenerate root filtration space. Assume $(x, y) \in E_0$ and $u \in E_{\leq -1}(x, y)$. There exists $v \in E_{\leq -1}(x, y)$ such that v is not collinear with u. In particular, every polar pair (x, y) is contained in a quadrangle.*

Proof By Lemma 6.7.13 and nondegeneracy of (E, F), there exists $y' \in E_{-1}(y) \cap E_1(u)$. Then $y = [u, y']$ and Lemma 6.7.8(vii) gives $y' \in E_1(x)$. Set $v = [x, y']$. Then $v \in E_{-1}(x)$. Furthermore, because of the filtration around y, we also have $v = [x, y'] \in E_{\leq -1}(y)$. Thus $v \in E_{\leq -1}(x, y)$. Also, $y' \in E_1(u)$ and $y' \in E_{-1}(v)$ exclude the possibility of $v \in E_{\leq -1}(u)$. This proves the first assertion. The second one follows from Lemma 6.7.15. $\qquad\square$

Remark 6.7.17 The relations $(E_i)_{-2 \leq i \leq 2}$ and the map $[\cdot, \cdot] : E_1 \to E$ of a nondegenerate root filtration space (E, F) are fully determined by the space (E, F) itself. For, thanks to Lemma 6.7.16, E_{-1}, $E_0 \cup E_1$, and E_2 are the relations of having distance 1, 2, and 3 in the collinearity graph of (E, F), and, for x, y in E at mutual distance two, we have $x \in E_1(y)$ if and only if x and y have a unique common neighbor (which coincides with $[x, y]$). Therefore, we will often introduce a nondegenerate root filtration space without explicit mention of the system $(E_i)_{-2 \leq i \leq 2}$.

Lemma 6.7.18 *Let (E, F) be a nondegenerate root filtration space. For $(x, y) \in E_0$ the following two statements hold.*

(i) $E_{-1}(x) \cap E_1(y) \neq \emptyset$.
(ii) $E_2(x) \cap E_0(y) \neq \emptyset$.

Proof (i) By Lemma 6.7.15 and (8), there is $u \in E_{-1}(x, y)$. By Lemma 6.7.13, there exists $v \in E_{-1}(x) \cap E_1(u)$. By Lemma 6.7.8(vii), $y \in E_1(v)$.

(ii) By (i), there is $u \in E_{-1}(y) \cap E_1(x)$. Set $v = [x, u]$. Parts (ii) and (iv) of Lemma 6.7.8 give $v \in E_{-1}(x, y, u)$. By (7) and Lemma 6.7.13, there exists $w \in E_2(v) \cap E_1(y)$. By (6) there is a point u' on the line vu such that $(u', w) \in E_1$. Set $z = [w, u']$. We are going to show that $z \in E_2(x) \cap E_0(y)$. By filtration around y, we must have $z \in E_{\leq 0}(y)$. The triangle condition for v, z, w shows that $v \notin E_{\leq 0}(z)$. Since $u' \in E_{\leq -1}(v, z)$, we find $(v, z) \in E_1$ and hence $u' = [v, z]$. By Lemma 6.7.8(v) applied to the pairs (x, u') and (v, z) of E_1, we obtain $z \in E_2(x)$. From $y \in E_{\leq 0}(z)$ and the triangle condition for x, y, z, we conclude $y \in E_0(z)$. $\square$

We give an important consequence of Lemma 6.7.12 for the nondegenerate case.

Lemma 6.7.19 *Let (E, F) be a nondegenerate root filtration space. For every pair $(v, w) \in E_{-1}$, the set $E_{\leq -1}(v) \cap E_{\leq 0}(w)$ is a geometric hyperplane of $E_{\leq -1}(v)$.*

Proof By Lemma 6.7.13(ii), the intersection $E_{-1}(v) \cap E_1(w)$ is nonempty and hence $E_{\leq -1}(v) \cap E_{\leq 0}(w)$ is a proper subspace of $E_{\leq -1}(v)$. To see that it is a geometric hyperplane let xu be a line in $E_{\leq -1}(v)$. If x does not lie in $E_{\leq 0}(w)$, then $x \in E_1(w)$ and $v = [x, w]$. Use Lemma 6.7.13(ii) and pick $z \in E_{-1}(w) \cap E_1(v)$. By Lemma 6.7.8(v), $(x, z) \in E_2$ and, by (6), there is a point s on the line xu such that $(z, s) \in E_1$. Now $w = [z, v]$ and Lemma 6.7.12 gives $(s, w) \in E_0$. We conclude $s \in xu \cap E_{\leq -1}(v) \cap E_{\leq 0}(w)$. $\square$

Example 6.7.20 Let $(\mathbb{P}, \mathbb{H})$ be a pair of thick projective spaces in duality. The root filtration space $E(\mathbb{P}, \mathbb{H})$ of Definition 6.7.5 satisfies two more remarkable properties.

(1) If $l \times \{H\}$ is a line of $E(\mathbb{P}, \mathbb{H})$, then $\mathbb{P} \times \{H\}$ is the unique maximal singular subspace containing this line.
(2) If (p, H) is a point of $E(\mathbb{P}, \mathbb{H})$, then $\mathbb{P} \times \{H\}$ and $\{p\} \times \mathbb{H}$ are the only maximal singular subspaces containing this point.

In order to characterize the root filtration spaces coming from a pair of projective spaces in duality, we will work with these properties.

Notation 6.7.21 For a set M of points and a point x of a root filtration space, we write $C_M(x) = M \cap E_{\leq 0}(x)$. Moreover, if X is also a set of points, we write $C_M(X) = \bigcap_{x \in X} C_M(x)$.

Lemma 6.7.22 *Let (E, F) be a nondegenerate root filtration space in which every line is contained in a unique maximal singular subspace. For each maximal singular subspace M of (E, F) and each $v \in E \setminus M$, exactly one of the following holds.*

(i) $M \cap E_1(v)$ *is a geometric hyperplane of M and $M \subseteq E_1(v) \cup E_2(v)$.*

(ii) $C_M(v)$ *is a geometric hyperplane of M, there is a unique point u of M such that $E_{-1}(v) \cap M = \{u\}$ and $M \subseteq E_{\leq 1}(v)$.*

Proof Obviously, case (i) holds if and only if $E_2(v) \cap M \neq \emptyset$. Assume, therefore, $M \subseteq E_{\leq 1}(v)$.

We claim that if ab is a line inside M, then $C_{ab}(v) \neq \emptyset$. To see this, assume the contrary, so $ab \subseteq E_1(v)$, and set $a' = [a, v]$ and $b' = [b, v]$. Then $(a', b) \in E_{\leq 0}$ by the filtration around b and either $b' \sim a'$ or $b' = a'$ by the filtration around a'. If $b' = a'$ then by Lemma 6.7.19 there is a point in $C_{ab}(v)$, contrary to assumption. Thus $a' \sim b'$. We also have $(a', b) \in E_0$ since $a' \sim b$ would lead to $a' = b'$ due to Lemma 6.7.8(ii). By Lemma 6.7.18(ii), there exists $z \in E_2(a') \cap E_0(b)$. Then $(z, a) \in E_1$. Set $x = [z, a]$. By the filtration around b, we have $x \sim b$. As also $x \sim a$, the point x belongs to M, the unique maximal singular subspace containing ab. The point x cannot commute with a' (by the triangle condition for z, x, a') and $(x, a') \in E_2$ is also excluded (by the triangle condition for a, x, a'). Thus $(a', x) \in E_1$ and $a = [a', x]$. Finally, the pair (x, v) is hyperbolic by Lemma 6.7.8(v) applied to the path x, a, a', v, a contradiction to $M \subseteq E_{\leq 1}(v)$. This settles the claim.

By the claim and Definition 6.7.2(5), $C_M(v)$ is a geometric hyperplane of M or all of M. Assume that there exists a point $u \in M$ such that $u \sim v$. (By uniqueness of M on any line, there is at most one such point.) By Lemma 6.7.13(ii), there is $v' \in E_{-1}(v) \cap E_1(u)$ (and hence $v = [u, v']$). If there is $w \in C_M(v')$, then the filtration around w gives $w \sim [u, v'] = v$, whence $w = u$ and $w \notin E_{\leq 0}(v')$, a contradiction. Thus v' does not commute with any point of M and hence, by the claim, there exists $x \in M \cap E_2(v')$. By the triangle condition for x, v, v', the points x and v cannot commute. Consequently, $C_M(v)$ is indeed a geometric hyperplane of M.

As $C_M(v)$ contains a hyperplane of M and M contains a line, there exists $y \in C_M(v)$. In view of the above it is sufficient to show that M also contains a point collinear with v. Pick $w \in E_{\leq -1}(y, v)$. Then, by the previous discussion, there exists a point $x \in M$ such that $(x, w) \in E_1$ and $y = [x, w]$. If v and x commute then, by the filtration around v, we have $v \sim [w, x] = y$, contrary to assumption. Thus $(x, v) \in E_1$. Also, by the filtration around y, we have $y \sim [x, v]$. But $x \sim [x, v]$ as well, so $[x, v]$ is in the unique maximal singular subspace containing yx, that is, M. As $[x, v]$ is also collinear with v, it is a point as required. $\qquad\square$

Lemma 6.7.23 *Suppose that each line is contained in a unique maximal singular subspace of the nondegenerate root filtration space (E, F). If $(u, v) \in E_0$ and $x \in E_{\leq -1}(u, v)$, then $E_{\leq 0}(u, v) \subseteq E_{\leq 0}(x)$. If, moreover, $y \in E_{\leq -1}(u, v)$ is distinct from x, then $(x, y) \in E_0$ and $E_{\leq 0}(u, v) = E_{\leq 0}(x, y)$.*

Proof Assume $w \in E_{\leq 0}(u, v) \setminus E_{\leq 0}(x)$. The only possibility for the relation of x and w is that they are special (use the triangle condition on x, v, w). Set $z = [x, w]$. The filtration around u gives $z \in E_{\leq -1}(u)$. Similarly, $z \in E_{\leq -1}(v)$. As (u, v) is a polar pair, the equalities $z = u$ and $z = v$ are impossible, so $z \in E_{-1}(u, v)$. It follows that the line xz is contained in the planes $\langle x, z, u \rangle$ and $\langle x, z, v \rangle$. Since the union of these planes does not generate a singular subspace, we have reached a contradiction. This establishes the first assertion.

Suppose now that y satisfies the hypotheses stated. By the first assertion, $E_{\leq 0}(u, v) \subseteq E_{\leq 0}(x, y)$. Clearly, $(x, y) \in E_{\leq 0}$, so the assumptions on y and on the uniqueness of the maximal singular subspace on a line imply $(x, y) \in E_0$. We can therefore apply the first assertion to x and y instead of u and v to obtain $E_{\leq 0}(x, y) \subseteq E_{\leq 0}(u, v)$. $\qquad\qquad\square$

Lemma 6.7.24 *Suppose that each line is contained in a unique maximal singular subspace of the nondegenerate root filtration space (E, F). Assume further $(x, y) \in E_0$ and $u, v \in E_{\leq -1}(x, y)$ with $u \neq v$. Let u' be a point of the line yu. There exists a (unique) point v' on the line vx such that $u' \sim v'$.*

The line yu exists as the points u and y are collinear by assumption and distinct as $u \in E_{\leq -1}(x)$ and $y \in E_0(x)$.

Proof Let M_1 be the unique maximal singular subspace of (E, F) containing xu and let M_2 be the analogue for xv. By Lemma 6.7.23, $C_{M_2}(\{u, v\}) = C_{M_2}(\{x, y\})$, so $C_{M_2}(u) = C_{M_2}(\{u, v\}) = C_{M_2}(\{x, y\}) = C_{M_2}(y)$.

For $i = 1, 2$, consider the map $\sigma_i : M_i/\{x\} \to (M_{3-i}/\{x\})^\star$ defined by $\sigma_i(z_i x) = C_{M_{3-i}}(z_i)/\{x\}$ for $z_i \in M_i$. Clearly, $z_{3-i} x \in \sigma_{3-i}(z_i x)$ if and only if $z_i x \in \sigma_i(z_{3-i} x)$. We claim that σ_i is injective. Suppose $u, u' \in M_i \setminus \{x\}$ are such that ux and $u'x$ are distinct and $\sigma_i(ux) = \sigma_i(u'x)$. Thus, $E_1(u) \cap M_{3-i} = E_1(u') \cap M_{3-i}$, which contains a point z in view of Lemma 6.7.22(ii). By Lemma 6.7.13(ii) there is $h \in E_1(x) \cap E_{-1}(z)$. Lemma 6.7.8(v) shows $(u_i, h) \in E_2$. In particular, the line $u_1 u_2$ meets the geometric hyperplane $E_{\leq 1}(h)$ in a single point $a \in E_1(h)$. Put $b = [a, h]$. The 5-circuit a, b, h, z, x contains the pair (a, h) from E_1, so by Lemma 6.7.8(vi), two of its points are collinear. The only possibility is $z \sim h$ (for instance, $a \sim z$ would lead to the contradiction $x = [u_1, z] = a$). But then, $C_{M_{3-i}}(a)$ would contain $\langle C_{M_{3-i}}(u_1) \cup \{z\} \rangle = M_{3-i}$, contradicting Lemma 6.7.22(ii). This establishes the claim that σ_i is injective.

As a consequence, we can apply Exercise 6.8.22 to the spaces $M_1/\{x\}$, $M_2/\{x\}$ and maps σ_1, σ_2. We find $xu = C_{M_1}(C_{M_2}(u)) = C_{M_1}(C_{M_2}(y))$. By symmetry, we also have $xv = C_{M_2}(C_{M_1}(y))$.

If $u' = u$, then $v' := x$ is as required. If $u' \neq u$, then $C_{M_1}(u') = C_{M_1}(y)$. Let v' be the unique point of M_2 collinear with u', so $v' \neq x$. The preceding argument, applied to x, u', u, and v' shows that $xv' = C_{M_2}(C_{M_1}(u')) = C_{M_2}(C_{M_1}(y)) = xv$. Therefore v' must be on the line xv. $\qquad\qquad\square$

Definition 6.7.25 According to Definition 5.3.1, the dimension of a line space is two less than the length of a maximal chain of subspaces. The *singular dimension* of a partial linear space is the supremum of all dimensions of its singular subspaces.

Theorem 6.7.26 *Suppose that (E, F) is a nondegenerate root filtration space of singular dimension at least two such that*

(1) *each line is on a unique maximal singular subspace and*
(2) *every point belongs to precisely two maximal singular subspaces.*

The collection of maximal singular subspaces can be partitioned into two sets $\mathcal{M}_1$ and $\mathcal{M}_2$ such that any two members of the same part are disjoint. Each $\mathcal{M}_i$ carries the structure of a projective space whose lines are the members of the class containing exactly one point each of a line in F. Moreover, the spaces $\mathcal{M}_2$ and $\mathcal{M}_1$ are in duality via $M \mapsto \{K \in \mathcal{M}_{3-i} \mid |K \cap M| = 1\}$ $(M \in \mathcal{M}_i)$, and (E, F) is isomorphic to $E(\mathcal{M}_1, \mathcal{M}_2)$.

In particular, if the singular dimension of (E, F) is finite, say $n - 1$ (so $n \geq 3$), and all lines are thick, then (E, F) is the shadow space of an $[n]$-geometry of type A_n on $\{1, n\}$.

Proof We say that two maximal singular subspaces of E are *parallel* if either $M_1 = M_2$ or $M_1 \cap M_2 = \emptyset$ and there exist points $x_1 \in M_1$ and $x_2 \in M_2$ with x_1 collinear with x_2.

STEP 1. *If M_1 and M_2 are distinct parallel maximal singular subspaces of E, then there exist geometric hyperplanes H_i of M_i such that for every point $z_1 \in H_1$ there exists a (unique) point $z_2 \in H_2$ such that $z_1 \sim z_2$ and vice versa.*

Let x_1, x_2 be points of M_1 and M_2, respectively, such that $x_1 \sim x_2$. Set $H_1 = C_{M_1}(x_2)$ and $H_2 = C_{M_2}(x_1)$. For $z_1 \in H_1 \setminus \{x_1\}$, the pair (x_2, z_1) is polar, so by Lemma 6.7.22 there exists a (unique) point $z_2 \in M_2 \cap E_{-1}(z_1)$. As $z_1, x_2 \in E_{\leq -1}(x_1, z_2)$, the pair x_1, z_2 must be polar, whence $z_2 \in C_{M_2}(x_1) = H_2$.

STEP 2. *Being parallel is a transitive relation.*

Let M_1, M_2, M_3 be distinct maximal singular subspaces of (E, F) such that M_1 is parallel to M_2 and M_2 is parallel to M_3. By Lemma 6.7.24 and dimension considerations there exist points $x_i \in M_i$ such that $x_1 \sim x_2$ and $x_2 \sim x_3$. As $M_2 \cap M_1 = \emptyset$ the line $x_1 x_2$ is not contained in M_2. The same holds for the line $x_2 x_3$. Therefore the points x_1 and x_3 are contained in the other maximal singular subspace on x_2. In particular, x_1 and x_3 are either equal or collinear. Equality cannot happen: then M_1 and M_3 would be two distinct maximal singular subspaces on x_1 not containing the line $x_1 x_2$, and M would be a third, contradicting (2). Thus we have $x_1 \sim x_3$. Assume that $M_1 \cap M_3$ is not empty. Then it must be a singleton $z \notin \{x_1, x_3\}$. Then x_1, x_3, z are pairwise collinear, so they are contained in a maximal singular subspace M. As M contains the lines zx_1 and zx_3, it must be equal to both M_1 and M_3, contrary to assumption. Thus M_1 and M_3 are disjoint, and therefore parallel.

STEP 3. *Let M be a maximal singular subspace of (E, F) and $x \in E \setminus M$. There is a maximal singular subspace parallel to M through x.*

Assume first that x is in hyperbolic relation with a point of M. Then, by Lemma 6.7.22, none of the maximal singular subspaces through x intersects M and there is a point $x_1 \in M \cap E_1(x)$. Set $x_2 = [x, x_1]$ and let N be the maximal singular subspace containing the line xx_2. Then N is as required.

Next assume that no point of M is hyperbolic with x. Again by Lemma 6.7.22 there is a point x_1 of M collinear with x. The maximal singular subspace through x not containing the line $x_1 x$ is as required.

We have shown that there are exactly two classes of parallel maximal singular subspaces: Fix a point $y \in E$, let M_1 and M_2 be the maximal singular subspaces on y, and let $\mathcal{M}_1$ be the class of M_1 and $\mathcal{M}_2$ the class of M_2. Then for every point $x \in E$, one of the maximal singulars containing x belongs to $\mathcal{M}_1$ while the other belongs to $\mathcal{M}_2$.

STEP 4. *Let $i \in \{1, 2\}$, and let M and N be distinct members of $\mathcal{M}_i$. For collinear points $x_1 \in M$ and $x_2 \in N$, define the line MN of $\mathcal{M}_i$ as the set of the maximal singular subspaces in $\mathcal{M}_i$ intersecting the line $x_1 x_2$. This definition is independent of the choice of x_1 and x_2. Moreover, with this definition of lines, $\mathcal{M}_i$ is a projective space.*

Suppose that $z_1 \in M$ and $z_2 \in N$ are also collinear. Assume that K is a member of $\mathcal{M}_i$ distinct from M and N and intersecting the line $x_1 x_2$ in the point x_3. We will show that K also meets $z_1 z_2$. This is obvious if $z_1 z_2 = x_1 x_2$, so suppose this is not the case. If $x_1 = z_1$ and $x_2 \neq z_2$, then x_1, x_2, and z_2 span a singular plane containing the line $x_2 z_2$ in N, so the plane must be contained in N, contradicting $M \cap N = \emptyset$. Therefore, $x_1 \neq z_1$ and, similarly, $x_2 \neq z_2$.

As x_1, z_2 is a polar pair, by Lemma 6.7.24 there is a point z_3 on the line $z_1 z_2$ collinear with x_3. Also, z_3 must be polar to x_1, so the maximal singular subspace containing the line $x_3 z_3$ must be K. We conclude that the definition of MN does not depend on the choice of x_1, x_2.

Let X_1, X_2, X_3 be three members of $\mathcal{M}_i$ not all on one line. As in Step 2, there exist $x_i \in X_i$ ($i \subset [3]$) such that $x_1 \frown x_2$ and $x_2 \frown x_3$. Again, as the lines $x_1 x_2$ and $x_2 x_3$ do not lie in X_2, they must be contained in the other maximal singular subspace, Y say, through x_2 (in particular, $x_1 \sim x_3$ as well). Now, by taking intersections with Y, we obtain an isomorphism between the plane of maximal singular subspaces spanned by X_1, X_2, X_3 and the plane in Y spanned by x_1, x_2, x_3.

STEP 5. *Let $i \in \{1, 2\}$, and let M be in $\mathcal{M}_i$. The set $\phi(M)$ of the maximal singular subspaces intersecting M is a geometric hyperplane of $\mathcal{M}_{3-i}$. The resulting map $\phi : \mathcal{M}_i \to \mathcal{M}_{3-i}{}^{\vee}$ is an injective homomorphism of projective spaces and the pair $(\mathcal{M}_i, \mathcal{M}_{3-i})$ is in duality.*

If M intersects the maximal singular subspaces X_1 and X_2 in the point x_1 and x_2, respectively, then $x_1 x_2$ is a line of M and the line $X_1 X_2$ consists of the maximal singular subspaces in $\mathcal{M}_{3-i}$ through a point of $x_1 x_2$. This shows that the set in question forms a subspace. Let $x \in E$ be a point which is hyperbolic to some point of M. Then x cannot be collinear with a point of M, so the maximal singular subspace on x in $\mathcal{M}_{3-i}$ does not intersect M. This shows that the subspace is proper.

To see that it is indeed a hyperplane of $\mathcal{M}_{3-i}$, assume that $X_1 X_2$ is a line in $\mathcal{M}_{3-i}$. Again, this line is given by a line $x_1 x_2$ where $x_i \in X_i$ and $x_1 \sim x_2$. Consider the maximal singular subspace Y containing the line $x_1 x_2$. As Y intersects X_1, it must be in $\mathcal{M}_i$, so, by Step 1, the points of Y collinear with some point of M form a geometric hyperplane of M. In particular, there is a point x_3 on the line x_1x_2 which is collinear with a point x of M. Now M intersects the unique maximal singular subspace containing x_3x, which lies on the line X_1X_2.

The final assertion follows easily from the previous ones.

STEP 6. *Conclusion.*

After the identification of $\mathcal{M}_2$ with a subspace of the hyperplane dual of $\mathcal{M}_2$, the pair of projective spaces $\mathcal{M}_1$ and $\mathcal{M}_2$ is in duality by Step 5, so $E(\mathcal{M}_1, \mathcal{M}_2)$ is well defined.

Since each point of E lies on a unique member of each $\mathcal{M}_i$ ($i = 1, 2$), there is a well-defined map $E \to \mathcal{M}_1 \times \mathcal{M}_2$, assigning to a point in E the pair in $\mathcal{M}_1 \times \mathcal{M}_2$ of maximal singular subspaces of (E, F) containing it. Its image can be identified with the set of incident pairs of $\mathcal{M}_1 \times \mathcal{M}_2$, that is, the point set of $E(\mathcal{M}_1, \mathcal{M}_2)$. It is straightforward to verify that this map is an isomorphism of spaces $(E, F) \to E(\mathcal{M}_1, \mathcal{M}_2)$.

If (E, F) has finite dimension, then $\mathcal{M}_2$ must be the hyperplane dual of $\mathcal{M}_1$. If, moreover, $\mathcal{M}_1$ has thick lines only, then, by Corollary 6.3.2, there is a natural number $n \geq 3$ and a division ring $\mathbb{D}$ such that $\mathcal{M}_1 \cong \mathbb{P}(\mathbb{D}^{n+1})$. As discussed in Remark 6.7.6, this space is isomorphic to the shadow space of the geometry of type A_n on $\{1, n\}$. $\qquad\square$

6.8 Exercises

Section 6.1

Exercise 6.8.1 Let Z be a thick projective space and let α be an automorphism of Z admitting a center c, i.e., leaving each line on c invariant. Show that α must be a perspectivity.

Exercise 6.8.2 Give non-thick counterexamples to Parts (ii) and (iii) of Lemma 6.1.3.
(*Hint*: Take the direct sum of a line having at least five points and a thick projective space.)

Exercise 6.8.3 (The non-Desarguesian Moulton plane) Fix a parameter $c \in \mathbb{R}$, $c > 1$. Let Moul_c be the line space whose points are those of $\mathbb{E}^2$ (cf. Example 1.1.2) and whose lines are of the following form, where $a, b \in \mathbb{R}$.

$$\{(x, y) \in \mathbb{E}^2 \,|\, x = b\}, \quad \text{or}$$

$$\{(x, y) \in \mathbb{E}^2 \,|\, y = ax + b\} \quad \text{with } a \leq 0, \quad \text{or}$$

$$\{(x, y) \in \mathbb{E}^2 \,|\, y = ax + b \text{ for } x \leq 0, \ y = cax + b \text{ for } x \geq 0\} \quad \text{with } a > 0.$$

(a) Show that Moul_c is an affine plane.
(b) Show that the *Moulton plane*, that is, the projective plane $\mathbb{P}(\mathrm{Moul}_c)$ obtained from Moul_c, is non-Desarguesian by using a configuration of Desargues in $\mathbb{E}^2$ of which only one of the ten lines is 'broken'.
(c) Show that no automorphism of $\mathbb{P}(\mathrm{Moul}_c)$ moves the projective line $\{0\} \times \mathbb{R}$ to any other line.
(d) Show that two Moulton planes Moul_c for distinct constants $c > 1$ are non-isomorphic.
   (*Hint*: Use (c).)

Section 6.2

Exercise 6.8.4 Show that Example 6.2.7 is indeed a projective line as defined in Definition 6.2.1.

Exercise 6.8.5 Let $(X, \{U_x \mid x \in X\})$ be a Moufang set whose projective group G acts sharply 3-transitively on X. Prove that, if, for distinct $x, y \in X$, the point-wise stabilizer $G_{x,y}$ is Abelian, then (X, π), where $\pi(x, x) = U_x$ and $\pi(x, y) = G_{x,y}$ for all distinct $x, y \in X$, is a Desarguesian projective line.

Exercise 6.8.6 Let $P = [4]$ and put $\pi(i, j) = \{\mathrm{id}\}$ for $i, j \in P$ whenever $i = j$ or $j \neq 4$. For $i < 4$, let $\pi(i, 4)$ consist of the identity and the transposition (j, k), where $\{i, j, k\} = [3]$. Show that (P, π) satisfies all properties of a projective line except for the part of Definition 6.2.4(4) stating that $\pi^{\vee}(4)$ is a subgroup of $\mathrm{Sym}(P)$. Conclude that this condition is not superfluous and that omission of it would lead to a notion of projective lines whose duals need not be projective lines.

Exercise 6.8.7 Suppose that $Z_1 = (P_1, L_2)$ and $Z_2 = (P_1, L_2)$ are two line spaces. Define the *direct product* $Z_1 \times Z_2$ of Z_1 and Z_2 as the space whose point set is the Cartesian product $P_1 \times P_2$ of the two point sets and whose lines are the subsets of $P_1 \times P_2$ of the form $\{x\} \times m$ or $l \times \{y\}$ for $x \in P_1$, $y \in P_2$, $l \in L_1$, and $m \in L_2$.

(a) Show that if both Z_1 and Z_2 are partial linear spaces, then so is $Z_1 \times Z_2$.
(b) For $i \in [2]$, let Γ_i be the geometry of $(P_i, L_i, *)$ as in Definition 2.5.12; denote the types points and lines for Γ_i by p_i and l_i, respectively. Consider the direct sum geometry Γ of Γ_1 and Γ_2; it is a geometry over $\{\mathrm{p}_1, \mathrm{p}_2, \mathrm{l}_1, \mathrm{l}_2\}$ (cf. Definition 2.1.8). Show that the direct product $Z_1 \times Z_2$ is isomorphic to the shadow space of Γ on $\{\mathrm{p}_1, \mathrm{p}_2\}$.
(c) Show that Γ is a generalized quadrangle if Z_1 and Z_2 each consist of a single line.

Exercise 6.8.8 Consider the Desarguesian projective line over the field $\mathbb{F}$. We will identify its point set with $\mathbb{F} \cup \{\infty\}$. If $a \in \mathbb{F}$, then the equation for the hyperplane

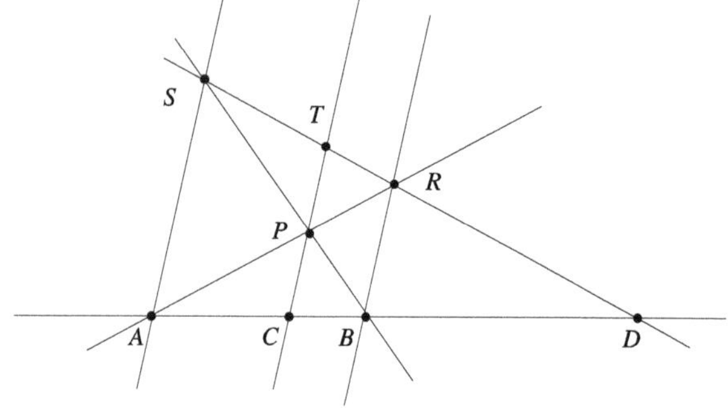

Fig. 6.6 Construction of the quadruple (A, B, C, D) from the triple (A, B, C) with ∞ at infinity

$\mathbb{F}(a, 1)$ is given by the matrix

$$J = \begin{pmatrix} 0 & 1 \\ -1 & 0 \end{pmatrix}$$

in the sense that $x \in \mathbb{F}(a, 1)$ if and only if x is a solution to the equation $x J (a, 1)^\top = 0$. The element a determines the rank 1 matrix

$$X(a) := \left(J(a, 1)^\top\right)(a, 1) = \begin{pmatrix} a & 1 \\ -a^2 & -a \end{pmatrix}$$

which squares to zero and represents the pair consisting of the point a and the hyperplane $\{a\}$ incident with a, as in Exercise 6.8.20. Show that the quadruple (a, b, c, d) (with distinct components from $\mathbb{F}$) is harmonic if and only if

$$\mathrm{Tr}\left(\left[X(a), X(b)\right]\left[X(c), X(d)\right]\right) = 0,$$

where $[A, B]$ is the commutator, that is, $AB - BA$, and Tr is the usual matrix trace function.

Exercise 6.8.9 Suppose that we are given the three points A, B, and C on a projective line l in a Desarguesian plane π. Consider the following construction of a fourth point D on l, depicted in Fig. 6.6.

(1) Pick points ∞ and P distinct from C on a line m on C distinct from l.
(2) Let R be the point of intersection in π of the lines AP and $B\infty$.
(3) Let S be the point of intersection in π of the lines BP and $A\infty$.
(4) Let D be the point of intersection of the line RS with l.

Show that (A, B, C, D) is a harmonic quadruple.

Section 6.3

Exercise 6.8.10 Theorem 6.3.1 also applies when all lines have exactly three points. Then there is no nontrivial perspectivity whose center is outside its axis. Give a more direct proof for this case.

Exercise 6.8.11 (This exercise is used in Theorem 6.3.8) Suppose that Z is a projective space all of whose lines are thick. Show that Z is a singleton, consists of a single line, or is thick (cf. Definition 2.5.12).

Section 6.4

Exercise 6.8.12 Let M be the vector space $\mathbb{F}_3^4$ of dimension 4 over $\mathbb{F}_3$ with standard basis $\varepsilon_1, \ldots, \varepsilon_4$. Define the following multiplication $\star$ on M:

$$a \star b = \varepsilon_4\big(a_4 + b_4 + (a_3 - b_3)(a_1 b_2 - b_1 a_2)\big) + \sum_{i=1}^{3} \varepsilon_i (a_i + b_i),$$

where $a = \sum_i \varepsilon_i a_i$ and $b = \sum_i \varepsilon_i b_i$. Show that $(M, \star)$ is a commutative Moufang loop on 81 elements that is not a group.

Section 6.5

Exercise 6.8.13 Let $n \in \mathbb{N}$, $n \geq 2$, and let $\mathcal{C}$ be a residually connected chamber system of type A_n and set $\Gamma = \Gamma(\mathcal{C})$. Take (W, S) to be the Coxeter system where $W = \mathrm{Sym}_{n+1}$ and $S = \{(i, i+1) \mid i \in [n]\}$. Fix two chambers c, d of $\mathcal{C}$ and set $w = w_{c,d}$ in the notation of Theorem 6.5.5. Prove that, if $w = r_1 \cdots r_q$ is a minimal expression for w with $r_i \in S$, then there is a minimal gallery from c to d of type $r_1 \cdots r_q \in S^*$ (in the notation of Remark 4.2.15).

Section 6.6

♮ **Exercise 6.8.14** Let Γ be a residually connected geometry over the diagram
$\overset{1}{\underset{\circ}{}} \xrightarrow{\; L^{\vee} \;} \overset{2}{\underset{\circ}{}} \xrightarrow{\; L \;} \overset{3}{\underset{\circ}{}}$ in which the following *intersection property* holds: for any type i, flag v, and element w of Γ with $\mathrm{Sh}_i(v) \cap \mathrm{Sh}_i(w) \neq \emptyset$, there exists a flag x incident with v and w such that $\mathrm{Sh}_i(v) \cap \mathrm{Sh}_i(w) = \mathrm{Sh}_i(x)$. Assume that there is a 1-element (respectively, 2-element) that is incident with a finite number of 2-elements (respectively, 1-elements). Prove that there is a projective space $\mathbb{P}$ and a

natural number i such that either the elements of type 1, 2, 3 of Γ are the subspaces of respective dimension $(i + 1)$, i, $(i - 1)$ in $\mathbb{P}$ and the incidence of Γ is symmetrized inclusion in $\mathbb{P}$, or the elements 1, 2, 3 of Γ are the subspaces of respective dimension $(i - 1)$, i, $(i + 1)$ in $\mathbb{P}$ with symmetrized inclusion as incidence.

Exercise 6.8.15 Let G be the group $\mathrm{PGL}(V)$.

(a) Prove that G is transitive on the set of pairs (c, Δ) consisting of a chamber c and an apartment Δ containing c (or, which amounts to the same, the set of ordered projective bases of $\mathbb{P}(V)$).
(b) Show that, for each $w \in \mathrm{Sym}_{n+1}$, there is a unique G-orbit of pairs of chambers c, d with $w = w_{c,d}$.
(*Hint*: In view of (a), we may assume that c is a fixed chamber and that d belongs to a fixed apartment Δ containing c. But then, by Corollary 4.2.11, the chamber d is uniquely determined by the fact that the type of a minimal gallery from c to d maps to $w_{c,d}$.)
(c) Conclude that there is a bijective correspondence between the G-orbits of pairs of chambers and the chambers of a fixed apartment Δ, given by

$$G(c_1, c_2) \leftrightarrow \text{the second component of } \{c\} \times \Delta \cap G(c_1, c_2).$$

Section 6.7

Exercise 6.8.16 (This exercise is used in Remark 6.7.6) Let $\mathbb{P}$ and $\mathbb{H}$ be a pair of projective spaces in duality. Prove that $\mathbb{H}$ and $\mathbb{P}$ also form a pair in duality and that $E(\mathbb{P}, \mathbb{H})$ is isomorphic to $E(\mathbb{H}, \mathbb{P})$.

Exercise 6.8.17 (This exercise is used in Remark 6.7.3) Prove that a root filtration space satisfies the following stronger version of Condition (4):

For each $(x, y) \in E_2$, we have $E_{\leq i}(x) \cap E_{\leq j}(y) = \emptyset$ whenever $i + j < 0$.

Exercise 6.8.18 Verify that no partition of the pairs of points of a thick generalized octagon into five classes E_i $(-2 \leq i \leq 2)$ corresponding to the five possible distances leads to a root filtration space with respect to $(E_i)_{-2 \leq i \leq 2}$.

Exercise 6.8.19 (This exercise is used in Example 7.10.1(ii)) Let $\mathbb{P}$ and $\mathbb{H}$ be two thick projective spaces in duality. Consider the space $D(\mathbb{P}, \mathbb{H})$ whose point set is the disjoint union of $\mathbb{P}$ and $\mathbb{H}$ and whose line set is the union of the line set of $\mathbb{P}$, the line set of $\mathbb{H}$, and the set of all unordered pairs $\{x, H\}$ with $x \in \mathbb{P}$ and $H \in \mathbb{H}$ such that $x \in H$. This space is called the *dualized projective space* of $\mathbb{P}$ and $\mathbb{H}$. Prove that the root filtration space $E(\mathbb{P}, \mathbb{H})$ defined in Example 6.7.5 is isomorphic to the subspace of the Grassmannian on $D(\mathbb{P}, \mathbb{H})$ whose points are the thin lines of $D(\mathbb{P}, \mathbb{H})$.

Exercise 6.8.20 Let V be a vector space over a field $\mathbb{F}$. We call a linear transformation X in $\mathrm{End}(V)$, the algebra of all endomorphisms of V, *extremal* if it has left and right rank 1 and satisfies $X^2 = 0$.

(a) Show that $\langle X \rangle \mapsto (X(V), \mathrm{Ker}(X))$ is an injective map from the subset of $\mathbb{P}(\mathrm{End}(V))$ of projective points of the form $\langle X \rangle$ with X extremal to $E(\mathbb{P}(V), \mathbb{P}(V^{\vee}))$.
(b) Describe the relations E_i on $E(\mathbb{P}(V), \mathbb{P}(V^{\vee}))$ in terms of matrix properties.

Exercise 6.8.21 Assume that (E, F) is a root filtration space with respect to $(E_i)_i$.

(a) Prove the following two statements.
  (i) If (E, E_2) is connected and $(E, \sim)$ has no isolated vertices, then $(E, \sim)$ is connected.
  (ii) Suppose that (E, F) is thick and satisfies the equivalent conditions of Lemma 6.7.13. If $(E, \sim)$ is connected, then so is (E, E_2).
(b) Let (E, F) be the ordinary hexagon. As stated in Example 6.7.10, this is a root filtration space with $E_0 = \varnothing$, and $(x, y) \in E_i$ if and only if x and y are at mutual distance $i + 1$ for $i \in \{2, 3\}$. Verify that (E, F) is an example showing that thickness is needed for (ii) of (a) to hold.

Exercise 6.8.22 (This exercise is used in Lemma 6.7.24) We will use the hyperplane dual $Z^{\star}$ of a linear space Z as defined in Exercise 5.7.12. Suppose that Z and Z' are linear spaces, and $\sigma : Z \to Z'^{\star}$ and $\sigma' : Z' \to Z^{\star}$ are maps such that, for all $x \in Z$ and $y \in Z'$,

$$y \in \sigma(x) \quad \Leftrightarrow \quad x \in \sigma'(y).$$

Extend σ to subsets X of Z by decreeing $\sigma(X) = \bigcap_{x \in X} \sigma(x)$, and similarly for σ'. Prove that, if σ is injective, then $\sigma'(\sigma(x)) = \{x\}$ for each $x \in Z$.

6.9 Notes

The topics of this chapter have a rich and intricate history. The point of view on thick projective spaces of dimension at least three is very close to Veblen & Young [296] in 1910. Pasch's Axiom (Definition 5.2.4) is also known as the Veblen and Young Axiom. Many of their ideas go back to Von Staudt (1847). Since then a lot of progress has been made. Veblen & Young obtained a geometric construction of a number system that is exactly a division ring. The reader is referred to [141] for an account of the 19th century geometry. A recent introduction into projective geometry is [292]. For many results on substructures of finite projective spaces, see [162].

Section 6.1

Perspectivities came to the foreground around 1940, with work of Artin [3] and Baer [10]. In [11] the basic equivalence between projective spaces and vector spaces is clearly discussed and explained. A new proof of this fact was given by [207, 208]. Our presentation, though clearly inspired from this classical tradition, differs from this work. The classification of non-thick projective spaces is implicit in the work of Birkhoff [20] and explicit in [251].

Section 6.2

The theory of projective lines in Sect. 6.2 is due to Buekenhout [41].

The idea of the Desarguesian line is closely connected to Tits systems of rank one.

The notion of a Moufang set, defined in Remark 6.2.3 was brought forward by Tits to describe (saturated split) Tits systems of rank one, and received much attention since. We refer the reader to [111] for a good introduction into the topic. Many examples come from algebraic groups of relative rank one. The finite examples are classified; the infinite case is still open.

Some examples of Moufang sets that are not Desarguesian lines arise from polarities on certain generalized hexagons defined over certain fields of characteristic three. The rank of the absolute geometry with respect to the polarity is one. So, it is just a set of points, albeit an interesting set of points when viewed as a subset of $\mathbb{P}(\mathbb{F}^7)$, because its normalizer in the full special linear group is the twisted Chevalley group ${}^2G_2(\mathbb{F})$ (cf. [62, 133]). In view of some special properties of this group, the set is a Moufang set.

We owe part of the proof of Step 14 of Theorem 6.2.11 to Rob Eggermont and Maxim Hendriks.

In case the skew field is a field $\mathbb{F}$, the *cross-ratio* of the four points a, b, c, d in $\mathbb{F}$ (viewed as affine points given by their first coordinates) is defined as

$$\frac{(c-a)(d-b)}{(d-a)(c-b)}.$$

This number is -1 if and only if the four points form a harmonic quadruple. The definition extends to a cross-ratio for any four points on the projective line $\mathbb{P}^1(\mathbb{F})$. For $a = 0$ and $b = \infty$ (so that the corresponding point is $(1 : 0) = \infty$), the cross-ratio is the scalar λ by which c needs to be multiplied to obtain d as the image of c under a perspectivity with center a and hyperplane b (so the unique element of $\pi(a, b)$ mapping c to d is identified with multiplication by λ). This theory is classical; see [4, 167].

The harmonic quadruples determine the full line structure provided that -1 is different from 1, so the definition of a projective line could have been given in terms of harmonic quadruples rather than regular groups.

Section 6.3

The proof of Theorem 6.3.1 is classical (cf. [4, Chap. 2]); its strength lies in the fact that it shows that a plane is Desarguesian in the presence of a limited set of perspectivities. The suggested proof using Theorem 6.2.11 on the Desarguesian projective line is closer to the classification of Moufang planes (cf. [291]). A detailed analysis of the corresponding ternary ring (cf. Example 2.3.4) leads to other classical proofs; see [240]. An elegant proof using Schur's Lemma can be found in [260, Sect. 5.6].

In [121] it is shown that, if Z is a linear line space that is locally projective of dimension at least four, then Z is either a projective or an affine space.

Section 6.4

The development and the proof of Theorem 6.3.1 follow [4]. The material of Theorem 6.4.8 is taken from [40].

The identity of Example 6.4.5, which, written multiplicatively, reads $(ab)(ca) = a((bc)a)$, is called a Moufang identity, after Moufang [222]. The multiplicative structure of the Cayley division ring of Example 2.3.4 and of any alternative ring (cf. Example 2.8.11) is also a Moufang loop. Basic surveys of loops are [38, 65, 239].

Section 6.5

Apartments play an important role in buildings, as will become clear in Chap. 11. Theorem 6.5.5(ii) gives the type of a gallery in a building of type A_n and Proposition 6.5.6 proves strong transitivity in the terminology of Definition 11.3.1.

Section 6.6

Characterizations of isomorphisms between Grassmannians are given in [68].

The geometric hyperplanes of Grassmannians over fields are determined by Shult [255]. Each geometric hyperplane arises from the universal embedding of the Grassmannian of k-dimensional subspaces of a vector space V into the projective space of the k-fold exterior power of V.

The shadow space on j of an apartment of $\mathrm{PG}(V)$, where V is a vector space of finite dimension $n+1$ and $j \in \{2, \ldots, n-1\}$, is a Johnson graph (cf. Exercise 2.8.19) that is isometrically embedded in $\mathrm{ShSp}(\mathrm{PG}(V), j)$. By [231], there are other examples of isometrically embedded Johnson graphs in $\mathrm{ShSp}(\mathrm{PG}(V), j)$ when $n \neq 2j - 1$.

Section 6.7

Root filtration spaces were introduced in [79, 80]. Most of the proofs of this section are from these two papers. The name triangle condition for Condition (4) of

Definition 6.7.2 on triples x, y, z stems from [279]. The abstract root subgroups defined in this book are examples of root filtration spaces.

The name filtration for Condition (3) of Definition 6.7.2 is related to a filtration in Lie algebras giving rise to examples of root filtration spaces.

The names polar, hyperbolic, and special for pairs of points in Definition 6.7.2 remind us that

(1) the convex closure (cf. Definition 11.4.16) of a polar pair is isomorphic to a polar space (symplecton; see Proposition 7.9.3 and Theorem 11.5.9),
(2) a hyperbolic pair in a Lie algebra spans a 3-dimensional linear subspace of the Lie algebra on which a certain symmetric bilinear form does not vanish and whose singular points are the points of the geometry, and
(3) a special pair of abstract root subgroups as in [279] generates a special group.

The classification of root filtration spaces will be continued in Sects. 7.9, 11.5, and 11.7; the parts not fully treated in this book are described in Sect. 11.9.

Section 6.8

The Moufang loop of Exercise 6.8.12, which is taken from [27], is not unique: according to [229] there are five non-associative Moufang loops of order 81 (up to isomorphism; besides the 15 groups of that order), two of which are commutative.

The non-Desarguesian Moulton plane, appearing in Exercise 6.8.3, dates back to 1902; see [223].

Exercise 6.8.6 was suggested to us by Pasini.

Exercise 6.8.14 is taken from Sprague's paper [263]. The variation

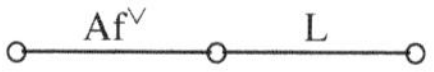

$$\circ \overset{\mathrm{Af}^{\vee}}{\rule{3cm}{0.4pt}} \circ \overset{\mathrm{L}}{\rule{3cm}{0.4pt}} \circ$$

of Sprague's diagram is used by Cuypers [103] to characterize affine Grassmannians.

Chapter 7
Polar Spaces

We come to one of the central themes in the theory of geometries of spherical Coxeter type. It can be described in various intertwining ways, such as: polarities in projective spaces and their absolutes (cf. Definition 1.4.12), (both algebraic and geometric) quadrics in projective spaces, geometries belonging to a diagram of type B_n, or polar spaces: line spaces satisfying the property that, for each line of the space, each point is collinear with either one or all points of that line.

The basic geometric theory of polar spaces is the main goal of this chapter. In Chaps. 8, 9, 10 attention will be given to the classification of polar spaces, a very powerful result. The major results, to be found in Theorem 10.3.13 and Corollary 10.3.14, concern spaces as well as geometries. The polar spaces that do not have subspaces isomorphic to Desarguesian planes will be left out of our classification.

In the first three sections of this chapter, we deal with the properties of polarities on geometries, especially projective geometries. The point shadows of their absolutes line spaces with the property described above, and so are polar spaces, which are introduced in Sect. 7.4. Such spaces are called nondegenerate if there are no points collinear with all other points.

In Sect. 7.4, it is shown that singular subspaces of nondegenerate polar spaces are projective spaces (Theorem 7.4.13) and in Sect. 7.5 that nondegenerate polar spaces whose maximal singular subspaces are projective spaces of dimension $n - 1$ provide geometries of type B_n (Theorem 7.5.8). We already know from Chap. 4 that the finite Coxeter groups of type B_n and the related hyperoctahedra lead to thin geometries of type B_n. Polarities of projective spaces, quadrics, and pseudo-quadrics in projective spaces provide many thick geometries of this type. A partial converse appears in Sect. 7.6. Under mild conditions, the shadow spaces on 1 of geometries of type B_n are shown to be polar spaces.

In Sect. 7.7, it is shown that singular planes of a polar space are Moufang planes. The geometries of type B_n associated with certain quadrics are not thick. By adjustment of the type function for these examples so-called oriflamme geometries (cf. Definition 7.8.5) emerge, which turn out to be of type D_n. This happens in Sect. 7.8, where we also look at other shadow spaces than the point shadow spaces

F. Buekenhout, A.M. Cohen, *Diagram Geometry*,
Ergebnisse der Mathematik und ihrer Grenzgebiete. 3. Folge / A Series of
Modern Surveys in Mathematics 57, DOI 10.1007/978-3-642-34453-4_7,

of geometries of type B_n. In Sect. 7.9, we show that the shadow space on 2, a space on the set of lines of a polar space, is a root filtration space (as introduced in Definition 6.7.2). An axiomatic characterization of this shadow space is treated in full for the case of rank three (Theorem 7.9.18). Finally, Sect. 7.10 is concerned with polar spaces having thin as well as thick lines.

7.1 Duality for Geometries over a Linear Diagram

In this section $\Gamma = (X, *, \tau)$ will be a geometry over an index set I.

An auto-correlation α of Γ (see Definition 1.3.1) induces a permutation α_I on I and an automorphism of the digon diagram of Γ. Recall from Definition 1.3.8 that a duality of Γ is an auto-correlation α of Γ such that α_I has order two. Also, if α is a duality such that both α and α_I have order two, then α is called a polarity.

Example 7.1.1 Here are some other polarities than those exhibited in Examples 1.3.9, 1.3.10, and 1.3.11.

 (i) Let T be the tetrahedron viewed as a geometry over {vertex, edge, face}. The map interchanging each vertex of T with the face opposite to that vertex and interchanging all pairs of opposite edges, is a polarity.
 (ii) The ordinary m-gon geometry Π is endowed with polarities. If m is even, these are all conjugate in the automorphism group of Π. If m is odd, then there is a unique polarity π on Π without points x incident with $\pi(x)$; all others are again conjugate in $\mathrm{Aut}(\Pi)$.
 (iii) Consider the real vector space $V = \mathbb{R}^{n+1}$, endowed with the standard inner product $f(x, y) = x_1 y_1 + x_2 y_2 + \cdots + x_{n+1} y_{n+1}$. In the $[n]$-geometry $\mathrm{PG}(V)$, the map sending each subspace U of V to $\mathbb{P}(U^\perp)$, where $U^\perp = \{x \in V \mid f(x, U) = \{0\}\}$, is a polarity. The same observation holds for $\mathbb{C}$ instead of $\mathbb{R}$. Later in this chapter (Theorem 7.2.12), we will have a closer look at constructions of this kind.

Remark 7.1.2 Every product of a polarity with an automorphism is a duality.

A symmetry σ of order two on a diagram may suggest that geometries belonging to it possess dualities and/or polarities inducing σ on the diagram, but this is not necessarily the case. By Exercise 7.11.31, the generalized quadrangle $\mathbb{P}(\mathbb{F}_q^4)_f$ (cf. Notation 7.2.13) for q a power of an odd prime and f a nondegenerate alternating form on $\mathbb{F}_q^4$ (cf. Definition 7.3.10), for instance, is a geometry of the symmetric type B_2 that is not isomorphic to its dual, which is $\mathbb{P}(\mathbb{F}_q^5)_\kappa$, where κ is a nondegenerate quadratic form of Witt index two on $\mathbb{F}_q^5$ (cf. Example 7.8.1). Also, Example 7.2.2 will give division rings that are not isomorphic to their opposites. The corresponding Desarguesian projective planes have no dualities although they belong to the symmetric diagram A_2.

It may also happen that a geometry possesses a duality but no polarity. A classical example is the generalized quadrangle of order $(4, 4)$ associated with the symplectic geometry on $\mathbb{P}(\mathbb{F}_4^4)$ (that is, the absolute of a symplectic polarity as in Definition 10.1.10). The details are treated in Exercise 7.11.30.

Conversely, a diagram may reflect the presence of dualities for geometries. But there may very well be dualities, even if they are not suggested by the diagram. For instance, the rank three tetrahedron geometry belongs to the symmetric diagram A_3, but also to the non-symmetric diagram

$$\overset{\text{C}}{\circ\!\!-\!\!\!-\!\!\!-\!\!\!-\!\!\!-\!\!\!-\!\!\circ\!\!-\!\!\!-\!\!\!-\!\!\!-\!\!\!-\!\!\!-\!\!\!-\!\!\!-\!\!\circ}.$$

Definition 7.1.3 Let $n \in \mathbb{N}$ and let $\Gamma = (X, *, \tau)$ be a geometry with a linear diagram over $[n]$. The *dual geometry*, denoted $\Gamma^\vee$, is the geometry whose elements and incidence are those of Γ while its type map $\tau^\vee$ is defined by $\tau^\vee(x) = n + 1 - \tau(x)$ for each $x \in X$.

Remark 7.1.4 Definition 7.1.3 extends Definition 2.2.5 beyond rank two.

Suppose that Γ is an $[n]$-geometry belonging to the diagram L_n (cf. Theorem 5.3.11) and possessing a duality δ. Obviously, δ can be viewed as an isomorphism $\Gamma \to \Gamma^\vee$. As the type L_2 residues cannot be generalized digons, the digon diagram of the geometry is a path, and so the duality induces either the identity or the permutation $\pi = (1, n)(2, n - 1) \cdots$ on $[n]$. Assume the latter. For $i \in [n - 1]$, the residues of Γ of type $\{i, i + 1\}$ are $(3, 3, 4)$-gons over $(i, i + 1)$ (see Theorem 2.5.15). But due to the existence of δ, residues of type $\{i, i + 1\}$ must be isomorphic to rank two residues of Γ of type $\{n + 1 - i, n - i\}$ which are $(3, 3, 4)$-gons over $(n + 1 - i, n - i)$ by a weak homomorphism sending type i to $n + 1 - i$ and type $i + 1$ to $n - i$. Therefore, the residues of Γ of type $\{i, i + 1\}$ are also $(3, 4, 3)$-gons over $(i, i + 1)$. This implies that the edges of the diagram can be labelled by $(3, 3, 3)$-gons (cf. Definition 2.2.6); in other words, by A_2, so that we are actually dealing with a projective geometry (cf. Corollary 5.4.4). So it makes sense to focus on projective geometries. In this case, $\Gamma^\vee$ is also a projective geometry.

Notation 7.1.5 Recall from Corollary 5.4.5 that there is a geometry of type A_n associated with each projective space $\mathbb{P}$ of finite dimension n, whose elements of type i are the subspaces of $\mathbb{P}$ of dimension $i - 1$. We write $\Gamma(\mathbb{P})$ for this geometry.

If $\mathbb{P} = \mathbb{P}(V)$ for some finite-dimensional vector space V, then $\Gamma(\mathbb{P}) = \mathrm{PG}(V)$.

Theorem 7.1.6 *Let $\mathbb{P}$ be a projective space of finite dimension and δ a duality of $\Gamma(\mathbb{P})$. Then, for every non-empty subspace X of $\mathbb{P}$, we have*

$$\delta(X) = \bigcap_{p \in X} \delta(p). \tag{7.1}$$

Proof If p is a point of X, then $\delta(X)$ is incident with $\delta(p)$ and so $\delta(X)$ is contained in the hyperplane $\delta(p)$ of $\mathbb{P}$, which proves $\delta(X) \subseteq \bigcap_{p \in X} \delta(p)$.

In order to prove the other inclusion, we proceed by induction on $i = \dim(X)$. For $i \leq 0$, the property is obvious. Let $i > 0$ and consider a hyperplane H of X and a point $q \in X \setminus H$. Then $\delta(H) = \bigcap_{p \in H} \delta(p)$ by the induction hypothesis and $\delta(X)$ must be a hyperplane of $\delta(H)$. Moreover $\delta(H)$ is not contained in $\delta(q)$ because q is not in H, hence $\delta(q) \cap \delta(H)$ is a proper hyperplane of $\delta(H)$. However, $\delta(X) \subseteq \delta(q) \cap \delta(H)$. By Proposition 5.2.9, $\delta(X)$ is a maximal subspace of $\delta(H)$, so $\delta(X) = \delta(q) \cap \delta(H)$. We conclude $\delta(X) \supseteq \bigcap_{p \in X} \delta(p)$. $\qquad\square$

Definition 7.1.7 Let $\mathbb{P}$ be a projective space of finite dimension. The hyperplane dual $\mathbb{P}^\star$ of $\mathbb{P}$ was introduced in Exercise 5.7.12. A *duality* of $\mathbb{P}$ is a map $\mathbb{P} \to \mathbb{P}^\star$ sending points of $\mathbb{P}$ to hyperplanes of $\mathbb{P}$, such that the corresponding weak homomorphism $\Gamma(\mathbb{P}) \to \Gamma(\mathbb{P})$ defined by (7.1) is a duality of $\Gamma(\mathbb{P})$. Similarly, a *polarity* of $\mathbb{P}$ is a map $\mathbb{P} \to \mathbb{P}^\star$, sending points of $\mathbb{P}$ to hyperplanes of $\mathbb{P}$ such that the corresponding weak homomorphism $\Gamma(\mathbb{P}) \to \Gamma(\mathbb{P})$ is a polarity of $\Gamma(\mathbb{P})$.

Lemma 7.1.8 *Let $\mathbb{P}$ be a projective space of finite dimension. A bijective map $\delta : \mathbb{P} \to \mathbb{P}^\star$ is a duality of $\mathbb{P}$ if and only if it satisfies the following property.*

Each triple of points a, b, c of $\mathbb{P}$ is collinear if and only if $\delta(a)$, $\delta(b)$, $\delta(c)$ are hyperplanes any two of which have the same intersection.

Proof A duality clearly satisfies the stated property, so we assume the property and derive that δ is a polarity of $\mathbb{P}$. By definition, $\Gamma((\mathbb{P})^\star)$ is isomorphic to $\Gamma(\mathbb{P})^\vee$. The property readily implies that $\mathbb{P}$ is isomorphic to $\mathbb{P}^\star$, from which we infer that $\Gamma(\mathbb{P})$ is isomorphic to $\Gamma(\mathbb{P})^\vee$. It follows that δ establishes an isomorphism $\Gamma(\mathbb{P}) \to \Gamma(\mathbb{P})^\vee$. $\qquad\square$

This criterion, which is useful when checking the examples derived from sesquilinear forms in Chap. 9, is completed now by a characterization of polarities. In order to incorporate the case where V is infinite-dimensional, we need a more general set-up for polarities. The definition below is geared to cope with this.

Definition 7.1.9 A map π on a projective space $\mathbb{P}$ sending each point x of $\mathbb{P}$ to either a hyperplane or $\mathbb{P}$ is called a *quasi-polarity* if it satisfies the following property for all points x, y of $\mathbb{P}$.

If $x \in \pi(y)$, then $y \in \pi(x)$.

The set $\{x \in \mathbb{P} \mid \pi(x) = \mathbb{P}\}$ is called the *kernel* of π and denoted $\mathrm{Ker}(\pi)$. If the kernel is empty, then we say that π is *nondegenerate*.

The *absolute (space)* with respect to a quasi-polarity π, denoted $\mathbb{P}_\pi$, is the line space whose points are projective points x with $x \in \pi(x)$ and whose lines are the projective lines l such that $l \subseteq \pi(x)$ for each $x \in l$.

Proposition 7.1.10 *Let $\mathbb{P}$ be a projective space of finite dimension. Every injective nondegenerate quasi-polarity of $\mathbb{P}$ is a polarity of $\mathbb{P}$.*

Proof Let π be an injective nondegenerate quasi-polarity of $\mathbb{P}$. First, using Lemma 7.1.8, we check that π determines a duality $\overline{\pi}$ of the corresponding geometry. Let a, b, c be distinct collinear points of $\mathbb{P}$. Consider $x \in \pi(a) \cap \pi(b)$. Then $a \in \pi(x)$ and $b \in \pi(x)$ so $ab \subseteq \pi(x)$, where ab stands for the projective line generated by a and b, and so $c \in \pi(x)$. Hence $x \in \pi(c)$. Thus $\pi(a) \cap \pi(b) \subseteq \pi(a) \cap \pi(c)$. In view of the symmetric role played by a, b, c, this forces $\pi(a) \cap \pi(b) = \pi(b) \cap \pi(c) = \pi(c) \cap \pi(a)$.

Conversely, suppose that a, b, c are points such that $\pi(a) \cap \pi(b) = \pi(b) \cap \pi(c) = \pi(c) \cap \pi(a)$. Let $c' \in \pi(c) \setminus \pi(a)$. Then $\pi(c')$ intersects ab in some point c'' and $\pi(c'')$ contains $\pi(a) \cap \pi(b)$ by the first argument. Moreover $\pi(c'')$ contains c', so $\pi(c'') = \langle \pi(a) \cap \pi(b), c' \rangle = \pi(c)$. Injectivity of π gives $c = c''$, forcing a, b, c to be collinear.

Next we verify that π is surjective. We use induction on the dimension, say n, of $\mathbb{P}$. If $n = 0$, the statement is obviously true. If $n = 1$, the map π can be viewed as a permutation of the point set of $\mathbb{P}$ of order 1 or 2, and so must be surjective. Assume now $n > 1$ and suppose that H is a hyperplane of $\mathbb{P}$ that is not the image of a point of $\mathbb{P}$. As $\pi|_H : x \mapsto \pi(x) \cap H$ is a bijective map from the points of H to the hyperplanes of H, there are points $x \in H$ and $y \in \mathbb{P} \setminus H$ such that $\pi(x) \cap H = \pi(y) \cap H = \pi(x) \cap \pi(y)$. Pick $u \in H \setminus \pi(x)$ and let z be a point in $xy \cap \pi(u)$. Then $u \in \pi(z)$ and $\pi(x) \cap H = \pi(x) \cap \pi(y) \subseteq \pi(z)$, so $\pi(z)$ contains $\langle u, \pi(x) \cap H \rangle_{\mathbb{P}} = H$. We conclude that $H = \pi(z)$, a contradiction.

Thus, the conditions of Lemma 7.1.8 are satisfied, so there is a duality $\overline{\pi}$ of the geometry $\Gamma(\mathbb{P})$ coinciding with π on the point shadow.

It remains to show that the duality $\overline{\pi}$ has order two if $n > 1$. If a is a point, the equality $\overline{\pi}(\overline{\pi}(a)) = x$, where x is a point of $\mathbb{P}$, means that for all $b \in \pi(a)$, one has $x \in \pi(b)$. Now $x = a$ is a solution. This solution is unique, for $\pi(x) \supseteq \pi(a)$ (as $\pi(x)$ contains each element of $\pi(a)$), whence $\pi(x) = \pi(a)$ and so $x = a$ (as π is injective). Therefore $\overline{\pi}(\overline{\pi}(a)) = a$ for any point a and so $\overline{\pi}$ is of order two on the set of all elements of $\mathbb{P}$. $\qquad\square$

Corollary 7.1.11 *If x, y, z are distinct collinear points of a projective space with quasi-polarity π, then $\pi(x) \cap \pi(y) = \pi(x) \cap \pi(z)$.*

Proof This is derived in the beginning of the proof of Proposition 7.1.10 (without recourse to the assumptions of injectivity and nondegeneracy of π). $\qquad\square$

7.2 Duality and Sesquilinear Forms

As we have seen in Exercise 1.9.11, for every division ring $\mathbb{D}$ there is an *opposite division ring* $\mathbb{D}^{\mathrm{op}}$ whose elements and whose addition are those of $\mathbb{D}$ and whose multiplication $*$ is given by $a * b = ba$ for $a, b \in \mathbb{D}$. Clearly, $(\mathbb{D}^{\mathrm{op}})^{\mathrm{op}} = \mathbb{D}$, while $\mathbb{D}^{\mathrm{op}} = \mathbb{D}$ if and only if $\mathbb{D}$ is commutative (i.e., a field).

Recall also from Exercise 1.9.11 that an anti-automorphism σ of $\mathbb{D}$ is an isomorphism of $\mathbb{D}$ onto $\mathbb{D}^{\mathrm{op}}$, i.e., a bijection $\sigma : \mathbb{D} \to \mathbb{D}^{\mathrm{op}}$ such that $\sigma(a+b) = \sigma(a) + \sigma(b)$ and $\sigma(ab) = \sigma(b)\sigma(a) = \sigma(a) * \sigma(b)$ for all $a, b \in \mathbb{D}$. We will see that the existence of a duality of $\mathrm{PG}(\mathbb{D}^{n+1})$ for $n > 1$ is equivalent to the existence of an anti-automorphism of $\mathbb{D}$.

Throughout this section, V is a right vector space over the division ring $\mathbb{D}$.

Example 7.2.1 Here are two familiar anti-automorphisms.

(i) If $\mathbb{D}$ is a field, anti-automorphisms of $\mathbb{D}$ coincide with automorphisms. This is the only case where the identity map is an anti-automorphism.

(ii) The division ring $\mathbb{H}$ of the *real quaternions* has elements $x_1 + x_2 i + x_3 j + x_4 k$, where x_1, x_2, x_3, x_4 are reals with the usual vector space structure of $\mathbb{R}^4$; its multiplication is determined by the rules $i^2 = j^2 = k^2 = -1$, $ij = k = -ji$, $jk = i = -kj$, $ki = j = -ik$ (cf. Exercise 1.9.11, where the rational quaternions were introduced). The map $x_1 + x_2 i + x_3 j + x_4 k \mapsto x_1 - x_2 i - x_3 j - x_4 k$ is an anti-automorphism, usually called *conjugation* and denoted by $x \mapsto \overline{x}$ ($x \in \mathbb{H}$). The composition of an anti-automorphism and an automorphism (e.g. $x \mapsto axa^{-1}$ for some nonzero $a \in \mathbb{H}$) is again an anti-automorphism. The element $\overline{x}x$ belongs to $\mathbb{R} \cdot 1$, and is called the *norm* of x and denoted by $N(x)$. The inverse of $x \in \mathbb{H} \setminus \{0\}$ can be conveniently expressed in terms of x itself: $x^{-1} = N(x)^{-1} \cdot \overline{x}$.

Example 7.2.2 In order to show that there exist non-commutative division rings having no anti-automorphisms, we revisit Example 1.4.7. We let $\mathbb{K} = \mathbb{F}(u)$ be a field admitting an automorphism σ of order three. The fixed points of σ form a subfield $\mathbb{F}$ of $\mathbb{K}$. We suppose that $a \in \mathbb{F}$ is not in the image of the *norm map* $N : \mathbb{K} \to \mathbb{F}$, given by $N(x) = x\sigma(x)\sigma^2(x)$ ($x \in \mathbb{K}$) and put $\mathbb{D} = \mathbb{K} \oplus \mathbb{K}j \oplus \mathbb{K}j^2$ with $jx = \sigma(x)j$ for $x \in \mathbb{K}$ and $j^3 = a$. Example 1.4.7, Step 2, shows that $\mathbb{D}$ is indeed a division ring. In addition, we suppose that $\mathrm{Aut}(\mathbb{F}) = 1$. Example 1.4.7, Step 3, satisfies this condition as $\mathrm{Aut}(\mathbb{Q}) = 1$.

We prove that $\mathbb{D}$ has no anti-automorphisms. Suppose that ϕ is one. Then ϕ fixes all elements of $\mathbb{F}$, as, by Example 1.4.7, Step 1, $\mathbb{F}$ is the center of $\mathbb{D}$ (and hence invariant under ϕ, so ϕ fixes $\mathbb{F}$ element-wise as $\mathrm{Aut}(\mathbb{F}) = 1$). We show that we can choose ϕ so that it fixes $\mathbb{K}$ element-wise. Put $v = \phi(u)$.

Right multiplication by an element $b = b_0 + b_1 j + b_2 j^2 \in \mathbb{D}$ is a $\mathbb{K}$-linear transformation with matrix

$$M_b = \begin{pmatrix} b_0 & b_1 & b_2 \\ \sigma(b_2)a & \sigma(b_0) & \sigma(b_1) \\ \sigma^2(b_1)a & \sigma^2(b_2)a & \sigma^2(b_0) \end{pmatrix}. \tag{7.2}$$

with respect to the basis $1, j, j^2$. Recall that we view $\mathbb{D}$ as a left vector space over $\mathbb{K}$. Accordingly, we view vectors as row vectors and have matrices acting on the right, so the i-th row of the matrix is the image of the i-th basis vector. In fact, this gives another representation of $\mathbb{D}$, namely as the set of matrices over $\mathbb{K}$ of the form (7.2), with the usual (matrix) multiplication and addition as ring operations. The map $b \mapsto M_b$ establishes the ring isomorphism. Applying the Cayley-Hamilton Theorem to M_b and translating the result to b by means of this isomorphism, we have

$$b^3 = \alpha_0(b) + \alpha_1(b)b + \alpha_2(b)b^2, \tag{7.3}$$

where $\alpha_i(b)$ are the coefficients of the polynomial $\det(XI_3 - M_b)$ in the indeterminate X (here I_3 stands for the identity matrix of dimension 3), so

$$\alpha_0(b) = \det(M_b),$$

$$\alpha_1(b) = -\det \begin{pmatrix} \sigma(b_0) & \sigma(b_1) \\ \sigma^2(b_2)a & \sigma^2(b_0) \end{pmatrix} - \det \begin{pmatrix} b_0 & b_1 \\ \sigma(b_2)a & \sigma(b_0) \end{pmatrix}$$

$$-\det \begin{pmatrix} b_0 & b_2 \\ \sigma^2(b_1)a & \sigma^2(b_0) \end{pmatrix},$$

$$\alpha_2(b) = \mathrm{Tr}(M_b) = b_0 + \sigma(b_0) + \sigma^2(b_0).$$

Observe that $\alpha_i(b) \in \mathbb{F}$, as they are fixed by σ.

In the notation of (7.3), we have $\alpha_2(v)v^2 + \alpha_1(v)v + \alpha_0(v) = v^3 = \phi(u)^3 = \phi(u^3) = \phi(\alpha_2(u)u^2 + \alpha_1(u)u + \alpha_0(u)) = \alpha_2(u)v^2 + \alpha_1(u)v + \alpha_0(u)$, because the coefficients $\alpha_i(u)$ belong to $\mathbb{F}$. Since $1, u, u^2$ are linearly independent over $\mathbb{F}$, so are $1, v, v^2$. Hence we must have $\alpha_i(u) = \alpha_i(v)$ for $i = 0, 1, 2$. In particular, (7.3) yields

$$u^3 - \alpha_2(v)u^2 - \alpha_1(v)u - \alpha_0(v) = 0.$$

By definition of the α_i, this equation for $v = v_0 + v_1 j + v_2 j^2$ with $v_j \in \mathbb{K}$ can be rewritten as $\det(uI_3 - M_v) = 0$. Consequently, there is a nonzero solution $(c_0, c_1, c_2) \in \mathbb{K}^3$ of the equation $(c_0, c_1, c_2)(uI_3 - M_v) = 0$. For such a solution, the element $c = c_0 + c_1 j + c_2 j^2$ satisfies $uc - cv = 0$. Since c is nonzero, it follows that $u = cvc^{-1}$. Hence the map $x \mapsto c\phi(x)c^{-1}$ is an anti-automorphism of $\mathbb{D}$ fixing u.

Consequently, we (may) suppose that ϕ fixes u. As it also fixes $\mathbb{F}$, it follows that ϕ fixes all of $\mathbb{K}$, so $juj^{-1} = \sigma(u) = \phi(\sigma(u)) = \phi(juj^{-1}) = \phi(j)^{-1}u\phi(j)$, giving $\phi(j)j \in C_\mathbb{D}(u) = C_\mathbb{D}(\mathbb{K}) = \mathbb{K}$ by use of Example 1.4.7, Step 1. Hence $\phi(j) = kj^2$ for some nonzero $k \in \mathbb{K}$. Now $a = \phi(a) = \phi(j^3) = (\phi(j))^3 = (kj^2)^3 = k\sigma^2(k)\sigma(k)a^2 = N(k)a^2$, which leads to $a = N(k^{-1})$, a contradiction with the choice of a. It follows that $\mathbb{D}$ admits no anti-automorphisms.

In view of Corollary 6.3.4 and Exercise 5.7.13, we conclude that the Desarguesian projective plane $\mathbb{P}(\mathbb{D}^3)$ is not isomorphic to its dual.

Definition 7.2.3 Let σ be an anti-automorphism of $\mathbb{D}$. A σ-*sesquilinear form* on V, or *sesquilinear form relative to* σ, is a map $f : V \times V \to \mathbb{D}$ such that, for all $a, b, c, d \in V$ and all $\lambda, \mu \in \mathbb{D}$,

$$f(a + b, c + d) = f(a, c) + f(b, c) + f(a, d) + f(b, d),$$

$$f(a\lambda, b\mu) = \sigma(\lambda)f(a, b)\mu.$$

The form f determines σ uniquely except when $f = 0$. Indeed, if $f(a, b) \neq 0$ for certain $a, b \in V$, then $\sigma(\lambda) = f(a\lambda, b)f(a, b)^{-1}$.

Example 7.2.4 Let I be an index set and V a vector space with basis $(e_i)_{i \in I}$. The prototype of a σ-sesquilinear form is obtained from a choice of a matrix $M = (f_{ij})_{i,j \in I}$ with entries in $\mathbb{D}$. For vectors $x = \sum_i e_i x_i$ and $y = \sum_i e_i y_i$ of V, put $f(x, y) = \sum_{i,j} \sigma(x_i) f_{ij} y_j = \sigma(x)^\top M y$. Here, $\sigma(x) = \sum_i e_i \sigma(x_i)$ and y are viewed as column vectors. The resulting map f is a sesquilinear form relative to σ.

If $\mathbb{D}$ is a field and σ is the identity, then a σ-sesquilinear form on V is just a bilinear form. The matrix M is the usual Gram matrix.

Recall the conjugation $x \mapsto \overline{x}$ of Example 7.2.1(ii) on $\mathbb{H}$, the division ring of real quaternions. The map $(x, y) \mapsto \overline{x}y$ is a sesquilinear form of the one-dimensional right vector space $\mathbb{H}$ over itself relative to $\sigma : x \mapsto \overline{x}$. The matrix M here is the 1×1 identity matrix. The map $(x, y) \mapsto (\overline{x}y + \overline{y}x)$ is a bilinear form of the four-dimensional right vector space $\mathbb{H}$ over $\mathbb{R}$. The corresponding matrix M with respect to the basis $1, i, j, k$ is a diagonal matrix with entries $2, -2, -2, -2$.

Lemma 7.2.5 *Let V be a right vector space over $\mathbb{D}$ and σ an anti-automorphism of $\mathbb{D}$.*

(i) *If f is a σ-sesquilinear form on V and $\lambda \in \mathbb{D} \setminus \{0\}$, then the map λf is a sesquilinear form relative to the anti-automorphism ρ given by $\rho(\alpha) = \lambda \sigma(\alpha) \lambda^{-1}$ ($\alpha \in \mathbb{D}$).*
(ii) *The σ-sesquilinear forms on V constitute an additive group under the usual addition of maps.*

Proof (i) For $a, b \in V$ and $\alpha, \beta \in \mathbb{D}$, we have $(\lambda f)(a\alpha, b\beta) = \lambda f(a\alpha, b\beta) = \lambda \sigma(\alpha) f(a, b)\beta = (\lambda \sigma(\alpha)\lambda^{-1})\lambda f(a, b)\beta = \rho(\alpha)(\lambda f)(a, b)\beta$.
(ii) Obvious. $\qquad\qquad\qquad\qquad\qquad\qquad\qquad\qquad\qquad\qquad\qquad\square$

Definition 7.2.6 The forms f and λf are called *proportional*. The set $\{\lambda f \mid \lambda \in \mathbb{D}\}$ is called the *proportionality class* of f.

The roles of proportional sesquilinear forms will be indistinguishable from the projective point of view. Later on, in Remark 7.2.9, we will make this observation more precise.

Using a basis of V, we see below that Example 7.2.4 comprises all σ-sesquilinear forms on V.

Proposition 7.2.7 *If $(e_i)_{i \in I}$ is a basis of V and if σ is an anti-automorphism of $\mathbb{D}$, then every σ-sesquilinear form on V is a map $f : V \times V \to \mathbb{D}$ such that, for certain $f_{ij} \in \mathbb{D}$,*

$$f(x, y) = \sum_{i,j \in I} \sigma(x_i) f_{ij} y_j.$$

Conversely, such maps are σ-sesquilinear forms.

Proof The last assertion was already made in Example 7.2.4 and is easily verified. Putting $f_{ij} = f(e_i, e_j)$ and letting $x = \sum_i e_i x_i$, $y = \sum_i e_i y_i$ be vectors of V, we have

$$f(x, y) = f\left(\sum_i e_i x_i, \sum_i e_i y_i \right) = \sum_{i,j} f(e_i x_i, e_j y_j) = \sum_{i,j} \sigma(x_i) f(e_i, e_j) y_j$$

$$= \sum_{i,j} \sigma(x_i) f_{ij} y_j. \qquad \qquad \square$$

We will now derive dualities from sesquilinear forms.

Notation 7.2.8 Let f be a sesquilinear form on V and $X \subseteq V$. By $f(v, X)$, for $v \in V$, we mean $\{ f(v, x) \mid x \in X \}$. We write $X^{\perp_f}$, or, if f is clear from the context, just $X^{\perp}$ to denote the linear subspace $\{ x \in V \mid f(X, x) = \{0\} \}$ of V. By $X^{\perp\perp}$ we mean $(X^{\perp})^{\perp}$ and by $a^{\perp}$ we mean $\{a\}^{\perp}$.

Remark 7.2.9 Given a σ-sesquilinear form f on V and a vector a in V, the map $f_a : V \to \mathbb{D}$ given by $x \mapsto f(a, x)$ $(x \in V)$ is a linear form on V, hence its kernel $a^{\perp}$ is either a hyperplane of V (and of $\mathbb{P}(V)$) or V itself.

For $\lambda \neq 0$, we have $(a\lambda)^{\perp} = a^{\perp}$ so $a^{\perp}$ depends only on the choice of $\langle a \rangle \in \mathbb{P}(V)$. Indeed, in the notation of Proposition 7.2.7, the equation in $x = (x_j)_{j \in I}$ of $a^{\perp}$ is $\sum_{i,j \in I} \sigma(a_i) f_{ij} x_j = 0$, which is equivalent to $\sum_{i,j \in I} \sigma(\lambda) \sigma(a_i) f_{ij} x_j = 0$, whence to $\sum_{i,j \in I} \sigma(a_i \lambda) f_{ij} x_j = 0$.

Also, $a^{\perp_f} = a^{\perp_{\lambda f}}$, as the map $(\lambda f)_a$ coincides with $f_{a \sigma^{-1}(\lambda)}$.

Definition 7.2.10 Let f be a σ-sesquilinear form on V. The *radical* of f, notation $\mathrm{Rad}(f)$, is the set of all $a \in V$ such that $a^{\perp} = V$. This is obviously a subspace of V. Notice that this definition extends the one given in Example 1.4.13 (and used in Definition 4.4.3).

The form f is called *nondegenerate* if $\mathrm{Rad}(f) = 0$ (equivalently, for $\dim(V) < \infty$, if the matrix $(f_{ij})_{i,j \in I}$ has maximal rank). We say that f is *reflexive* if, for each x and y in V, the relation $f(x, y) = 0$ implies $f(y, x) = 0$.

Notation 7.2.11 For a sesquilinear form f on V, we denote by δ_f the map $\mathbb{P}(V) \to \mathbb{P}(V)^{\star} \cup \{\mathbb{P}(V)\}$ given by $\delta_f(\langle x \rangle) = x^{\perp_f}$ for $x \in V \setminus \{0\}$.

Here is the promised generalization of Example 7.1.1(iii) to σ-sesquilinear forms f. (There $\sigma = \mathrm{id}$.)

Theorem 7.2.12 *Let f be a σ-sesquilinear form on V for some anti-automorphism σ of $\mathbb{D}$. Then δ_f does not depend on the choice of f from its proportionality class. Moreover, the following holds.*

(i) *If f is reflexive, then δ_f is a quasi-polarity of $\mathbb{P}(V)$ with kernel $\mathrm{Rad}(f)$.*
(ii) *If $\dim(V) < \infty$ and f is nondegenerate, then the map δ_f determines a duality of $\mathrm{PG}(V)$.*

Proof The independence of δ_f of the choice of f from its proportionality class is mentioned in Remark 7.2.9.

(i) By Remark 7.2.9, $\delta_f(\langle y \rangle)$ is either a hyperplane or all of $\mathbb{P}(V)$. Suppose $\langle x \rangle \in \delta_f(\langle y \rangle)$. In terms of the form, $f(y, x) = 0$. As f is reflexive, this is equivalent to $f(x, y) = 0$, which implies $\langle y \rangle \in \delta_f(\langle x \rangle)$. Hence δ_f is a quasi-polarity.

(ii) We show that δ_f is a bijection from the set of points of $\mathbb{P}(V)$ onto the set of hyperplanes of $\mathbb{P}(V)$. Take a hyperplane H in $\mathbb{P}(V)$. Then H is the kernel of some linear form g on V. There is a vector $a \in V$ such that $f(a, x) = g(x)$ for all $x \in V$; for, using a basis $(e_i)_{i \in I}$ as in Proposition 7.2.7, and putting $g_i = g(e_i)$ for all i, such an a would satisfy $\sum_i \sigma(a_i) f_{ij} = g_j$ for all j or $\sum_i \sigma^{-1}(f_{ij}) a_i = \sigma^{-1}(g_j)$ and this system of linear equations has a solution $a = \sum_i e_i a_i$ because the matrix $(f_{ij})_{ij}$ is of maximal rank. Hence $a^\perp = H$. Moreover, if $a^\perp = b^\perp$, then there is a nonzero scalar $\lambda \in \mathbb{D}$ such that $\sum_i \sigma(a_i) f_{ij} = \lambda(\sum_i \sigma(b_i) f_{ij})$ for all j. Again, as $(f_{ij})_{ij}$ has maximal rank, we find $\sigma(a_i) - \sigma(b_i \sigma^{-1}(\lambda)) = 0$ for all $i \in I$ and so $\langle a \rangle = \langle b \rangle$.

Now, let $a, b, c \in V \setminus \{0\}$ be such that $\langle a \rangle$, $\langle b \rangle$, $\langle c \rangle$ are distinct. Assume first that these points are collinear. Then $c = a\lambda + b\mu$ for some $\lambda, \mu \in \mathbb{D}$ and so $f(c, x) = 0$ if and only if $f(a\lambda, x) + f(b\mu, x) = 0$, forcing $c^\perp$ to contain $a^\perp \cap b^\perp$. In view of the symmetry of the roles of a, b, c it follows that $a^\perp \cap b^\perp = b^\perp \cap c^\perp = c^\perp \cap a^\perp$.

Conversely, assume that $a^\perp \cap b^\perp = b^\perp \cap c^\perp = c^\perp \cap a^\perp$. Take linearly independent vectors v_1, v_2 complementary to $a^\perp \cap b^\perp$. Since a and b are linearly independent, the matrix

$$\begin{pmatrix} f_a(v_1) & f_b(v_1) \\ f_a(v_2) & f_b(v_2) \end{pmatrix}$$

has rank two. Hence there are unique $\lambda, \mu \in \mathbb{D}$ with $f_c(v_i) = \lambda f_a(v_i) + \mu f_b(v_i)$ for $i = 1, 2$. Consequently $f_c = \lambda f_a + \mu f_b$ and, for $x \in V$, we have

$$f(c, x) = f_c(x) = \lambda f(a, x) + \mu f(b, x)$$
$$= f\left(a\sigma^{-1}(\lambda), x\right) + f\left(b\sigma^{-1}(\mu), x\right) = f\left(a\sigma^{-1}(\lambda) + b\sigma^{-1}(\mu), x\right).$$

Since $\mathrm{Rad}(f) = 0$, we obtain $c = a\sigma^{-1}(\lambda) + b\sigma^{-1}(\mu)$. Thus $\langle a \rangle$, $\langle b \rangle$, $\langle c \rangle$ are collinear; now Lemma 7.1.8 applies. $\qquad\square$

The above proof of (ii) is in terms of the projective space $\mathbb{P}(V)$. Exercise 7.11.4 deals with a proof in terms of the projective geometry.

Notation 7.2.13 For a sesquilinear reflexive form f on V, we also write $\mathbb{P}(V)_f$ rather than $\mathbb{P}(V)_{\delta_f}$ as in Definition 7.1.9.

Here is a converse to Theorem 7.2.12.

Theorem 7.2.14 *Let* $\mathbb{P} = \mathbb{P}(V)$ *and suppose that* π *is a quasi-polarity of* $\mathbb{P}$. *The kernel* K *of* π *is a subspace of* $\mathbb{P}$ *and* π *defines a homomorphism* $\mathbb{P}/K \to \mathbb{P}^\star$.

(i) *There is a reflexive* σ-*sesquilinear form* f *on* V, *relative to some anti-automorphism* σ *of* $\mathbb{D}$, *such that* $\pi = \delta_f$ *and* $\mathrm{Rad}(f) = K$.

(ii) *The quasi-polarity π is nondegenerate if and only if the sesquilinear form f is nondegenerate.*

Proof First observe that K is a subspace of $\mathbb{P}$. For, if x and y are distinct points of K and z belongs to the projective line xy on x and y, then, for each point u of $\mathbb{P}$, we have $u \in \mathbb{P} = \pi(x) \cap \pi(y)$, so $x, y \in \pi(u)$, which implies $z \in \pi(u)$ as $\pi(u)$ is a subspace of $\mathbb{P}$, and so $u \in \pi(z)$. This settles $\pi(z) = \mathbb{P}$, so $z \in K$.

If x and y are distinct points of $\mathbb{P}$ with $\langle K, x \rangle = \langle K, y \rangle$, then xy meets K is a point, say z, so $\pi(z) = \mathbb{P}$ and Corollary 7.1.11 gives $\pi(x) = \pi(x) \cap \pi(z) = \pi(y) \cap \pi(z) = \pi(y)$. Therefore, the map π naturally induces a map $\pi_K : \mathbb{P}/K \to \mathbb{P}^\star$ with $\pi_K(\langle K, x \rangle) = \pi(x)$. By Corollary 7.1.11 the map π_K is a homomorphism of projective spaces.

(i) Recall from Exercise 5.7.13 that $\mathbb{P}^\star = \mathbb{P}(V^\vee)$ and that $V^\vee$ is a right vector space over $\mathbb{D}^{\mathrm{op}}$ (equivalently, a left vector space over $\mathbb{D}$). By Corollary 6.3.4, there is an anti-automorphism σ of $\mathbb{D}$ and a σ-linear map $\psi_K : V/K \to V^\vee$, whose projectivization is π_K. Thus,

$$\pi_K(\langle x, K \rangle) = \left\{ y \in V \mid \psi_K(x + K)y = 0 \right\}.$$

Let ψ be the σ-linear map $V \to V^\vee$ with $\psi(x) = \psi_K(\langle x, K \rangle)$ for each $x \in V$. The map $f : V \times V \to \mathbb{D}$ given by $f(x, y) = \psi(x)(y)$ is a σ-sesquilinear form as required. The reflexivity of f follows from the equivalence of $x \in \pi(y)$ and $y \in \pi(x)$ for points x, y of $\mathbb{P}(V)$.

(ii) If $x \in \mathbb{P}(V)$, then $x \in K$ means $\pi(x) = \mathbb{P}(V)$, which is equivalent to $\psi(x) = 0$ and hence to $f(x, y) = 0$ for all $y \in \mathbb{P}(V)$. So $K = \emptyset$, which is equivalent to π being nondegenerate, implies $\mathrm{Rad}(f) = 0$, that is, f is nondegenerate. $\square$

Remark 7.2.15 For V and f as in Theorem 7.2.12, we determine the semi-linear transformation $f' : V \to V^\vee$ pertaining to δ_f. Write vectors of V out in coordinates with respect to a basis $(e_i)_{i \in I}$ of V, and vectors of $V^\vee$ with respect to a dual basis $(e_i^\vee)_{i \in I}$ of $V^\vee$. The point $(x_i)_i \mathbb{D}$ of $\mathbb{P}(V)$ is mapped by f' onto the point $(x_j')_j \mathbb{D}^{\mathrm{op}}$ of $\mathbb{P}(V^\vee)$ where $x_j' = \sum_i \sigma(x_i) f_{ij}$ for some matrix $(f_{ij})_{i,j \in I}$ of maximal rank. Indeed, with $f'(e_i) = \sum_{j \in I} e_j^\vee f_{ij}$, and using $*$ to denote multiplication in $\mathbb{D}^{\mathrm{op}}$, we have

$$f'(x) = \sum_{i \in I} f'(e_i x_i) = \sum_{i \in I} f'(e_i) \sigma(x_i) = \sum_{i,j \in I} e_j^\vee f_{ij} * \sigma(x_i)$$

$$= \sum_{j \in I} e_j^\vee \left(\sum_{i \in I} \sigma(x_i) f_{ij} \right) = \sum_{j \in I} e_j^\vee x_j'.$$

The hyperplane $\delta(x_i)$ consists of all points $(y_j)_j$ such that $\sum_{i,j} \sigma(x_i) f_{ij} y_j = 0$ and so $\delta = \delta_f$, where f is the form determined by $(f_{ij})_{i,j \in I}$.

Example 7.2.16 There exist projective spaces admitting a duality but no polarity. As a consequence of Corollary 6.3.4 and Exercise 5.7.13, an instance of such a space of big enough dimension boils down to a division ring with an anti-automorphism but no involutory anti-automorphisms. We exhibit such an example in five steps, starting again from the division rings constructed in Example 1.4.7.

Let $\mathbb{K}$ be a Galois field extension of $\mathbb{F}$ of degree three and let σ be a generator of its Galois group. Again, we take $a \in \mathbb{F}$ outside the image of the *norm map* $N : \mathbb{K} \to \mathbb{F}$ and let $\mathbb{D}$ be defined as in Example 1.4.7, so $\mathbb{D} = \mathbb{K} \oplus \mathbb{K}j \oplus \mathbb{K}j^2$ with the multiplication rules $jx = \sigma(x)j$ for $x \in \mathbb{K}$ and $j^3 = a$. We have seen that then $\mathbb{D}$ is a division algebra.

STEP 1. $\mathbb{D}$ *has no anti-automorphism* φ *such that* $\varphi(x) = x$ *for all* $x \in \mathbb{F}$.

By way of contradiction, suppose $\varphi \colon \mathbb{D} \to \mathbb{D}$ is an anti-automorphism that restricts to the identity on $\mathbb{F}$. We first show that we may assume $\varphi(x) = x$ for all $x \in \mathbb{K}$.

Let $k \in \mathbb{K} \setminus \mathbb{F}$, so $\mathbb{K} = \mathbb{F}(k)$, and let $X^3 + aX^2 + bX + c \in \mathbb{F}[X]$ be the minimal polynomial of k over $\mathbb{F}$ (that is, the minimal polynomial of M_k as defined in Example 7.2.2). The element $\varphi(k) \in \mathbb{D}$ is a root of this polynomial. Let

$$u = k^2 + k\varphi(k) + \varphi(k)^2 + a\big(k + \varphi(k)\big) + b \quad \text{and}$$
$$v = k^2 + \varphi(k)k + \varphi(k)^2 + a\big(k + \varphi(k)\big) + b.$$

The equation $k^3 + ak^2 + bk + c = \varphi(k)^3 + a\varphi(k)^2 + b\varphi(k) + c$ (expressing that k and $\varphi(k)$ have the same minimal polynomial) implies

$$ku = u\varphi(k) \quad \text{and} \quad vk = \varphi(k)v.$$

If $u = v$, then comparison of their definitions gives $k\varphi(k) = \varphi(k)k$, so, by Example 1.4.7 Step 1, $\varphi(k) \in C_{\mathbb{D}}(k) = \mathbb{K}$. Hence $\varphi(k) = k$, $\sigma(k)$, or $\sigma^2(k)$. If $\varphi(k) = \sigma(k)$ we replace φ by $\psi \colon x \mapsto j^{-1}\varphi(x)j$; if $\varphi(k) = \sigma^2(k)$ we replace φ by $\psi \colon x \mapsto j^{-2}\varphi(x)j^2$. Thus, in each case we obtain an anti-automorphism ψ of $\mathbb{D}$ such that $\psi(x) = x$ for all $x \in \mathbb{K}$. Therefore, we may assume that at least one of u, v is nonzero.

If $u \neq 0$, we replace φ by the map $\psi \colon x \mapsto u\varphi(x)u^{-1}$; if $v \neq 0$ we replace it by the map $\psi \colon x \mapsto v^{-1}\varphi(x)v$. In each case, ψ is an anti-automorphism which restricts to the identity on $\mathbb{F}$ and such that $\psi(k) = k$, hence $\psi(x) = x$ for all $x \in \mathbb{K}$.

Taking the image under ψ of the equation $jk = \sigma(k)j$, we find

$$k\psi(j) = \psi(j)\sigma(k),$$

so $k\psi(j)j = \psi(j)jk$, and again Example 1.4.7 Step 1 gives $\psi(j)j \in \mathbb{K}$. Let $\lambda = \psi(j)j$, so

$$\psi(j)^3 = \big(\lambda j^{-1}\big)^3 = \lambda\sigma(\lambda)\sigma^2(\lambda)j^{-3}.$$

Since $j^3 = a$ we have $\psi(j)^3 = a$; hence, with N the norm of $\mathbb{K}$ over $\mathbb{F}$

$$a = \lambda\sigma(\lambda)\sigma^2(\lambda)a^{-1} = N(\lambda)a^{-1}$$

and therefore $a^2 = N(\lambda)$, so $a = N(a\lambda^{-1})$, a contradiction.

We now provide a specific extension of the rationals.

STEP 2. *Let* $\mathbb{K} = \mathbb{Q}(u, v) \subset \mathbb{C}$, *where*

$$u^4 + 10u^2 + 20 = 0 \quad \textit{and} \quad v^3 - 18v + 18 = 0.$$

Then $\mathbb{K}$ is a Galois extension of the rationals of degree 12 with Galois group $\langle \sigma, \tau \rangle$, where

$$\tau(u) = \frac{2(u^2 + 5)}{u}, \qquad \tau(v) = v,$$

$$\sigma(u) = u, \qquad \sigma(v) = \frac{1}{2}\left(-v \pm \frac{1}{u^2 + 5}(-24 + 3v + 2v^2)\right).$$

These automorphisms satisfy $\sigma^3 = \tau^4 = I$ and $\tau\sigma = \sigma^2\tau$.

Eisenstein's criterion shows that the polynomials $X^4 + 10X^2 + 20$ and $X^3 - 18X + 18$ are irreducible over $\mathbb{Q}$, so $[\mathbb{K}:\mathbb{Q}] = 12$. Notice that $(u^2 + 5)^2 = 5$, so $\mathbb{Q}(u) \supset \mathbb{Q}(\sqrt{5})$ (and in fact $u^2 = -5 + \sqrt{5}$). The extension $\mathbb{Q}(u)$ of $\mathbb{Q}$ is Galois, with a cyclic Galois group generated by $\tau|_{\mathbb{Q}(u)}$.

On the other hand, the discriminant of $X^3 - 18X + 18$ is $5 \cdot (54)^2$, so all of the roots of this polynomial are in $\mathbb{Q}(\sqrt{5})(v)$. The roots are in fact

$$v \quad \text{and} \quad \frac{1}{2}\left(-v \pm \frac{1}{\sqrt{5}}(-24 + 3v + 2v^2)\right).$$

Therefore σ defines an automorphism of $\mathbb{K}$.

Finally, the relations between σ and τ are easy to verify by explicit computation. They imply that the group generated by σ and τ has order 12. Since $[\mathbb{K}:\mathbb{Q}] = 12$, this group must be the Galois group of $\mathbb{K}$ over $\mathbb{Q}$.

STEP 3. *The number 11 is not a norm of $\mathbb{Q}(v)$ over $\mathbb{Q}$.*

For, suppose $x_0, x_1, x_2 \in \mathbb{Q}$ are such that

$$11 = N_v\big(x_0 + x_1 v + x_2 v^2\big),$$

where N_v denotes the norm of $\mathbb{Q}(v)$ over $\mathbb{Q}$ (so $N_v(x) = x\sigma(x)\sigma^2(x)$ for $x \in \mathbb{Q}(v)$). Clearing denominators, we find integers y_0, y_1, y_2, z such that

$$N_v\big(y_0 + y_1 v + y_2 v^2\big) = 11z^3, \tag{7.4}$$

and we may assume that y_0, y_1, y_2 are not all divisible by 11. Now, observe that the polynomial $X^3 - 18X + 18$ is irreducible over the finite field $\mathbb{F}_{11}$. Reducing modulo 11 the polynomial

$$N_v\big(Y_0 + Y_1 v + Y_2 v^2\big) = \det\begin{pmatrix} Y_0 & -18Y_2 & -18Y_1 \\ Y_1 & Y_0 + 18Y_2 & 18(Y_1 - Y_2) \\ Y_2 & Y_1 & Y_0 + 18Y_2 \end{pmatrix} \in \mathbb{Z}[Y_0, Y_1, Y_2]$$

yields the polynomial $N_{\mathbb{F}_{11^3}/\mathbb{F}_{11}}(Y_0 + Y_1 w + Y_2 w^2) \in \mathbb{F}_{11}[Y_0, Y_1, Y_2]$, where $N_{\mathbb{F}_{11^3}/\mathbb{F}_{11}}$ denotes the norm of $\mathbb{F}_{11^3}$ over $\mathbb{F}_{11}$ and $w \in \mathbb{F}_{11^3}$ is a root of $X^3 - 18X + 18$. Since the norm polynomial has no non-trivial zero, it follows from (7.4) that y_0, y_1, and y_2 are all divisible by 11, a contradiction.

STEP 4. *Let $\mathbb{F} = \mathbb{Q}(u)$ and $a = 11$. Then $\mathbb{F}$ is the subfield of $\mathbb{K}$ fixed under σ, so $\mathbb{K}$ is a Galois extension of $\mathbb{F}$ of degree three and $a \in \mathbb{F} \setminus N(\mathbb{K})$.*

For, otherwise, $11 = N(x)$ for some $x \in \mathbb{K}$, and, taking the norm of $\mathbb{F}$ over $\mathbb{Q}$ on each side, we obtain

$$11^4 = N_{\mathbb{K}/\mathbb{Q}}(x),$$

where $N_{\mathbb{K}/\mathbb{Q}}$ is the norm of $\mathbb{K}$ over $\mathbb{Q}$, so $N_{\mathbb{K}/\mathbb{Q}}(x) = \prod_{\gamma \in \langle \sigma, \tau \rangle} \gamma(x) = N_v(N(x))$ for $x \in \mathbb{K}$. Consequently, $11 = N_v(N(x)/11)$. By Step 3, this is impossible.

STEP 5. *Take $\mathbb{K}$ and σ as in Step 2 and $\mathbb{F}$ and a as in Step 4. Then $\mathbb{D}$ is a division ring. Moreover, there exists an anti-automorphism φ of $\mathbb{D}$ such that*

$$\varphi(x) = \tau(x) \quad \text{for } x \in \mathbb{K} \quad \text{and} \quad \varphi(j) = j,$$

but $\mathbb{D}$ admits no involutory anti-automorphism.

By Step 4 and Example 1.4.7, $\mathbb{D}$ is a division ring. Straightforward computations show that φ is an anti-automorphism.

By Steps 1 and 4, $\mathbb{D}$ has no anti-automorphism whose restriction to $\mathbb{F}$ is the identity. Suppose $\psi \colon \mathbb{D} \to \mathbb{D}$ is an anti-automorphism such that $\psi^2 = \mathrm{id}$. Since $\psi|_{\mathbb{F}}^2 = \mathrm{id}$ and the Galois group of $\mathbb{F}$ over $\mathbb{Q}$ is $\langle \tau \rangle$, we have either $\psi|_{\mathbb{F}} = \mathrm{id}$ or $\psi|_{\mathbb{F}} = \tau^2$. Then either ψ or $\varphi^2 \circ \psi$ is an anti-automorphism whose restriction to $\mathbb{F}$ is the identity, a contradiction.

7.3   Absolutes and Reflexive Sesquilinear Forms

The absolute geometries (cf. Definition 1.4.12) corresponding to polarities of projective geometries are very interesting subgeometries; they are related to the so-called classical (simple) groups. The corresponding spaces satisfy the polar space axiom, to be introduced later, in Definition 7.4.1.

Definition 7.3.1 Let $\Gamma = (X, *, \tau)$ be a geometry over $[n]$ belonging to the linear digon diagram $\mathcal{I}(\Gamma)$ of rank n. As we discussed above, a polarity π of Γ maps an element of type 1 to an element of type n and, more generally, an element of type i to an element of type $n + 1 - i$. An element $x \in X$ is said to be *absolute* with respect to π if $x * \pi(x)$ and $\tau(x) \le \tau(\pi(x))$.

Notation 7.3.2 We write $\mathrm{Abs}(\Gamma, \pi)$ for the subgeometry of Γ over $[\lfloor n/2 \rfloor]$ induced on the set of all elements of Γ that are absolute with respect to π.

The absolute geometry with respect to $\langle \pi \rangle$ in Γ is isomorphic to the subgeometry of Γ over $[\lfloor n/2 \rfloor]$ induced on the set of all absolute elements with respect to π. We will often identify the two 'absolute' geometries.

Remark 7.3.3 The absolute geometry $\mathrm{Abs}(\Gamma, \pi)$ may be empty, as is the case for the polarity $\pi = \delta_f$ (cf. Theorem 7.2.12) on $\Gamma = \mathrm{PG}(\mathbb{R}^{n+1})$ where $f(x, y) = x_1 y_1 + x_2 y_2 + \cdots + x_{n+1} y_{n+1}$ (cf. Example 7.1.1(iii)). The same example with $\mathbb{C}$ instead of

$\mathbb{R}$ allows for absolute subspaces of projective dimension $m - 1$, where $m = \lfloor (n + 1)/2 \rfloor$. An example of an absolute element of maximal dimension is

$$\langle \varepsilon_1 + i\varepsilon_2, \varepsilon_3 + i\varepsilon_4, \ldots, \varepsilon_{2m-1} + i\varepsilon_{2m} \rangle.$$

Remark 7.3.4 The terminology of isotropic elements is classical in projective geometries. Here, an element x is *isotropic* with respect to a polarity π whenever x and $\pi(x)$ are both incident with some point. We will refrain from generalizing this to the situation under discussion.

If we reverse the natural ordering of the diagram we obtain an absolute geometry which is canonically isomorphic to $\mathrm{Abs}(\Gamma, \pi)$ by the map sending x to $\pi(x)$.

Theorem 7.3.5 *Let $\Gamma = (X, *, \tau)$ be an $[n]$-geometry ($n \geq 2$) with linear digon diagram, whose types are labelled from left to right by $1, \ldots, n$. Then the following hold for every polarity π of Γ.*

(i) *If v is an absolute element with respect to π and x is an element of Γ incident with v such that $\tau(x) \leq \tau(v)$, then x is an absolute element with respect to π.*

(ii) *If a is an absolute element of type 1 with respect to π, then the restriction π_a of π to the residue geometry $\Gamma_{a,\pi(a)}$ is a polarity.*

(iii) *If a is an element of type 1 which is absolute with respect to π, then $\mathrm{Abs}(\Gamma_{a,\pi(a)}, \pi_a) = \mathrm{Abs}(\Gamma, \pi)_a$.*

Proof (i) From $x * v$ and $\tau(x) \leq \tau(v)$ we obtain $\pi(x) * \pi(v)$ and $\tau(\pi(v)) \leq \tau(\pi(x))$, so the Direct Sum Theorem 2.1.6 applied to the residue $\Gamma_{\pi(v)}$ gives $v * \pi(x)$. Next, in the residue Γ_v, we find $x * \pi(x)$ for the same reason.

(ii) It suffices to show that π maps every element v of $\Gamma_{a,\pi(a)}$ to an element of $\Gamma_{a,\pi(a)}$. First, $a * v$ forces $\pi(v) * \pi(a)$. Second, $\pi(a) * v$ forces $\pi(\pi(a)) * \pi(v)$ whence $a * \pi(v)$.

(iii) If x is an absolute element with respect to π which is incident with a, then $x * \pi(x)$ and, since $1 \leq \tau(x) \leq \tau(\pi(x)) \leq n$, we see in $\Gamma_{\pi(x)}$ that $x * \pi(a)$. Consequently, x belongs to $\Gamma_{a,\pi(a)}$ and is absolute with respect to π_a. Conversely, if x is absolute with respect to π_a in $\Gamma_{a,\pi(a)}$, then x is absolute with respect to π and it is in Γ_a, hence in $\mathrm{Abs}(\Gamma, \pi)_a$. $\square$

Theorem 7.3.6 *Let Γ be a projective geometry of finite dimension $n \geq 2$ and $\mathbb{P}$ one of the two corresponding projective spaces. The absolute geometry $\mathrm{Abs}(\Gamma, \pi)$ of the projective geometry Γ with respect to a polarity π satisfies the following properties.*

(i) *A line ab of $\mathbb{P}$ is absolute if and only if a and b belong to $\pi(a) \cap \pi(b)$.*

(ii) *A subspace x of $\mathbb{P}$ is absolute if and only if all points and lines contained in x are absolute.*

(iii) *Given a line l of $\mathrm{Abs}(\Gamma, \pi)$ and a point $p \in \mathrm{Abs}(\Gamma, \pi)$ with p not on l, either a unique point of l or all points of l are collinear with p in $\mathrm{Abs}(\Gamma, \pi)$.*

(iv) *If the points of $\mathbb{P}_\pi$ generate $\mathbb{P}$, then there is no point of $\mathrm{Abs}(\Gamma, \pi)$ which is collinear with all points of $\mathrm{Abs}(\Gamma, \pi)$.*

Proof (i) Since the image of a line ab under π is the subspace $\pi(a) \cap \pi(b)$ (see Theorem 7.1.6), the statement follows.

(ii) If x is absolute, then Theorem 7.3.5 shows that all points and lines in x are absolute. Conversely, let all points and lines in x be absolute. Take a point $a \in x$. For any other point $y \in x$, the line ay is absolute so, by (i), $\pi(a)$ contains y, hence $\pi(a)$ contains x for all $a \in x$. Now Theorem 7.1.6 shows that $\pi(x) = \bigcap_{a \in x} \pi(a)$, hence $x \subseteq \pi(x)$ and so x is absolute.

(iii) By (i) the set of points of l which are collinear with p in $\mathrm{Abs}(\Gamma, \pi)$ coincides with $l \cap \pi(p)$. Since $\pi(p)$ is a hyperplane of $\mathbb{P}$, the result follows.

(iv) If p is collinear with all other points in $\mathrm{Abs}(\Gamma, \pi)$, then $\pi(p)$ contains all points of $\mathbb{P}$, a contradiction. $\square$

Assertions (iii) and (iv) of the theorem will recur in the definition of (nondegenerate) polar spaces in the next section.

For the remainder of this section, let V be a finite-dimensional right vector space over the division ring $\mathbb{D}$, and let σ be an anti-automorphism of $\mathbb{D}$.

According to Theorems 7.2.12 and 7.2.14, the set of absolute points of a polarity π of $\mathrm{PG}(V)$ coincides with $\{\langle x \rangle \in \mathbb{P}(V) \mid f(x, x) = 0\}$ for a nondegenerate σ-sesquilinear form f on V such that $\pi = \delta_f$. Recall from Theorem 7.1.6 that π can be viewed as acting on the geometry $\mathrm{PG}(V)$ of all subspaces of V via $\pi(X) = \bigcap_{x \in X} \pi(x)$ for each subspace X. Since π is a polarity, for $\langle x \rangle, \langle y \rangle \in \mathbb{P}(V)$, incidence of $\langle x \rangle$ and $\pi(\langle y \rangle)$ implies incidence of $\pi(\langle x \rangle)$ and $\langle y \rangle$. In other words, $f(x, y) = 0$ implies $f(y, x) = 0$. We analyze this symmetry in greater detail.

Lemma 7.3.7 *Let f be a nondegenerate σ-sesquilinear form on V. Then δ_f is a polarity of $\mathrm{PG}(V)$ if and only if f is reflexive.*

Proof In view of the above, we need only establish the 'if' part. If f is reflexive, then, for $x \in V \setminus \{0\}$,

$$\delta_f^2 \langle x \rangle = \bigcap_{f(x,y)=0} \delta_f(y) = \bigcap_{x \in \delta_f(y)} \delta_f(y) = \langle x \rangle,$$

which shows that δ_f is a polarity. $\square$

Proposition 7.3.8 *Suppose that V is a vector space with $\dim(V) \geq 2$. A nondegenerate σ-sesquilinear form f on V is reflexive if and only if there is a scalar $\varepsilon \in \mathbb{D} \setminus \{0\}$ such that $f(y, x) = \sigma(f(x, y))\varepsilon$ for all $x, y \in V$.*

Proof If $f(y, x) = \sigma(f(x, y))\varepsilon$ for all $x, y \in V$, then f is clearly reflexive. As for the converse, assume that f is reflexive. For every $a \in V \setminus \{0\}$, the equations $f(a, x) = 0$ and $\sigma^{-1}(f(x, a)) = 0$ in x define the same hyperplane of V (or V itself). Therefore, there is a scalar $\varepsilon_a \in \mathbb{D} \setminus \{0\}$ depending on a, such that $\sigma^{-1}(f(x, a)) = \sigma^{-1}(\varepsilon_a) f(a, x)$. Applying σ to both sides, we find $f(x, a) = \sigma(f(a, x))\varepsilon_a$. We must show that $\varepsilon_a = \varepsilon_b$ for every $b \in V \setminus \{0\}$. Clearly, it suffices to prove this for a, b linearly independent. Then

$$f(x, a + b) = f(x, a) + f(x, b) = \sigma\big(f(a, x)\big)\varepsilon_a + \sigma\big(f(b, x)\big)\varepsilon_b,$$

but also

$$f(x, a + b) = \sigma\big(f(a + b, x)\big)\varepsilon_{a+b} = \sigma\big(f(a, x)\big)\varepsilon_{a+b} + \sigma\big(f(b, x)\big)\varepsilon_{a+b},$$

whence

$$\sigma\big(f(a, x)\big)(\varepsilon_{a+b} - \varepsilon_a) = \sigma\big(f(b, x)\big)(\varepsilon_b - \varepsilon_{a+b}) \qquad (7.5)$$

for $x \in V$. Because f is nondegenerate, we can choose $x \in V$ such that $f(a, x) = 1$ and $f(b, x) = 0$. Then (7.5) forces $\varepsilon_a = \varepsilon_{a+b}$ and, by the same argument with a, b interchanged, $\varepsilon_b = \varepsilon_{a+b}$. The conclusion is $\varepsilon_a = \varepsilon_b$, as required. $\qquad \square$

Corollary 7.3.9 *Let V be a vector space of dimension at least two. Suppose that f is a nonzero reflexive σ-sesquilinear form on V for some anti-automorphism σ of $\mathbb{D}$. Then there is a nonzero element ε of $\mathbb{D}$ with*

$$\sigma(\varepsilon) = \varepsilon^{-1} \quad \text{and} \quad \sigma^2(t) = \varepsilon t \varepsilon^{-1} \quad \text{for all } t \in \mathbb{D} \qquad (7.6)$$

such that $f(y, x) = \sigma(f(x, y))\varepsilon$ for all $x, y \in V$.

Proof By Proposition 7.3.8 there is a nonzero element ε of $\mathbb{D}$ satisfying $f(y, x) = \sigma(f(x, y))\varepsilon$ for all $x, y \in V$. Let $t \in \mathbb{D}$. As f is nonzero, there exist vectors $x, y \in V$ such that $f(x, y) = t$. Then

$$t = f(x, y) = \sigma\big(f(y, x)\big)\varepsilon = \sigma\big(\sigma\big(f(x, y)\big)\varepsilon\big)\varepsilon = \sigma(\varepsilon)\sigma^2(t)\varepsilon.$$

Taking $t = 1$, we see $1 = \sigma(\varepsilon)\varepsilon$. $\qquad \square$

Definition 7.3.10 A pair (σ, ε) of an anti-automorphism σ of $\mathbb{D}$ and an element $\varepsilon \in \mathbb{D}$ satisfying the relations (7.6) of Corollary 7.3.9 will be called *admissible* (for $\mathbb{D}$).

Let (σ, ε) be an admissible pair for $\mathbb{D}$. A σ-sesquilinear form f such that $f(y, x) = \sigma(f(x, y))\varepsilon$ for all $x, y \in V$, is called (σ, ε)-*hermitian*. Moreover,

(1) if $\varepsilon - 1$, then f is called σ-*hermitian* (then $\sigma^2 = \mathrm{id}$);
(2) if $\varepsilon = -1$, then f is called σ-*anti-Hermitian* (then $\sigma^2 = \mathrm{id}$);
(3) if $\sigma = \mathrm{id}$ and $\varepsilon = 1$, then f is called *symmetric*;
(4) if $\sigma = \mathrm{id}$ and $\varepsilon = -1$, then f is called *antisymmetric*.

Furthermore, *hermitian* means σ-hermitian for some anti-automorphism σ, and similarly for *anti-Hermitian*. Finally, a σ-sesquilinear form f on V such that $f(x, x) = 0$ for all $x \in V$ is called an *alternating form*.

Let us rephrase in this terminology what we have achieved so far regarding quasi-polarities. Recall the duality δ_f from Theorem 7.2.12.

Theorem 7.3.11 *Every quasi-polarity on the projective space $\mathbb{P}(V)$ of a right vector space V of finite dimension at least two over $\mathbb{D}$ is of the form δ_f for a (σ, ε)-hermitian form f on V with respect to an admissible pair (σ, ε) for $\mathbb{D}$. The quasi-polarity is the same for any form proportional to f.*

Proof Put together Theorems 7.2.12 and 7.2.14, Lemma 7.3.7, Proposition 7.3.8, and Corollary 7.3.9.   □

Remark 7.3.12 Let V be a right vector space over $\mathbb{D}$ with basis $(e_i)_{i \in I}$ and let f be a (σ, ε)-hermitian form on V. For all $i, j \in I$, let $f_{ij} \in \mathbb{D}$ be such that $f(e_i, e_j) = f_{ij}$. Similarly to what we have seen in the proof of Theorem 7.2.14, f can be described by the matrix $(f_{ij})_{i,j \in I}$ satisfying $f(x, y) = \sum_{i,j \in I} \sigma(x_i) f_{ij} y_j$. Observe that this sum is well defined as only a finite number of x_i and y_j are nonzero. Now

  (i) f is σ-hermitian if and only if $\sigma(f_{ij}) = f_{ji}$ for each i, j;
  (i) f is σ-anti-Hermitian if and only if $\sigma(f_{ij}) = -f_{ji}$ for each i, j;
 (iii) f is symmetric if and only if $f_{ij} = f_{ji}$ for each i, j;
 (iv) f is antisymmetric if and only if $f_{ij} = -f_{ji}$ for each i, j;
  (v) f is alternating if and only if f is antisymmetric and $f_{ii} = 0$ for each i.

In particular, if f is a symmetric, respectively, alternating form, then so is the corresponding matrix $(f_{ij})_{i,j \in I}$, and conversely.

We next describe how an admissible pair transforms under scalar multiplication of a (σ, ε)-hermitian form.

Definition 7.3.13 Two admissible pairs (σ, ε) and (ρ, δ) are called *proportional* if there exists $\lambda \in \mathbb{D}$, $\lambda \neq 0$, such that $\delta = \lambda \sigma(\lambda)^{-1} \varepsilon$ and, $\rho(x) = \lambda \sigma(x) \lambda^{-1}$ for each $x \in \mathbb{D}$. In order to specify λ, we also say that the pair (ρ, δ) is *proportional* to (σ, ε) by λ.

Proportionality is an equivalence relation on the set of admissible pairs for $\mathbb{D}$.

Lemma 7.3.14 *If f is a (σ, ε)-hermitian form and $\lambda \in \mathbb{D}$, $\lambda \neq 0$, then, λf is a (ρ, δ)-hermitian form, where (ρ, δ) is proportional to (σ, ε) by λ.*

Proof This is a straightforward extension of the proof of Lemma 7.2.5. Having dealt with ρ in the proof cited, we now verify δ.

$$\lambda f(y, x) = \lambda \sigma\big(f(x, y)\big)\varepsilon = \lambda \sigma\big(f(x, y)\big)\sigma(\lambda)\lambda^{-1}\delta = \lambda \sigma\big(\lambda f(x, y)\big)\lambda^{-1}\delta$$
$$= \rho\big(\lambda f(x, y)\big)\delta.$$
□

In other words, proportional admissible pairs come from proportional reflexive sesquilinear forms.

Theorem 7.3.15 *Every nondegenerate reflexive sesquilinear form on a vector space of dimension at least two is proportional to*

(1) *a hermitian form if it is not antisymmetric;*
(2) *an anti-Hermitian form if it is not symmetric.*

Fig. 7.1 The polar space axiom

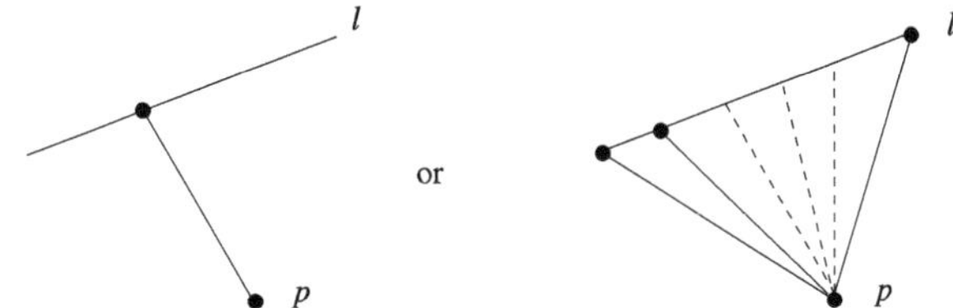

Proof In view of Proposition 7.3.8 and its corollary, our nondegenerate reflexive sesquilinear form f is (σ, ε)-hermitian for some admissible pair (σ, ε).

Put $\delta = 1$ or -1 according to whether f is not antisymmetric or not symmetric. Then there exists $s \in \mathbb{D}$ such that $\lambda := \sigma(s) + \delta s \varepsilon^{-1} \neq 0$; for otherwise $1 + \delta \varepsilon^{-1} = 0$, so $\varepsilon = -\delta \in Z(\mathbb{D})$, the center of $\mathbb{D}$, and $\sigma(s) = s$ for all $s \in \mathbb{D}$, so that f is antisymmetric, respectively symmetric.

Now $\delta = \lambda \sigma(\lambda)^{-1} \varepsilon$, so, by Lemma 7.3.14, λf is a (ρ, δ)-hermitian form. $\qquad\square$

Corollary 7.3.16 *A nondegenerate reflexive sesquilinear form is antisymmetric or proportional to a non-alternating hermitian form.*

Proof If f is an alternating form on V, then, for all $x, y \in V$, we have $0 = f(x + y, x + y) = f(x, y) + f(y, x)$, and so f is antisymmetric. Therefore, the assertion is a consequence of the above proof. $\qquad\square$

Example 7.3.17 If f is antisymmetric then $f(x, x) = -f(x, x)$ for each $x \in V$, so $\mathbb{D}$ has characteristic two or f is alternating. Thus, it is conceivable only in characteristic two that non-alternating antisymmetric forms exist.

Let $\mathbb{D} = \mathbb{F}_2$ be the field of two elements, viewed as a right vector space, and let $f : \mathbb{D} \times \mathbb{D} \to \mathbb{D}$ be the map given by $f(x, y) = x^2 + xy + y^2$. Then f is both antisymmetric and non-alternating hermitian (even symmetric).

7.4 Polar Spaces

We introduce the abstract notion of polar space and derive some basic properties. As we will see later, this provides an intrinsic characterization of the shadow space on points and lines of the absolute geometries. In the subsequent section, the geometry will be reconstructed from the polar space.

Definition 7.4.1 A line space (P, L) is a *polar space* if, for each point $p \in P$ and each line $l \in L$, the set of points of l which are collinear with p, is either a singleton or l (cf. Fig. 7.1).

Example 7.4.2 Each projective space is a polar space in which, for each point p and each line l, the set of points of l collinear with p, coincides with l. At the other extreme, a generalized quadrangle is a polar space in which, for each point p and each line l not on p, the set of points of l collinear with p, is a singleton.

By Theorem 7.3.6(iii), the point shadow space of the absolute of a polarity on a projective geometry is a polar space.

In the next section we will show that, under certain finiteness conditions, a polar space gives rise to a geometry with a diagram of type B_n (cf. Table 4.2).

Since the collinearity relation will play an important role, we recall from Definition 2.2.1 that, for a point x, we denote by $x^\perp$ the set of points collinear with x. (Thus it contains x.) If $y \in x^\perp$ we write $x \perp y$. For a set X of points, we put $X^\perp = \bigcap_{x \in X} x^\perp$. Recall from Definition 2.5.8 that a subset X of points in a line space is a subspace if any line having two distinct points in X is entirely contained in X.

Some of the properties for polar spaces can be stated for the more general class of gamma spaces (cf. Remark 6.7.3). A gamma space fails to be a polar space whenever there are a point p and a line l with $p^\perp \cap l = \emptyset$.

Definition 7.4.3 Let X be a set of points in a gamma space $Z = (P, L)$. We put $\mathrm{Rad}(X) = X \cap X^\perp$ and call it the *radical* of X. The set $\mathrm{Rad}(P)$, which coincides with $P^\perp$, is called the *radical* of the gamma space Z and also denoted by $\mathrm{Rad}(Z)$. The gamma space Z is called *degenerate* if $\mathrm{Rad}(Z)$ is non-empty, and *nondegenerate* otherwise.

In view of Theorem 7.3.6(iii), (iv), the absolute points and lines of a polarity π form a nondegenerate polar space if the point set generates the whole projective space.

The following result shows that quasi-polarities also lead to polar spaces.

Proposition 7.4.4 *The absolute $\mathbb{P}_\pi$ with respect to a quasi-polarity is a polar space. Moreover, $\mathrm{Ker}(\pi) \subseteq \mathrm{Rad}(\mathbb{P}_\pi)$, with equality if the points of $\mathbb{P}_\pi$ generate $\mathbb{P}$.*

Proof Write $\mathbb{P}_\pi = (P, L)$. Suppose $p \in P$ and $l \in L$. Then $l \cap \pi(p)$ is either all of l or a singleton contained in P. This establishes that $\mathbb{P}_\pi$ is a polar space.

The second assertion follows from $\{x \in \mathbb{P} \mid \pi(x) = \mathbb{P}\} = \{x \in P \mid \pi(x) = \mathbb{P}\} \subseteq \{x \in P \mid P \subseteq \pi(x)\} = \mathrm{Rad}(\mathbb{P}_\pi)$. As $P \subseteq \pi(x)$ is equivalent to $\langle P \rangle_\mathbb{P} \subseteq \pi(x)$, the inclusion can be replaced by an equality if $\langle P \rangle_\mathbb{P} = \mathbb{P}$. $\qquad\square$

Example 7.4.5 Here are some further examples of polar spaces.

(i) Every projective space and, more generally, every linear space is a polar space whose radical is the full set of points.

(ii) A generalized quadrangle is a polar space in which $|x^\perp \cap l| = 1$ holds whenever x is a point not on the line l.

(iii) Let (P, L) be a polar space. Then the pair

$$\left(P \cup \{\infty\}, \ L \cup \{\{x, \infty\} \mid x \in P\}\right)$$

is again a polar space. It has lines of length two if $P \neq \emptyset$, and ∞ is contained in its radical. In Sect. 7.10 we will deal with a more general construction of this kind, and show how a polar space with lines of size two can be decomposed into subspaces all of whose lines have at least three points.

Definition 7.4.6 A set of points X of a line space is *singular* if $x \perp y$ for all x and y in X.

The difference between a singular space and a linear space (cf. Definition 2.5.13) is that, in a singular space it may happen that a pair of points is on more than one line. We are primarily interested in the singular subspaces of a polar space.

The next lemma states some basic properties of gamma spaces. Recall from Definition 2.5.11 that, for a subset X of points of a line space Z, we denote by $\langle X \rangle$ the subspace of Z generated by X.

Lemma 7.4.7 *Let $Z = (P, L)$ be a gamma space. Then the following hold.*

 (i) *Every subspace of Z is a gamma space.*
 (ii) *If X and Y are subsets of P, then $(X \cup Y)^{\perp} = X^{\perp} \cap Y^{\perp}$. In particular, if $X \subseteq Y$, then $Y^{\perp} \subseteq X^{\perp}$ and $X \subseteq X^{\perp\perp} \subseteq Y^{\perp\perp}$.*
 (iii) *If X is a subset of P, then $X^{\perp}$ is a subspace; moreover, $X^{\perp} = \langle X \rangle^{\perp}$ and $X^{\perp} = X^{\perp\perp\perp}$.*
 (iv) *Every maximal singular subset of P is a singular subspace of Z.*
 (v) *If X is a singular subset of P, then $\langle X \rangle$ is a singular subspace.*

Proof (i), (ii), (iii) are straightforward and left to the reader.

(iv) Fix a maximal singular subset M of P. Let a, b be distinct points in M and let l be a line on a and b. For $x \in M$, we have $x \perp a$ and $x \perp b$, whence $l \subseteq x^{\perp}$. Therefore $M \cup l$ is a singular subset and since M is maximal, $l \subseteq M$, which proves that M is a subspace.

(v) X is contained in some maximal singular subset M by Zorn's lemma. Now $\langle X \rangle \subseteq M$ by (iv), so $\langle X \rangle$ is a singular subspace. $\square$

Lemma 7.4.8 *Let Z be a polar space.*

 (i) *Every subspace of Z is a polar space.*
 (ii) *If x and y are non-collinear points of Z, then $\mathrm{Rad}(x^{\perp} \cap y^{\perp}) \subseteq \mathrm{Rad}(Z)$.*
 (iii) *If $l \in L$ and $x \in l$ satisfy $x^{\perp} \subseteq l^{\perp}$, then $l \cap \mathrm{Rad}(Z) \neq \emptyset$; in particular, Z is degenerate.*
 (iv) *If x, y are distinct points of Z with $x^{\perp} \subseteq y^{\perp}$, then $\mathrm{Rad}(Z) \neq \emptyset$.*

Proof (i) is straightforward.

(ii) We show first that $\mathrm{Rad}(x^{\perp} \cap y^{\perp}) \subseteq \mathrm{Rad}(x^{\perp})$. Let $p \in \mathrm{Rad}(x^{\perp} \cap y^{\perp})$ and take $q \in x^{\perp} \setminus \{p\}$. We must show that $p \perp q$ and we may assume $q \neq x$. Then there is a line l on q and x. On l there is a point r such that $r \perp y$, by the definition of a polar space. Now $r \in x^{\perp} \cap y^{\perp}$, hence $r \perp p$. Also, $r \neq x$ (as $y \in r^{\perp} \setminus x^{\perp}$), so p

is collinear with (two and hence) all points of l, and $q \perp p$ as required. Therefore $\mathrm{Rad}(x^\perp \cap y^\perp)$ is contained in $\mathrm{Rad}(x^\perp)$ and, by symmetry, it is also contained in $\mathrm{Rad}(y^\perp)$.

Now, let $z \in P$, where P is the point set of Z, and, again, $p \in \mathrm{Rad}(x^\perp \cap y^\perp)$. We show that $p \perp z$. In view of the preceding argument, we may assume that $z \notin x^\perp \cup y^\perp$. Take a line m on p and x and consider a point $u \in z^\perp \cap m$. If $u = p$ we have finished, so we may assume $u \neq p$. Since y is not in $x^\perp$, it can be collinear with only one point on m namely p; in particular, $y \notin u^\perp$. Observing that $p \in \mathrm{Rad}(y^\perp)$ by the previous paragraph, we see $p \in \mathrm{Rad}(\{y, u\}^\perp)$. Applying our first argument to $\{u, y\}$ instead of $\{x, y\}$, we derive $p \in \mathrm{Rad}(u^\perp)$ and since $z \in u^\perp$ we obtain $p \perp z$ as wanted. Therefore, $p \in P^\perp = \mathrm{Rad}(Z)$.

(iii) Suppose that $l \in L$ and $x \in L$ satisfy $x^\perp \subseteq l^\perp$. Since $p^\perp \cap l \neq \emptyset$ for each $p \in P$, the point set P is partitioned by $l^\perp$ and the sets $y^\perp \setminus l^\perp$ with $y \in l \setminus \{x\}$. Suppose $l \cap \mathrm{Rad}(Z) = \emptyset$. Then there must be distinct y_1, y_2 on $l \setminus \{x\}$ such that $y_i^\perp \setminus l^\perp$ is non-empty for each $i = 1, 2$.

We claim that no point of $y_1^\perp \setminus l^\perp$ is collinear with a point of $y_2^\perp \setminus l^\perp$. Indeed, suppose that there exist $x_i \in y_i^\perp \setminus l^\perp$ $(i = 1, 2)$ with $x_1 \perp x_2$ and let m be a line on x_1, x_2. If $m \cap l^\perp \neq \emptyset$ then m has two of its points collinear with y_1, respectively, y_2, so m must be in $l^\perp$, contradicting $x_1 \in y_1^\perp \setminus l^\perp$. Hence $m \cap l^\perp = \emptyset$. Now, there is a point $x_3 \in m \cap x^\perp$; but $x_3 \in x^\perp \setminus l^\perp$ contradicts $x^\perp \subseteq l^\perp$. Therefore, indeed, no point of $y_1^\perp \setminus l^\perp$ is collinear with a point of $y_2^\perp \setminus l^\perp$.

Take $z_i \in y_i^\perp \setminus l^\perp$ $(i = 1, 2)$ and let l_1 be a line on z_1 and y_1. Then $z_2^\perp \cap l_1 \neq \emptyset$. Moreover, $z_2^\perp \cap (l_1 \setminus y_1) = \emptyset$ by our claim, so $z_2 \perp y_1$. But we also have $z_2 \perp y_2$, whence $z_2 \in l^\perp$ contradicting $z_2 \in y_2^\perp \setminus l^\perp$. We conclude that $l \cap \mathrm{Rad}(Z)$ is non-empty.

(iv) Suppose $x^\perp \subseteq y^\perp$ and $x \neq y$. Then $x \perp y$, so there is a line l on x and y. Now $x^\perp \subseteq \{x, y\}^\perp = l^\perp$, so (iii) applies. $\qquad\square$

Each gamma space has a natural quotient which is a nondegenerate gamma space.

Definition 7.4.9 The *nondegenerate quotient space* of a gamma space $Z = (P, L)$ with radical $R = \mathrm{Rad}(Z)$ is the line space $\rho(Z) = (\rho(P), \rho(L))$ whose points are the sets $\rho(x) := \langle R, x \rangle$ for $x \in P \setminus R$, and whose lines are the sets $\rho(l) := \{\rho(x) \mid x \in l\}$ of size at least two for $l \in L$ (so $l \cap R = \emptyset$). We also write Z/R or $(P/R, L/R)$ for $\rho(Z)$.

Proposition 7.4.10 *Suppose that Z is a non-empty gamma space with radical R. Its nondegenerate quotient $\rho(Z)$ is a nondegenerate gamma space. Moreover, it is a polar space if and only if Z is a polar space.*

Proof Write $Z = (P, L)$. We first show that, for any two points $x, y \in P$, their images $\rho(x)$ and $\rho(y)$ are collinear if and only if x and y are collinear. For, suppose that $\rho(x)$ and $\rho(y)$ are collinear in $\rho(Z)$. Then there is a line l of Z on points x' and y' of Z such that $\langle x, R \rangle = \langle x', R \rangle$ and $\langle y, R \rangle = \langle y', R \rangle$. By Lemma 7.4.7(iii),

$x' \perp (R \cup \{y'\})$ implies $x' \perp \rho(y)$, whence $x' \perp y$. Thus $y \perp (R \cup \{x'\})$, so by the same lemma, $y \perp \rho(x)$, proving $y \perp x$.

Now suppose that z is a point in $P \setminus R$ and m is a line of Z such that $\rho(m)$ is a line of $\rho(Z)$ with $|(\rho(z))^\perp \cap \rho(m)| > 2$. By the above, there are two points of m collinear with z, so, by the definition of gamma space, $z \in m^\perp$, proving $\rho(z) \in (\rho(m))^\perp$. Hence $\rho(Z)$ is a gamma space.

Suppose that w is a point in $P \setminus R$ such that $\rho(w)$ belongs to $\mathrm{Rad}(\rho(Z))$. By the above, w is collinear with all of $P \setminus R$, whence with all of P, so $w \in R$, a contradiction. Hence $\rho(Z)$ is nondegenerate.

Finally, it also follows from the above that $\rho(Z)$ is a polar space if and only if Z is a polar space. $\qquad\square$

We now prove that a nondegenerate polar space is partial linear (cf. Definition 2.5.13).

Theorem 7.4.11 *Each nondegenerate polar space is a partial linear space.*

Proof Let Z be a nondegenerate polar space. Suppose that l_1 and l_2 are distinct lines on two distinct points a, b of Z. Then $l_j \setminus (l_1 \cap l_2) \neq \emptyset$ for some $j \in [2]$, so Lemma 7.4.8(iii) shows that there exists a point $c \in Z \setminus \bigcup_{y \in l_1 \cap l_2} y^\perp$. For each $i \in [2]$, the point c is collinear with a point y_i on l_i; it follows that $y_i \in c^\perp \cap l_i \setminus (l_1 \cap l_2)$. Since y_1 is collinear with a and b, we have $y_1 \perp y_2$. Let l be a line on y_1 and y_2. We clearly have $l \subseteq \{a, b, c\}^\perp$. We will establish

$$y_1^\perp \cap \{a, c\}^\perp \subseteq l^\perp \cap \{a, c\}^\perp. \tag{7.7}$$

Let $u \in y_1^\perp \cap \{a, c\}^\perp$. Then u is collinear with the distinct points y_1 and a of l_1, hence with each point of l_1 and in particular with b. Therefore, u is collinear with each point of the line l_2 on a and b, hence with y_2. Finally, u is collinear with the two distinct points y_1 and y_2 of l, so with each point of l, proving (7.7). Now, by Lemma 7.4.8(iii), $\{u, c\}^\perp$ is a degenerate polar space (containing l) and, by Lemma 7.4.8(ii), Z is degenerate, a contradiction. $\qquad\square$

Corollary 7.4.12 *If x and y are distinct collinear points of a nondegenerate polar space Z, then $l = \{x, y\}^{\perp\perp}$ is the unique line of Z on x and y.*

Proof By Theorem 7.4.11, there is a unique line l on x and y, so $l = \langle x, y \rangle$. The polar axiom (Definition 7.4.1) forces $l^\perp = \{x, y\}^\perp$. If $w \in l$ and $v \in \{x, y\}^\perp$, then $v \in l^\perp \subseteq w^\perp$, so $w \in v^\perp$. This shows $l \subseteq \{x, y\}^{\perp\perp}$. It remains to establish $\{x, y\}^{\perp\perp} \subseteq l$.

Suppose $z \in \{x, y\}^{\perp\perp} \setminus l$. Parts (ii) and (iii) of Lemma 7.4.7 give $l^\perp = \{x, y\}^{\perp\perp\perp} \subseteq z^\perp$. As $x, y \in l^\perp$, we find $z \perp \{x, y\}$. As $z \notin l$, we have $z \neq y$, so Lemma 7.4.8(iii), applied to the line on z and y gives a point, u say, in $z^\perp \setminus y^\perp$. Then $u^\perp \cap l$ is a singleton, which, without loss of generality may be taken to be $\{x\}$. Thus, x, z are collinear points of $\{u, y\}^\perp$, say on the line m. If $w \in \{u, y\}^\perp \cap x^\perp$, then $w \in l^\perp \subseteq z^\perp$, whence $\{u, y\}^\perp \cap x^\perp \subseteq \{u, y\}^\perp \cap m^\perp$. Thus, by Lemma 7.4.8(iii),

$\text{Rad}(\{u, y\}^{\perp}) \neq \emptyset$, whence, by Lemma 7.4.8(ii), $\text{Rad}(Z) \neq \emptyset$, contradicting the non-degeneracy assumption. Hence $\{x, y\}^{\perp\perp} \setminus l = \emptyset$, proving $\{x, y\}^{\perp\perp} = l$. $\qquad\square$

The corollary shows that the nondegenerate polar space $Z = (P, L)$ is fully determined by the graph $(P, \perp)$. This is no longer the case in a degenerate polar space, see Exercise 7.11.17.

We now derive that singular subspaces of nondegenerate polar spaces are projective spaces.

Recall from Definition 5.2.7 that a geometric hyperplane of a line space Z is a proper subspace X of Z such that $l \cap X \neq \emptyset$ for each line l in Z.

Theorem 7.4.13 *Let Z be a polar space and X a singular subspace of Z.*

 (i) *If $p \in Z$, then either $X \subseteq p^{\perp}$ or $p^{\perp} \cap X$ is a geometric hyperplane of X.*

 (ii) *If X is a maximal singular subspace of Z and $p \in Z \setminus X$, then $p^{\perp} \cap X$ is a geometric hyperplane of both X and $\langle p, p^{\perp} \cap X \rangle$; moreover, the latter is also a maximal singular subspace of Z.*

(iii) *Every geometric hyperplane Y of X is a maximal subspace of X and X is the union of all lines containing p and a point of Y, where p is any point of $X \setminus Y$.*

(iv) *If Z is nondegenerate, then X is a projective space.*

Proof (i) is straightforward.

(ii) Put $Y = \langle p, p^{\perp} \cap X \rangle$. This is a singular subspace by Lemma 7.4.7(v). Since X is maximal and $p \notin X$, we cannot have $X \subseteq p^{\perp}$. Consequently, by (i), $p^{\perp} \cap X$ is a geometric hyperplane of X. Let $q \in X \setminus p^{\perp}$. Then $q \notin Y$ and $p^{\perp} \cap X \subseteq q^{\perp} \cap Y$. We will prove that these sets are equal. Suppose that there is a point $x \in q^{\perp} \cap Y \setminus p^{\perp} \cap X$. Then $x^{\perp} \cap X$ contains $p^{\perp} \cap X$ and q, so it contains $\langle p^{\perp} \cap X, q \rangle = X$ (recall that $p^{\perp} \cap X$ is a geometric hyperplane of X, hence maximal in X by Lemma 5.2.8). Now $X \cup \{x\}$ is a singular set of points of Z and so $x \in X$. Since $x \notin p^{\perp} \cap X$, we have $x \notin p^{\perp}$. But, by Lemma 7.4.8(iii), $x \in Y = \langle p, p^{\perp} \cap X \rangle \subseteq p^{\perp}$, a contradiction. Therefore $p^{\perp} \cap X = q^{\perp} \cap Y$ and, by (i), $p^{\perp} \cap X$ is a geometric hyperplane of Y while Y is the union of all lines on p and on a point of $p^{\perp} \cap X$.

We next show that Y is maximal. Suppose, instead, that Y is properly contained in a maximal singular subspace T of Z. Then T has a line l on p and on a point of $T \setminus Y$. On l we have a point $d \in Z \setminus (p^{\perp} \cap X)$ with $d \in q^{\perp}$. Then $X \cup \{d\}$ is a singular subset and, since X is maximal, $d \in X$. But then, as l is a line on p and d, we have $d \in p^{\perp} \cap X \subseteq Y$, a contradiction. So Y is a maximal singular subspace of Z.

(iii) Let p be a point of $X \setminus Y$ and set $Y' = \bigcup_{y \in Y} py$. If $x \in X \setminus \{p\}$, then x is collinear with p (as X is singular) and xp is a line on p meeting Y in a point distinct from p (as Y is a geometric hyperplane and $p \notin Y$). Hence $x \in Y'$, which proves $Y' = X$ as clearly $p \in Y'$. This settles (iii).

(iv) We need only verify Pasch's axiom. Suppose that a, b, c, p, q are distinct points such that bc and pq are distinct lines in X meeting in a. We will derive that bp and cq intersect. By Corollary 7.4.12, there is $v \in \{c, q\}^{\perp} \setminus a^{\perp}$ (because $a \notin cq$). Clearly $a \not\perp v$ implies $v \not\perp b$ and $v \not\perp p$. Thus, there is a unique point

$u \in b\,p \cap v^\perp$. We claim that u is the point of intersection of $b\,p$ and $c\,q$. To show this, it suffices to establish that $u \in c\,q$. Observe that all three points u, c and q lie inside $\{a, v\}^\perp$, which is a nondegenerate polar space by Lemma 7.4.8(ii). Consider $z \in \{c, q\}^\perp \cap \{a, v\}^\perp$. We have $z \perp a\,c$ and $z \perp a\,q$ so $z \perp \langle b, p \rangle = b\,p$, whence $z \perp u$. This shows that $u \in (\{c, q\}^\perp \cap \{a, v\}^\perp)^\perp$, which, as $\{a, v\}^\perp$ is nondegenerate, yields $u \in c\,q$ by Corollary 7.4.12. Thus $u \in b\,p \cap c\,q$, as claimed. $\square$

7.5 The Diagram of a Polar Space

The topic of this section is that nondegenerate polar spaces of rank n lead to geometries of type B_n. As pointed out in the proof of Lemma 7.4.7(v), by Zorn's lemma, there exist maximal singular subsets of a gamma space, and so there exist maximal singular subspaces. However, in order to proceed with induction, we need a much stronger property which corresponds to the finite rank in the context of geometries.

Definition 7.5.1 The singular dimension of a line space was introduced in Definition 6.7.25. The *nondegenerate singular dimension* of a gamma space Z is the singular dimension of the nondegenerate quotient $\rho(Z)$ of Proposition 7.4.10. The *rank* and the *nondegenerate rank* of a polar space are understood to be one more than its singular dimension and its nondegenerate singular dimension, respectively. We sometimes write $\mathrm{rk}(Z)$ to denote the rank of Z.

Example 7.5.2 A nondegenerate polar space of rank two is the shadow space $\mathrm{Sh}_1(\varGamma)$ of a generalized quadrangle $\varGamma$ over [2], and conversely. The elements of type 1 have shadows on 1 of dimension 0 while the elements of type 2 have shadows on 1 of dimension 1.

Proposition 7.5.3 *Let Z be a nondegenerate polar space of finite rank and let X be a singular subspace of Z.*

  (i) *There is a maximal singular subspace M disjoint from X.*
 (ii) *If Y is a singular subspace containing X, then there exists a maximal singular subspace M such that $M \cap Y = X$.*
(iii) $X^{\perp\perp} = X$.

Proof (i) Let M' be a maximal singular subspace of Z. If $M' \cap X = \emptyset$ we are done. Otherwise, the finite rank assumption shows that it suffices to prove the existence of a maximal singular subspace M such that $M \cap X$ is a proper subspace of $M' \cap X$. Let $q \in M' \cap X$. There exists a point $p \in Z \setminus q^\perp$ since Z is nondegenerate. Then $p \in Z \setminus M'$, so, by Theorem 7.4.13(ii), $M = \langle p^\perp \cap M', p \rangle$ is a maximal singular subspace of Z and $p^\perp \cap M'$ is a geometric hyperplane of M'. In particular, by Theorem 7.4.13(iii), $M \cap M'$, being a proper subspace of M' (it does not contain q) containing $p^\perp \cap M'$, must coincide with $p^\perp \cap M'$. If $x \in M \cap X$, then $q \in x^\perp$ (as both are in X) and $M \cap M' \subseteq x^\perp$, so $M' = \langle q, p^\perp \cap M' \rangle \subseteq x^\perp$. As M' is a maximal singular subspace of Z, we conclude $x \in M'$. Therefore $M \cap X \subseteq M' \cap X$ and, since $q \in (M' \cap X) \setminus (M \cap X)$, the inclusion is proper.

(ii) We proceed by induction on $\dim(X)$. If $\dim(X) = -1$, then X is empty and the property follows from (i). Assume $\dim(X) \geq 0$. There exists a geometric hyperplane X' in X (take $X' = p^{\perp} \cap X$ for some $p \in Z \setminus X^{\perp}$). We have $\dim(X') < \dim(X)$ so the induction hypothesis applies and there exists a maximal singular subspace M' such that $M' \cap Y = X'$. Let $p \in X \setminus X'$. Then $p \notin M'$ and, by the lemma, $M = \langle M' \cap p^{\perp}, p \rangle$ is a maximal singular subspace, while M contains $\langle X', p \rangle = X$. It remains to show that $M \cap Y \subseteq X$. Assume to the contrary that $x \in M \cap Y \setminus X$. Then $x \neq p$. Let l be a line on p and x. Clearly, $l \subseteq Y$. Moreover, l intersects the geometric hyperplane $M' \cap p^{\perp}$ of M in some point a. Notice that $a \neq p$ as $p \notin M'$. Also, $a \in M' \cap Y = X' \subseteq X$. Since l contains the two distinct points p and a of X, it must be contained in X. So $x \in X$, proving $M \cap Y \subseteq X$, as required.

(iii) If X is a maximal singular subspace, then $X^{\perp} = X$, whence $X^{\perp\perp} = X^{\perp} = X$. Otherwise let Y be a singular subspace properly containing X; by induction downwards on the dimension, we have $Y^{\perp\perp} = Y$. Due to (ii) there is a maximal singular subspace M such that $Y \cap M = X$. Now $X^{\perp\perp} \subseteq Y^{\perp\perp} \cap M^{\perp\perp} = Y \cap M = X$. From Lemma 7.4.7(ii), we conclude $X = X^{\perp\perp}$. $\square$

In view of Exercise 7.11.13, the last assertion also holds for singular subspaces of finite dimension in nondegenerate polar spaces of arbitrary rank. Remark 7.7.4 shows that it does not hold in general.

In order to state the main result of this section, we need one more definition.

Definition 7.5.4 In a polar space Z, two maximal singular subspaces A, B are called *adjacent* if they have a common geometric hyperplane. Notation: $A \sim B$. The resulting graph is called the *dual polar graph* of Z.

The *dual polar space* of Z is the line space Z^{δ} whose points are the maximal singular subspaces of Z and whose lines are the maximal cliques of the dual polar graph.

Clearly, the dual polar graph is the collinearity graph of the dual polar space. The term dual polar is historical and might be confused with dual as in dual line space of Definition 2.5.8 or hyperplane dual as in Exercise 5.7.12. Later, in Definition 7.7.1, we will see that the dual polar space is in fact the shadow space of a geometry with the linear diagram B_n on the node opposite the node representing the points of the polar space. This is the (unorthodox) reason for naming this graph dual polar.

Theorem 7.5.5 *Let Z be a nondegenerate polar space of finite rank n.*

(i) *If A and B are maximal singular subspaces of Z, there exists a path of maximal singular subspaces $A \sim M_1 \sim M_2 \sim \cdots \sim M_{k-1} \sim B$ of length $k \leq \dim(A) - \dim(A \cap B)$ in the dual polar graph of Z.*

(ii) *For every maximal singular subspace M of Z we have $\dim(M) = n - 1$.*

Proof By Theorem 7.4.13(iv), all singular subspaces of Z are projective.

(i) We have $\dim(A) = n - 1$ and we proceed by induction on $i = \dim(A) - \dim(A \cap B)$. If $i = 0$, we have $A \cap B = A$, whence $A \subseteq B$ proving $A = B$, as required. If $i > 0$, let $a \in A \setminus B$ and consider $B' = \langle a, a^{\perp} \cap B \rangle$ which is a maximal singular subspace such that $a^{\perp} \cap B$ is a common geometric hyperplane of B and B' (see Theorem 7.4.13(ii)). Hence $B' \sim B$. Moreover, $j := n - 1 - \dim(A \cap B') < i$ since B' contains $A \cap B$ as well as a. So, the induction hypothesis applies to A and B', and there is a path $A \sim M_1 \sim M_2 \sim \cdots \sim M_{j-1} \sim B'$ where each M_i is a maximal singular subspace.

(ii) It suffices to show that any two maximal singular subspaces of Z have the same dimension. This follows immediately from (i) and from the fact that adjacent maximal singular subspaces have the same dimension. $\square$

Example 7.5.6 We exhibit nondegenerate polar spaces of infinite rank for which Theorem 7.5.5 does not hold. Let V be a vector space, and set $U = V \oplus V^{\vee}$, where $V^{\vee}$ is the dual of V. Define the bilinear form f on U by

$$ f(v + w, v' + w') = w(v') - w'(v) \quad (v, v' \in V; w, w' \in V^{\vee}). $$

As f is reflexive, Theorem 7.2.12 gives that δ_f is a quasipolarity, and we can construct a polar space $Z = (P, L)$ from δ_f as in Proposition 7.4.4. This means that P is the set of points x of $\mathbb{P}(U)$ with $f(x, x) = 0$ and L is the collection of all projective lines l of $\mathbb{P}(U)$ with $f(x, y) = 0$ for all $x, y \in l$. The polar space Z is nondegenerate as $\mathrm{Rad}(f) = \{0\}$. The subspaces $\mathbb{P}(V)$ and $\mathbb{P}(V^{\vee})$ of Z are maximal singular. If $\alpha := \dim(V)$ is infinite, then $V \not\cong V^{\vee}$. As $\dim(V^{\vee}) \geq 2^{\alpha} > \alpha$, the projective spaces $\mathbb{P}(V)$ and $\mathbb{P}(V^{\vee})$ do not even have the same dimension.

Now we build a geometry of type B_n from a nondegenerate polar space of rank n. The construction is quite natural.

Definition 7.5.7 Let Z be a nondegenerate polar space of finite rank $n \geq 2$. The *polar geometry* of Z, notation $\Gamma(Z)$, is an incidence system over $[n]$, whose elements of type $i \in [n]$ are the singular subspaces of Z of dimension $i - 1$, and in which two elements X, Y are incident if either $X \subseteq Y$ or $Y \subseteq X$.

Below is the diagram for the polar geometry of a nondegenerate polar space of rank n. Near each node, we have written the set of elements of the corresponding type.

o———————o · · · · · o———————o——4——o
P L M_{n-3} M_{n-2} M_{n-1}

By M_i, we denote the set of singular subspaces of Z of dimension i. This is the type B_n introduced in Table 4.2. The elements of type $n - 1$ in $\Gamma(Z)$ are the maximal singular subspaces of Z. Of course, the terminology is chosen so that the polar geometry of a polar space is indeed a geometry.

Theorem 7.5.8 *Let Z be a nondegenerate polar space of finite rank $n \geq 2$. Its polar geometry $\Gamma := \Gamma(Z)$ is a firm and residually connected geometry, which belongs to the diagram B_n and satisfies the following two additional properties.*

(i) *Distinct elements of Γ have distinct 1-shadows.*
(ii) *Two distinct lines have at most one common point, that is, the shadow space* ShSp$(\Gamma, 1)$ *is a partial linear space.*

Proof Property (ii) follows from Theorem 7.4.11 and (i) is obvious from the construction of $\Gamma(Z)$.

Firmness of Γ is due to two facts. First of all, singular subspaces are projective spaces by Theorem 7.4.13(iv) and the geometry of all subspaces of such a space is firm, while it corresponds to some residue of Γ. Secondly, every singular subspace V of Z of dimension $n-2$ is in at least two singular subspaces of dimension $n-1$. For, by Lemma 7.4.7(iv) and Theorem 7.5.8(ii), there is at least one, say M, and then, by Proposition 7.5.3(ii), there is a maximal subspace M' such that $M' \cap M = V$.

In order to obtain the required diagram, observe first of all that, in view of the definition of incidence in Γ, the digon diagram is linear. We proceed by induction on n. Assume that $n = 2$. We need to show that Z is a generalized quadrangle (cf. Definition 2.2.7). If p is a point of Z, then there is a point a of Z not collinear with p, whence $d(p, a) = 4$ in the incidence graph (for, there is a line l on p and $a^\perp$ contains a point of l). For the same reason, $d(p, l) \le 3$ for each line l. Hence the point-diameter $d_1 = 4$ in Γ equals 4. Similarly, the line-diameter d_2 equals 4 (there are no circuits of length 6 in the incidence graph). As the girth g satisfies $g > 3$ for the same reason and $g \le d_1$, we see that $g = d_1 = d_2 = 4$; so Γ is indeed a generalized quadrangle.

For $n > 2$, observe first that if M is a maximal singular subspace of Z, then its residue Γ_M is a projective geometry of rank $n-1$ and so the restriction of the diagram to $[n-2]$ is as required. Consider a point a in Z and let b be a point such that b is not collinear with a. The residue Γ_a is isomorphic to $\Gamma(\{a, b\}^\perp)$. According to Lemma 7.4.8(i), (ii), the subspace $\{a, b\}^\perp$ of Z is a nondegenerate polar space, which obviously has rank $n-1$. Hence the induction hypothesis applies to $\{a, b\}^\perp$, giving that $\Gamma(\{a, b\}^\perp)$ is a geometry belonging to B_{n-1}. This forces Γ to be of type B_n.

We next prove that Γ is residually connected by induction on n. It holds for $n = 2$, thanks to the above. For $n > 2$, consider first the residue of a non-empty flag. By the above treatment of the case $n = 2$ and the Direct Sum Theorem, this residue is a direct sum of projective geometries and (possibly) a geometry $\Gamma(Y)$ where Y is a polar space to which induction applies. Hence, this residue has a connected incidence graph whenever it is of rank at least two. It remains to show that the incidence graph of Γ is connected. If p is a point and X is an element of Γ, then there is some point q incident to X and there is at least one point a in $p^\perp \cap q^\perp$. This implies $p * pa * a * aq * q * X$, so p and X are in the same connected component of the incidence graph. $\qquad\square$

Example 7.5.9 If π is a quasi-polarity on the projective space $\mathbb{P}(V)$, then the polar geometry $\Gamma(\mathbb{P}(V)_\pi)$ of $\mathbb{P}(V)_\pi$ as in Theorem 7.5.8 is isomorphic to the absolute geometry $\mathrm{Abs}(\mathrm{PG}(V), \pi)$.

7.6 From Diagram to Space

Let $n \in \mathbb{N}$, $n > 0$. In this section, we study geometries over the diagram B_n. We have seen that nondegenerate polar spaces are examples. As usual, we assume that our geometries are firm and residually connected. This, however, will not suffice to derive that the shadow on 1 is a nondegenerate polar space. Therefore, the notion of a polar geometry will be introduced. In Corollary 7.6.8, these will turn out to correspond nicely to nondegenerate polar spaces.

Definition 7.6.1 Let Γ be a residually connected geometry with linear digon diagram over $[n]$, with ordering of the nodes from left to right, by $1, 2, \ldots, n$. Elements of type 1, 2, and 3, are called *points lines*, and *planes*, respectively. If a point p is incident with a line l, we will freely say that p is on l. Sets and sequences of points on the same line are called *collinear*.

Remark 7.6.2 The geometry Γ/A of Example 1.3.6 with $n = 3$ shows that all points of a geometry Γ of type B_3 may be pairwise collinear while the more classical examples like the absolutes with respect to polarities and the hyperoctahedron have no points collinear with all other points. To overcome these difficulties, we employ the additional properties derived in Theorem 7.5.8.

Definition 7.6.3 An $[n]$-geometry Γ of type B_n satisfying Conditions (1) and (2) below is called a *polar geometry* of rank n.

(1) Distinct elements of Γ have distinct 1-shadows.
(2) Distinct lines have at most one common point, i.e., $\mathrm{ShSp}(\Gamma, 1)$ is a partial linear space.

Example 7.6.4 As suggested by the name, according to Theorem 7.5.8, the polar geometries of nondegenerate polar spaces are examples of polar geometries.

The Alt_7-geometry of type B_3 of Example 2.4.11 satisfies neither (i) nor (ii) of Theorem 7.5.8.

We first establish that (1) and (2) are inductive in the sense that residues of type B_{n-1} inherit (1) and (2) from the original geometry.

Lemma 7.6.5 *Let Γ be a polar geometry of rank n. For every point p of Γ, the residue Γ_p is a polar geometry as well.*

Proof We verify the conditions (1) and (2) for Γ_p; the rest is obvious.

Ad (1) Assume that v and w are elements of Γ_p having the same point-shadow (that is, type 2 in Γ). We show that $v = w$. It suffices to establish that v and w have the same 1-shadow in Γ and to apply (1) in Γ. For $q \in \mathrm{Sh}_1(v)$ with $q \neq p$, there is some line L in Γ_v which is incident with p and q. Now $L \in \mathrm{Sh}_2(v)$ is in Γ_p, hence $L \in \mathrm{Sh}_2(w)$ in Γ_p, and so $L * w$. The Direct Sum Theorem 2.1.6 for Γ_L gives $q * w$.

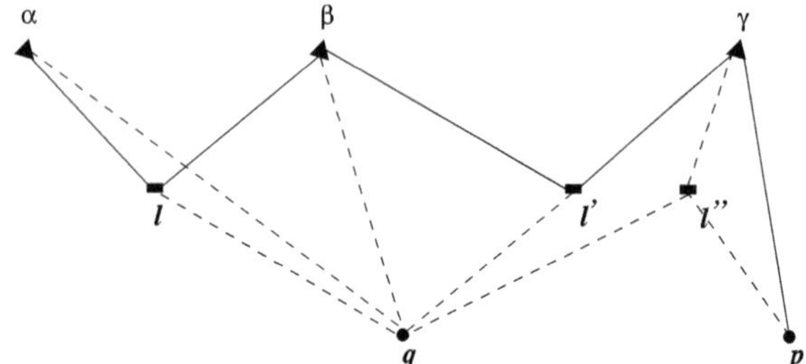

Fig. 7.2 Configuration in Γ of Step 1. The *interrupted lines* represent incidences derived by reasoning

Therefore $\mathrm{Sh}_1(v) \subseteq \mathrm{Sh}_1(w)$ and, by the symmetry of the roles of v and w, we obtain $\mathrm{Sh}_1(v) = \mathrm{Sh}_1(w)$. By (1) for Γ, we conclude that $v = w$.

Ad (2) Assume that α and β are planes in Γ incident with p, having two distinct lines L, L' in common. Let $q \neq p$, be a point in Γ incident with α. In Γ_α, there is a line L'' on q, intersecting L, L', respectively, in points a, a' distinct from p. Now a and a' are also in Γ_β, so L'' must be in Γ_β by (2) and this forces q in Γ_β. So $\mathrm{Sh}_1(\alpha) \subseteq \mathrm{Sh}_1(\beta)$ and in fact $\mathrm{Sh}_1(\alpha) = \mathrm{Sh}_1(\beta)$. By (1), we obtain $\alpha = \beta$. Therefore, (2) holds in Γ_p. $\qquad\square$

Proposition 7.6.6 *Let $n \geq 2$. Each $[n]$-geometry Γ with diagram B_n satisfies the following properties.*

(i) *For each point p and each line l, there is a point on l collinear with p.*

(ii) *If Γ is a polar geometry, then, for any line l and any point p, either one or all points on l are collinear with p.*

(iii) *If Γ is a polar geometry, then, for any line l and point p off l but collinear with two points of l, there is a plane incident with l and p, while $n \geq 3$.*

Proof In (ii) we may assume that p is not on l, for otherwise the assertion is trivially satisfied. We proceed by induction on n. For $n = 2$, the geometry Γ is a generalized quadrangle and so (i), (ii), (iii) hold trivially. So, from here on, we assume that $n \geq 3$.

STEP 1. *For every plane α and point p not on α, there exists a plane β incident with p such that α and β are both incident with some line.*

Since Γ is residually connected, Corollary 1.6.6 shows that there exists a chain from α to p consisting of planes and lines only. By induction on the length of that chain it suffices to treat the case where the chain has length six, say $\alpha, l, \beta, l', \gamma, p$ (cf. Fig. 7.2), where γ, β are planes and l, l' are lines.

The lines in Γ_β are those of a projective plane, so there is a point q incident with l', β, and l. In $\Gamma_{l'}$ the Direct Sum Theorem 2.1.6 gives $\gamma * q$ and in Γ_l, it gives $\alpha * q$. In Γ_γ, the points and lines constitute a projective plane, so there is a line l'' in Γ_γ, incident with p and q; cf. Fig. 7.2. Moreover, in Γ_q, the induction hypothesis applies and so (i) holds, forcing the existence of a line l''' and a plane δ with $\alpha * l''' * \delta * l''$. Finally, in $\Gamma_{l''}$ the Direct Sum Theorem 2.1.6 gives $p * \delta$ and so we find the chain p, δ, l''', α, as required.

Fig. 7.3 The configuration in Γ of Step 2

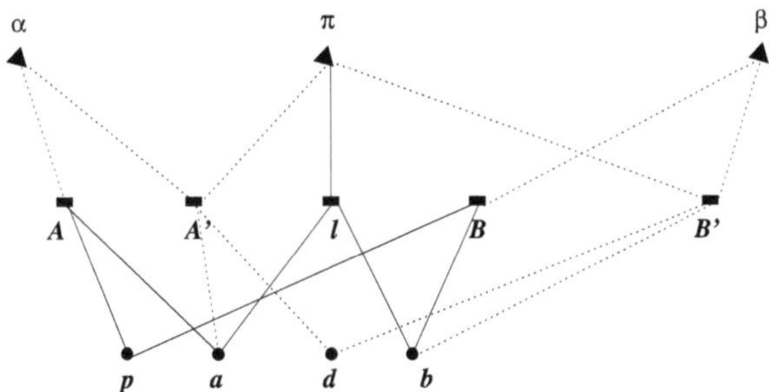

STEP 2. We prove (i). Let p and l be an arbitrary point and line. Let α be a plane incident with l. Then α is incident with all points on l, by the Direct Sum Theorem 2.1.6. If $p * \alpha$, then Γ_α shows that p is collinear with all points of l. So we assume that p and α are not incident. By Step 1 we find a chain p, δ, l', α, l where δ is a plane and l' a line. In Γ_α, the lines l and l' are incident with some point q and in $\Gamma_{l'}$ the Direct Sum Theorem gives $\delta * q$. Finally, in Γ_δ, the points p and q are incident with some line. Thus we have proved that there is at least one point on l which is collinear with p and so (i) holds.

STEP 3. As for the proof of (ii) and (iii), suppose that Γ is a polar geometry. We assume that p is collinear with two distinct points a, b on l. In view of (i) it suffices to show that there is a plane incident with both p and l. Let π be a plane incident with l. We may assume that p is not incident with π (otherwise (ii) and (iii) hold). Let A be a line incident with a and p and let B be a line incident with b and p. By the Direct Sum Theorem 2.1.6, $a * \pi * b$. In Γ_a, the induction hypothesis on (i) applies to π and A and gives a plane α of Γ incident with A and a line A' of Γ incident with π and α. Similarly, in Γ_b, we obtain a plane β and a line B' such that $B * \beta * B' * \pi$; see Fig. 7.3.

In Γ_A, we have $p * \alpha$, and in Γ_B, we have $p * \beta$. In Γ_π, the lines A' and B' have a point d in common, from which we deduce $\alpha * d * \beta$. Let D be a line on p and d. There is such a line in Γ_α and also in Γ_β and, by axiom (ii) for polar geometries, D is this line. Hence $\alpha * D * \beta$.

Now consider Γ_d. In there, D is a point off π (as p is not incident with π) whereas A' and B' are points incident with π and collinear with D via α and β, respectively. If $n = 3$, then A' and B' are points on the line π in the generalized quadrangle Γ_d which are both collinear with D. Therefore, they coincide, and so do α and β, which is then a plane as required for (iii). So assume $n \geq 4$. In view of Lemma 7.6.5, the induction hypothesis on (iii) for Γ_d provides an element v of type 4, incident with π and D. In Γ_v, the points, lines, and planes constitute a projective geometry. As p and l are in it, there is a plane in Γ_v, incident with l and p, and we are done.　　　　　　　　　　　　　　　　$\square$

Theorem 7.6.7 *Let Γ be a polar geometry of rank at least two and set $Z = \mathrm{ShSp}(\Gamma, 1)$.*

(i) *No point of Z is collinear with all other points*: $\mathrm{Rad}(Z) = \emptyset$.

(ii) *Every singular subspace of Z is the* 1-*shadow of some element of Γ and is a projective space.*

Proof We proceed by induction on $n := \mathrm{rk}(\Gamma)$. The properties are trivial for $n = 2$, so let $n \geq 3$.

(i) Let p be a point of Z and let l be a line of Z on p. There exists a point $q \neq p$ on l since Γ is firm. In Γ_q, the induction hypothesis applies thanks to Lemma 7.6.5 and gives some line l' on q with the property that l, l' are not both incident with the same plane. There exists a point $p' \neq q$ on l'. We claim that p and p' are not collinear. Assume the contrary. By Proposition 7.6.6(iii), there is a plane α incident to l and p'. In Γ_α, the points q and p' are incident with a line l''. Condition (2) of Definition 7.6.3 forces $l'' = l'$, so l and l' are incident with the plane α in Γ_q, a contradiction.

(ii) Let X be a singular subspace of Z. If l is a line of Z contained in X and p a point of $X \setminus l$, then Proposition 7.6.6(iii) shows that there is a plane α incident with l and p. Now $\mathrm{Sh}_1(\alpha)$, endowed with the lines contained in it, is a projective plane by Condition (2) of Definition 7.6.3 and the diagram. This proves that X is a projective space.

Let $p \in X$ and let X_p be the set of lines in X on p. In Γ_p, the set X_p consists of lines which are pairwise in a plane and the plane on any two of these is contained in X_p. So, in view of Lemma 7.6.5 again, induction applies and there exists an element x of Γ_p with $\mathrm{Sh}_2(x) = X_p$. Now $\mathrm{Sh}_1(x) = X$ by the Direct Sum Theorem 2.1.6, and we have finished. $\qquad\square$

Below, we summarize the correspondence between polar geometries and polar spaces. Recall the notion of the polar geometry of a polar space from Definition 7.5.7.

Corollary 7.6.8 *If Γ is a polar geometry of rank n, then its shadow space $Z := \mathrm{ShSp}(\Gamma, 1)$ on 1 is a nondegenerate polar space of rank n whose polar geometry $\Gamma(Z)$ is isomorphic to Γ.*

Proof Immediate from Proposition 7.6.6 and Theorem 7.6.7. $\qquad\square$

7.7 Singular Subspaces

We continue to investigate the structure of a nondegenerate polar space of finite rank n. Given the polar geometry viewpoint, we now focus on the 'other end' of the linear diagram of type B_n, and so we are interested in maximal singular subspaces. In this section we derive some properties of singular subspaces of polar spaces in general. The highlight is Theorem 7.7.10, which shows that all singular planes in a polar space of rank three are Moufang.

The first lemma is a direct consequence of the polar geometry (cf. Definition 7.5.7) being of type B_n.

Lemma 7.7.1 *Let Z be a nondegenerate polar space of finite rank n. Its dual polar space Z^δ is isomorphic to the shadow space on n of the polar geometry of Z over $[n]$.*

The points of Z^δ are the maximal singular subspaces of Z. For every singular subspace U of dimension $n-2$ of Z, the set U^δ of all maximal singular subspaces of Z containing U is a line of Z^δ. So two distinct points of Z^δ are collinear if and only if they meet in an $(n-2)$-dimensional singular subspace of Z.

Theorem 7.7.2 *Let M_1 and M_2 be maximal singular subspaces of a nondegenerate polar space Z of finite rank $n \geq 2$.*

  (i)   *The distance between M_1 and M_2 in Z^δ is equal to $n - 1 - \dim(M_1 \cap M_2)$.*
 (ii)   *Each member of a shortest path from M_1 to M_2 in Z^δ contains $M_1 \cap M_2$.*
(iii)   *If U^δ is a line of Z^δ, then there is a unique point of U^δ nearest to M_1.*
(iv)   *The diameter of Z^δ is n.*

Proof We proceed by induction on n. For $n = 2$, everything is obvious. So, take $n \geq 3$.

(i) and (ii) We proceed by induction on the distance d between M_1 and M_2. If $d = 0$, then $M_1 = M_2$ and we have finished. So, let $d \geq 1$ and pick a point M' of Z^δ with $d(M_1, M') = 1$ and $d(M', M_2) = d - 1$. Then $\dim(M' \cap M_2) = n - 1 - (d-1)$ by the induction hypothesis and $\dim(M_1 \cap M') = n - 2$ by the definition of collinearity. In the projective space M' of dimension $n - 1$, the subspace $M_1 \cap M'$ is a geometric hyperplane and $M' \cap M_2$ intersects this in a subspace V of dimension $n - 1 - (d - 1) - 1$ or $n - 1 - (d - 1)$. In the first case, $\dim(M_1 \cap M_2) \geq n - 1 - d$. If $\dim(M_1 \cap M_2) = i$ and if $a_{i+1}, \ldots, a_{n-1}$ are points of M_2 such that $\langle M_1 \cap M_2, a_{i+1}, \ldots, a_{n-1} \rangle = M_2$, put

$$X_1 = \langle M_1 \cap a_{i+1}^\perp, a_{i+1} \rangle, \qquad X_2 = \langle X_1 \cap a_{i+2}^\perp, a_{i+2} \rangle, \ldots,$$
$$X_{n-1-i} = \langle X_{n-2-i} \cap a_{n-1}^\perp, a_{n-1} \rangle.$$

Then $M_1, X_1, \ldots, X_{n-1-i} = M_2$ is a path of the collinearity graph of Z^δ, all of whose members contain $M_1 \cap M_2$ which shows that $d \leq n - 1 - i$. Hence we reach $d = n - 1 - i$, (i) and (ii) in that case. In the second case, $\dim(M_1 \cap M_2) = i \geq n - 1 - (d-1)$ and so by the preceding argument $d \leq n - 1 - i \leq n - 1 - (n - d) \leq d - 1$, a contradiction.

(iii) Assume first that there is a point $x \in U \cap M_1$ in Z. Consider the residue Z_x of x; using Proposition 7.4.10 and Proposition 7.6.6, we obtain that this is a nondegenerate polar space of rank $n - 1$ in which U and M_1 are again a line and a point of Z_x^δ, respectively. So induction applies and, by (ii), a point of U^δ that is nearest to M_1 in Z_x^δ is also nearest to M_1 in Z^δ.

Assume next that $U \cap M_1 = \emptyset$. Let $a_1, \ldots, a_{n-1}$ be a minimal generating set of the projective space U. Then $\bigcap_{i=1}^{n-1} a_i^\perp \cap M$ consists of a unique point q. Hence $\langle U, q \rangle$ is the unique point of U^δ that is nearest M.

(iv) By Proposition 7.5.3, there are two maximal singular subspaces M_1, M_2 in Z, such that $M_1 \cap M_2 = \emptyset$. It suffices to apply (i) to that pair in order to see that their distance is n and again (i) to observe that the diameter cannot be larger. $\square$

As a first application, we show (in Theorem 7.7.10) that the singular planes of nondegenerate polar spaces have many nontrivial automorphisms. We begin with a lemma that is closely connected with Exercise 7.11.22.

Lemma 7.7.3 *Suppose that Z is a nondegenerate polar space of rank at least three. If M and N are disjoint maximal singular subspaces of Z, then the map $\pi_{M,N}$: $M \to N^\vee$ given by $x \mapsto x^\perp \cap M$ is an injective homomorphism of projective spaces, where $N^\vee$ is the dual projective space of N.*

Proof Let $x \in M$ by Theorem 7.4.13(ii), $x^\perp \cap N$ is a hyperplane of N. Suppose x, y, z are distinct collinear points of M. Then, by the polar space property $\{x, y\}^\perp \cap N = \{z, y\}^\perp \cap N$, so the image of the line xy is a line and $\pi_{M,N}$ is a homomorphism.

We next show that $\pi_{M,N}$ is injective. Assume, by way of contradiction, that p and q are distinct points of M such that $\pi_{M,N}(p) = \pi_{M,N}(q)$. Let $x \in N \setminus \pi_{M,N}(p)$. Then the line pq of M intersects $x^\perp$ in some point r. Now $r^\perp$ contains x and $\pi_{M,N}(p)$, whence also N, contradicting the fact that N is a maximal singular subspace. Therefore, $\pi_{M,N}$ is injective. $\square$

Remark 7.7.4 A consequence of Lemma 7.7.3 is $\dim(M) \leq \dim(N^\vee)$ and, similarly, $\dim(N) \leq \dim(M^\vee)$. If $\mathrm{rk}(Z) = \infty$, then, up to an interchange of the two maximal singular subspaces, we have $\dim(M) \leq \dim(N) < \dim(N^\vee)$. This shows that there exists a hyperplane H of N which is not of the form $\pi_{M,N}(p)$ for $p \in M$.

In Example 7.5.6, let $(e_i)_{i \in I}$ be a basis of V and consider the linear span H of the functionals $(e_i^\vee)_{i \in I}$, determined by $e_i^\vee(e_j) = \delta_{ij}$. Then H is a proper subspace of $N := V^\vee$, but $H^\perp = N$. Therefore, Proposition 7.5.3(ii) does not hold for every nondegenerate polar space.

If Z is a nondegenerate polar space of rank at least four, then its maximal singular subspaces are projective spaces of dimension at least three, and hence Desarguesian. If the rank is three, this need not be the case. On the other hand, not every projective plane occurs as a singular subspace of a nondegenerate polar space, as we will see in Theorem 7.7.10. First we introduce the necessary terminology.

Definition 7.7.5 Let Z be a nondegenerate polar space. A pair $\{U, V\}$ of two singular subspaces of Z of the same dimension is called an *opposite pair* if, for each $x \in U \cup V$, the subspace $x^\perp$ does not contain $U \cup V$.

For an opposite pair $\{U, V\}$ of singular subspaces of Z as above, we have $U \cap V = \emptyset$ in view of Theorem 7.4.13(ii).

Proposition 7.7.6 *Suppose that Z is a nondegenerate polar space of finite rank n whose polar geometry is thick. If U_1 and U_2 are singular subspaces of Z of the* *same dimension, then there is a singular subspace V of the same dimension such that both $\{U_1, V\}$ and $\{U_2, V\}$ are opposite pairs.*

Proof For $n = 1$, the proposition is immediate from the thickness assumption. We proceed by induction on n. Suppose $n \geq 2$. We claim that for any two points U_1, U_2, there is a point in $Z \setminus (U_1^\perp \cup U_2^\perp)$. If $U_1 = U_2$, then the claim is immediate from nondegeneracy of Z. Assume that U_1 and U_2 are distinct points.

If they are collinear, there is a third point U_3 on the line $U_1 U_2$. By Lemma 7.4.8(iii) there is a line l of Z on U_3 not contained in $(U_1 U_2)^\perp$. Every point V on $l \setminus U_3$ is as required.

If U_1 and U_2 are non-collinear, there is a point b in $U_2^\perp \setminus U_1^\perp$ with $b \neq U_2$. By Lemma 7.4.8(ii), $U_1^\perp \cap U_2^\perp$ is not contained in $b^\perp$ since otherwise the point $bU_2 \cap (U_1^\perp \cap U_2^\perp)$ is in $\mathrm{Rad}(U_1^\perp \cap U_2^\perp)$. Therefore, there is a point $c \in U_1^\perp \setminus (U_1^\perp \cap U_2^\perp)$ with $b \perp c$. On the line bc, any third point V is as required. This establishes the claim.

Next, let U_1 and U_2 be as in the hypotheses. Take $u_1 \in U_1$, $u_2 \in U_2$ and let v be a point of Z such that $\{u_1, v\}$ and $\{u_2, v\}$ are opposite pairs. Set $W_1 = \langle v, v^\perp \cap U_1 \rangle$ and $W_2 = \langle v, v^\perp \cap U_2 \rangle$. We apply Proposition 7.4.10 and consider the polar space $v^\perp/\{v\}$ in which $W_1/\{v\}$ and $W_2/\{v\}$ are singular subspaces of the same dimension. As the polar geometry of Z is thick, so is the polar geometry of the polar space $v^\perp/\{v\}$. Now the induction hypothesis yields a singular subspace V of Z of dimension $\dim(U_1)$ containing v such that $\{W_1/\{v\}, V/\{v\}\}$ and $\{W_2/\{v\}, V/\{v\}\}$ are opposite pairs. Then $\{U_1, V\}$ and $\{U_2, V\}$ are opposite pairs as well, proving the proposition. $\qquad\square$

Corollary 7.7.7 *Let Z be a nondegenerate polar space of finite rank whose polar geometry is thick. Suppose that U_1 and U_2 are singular subspaces of the same finite dimension. Then there is an isomorphism $U_1 \to U_2$ which is the identity on $U_1 \cap U_2$. In fact, the map $\pi_{U_2,V}^{-1}\pi_{U_1,V}$, for any singular subspace V of dimension $\dim(U_1)$ such that $\{U_1, V\}$ and $\{U_2, V\}$ are opposite, establishes such an isomorphism. In particular, all maximal singular subspaces of Z are selfdual.*

Proof Take V as in Proposition 7.7.6. Then the maps $\pi_{U_1,V}$ and $\pi_{U_2,V}$ are dualities, so $\pi_{U_2,V}^{-1}\pi_{U_1,V}$ is an isomorphism $U_1 \to U_2$. It is readily verified that it is the identity on $U_1 \cap U_2$. Applying this observation to the case where U_1 and U_2 have dimension two, we find that all planes of Z are isomorphic. In particular, U_1 and V are isomorphic. On the other hand, they are each other's duals (by $\pi_{U_1,V}$), so V, and hence every maximal singular subspace, must be selfdual. $\qquad\square$

Definition 7.7.8 Let $\mathbb{P}$ be a thick projective plane. We call $\mathbb{P}$ *Moufang* if, for every line l of $\mathbb{P}$, each point $p \in l$ and points a, b of $\mathbb{P}$ collinear with p and not on l, there is a perspectivity of $\mathbb{P}$ with center p and axis l mapping a onto b.

Remark 7.7.9 Desarguesian planes are Moufang. Recall from Corollary 6.3.2 that a Desarguesian plane is of the form $\mathbb{P}(\mathbb{D}^3)$ for some division ring $\mathbb{D}$. If $\mathbb{D}$ has only two elements, then the Moufang property is verified by analysis of the Fano plane:

the permutation in $(3, 7)(5, 6)$ is a perspectivity in Fig. 1.21 with center 1 and axis $\{1, 2, 4\}$, whereas transitivity of the automorphism group of the Fano plane on the set of incident point-line pairs gives that it suffices to check only this pair of center and axis. If the underlying division ring $\mathbb{D}$ has at least three points, then nontrivial perspectivities exist whose centers do not lie on the axes; see Definition 6.1.1 and Theorem 6.1.8. To see that $\mathbb{P}(\mathbb{D}^3)$ has perspectivities, consider the matrix

$$\begin{pmatrix} 1 & \lambda & 0 \\ 0 & 1 & 0 \\ 0 & 0 & 1 \end{pmatrix}$$

for arbitrary $\lambda \in \mathbb{D}$. Left multiplication by this matrix represents an arbitrary perspectivity with center $e_1\mathbb{D}$ and axis $\langle e_1, e_3 \rangle$. Since $\mathrm{GL}(\mathbb{D}^3)$ is transitive on bases and induces automorphisms of $\mathbb{P}(\mathbb{D}^3)$, it follows that $\mathbb{P}(\mathbb{D}^3)$ is Moufang.

Not all Moufang planes are Desarguesian; see Exercise 7.11.26.

Theorem 7.7.10 *If Z is a nondegenerate polar space of finite rank at least three whose polar geometry is thick, then the singular planes of Z are Moufang.*

Proof Suppose that M is a singular plane of Z and that l is a line of M. Let $p \in l$ and b_1, $b_2 \in M \setminus l$ be such that p, b_1, b_2 are collinear. As $\mathrm{Rad}(p^\perp) = \{p\}$ (cf. Lemma 7.4.8(iii)), there is a plane, A say, on l such that $\{M/l, A/l\}$ is an opposite pair in $l^\perp/l$. Then $M \cap A = l$. Also, by Proposition 7.7.6, there is a singular plane B of Z containing p such that $(M/p, B/p)$ and $\{A/p, B/p\}$ are opposite pairs of $p^\perp/p$. In particular, $M \cap B = A \cap B = \{p\}$. Let c be a point distinct from p on the line $b_1^\perp \cap B = b_2^\perp \cap B$, and choose d to be a point distinct from p on the line $c^\perp \cap A$. Then d is not on l because of the above opposite pairs. Put $m = cd$. Let C be a plane on the line m such that $\{C/m, \langle\{p\} \cup m\rangle/m\}$ is an opposite pair in $m^\perp/m$. Let $i \in \{1, 2\}$. Since $\{A/l, M/l\}$ is an opposite pair in $l^\perp/l$, we have $b_i \not\perp d$ so b_i is not in C. The line $b_i^\perp \cap C$ contains c but is not contained in B, for otherwise it would contain p, which is not in C. Let D_i be the unique plane containing $b_i^\perp \cap C$ and intersecting B along a line. Then D_i is disjoint from M and A. See Fig. 7.4.

Now, set $\alpha_i = \pi_{A,D_i}^{-1} \pi_{M,D_i}$ and write $\alpha = \alpha_2^{-1}\alpha_1$. By Corollary 7.7.7, this is an isomorphism $M \to M$ fixing l point-wise. Moreover, $\alpha_i(b_i) = d$, whence $\alpha(b_1) = b_2$. Finally, for every line X of M containing p, we see that $\alpha_1(X)$ and $\alpha_2(X)$ are the same line of Z on p, so $\alpha(X) = X$. Therefore, α is a perspectivity of M with center p and axis l, mapping b_1 to b_2.　　　　　　　　　　□

7.8 Other Shadow Spaces of Polar Geometries

Corollary 7.6.8 shows that the study of a polar geometry of rank n is equivalent to the study of a nondegenerate polar space Z of rank n. The polar space is the shadow space on the left most node of the linear diagram B_n associated with the polar geometry. In this section we derive some properties of the shadow spaces on n (the right most node) and on 2 (the one but left most node) of a polar geometry.

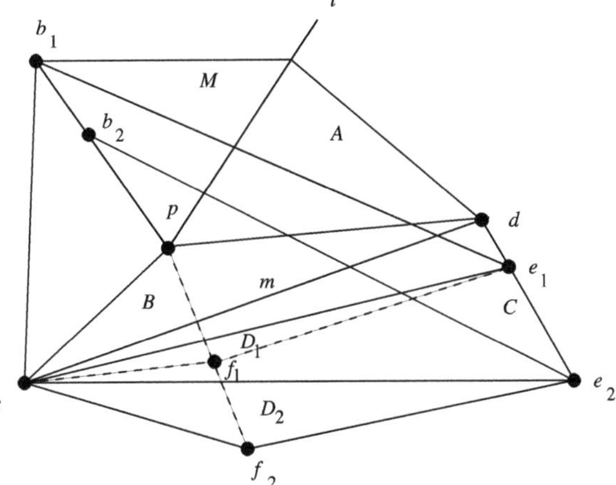

Fig. 7.4 The configuration of planes involved in the proof that planes of polar spaces are Moufang. We have $b_i^\perp \cap C = ce_i$, $D_i = \langle c, e_i, f_i \rangle$ and $D_i \cap B = cf_i$

The first is interesting because it also has a very nice and natural graph-theoretic description. The second is significant because it is a root filtration space.

It is immediate from Definition 7.5.4 that the points of a dual polar space of a nondegenerate polar space Z of finite rank are the maximal singular subspaces and that two such, say A and B, are collinear whenever $\dim(A \cap B) = \dim(A) - 1$. Its lines are the maximal cliques in this graph.

In a nondegenerate polar space of rank n whose dual polar spaces have thin lines, each singular subspace of dimension $n - 2$ is contained in exactly two maximal singular subspaces; that is, the n-order of its polar geometry is equal to one. Such polar spaces lead to oriflamme geometries, which have diagram D_n. Examples are a quadric of rank n in $\mathbb{P}(\mathbb{F}^{2n})$ (see Example 7.8.7) and the hyperoctahedron (see Example 4.1.13(iii)). By the way, the lines of a dual polar space of a nondegenerate polar space with thick lines are either all thick or all thin; see Exercise 7.11.32.

Example 7.8.1 Let V be a vector space over a field $\mathbb{F}$ and consider a quadratic form κ on V (cf. Definition 4.4.2). The *quadric* corresponding to κ is the line space (Q, M) where

$$Q := \big\{ \langle x \rangle \in \mathbb{P}(V) \mid \kappa(x) = 0 \big\},$$

and M is the set of lines of $\mathbb{P}(V)$ entirely contained in Q. We denote (Q, M) by $\mathbb{P}(V)_\kappa$.

We claim that $\mathbb{P}(V)_\kappa$ is a polar space embedded in $\mathbb{P}(V)$. To verify this, we use the symmetric bilinear form f of κ. The map $\pi = \delta_f$ (cf. Notation 7.2.11) is a quasi-polarity of $\mathbb{P}(V)$. If $p = \langle x \rangle \in Q$, then $f(x, x) = \kappa(2x) - 2\kappa(x) = 2\kappa(x) = 0$, so $p \in \mathbb{P}(V)_f$. In particular, $\mathbb{P}(V)_\kappa$ is embedded in $\mathbb{P}(V)_f$ (cf. Notation 7.2.13). Moreover, if l is a line of Q, then there is a point $q \in l$ of the form $q = \langle y \rangle$ with $f(x, y) = 0$. But then, for an arbitrary point $\langle \alpha x + \beta y \rangle$ on the line $pq = x\mathbb{F} + y\mathbb{F}$ of $\mathbb{P}$, we have $\kappa(x\alpha + y\beta) = f(x\alpha, y\beta) + \kappa(x\alpha) + \kappa(y\beta) = \alpha f(x, y)\beta + \alpha^2 \kappa(x) + \beta^2 \kappa(y) = 0$, so $pq \in M$. We conclude that indeed $\mathbb{P}(V)_\kappa$ is a polar space embedded in $\mathbb{P}(V)_f$. In particular, $\mathbb{P}(V)_\kappa$ is a polar space embedded in $\mathbb{P}(V)$.

Observe that Q can be empty, for instance if we take $\mathbb{F} = \mathbb{R}$ and κ positive definite.

If the characteristic of $\mathbb{F}$ is distinct from two, then $\kappa(x) = \frac{1}{2} f(x, x)$ for $x \in V$, and so $\mathbb{P}(V)_\kappa$ coincides with $\mathbb{P}(V)_f$. As a consequence, new polar spaces of this kind in $\mathbb{P}(V)$ are only to be expected for fields of characteristic two.

The *Witt index* of a quadratic form κ on V is the maximum dimension of a linear subspace of V on which κ vanishes. Such a form κ is called *anisotropic* if its Witt index is zero. Compare this terminology with Remark 7.3.4.

There is yet another important example.

Example 7.8.2 Let $\mathbb{D}$ be a division ring. Put $\mathbb{P} = \mathrm{PG}(\mathbb{D}^4)$. By Proposition 2.4.7, this is a geometry with diagram A_3. Let P be the set of lines of $\mathbb{P}$. For $l, m \in P$, write $l \perp m$ to denote that l and m have a non-empty intersection in $\mathbb{P}$. Then, for two distinct such l and m, the set $\{l, m\}^{\perp\perp}$ consists of all projective lines containing the point $l \cap m$ of $\mathbb{P}$ and contained in the plane in $\mathbb{P}$ generated by l and m. Such a set of projective lines is called a *pencil*. Write L for the collection of all such 'lines' (or pencils) $\{l, m\}^{\perp\perp}$. By construction, (P, L) is the shadow space $\mathrm{ShSp}(\mathrm{PG}(\mathbb{D}^4), 2)$ as introduced in Definition 2.5.1. It is the Grassmannian of lines as introduced in Definition 6.6.1. The space (P, L) is a nondegenerate polar space of rank three (proving this statement is Exercise 7.11.28). There are two kinds of maximal singular subspace: those consisting of all lines on a given projective point of $\mathbb{P}$ and those consisting of all lines contained in a given projective plane of $\mathbb{P}$. Each pencil belongs to exactly one of each kind.

The two previous examples are related by means of the so-called Klein correspondence.

Theorem 7.8.3 (Klein Correspondence) *If $\mathbb{F}$ is a field, then the Grassmannian of lines in $\mathrm{PG}(\mathbb{F}^4)$ is isomorphic to the quadric corresponding to the quadratic form κ on $\mathbb{F}^6$ given by*

$$\kappa(x) = x_1 x_2 - x_3 x_4 + x_5 x_6 \quad \left(x = \varepsilon_1 x_1 + \cdots + \varepsilon_6 x_6 \in \mathbb{F}^6 \right).$$

Proof Write (P, L) for the Grassmannian of lines in $\mathrm{PG}(\mathbb{F}^4)$ and (Q, M) for the quadric corresponding to κ. Consider the bilinear map $g : \mathbb{F}^4 \times \mathbb{F}^4 \to \mathbb{F}^6$ determined by

$$g(x, y) = (x_1 y_2 - x_2 y_1, x_3 y_4 - x_4 y_3, x_3 y_1 - x_1 y_3,$$

$$x_4 y_2 - x_2 y_4, x_1 y_4 - x_4 y_1, x_2 y_3 - x_3 y_2). \tag{7.8}$$

It induces a map $k : P \to Q$ by $k(l) = \langle g(x, y) \rangle$ whenever $l = \langle x, y \rangle \in P$. For, the expression $g(x, y)$ does not depend on the choice of distinct projective points $\langle x \rangle$, $\langle y \rangle$ inside l, and a straightforward computation shows that $\kappa(g(x, y)) = 0$.

If l, m, n is a collinear triple in (P, L), then there are nonzero vectors $x, a, b \in \mathbb{F}^4$ such that $l = \langle x, a \rangle$, $m = \langle x, b \rangle$ and $n = \langle x, a + b \rangle$. Then, by linearity of g in the second argument, we have $g(x, a + b) = g(x, a) + g(x, b)$, and so $k(n)$ is collinear with $k(l)$ and $k(m)$. This establishes that k is in fact a homomorphism of line spaces. It is readily seen to be an isomorphism. $\qquad\square$

Recall from Definition 1.2.1 what it means for a graph to be bipartite.

Lemma 7.8.4 *Let Z be a nondegenerate polar space of finite rank n whose polar geometry has n-order equal to one.*

(i) *The dual polar graph is bipartite (i.e., has no circuits of odd length, or, equivalently, the vertex set is the disjoint union of two cocliques C_1, C_2).*
(ii) *For all M, $M' \in C_1 \cup C_2$, the number $n - 1 - \dim(M \cap M')$ is even if and only if M and M' are in the same constituent C_1 or C_2.*

Proof (i) We need only show that there is no circuit of odd length in the dual polar graph of Z. Assume, by way of contradiction, that E is such a circuit of odd length. Let it have smallest possible cardinality, say $2m + 1$. Let a be a vertex of E and let b, c be the two vertices of E with the property $d(a, b) = m = d(a, c)$, where we take distances in E. Then $d(b, c) = 1$. By our hypothesis, the line bc of the dual polar space of Z has only two points namely b and c. Theorem 7.7.2 gives that one of b and c, say c, is nearest to a. Then there is a path P from a to c whose length is smaller than m. Combining this path P with one of the two paths from a to c, in E (one of length m and one of length $m + 1$) we find a circuit of odd length which is shorter than $2m + 1$, a contradiction.

(ii) follows immediately from (i) and Theorem 7.7.2. $\qquad\square$

For Z as in the lemma, we construct a geometry of type D_n.

Definition 7.8.5 Let Z be a nondegenerate polar space of finite rank n whose polar geometry has n-order equal to one. The *oriflamme geometry* of Z is the incidence system, denoted by $\Delta(Z)$, over $[n]$ whose elements of type $i = 1, \ldots, n - 2$ are the elements of type i of the polar geometry $\Gamma(Z)$ of Z and in which incidence for these elements is as in $\Gamma(Z)$. The elements of type $n - 1$ (respectively, n) of $\Delta(Z)$ are the members of C_1 (respectively, C_2) as defined in Lemma 7.8.4. Elements in $C_1 \cup C_2$ are incident in $\Delta(Z)$ if and only if they are adjacent in Z^δ.

Theorem 7.8.6 *Let Z be a nondegenerate polar space of finite rank $n \geq 2$ whose polar geometry has n-order one. The oriflamme geometry $\Delta(Z)$ is an $[n]$-geometry over D_n whose point shadow space is isomorphic to Z.*

Proof We proceed by induction on n. For $n = 2$, let $x \in C_1$ and $y \in C_2$. If p is a point of y, then there is a line l of Z on p that has a nontrivial intersection with x, because Z is a polar space. Now $\dim(l \cap x) = 1$, so $l \cap x$ is a point q. As q is on exactly two lines (for the 2-order of the polar geometry is equal to 1) and as these are in distinct constituents C_1, C_2, we conclude from $x \in C_1$ that $l \in C_2$. But then l and y are both lines from C_2 on p, so $l = y$. Therefore $x * y$ in $\Delta(Z)$ and so the latter is a generalized digon and all statements hold. Let $n > 2$. If p is a point of Z, then the residue Z_p is a nondegenerate polar space of rank $n - 1 \geq 2$, whose polar geometry has $n - 1$-order equal to one, so induction yields that $\Delta(Z_p)$ belongs to the diagram D_{n-1} while $\Delta(Z_p)$ is firm and residually connected. If M is a maximal singular subspace of Z then the residue of M in $\Delta(Z)$ is clearly a projective geometry of rank $n - 1$. Hence $\Delta(Z)$ belongs to D_n and $\Delta(Z)$ is firm. Residual connectedness of $\Delta(Z)$ readily follows. $\square$

Example 7.8.7 Let $\kappa : V \to \mathbb{F}$ be a quadratic form on a vector space of even dimension $2t$ such that the bilinear form associated with κ is nondegenerate of Witt index t. Write $\mathbb{P} = \mathbb{P}(V)$. The polar space $\mathbb{P}_\kappa$ of Example 7.8.1 leads to an oriflamme geometry. For, the geometry of $\mathbb{P}_\kappa$ is of rank t and, fixing a subspace, U say, of V of dimension $t - 1$ such that $\mathbb{P}(U)$ is a singular subspace of $\mathbb{P}_\kappa$ of dimension $t - 2$, we can compute the t-order of the polar geometry of $\mathbb{P}_\kappa$ as the number of projective points in the residue $\mathbb{P}(U^\perp / U)$, which is isomorphic to $\mathbb{P}(W)_{\kappa'}$ for κ' the restriction of κ to a complement W of U in $U^\perp$. Observe that $\dim(W) = 2$ (here we are using vector space dimensions) as $\dim(U) = t - 1$ and $\dim(U^\perp) = t + 1$. So the t-order is the number of points $\langle x \rangle \in \mathbb{P}(W)$ such that $\kappa(x) = 0$. Choose a basis w_1, w_2 of W. Now there are $a, b, c \in \mathbb{F}$ such that $\kappa(x) = ax_1^2 + bx_1x_2 + cx_2^2$, where $x = x_1 w_1 + x_2 w_2$. As the Witt index of κ is equal to t, the Witt index of κ' is equal to one, and the set of solutions is neither empty nor $\mathbb{P}(W)$. This implies that the quadratic equation $ax_1^2 + bx_1x_2 + cx_2^2 = 0$ has exactly two projective solutions. Therefore, the t-order is equal to one, and Theorem 7.8.6 applies.

The theorem admits a converse.

Notation 7.8.8 Let $n \geq 3$ and let Γ be an $[n]$-geometry of type D_n. Denote by $\widetilde{\Gamma}$ the incidence system over $[n]$ constructed from Γ as follows. For $i \in [n - 2]$, the elements of type i of $\widetilde{\Gamma}$ are the elements of type i of Γ. Elements of type $n - 1$ of $\widetilde{\Gamma}$ are flags of Γ of type $\{n - 1, n\}$. Finally, the set of elements of type n of $\widetilde{\Gamma}$ is the union of the sets of elements of type $n - 1$ and n in Γ. Incidence in Γ determines incidence in $\widetilde{\Gamma}$ except that incident elements of distinct types $n - 1$, n, respectively, in Γ are no longer incident in $\widetilde{\Gamma}$.

Proposition 7.8.9 *Let $n \geq 3$ and let Γ be an $[n]$-geometry of type D_n.*

(i) *$\widetilde{\Gamma}$ is an $[n]$-geometry of type B_n whose n-order is equal to one.*
(ii) *$\widetilde{\Gamma}$ is a polar geometry.*
(iii) *If Z is the polar space associated with $\widetilde{\Gamma}$, then $\Delta(Z) \cong \Gamma$.*

Proof (i) is straightforward.

(ii) We need to prove Conditions (1) and (2) of Definition 7.6.3 for $\widetilde{\Gamma}$. We proceed in four of steps.

STEP 1. *Given a point q and an element v_{n-1} of Γ of type $n-1$, there exists an element v_n of Γ of type n such that $q * v_n * v_{n-1}$.*

We use induction on n. For $n = 3$, Γ is a projective geometry and so, by Corollary 5.4.4 and Theorem 6.3.1, the statement follows from a study of $\mathbb{P}(\mathbb{D}^4)$, where $\mathbb{D}$ is a division ring. In this interpretation, q is a line and v_2 is a point of $\mathbb{P}(\mathbb{D}^4)$, so we can take v_3 to be the plane generated by q and v_2 in $\mathbb{P}(\mathbb{D}^4)$.

For $n > 3$, let l be a line incident with v_{n-1}. As q and l have types at most two, they are also elements of $\widetilde{\Gamma}$. By (i) and Proposition 7.6.6(ii) applied to $\widetilde{\Gamma}$, there is a line l' incident with q, intersecting l in some point p. Both l' and p are also elements of Γ. The induction hypothesis applied to the residue Γ_p, provides an element v_n of Γ incident with v_{n-1} and l', and therefore also with q.

STEP 2. *Given two points p, q there exists a chain $p * v_{n-1} * v_n * q$ where v_{n-1} and v_n are elements of respective types $n-1$ and n.*

Indeed, take an element v_{n-1} of type $n-1$ of Γ, incident with p, and apply Step 1.

STEP 3. *Condition (2) of Definition 7.6.3 holds for $\widetilde{\Gamma}$.*

Let p and q be distinct points and let l and l' be lines of $\widetilde{\Gamma}$ incident with p and q. We need to show $l = l'$. This is obvious for $n = 3$. We use induction on n and assume $n > 3$. Step 2, applied to Γ_p, gives a chain l, v_{n-1}, v_n, l' in Γ. By the Direct Sum Theorem 2.1.6, $p, q \in \{v_{n-1}, v_n\}^*$. In the projective geometry $\Gamma_{\{v_{n-1}, v_n\}}$, there is a unique line l'' incident with p and q. This is also the case, in the projective geometry $\Gamma_{v_{n-1}}$ and so $l'' = l$. Similarly, in the projective geometry Γ_{v_n}, we find $l'' = l'$. Hence, $l = l'$.

STEP 4. *Condition (1) of Definition 7.6.3 holds for $\widetilde{\Gamma}$.*

Suppose that x and y are elements of $\widetilde{\Gamma}$ with $p \in \mathrm{Sh}_1(\{x\}) = \mathrm{Sh}_1(\{y\})$ (notation of Definition 2.5.1). We need to show $x = y$. By Step 2, the shadow of an element of type at least two contains a line, so, without loss of generality, we may take the types of both x and y to be at least two.

First, let $n = 3$. Then, as in Step 1, we may assume that, for some division ring $\mathbb{D}$, the elements of Γ of type 1, 2, and 3 are the lines, points, and planes of $\mathbb{P}(\mathbb{D}^4)$, respectively, so the elements of $\widetilde{\Gamma}$ of type 1, 2, and 3 correspond to lines, the flags consisting of points and planes (in terms of 1-shadows: the pencils of lines), and points or planes (in terms of 1-shadows, all lines on a point or all lines in a plane), respectively. In these cases, clearly, the 1-shadows uniquely determine the element, so $x = y$ follows.

Proceed again by induction on n. Assume $n > 3$. Let l be a line on p with $l * x$ and let q be a point on l distinct from p. Then p and q both belong to $\mathrm{Sh}_1(y)$; but Γ_y is a projective geometry, so there is a line of Γ incident with y, p, and q, which must coincide with l by Step 3. Therefore x and y have equal 2-shadows in Γ_p and the induction hypothesis forces $x = y$.

(iii) By (ii) and Corollary 7.6.8 there is a nondegenerate polar space Z whose polar geometry is isomorphic to $\widetilde{\Gamma}$. By Theorem 7.8.6, $\Delta(Z)$ is as stated. $\square$

Apparently, geometries of type D_n behave better than geometries of type B_n in that no other examples than those arising from polar spaces occur.

The final result of this section revisits the Grassmannian of lines introduced in Example 7.8.2. It restates results found in the proof of Proposition 7.8.9 in terms of spaces.

Corollary 7.8.10 *Suppose that Z is a nondegenerate polar space of rank three all of whose lines are thick. If there is a line in Z on which there are precisely two maximal singular subspaces (planes), then Z is isomorphic to the Grassmannian of lines of $\mathbb{P}(\mathbb{D}^4)$ for some division ring $\mathbb{D}$.*

Proof By Exercise 7.11.32 every line of the dual polar space of Z is thin. Therefore, Lemma 7.8.4 applies, so there is a partition of the maximal singular subspaces into two collections, say M_0 and M_1, such that each member of M_0 meets each member of M_1 either not at all or in a line. Now consider the geometry $(M_0, L, M_1, *)$ where L is the set of lines of Z and $*$ is defined as incidence on $(M_i \times L) \cup (L \times M_i)$ for $i \in \{0, 1\}$ and as meeting in a line on $(M_0 \times M_1) \cup (M_1 \times M_0)$. This geometry is readily seen to be thick, residually connected, and to belong to the diagram A_3, and so, by Corollary 5.4.4 and Theorem 6.3.1, (M_0, L) is isomorphic to a projective space $\mathbb{P}(\mathbb{D}^4)$ for some division ring $\mathbb{D}$. The result follows as in the proof of Proposition 7.8.9. $\qquad\square$

7.9 Root Filtration Spaces

In Sect. 6.7, we introduced the notion of a root filtration space and in Theorem 6.7.7 we found that in each projective space there is a root filtration space whose point set consists of all pairs of an incident point and hyperplane. Polar spaces, nondegenerate or not, are degenerate root filtration spaces (cf. Exercise 7.11.35). But the Grassmannian of lines in a nondegenerate polar space of rank at least three is always a nondegenerate root filtration space; see Proposition 7.9.2. This observation leads to the question how to recognize these Grassmannians among all root filtration spaces. In this section, we give a partial answer in that we characterize the smallest nontrivial case, where the rank of the polar space is three; see Theorem 7.9.18. There is a recognition theorem for the general case of finite rank, but this is more involved and will not be treated here (the result is stated in Theorem 11.7.11). Some of the early results of this section, however, are applicable to all nondegenerate root filtration spaces and provide a beginning to the overall classification.

Definition 7.9.1 Let Z be a polar space of rank at least three. The *Grassmannian* of lines of Z is the line space whose points are the lines of Z and whose lines are the sets of lines of Z containing a point x of Z and contained in a singular plane π of Z with $x \in \pi$.

Let Γ be a polar geometry of rank $n \geq 3$. For $j \in [n]$, the Grassmannian of lines of the polar space $\mathrm{ShSp}(\Gamma, 1)$ is isomorphic to $\mathrm{ShSp}(\Gamma, 2)$. This gives the connection with the notion of a Grassmannian for geometries of type A_n given in Definition 6.6.1.

Proposition 7.9.2 *Let Z be a nondegenerate polar space of rank at least three. The Grassmannian (E, F) of lines of Z is a nondegenerate root filtration space.*

Proof Clearly, (E, F) is a partial linear space. We give a system of symmetric relations $(E_i)_{-2 \leq i \leq 2}$ on E satisfying the conditions of Definition 6.7.2. These are defined as follows, where $l, m \in E$. As always, E_{-2} is the diagonal, and $(l, m) \in E_{-1}$ if and only if l and m are distinct and collinear in (E, F), which means that they generate a singular plane of Z.

(0) $l \in E_0(m)$ if and only if either l and m generate a singular subspace not contained in a plane, or l and m intersect nontrivially but do not generate a singular plane,
(1) $l \in E_1(m)$ if and only if l and m do not intersect nontrivially and do not generate a singular subspace and there is a unique line n such that both the subspace generated by l and n and the subspace generated by n and m are singular planes, in which case $n = [l, m]$.
(2) E_2 is the complement of $E_{\leq 1}$ in $E \times E$.

The verification that this system satisfies Definition 6.7.2(1)–(8) is left to the reader. $\qquad\square$

The same construction for a projective space instead of a polar space leads to a root filtration space with $E_1 = E_2 = \emptyset$.

In order to characterize the case where the polar space has rank three, we begin with a study of root filtration spaces in general, building upon the results of Sect. 6.7. The adjective 'polar' introduced there for pairs of points in E_0 is justified by the following proposition.

Proposition 7.9.3 *Let $(x, y) \in E_0$ inside a nondegenerate root filtration space (E, F). The set $E_{-1}(x, y)$ is a subspace of (E, F) which is a nondegenerate polar space (possibly with an empty set of lines).*

Proof By Definition 6.7.2(8) and Lemma 6.7.15, $E_{-1}(x, y)$ is not empty; it clearly coincides with $E_{\leq -1}(x, y)$, so by Definition 6.7.2(5) it is a subspace of (E, F). We verify the polar space axiom, Definition 7.4.1, for this subspace. Let $u, v, w \in E_{\leq -1}(x, y)$ be such that $u \sim v$ (recall that this means $(u, v) \in E_{-1}$). By Lemma 6.7.18(ii), there exists $z \in E_2(x) \cap E_0(y)$. By the triangle condition on w, z, y, we must have $w \in E_{\leq 1}(z)$. The triangle condition on x, z, w, then forces $(w, z) \in E_1$. Hence $w' = [w, z]$ exists; it satisfies $(w', x) \in E_1$ and $w = [x, w']$. Furthermore, by the filtration around y, we have $y \sim w'$. By Lemma 6.7.19 (applied with y and w' in the roles of v and w, respectively), there is a point

$s \in uv \cap E_{\leq 0}(w')$. By the filtration around s and by using $x \sim s$, we have $s \sim [x, w'] = w$. As (E, F) is a gamma space (cf. Remark 6.7.3), this establishes that $E_{\leq -1}(x, y)$ is a polar space. By Lemma 6.7.16, this polar space is nondegenerate. $\qquad\square$

Proposition 7.9.4 *Let (E, F) be a nondegenerate root filtration space. If $y \in E$ and $l \in F$ satisfy $y \in E_0(l)$, then $E_{\leq -1}(y, l)$ is a non-empty singular subspace of (E, F).*

Proof Suppose that a and b are non-collinear points of $E_{\leq -1}(y, l)$. As $E_{\leq -1}(a, b)$ has at least two points, $(a, b) \in E_0$ (cf. Remark 6.7.17), so $E_{\leq -1}(a, b)$, which by Proposition 7.9.3 is a polar space containing both y and l, has a point in l collinear with y, contradicting the assumption $y \in E_0(l)$. Hence, $E_{\leq -1}(y, l)$ is a singular subspace.

It remains to show that $E_{\leq -1}(y, l)$ is non-empty. Pick $x \in l$. By Lemma 6.7.18(ii), there exists $z \in E_2(x) \cap E_0(y)$. Let v be the point of the line l such that $(v, z) \in E_1$, which exists by Condition (6) of Definition 6.7.2, and set $v' = [v, z]$. Then $v \sim v'$, $(v', x) \in E_1$ (by the triangle condition) and $v = [v', u]$ for every point u on the line l distinct from v. Furthermore, the y and v' commute (by filtration around y, as $v, z \in E_{\leq 0}(y)$). Assume $y \sim v'$. Then, by the filtration around y we have $y \sim v = [x, v']$, contradicting $y \in E_0(l)$. Thus the pair (y, v') is actually polar.

Using Lemma 6.7.18(ii) again, pick $z' \in E_2(v) \cap E_0(y)$, let u be the point on the line l such that $(z', u) \in E_1$, and put $u' = [z', u]$. As for v above, we see that $u' \in E_{-1}(u) \cap E_1(v) \cap E_0(y)$. Observe that $u \neq v$, as $v \in E_2(z')$ and $u \in E_1(z')$. We also have $(u', v') \in E_2$ by Lemma 6.7.8(v) and $E_{\leq -1}(y, l) = E_{\leq -1}(y, u, v)$ by Definition 6.7.2(5). Furthermore, from the fact that v is the only point of the line uv which is also in $E_{\leq 1}(z)$ we infer $(z, u) \in E_2$. For a point $w \in E_{\leq -1}(u, y)$ the triangle condition on z, u, w gives $z \in E_{\geq 1}(w)$ and the triangle condition on y, z, w gives $z \in E_{\leq 1}(w)$, so the pair (w, z) is special. Hence, by Lemma 6.7.12, w is in $E_{\leq -1}(v)$ if and only if $(w, v') \in E_0$, and otherwise $(w, v') \in E_1$.

By the argument of the first paragraph applied to uu' instead of the line l, and by using Definition 6.7.2(5) to see that uu' is contained in $E_{\leq 0}(y)$, we find that $E_{\leq -1}(y, u, u')$ is a singular subspace. Since $E_{\leq -1}(y, u)$ is a nondegenerate polar space, there exists $w \in E_{\leq -1}(y, u) \setminus E_{\leq -1}(u')$. If $(w, v') \in E_0$, then by Lemma 6.7.8(iv) applied to v', w and u we have $v = [v', u] \in E_{-1}(v', w, u)$, so $w \in E_{\leq -1}(y, u, v)$, as required. Therefore, we may assume $(w, v') \in E_1$.

Assume $E_{\leq -1}(y, u, v) = \emptyset$. For $w' = [w, v']$, by the filtration around y, we have $y \sim w'$. Also, if $(v, w) \in E_1$ then $u = [v, w]$, and the filtration around y would imply $y \sim u = [v, w]$, a contradiction. Therefore, $(v, w) \in E_0$ and hence, by the filtration around v, we find $v \sim w'$. We thus have $w' \in E_{\leq -1}(y, v, v')$.

Similarly, we find that $(u', w') \in E_1$ and $t := [w', u'] \in E_{\leq -1}(y, u, u')$. Since $w \notin E_{-1}(u')$, the points t and w are distinct. Obviously, $w' \sim t$ and $w \sim w'$. If $w \sim t$ as well, then by Lemma 6.7.19 there is a point s on the line wt such that $(s, v') \in E_0$ whence, by filtration around s, the point s is collinear with $[u, v'] = v$, contrary to assumption. As u and w' are distinct points of $E_{\leq -1}(w, t)$, the pair (w, t) is polar. Now u is a point and yw' is a line of the polar space $E_{\leq -1}(w, t)$, so there is a point p on the line yw' collinear with u. As p lies on the line yw' of the subspace $E_{\leq 0}(v')$ (cf. Definition 6.7.2(5)), we have $p \in E_{\leq -1}(u) \cap E_{\leq 0}(v')$, so the filtration around p gives $p \sim v$. This implies $p \in E_{\leq -1}(y, u, v)$, a final contradiction. $\square$

Theorem 7.9.5 *Let (E, F) be a nondegenerate root filtration space. If some line in F is contained in a unique maximal singular subspace, then so is every line in F. Otherwise, $\mathrm{rk}(E_{-1}(x, y)) \geq 2$ for any polar pair (x, y).*

Proof Assume that the line m is contained in two planes P and Q that do not generate a singular subspace. Let x be a point of m.

We first show that there is $y \in E_0(x)$ such that $E_{\leq -1}(x, y)$ is a polar space of rank at least 2. Pick $z \in E_2(x)$ and $u \in P \setminus m$ and $v \in Q \setminus m$ such that $z \in E_1(u, v)$. We are in the case $(u, v) \in E_0$ (because $m \subseteq E_{\leq -1}(u, v)$) and hence, by Lemma 6.7.12, $(u, [z, v]) \in E_1$. Setting $y = [u, [z, v]]$, we find $y \in E_{\leq -1}(u, v)$. Notice that $y \sim x$ would imply that both u and y belong to $E_{\leq -1}([u, z], x)$, so $u = y$, whence $u \sim v$, a contradiction. Therefore, $(x, y) \in E_0$. By Proposition 7.9.3, the subset $E_{-1}(x, y)$ is a polar space. It contains m and y, so there is $w \in m \cap E_{\leq -1}(y)$; now uw is a line contained in $E_{\leq -1}(x, y)$, so the latter is a polar space of rank at least 2, as claimed.

Let l be a line that intersects m in the point x. If l intersects $E_{-1}(x, y)$, then, as $E_{-1}(x, y)$ is a nondegenerate polar space of rank at least two, the line l is contained in two planes that do not generate a singular subspace.

If l does not intersect $E_{-1}(x, y)$ then either $l \subseteq E_0(y)$ and, by Proposition 7.9.4, $E_{\leq -1}(y, l)$ must be a non-empty singular subspace, or there is a point $p \in l \cap E_1(y)$ with $[p, y] \in E_{\leq -1}(y, l)$ by Lemma 6.7.8(iv). In each case, there exists $t \in E_{\leq -1}(y, l)$. As $E_{-1}(x, y)$ is a polar space of rank at least 2, the point t is contained in a quadrangle of $E_{-1}(x, y)$. But $E_{\leq -1}(y, l)$ is a singular subspace, so there is a point s of $E_{\leq -1}(x, y, t)$ which is not in $E_{\leq -1}(l)$. Now we replace m with the line xt, P with the plane lt and Q with xts and apply the first part of the proof. To be more specific, in this case $z \in E_2(x)$ as above, the role of u is played by some point u' on l, and the role of v by some point v' on xs such that $z \in E_1(u', v')$. We obtain that $y' := [u', [z, v']]$ is a point in $E_0(x)$ such that $E_{-1}(x, y')$ is a polar space of rank at least 2 and now l intersects $E_{-1}(x, y')$ in u' (and l still intersects m in x). It follows that l is contained in two planes that do not generate a singular subspace.

Thus, the property of m (being contained in two planes) is inherited by any line intersecting m and consequently by every line. We have seen that if there is a line contained in two planes not in a common singular subspace then every line is. Finally, if there are two distinct maximal subspaces containing a line l, then we can choose non-collinear points a and b of these subspaces and these points together with the line l generate a subspace containing two planes on the line l, which is not singular.

An even stronger statement can be proved in this case: Assume that the point $x \in E$ and the line $l \in F$ generate a singular plane. There exists a point $y \in E_0(x)$ such that y and l also generate a singular plane. Indeed, pick a and b as above. As $l \subseteq E_{-1}(a, b, x)$, the points a and b belong to $E_{\leq 0}(x)$. If $x \notin E_{-1}(a, b)$ then either a or b will be a good choice for y. Otherwise as l is a line in the nondegenerate polar space $E_{-1}(a, b)$, it coincides with the set of points of $E_{-1}(a, b)$ that are collinear with every point collinear with all of the points of l. This implies the existence of a point $y \in E_{-1}(a, b) \setminus E_{\leq -1}(x)$.

Suppose now that every line is contained in at least two distinct maximal singular subspaces of a root filtration space. If x and y are two points of E at mutual distance two, then, by Lemma 6.7.8(ii) and Proposition 7.9.3, either $(x, y) \in E_1$ and $E_{-1}(x, y) = \{[x, y]\}$, or $(x, y) \in E_0$ and $E_{-1}(x, y)$ is a nondegenerate polar space.

To finish the proposition, we need to show that, in the polar case, the polar space $E_{-1}(x, y)$ has rank at least two. To see this, pick $z \in E_2(x) \cap E_0(y)$ (it exists by Lemma 6.7.18). By Lemma 6.7.16 there are $u, v \in E_{-1}(x, y)$ with $(u, v) \in E_0$. By the triangle condition, $u, v \in E_1(z)$ and, by Lemma 6.7.12, $v' = [v, z] \in E_1(u)$. Also, $v' \in E_{-1}(y)$ by the filtration around y. Together with $y \in E_{-1}(u)$ this implies $y = [u, v']$. By the strong property proved two paragraphs above there exists a point $w' \in E_{-1}(v, v') \cap E_0(y)$. Lemma 6.7.19 implies that there is a point w on the line $v'w'$ which is also in $E_0(x)$. Since $(v', x) \in E_1$, the point w is distinct from v' and hence $w \in E_0(y)$. In view of the triangle condition on w, y, u, we have $w \in E_{\leq 1}(u)$. If w were in $E_{\leq 0}(u)$ then the filtration around w would give $y = [u, v'] \in E_{\leq -1}(w)$, a contradiction with $y \in E_0(w)$. Thus $w \in E_1(u)$. By the filtration around x and y, respectively, we have $[w, u] \in E_{-1}(x, y)$. As the subspace $E_{-1}(x, y)$ contains the line $u[w, u]$, it is a polar space of rank at least two. This proves the proposition. $\square$

In the proposition below we will use a Galois connection between two maximal singular subspaces.

Definition 7.9.6 A *Galois connection* between two partially ordered sets is a pair of order reversing maps between these sets, one way each, with the property that each element is smaller than or equal to its image under the composition of the two maps (in the right order). The elements equal to their images under these compositions are called *closed*.

Recall from Exercise 3.7.7 that, for a set X, we denote by 2^X the power set of X, that is, the collection of all subsets of X. This set is partially ordered by inclusion.

Recall from Notation 6.7.21 that $C_M(x) = M \cap E_{\leq 0}(x)$ whenever x is a point and M a subset of the point set E of a root filtration space. Moreover, $C_M(X) = \bigcap_{x \in X} C_M(x)$ for $X \in 2^E$.

Proposition 7.9.7 *Suppose that (E, F) is a nondegenerate root filtration space in which each line is contained in exactly one maximal singular subspace. Let M_1 and M_2 be two maximal singular subspaces of (E, F) containing a point z. The maps $C_{M_2} : 2^{M_1} \to 2^{M_2}$ and $C_{M_1} : 2^{M_2} \to 2^{M_1}$ give a Galois connection between 2^{M_1} and 2^{M_2}. For $i \in [2]$, each closed subset of M_i is a subspace containing z and each subspace of M_i of finite dimension containing z is closed. In particular, the dimensions of maximal singular subspaces of (E, F) are equal.*

Proof As the relation $E_{\leq 0}$ is symmetric and inclusion reversing, the maps C_{M_i} $(i = 1, 2)$ provide a Galois connection between 2^{M_1} and 2^{M_2}. By Definition 6.7.2(5), closed subsets are indeed subspaces containing z. Let $P = M_1 \cup M_2$ and $L = (F \cap (2^{M_1} \cup 2^{M_2})) \cup \{\{u, v\} \mid u \in M_1 \setminus \{z\}, v \in C_{M_2}(u) \setminus \{z\}\}$. By Lemma 6.7.22, for every $u \in M_1 \setminus \{z\}$, the subspace $C_{M_2}(u)$ is a hyperplane of M_2 and $C_{M_1}(v)$ is a hyperplane of M_1 for every $v \in M_2 \setminus \{z\}$, so the point-line space (P, L) is a polar space with radical $\{z\}$. By Exercise 7.11.13, every singular subspace X of finite dimension in the nondegenerate polar space $(P, L)/\{z\}$ satisfies $X = X^{\perp\perp}$. For every subspace X of M_i ($i \in [2]$) of finite dimension containing z, this implies $X = C_P(C_P(X)) = C_{M_i}(C_{M_{3-i}}(X))$, proving that X is closed indeed.

In view of the Galois connection, two intersecting maximal singular subspaces have equal dimension. As (E, F) is connected, this implies that all maximal singular subspaces of (E, F) have the same dimension. $\qquad\square$

Corollary 7.9.8 *If (E, F) is a nondegenerate root filtration space having a line that is a maximal singular subspace of (E, F), then (E, F) is a generalized hexagon.*

Proof According to Theorem 7.9.5 and Proposition 7.9.7, the singular dimension of (E, F) equals 1. Emptiness of E_0 follows from Lemma 6.7.22. As stated in Example 6.7.10, this suffices for the proof of the corollary. $\qquad\square$

From now on we assume that the singular dimension of (E, F) is at least two.

Lemma 7.9.9 *Suppose that (E, F) is a nondegenerate root filtration space of singular dimension at least two in which each line is contained in exactly one maximal singular subspace. For any three points x, y, z of a singular subspace of E not all on one line, the following assertions hold.*

(i) *$E_1(x) \cap E_0(y) \cap E_{-1}(z) \neq \emptyset$.*
(ii) *The union of all lines zv for $v \in xy$ is a subspace of E.*
(iii) *Every singular subspace of E is projective.*

Proof Let M_1 be the maximal singular subspace containing x, y, z and let M_2 be another one through z.

(i) By Proposition 7.9.7, we have $yz = C_{M_1}(C_{M_2}(y))$, which shows (i).

(ii) Proposition 7.9.7 also gives $\langle x, y, z \rangle = C_{M_1}(C_{M_2}(xy))$. Clearly, $\bigcup_{v \in xy} zv \subseteq \langle x, y, z \rangle$. Suppose that $u \in \langle x, y, z \rangle$ does not lie on the line xz. As $C_{M_2}(x)$ and $C_{M_2}(u)$ are distinct geometric hyperplanes of M_2, and geometric hyperplanes of M_2 are maximal (cf. Lemma 5.2.8), there is a point $v \in C_{M_2}(u) \setminus C_{M_2}(x)$. Consequently, x does not lie in the geometric hyperplane $C_{M_1}(v)$ of M_1, which meets the line xy in a point, say q. Now $C_{M_2}(xy) \subset \langle v, C_{M_2}(xy) \rangle \subseteq C_{M_2}(q)$ and $C_{M_2}(xy) = C_{M_2}(xq)$ is a geometric hyperplane of $C_{M_2}(q)$, so Lemma 5.2.8 forces $\langle v, C_{M_2}(xy) \rangle = C_{M_2}(q)$. This implies $C_{M_2}(q) = \langle v, C_{M_2}(xy) \rangle \subseteq C_{M_2}(u)$, and so $C_{M_2}(q) = C_{M_2}(u)$, giving $qz = C_{M_1}(C_{M_2}(q)) = C_{M_1}(C_{M_2}(u)) = uz$. We conclude that u is on the line qz containing the point q of xy.

(iii) Let $u \in xy \setminus \{y\}$ and $v \in xz \setminus \{z\}$ with $u \neq v$. Clearly, $\langle x, y, z \rangle = \langle y, u, v \rangle$. By (ii) applied to y, u, v, the point z lies on a line through y meeting uv in a point. This establishes Pasch's axiom, so we finish by invoking Theorem 5.2.6. $\qquad\square$

Lemma 7.9.10 *Suppose that (E, F) is a nondegenerate root filtration space in which each line is contained in exactly one maximal singular subspace. There is no quadruple $x, u, v, w \in E$ such that $x \in E_{\leq -1}(u, v, w)$, $(u, v) \in E_1$, and $w \in E_0(u, v)$.*

Proof Suppose the contrary. By Lemma 6.7.16 we can pick $y \in E_{\leq -1}(u, w) \cap E_0(x)$. Then $(y, v) \in E_2$ is excluded by the triangle y, v, w. On the other hand, in the case $(y, v) \in E_0$, we infer from the filtration around y that $x = [u, v] \sim y$, a contradiction with $y \in E_0(x)$. Thus we have $(v, y) \in E_1$. Let $z = [v, y]$. By the filtration around x we have $x \sim z$ or $x = z$. The second case cannot happen as x is not collinear with y, so $x \sim z$. A similar argument shows $z \sim w$. But then the line xz is in $E_{\leq -1}(w, v)$, a contradiction with the assumption that there is only one maximal singular subspace on it, as w and v are not collinear. $\qquad\square$

Proposition 7.9.11 *Suppose that (E, F) is a nondegenerate root filtration space in which each line is contained in exactly one maximal singular subspace, which is of dimension at least two. If there is a point of E that is contained in at least three maximal singular subspaces, then (E, F) has singular dimension two.*

Proof Let x be a point belonging to at least 3 maximal singular subspaces. Assume that l is a line on x that is not a hyperplane of the unique maximal singular subspace M on l. By Lemma 7.9.9(i), we can find $u \in E_{-1}(x) \cap E_0(l \setminus \{x\})$. Let M' be a maximal singular subspace containing x not containing l and xu. By Lemma 6.7.22 there exists a point $v \in M'$ such that $(u, v) \in E_1$. The set $C_M(u)$ is a proper hyperplane of M containing l. Therefore there exists a point $z \in M \setminus l$ such that $(u, z) \in E_0$. Pick $y \in l \setminus \{x\}$. For every point w of the line yz either $(v, w) \in E_0$ or $(v, w) \in E_1$ holds. By Lemma 6.7.19, there exists at least one point $w \in yz$ such that $(v, w) \in E_0$. Now we have a contradiction with Lemma 7.9.10. $\qquad\square$

In view of Proposition 7.9.11, we can distinguish two cases: either the singular dimension of (E, F) equals 2, or each point in E is on precisely 2 maximal singular subspaces. But the latter case has been dealt with in Theorem 6.7.26. From now on, we assume that (E, F) has singular dimension two.

For a line $l \in F$ we define

$$S(l) = l \cup \left\{ z \in E_{\leq 0}(l) \mid \left| E_{\leq -1}(z) \cap l \right| = 1 \right\}.$$

Lemma 7.9.12 *Suppose that (E, F) is a non-degenerate root filtration space of singular dimension two in which each line is on a unique maximal singular subspace. For $l \in F$, $(u, v) \in E_0$, and $x \in E_{-1}(u, v)$, the following properties hold, where M is the maximal singular subspace on l.*

(i) $S(l) = E_{\leq 0}(l) \setminus (M \setminus l)$.
(ii) $S(ux) = S(vx)$.
(iii) $S(vx) = E_{\leq 0}(E_{\leq 0}(u, v))$.

Proof (i) Suppose $x \in E_{\leq 0}(l) \setminus M$. By Lemma 6.7.22, $C_M(x)$ is a proper hyperplane of M. As it contains l and M has dimension two, this forces $C_M(x) = l$. The unique point of $M \cap E_{-1}(x)$ belongs to this hyperplane, and hence to l. Therefore, $x \in S(l)$. The other inclusions are trivial.

(ii) By symmetry it is sufficient to show that $S(vx) \subseteq S(ux)$. It is obvious that $vx \subseteq S(ux)$. Pick $w \in S(vx) \setminus (vx \cup ux)$. Write N for the unique maximal singular subspace on xu. Notice that $w \notin N$, for otherwise $C_N(v)$ would contain xuw, whence all of N, contradicting Lemma 6.7.22.

Assume that w does not commute with u. The triangle condition on x, u, w implies $(u, w) \in E_1$. Set $z = [u, w]$. If $x \sim w$, then $x = z$ and we obtain a contradiction to Lemma 7.9.10 with x, u, w and some point of xv; therefore $(x, w) \in E_0$. By the filtration around x, we have $x \sim z$ and therefore z is on the singular plane P on ux. Furthermore, z is not on the line xu for otherwise each point of $xu \setminus \{z\}$ would be special with w, contradicting $(x, w) \in E_0$. By the filtration around v we have $z \in E_{\leq 0}(v)$. This means that v commutes with the whole plane P, a contradiction to Lemma 6.7.22. We have shown $w \in E_{\leq 0}(u, x) \setminus N$, so, by (i), $w \in S(xu)$.

(iii) To see the inclusion $E_{\leq 0}(E_{\leq 0}(u, v)) \subseteq S(vx)$ note that $E_{\leq 0}(E_{\leq 0}(u, v)) \subseteq E_{\leq 0}(u, v)$, since $u, v \in E_{\leq 0}(u, v)$, and that $E_{\leq 0}(u, v) \subseteq S(vx)$ by Lemmas 6.7.22, 6.7.23, and (i). To see the reverse inclusion, let $w \in S(vx) \setminus (vx \cup ux) = S(ux) \setminus (vx \cup ux)$ and $z \in E_{\leq 0}(u, v)$. We need to show $w \in E_{\leq 0}(z)$. By Lemma 6.7.23, $E_{\leq 0}(u, v) = E_{\leq 0}(ux \cup vx)$.

If z is not collinear with x then, there are noncollinear $u' \in ux$ and $v' \in vx$ such that $z \in E_{-1}(u', v')$, and, since $w \in E_{\leq 0}(u', v')$, Lemma 6.7.23 gives $w \in E_{\leq 0}(z)$.

Arguing as in the previous paragraph, but with w and z interchanged, we find $z \in E_{\leq 0}(w)$ if w is not collinear with x. Therefore, we are left with the case where both $z \sim x$ and $w \sim x$. But then we can use Lemma 7.9.10 to exclude the possibility of $(z, w) \in E_1$. $\qquad\square$

Definition 7.9.13 A subset of a line space is called *convex* if each point on a geodesic (cf. Definition 1.6.1) between two points of the subset belongs to it.

A subspace of a root filtration space (E, F) of the form $E_{\leq 0}(E_{\leq 0}(x, y))$, where $(x, y) \in E_0$, will be called a *symplecton* (plural: symplecta).

The following consequence of Lemma 7.9.12 concerns symplecta.

Corollary 7.9.14 *Suppose that (E, F) is a non-degenerate root filtration space of singular dimension two in which each line is on a unique maximal singular subspace. If $(x, y) \in E_0$, then $E_{\leq 0}(E_{\leq 0}(x, y))$ is the unique symplecton containing x and y. These symplecta are convex. The intersection of two symplecta is either empty or a singleton. The intersection of a symplecton and a singular plane of (E, F) is either empty or a line.*

Proof Let $S = E_{\leq 0}(E_{\leq 0}(x, y))$ where $(x, y) \in E_0$. Then, by Definition 6.7.2(5), S is a subspace. It is a gamma space as (E, F) is. Assume that l is a line of S and $w \in S$. Applying E_0 twice to both sides of the inclusion $\{u, v\} \subseteq E_{\leq 0}(u, v)$, we find

$S \subseteq E_{\leq 0}(S)$. Consequently, $w \in E_{\leq 0}(S) \subseteq E_{\leq 0}(l)$ and hence, by Lemma 6.7.22, either w belongs to the maximal singular subspace M, say, containing l or there is exactly one point of l collinear with w. Therefore S is a polar space. To see that the polar rank is at least 2, we need to show that S contains at least one line. To this end, recall that, by Proposition 7.9.3, $E_{\leq -1}(x, y)$ is a nondegenerate polar space (in the present case with no lines). Therefore there are $u, v \in E_{\leq -1}(x, y)$ such that $(u, v) \in E_0$. Repeated applications of Lemma 7.9.12 give $S = E_{\leq 0}(E_{\leq 0}(x, y)) = S(ux) = E_{\leq 0}(E_{\leq 0}(u, v))$. In particular, $u \in S$ and ux is a line of S. Thus S is indeed a polar space. Non-degeneracy of S follows from non-degeneracy of $E_{\leq -1}(x, y)$ which is a subset of S by the preceding argument.

Let l be an arbitrary line of S on x. There exists a unique point u of l collinear with y. By non-degeneracy of the polar space $E_{\leq -1}(x, y)$ there is an element $v \in E_{\leq -1}(x, y)$ such that $(u, v) \in E_0$. Then, as above, repeated applications of Lemma 7.9.12 give $S = S(l)$. Let m be a line of S intersecting l in a point u' and let $v' \in S$ be polar to u'. Then we can choose $x' \in l$ and $y' \in m$ collinear with v'. Again from repeated applications of Lemma 7.9.12 we infer that $S(m) = S(l) = S$. Finally let m' be a line of S not intersecting l. Then we can find a line m of S connecting a point of l with one of m'. An argument like the one just given shows that $S(m') = S(m)$. We obtain that $S = S(l)$ for an arbitrary line l of S.

Now let (x', y') be an arbitrary polar pair of S. As S is a nondegenerate polar space we can embed x', y' into a quadrangle of S. We can use again Lemma 7.9.12 to show that $E_{\leq 0}(E_{\leq 0}(x', y')) = S(l) = S$ for some line l of S on x' and also $E_{-1}(x', y') \subseteq S$. As a consequence, S is convex.

In view of the above, every line n in F is contained in at most one symplecton, namely in $S(n)$. To see that $S(n)$ is indeed a symplecton, observe that by connectedness of (E, F), there is a line $m \in F$ intersecting the maximal singular subspace N on n in a point x of n. The maximal singular subspace M on m intersects N in x. Pick $v \in n \setminus \{x\}$. By Lemma 6.7.22, $m' = C_M(v)$ is a line containing x and any point u of m' different from x is in $E_0(v)$. Lemma 7.9.12 gives that $S(n) = E_{\leq 0}(E_{\leq 0}(u, v))$, a symplecton. Thus every line $n \in F$ is contained in the unique symplecton $S(n)$. As $N \not\subseteq S(n)$, the rank of the polar space $S(n)$ must be two.

The assertions regarding intersections follow from convexity of symplecta. □

Lemma 7.9.15 *Suppose that (E, F) is a non-degenerate root filtration space of singular dimension two in which each line is on a unique maximal singular subspace. Let $x, y, z \in E$ be three distinct points with $x \in E_{\leq 0}(y, z)$ and $y \in E_{\leq 0}(z)$, not lying in a singular plane. Then x, y, and z are contained in a symplecton.*

Proof By Lemma 7.9.12(iii) it suffices to find a line l with $x, y, z \in S(l)$. Assume first that at least two of the three points are collinear, say $x \sim y$. Then, by Lemma 7.9.12(i), $x, y, z \in S(xy)$. If none of the pairs are collinear, choose $u \in E_{\leq -1}(x, y)$. By Corollary 7.9.14, the symplecton containing x and y is $S(xu)$. As $z \in E_{\leq 0}(x, y)$, Lemma 6.7.23 gives $z \in E_{\leq 0}(xu)$. Since z is not collinear with x, we even have $z \in S(xu)$. □

Lemma 7.9.16 *Suppose that Z is a non-degenerate root filtration space of singular dimension two in which each line is on a unique maximal singular subspace. If three distinct symplecta pairwise intersect each other, then there is a singular plane of Z containing the intersections.*

Proof By Corollary 7.9.14 the intersections are the singletons $\{x\}$, $\{y\}$, $\{z\}$. If these points are distinct, then Lemma 7.9.15 implies that they lie on a singular plane of Z. $\square$

Lemma 7.9.17 *Suppose that (E, F) is a non-degenerate root filtration space of singular dimension two in which each line is on a unique maximal singular subspace. Let S be a symplecton of (E, F) and $z \in E \setminus S$.*

  (i) *If S intersects two distinct symplecta S_1 and S_2 containing z, then S intersects every symplecton containing z.*
 (ii) $E_{\leq 0}(z) \cap S \neq \emptyset$.

Proof (i) Let $\{y\} = S \cap S_1$ and $\{x\} = S \cap S_2$. Then x, y, z lie in a singular plane P. Any further symplecton containing z intersects P in a line l on z. As l is a hyperplane of P, it intersects the line $xy = S \cap P$.

   (ii) Assume that $S \subseteq E_{\geq 1}(z)$. First we show that the set $S \cap E_1(z)$ contains a polar pair. This is trivial if $S \subseteq E_1(z)$. Otherwise S contains a point $x \in E_2(z)$ and there exist two distinct lines l and m of S on x (choose two singular planes of E through x and take the intersections with S). Furthermore, by Condition (6) of Definition 6.7.2 there exist $u \in l$ such that $(z, u) \in E_1$ and $v \in m$ such that $(v, z) \in E_1$. As $u \neq x \neq v$, we have $(u, v) \in E_0$.

   Assume that (u, v) is a polar pair from $S \cap E_1(z)$. By filtration around v, we have $[z, u] \in E_{\leq 1}(v)$. If $(v, [z, u]) \in E_{\leq 0}$ then $v \in S(u[z, u])$, whence $S(u[z, u]) = E_{\leq 0}(E_{\leq 0}(u, v)) = S$ and $[z, u]$ is a point of $S \cap E_{\leq 0}(z)$. Otherwise put $w = [v, [z, u]]$. Then $v \sim w$ (by definition of w), $u \sim w$ (by filtration around u), and $(z, w) \in E_{\leq 0}$ (by filtration around z). Thus w is a point of $E_{\leq -1}(u, v) \subseteq S$ belonging to $E_{\leq 0}(z)$. $\square$

   We conclude that the space is the Grassmannian of lines in a nondegenerate polar space of rank three.

Theorem 7.9.18 *Every nondegenerate root filtration space of singular dimension two such that each line is on a unique maximal singular subspace and each point is on at least three maximal singular subspaces, is the shadow space on 2 of a polar geometry of type B_3.*

Proof Let (E, F) be a nondegenerate root filtration space as in the hypotheses and write $\mathcal{S}$ for the collection of symplecta of this space. By the well-known characterization of shadow spaces of type $B_{n,1}$ as nondegenerate polar spaces of rank n, it suffices to show that the pair $(\mathcal{S}, E)$ is a nondegenerate polar space of (polar) rank three. Here, a member of E is interpreted as a line by identifying it with the collection of symplecta it contains. By Lemma 7.9.17, $(\mathcal{S}, E)$ satisfies the polar space property.

Suppose that S is a symplecton. Choose $x \in S$ and let T be another symplecton on x. Finally, choose $y \in T \cap E_0(x)$ and let U be a symplecton on y distinct from T. If $z \in S \cap U$, then, by Lemma 7.9.16, z, x, y are in a singular subspace of (E, F), contradicting $(x, y) \in E_0$. Therefore, U is a point of $(\mathcal{S}, E)$ non-collinear with S. It follows that $(\mathcal{S}, E)$ is a nondegenerate polar space.

Let M be a maximal singular subspace of (E, F), so its dimension is two. The image of the homomorphism $l \mapsto S(l)$ from the dual of M to $\mathcal{S}$ is a maximal singular subspace of $(\mathcal{S}, E)$, so the latter is a nondegenerate polar space of rank three. $\qquad \square$

We summarize the conclusions of Theorems 7.9.5, 6.7.26, 7.9.18, and Corollary 7.9.8.

Theorem 7.9.19 *Suppose that Z is a nondegenerate root filtration space having a line that is contained in a unique maximal singular subspace. Then either Z is the shadow space on 2 of a polar geometry of rank three or a generalized hexagon, or there is a projective space $\mathbb{P}$ and a collection of hyperplanes $\mathbb{H}$ of $\mathbb{P}$ forming a subspace of the dual of $\mathbb{P}$ and annihilating $\mathbb{P}$ such that Z is isomorphic to $E(\mathbb{P}, \mathbb{H})$.*

In particular, if the singular dimension of Z is finite, then Z is the shadow space on $\{1, n\}$ of a projective geometry of type A_n, a shadow space on 2 of a polar geometry of type B_3, or a generalized hexagon, that is, a shadow on 1 of a geometry of type G_2.

7.10 Polar Spaces with Thin Lines

In Theorem 6.3.8 we saw that the theory of projective spaces reduces essentially to the study of projective spaces all of whose lines are thick (i.e., have at least three points). There is a similar though slightly different theory for polar spaces. The difference between the two cases is easily understood. In a projective space, the fact that all lines are thick forces the corresponding projective geometry to be thick. This is not the case for polar spaces. If Γ is a geometry of type B_n all of whose lines are thick, then the i-orders s_i ($i \in [n-1]$) of Γ are all equal (and at least two). However, the n-order s_n may still be equal to one; it happens for quadrics of maximal index in projective spaces. As we saw in Theorem 7.8.6, this phenomenon gives rise to geometries of type D_n. Therefore, rather than considering polar spaces whose polar geometries are not thick in general, we study polar spaces having a thin line (cf. Definition 2.5.8).

Example 7.10.1 We provide two constructions.

(i) Let $(Z_i)_{i \in I}$ be a family of non-empty nondegenerate polar spaces $Z_i = (P_i, L_i)$ indexed by the set I. We assume that the sets P_i are pairwise disjoint. Let

$Z = (P, L)$ be the direct sum of the Z_i as introduced in Definition 6.3.6. Clearly, Z is a nondegenerate polar space and it is of finite rank $n = \sum_{i \in I} n_i$ if and only if I is finite and each rank n_i of Z_i is finite.

(ii) Let Z be a thick projective space of dimension $d \geq 1$ (where d need not be finite) and let $Z^\star$ be its hyperplane dual as defined in Exercise 5.7.12. The dualized projective space $D(Z, Z^\star)$ (defined in Exercise 6.8.19) is a nondegenerate polar space of rank $d + 1$.

Definition 7.10.2 A polar space is called *irreducible* if it is not isomorphic to the direct sum of at least two non-empty polar spaces.

Our purpose in this section is to characterize irreducible polar spaces by their collinearity graphs and to classify the irreducible polar spaces having some thin line.

Proposition 7.10.3 *Let Z be a nondegenerate polar space of finite rank at least two. The space Z is the direct sum of at least two irreducible nondegenerate polar spaces Z_i $(i \in I)$ of positive rank if and only if the complement of the collinearity graph of Z is not connected. In that case, the subsets Z_i of Z are the connected components of that graph.*

Proof Let Z be a nondegenerate polar space with point set P. Write $a \approx b$ for two points $a, b \in P$ if and only if $a \notin b^\perp$. We consider the collection $\{P_i \mid i \in I\}$ of all connected components of the graph $(P, \approx)$.

Let $a \in P_i, b \in P_j$ with $i \neq j$. Then $a \perp b$ and so there is a line ab on a and b. Assume there exists $c \in ab \setminus \{a, b\}$. Since $\mathrm{Rad}(Z)$ is empty, there is a point $d \in b^\perp \setminus a^\perp$ and so $d^\perp \cap ab = \{b\}$. This gives $c \approx d \approx a$, so $c \in P_i$. For the same reason, $c \in P_j$, which is a contradiction. Therefore ab is a thin line. In particular, a line having two points in P_i must be contained in P_i, so P_i is a subspace of Z. We find that Z is the direct sum of the polar spaces P_i $(i \in I)$, which are readily seen to be irreducible, nondegenerate, and of positive rank.

Finally, if Z is the direct sum of at least two irreducible nondegenerate polar spaces Z_i $(i \in I)$ of positive rank, then the graph $(P, \approx)$ is not connected and the connected component of a point of Z_i is contained in Z_i. It follows from the above argument that this component coincides with Z_i.   □

Theorem 7.10.4 *If S is an irreducible nondegenerate polar space of finite rank $n \geq 2$ having a thin line, then S is the dualized projective space of some thick projective space of dimension $n - 1$.*

Proof Put $S = (P, L)$. We proceed in four steps. Maximality is taken with respect to the ordering by inclusion.

STEP 1. *Every maximal thick singular subspace of S is a maximal singular subspace.*

Assume the contrary: let M be a maximal thick singular subspace of S that is not maximal singular. Then $N = M^\perp \setminus M$ is not empty and all lines joining a point of M to a point of N are thin lines. Also, M is not empty since a point would be a thick singular subspace anyway. If a is a point not in $N \cup M$, then there is some $x \in M$ with $a \notin x^\perp$ and therefore $a \in y^\perp$ for all $y \in N$ since $\{x, y\}$ is a thin line. Hence $P = N \cup N^\perp$. Now $N \cap N^\perp$ is empty since it is contained in $\mathrm{Rad}(S)$. We observe that S is the direct sum of the polar spaces N and $N^\perp$ which are subspaces of S. Since S is irreducible, this forces N to be empty and M to be a maximal singular subspace.

For the remainder of the proof, fix a maximal thick singular subspace M of S. Moreover, fix a maximal singular subspace N disjoint from M; it exists by Proposition 7.5.3.

STEP 2. *Every thin line with some point in $M \cup N$ is contained in $M \cup N$.*

Every point $n \in N$ determines a geometric hyperplane $n^\perp \cap M$ in M and this forces an isomorphism of N with the hyperplane dual $M^\star$ (cf. Exercise 5.7.12), implying that N is also thick. So the roles of M and N are symmetric.

Let $a \in M$ and let ab be a thin line of S. Then b is not in M and M is a maximal singular subspace by Step 1. Next, we see that $a^\perp \cap N$ is a geometric hyperplane of N and, since every point of N is collinear with either a or b, the subspace $b^\perp \cap N$ must contain at least the complement in N of some geometric hyperplane. In view of the thickness of N this shows that $b^\perp$ contains N and so that $b \in N$. The argument for $a \in N$ instead of $a \in M$ is now immediate from the symmetric roles of M and N.

STEP 3. *Every line mn with $m \in M$ and $n \in N$, is a thin line.*

Assume first that $m = a, n \neq b$ where a, b are as in Step 2. Then a, b, n generate a projective plane containing the thin line ab and the line bn which has at least 3 points. Therefore, na is a thin line. Now we assume $m \neq a, n \neq b$. If $m \notin b^\perp$ and $n \notin a^\perp$, then each point of mn is collinear with a unique point of ab and conversely. Therefore mn is a thin line. Assume then that $n \in a^\perp$. By our first argument na is a thin line and so, replacing $\{a, b\}$ by $\{n, a\}$ we see that mn is a thin line.

STEP 4. $P = M \cup N$.

Otherwise, we may assume that there is some line bp with p not in $M \cup N$ but $b \in N$. Then bp has at least one more point say q, in view of Step 2. Now $p^\perp \cap N = q^\perp \cap N$ is a geometric hyperplane in N. Let $b' \in N \setminus p^\perp$ and consider $b'^\perp \cap M$ which is a geometric hyperplane in M. For every $x \in b'^\perp \cap M$, the line $b'x$ has only two points by Step 3. Hence $x \in p^\perp \cap q^\perp$ and so $x \in y^\perp$ for every point y on bp. If $a' \in M \setminus b'^\perp$, then a' is collinear with some point of bp say y, hence $y^\perp$ contains the geometric hyperplane $b'^\perp \cap M$ of M and an additional point a', forcing M to be in $y^\perp$, a contradiction. Therefore $P = M \cup N$.

Now it is clear that S is a dualized thick projective space. $\square$

7.11 Exercises

Section 7.1

Exercise 7.11.1 Consider the simplex on the vertex set $[n]$ as a geometry of rank $n-1$ and let $\sigma \in \mathrm{Sym}_n$ be a permutation of $[n]$. Prove that $\pi : [n] \to [n]$ determined by $\pi(x) = [n] \setminus \{\sigma(x)\}$ is a quasi-polarity if and only if $\sigma^2 = \mathrm{id}$.

Exercise 7.11.2 Let Z be the generalized quadrangle of order $(2, 2)$. See Example 2.2.10 for a construction and Theorem 2.2.11 for its uniqueness.

(a) Prove that $\mathrm{Aut}(Z)$ is isomorphic to Sym_6.
(b) In Exercise 2.8.7, the dualities (and polarities) of Z were made explicit as those elements (involutions) of $G/Z(G)$ that do not belong to $H/Z(H)$. Derive from Theorem 2.2.11 that there exists a duality α of Z, corresponding to an outer automorphism of Sym_6 sending $(1, 2)$ to $(1, 2)(3, 4)(5, 6)$.

Exercise 7.11.3 Prove that a nondegenerate quasi-polarity of a projective line l can be viewed as a permutation of l of order at most two and determine the degenerate quasi-polarities of l.

Section 7.2

Exercise 7.11.4 Prove Theorem 7.2.12 in terms of projective geometries without recourse to Lemma 7.1.8. That is, assume that f is a nondegenerate σ-sesquilinear form on a vector space V of finite dimension. Show that the map $X \mapsto X^{\perp f}$ is an isomorphism of geometries $\mathrm{PG}(V) \to \mathrm{PG}(V^\vee)$.
(*Hint:* Use the properties of sesquilinear forms given in Definition 7.2.3.)

Exercise 7.11.5 Suppose that f is a symmetric or antisymmetric bilinear form on a vector space V.

(a) Suppose that U is a linear subspace of $\mathrm{Rad}(f)$. Show that f induces a unique symmetric or antisymmetric bilinear form, $\overline{f}$ say, on the quotient vector space V/U.
(b) Prove that $\mathrm{Rad}(\overline{f}) = \mathrm{Rad}(f)/U$.
(c) Conclude that f induces a nondegenerate bilinear form on $V/\mathrm{Rad}(f)$.

Exercise 7.11.6 Prove that, for every quasi-polarity π of a projective space $\mathbb{P}$, we have

$$\mathrm{Ker}(\pi) = \bigcap_{x \in \mathbb{P}} \pi(x).$$

Exercise 7.11.7 Let $k \leq n/2$. For a set Y, we denote by $\binom{Y}{k}$ the collection of subsets of Y of size k (as in Example 1.2.6). Prove the existence of a bijective map $f : \binom{[n]}{k} \to \binom{[n]}{n-k}$ such that $X \subseteq f(X)$ for each $X \in \binom{[n]}{k}$.

Section 7.3

Exercise 7.11.8 The next two questions regard Proposition 7.3.8.

(a) Show that the condition $\dim(V) \geq 2$ is necessary.
(b) Generalize the proposition to the case where f need not be nondegenerate but the codimension of $\mathrm{Rad}(f)$ is at least two.

Exercise 7.11.9 Let V be the real vector space with basis $(\varepsilon_i)_{i \in \mathbb{N}}$ and let f be the symmetric bilinear form on V given by

$$f(x, y) = x_1 y_2 + x_2 y_1 + \cdots + x_{2n-1} y_{2n} + x_{2n} y_{2n-1} + \sum_{i>n} x_i y_i \quad (x, y \in V)$$

where $x = \sum_{i \geq 1} x_i \varepsilon_i$ and $y = \sum_{i \geq 1} y_i \varepsilon_i$. Verify that $\mathbb{P}(V)_f$ is a nondegenerate polar space of rank n that does not embed in a proper subspace of $\mathbb{P}(V)$. (Note: in Exercise 9.6.13 it will become clear that this polar space does not have an embedding in a projective space of finite dimension.)

Section 7.4

Exercise 7.11.10 Let Z be a line space. Prove that Z is a polar space if and only if, for each point p of Z, the set $p^\perp$ of points collinear with p is either the whole point set of Z or a geometric hyperplane of Z.

Exercise 7.11.11 We continue the study of quotient spaces in Exercise 2.8.24. Let $Z = (P, L)$ be a gamma space. Suppose that A is a group of automorphisms of Z. Prove that, if $d(x, \sigma(x)) \geq 4$ for each $x \in P$ and $\sigma \in A \setminus \{1\}$, then Z/A is a gamma space.

Exercise 7.11.12 Let $Z = (P, L)$ be a gamma space. Consider the equivalence relation $\equiv$ on P given by $x \equiv y$ if and only if $x^\perp = y^\perp$ in the collinearity graph of Z; cf. Definition 2.2.1. Prove that it is a standard equivalence as defined in Exercise 2.8.24. Compare $\rho(Z/\equiv)$ and $\rho(Z)/\equiv$, where ρ is as in Proposition 7.4.10.

Exercise 7.11.13 (This exercise is used in Theorem 8.2.9 and Proposition 7.9.7) Suppose that Z is a polar space. Prove the following two assertions.

(a) If Z is nondegenerate and contains a singular subspace X of finite dimension, then $X^{\perp\perp} = X$.

(*Hint:* Use induction on the dimension of X, and Lemma 7.4.8(iv). Corollary 7.4.12 helps to start the induction.)

(b) If Z is of finite rank, then, for every singular subspace X, we have $X^{\perp\perp} = \langle \mathrm{Rad}(Z), X \rangle$.

Exercise 7.11.14 Use Exercise 5.7.19 to give an alternative proof of Theorem 7.4.13(iv).

(*Hint:* Fix a maximal singular subspace M of the nondegenerate polar space Z and use the collection of geometric hyperplanes $x^\perp \cap M$ of it for x ranging over the points of Z off M.)

Exercise 7.11.15 Let V be the real vector space of all sequences with entries in $\mathbb{R}$ that are zero in all but a finite number of places. For $a = (u_i)_{i \in \mathbb{N}}$ in V, put $\pi(\langle a \rangle) = \{\langle (v_i)_{i \in \mathbb{N}} \rangle \in \mathbb{P}(V) \mid \sum_{i \in \mathbb{N}} (-1)^i u_i v_i = 0\}$. Prove that π is a quasi-polarity whose absolute $\mathbb{P}(V)_\pi$ is a nondegenerate polar space with singular subspaces of infinite dimension.

(*Hint:* Consider the set of points $\langle (u_i)_{i \in \mathbb{N}} \rangle$ with $u_{2i-1} = u_{2i}$ for $i = 1, 2, \ldots$.)

Exercise 7.11.16 (This exercise is used in Proposition 10.3.4) Suppose that X is a set of points of a polar space Z with $\mathrm{Rad}(X) = \emptyset$ whose induced lines (that is, lines of the line space induced on X as in Definition 2.5.8) have size at least three. Prove that the non-collinearity graph on X (with point set X and in which two points are adjacent if they are distinct and not collinear in Z) is connected.

(*Hint:* Suppose that $x, y \in X$ are collinear in Z. As the radical of X is empty, there are points $v, w \in X$ with $x \not\perp w$ and $y \not\perp v$. So we are done unless $x \perp v \perp w \perp y$. But then there is a point $u \in X \cap xy \setminus \{x, y\}$ satisfying $u \not\perp v$ and $u \not\perp w$.)

Exercise 7.11.17 Let $\mathbb{P}$ be a projective space and let A, B be disjoint non-empty subspaces of $\mathbb{P}$. Assume that Z is a non-empty and nondegenerate polar space whose points and lines are also points and lines of A. (In terms of Definition 9.1.1, the space Z is embedded in A.) Consider the space (X, Y) whose point set X is the union of all lines joining a point of Z to a point of B and whose line set Y consists of the lines of B and those lines of $\mathbb{P}$ that are contained in a subspace of $\mathbb{P}$ generated by B and a point or a line of Z. Prove that (X, Y) is a polar space whose radical is B.

Exercise 7.11.18 Consider the generalized quadrangle arising from a symplectic polarity in $\mathbb{P}(\mathbb{F}_2^4)$ (that is, δ_f for an alternating form f; cf. Definition 7.3.10). Prove that it is isomorphic to the generalized quadrangle described in Example 2.2.10 and that it is isomorphic to a quadric in $\mathbb{P}(\mathbb{F}_2^5)$.

Exercise 7.11.19 Let $\mathbb{F}$ be a field of characteristic two in which each element is a square (e.g., $\mathbb{F}_2$; this property is also known as the field being *perfect*). Fix $m \in \mathbb{N}$, $m \geq 1$. Suppose that f is a nondegenerate alternating form on $\mathbb{F}^{2m}$. Prove the following assertions, where $\varepsilon_1, \ldots, \varepsilon_{2m+1}$ is the standard basis of $\mathbb{F}^{2m+1}$.

(a) There is a quadratic form κ on $\mathbb{F}^{2m}$ such that $f(x, y) = \kappa(x + y) + \kappa(x) + \kappa(y)$ $(x, y \in \mathbb{F}^{2m})$.
(b) There is a quadratic form κ_0 on $\mathbb{F}^{2m+1}$ such that $\kappa_0(\varepsilon_{2m+1}) = 1$ and $\kappa_0(x) = \kappa(x)$ and $\kappa(x) = 1 + \kappa_0(x + \varepsilon_{2m+1})$ for $x \in \mathbb{F}^{2m}$.
(c) The polar space $\mathbb{P}(\mathbb{F}^{2m})_{\delta_f}$ is isomorphic to the polar space $\mathbb{P}(\mathbb{F}^{2m+1})_{\kappa_0}$.

Section 7.5

Exercise 7.11.20 (This exercise is used in Proposition 10.5.1) Let Z be a nondegenerate polar space with a point p on a line l. Show that Z has a point q such that $q^{\perp} \cap l = \{p\}$.

Exercise 7.11.21 Let Γ be the polar geometry of a nondegenerate polar space of finite rank $n \geq 2$. Show that each line of $\mathrm{ShSp}(\Gamma, n-1)$, the shadow space on $\{n-1\}$ of Γ, is contained in a unique maximal singular subspace.

Exercise 7.11.22 Suppose that Z is a polar space having a maximal singular subspace of finite dimension. Prove that the rank of Z is finite.
(*Hint:* Establish first that, if M and N are disjoint maximal singular subspaces of Z, then the map $M \to N$ sending p to $p^{\perp} \cap N$ is an injective homomorphism $M \to N^{\star}$.)

Section 7.6

Exercise 7.11.23 Let $n \in \mathbb{N}$, $n \geq 2$, and let Γ be the polar geometry of a nondegenerate polar space Z of finite rank n. Show that the collinearity graph of the shadow space $\mathrm{ShSp}(\Gamma, 2)$ can be identified with the space of lines of Z in which two distinct lines are adjacent whenever they generate a singular plane. Prove that the diameter of this collinearity graph is three.

Section 7.7

Exercise 7.11.24 Let l be a line of a Moufang projective plane Z. Show that the point-wise stabilizer of l in $\mathrm{Aut}(Z)$ is transitive on the set of points of Z off l.

Exercise 7.11.25 (This exercise is used in Lemma 9.3.4) Let Z be a nondegenerate polar space of rank at least two. Suppose that H is a geometric hyperplane of Z. Prove that either $H = x^{\perp}$ for some point x of Z or H is nondegenerate.

Exercise 7.11.26 Let C be the Cayley division ring and Z the projective plane corresponding to the affine plane defined by means of C in Example 2.3.4. Prove that Z is Moufang but not Desarguesian.

Exercise 7.11.27 Prove that removal of the thickness condition of Proposition 7.7.6 results in a false statement. Same for Corollary 7.7.7.

Section 7.8

Exercise 7.11.28 Prove the statement of Example 7.8.2 that (P, L) is a nondegenerate polar space of rank three.

♯ **Exercise 7.11.29** Let $\mathbb{F}$ be a field of characteristic two. We will exhibit a duality on the generalized quadrangle that is the absolute of the polarity associated with the bilinear form f on $V := \mathbb{F}^4$ defined by $f(x, y) = x_1 y_2 + x_2 y_1 + x_3 y_4 + x_4 y_3$ $(x, y \in V)$. (The polarity is symplectic in the terminology of Definition 10.1.10.)

(a) In the Klein Correspondence 7.8.3, an isomorphism $k : (P, L) \to (Q, M)$ was established between the Grassmannian (P, L) of lines of $\mathbb{P}(V)$ and the quadric (Q, M) determined by the quadratic form κ on $\mathbb{F}^6$ given by $\kappa(x) = x_1 x_2 - x_3 x_4 + x_5 x_6$. In particular, L consists of the pairs (p, π) of a point p and plane π of $\mathbb{P}(V)$ with $p \in \pi$. Establish that the map k is determined on L by the rule: if $x, a, b \in V$ span a 3-dimensional subspace of V, then $k(\langle x \rangle, \langle x, a, b \rangle) = \langle g(x, a), g(x, b) \rangle$, where g is as in (7.8).
(b) Let L_f be the subset of L consisting of the pairs (p, π) such that $\pi = p^\perp$, where $\perp = \perp_f$, and let P_f be the subset of P of all lines $\langle x, y \rangle$ of $\mathbb{P}(V)$ with $f(x, y) = 0$. Prove that (P_f, L_f) is a subspace of (P, L) isomorphic to the dual of the generalized quadrangle $\mathbb{P}(V)_f$.
(c) Verify that the image of P_f under k coincides with $Q \cap (\varepsilon_1 + \varepsilon_2)^\perp$.
(d) Let W be the 4-dimensional vector space $(\varepsilon_1 + \varepsilon_2)^\perp / \langle \varepsilon_1 + \varepsilon_2 \rangle$. We will identify it with the subspace of $\mathbb{F}^6$ spanned by $\varepsilon_3, \ldots, \varepsilon_6$. Let f_W be the restriction to W of the bilinear form associated with κ. Show that the linear map $\phi : W \to V$ determined by $\phi(\varepsilon_{i+2}) = \varepsilon_i$ $(i \in [4])$ is an isomorphism of vector spaces with $f_W(x, y) = f(\phi(x), \phi(y))$.
(e) For $x, y \in V$ with $f(x, y) = 0$, write $\overline{g}(x, y)$ for the image in W of $g(x, y)$ under the natural projection $(\varepsilon_1 + \varepsilon_2)^\perp \to W$. In view of (b), we may interpret L_f as a set of points of $\mathbb{P}(V)$ by identifying $\langle x \rangle$ with $(\langle x \rangle, x^\perp)$. Let $\delta : (P_f, L_f) \to (L_f, P_f)$ be the map given by

$$\delta(l) = \langle \phi(\overline{g}(x, y)) \rangle \quad \text{whenever } l = \langle x, y \rangle,$$

$$\delta(\langle x \rangle) = \langle \phi(\overline{g}(x, a)), \phi(\overline{g}(x, b)) \rangle \quad \text{whenever } x^\perp = \langle x, a, b \rangle.$$

Derive that δ is a duality of (P_f, L_f).
(f) Show that $\delta^2(\langle x \rangle) = \langle \varepsilon_1 x_1^2 + \cdots + \varepsilon_4 x_4^2 \rangle$ for each $x \in V \setminus \{0\}$.

♯ **Exercise 7.11.30** (Cited in Remark 7.1.2) This exercise builds upon Exercise 7.11.29. Fix $n \in \mathbb{N}$ with $n > 0$, and let $\mathbb{F} := \mathbb{F}_{2^n}$ be the field of order 2^n. On $V := \mathbb{F}^4$, we define the bilinear form f by $f(x, y) = x_1 y_2 + x_2 y_1 + x_3 y_4 + x_4 y_3$ $(x, y \in V)$. We will study the absolute Z of the polarity δ_f determined by f. Thus, $Z = (L_f, P_f)$ in the notation of Exercise 7.11.29. It is a generalized quadrangle of order $(2^n, 2^n)$. In particular, we will look for dualities that are not polarities of the rank two geometry of Z.

(a) Prove that the group $\mathrm{Sp}(V, f)$ of elements of $\mathrm{GL}(V)$ preserving f (so an element g of $\mathrm{GL}(V)$ belongs to $\mathrm{Sp}(V, f)$ if and only if $f(gx, gy) = f(x, y)$ for all $x, y \in V$) induces a group of automorphisms of Z. (This group, the symplectic group, will make its full introduction in Definition 10.1.10.)

(b) Verify that, for each nonzero $a \in V$, the map $r_{a,\phi}$ of Exercise 1.9.30, where $\phi(x) = f(a, x)$ $(x \in V)$, belongs to $\mathrm{Sp}(V, f)$; it is a transvection with center $\langle a \rangle$. Verify also that the set $T_a := \{r_{a\lambda,\phi} \mid \lambda \in \mathbb{F} \setminus \{0\}\}$ of transvections with fixed center $\langle a \rangle$, together with the identity, is a subgroup of $\mathrm{Sp}(V, f)$ isomorphic to the additive group of $\mathbb{F}$.

(c) Consider the map $\langle a \rangle \mapsto T_a$ from Z to the collection of transvection subgroups of $\mathrm{Sp}(V, f)$. Show that it is an isomorphism from the collinearity graph of Z to the graph on the transvection subgroups in which two vertices are adjacent whenever they commute. Conclude that Z can be fully described in terms of subgroups T_a of $\mathrm{Sp}(V, f)$.

(d) Establish that, by adding the regular lines (that is, the subsets of Z of the form $\{x, y\}^{\perp\perp}$ for noncollinear points x, y), we obtain the projective space $\mathbb{P}(V)$ from Z. Derive from this fact that the automorphism group of Z embeds in $\mathrm{P\Gamma L}(V)$, where it normalizes $\mathrm{Sp}(V, f)$. Conclude that each duality of Z is of the form δt where δ is the duality found in Exercise 7.11.29 and $t \in \mathrm{P\Gamma L}(V)$ normalizes $\mathrm{Sp}(V, f)$.

(e) Suppose that $n = 2m + 1$ is odd. Let $t \in \mathrm{P\Gamma L}(V)$ be induced from the semilinear transformation on V mapping each coordinate of $x \in V$ to its 2^m-th power. Verify that δt is a polarity of Z.

(f) Show that, for even n, the rank two geometry of Z admits a duality but no polarity.

Exercise 7.11.31 (Cited in Remark 7.1.2) Let $\mathbb{F}$ be a field of odd characteristic and let f be a nondegenerate alternating form on $\mathbb{F}^4$. Use the methods of Exercise 7.11.29 to show that there is a nondegenerate quadratic form κ of Witt index two on $\mathbb{F}^5$ such that the generalized quadrangle $\mathbb{P}(\mathbb{F}^4)_f$ is isomorphic to the dual of $\mathbb{P}(\mathbb{F}^5)_\kappa$. Show also, at least in case $\mathbb{F}$ is finite, that these generalized quadrangles are not self-dual.

Exercise 7.11.32 (This exercise is used in Corollary 7.8.10) Let Z be a nondegenerate polar space of finite rank at least three all of whose lines are thick. Prove that either every line of the dual polar space of Z is thin, or every line is thick.

Exercise 7.11.33 A line space is called a *near-polygon* if it is partial linear and if, for each point p and line l, there is a unique point on l nearest p.

(a) Prove that every dual polar space is a near-polygon.
(b) Show that the Shult-Yanushka space of Definition 5.6.8 is a near-hexagon but not a dual polar space.

Exercise 7.11.34 (Cited in Sect. 11.9) Let $\Gamma = (X_1, X_2, X_3, X_4, *)$ be a [4]-geometry of type D_4. Construct an incidence system $\widehat{\Gamma} = (Y_1, Y_2, Y_3, Y_4, \widehat{*})$ over [4] as follows.

(1) $Y_1 = X_2$; Y_2 is the set of flags of Γ of type $\{1, 3, 4\}$; Y_3 is the set of flags of Γ of type $\{1, 3\}$, $\{1, 4\}$, or $\{3, 4\}$; $Y_4 = X_1 \cup X_3 \cup X_4$;
(2) $x \widehat{*} y$ if and only if $x * y$ for elements x and y of $\widehat{\Gamma}$ of different type.

Prove that $\widehat{\Gamma}$ is a [4]-geometry of type F_4 whose 3-order and 4-order are 1.

Section 7.9

Exercise 7.11.35 Let Z be a polar space with point set E. As usual, we take E_{-2} to be the identity relation. Furthermore, we set $E_1 = E_{-1} = \emptyset$ and let $(x, y) \in E_0$ if and only if x and y are distinct and collinear points of Z. Finally, we take E_2 to be $(E \times E) \setminus E_{\leq 1}$. Show that $(E, \emptyset)$ is a degenerate root filtration space with respect to $(E_i)_{-2 \leq i \leq 2}$.

Exercise 7.11.36 Let V be a vector space over a field $\mathbb{F}$ supplied with a nondegenerate bilinear alternating form f. Let E be the set of projective points of $\mathrm{End}(V)$ (the space of linear maps $V \to V$) of the form $\mathbb{F}(r_{a,\phi} - \mathrm{id})$ where $a \in V \setminus \{0\}$ and $\phi(x) = f(a, x)$ for $x \in V$. (See Exercise 1.9.30 for $r_{a,\phi}$.)

(a) Prove that $(E, \emptyset)$ is a degenerate root filtration space with $(\mathbb{F}x, \mathbb{F}y) \in E_0$ if and only if x and y commute in $\mathrm{End}(V)$.
(b) Verify that the root filtration space of (a) is a subspace of the root filtration space of Exercise 6.8.20.
(c) Show that this root filtration space is isomorphic to a root filtration space as constructed in Exercise 7.11.35.

Section 7.10

Exercise 7.11.37 Let $\mathbb{P}$ be a thick projective space of infinite dimension. Consider its dualized projective space $Z := D(\mathbb{P}, \mathbb{P}^\star)$, defined in Exercise 6.8.19. Let M be the point set of $\mathbb{P}^\star$. Prove that M is a maximal singular subspace of the polar space Z. Take a hyperplane H of M lying outside the image of $\mathbb{P}$ under its canonical embedding in $(\mathbb{P}^\star)^\vee$. Show that H is in no maximal singular subspace of Z

other than M. Conclude that the condition that the polar space have finite rank is necessary in Theorem 7.5.8.

7.12 Notes

Duality and polarities in the real and complex projective plane were studied around 1825 by Gergonne [131] and Poncelet [241]. Since then there have been many developments. Several ideas occurring in the first four sections are inspired by work of Tits. Our approach will follow Veldkamp [297]. As it uses an embedding of the polar space in a projective space (the topic of Chap. 8), it is only valid for polar spaces whose singular planes are Desarguesian. A full classification of nondegenerate polar spaces of rank at least three is given by Tits in [286]. The fact that one can start with axioms as simple as those of Definition 7.4.1 stems from [55].

General textbook introductions to polar spaces appear in [14, 260, 292].

Section 7.2

The fundamentals on sesquilinear forms can also be found in, e.g., [143], where reflexive forms are called orthosymmetric.

The material for Example 7.2.2 is from [23] and [194]. The counterexample in Example 7.2.16 is due to Tignol. When asked by us, he actually came up with a general construction, which led to the paper [219]. The division algebra introduced in the example is a cyclic algebra of degree three, and can be found in textbooks like [197]. Eisenstein's criterion states that the polynomial $a_n X^n + a_{n-1} X^{n-1} + \cdots + a_1 X + a_0 \in \mathbb{Z}[X]$ is irreducible over $\mathbb{Q}$ if there exists a prime $p \in \mathbb{N}$ such that

(1) $a_i \equiv 0 \pmod{p}$ if and only if $i \neq n$,
(2) $a_0 \not\equiv 0 \pmod{p^2}$.

It is treated in many textbooks on algebra, such as [271].

Section 7.3

The notion of quasi-polarity (cf. Definition 7.1.9) differs a little from the definition given by Lenz [204]. In fact, what he called quasi-polarity is an injective nondegenerate quasi-polarity in our terms.

The Cayley-Hamilton theorem, used in Example 7.2.2, is standard in linear algebra and can be found in most standard introductions to algebra, such as [87, Theorem 3].

Section 7.4

The basic axiomatic theory of polar spaces appearing here was first developed by Veldkamp [297], who went far towards a classification. Some of his ideas will also appear in the next chapter. His theory was simplified and completed by use of new ideas by Tits [286]. Much of the present theory, including Definition 7.4.1 and Theorem 7.4.13, stems from Buekenhout and Shult [55]. We have also made use of more general treatments following this paper, by Buekenhout [49], Percsy [237], and Johnson [177]. Johnson and Shult have given even weaker axioms for polar spaces in [180].

For extensions of polar spaces, see [43, 51, 276]. For a connection with Zara graphs, see [26].

Section 7.5

Dual polar spaces have been studied from an axiomatic point of view by Cameron [57] and Shult and Yanushka [262]. See also Brouwer and Cohen [34] for a weakening of the axioms.

Section 7.6

Conditions (1) and (2) of Definition 7.6.3 can be found in [287, Proposition 9]

Section 7.7

The fact that singular planes of polar spaces are Moufang, Theorem 7.7.10, is a special instance of the results on buildings in [286]. In particular, all thick buildings of spherical irreducible type and rank at least three are Moufang (cf. Exercise 11.8.13) and if a building is Moufang, then so are its residues. Applying this result to polar geometries of rank three (which are buildings by Proposition 11.1.9), we find that its singular planes (which are residues by Proposition 11.5.14) are Moufang as well.

Examples of nondegenerate rank three polar spaces whose singular planes are not Desarguesian come from algebraic groups and can be constructed as fixed point geometries of geometries of type E_7; see [224]. Each Moufang plane occurs (up to isomorphism) as a singular plane of a nondegenerate polar space of rank three (cf. [286]). This will be discussed in [75].

The more general fact behind Remark 7.7.9 is that a projective plane is known to be Moufang if and only if the corresponding ternary ring T is alternative (see Exercise 2.8.11); see, for instance, [167].

Section 7.8

The oriflamme geometry of Theorem 7.8.6 was given its name by Tits in [286]. The corresponding Coxeter diagram D_n has a symmetry of order two. In Corollary 10.5.7 we will see that it is realized by a duality of the underlying geometry. If $n = 4$, the Coxeter diagram D_4 has an automorphism group of order 6. The corresponding correlations of order three were first studied in [281]. The Klein Correspondence 7.8.3 is related to the isomorphism of Coxeter diagrams D_3 and A_3; it is expressed here in terms of the shadow spaces of $PG(\mathbb{F}^4)$ on 2 and of $PG(\mathbb{F}^6)_\kappa$ on 1. A more elaborate treatment can be found in [272].

Section 7.9

The bulk of this section is from [80]. The general classification is described in this paper and [79]. It depends on the work of Cooperstein, Kasikova, and Shult. See [76] for a survey.

   The notion convex, introduced in Definition 7.9.13, is also known as *geodesically closed*.

Section 7.10

Polar spaces with thin lines were treated by Buekenhout and Sprague [56].

Section 7.11

Exercises 7.11.29 and 7.11.30 barely scratch the surface of a deep subject related to ovoids, Suzuki groups, Moufang sets, and twisted Chevalley groups; the interested reader is recommended to read further in [62, 272, 281].

   Near-polygons, introduced in Exercise 7.11.33, are studied in many papers; see [110] for an extensive treatment. Dual polar spaces are special near-polygons, but the converse does not hold. The finite examples among the former are classified, but the regular finite near-polygons are not, although these give rise to distance-regular graphs. A significant step in this direction is [13]. Often, dual polar spaces give rise to distance-transitive graphs; see [36]. In [58], Cameron proved that there are a finite number of distance-transitive graphs of given valency at least three.

Chapter 8
Projective Embeddings of Polar Spaces

This is the second chapter devoted to polar spaces. The motivation for studying these spaces is their occurrence as partial subspaces—in fact, as absolutes of polarities—in projective spaces, whose lines are projective lines. This chapter shows that the following converse holds: a polar space with a thick polar geometry is isomorphic to a partial subspace of a projective space if it is nondegenerate of rank at least three and all of its singular subspaces are Desarguesian; this is the content of Corollary 8.4.26. The conditions mentioned in this statement are, in a sense, best possible. For, if a space is a partial subspace of a projective space, its linear subspaces must be projective as well. Thus, if Z is a degenerate polar space whose radical is not a generalized projective space, it will not be embeddable. There are numerous examples of thick generalized quadrangles that cannot be embedded in a projective space; for instance there are finite examples whose point orders are not prime powers; see Example 8.3.17. Finally, there are examples of nondegenerate polar spaces of rank three whose planes are Moufang but not Desarguesian. These are most easily defined in terms of algebraic groups and will not be discussed here.

The construction of embeddings of polar spaces of rank at least three comes in three main steps. The first step, taken in Sect. 8.1, is the study of geometric hyperplanes of polar spaces. For every point x of a polar space, the set $x^\perp$ of all points collinear with x is a geometric hyperplane. The notion of a geometric hyperplane was introduced in Definition 5.2.7 (see also Exercise 5.7.12). The second step is the embedding of the polar space in a space whose points are geometric hyperplanes by means of the map $x \mapsto x^\perp$, which takes place in Sect. 8.2. The final and hardest step, namely the proof that the space of geometric hyperplanes is actually a projective space is carried out in Sects. 8.3 and 8.4. The proof of the embedding result for rank three, given in Sect. 8.4, is long.

If the rank is four or more, there is a more elementary proof, given in Sect. 8.3. By Corollary 6.3.2, the singular subspaces of these polar spaces are Desarguesian.

The chapter ends with Sect. 8.5 in which automorphisms of polar spaces of rank at least three are found.

F. Buekenhout, A.M. Cohen, *Diagram Geometry*,
Ergebnisse der Mathematik und ihrer Grenzgebiete. 3. Folge / A Series of
Modern Surveys in Mathematics 57, DOI 10.1007/978-3-642-34453-4_8,
© Springer-Verlag Berlin Heidelberg 2013

8.1 Geometric Hyperplanes and Ample Connectedness

Recall from Definition 5.2.7 that a geometric hyperplane of a line space Z is a proper subspace of Z with the property that every line of Z meets it in at least a point. The study of geometric hyperplanes will be of use both for the embedding of polar spaces in linear spaces, the main goal of this chapter, and for the construction of automorphisms of polar spaces, to be discussed at the end of this chapter, in Sect. 8.5. For our arguments to work, we need that the complement of each geometric hyperplane is connected.

As described in Definition 2.5.8, for a line space Z and a set of points H, we will write $Z \setminus H$ to denote the set of points of Z outside H.

Definition 8.1.1 We call the line space Z *amply connected* if, for every proper subspace X of Z, the subgraph of the collinearity graph of Z induced on $Z \setminus X$, is connected.

Lemma 8.1.2 *Suppose that Z is a polar space.*

  (i) *For each point p of Z, the subspace $p^\perp$ is either the whole space (in which case $p \in \mathrm{Rad}(Z)$) or a geometric hyperplane of Z.*
 (ii) *If Z is nondegenerate and H is a degenerate geometric hyperplane of Z, then there is a unique point p of Z such that $H = p^\perp$.*
(iii) *If r is the nondegenerate rank of Z, then the nondegenerate rank of a geometric hyperplane of Z is at least $r - 1$.*

Proof (i) is obvious: clearly, $p \in \mathrm{Rad}(Z)$ if and only if $p^\perp$ is the whole space and $p^\perp$ is a proper subspace otherwise, in which case each line not contained in $p^\perp$ has a unique point of $p^\perp$ by the polar space axiom.

(ii) Suppose $p \in \mathrm{Rad}(H)$, so $H \subseteq p^\perp$. If l is a line of Z on p not contained in H, then, by nondegeneracy of Z, there is a point q of Z non-collinear with p and collinear with a point $s \in l \setminus H$. Then $qs = \{q, s\}^{\perp\perp}$ is a line disjoint from H, a contradiction. Hence every line of Z on p is in H, so $H = p^\perp$. If $q \in H$ also satisfies $H = q^\perp$, then $p^\perp = q^\perp$, contradicting that Z is nondegenerate in view of Lemma 7.4.8(iv).

(iii) Fix a geometric hyperplane H of Z. If $\langle H, \mathrm{Rad}(Z)\rangle = Z$, then the nondegenerate rank of H is equal to r, so we may assume that $\langle H, \mathrm{Rad}(Z)\rangle/\mathrm{Rad}(Z)$ is a geometric hyperplane of $Z/\mathrm{Rad}(Z)$. By going over to the nondegenerate quotient of Z (as described in Proposition 7.4.10), we may assume $\mathrm{Rad}(Z) = \emptyset$. Let M be a maximal singular subspace of Z (so $\dim(M) = r - 1$) and write $N = H \cap M$. Either $M = N$ or N is a geometric hyperplane of M and hence, by Theorem 7.4.13(iii), a maximal subspace of M. In particular, the dimension of N is at least $r - 2$. If H is a nondegenerate polar space, we are done. Otherwise, the result follows from (ii). $\qquad\square$

We describe a class of polar spaces which are clearly not amply connected.

Definition 8.1.3 A *rosette* is the union of any collection of at least two linear spaces having a geometric hyperplane in common.

A rosette is obviously a polar space whose radical is the geometric hyperplane in the intersection of the linear spaces of which it is built up. In other words, a rosette is a polar space of nondegenerate rank one. Moreover, the complement of this geometric hyperplane is disconnected and so the rosette is not amply connected.

Recall from Definition 2.5.8 that a thick line is a line having at least three points.

Theorem 8.1.4 *Each polar space of nondegenerate rank at least two whose lines are thick, is amply connected.*

Proof Suppose that Z is a polar space of nondegenerate rank at least two whose lines are thick. Let U be a proper subspace of Z and let x and y be points in $Z \setminus U$. We need to show that there is a path from x to y in $Z \setminus U$, and so we may assume that x and y are not collinear. If there is a point $z \in \mathrm{Rad}(Z) \setminus U$ then x, z, y form a path from x to y in $Z \setminus U$. Therefore, we (may) assume that U contains $\mathrm{Rad}(Z)$.

Consider the nondegenerate quotient $Z/\mathrm{Rad}(Z)$ of Z by $\mathrm{Rad}(Z)$. For two points u, v in $Z \setminus \mathrm{Rad}(Z)$, we have $u \perp v$ if and only if $\langle u, \mathrm{Rad}(Z)\rangle \perp \langle v, \mathrm{Rad}(Z)\rangle$. This implies that x and y are connected by a path in $Z \setminus U$ if and only if $\langle x, \mathrm{Rad}(Z)\rangle$ and $\langle y, \mathrm{Rad}(Z)\rangle$ are connected by a path in $(Z/\mathrm{Rad}(Z)) \setminus (U/\mathrm{Rad}(Z))$. As the nondegenerate ranks of Z and $Z/\mathrm{Rad}(Z)$ are equal, for the proof of the theorem, we may pass over to $Z/\mathrm{Rad}(Z)$. Therefore, we (may) assume that Z is nondegenerate.

Lemma 7.4.8(ii) gives $\mathrm{Rad}(x^\perp \cap y^\perp) = \emptyset$. By the assumptions that Z is nondegenerate and that its (nondegenerate) rank is at least two, there exist non-collinear points a, b in $x^\perp \cap y^\perp$. If a or b is outside U, we are done, so we will assume both are in U. Let x' be a third point on a line m containing a and x (it exists as lines are thick). Now x' is not in U, for otherwise m would lie in U and so $x \in U$, a contradiction. On a line containing b and y, there is a point y' collinear with x'. Here, $y' \neq y$, for otherwise y would be collinear with each point of m and hence with x, a contradiction. Similarly, $y' \neq b$. Moreover y' is not in U. We conclude that x, x', y', y is a path from x to y inside $Z \setminus U$. $\square$

If $x \in Q$, then $U = x^\perp$ is a subspace of $\mathrm{K}_{m,n}$ whose complement is the subset $Q \setminus \{x\}$ of the part Q. So, if $m > 2$, then this complement is not connected and so $\mathrm{K}_{m,n}$ is neither amply connected nor a rosette. Exercise 8.6.2 shows that the condition that lines are thick is essential in Theorem 8.1.4.

The line spaces appearing in the next lemma and subsequent proposition are not necessarily polar spaces.

Lemma 8.1.5 *Let H be a geometric hyperplane of a space Z. If U is a subspace of Z not contained in H, then $H \cap U$ is a geometric hyperplane in U.*

Proof Let l be a line contained in U. It intersects H in some point p, so l intersects $H \cap U$ in p. Moreover, $H \cap U$ is a proper subspace of U since U is not contained in H. $\square$

The proposition below obviously applies to amply connected spaces, for which it will be used primarily.

Proposition 8.1.6 *If H a geometric hyperplane of a space Z such that $Z \setminus H$ is connected, then H is a maximal subspace of Z.*

Proof For distinct points x, y in $Z \setminus H$, we must show that $\langle H, x \rangle$ contains y. Consider a chain $x = x_1, \ldots, x_n = y$ from x to y in which consecutive members are distinct, collinear, and not in H. A line l on x_1 and x_2 intersects H in a point p and so $\langle H, x \rangle$ contains p and x, hence l, and x_2. The same argument applies inductively to lines containing x_i and x_{i+1} for $i = 2, \ldots, n - 1$. Therefore, $y \in \langle H, x_n \rangle = \cdots = \langle H, x_2 \rangle = \langle H, x \rangle$. $\qquad\square$

Observe that Proposition 8.1.6 need not hold if Z is a rosette.

Corollary 8.1.7 *Let Z be a polar space of rank at least two whose lines are thick, but not a rosette. If H is a geometric hyperplane of Z, then $\langle Z \setminus H \rangle = Z$.*

Proof Since Z is not a rosette, Theorem 8.1.4 shows that Z is amply connected and so $Z \setminus H$ is connected. Take $p \in H$. If there is a line l of Z with $l \cap H = \{p\}$, then, since l has at least two points in $Z \setminus H$, the point p belongs to $\langle Z \setminus H \rangle$. Suppose, therefore, that p is not collinear with any point of $Z \setminus H$. Then $p^\perp \subseteq H$. By Lemma 8.1.2(i) and Proposition 8.1.6 this implies $p^\perp = H$, so $p \in \mathrm{Rad}(H)$. Clearly, there is a line m on p and a point of $H \setminus \mathrm{Rad}(H)$. Then $m \cap \mathrm{Rad}(H) = \{p\}$, so p is the only point of H with $p^\perp \subseteq H$. It follows that $\langle Z \setminus H \rangle$ contains $m \setminus \{p\}$. Since m is thick, $p \in \langle m \setminus \{p\} \rangle$, whence $p \in \langle Z \setminus H \rangle$. $\qquad\square$

Corollary 8.1.8 *If Z is a nondegenerate polar space of rank at least two with thick lines and H is a geometric hyperplane of Z, then $\langle Z \setminus H \rangle = Z$ and H is a maximal subspace of Z.*

Proof By Theorem 8.1.4, Z is amply connected, so, by Proposition 8.1.6, every geometric hyperplane of Z is a maximal subspace of Z. The other assertion follows from Corollary 8.1.7. $\qquad\square$

Corollary 8.1.9 *Let Z be a polar space of rank at least two whose lines are thick and let H be a geometric hyperplane of Z. If Z is not a rosette, then $H \cap \mathrm{Rad}(Z)$ is either a geometric hyperplane of $\mathrm{Rad}(H)$ or $\mathrm{Rad}(H)$ itself.*

Proof By Theorem 8.1.4, Z is amply connected. Hence, by Proposition 8.1.6, H is a maximal subspace of Z. Let l be a line in $\mathrm{Rad}(H)$ and consider a point $p \in Z \setminus H$. Then $p^\perp$ intersects l in some point a. Now $a^\perp$ contains H and p, so it contains all points of Z, whence $a \in \mathrm{Rad}(Z)$. This shows that each line of $\mathrm{Rad}(H)$ contains a point of $H \cap \mathrm{Rad}(Z)$, so the corollary follows. $\qquad\square$

Later on in this chapter, in Proposition 8.5.9, we show that nondegenerate polar spaces of rank at least three all of whose lines are thick possess automorphisms. The proof is based on a construction that is quite similar to the way perspectivities were obtained for projective spaces. The following result plays a crucial role.

Observe that, if Z is a nondegenerate polar space of rank at least three all of whose lines are thick, then each point lies on a thick projective plane and so belongs to at least three lines. Therefore, it is a thick polar space.

Proposition 8.1.10 *Let Z be a thick nondegenerate polar space of rank at least three. Let p and q be non-collinear points, $a \in \{p, q\}^{\perp}$, and H a geometric hyperplane of $p^{\perp}$ such that $H \cap q^{\perp} \subseteq a^{\perp}$. The line aq contains a unique point w such that $\{w, p\}^{\perp} = H$.*

Proof Observe that $p^{\perp}$ cannot be a rosette, for otherwise $\mathrm{Rad}(Z) \neq \emptyset$ by Lemma 8.1.2. Hence $p^{\perp}$ is amply connected by Theorem 8.1.4.

Suppose $H \subseteq q^{\perp}$. Then $H \subseteq \{a, p, q\}^{\perp}$, so, by Proposition 8.1.6, applied to $p^{\perp}$, we have $H = \{a, p, q\}^{\perp}$. Taking a plane A on p not contained in $a^{\perp}$, we see that $A \cap H = \{a, q\}^{\perp} \cap A$ is a point of A, contradicting the fact that H is a geometric hyperplane of $p^{\perp}$.

Hence there exists $b \in H \setminus q^{\perp}$. Then $b^{\perp} \cap aq$ has a unique point w. As $p^{\perp}$ is amply connected, Proposition 8.1.6 gives that H is a maximal subspace of $p^{\perp}$. Now, each point of $H \cap q^{\perp}$ lies in $H \cap (aq)^{\perp}$, so $w^{\perp}$ contains both b and the geometric hyperplane $H \cap q^{\perp}$ of the polar space H. If H is not a rosette, then, by Theorem 8.1.4, it is amply connected. In this case, Proposition 8.1.6 yields that $H \cap q^{\perp}$ is a maximal subspace of it, so $\langle H \cap q^{\perp}, b \rangle = H$, whence $H \subseteq \{w, p\}^{\perp}$. The same proposition applied to $p^{\perp}$, shows $\{w, p\}^{\perp} = H$. If H is a rosette, then we must have $H = \{a, p\}^{\perp}$ and so $w = a$ is as required. $\square$

To finish this section, we show that on the complement of a hyperplane, we can find a standard equivalence.

Remark 8.1.11 Let Z be a thick nondegenerate polar space of rank at least three having a geometric hyperplane H. Consider the restriction of Z to the complement $A = Z \setminus H$ of H; cf. Definition 2.5.8. Its lines are the intersections with A of lines of Z that are not contained in H. Having distance 0 or 3 in the collinearity graph of this line space on A is a standard equivalence relation, in the sense of Exercise 2.8.24. It is an equivalence relation because the classes consist of points of A having the same sets of collinear points in H (with respect to Z). To see that this equivalence relation is standard, let x and x' be distinct equivalent points of A, and let $z \neq x$ be collinear with x. We need to prove the existence of a point z' equivalent to z collinear with x'. Because Z is nondegenerate, Lemma 7.4.8(iii) shows that there is a line m of Z on z not contained $x^{\perp}$. This line has a point $q \in z^{\perp} \cap H \setminus x^{\perp}$. As q is not collinear with the intersection p of the line xz with H and p is collinear with x' (for x and x' are equivalent), there is a single point z' in $q^{\perp} \cap x'p \setminus H$. Notice that p and q are not collinear, for otherwise $q \in \{p, z\}^{\perp} \subseteq x^{\perp}$, contradicting the choice of q.

It remains to show that z' is equivalent to z. Suppose not; then there is a common neighbor r of z and z' outside H. As the rank of Z is at least three, the subspace $\{z, z'\}^\perp$ is a nondegenerate polar space of rank at least two, and so there are lines l and l' on r inside $\{z, z'\}^\perp$ with $l' \not\subseteq l^\perp$. Now $p^\perp \cap l$ contains a point v of $\{x, x'\}^\perp$ which is in $\{z, z'\}^\perp \cap H$, and, similarly, $p^\perp \cap l'$ contains a point v' of $\{z, z'\}^\perp \cap H$. Notice $v \not\perp v'$. If $\{z, z'\}^\perp \cap H$ were a rosette, it would be a degenerate geometric hyperplane of $\{z, z'\}^\perp$, so Lemma 8.1.2(ii) shows $\{z, z'\}^\perp \cap H = \{t, z, z'\}^\perp$ for some $t \in H$. But then the rank of $\{z, z'\}$ must be two and $\mathrm{Rad}(\{t, z, z'\}^\perp) = \{t\}$. As $v'p$ and vp are lines of $\{t, z, z'\}^\perp$, it follows that $t = p$, a contradiction with $q \in \{z, z'\}^\perp \cap H \setminus p^\perp$. Therefore, $\{z, z'\}^\perp \cap H$ is not a rosette.

By Theorem 8.1.4, the space $\{z, z'\}^\perp \cap H$ is amply connected, so Proposition 8.1.6 gives that $\{r, z, z'\}^\perp \cap H$ is a maximal subspace of $\{z, z'\}^\perp \cap H$. If $u \in \{r, z, z'\}^\perp \cap H$, then u is distinct from r (which is not in H) and the unique point of $p^\perp \cap ur$ also lies in $\{x, x'\}^\perp$ and hence in $ur \cap H = \{u\}$, so it coincides with u. In particular $u \in p^\perp$, proving $\{r, z, z'\}^\perp \cap H \subseteq \{p, z, z'\}^\perp \cap H$. As $\{r, z, z'\}^\perp \cap H$ is a maximal subspace of $\{z, z'\}^\perp \cap H$, we have either $\{r, z, z'\}^\perp \cap H = \{p, z, z'\}^\perp \cap H$ or $\{p, z, z'\}^\perp \cap H = \{z, z'\}^\perp \cap H$. In the former case, $p \in r^\perp$, so $r \in \{x, x'\}^\perp$, contrary to the assumption. In the latter case, $q \in \{p, z, z'\}^\perp \cap H$, so $q \perp p$, a contradiction. We conclude that $\{z, z'\}^\perp$ is entirely contained in H, which settles that z and z' are equivalent.

8.2 The Veldkamp Space

We will develop a theory of geometric hyperplanes in amply connected spaces. These hyperplanes will be the points of the Veldkamp space to be defined below. Recall the notion of homomorphism for spaces from Definition 2.5.8, where an injective homomorphism mapping lines to lines is called an embedding. The main result of this section, Theorem 8.2.9, establishes that a thick nondegenerate polar space of rank at least three embeds in its Veldkamp space. We begin with a lemma in which ample connectedness comes really into play.

Lemma 8.2.1 *Let Z be an amply connected space all of whose geometric hyperplanes are amply connected and let H_1, H_2, H_3 be distinct geometric hyperplanes of Z such that $H_3 \supseteq H_1 \cap H_2$. Then $H_1 \cap H_2 = H_2 \cap H_3 = H_3 \cap H_1$.*

Proof We have the inclusions $H_2 \supseteq H_3 \cap H_2 \supseteq H_1 \cap H_2$ and, by Lemma 8.1.5, $H_3 \cap H_2$ and $H_1 \cap H_2$ are both geometric hyperplanes of H_2. Hence, by the hypotheses and Proposition 8.1.6, $H_3 \cap H_2 = H_1 \cap H_2$. The lemma follows. $\qquad\square$

Proposition 8.2.2 *Let Z be an amply connected space all of whose geometric hyperplanes are amply connected. If H_1, H_2 are distinct geometric hyperplanes and $p \in Z \setminus (H_1 \cap H_2)$, then there is at most one geometric hyperplane containing p and $H_1 \cap H_2$.*

Proof Assume, by way of contradiction, that H_3 and H_4 are distinct geometric hyperplanes containing p and $H_1 \cap H_2$. As $H_1 \cap H_2$ is a geometric hyperplane of H_1 and of H_2, which are amply connected, Proposition 8.1.6 shows that at most one of H_1, H_2 contains $\langle H_1 \cap H_2, p \rangle$. Therefore, without loss of generality, we can assume that $H_3 \neq H_1, H_2$ and $H_4 \neq H_1$. Applying Lemma 8.2.1 to the triples $\{H_1, H_2, H_3\}$ and $\{H_1, H_3, H_4\}$, we find $H_1 \cap H_2 = H_1 \cap H_3 = H_3 \cap H_4$, a contradiction as $p \in H_3 \cap H_4 \setminus (H_1 \cap H_2)$. $\square$

Definition 8.2.3 The *Veldkamp space* $\mathcal{V}(Z)$ of a line space Z is the space in which

(1) a *point* is a geometric hyperplane of Z;
(2) a *line* is the collection $H_1 H_2$ of all geometric hyperplanes H of Z such that $H_1 \cap H_2 = H_1 \cap H = H_2 \cap H$ or $H = H_i$ ($i = 1$ or 2), where H_1, H_2 are distinct points of $\mathcal{V}(Z)$.

If Z is a projective space, then $\mathcal{V}(Z)$ coincides with the hyperplane dual $Z^\star$ defined in Exercise 5.7.12.

Theorem 8.2.4 *For every amply connected space Z all of whose geometric hyperplanes are amply connected, the Veldkamp space $\mathcal{V}(Z)$ is a linear space. Each line is of the form $\{H \in \mathcal{V}(Z) \mid H \supseteq H_1 \cap H_2\}$ for any two distinct points H_1, H_2 of $\mathcal{V}(Z)$ on that line.*

Proof By definition, for any two distinct points H_1, H_2 of $\mathcal{V}(Z)$, there is a line $H_1 H_2$ joining them. From the definition of Veldkamp line it is immediate that $H_1 H_2$ is contained in $\{H \in \mathcal{V}(Z) \mid H \supseteq H_1 \cap H_2\}$. Suppose now $H \in \mathcal{V}(Z)$ contains $H_1 \cap H_2$. Lemma 8.2.1 shows that H belongs to $H_1 H_2$.

For any pair of distinct geometric hyperplanes H_3, H_4 on the line $H_1 H_2$ of $\mathcal{V}(Z)$, we show that $H_1 H_2 = H_3 H_4$. If H_3 is distinct from H_2, then, by the definition of a line in $\mathcal{V}(Z)$, we have $H_1 \cap H_2 = H_2 \cap H_3$, whence, by Lemma 8.2.1, $H_1 H_2 = H_2 H_3$. Similarly, we obtain $H_2 H_3 = H_3 H_4$. If $H_3 = H_2$, then $H_1 \cap H_2 = H_2 \cap H_4$ whence $H_1 H_2 = H_2 H_4 = H_3 H_4$. $\square$

Corollary 8.2.5 *If Z is a thick nondegenerate polar space of rank at least three, then Z and all of its geometric hyperplanes are amply connected. In particular, $\mathcal{V}(Z)$ is a linear space.*

Proof The second statement follows from the first statement and Theorem 8.2.4. As for the first statement, Theorem 8.1.4 and the fact that a rosette has a non-empty radical give that Z is amply connected. Suppose that H is a geometric hyperplane of Z that is not amply connected. As H is a polar space of rank at least two whose lines are thick, Theorem 8.1.4 gives that H is a rosette. By Lemma 8.1.2(ii), $H = p^\perp$ for some point p of Z, so $p^\perp$ is a rosette. This contradicts the fact that $p^\perp$ has nondegenerate rank at least two. $\square$

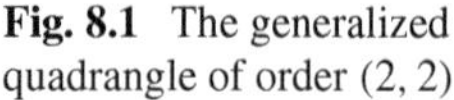

Fig. 8.1 The generalized
quadrangle of order $(2, 2)$

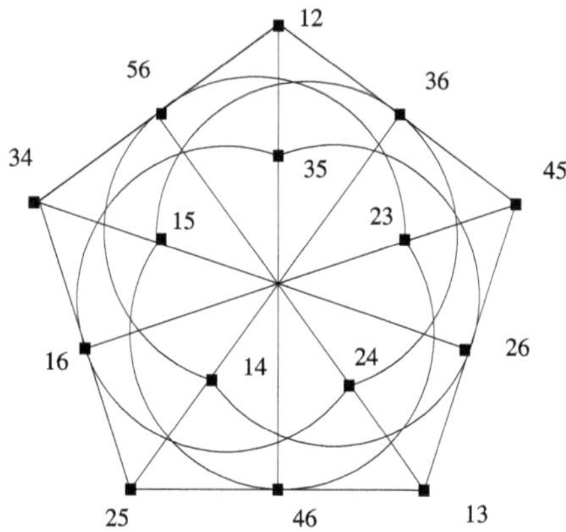

Example 8.2.6 The smallest model of a nondegenerate polar space with rank $r \geq 2$
and thick lines is the generalized quadrangle Z of 15 points and 15 lines described
in Example 2.2.10 (see also Exercises 7.11.18 and 7.11.19). It is pictured in Fig. 2.3,
which reappears as Fig. 8.1. Let H be a geometric hyperplane. It is easy to check
that H is of one of the following three kinds.

(1) A set $p^{\perp}$ for some point p. There are 15 such hyperplanes.
(2) A grid, consisting of 9 points and 6 lines. There are 10 such hyperplanes.
(3) An ovoid, consisting of 5 points containing no line. There are 6 such hyper-
     planes.

Hence $\mathcal{V}(Z)$ has $15 + 10 + 6 = 31$ points and all lines of $\mathcal{V}(Z)$ have cardinality 3.
In fact, $\mathcal{V}(Z)$ is the projective space $\mathbb{P}(\mathbb{F}_2^5)$. Here, the sets $\{H \in \mathcal{V}(Z) \mid H \supseteq H_1 \cap H_2\}$, for H_1, H_2 distinct points of $\mathcal{V}(Z)$, do not always represent lines of $\mathcal{V}(Z)$:
consider the intersection of $a^{\perp}$ and $b^{\perp}$ for two distinct collinear points a, b of Z.
It is the line ab. The line on $a^{\perp}$ and $b^{\perp}$ in $\mathcal{V}(Z)$ has a unique third point, namely
$c^{\perp}$, where $a\,b = \{a, b, c\}$. But there exists a hyperplane H containing ab that is a
generalized subquadrangle of order $(2,1)$. Thus, $H \supseteq a^{\perp} \cap b^{\perp}$, but H is not on the
line in the Veldkamp space generated by $a^{\perp}$ and $b^{\perp}$.

Pushing Theorem 8.2.4 further, one might ask for fairly general conditions under
which the Veldkamp space $\mathcal{V}(Z)$ is a projective space. The following result indicates
how this may be achieved.

Theorem 8.2.7 *Let Z be an amply connected space all of whose geometric hyper-*
planes are amply connected. Assume also that all geometric hyperplanes of geo-
metric hyperplanes of Z are amply connected, and that, for all distinct geometric
hyperplanes H_1, H_2 and every point p of Z outside $H_1 \cap H_2$, there is a geometric
hyperplane containing p and $H_1 \cap H_2$. Then $\mathcal{V}(Z)$ is a projective space.

Proof By Theorem 8.2.4, $\mathcal{V}(Z)$ is a linear space. We check Pasch's axiom 5.2.4 for
projective spaces. To this end, suppose that $\{A, B, P\}$ and $\{A, C, Q\}$ are collinear

triples on distinct lines of $\mathcal{V}(Z)$. We need to show that the lines BC and PQ of $\mathcal{V}(Z)$ have a common point R.

Suppose $P \cap Q \subseteq B \cap C$. This implies $P \cap Q \subseteq B \cap C \cap Q = A \cap B \cap C$ (the equality holds by Proposition 8.1.6 as $A \cap B \cap C$ is a geometric hyperplane of $A \cap C = C \cap Q$ contained in $B \cap C \cap Q$ and $A \cap C$ is amply connected). In particular, $P \cap Q \subseteq C$, so $P \cap Q = P \cap C$ is a geometric hyperplane of C contained in $B \cap C$. Therefore, $B \cap C = P \cap Q$, contradicting that BC and PQ are distinct lines.

We conclude that Z has a point p in $P \cap Q \setminus B \cap C$. By assumption, there is a geometric hyperplane R containing $B \cap C$ and p. Observe that, by Lemma 8.1.5, $A \cap B \cap C = A \cap P \cap Q$ is a geometric hyperplane of $P \cap Q$. Since $P \cap Q$ is amply connected, Proposition 8.1.6 yields that $A \cap B \cap C$ is a geometric hyperplane of $P \cap Q$ (observe that $P \cap Q \nsubseteq A$ for otherwise A, P, Q would be collinear by Theorem 8.2.4). Thus R contains $P \cap Q$, so R is on the line PQ. We are done, as R has been chosen so as to lie on BC. $\qquad\square$

In the special case where the lines of a space have size three, its Veldkamp space is often a projective space.

Theorem 8.2.8 *Let Z be an amply connected space all of whose lines have exactly three points. If every geometric hyperplane is amply connected as well, then $\mathcal{V}(Z)$ is a projective space all of whose lines have size three.*

Proof Let A and B be distinct geometric hyperplanes of Z. The set

$$C = A * B := (A \cap B) \cup \big(Z \setminus (A \cup B)\big)$$

is a geometric hyperplane with $A \cap C = A \cap B = B \cap C$. Hence lines of $\mathcal{V}(Z)$ are thick.

Assume now that D is a geometric hyperplane of Z containing $A \cap B$, which is distinct from A and B. There is a point $p \in D \setminus (A \cup B)$. It also belongs to C, so both C and D contain $\{p\} \cup (A \cap B)$. Now Proposition 8.2.2 forces $C = D$, so lines of $\mathcal{V}(Z)$ have cardinality three.

We next check that $\mathcal{V}(Z)$ satisfies Pasch's axiom 5.2.4. Suppose that A, B, C, P, Q are geometric hyperplanes of Z such that $\{A, B, P\}$ and $\{A, C, Q\}$ are collinear triples on distinct lines of $\mathcal{V}(Z)$. Then $P = A * B$ and $Q = A * C$. Straightforward set operations show

$$P \cap Q = (A \cap B \cap C) \cup \big(Z \setminus (A \cup B \cup C)\big), \quad \text{and}$$

$$Z \setminus (P \cup Q) = \big((B \cap C) \setminus A\big) \cup \big(A \setminus (B \cup C)\big),$$

whence $P * Q = (B \cap C) \cup (Z \setminus (B \cup C)) = B * C$. Thus, the lines BC and PQ of $\mathcal{V}(Z)$ have a common point. So Pasch's axiom holds and Theorem 5.2.6 gives that $\mathcal{V}(Z)$ is a projective space. $\qquad\square$

Theorem 8.2.9 *Let Z be a thick nondegenerate polar space of rank at least three. The map $x \mapsto x^{\perp}$ is an embedding of Z in its Veldkamp space $\mathcal{V}(Z)$.*

Proof Let x, y be points of Z. If $x^\perp = y^\perp$, then $x = y$ since Z is nondegenerate. This shows injectivity of the map $x \mapsto x^\perp$.

Suppose that x and y are both on the line l of Z and take H to be a geometric hyperplane distinct from $x^\perp$ and containing $x^\perp \cap y^\perp = l^\perp$. By Corollary 8.2.5, every hyperplane of Z is amply connected, so by Lemma 8.2.1, we have $H \cap x^\perp = x^\perp \cap y^\perp$.

There is a point $z \in H \setminus l^\perp$. Now $z^\perp \cap l$ consists of a single point u, say, and $u^\perp$ is a geometric hyperplane of Z containing $l^\perp$ and z. Suppose $u^\perp \neq H$. Lemma 8.2.1 with $H_1 = u^\perp$, $H_2 = x^\perp$, and $H_3 = H$ gives $u^\perp \cap H = x^\perp \cap H = l^\perp$, contrary to $z \in (u^\perp \cap H) \setminus l^\perp$. Therefore, $u^\perp = H$; we find that the image of l under $x \mapsto x^\perp$ is the line $\{w^\perp \mid w^\perp \supseteq l^\perp\}$ of $\mathcal{V}(Z)$. By Lemma 7.4.7(ii) and (iii) we have $w^\perp \supseteq l^\perp$ if and only if $w^{\perp\perp} \subseteq l^{\perp\perp}$, which, by Exercise 7.11.13, amounts to $w \in l$. We conclude that the line of $\mathcal{V}(Z)$ on $x^\perp$ and $y^\perp$ is equal to $\{w^\perp \mid w \in l\}$, the image of l under the map $x \mapsto x^\perp$. $\qquad\square$

8.3 Projective Embedding for Rank at Least Four

Recall from Definition 8.2.3 that $\mathcal{V}(Z)$ stands for the Veldkamp space of a line space Z. In Theorem 8.2.9, we have seen that each thick nondegenerate polar space Z of rank at least three can be embedded in its Veldkamp space $\mathcal{V}(Z)$. Here, our purpose is to show that $\mathcal{V}(Z)$ is a projective space. In this section, culminating in Theorem 8.3.16, we handle the case where Z has rank $r \geq 4$; infinite rank, $r = \infty$, is allowed. The next section uses a different approach for $r = 3$. The proof for the case of rank at least four is not easy, but it is much simpler than the proof for rank three.

By Theorem 7.4.13(iv) all maximal singular subspaces of Z are projective spaces. In view of Theorem 7.5.5 (for $r < \infty$) and Exercise 7.11.22 (for $r = \infty$), they all have dimension $r - 1 \geq 3$, so, by Theorem 6.1.8, all singular subspaces of Z are Desarguesian.

Notation 8.3.1 Throughout this section, Z is a nondegenerate polar space with thick lines of rank at least four. Moreover, A, B, C, P, Q are distinct points of $\mathcal{V}(Z)$ such that $\{A, B, P\}$ and $\{A, C, Q\}$ are collinear triples spanning distinct lines (cf. Fig. 8.2).

Figure 8.2 is another incarnation of the Pasch configuration displayed in Fig. 5.2. In view of Corollary 8.2.5 and Theorem 5.2.6 we need only show (Pasch's axiom 5.2.4) that the lines PQ and BC of $\mathcal{V}(Z)$ have a common point R. This will be our goal for the greater part of the section.

Lemma 8.3.2 $A \cap P \cap Q = A \cap B \cap C = A \cap B \cap C \cap P \cap Q = (B \cap C) \cap (P \cap Q)$.

Proof This follows from $A \cap P \subseteq B$, $A \cap Q \subseteq C$, etc. $\qquad\square$

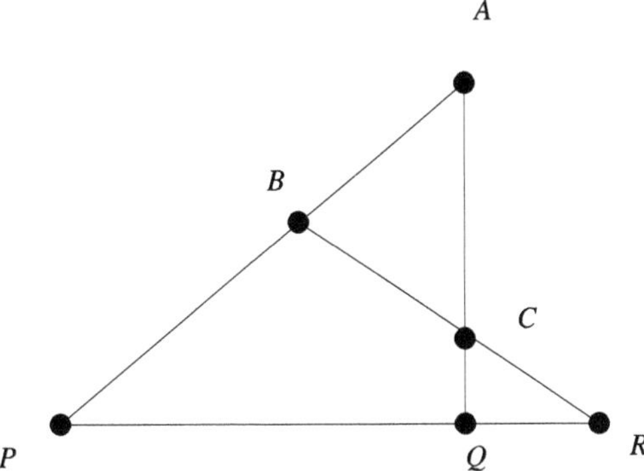

Fig. 8.2 The Pasch configuration

The next lemma deals with degeneracies.

Lemma 8.3.3 *Neither $B \cap C$ nor $P \cap Q$ is a rosette. If there is a point x of Z such that $(B \cap C) \cup (P \cap Q) \subseteq x^\perp$, then $R = x^\perp$ is a geometric hyperplane of Z on the Veldkamp lines BC and PQ.*

Proof Assume that $B \cap C$ is a rosette with radical X. Then $B \cap C \subseteq X^\perp$. Since Z is nondegenerate of rank at least four, the subspace $X^\perp$ is not a rosette. In particular, $B \cap C$ is a proper subspace of $X^\perp$. Now, for $x \in X$, the space $x^\perp$ is a polar space of rank at least three which is not a rosette, so by Lemma 8.1.2(ii) applied to $B \cap C \subset X^\perp \subseteq x^\perp$, we obtain $X^\perp = x^\perp$, proving $X = \{x\}$. Now $x^\perp \neq B$ or $x^\perp \neq C$, and without loss of generality, we (may) assume the former. Then $B \cap C = x^\perp \cap B$ is a polar space of rank at least three with radical $\{x\}$ which is a rosette, a contradiction.

The second statement of the lemma is immediate as $x^\perp$ is a geometric hyperplane of Z belonging to the lines BC and PQ of $\mathcal{V}(Z)$. $\qquad\square$

Thus, from now on, we work in the following setting.

Hypothesis 8.3.4 *There is no point x of Z such that $x^\perp$ contains $B \cap C$ and $P \cap Q$.*

We will construct a geometric hyperplane R of Z containing $P \cap Q$ as well as $B \cap C$, as required for Pasch's axiom. In order to motivate the construction of R in Definition 8.3.6 below, we first derive some properties that R should have.

Proposition 8.3.5 *If R is a geometric hyperplane of Z containing $B \cap C$ and $P \cap Q$ that is not of the form $x^\perp$ for some $x \in Z$, then it satisfies the following properties.*

(i) *R is a nondegenerate polar space of rank at least three.*
(ii) *R is the disjoint union of sets R^0, R^1, R^2, R^3 where*
    (0) *$R^0 = A \cap B \cap C$,*
    (1) *$R^1 = (B \cap C \setminus R^0) \cup (P \cap Q \setminus R^0)$,*
    (2) *R^2 is the set of points outside $R^0 \cup R^1$ lying on a line bp with $b \in B \cap C \setminus A$ and $p \in P \cap Q \setminus A$, and*

(3) R^3 *consists of all* $y \in (R^0)^\perp$ *with* $y^\perp \cap B \cap C = y^\perp \cap P \cap Q = R^0$ *and with the property that, for each* $a \in R^0$ *and each* $y' \in ay \setminus \{a, y\}$, *we have* $y' \in R^2$.

(iii) *If* R^3 *is non-empty, then* R^0 *is a nondegenerate polar space of rank at least two.*

Proof Assertion (i) is easy. The crucial part of (ii) and (iii) is to prove that every point $y \in R \setminus (R^0 \cup R^1 \cup R^2)$ belongs to R^3. To this end, we observe that $y^\perp \cap B \cap C$ is a geometric hyperplane of $B \cap C$ contained in R^0 (for otherwise there is a line of the form bp with $b \in B \cap C \setminus A$ and $p \in P \cap Q \setminus A$ containing y, leading to the contradiction $y \in R^2$). This implies $y^\perp \cap B \cap C = R^0$. Similarly, $y^\perp \cap P \cap Q = R^0$, and $y \in (R^0)^\perp$ follows.

(iii) A singular subspace of Z on y and on a point of R^0 of dimension three meets $B \cap C$ and hence R^0 in a line, so $\mathrm{rk}(R^0) \geq 2$. Next assume $a \in \mathrm{Rad}(R^0)$. All points $y' \in ay \setminus \{a\}$ satisfy $(y')^\perp \cap B \cap C = (y')^\perp \cap P \cap Q = R^0$. Therefore, if $b \in B \cap C \setminus A$, the (non-empty) space $b^\perp \cap ay$ must coincide with $\{a\}$. This gives $B \cap C \setminus A \subseteq a^\perp$, whence $B \cap C \subseteq a^\perp$. Since the same can be said about $P \cap Q$, we derive that $(B \cap C) \cup (P \cap Q) \subseteq a^\perp$, contrary to the hypotheses. Therefore, R^0 is nondegenerate, proving (iii).

(ii) Let $a \in R^0$ and $y' \in ay \setminus \{a, y\}$. We need to show $y' \in R^2$. Pick a point $a' \in R^0$ collinear with a point c in $B \cap C \setminus A$ but not collinear with a. Then y' is not collinear with a', so $(y')^\perp \cap ca'$ consists of a single point in $ca' \setminus \{a\} \subseteq B \cap C \setminus A$. In particular, $(y')^\perp \cap B \cap C \neq R^0$. By the first paragraph of this proof with y' instead of y, we find $y' \in R^2$. $\qquad \square$

The proof of Proposition 8.3.5 helps us understand how R should be constructed.

Definition 8.3.6 We take R to be the disjoint union of the sets R^0, R^1, R^2, and R^3, defined as in Proposition 8.3.5(ii).

(1) A line bp joining a point $b \in B \cap C \setminus A$ and a point $p \in P \cap Q \setminus A$ is called *generic*.
(2) A line contained in $R^0 \cup R^2$ with a unique point in R^0 will be called *special*.
(3) A line ay joining a point $a \in R^0$ with a point $y \in R^3$ will be called *extraspecial*.

In Fig. 8.3, a special and a generic line are drawn. Clearly, $R \neq Z$ because B is not contained in R. In order to establish that R is a point of $\mathcal{V}(Z)$, it remains to derive that R is a subspace of Z and that every line of Z intersects R in some point. The proof of the former will be our first and by far the biggest enterprise.

Let x, y be distinct collinear points in R. We have to show $xy \subseteq R$. If x, y are both in $R^0 \cup R^1$, then this is obvious. Let $i, j \in \{0, 1, 2, 3\}$ be such that $x \in R^i$, $y \in R^j$. We distinguish the cases $(i, j) = (1, 2), (0, 2), (2, 2), (0, 3), (1, 3), (2, 3)$, and $(3, 3)$. In each case we will show that xy is generic, special, or extraspecial. Since, clearly, each such line is contained in R, this suffices for establishing $xy \subseteq R$.

Lemma 8.3.7 *If* $(i, j) = (1, 2)$, *then* xy *is a generic line.*

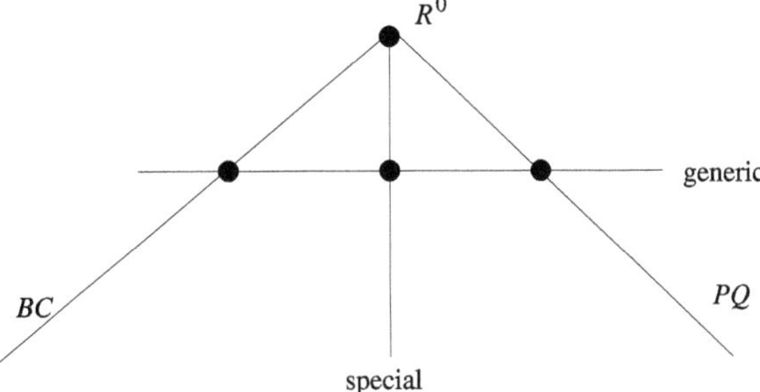

Fig. 8.3 A special and a generic line

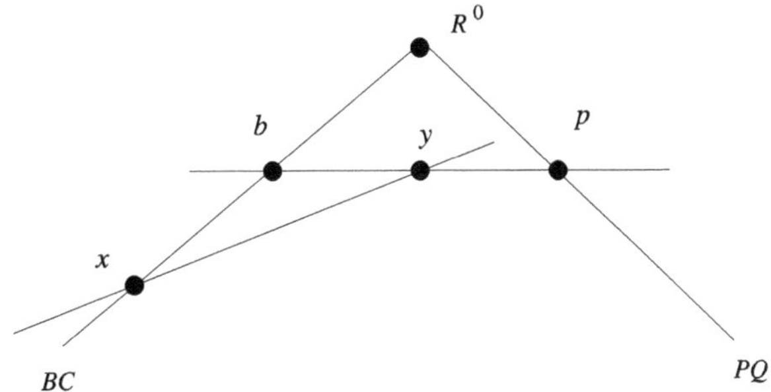

Fig. 8.4 More generic lines

Proof Let y be on the generic line bp with $b \in B \cap C \setminus R^0$ and $p \in P \cap Q \setminus R^0$ and assume $y \neq b, p$. Without loss of generality, we may assume $x \in B \cap C$ and $b \neq x$.

Suppose $x \perp b$. Now bx intersects P in a point, a say, of $P \cap B \cap C = R^0$. Moreover, $x \perp p$, so $bx \perp p$, whence $a \perp p$. As $p \notin A$, we have $a \neq p$. Now ap intersects the line xy in a point of $P \cap Q \setminus R^0$, so xy is a generic line; cf. Fig. 8.4.

Let $x \not\perp b$, so $x \not\perp p$. We investigate $y^\perp \cap B \cap C$. We know that y is not in B (for otherwise, $p \in B \cap P \cap Q = R^0$).

Since $y^\perp$ contains a singular plane and since it has a unique point in its radical by Lemma 7.4.8(iii), $y^\perp$ is not a rosette. Hence it is amply connected by Theorem 8.1.4. Then $y^\perp \cap B$ is a geometric hyperplane of $y^\perp$ not containing y, of rank at least $r - 1$. Moreover $\mathrm{Rad}(y^\perp \cap B) = \emptyset$ for if $a \in \mathrm{Rad}(y^\perp \cap B)$ then $a^\perp \supseteq y^\perp$, hence $a = y$ and $y \in B$, a contradiction.

By Lemma 8.1.5, $y^\perp \cap B \cap C$ is a geometric hyperplane of $y^\perp \cap B$ or coincides with it. As $y^\perp \cap B$ is nondegenerate of rank at least three, Corollary 8.2.5 yields that $y^\perp \cap B \cap C$ is amply connected. In particular, there is a path from b to x inside $(y^\perp \cap B \cap C) \setminus A$. Now, repeated application of the argument for the case $x \perp b$ leads to the desired result. This ends the proof of the lemma. $\qquad\square$

Lemma 8.3.8 *If Hypothesis 8.3.4 holds and* $y \in R^2$, *then* $y^\perp \cap B \cap C$ *is a geometric hyperplane of* $B \cap C$ *that is a polar space of rank at least two.*

Proof As $y \in R^2$, there are $b \in B \cap C \setminus R^0$ and $p \in P \cap Q \setminus R^0$ distinct from y such that bp is a line of Z containing y.

Suppose $B \cap C \subseteq y^{\perp}$. Then also $\langle R^0, p \rangle \subseteq y^{\perp}$. But R^0 is a geometric hyperplane of $P \cap Q$, and $P \cap Q$ is amply connected (by Corollary 8.2.5 and Lemma 8.3.3), so R^0 is a maximal subspace of $P \cap Q$ (cf. Proposition 8.1.6). As $p \notin R^0$, this implies $P \cap Q = \langle R^0, p \rangle \subseteq y^{\perp}$, so $(B \cap C) \cup (P \cap Q) \subseteq y^{\perp}$, contradicting Hypothesis 8.3.4. We conclude that $y^{\perp} \cap B \cap C$ is a geometric hyperplane of $B \cap C$.

Suppose $w \in \mathrm{Rad}(y^{\perp} \cap B \cap C)$. Then $(wy)^{\perp} \supseteq y^{\perp} \cap B \cap C$. There is $y' \in wy$ with $y'^{\perp} \supseteq B \cap C$. (For if $y' = w$ fails, there exists $q \in B \cap C \setminus w^{\perp}$, and we can take $y' \in wy$ such that $\{y'\} = wy \cap q^{\perp}$; then $y'^{\perp} \supseteq \langle y^{\perp} \cap B \cap C, v \rangle = B \cap C$ as $B \cap C$ is amply connected.) But then $y'^{\perp} \supseteq \langle B \cap C, p \rangle \supseteq B \cap C \cup \langle R^0, p \rangle = (B \cap C) \cup (P \cap Q)$ since $P \cap Q$ is amply connected and $p \in \langle b, y, w \rangle \subseteq y'^{\perp}$, and so y' is a point contradicting Hypothesis 8.3.4. Therefore, $\mathrm{Rad}(y^{\perp} \cap B \cap C) = \emptyset$. By the same argument as in the proof of Lemma 8.3.7, the subspace $y^{\perp} \cap B \cap C$ contains a line, so it has rank at least two. $\square$

By symmetry between $B \cap C$ and $P \cap Q$, we have the same conclusions with $B \cap C$ replaced by $P \cap Q$.

Lemma 8.3.9 *If $(i, j) = (0, 2)$ and Hypothesis 8.3.4 holds, then the line xy is special or extraspecial.*

Proof Let y be on the generic line bp with $b \in B \cap C \setminus R^0$ and $p \in P \cap Q \setminus R^0$ such that $y \neq b, p$. Notice that $b \neq x$.

Suppose that $x \perp b$. Then $\langle b, x, y \rangle$ is a singular plane containing p, so, for every $z \in xy$, the line bz intersects px, whence $xy \setminus \{x\} \subseteq R^2$, and xy is a special line.

Therefore, we (may) assume $x \not\perp b$ and so $x \not\perp p$. By Lemma 8.3.8, the subspace $y^{\perp} \cap B \cap C$ contains a line on b, which, by the polar space property for $B \cap C$, contains a point $u \in x^{\perp}$. Thus, $u \in \{y, x, b\}^{\perp} \cap B \cap C$. If $u \notin R^0$, then uy is a generic line by Lemma 8.3.7 and, by the previous paragraph, xy is a special line.

Thus we are led to the case where $\{y, x, b\}^{\perp} \cap B \cap C \subseteq R^0$. Now each line on x in $y^{\perp} \cap B \cap C$ has a second point (its intersection with $b^{\perp}$) in R^0, so $(xy)^{\perp} \cap B \cap C \subseteq y^{\perp} \cap R^0 \subset y^{\perp} \cap B \cap C$. (Here, the existence of b forces that the latter inclusion is proper.) By Theorem 8.1.4 and Lemma 8.3.8, $y^{\perp} \cap B \cap C$ is amply connected. But $(xy)^{\perp} \cap B \cap C$ is a geometric hyperplane in $y^{\perp} \cap B \cap C$, so Corollary 8.1.8 forces that it coincides with $y^{\perp} \cap R^0$. We have shown

$$(xy)^{\perp} \cap B \cap C = y^{\perp} \cap R^0. \tag{8.1}$$

By Lemma 8.3.8, the space $y^{\perp} \cap B \cap C$ is nondegenerate. We claim that this implies

$$\mathrm{Rad}\big(y^{\perp} \cap R^0\big) = \{x\}. \tag{8.2}$$

For, x belongs to this radical and if l is a line in $\mathrm{Rad}(y^{\perp} \cap R^0)$, then there is $u \in b^{\perp} \cap l$. So, by Corollary 8.1.8 again, $u^{\perp} \supseteq \langle b, (xy)^{\perp} \cap B \cap C \rangle = y^{\perp} \cap B \cap C$, contradicting that $y^{\perp} \cap B \cap C$ has empty radical.

We next claim that

$$\text{for each } v \in (xy)^{\perp} \cap B \cap C \setminus \{x\}, \quad \text{the line } vy \text{ is special.} \tag{8.3}$$

Indeed, if vy is not special, then, arguing as above with v in the role of x, we find $\{y, v, c\}^\perp \cap B \cap C \subseteq R^0$ for a suitable point $c \in y^\perp \cap B \cap C \setminus R^0$, and derive (8.2) with v replacing x, which implies that $v = x$, a contradiction with the choice of v.

Now fix $v \in (xy)^\perp \cap B \cap C \setminus \{x\}$ and let α be the plane $\langle x, y, v \rangle$. We will show that all points of $\alpha \setminus xv$, with a possible exception of a single point of xy, are in R^2. Let $y' \in \alpha \setminus (xy \cup xv)$. Then $y' \in R^2$ (by (8.3) applied to the unique point in $yy' \cap xv$ instead of v). For $a \in xv \setminus \{x\}$, it follows from (8.3), (8.1), and (8.2) that $ay' \setminus \{a\} \subseteq R^2$ with a possible exception if a is the unique point in $\mathrm{Rad}(y'^\perp \cap R^0)$. So all points of $yx \setminus \{x\}$ but at most one are in R^2.

Assume that xy is not special. There is a unique point $z \in xy \setminus R^2$ with $z \neq x$. In order to prove that xy is extraspecial, we need to derive $z \in R^3$.

We claim $z \in (R^0)^\perp$. For each $y' \in \alpha \setminus (xy \cup xv)$, we have $zy' \cap xv = \mathrm{Rad}(y'^\perp \cap R^0)$, which implies $y'^\perp \cap R^0 \subseteq (zy' \cap xv)^\perp \cap y'^\perp \cap R^0 \subseteq z^\perp \cap R^0$.

As every point of R^0 is collinear with a line of α and hence either with a point y' as described or with xv or xy, and as $(xy)^\perp \cap R^0 \subseteq z^\perp \cap R^0$, we have

$$R^0 \setminus (xv)^\perp \subseteq \left((xy)^\perp \cap R^0\right) \cup \bigcup_{y' \in \alpha \setminus (xv \cup xy)} y'^\perp \cap R^0 \subseteq z^\perp \cap R^0.$$

By Corollary 8.1.7, the claim can only fail to hold if R^0 is a rosette. Then $x \in \mathrm{Rad}(R^0)$ because $y^\perp \cap R^0$ is a subspace of R^0 with radical $\{x\}$ that contains lines. But now we can argue similarly with pairs x', y' such that $y' \in \alpha \setminus (xy \cup vx)$ and $\{x'\} = xv \cap zy'$ replacing x, y to obtain that every $x' \in xv$ also belongs to $\mathrm{Rad}(R^0)$. This means $xv \subseteq \mathrm{Rad}(R^0)$.

By (8.1) and (8.2), the subspace $y^\perp \cap R^0$ contains a point $v' \in (xy)^\perp \cap B \cap C$ not collinear with v, so we can argue with v' instead of v to derive $xv' \subseteq \mathrm{Rad}(R^0)$. But then v and v' are non-collinear points of $\mathrm{Rad}(R^0)$, a contradiction. We conclude $R^0 \subseteq z^\perp$. This settles the claim $z \in (R^0)^\perp$.

Let $a \in R^0$ and $w \in az \setminus \{a, z\}$. In order to prove $z \in R^3$, it remains to show $w \in R^2$. If $a \perp x$, then we see from the preceding arguments that $w \in R^2$, as required. If $a \not\perp x$, then choose $x' \in vx \cap a^\perp$. On $x'z \setminus \{x', z\}$ there is a point $y' \in R^2$. Replacing x, y by x', y' in the preceding arguments, we find $w \in R^2$, so $z \in R^3$. This ends the proof of the lemma. $\square$

Lemma 8.3.10 *If $\{i, j\} = \{2\}$ and Hypothesis 8.3.4 holds, then xy is generic, special, or extraspecial.*

Proof Let y be on the generic line bp, and x on the generic line $b'p'$, where $b, b' \in B \cap C \setminus R^0$ and $p, p' \in P \cap Q \setminus R^0$. Thus we have the configuration of Fig. 8.5. Observe that, by Lemma 8.3.7, every line on x or y meeting $B \cap C$ in a point outside R^0 is generic.

If $b = b'$, then $p \perp p'$ and, for $z \in xy$, the line bz intersects pp', hence bz is generic unless $z \in B \cap C$. In particular, $z \in R^0 \cup R^1 \cup R^2$, and so $xy \subseteq R^0 \cup R^1 \cup R^2$, which implies that xy is generic or special.

From now on, we (may) assume $b \neq b'$ and $p \neq p'$, and $(xy)^\perp \cap B \cap C \subseteq R^0$.

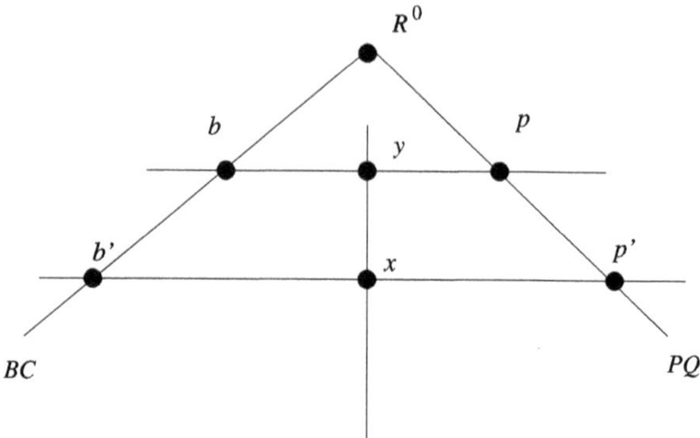

Fig. 8.5 Two generic lines

As $\mathrm{rk}(Z) \geq 4$, there is a line in $y^\perp \cap B \cap C$ on b, and so $(yxb)^\perp \cap B \cap C$ is non-empty. Let $u \in (yxb)^\perp \cap B \cap C$. Then $u \in R^0$ and Lemma 8.3.9 shows that ux and uy are special or extraspecial.

Assume that there exists $y' \in yu \setminus \{y, u\}$ such that $b' \perp y'$. If $(y')^\perp \supseteq R^0$, then $(y')^\perp$ contains R^0 and b', whence $B \cap C$. But $y' \perp b'x$, so $(y')^\perp$ also contains p', whence $P \cap Q$, so y' contradicts Hypothesis 8.3.4. Therefore, $y' \in R^2$. By the above case where $b = b'$, applied here to b', x, and y', the line xy' is generic or special. If xy' is generic with points $b'' \in B \cap C \setminus R^0$ and $p'' \in P \cap Q \setminus R^0$, then xy meets the lines up'' and ub'' of the singular plane $\langle ub'', p'' \rangle$, so xy is generic by Lemma 8.3.7. If xy' is special, then xy' has a point $a' \in R^0 \setminus u$ and xy has a point $a'' \in R^0$, on the line ua'. Now $xy = xa''$ is special or extraspecial by Lemma 8.3.9. As $b' \not\perp y$, the only case not dealt with is where u is the only point of uy collinear with b'.

But b' can be taken to vary over all points of $x^\perp \cap B \cap C \setminus R^0$. Therefore, it remains to consider the case where $x^\perp \cap B \cap C \setminus R^0 \subseteq u^\perp$. By Lemma 8.3.8 and Corollary 8.1.8 applied to the geometric hyperplane $x^\perp \cap R^0$ of $x^\perp \cap B \cap C$, we find $x^\perp \cap B \cap C \subseteq u^\perp$. But now $B \cap C = \langle x^\perp \cap B \cap C, b \rangle$ and $P \cap Q = \langle R^0, p \rangle$ also belong to $u^\perp$ (as, by Lemma 8.3.3, $B \cap C$ and $P \cap Q$ are not rosettes), contradicting Hypothesis 8.3.4. $\qquad\square$

Lemma 8.3.11 *If $z \in R^3$, then $z^\perp \cap B \cap C = R^0$. In particular, $(i, j) \neq (1, 3)$.*

Proof This is immediate from the definition of R^3. $\qquad\square$

Lemma 8.3.12 *If $(i, j) = (0, 3)$, then xy is extraspecial.*

Proof This is obvious from the definition of R^3. $\qquad\square$

Lemma 8.3.13 *If $(i, j) = (2, 3)$ and Hypothesis 8.3.4 holds, then xy is extraspecial.*

Proof Let $a \in R^0$, so ya is extraspecial. Suppose $x \perp a$. We have $x \perp ay$. Let $y' \in ay \setminus \{a, y\}$, so $y' \in R^2$. If $x = y'$, then $xy = ay$ is extraspecial and we are done, so we (may) assume $x \neq y'$. In view of Lemma 8.3.10, the line xy' is generic, special, or extraspecial. In the first case, $y^\perp$ contains R^0 and xy', whence it contains R (cf. Lemma 8.3.3), a contradiction with Hypothesis 8.3.4. In the second and third case, xy is extraspecial by Lemma 8.3.9.

Assume, therefore, $x \not\perp a$. By definition of R^3, there is a line in R^0 on a; this line contains a point $a' \in R^0$ collinear with x. Now ya' is extraspecial and $x \perp a'$, so, by the above paragraph with a' instead of a, we find that xy is extraspecial. $\square$

Lemma 8.3.14 *If Hypothesis* 8.3.4 *holds, then* $(i, j) \neq (3, 3)$.

Proof Assume $(i, j) = (3, 3)$ and let $a \in R^0$. Then $a \perp y$, and, by Lemma 8.3.12, ay is extraspecial. Take $x' \in ax \setminus \{x, a\}$. Then $x' \in R^2$ as $x \in R^3$. Also, $y \perp \{x, a\}$, so $y \perp x'$. As xy has more than one point of R^3, it does not meet R^0. This implies that y is not on the line xa; in particular, $y \neq x'$. According to Lemma 8.3.13, the line $x'y$ is extraspecial. Consequently, $x'y$ contains a point a' in R^0 distinct from a, which lies in the singular plane $\alpha = \langle x, y, a \rangle$. This plane meets R^0 in the line aa', so $xy \cap aa'$ is a point of R^0 on the line xy, which contradicts that xy does not meet R^0. $\square$

So far, we have established that R is a subspace.

Lemma 8.3.15 *The subspace* R *is a geometric hyperplane of* Z.

Proof Again, we (may) work under Hypothesis 8.3.4. Assume that $l \cap R = \emptyset$. Let α be a singular plane on l. Then $\alpha \cap B \cap C$ and $\alpha \cap P \cap Q$ must be points. If they were distinct, then the line in α joining them would be a generic line meeting l, contradicting the assumption on l. Thus, $\alpha \cap B \cap C = \alpha \cap P \cap Q = \alpha \cap R^0$ is a single point outside l.

For $x \in B \cap C \setminus A$, consider the plane $\beta = \langle x, x^\perp \cap \alpha \rangle$. If $x \not\perp \alpha \cap R^0$, then $\beta \cap \alpha$ is a line disjoint from R, so, by the first paragraph, $\{x\} = \beta \cap B \cap C = \beta \cap R^0$, contradicting $x \notin A$. Hence $x \in (\alpha \cap R^0)^\perp$, proving $B \cap C \setminus A \subseteq (\alpha \cap R^0)^\perp$. By Theorem 8.2.4 and Lemma 8.3.3 (the part that $B \cap C$ is not a rosette), we obtain $B \cap C \subseteq (\alpha \cap R^0)^\perp$. Similarly, $P \cap Q \subseteq (\alpha \cap R^0)^\perp$. This, however, contradicts Hypothesis 8.3.4.

We have shown that every line of Z either has exactly one point in R or is entirely contained in R. Suppose $R = Z$. Every line of Z is either generic, special, or extraspecial. In particular, $B \cap C$ is a geometric hyperplane of Z, so $B = C$, contrary to the assumptions.

The conclusion is that R is a geometric hyperplane of Z, as required. $\square$

Theorem 8.3.16 *If* Z *is a thick nondegenerate polar space of rank at least four, then its Veldkamp space* $\mathcal{V}(Z)$ *is a projective space into which it embeds via the map* $x \mapsto x^\perp$.

Proof By Theorem 8.2.9 the map $x \mapsto x^\perp$ is an embedding of Z into $\mathcal{V}(Z)$. Given five distinct points A, B, C, P, Q as in Notation 8.3.1, the construction of Definition 8.3.6 gives a subset of Z, which is a geometric hyperplane by Lemma 8.3.15.

This shows that R is a point of $\mathcal{V}(Z)$ and so Pasch's axiom is satisfied. It follows from Theorem 5.2.6 that $\mathcal{V}(Z)$ is a projective space. $\square$

Example 8.3.17 Starting from the generalized quadrangle Z of absolute points and lines in $\mathbb{P}(\mathbb{F}^4)$ with respect to a nondegenerate symplectic polarity π, another generalized quadrangle, denoted Z_p, can be obtained by deleting all points collinear with some fixed point p of Z and adding all projective lines containing p but not contained in Z. More precisely, denote by P the point set $\mathbb{P}(\mathbb{F}^4) \setminus p^{\perp}$, where $\perp$ is the collinearity relation in Z, and by L the union of the set of lines of Z not containing p and the set of lines of $\mathbb{P}(\mathbb{F}^4)$ on p that do not belong to Z. Then $Z_p = (P, L)$ is the new generalized quadrangle. Proving this is Exercise 8.6.13.

If $\mathbb{F} = \mathbb{F}_q$, then the order of the new generalized quadrangle is $(q - 1, q + 1)$. In particular, if $q - 1$ is not a prime power (an example occurs for $q = 7$), then Z cannot be embedded in a thick projective space (cf. Theorem 6.3.1).

8.4 Projective Embedding for Rank Three

In this section, we prove the analogue of Theorem 8.3.16 for the case of rank three. It is stated in Theorem 8.4.25. Since every projective space in which a nondegenerate polar space is embedded has dimension at least three (cf. Exercise 8.6.9), the singular subspaces of the polar space are necessarily Desarguesian. If the singular planes of a polar space are non-Desarguesian and the polar geometry (cf. Theorem 7.5.8) is not thick, then Exercise 8.6.15 shows that we need not have a projective embedding.

Therefore, similarly to Notation 8.3.1, we adopt the following setting.

Notation 8.4.1 Throughout this section, we take Z to be a nondegenerate polar space of rank three all of whose singular planes are Desarguesian and whose corresponding polar geometry is thick. In particular, every line is on at least three planes.

Let A, B, C, P, Q be distinct points of $\mathcal{V}(Z)$ such that $\{A, P, B\}$ and $\{A, Q, C\}$ are collinear triples spanning distinct lines.

In this extremely long section, we show that the Veldkamp space $\mathcal{V}(Z)$ is again a projective space. If the lines of Z have size three, then Corollary 8.2.5 and Theorem 8.2.8 give the required result. Therefore, we (may) assume that all lines of Z have size at least four.

In view of Theorem 5.2.6 we need only show that the Veldkamp lines PQ and BC have a common point R (that is, Pasch's axiom 5.2.4). This will be done in a series of lemmas and a proposition leading up to Theorem 8.4.25.

The construction of R and the verification of its properties will be heavily based on the use of singular planes in Z, to which we will generally (in this section) refer as planes.

Notation 8.4.2 For planes α of Z, we abbreviate $\alpha \cap A$, $\alpha \cap B \cap C$, $\alpha \cap A \cap B \cap C$ to α_A, α_{BC}, α_{ABC}, respectively.

Lemma 8.4.3 *The points A, B, C, P, and Q of $V(Z)$ satisfy the following properties.*

(i) $A \cap P \cap Q = A \cap B \cap C = (B \cap C) \cap (P \cap Q)$.
(ii) *If α is a singular plane in Z, then $\alpha_{ABC} = \alpha_{APQ} = \alpha_{ABCPQ}$.*

Proof Statement (i) is due to Theorem 8.2.4, the fact that $A \cap P$ is contained in B, $A \cap Q$ in C, etc. Statement (ii) is a direct consequence of (i). $\qquad\square$

Lemma 8.4.4 *None of A, B, C, P, Q is a rosette. In particular, $A \setminus B$ generates A, the subset $P \setminus Q$ generates P, and so on.*

Proof The first statement immediately follows from Corollary 8.1.7. The second statement repeats Corollary 8.1.7. $\qquad\square$

For the remainder of this section, we will assume Hypothesis 8.3.4: there is no point x of Z such that $x^{\perp}$ contains $B \cap C$ and $P \cap Q$. This assumption is allowed, for otherwise we can take $R = x^{\perp}$.

Now we construct R as a set.

Notation 8.4.5 For any plane α of Z, define a sequence of subsets

$$\alpha_1 \subseteq \alpha_2 \subseteq \alpha_3 \subseteq \cdots \subseteq \alpha,$$

as follows.

(1) First, put $\alpha_1 = \alpha_{BC} \cup \alpha_{PQ}$. Observe that α_{BC} and α_{PQ} are non-empty but that α_1 may be a singleton.
(2) Next define α_2 as the union of α_1 and of all lines joining two points in α_1. By Lemma 8.4.3, if $x \in \alpha_2 \setminus \alpha_1$, then x is on a line bp with $b \in B \cap C \setminus A$ and $p \in P \cap Q \setminus A$. Observe that α_2 is a linear subspace of α, hence a point, a line, or α itself.
(3) Finally for $n \geq 2$, define α_{n+1} as the union of α_n and of all lines xy with $x \in \alpha_n$ and $y \in \alpha \cap \beta_n$ for some plane β of Z adjacent to α (i.e., such that $\alpha \cap \beta$ is a line).

The set R is the union of all α_n ($n \in \mathbb{N}$, $n \geq 1$) where α runs over all planes of Z.

Definition 8.4.6 For adjacent planes α and β we say that β *contributes* to α_n if $(\alpha \cap \beta_{n-1}) \setminus \alpha_{n-1}$ is non-empty.

So far, we cannot claim that each α_n is a linear subspace of α but later on this will turn out to be true for $n \geq 3$. On the other hand, an easy inductive argument shows that a geometric hyperplane of Z containing $B \cap C$ and $P \cap Q$ must indeed contain the set R.

Our first goal is to establish that R is a subspace of Z.

Lemma 8.4.7 *Let $d \geq 1$. If α and β are planes of Z at mutual distance d in the dual polar space and if $x \in \alpha \cap \beta_n$ for some $n \geq 2$, then $d \leq 2$ and $x \in \alpha_{n+d}$.*

Proof First of all, $d \leq 2$ because $x \in \alpha \cap \beta$. For $d = 1$, let a be a point in α_1. Then a is also in α_n and so $x \in xa \subseteq \alpha_{n+1}$. For $d = 2$, there is a plane γ adjacent to both α and β, with $x \in \alpha \cap \beta \cap \gamma$. Thus, $x \in \gamma_{n+1}$ by the preceding argument and, for the same reason, $x \in \alpha_{n+2}$. $\qquad\square$

Lemma 8.4.8 *The set R is a subspace of Z.*

Proof Let x, y be distinct points in R with $x \perp y$ and let α be a plane containing xy. By the construction of R, there is a plane γ and a positive integer n such that $x \in \gamma_n$. By Lemma 8.4.7, $x \in \alpha_{n+2}$. Similarly, $y \in \alpha_{m+2}$ for some m. We may assume that $m \leq n$ and so $x, y \in \alpha_{n+2}$. Let δ be a plane on x and y, adjacent to α. Lemma 8.4.7 shows that $x \in \delta_{n+3}$, whence xy is contained in α_{n+4} by Lemma 8.4.7. This shows $xy \subseteq R$, so R is a subspace of Z. $\qquad\square$

Our second goal is to show that every line intersects R non-emptily. This will be achieved in Proposition 8.4.11. We use the following partition of the planes of Z into four types.

Lemma 8.4.9 *Each plane α of Z has one of the following four types.*

type I: $\alpha_2 = \alpha$,
type II: $\alpha_2 = \alpha_1$ *is a line,*
type III: α_2 *is a line and α_1 consists of the two distinct points α_{BC} in $B \cap C \setminus A$ and α_{PQ} in $P \cap Q \setminus A$,*
type IV: $\alpha_2 = \alpha_1$ *is a point in α_{ABC}.*

If α is of type I, then both α_{BC} and α_{PQ} contain a line, while α_{ABC} is non-empty.
If α is of type II, then one of α_{BC}, α_{PQ} contains the other set and contains a line, while α_{ABC} is non-empty.

Proof Straightforward. $\qquad\square$

The next lemma not only prepares for our second goal but also for the crucial existence of planes of type III (in Lemma 8.4.13).

Lemma 8.4.10 *Let α be a plane of type IV. One of the following holds.*

(i) *There is a plane β of type III which is adjacent to α and which contributes a point to α_3 other than α_{ABC}.*
(ii) *There are planes β, γ with β adjacent to both α and γ, with β of type IV, γ of type III, and β contributing a point to α_4 other than α_{ABC}.*

Proof There exists a point x in $B \cap C \setminus A$, for otherwise $B \cap C \subseteq A \cap B \cap C \subseteq P \cap Q$, so P is common to the lines PQ and BC in $\mathcal{V}(Z)$. Assume first that x

Fig. 8.6 Argument of
Lemma 8.4.10

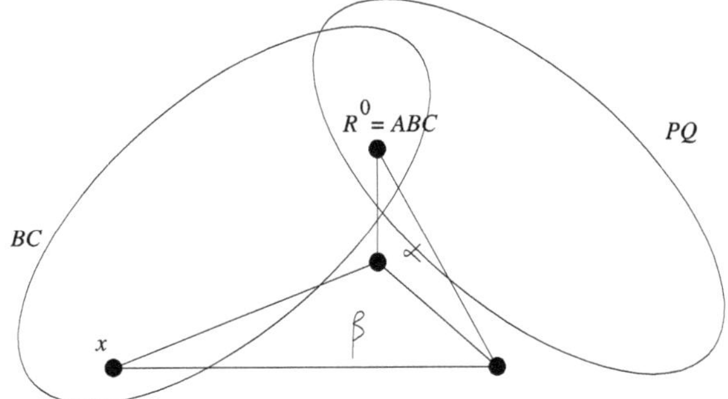

is not in $\alpha_{ABC}^{\perp}$ and let β be the plane $\langle x^{\perp} \cap \alpha, x \rangle$. (See Fig. 8.6.) Then $\beta_{BC} = x$,
for otherwise β_{BC} contains a line and so contains a point of α which is necessarily
α_{ABC}, a contradiction. Similarly there is a unique point in β_{PQ} which cannot be in
$B \cap C$, for otherwise it would be x, which would belong to $A \cap B \cap C$. Therefore,
β is of type III, β_2 intersects α in a point and (i) holds.

If x is in $\alpha_{ABC}^{\perp}$ for each $x \in B \cap C \setminus A$, then we may also assume, for reasons
of symmetry, that $\alpha_{ABC}^{\perp}$ contains $P \cap Q \setminus A$. By Hypothesis 8.3.4, we may further
assume that there is a point $y \in B \cap C$ with y not in $\alpha_{ABC}^{\perp}$ (for otherwise, after a
symmetry argument, $(B \cap C) \cup (P \cap Q) \subseteq \alpha_{ABC}^{\perp}$). Then $y \in A \cap B \cap C$.

Let β be the plane $\langle y^{\perp} \cap \alpha, y \rangle$, which is adjacent to α. The plane β is necessarily
of type IV. Take $x \in B \cap C \setminus A$. As lines are thick, the line $x\alpha_{ABC}$ has a point
$z \neq \alpha_{ABC}$ with z outside $y^{\perp}$. Let γ be the plane $\langle z^{\perp} \cap \beta, z \rangle$. Since $z \in B \cap C \setminus A$,
the plane γ must be of type III, γ_2 contributes a point to β_3, say t, and ty lies in β_3,
hence β contributes a point to α_4, and (ii) holds.   □

Proposition 8.4.11 *Every line of Z intersects R in some point.*

Proof Let l be a line and α a plane containing l in Z. It suffices to show that α
intersects R in at least one line. If α is of one of the types I, II, III, then this is
obvious and if α is of type IV, then it follows from Lemma 8.4.10.   □

Now that our second goal has been achieved, we come to the third and hardest
goal, namely to show that R is a proper subspace of Z. Once this is established, we
know that R is a geometric hyperplane of Z by Lemma 8.4.8 and Proposition 8.4.11.

The strategy will be to prove that for every plane α of type III, $\alpha_2 = \alpha_n$ for all
$n \geq 2$, which forces $\alpha \cap R$ to be a line by Lemma 8.4.9. This will end the proof,
provided we can show that there is at least some plane of type III. We will do this
first of all.

Lemma 8.4.12 *There exists a plane α intersecting P and B in a line and $P \cap Q$ in
precisely a point. Each such plane intersects Q in a line as well.*

Proof Let $p \in P \setminus Q$ which is a non-empty set (for otherwise $P = Q$ by Corollary 8.1.8). Actually, we can choose p in such a way that $p^{\perp} \neq P$ (for there is at

most one point p_0 of P such that $P = p_0^\perp$, in which case a point $p \in P \setminus (Q \cup \{p_0\})$ is easily found). Let $q \in p^\perp \cap Q$ be such that pq does not contain a point $p_0 \in P$ with $p_0^\perp = P$. At least one plane α containing pq is such that α_P is a line and α_{PQ} is a point. Assume, by way of contradiction, that each such plane is in B for each choice of p and q as above. Then the subspace generated by $P \setminus Q$ is in B, and so $P \subseteq B$ by Lemma 8.4.4, a contradiction. Therefore, the first statement holds. The second statement also holds, for otherwise α would intersect Q in a plane and so have a line in $P \cap Q$, contradicting the first statement. $\qquad\square$

Lemma 8.4.13 *There exists a plane of type III.*

Proof By Lemma 8.4.10, it suffices to show that there is either a plane of type III or a plane of type IV. Let α be a plane as in Lemma 8.4.12. We assume, by way of contradiction, that α is not of type III and not of type IV. Since α_{PQ} is a singleton, α is of type II, $\alpha_{ABC} = \alpha_{PQ}$, and α_{BC} is a line containing the point α_{PQ}. After an interchange of B and C, if needed, we can find a point $q \in B \setminus C$ that is not in $\alpha_{ABC}^\perp$ (for otherwise, $\alpha_{ABC}^\perp$ would contain B and C, and this would force $B = C = \alpha_{ABC}^\perp$, a contradiction). If $q \in P$, then q is adjacent to some point $p \neq \alpha_{ABC}$ on α_{BC} and the line pq contains some point q' distinct from q and p. Then q' is outside P because p is not in P. Moreover q' is not in $\alpha_{ABC}^\perp$ and $q' \in B \setminus C$. Therefore, we may assume that q is not in P. Furthermore q is outside α, since q is not in $\alpha_{ABC}^\perp$. Then $\beta = \langle q^\perp \cap \alpha, q \rangle$ is a plane adjacent to α and $\alpha \cap \beta$ intersects both B and P in a unique point (for otherwise $\alpha_{ABC} \subseteq \alpha_B \cap \alpha_P \subseteq \alpha \cap \beta$ and $q \perp \alpha_{ABC}$). Therefore, β intersects both P and B in exactly one line and $P \cap Q$ in a point. By Lemma 8.4.12, β intersects Q in a unique line. As β_B is a line and $q \notin C$, the subset β_{BC} is a unique point. Since β_{PQ} and β_{BC} are points, β is of type III or IV and we have finished. $\square$

Lemma 8.4.14 *If α is a plane of type III, then $\alpha_A, \alpha_B, \alpha_C, \alpha_P, \alpha_Q$ are lines and $\alpha_{ABP}, \alpha_{ACQ}, \alpha_{BC}, \alpha_{PQ}$ are points such that any three of them are non-collinear.*

Proof Since α_{BC} and α_{PQ} are distinct points, $\alpha_B, \alpha_C, \alpha_P, \alpha_Q$ are lines. If $\alpha_A = \alpha$, then α_{AB} is a line and α_{ABC} is non-empty, a contradiction. For the same reason, $\alpha_{ABP} = \alpha_{AB}$ and $\alpha_{ACQ} = \alpha_{AC}$ are points. If three of the given points are collinear we may assume, for reasons of symmetry, that either $\alpha_{ABP}, \alpha_{BC}, \alpha_{ACQ}$ are collinear or that $\alpha_{PQ}, \alpha_{ABP}, \alpha_{BC}$ are collinear. In the first case, $\alpha_{BC} \in A$ or $\alpha_{ABP} = \alpha_{ACQ}$ and in both cases, α_{ABC} is non-empty; a contradiction. In the second case we find similarly that α_{ABC} is non-empty. $\qquad\square$

For showing that $R \neq Z$, it suffices to prove that $\alpha \cap R$ is a line for planes α of type III. (Such planes exist by Lemma 8.4.13.) We proceed to show that this condition holds.

Recall that if α is a plane of type III, we want to show that $\alpha_n = \alpha_2$ for all $n \geq 2$. Assuming that $x \in \alpha_n \setminus \alpha_{n-1}$, we have a plane α^1 adjacent to α, with α_{n-1}^1 contributing to α_n, a plane α^2 adjacent to α^1, with α_{n-2}^2 contributing to α_{n-1}^1, and so

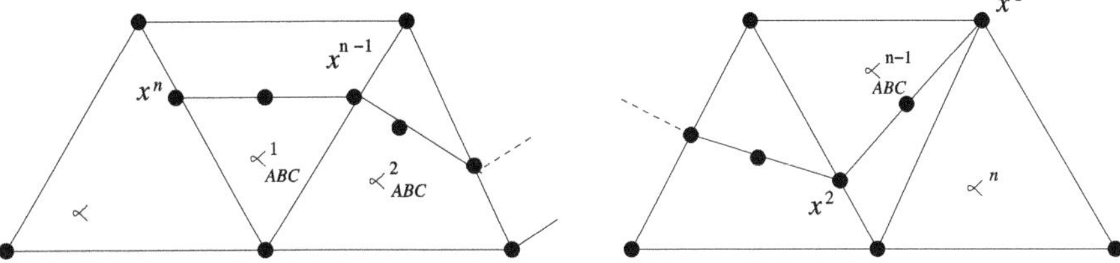

Fig. 8.7 A Veldkamp sequence

a sequence of planes in which any two consecutive terms are adjacent, ending with a contribution to α_3^{n-3} of α_2^{n-2}, where α^{n-2} is the last plane in the sequence. Among all possible configurations of such sequences, there is a particularly important one that we describe and study now.

Definition 8.4.15 Let the adjacency of planes be denoted by $\sim$. A *Veldkamp sequence* of length $n + 1$ is a sequence $\alpha, \alpha^1, \alpha^2, \ldots, \alpha^n$ of consecutively adjacent planes (so $\alpha^{i-1} \sim \alpha^i$ for $i \in [n]$, where $\alpha^0 = \alpha$) with $n \geq 1$, such that

(i) α and α^n are of type III;

(ii) α^i is of type IV for each $i \in [n - 1]$;

(iii) α_{ABP}, α_{BC}, α_{PQ}, and α_{ACQ} are not in α^1, and α_{ABC}^i is not in α^{i+1} for each $i \in [n - 1]$.

Notation 8.4.16 Given a Veldkamp sequence $\alpha, \alpha^1, \ldots, \alpha^n$, we construct a sequence of points as follows. In α^n, there is a unique point $x^1 \in \alpha_2^n \cap \alpha^{n-1}$ which contributes to α_3^{n-1}. Since $\{x^1\} \neq \alpha_{ABC}^{n-1}$, the line $x^1 \alpha_{ABC}^{n-1}$ intersects α^{n-2} in a unique point $x^2 \in \alpha_3^{n-1} \cap \alpha^{n-2}$. Recursing in this way, we find a unique sequence of points $x^1, x^2, \ldots, x^n$ such that, for each $i \in \int n - 1$,

(1) $x^n \in \alpha^1 \cap \alpha$ and $x^i \in \alpha^{n-i+1} \cap \alpha^{n-i}$;

(2) x^{i+1}, α_{ABC}^{n-i}, and x^i are collinear in α^{n-i};

(3) $x^i \in \alpha^{n-i} \cap \alpha_{i+1}^{n-i+1} \setminus \alpha_{i+1}^{n-i}$ and $x^n \in \alpha \cap \alpha_{n+1}^1 \setminus \alpha_{n+1}$.

The notion of a Veldkamp sequence and the associated sequence of points x^i ($i \in [n]$) is illustrated in Fig. 8.7.

We now come to the role of Desargues' axiom.

Lemma 8.4.17 *Let $\alpha, \alpha^1, \ldots, \alpha^n$ be a Veldkamp sequence. If $x^1, \ldots, x^n$ is the underlying sequence of points satisfying* (1), (2), (3), *then $x^n \in \alpha_2$.*

Proof We will construct an isomorphism from α^n to α, mapping α_{BC}^n, α_{PQ}^n, x^1 onto α_{BC}, α_{PQ}, x^n, respectively, from which we then conclude $x^n \in \alpha_2$. In order to simplify notation, put $\alpha^0 = \alpha$. We define π as a product $\pi_0 \circ \phi_1 \circ \ldots \circ \phi_{n-2} \circ \pi_{n-2} \circ \phi_{n-1} \circ \pi_{n-1}$ where π_i is an isomorphism from α^{i+1} to α^i for $i \in \{0, \ldots, n - 1\}$ which fixes $\alpha^{i+1} \cap \alpha^i$ point-wise and ϕ_i is a perspectivity of α^i with center α_{ABC}^i

mapping $\alpha^i \cap \alpha^{i+1}$ on $\alpha^i \cap \alpha^{i-1}$ for each $i = n - 1, \ldots, 1$. The π_i exist in view of Corollary 7.7.7 and the ϕ_i exist by Desargues' axiom. Clearly, π maps x^1 to x^n. Moreover, π maps the points $\alpha_A^n \cap \alpha^{n-1}$, $\alpha_B^n \cap \alpha^{n-1}$, $\alpha_C^n \cap \alpha^{n-1}$, $\alpha_P^n \cap \alpha^{n-1}$, $\alpha_Q^n \cap \alpha^{n-1}$ to $\alpha_A \cap \alpha^1$, $\alpha_B \cap \alpha^1$, $\alpha_C \cap \alpha^1$, $\alpha_P \cap \alpha^1$, $\alpha_Q \cap \alpha^1$, respectively. By Desargues' axiom we can adjust π_{n-1}, hence π, in such a way that π maps α_{BC}^n to α_{BC}. Then π maps α_B^n, α_C^n to α_B, α_C, respectively.

The points $\alpha_{ABP}^n, \alpha_{ACQ}^n$ are not both on α^{n-1}, for otherwise α_A^{n-1} contains the two points. By Condition (iii) of Definition 8.4.15, this forces $n \geq 2$ and then α^{n-1} is of type IV and α_A^{n-1} contains three non-collinear points namely $\alpha_{ABC}^{n-1}, \alpha_{ABP}^n, \alpha_{ACQ}^n$. Thus, α^{n-1} is in A and similarly, $\alpha^{n-2}, \ldots, \alpha^1$ are in A, forcing α_{ABP} and α_{ACQ} on $\alpha \cap \alpha^1$, a contradiction again with (iii) of Definition 8.4.15.

Without loss of generality, we (may) assume that α_{ABP}^n is not in α^{n-1}. Consequently, α_B^n and α_P^n intersect α^{n-1} in distinct points. Applying π, we see that α_B and α_P intersect α^1 in distinct points as well. Desargues' axiom allows us once more to adjust π_{n-1} in such a way that π maps $\alpha_{ABP}^n \in \alpha_B^n$ to $\alpha_{ABP} \in \alpha_B$. Now π maps α_A^n to α_A and α_P^n to α_P, hence α_{ACQ}^n to α_{ACQ}, α_{PQ}^n to α_{PQ}, and α_2^n to α_2; this forces $\pi(x^1) = x^n \in \alpha_2$. $\qquad\square$

Lemma 8.4.18 *If α is a plane of type III, then $\alpha_3 = \alpha_2$.*

Proof Assume, by way of contradiction, that there is a plane β with $\alpha \sim \beta$ and that there is a point p in $\beta_2 \cap \alpha$ which is not on α_2. Then $\beta_2 \neq \beta_1$, so β is not of type II or IV. If β is of type I, then β_{BC} and β_{PQ} contain lines. If these two sets intersect α in distinct points, then the latter two points are α_{BC}, α_{PQ} and $\alpha_2 = \alpha \cap \beta$, a contradiction. If the two sets intersect α in a unique point, this is in α_{ABC}, another contradiction. Therefore, β is of type III. If α_{ABP} and α_{ACQ} are not on $\alpha \cap \beta$, then α, β is a Veldkamp sequence and by Lemma 8.4.17, $\beta_2 \cap \alpha = \alpha_2 \cap \beta$. Hence we may assume, without loss of generality, that $\alpha_{ABP} \in \alpha \cap \beta$.

Observe that $\alpha_2 \cap \beta \neq \alpha_{ABP} \neq \beta_2 \cap \alpha$, for otherwise B would contain α_{PQ}, contradicting that α is of type III. By the assumption on line length, β has order at least three (cf. Definition 2.2.16), so there is a line l on p in β that contains no point among $\beta_{BC}, \beta_{PQ}, \beta_{ABP}, \beta_{ACQ}$. There exists a plane γ intersecting β along l. It is of type III or IV. We deal with these two cases separately.

Suppose that γ is of type III. Then β, γ is a Veldkamp sequence and so, by Lemma 8.4.17, γ_2 contains p. Let δ be the plane $\langle \gamma_{BC}, \gamma_{BC}^{\perp} \cap \alpha \rangle$. It is of type I or III. If it were of type I, then $\delta \cap \alpha$ would be α_2 and so $p \in \alpha_2$, a contradiction. We conclude that δ is of type III. See Fig. 8.8 for an illustration of the configuration of planes.

If α, δ were a Veldkamp sequence, then $p \in \alpha_2$, so we must have $\alpha_{ACQ} \in \alpha \cap \delta$. The current configuration forces the order of α to be at least four. Indeed, if the order were three, then on the line $k := \alpha \cap \delta$ two of the five points $\alpha_2 \cap k$, $B \cap k$, $P \cap k$, α_{ACQ}, p would coincide, which would contradict one of the assumptions $p \in \alpha_2$ or α of type III. This implies of course that the order of γ is also at least four, so there is a line m on p in γ distinct from l and γ_2 that misses γ_{ABP}, and

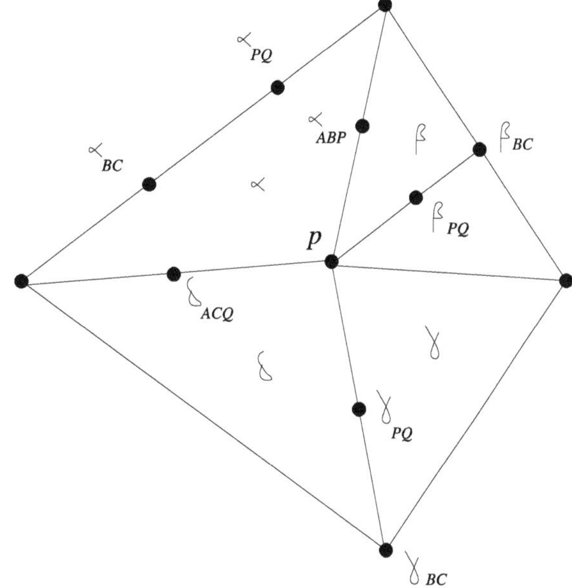

Fig. 8.8 The planes α, β, γ, δ in the proof of Lemma 8.4.18

γ_{ACQ} (and hence also misses γ_{BC} and γ_{PQ}). Set $\varepsilon = \langle m, m^{\perp} \cap \alpha \rangle$. Then $\alpha \cap \varepsilon$ contains neither α_{ACQ} nor α_{ABP}, and ε is of type III or IV. In the latter case, $\alpha, \varepsilon, \gamma$ is a Veldkamp sequence, and so $p \in \alpha_2$. Therefore, ε is of type III. Now γ, ε is a Veldkamp sequence, so $p \in \varepsilon_2$. If also α, ε is a Veldkamp sequence, then $p \in \alpha_2$, and we are done. So, assume it is not. As $\alpha \cap \varepsilon$ contains neither $\alpha \cap \beta = \alpha_{ABP}$ nor $\alpha \cap \delta = \alpha_{ACQ}$, we must have one of $\alpha_{BC} \in \varepsilon$ or $\alpha_{PQ} \in \varepsilon$. Without harming generality, we (may) assume $\alpha_{PQ} \in \varepsilon$. But then $\alpha_{PQ} = \varepsilon_{PQ}$ and $\varepsilon_{BC}\varepsilon_{PQ} = p\varepsilon_{PQ} = p\alpha_{PQ}$ forces $\varepsilon_{BC} \in \alpha \cap \varepsilon$, so $p \in \alpha_{BC}\alpha_{PQ} = \alpha_2$, a contradiction again.

We are left with the case where γ is of type IV. In γ there are at least two lines n and n' on p missing γ_{ABC} and distinct from l. Consider the planes $\varepsilon = \langle n, n^{\perp} \cap \alpha \rangle$ and $\varepsilon' = \langle n', n'^{\perp} \cap \alpha \rangle$. These are of type III or IV, again. If ε is of type IV, then $\beta, \gamma, \varepsilon, \alpha$ is a Veldkamp sequence, so, by Lemma 8.4.17, we have $p \in \alpha_2$, and similarly for ε'. This shows that both ε and ε' are of type III. Moreover, $\beta, \gamma, \varepsilon$ and $\beta, \gamma, \varepsilon'$ are Veldkamp sequences, showing that $p \in \varepsilon_2 \cap \varepsilon_2'$. But α_{ACQ} lies in at most one of $\varepsilon, \varepsilon'$, say $\alpha_{ACQ} \in \varepsilon'$. The cases where $\alpha_{BC} \in \varepsilon$ and where $\alpha_{PQ} \in \varepsilon$ can be excluded as above. Thus, without loss of generality, we (may) assume that α, ε is a Veldkamp sequence, which gives $p \in \alpha_2$, the final contradiction. This ends the proof of the lemma. $\square$

Lemma 8.4.19 *Suppose that α, β are adjacent planes with β contributing to α_3. One of the following holds.*

(i) *The type of α is IV, the type of β is III, and $\alpha_{ABC}, \beta_{BC}, \beta_{PQ}$ are distinct points lying off $\alpha \cap \beta$.*

(ii) *The type of α is IV, the type of β is I, and $\alpha \cap \beta_{PQ} = \alpha \cap \beta_{BC} = \beta_{ABC}$.*

(iii) *The type of α is II, the type of β is I, and $\alpha \cap \beta_{PQ} = \alpha \cap \beta_{BC} = \beta_{ABC}$.*

Proof Straightforward by use of Lemma 8.4.18 to eliminate the case where α is of type III. $\qquad\square$

Lemma 8.4.20 *Suppose that α, β are planes of type III with $\alpha_2 \cap \beta = \alpha \cap \beta = \{x\}$ for some point x. Then $\alpha \cap \beta_2 = \alpha \cap \beta = \{x\}$.*

Proof We may assume $x \notin (B \cap C) \cup (P \cap Q)$ (for otherwise the statement holds). We may assume that one of β_{BC}, β_{PQ}, say the former, is not in $\alpha_2^\perp$ (for otherwise, $x \in \langle \beta_{BC}, \beta_{PQ} \rangle = \beta_2$ and we are done). The plane $\gamma = \langle \beta_{BC}, \alpha \cap \beta_{BC}^\perp \rangle$ is necessarily of type III. By Lemma 8.4.18, we have $x \in \gamma_2$ (since $x \in \alpha_2$ forces $x \in \gamma_3$). Therefore, $\gamma_2 = \gamma \cap \beta$ and $\gamma_2 = \beta_2$. Thus, $x \in \beta_2 \cap \alpha$. $\qquad\square$

Lemma 8.4.21 *Suppose $\mathrm{rk}(B \cap C) \geq 2$. For every point $a \in A \cap B \cap C$, the set $((P \cap Q) \cup (B \cap C)) \setminus (A \cup a^\perp)$ is non-empty.*

Proof Suppose the contrary. Then $P \cap Q \setminus A$ and $B \cap C \setminus A$ are subsets of $a^\perp$. Hypothesis 8.3.4 shows that we may assume $A \cap B \cap C \not\subseteq a^\perp$. Corollary 8.1.7, applied to the polar space $B \cap C$ and its geometric hyperplane $A \cap B \cap C$, shows that $B \cap C$ is a rosette. Denote its radical by U. Clearly, $a \notin U$ (for otherwise $A \cap B \cap C \subseteq a^\perp$), and $U \subseteq A \cap B \cap C$ (for otherwise, there are $u \in U \setminus A$ and $a' \in A \cap B \cap C \setminus a^\perp$ so that each point $u' \in ua'$ distinct from u is not in $a^\perp$). Now $B \cap C = (B \cap C \cap a^\perp) \cup (A \cap B \cap C) = \langle U, a \rangle \cup (A \cap B \cap C) = A \cap B \cap C$, a contradiction. $\qquad\square$

Lemma 8.4.22 *For each plane α of type III we have $\alpha_4 = \alpha_2$.*

Proof By Lemma 8.4.18 we have $\alpha_3 = \alpha_2$. Let $x \in \alpha_4 \setminus \alpha_3$, so there are planes β, γ with $\alpha \sim \beta \sim \gamma$ such that β_3 contributes the point x to α_4 and γ_2 contributes to β_3. Lemma 8.4.19 describes the various possibilities for β, γ, which we now treat separately in order to derive a contradiction.

(i) *The type of β is IV and the type of γ is III.* Then $\beta \cap \gamma_{PQ} = \beta \cap \gamma_{BC} = \emptyset$, $x \in \langle \beta_{ABC}, y \rangle$, where $y \in \gamma_2 \cap \beta$. Let x' be the unique point of $\alpha_2 \cap \beta$. We have to show $x = x'$. Suppose $x \neq x'$. We are done if α, β, γ is a Veldkamp sequence, so suppose it is not. Without loss of generality, $\alpha_{ABP} \in \alpha \cap \beta$. In order to finish this case, we need two steps.

STEP 1. *Each plane δ on xy is of type IV.*

Suppose that δ is of type I. The plane $\varepsilon = \langle \alpha_{BC}, \alpha_{BC}^\perp \cap \delta \rangle$ does not pass through xy, so it has at least two distinct points in $\varepsilon \cap \delta$ from $(B \cap C) \cup (P \cap Q)$, and its type is I. But then $\varepsilon \cap \alpha = \alpha_2$, whence $x' \in \varepsilon \cap \alpha \cap \beta = \{x\}$, a contradiction.

Suppose, therefore, that δ is of type II. Without loss of generality, we may assume that $\delta \cap B \cap C$ is a line. Set $\gamma' = \langle \gamma_{BC}^\perp \cap \delta, \gamma_{BC} \rangle$. This is a plane of type I in which γ_1' is the union of two lines meeting in z, where $\{z\} = \delta_{BC} \cap \gamma_{BC}^\perp$.

Consider $\pi = \langle x', x'^{\perp} \cap \gamma' \rangle$. This is a plane in which $\pi \cap \gamma'$ is a line on y distinct from $\delta \cap \gamma'$ (for otherwise $x' \perp z$, so we find a 3-dimensional singular subspace in Z, a contradiction). As $(x'y)_{BC} = (x'y)_{PQ} = \emptyset$ and π_{BC}, π_{PQ} contain distinct points of γ'_1, the plane π is of type III. Now $\alpha_2 \cap \pi$ contains x', so Lemmas 8.4.18 and 8.4.20 give $x' \in \pi_2$, whence $\pi_2 = x'y$, contradicting $(x'y)_{BC} = \emptyset$.

STEP 2. *For each plane δ on xy, at least one of δ_{AB}, δ_{AC} is a line. If δ' is another such plane, and δ'_{AB} is a line, then δ_{AB} is a point.*

(Observe that the choice $\delta = \beta$ is allowed here.) Set $\rho = \langle \gamma_2, \gamma_2^{\perp} \cap \delta \rangle$ and $\sigma = \langle \alpha_{BC}, \alpha_{BC}^{\perp} \cap \delta \rangle$. Because both planes have a point in $BC \setminus PQ$ and a line in δ disjoint from $(B \cap C) \cup (P \cap Q)$, they are of type III. If δ_{AB} and δ_{AC} are points, then $\delta_{AB} = \delta_{AC} = \delta_{ABC} \notin \rho \cup \sigma$, so ρ, δ, σ is a Veldkamp sequence, showing that $x \in \alpha_2$, a contradiction. Thus, indeed at least one of δ_{AB}, δ_{AC} is a line.

Suppose now that δ' is another plane on xy. Set $\sigma' = \langle \alpha_{BC}, \alpha_{BC}^{\perp} \cap \delta' \rangle$, and $\{z'\} = \sigma'_2 \cap \delta'$, and $\pi = \langle z', (z')^{\perp} \cap \rho \rangle$. Now $\pi \cap \rho$ is a line on y distinct from $\delta \cap \rho$, for otherwise $z' \perp \delta$ or $\delta \cap \rho = xy$ which lead to contradictions. Also, $\pi \cap \rho \neq \gamma_2$, for otherwise π would have type III and by Lemmas 8.4.18 and 8.4.20, $\pi_2 = \gamma_2$ contains z', another contradiction. As $z'y$ has no points in $(B \cap C) \cup (P \cap Q)$, the plane π is of type III or IV. If its type is III, then Lemmas 8.4.18 and 8.4.20 give $z' \in \pi_2$, which is a contradiction as before. Hence π is of type IV, and $\pi_{ABC} \notin z'y$. If δ'_{AB} is a line, then $z'y$ contains a point of it, and so π_{AB} contains a line meeting ρ. As ρ is of type III, it has a unique point in $A \cap B$ by Lemma 8.4.14, so $\pi_{AB} \cap \rho = \rho_{AB}$. Suppose now that δ_{AB} is a line. It meets ρ in a point which must be ρ_{AB}, a point of $\pi \cap \delta \cap \rho = \{y\}$, forcing $\gamma_{BC}, \gamma_{PQ}, \gamma_{AB} = \{y\}$ to be collinear, in conflict with Lemma 8.4.14. This ends the proof of Step 2.

Now take three planes on xy. Each is of type IV and has a line in $A \cap B$ or in $A \cap C$, so at least two of them have a line from one of the sets $A \cap B$ or $A \cap C$, contradicting Step 2. Hence $x = x'$, so Case (i) is impossible.

(ii) *The type of β is IV and the type of γ is I.* We may assume $x \in \alpha \cap \beta \cap \gamma$. Set $\delta = \langle \alpha_{BC}, \alpha_{BC}^{\perp} \cap \gamma \rangle$, so $x \in \alpha \cap \delta$. Then δ is of type I and $\alpha_{BC}, \alpha_{PQ} \in \alpha \cap \delta$, so $x \in \alpha_2$, contradicting $x \in \alpha_4 \setminus \alpha_3$.

(iii) *The type of β is II and the type of γ is I.* Without loss of generality we may take $\alpha_{BC} \notin \gamma$. Now $x \in \alpha \cap \beta \cap \gamma$ and $\delta = \langle \alpha_{BC}, \alpha_{BC}^{\perp} \cap \gamma \rangle$ is a plane with a line in $B \cap C$ and a point in $P \cap Q$ not on that line, so it must be of type I. As $x \in \delta_2$, we find $x \in \alpha_3 = \alpha_2$, the final contradiction. $\square$

We continue with planes of type II.

Lemma 8.4.23 *For each plane α of type II we have $\alpha_4 = \alpha_3$.*

Proof Let x, β, γ be as in the first paragraph of the proof of Lemma 8.4.22. Again we distinguish the three possibilities for β, γ described in Lemma 8.4.19 and derive a contradiction. In Cases (i) and (ii), β is of type IV and $\beta_{ABC} = \alpha \cap \beta \cap \gamma$, which implies $\{x\} = \beta_{ABC}$, a contradiction. So case (iii) remains.

Fig. 8.9 The planes α, β, γ, α', β', γ', in the proof of Lemma 8.4.23

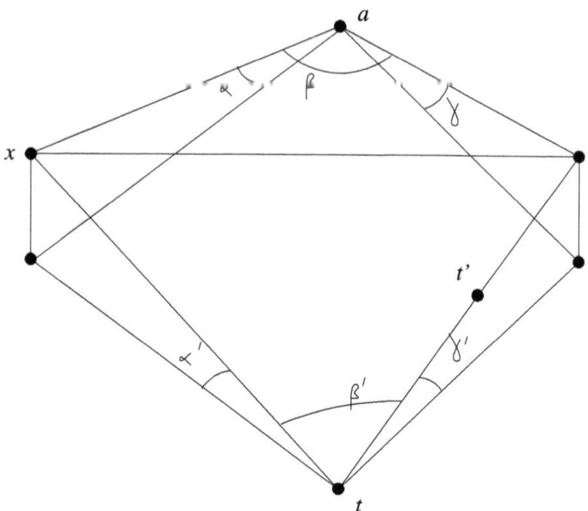

(iii) *β is of type II and γ type I.* Set $\{a\} = \alpha \cap \beta \cap \gamma$. Then $a \in A \cap B \cap C$ (for otherwise $\delta = \langle p^{\perp} \cap \alpha, p \rangle$ for $p \in \gamma_{BC} \setminus \beta$ is a plane of type I or III with $a \in \delta_2$ and $x \in \alpha \cap \delta_2 \setminus \alpha_2$). By Lemma 8.4.21, there exists $t \in ((B \cap C) \cup (P \cap Q)) \setminus (A \cup a^{\perp})$. Without loss of generality, we (may) assume $t \in B \cap C$. Observe that $\{a\} = \gamma_{BC} \cap \gamma_{PQ}$. Let α', β', γ' denote the plane on t adjacent to α, β, γ, respectively. Without loss of generality we may take $x \in \beta' \cap \alpha'$. Now $a \notin \gamma'$, so $\gamma' \cap \gamma$ contains a unique point from $B \cap C$ and a unique one from $P \cap Q$ (and they are distinct); see Fig. 8.9. Thus, γ' is of type I, and $\gamma' \cap \beta'$ contains a point, t' say, of $P \cap Q$. (Notice that $t \neq t'$ as $t \notin A \cap B \cap C$.) Now β', containing three non-collinear points (viz., t, t' and $\beta' \cap \beta_2$) of $(B \cap C) \cup (P \cap Q)$, must be of type I. In particular β'_{ABC} is the point $\beta_1 \cap \beta'$, which does not lie in α'. Then β'_1 meets α' in distinct points which also belong to α'_1, so $x \in \alpha'_2$, whence $x \in \alpha_3$, a contradiction. $\qquad\square$

Lemma 8.4.24 *For each plane α of type IV, we have $\alpha_4 = \alpha_3$.*

Proof Again, x, β, γ are chosen as in the first paragraph of the proof of Lemma 8.4.22. In particular, $x \in \alpha_4 \setminus \alpha_3$ and we derive a contradiction in each of the three cases for the types of β, γ described in Lemma 8.4.19.

(i) *β is of type IV, and γ is of type III.* At least one of the two points γ_{BC}, γ_{PQ}, say γ_{BC}, is non-collinear with α_{ABC}. Set $\delta = \langle \gamma_{BC}, \gamma_{BC}^{\perp} \cap \alpha \rangle$. Then δ has a line (namely $\alpha \cap \delta$) disjoint from $(B \cap C) \cup (P \cap Q)$ and a point in $(B \cap C) \setminus A$, so is of type III. Set $\{y\} = \beta \cap \gamma_2 \setminus \beta_1$, $\{y'\} = \alpha \cap \delta_2 \setminus \alpha_1$, and $\{x'\} = y'\alpha_{ABC} \cap \beta$. Since $x' \in \alpha_3$ (for $y' \in \alpha \cap \delta_2 \setminus \alpha_2$), we have $x \neq x'$. Symmetry allows us to interchange the roles of $x, y, \alpha, \beta, \gamma$ with $x', y', \beta, \alpha, \delta$. We will use this observation at the end of six steps.

STEP 1. *Every plane on xy is of type IV.*

Clearly, type III does not occur as xy contains the point β_{ABC}. Let ε be a plane on xy and let ζ be the plane on y' adjacent to ε. The type of β is IV, so we (may) assume $\varepsilon \neq \beta$. The type of ε cannot be I as both ε_{BC} and ε_{PQ} meet xy only in β_{ABC}. Suppose that ε is of type II. Consider the plane $\eta = \langle \gamma_2, \gamma_2^{\perp} \cap \varepsilon \rangle$. Since $\eta \cap \varepsilon$ contains a point of ε_1, the type of η must be I, and $\eta_{ABC} = \varepsilon_1 \cap \eta$. Hence the plane $\pi = \langle x', (x')^{\perp} \cap \eta \rangle$, meeting η in a line on y distinct from $\varepsilon \cap \eta$, has two points from $(B \cap C) \cup (P \cap Q)$ in η. As $\pi \cap \beta = x'y$ has no points of $(B \cap C) \cup (P \cap Q)$, the type of π is III and $\pi_1 \subseteq \eta$. Take $z \in \pi_{BC} \cup \pi_{PQ}$ such that $\rho = \langle z, z^{\perp} \cap \alpha \rangle$ does not pass through α_{ABC} (observe that this can be done as π_1 is disjoint from α). As for π, we obtain that ρ is of type III. Now $\rho_4 = \rho_2$ (cf. Lemma 8.4.22) and $x' \in \rho \cap \alpha_3$, so $x' \in \rho_2$. Therefore, $\rho_2 = x'z \subseteq \pi$, which forces $\rho_1 = \pi_1$, contradicting $\pi_1 \subseteq \eta$. The conclusion is that ε is of type IV.

STEP 2. *Every plane on xy has a line in $A \cap B$ or in $A \cap C$.*

Suppose, to the contrary, that ε is a plane on xy with $\varepsilon_{AB} = \varepsilon_{AC} = \varepsilon_{ABC}$. First we deal with the case where $\varepsilon \neq \beta$. Consider the planes $\gamma' = \langle \gamma_2^{\perp} \cap \varepsilon, \gamma_2 \rangle$, $\alpha' = \langle \alpha_{ABC}^{\perp} \cap \varepsilon, \alpha_{ABC} \rangle$ and $\delta' = \langle \delta_{PQ}^{\perp} \cap \alpha', \delta_{PQ} \rangle$. Then γ' is of type III, α' of type IV, and δ' of type III. Also $\alpha'_{AB}, \alpha'_{AC}$ are points as α' is adjacent to ε. Thus, $\gamma', \varepsilon, \alpha', \delta'$ is a Veldkamp sequence, forcing $\delta'_2 = \delta \cap \delta' = \delta_2$ (cf. Lemma 8.4.17), a contradiction with $y' \in \delta_2 \setminus \delta'$. Therefore, ε must have a line in $A \cap B$ or in $A \cap C$.

Now suppose $\varepsilon = \beta$. Choose ε' as ε above and repeat the above construction for ε' instead of ε, but now with $\alpha' = \langle xy', (xy')^{\perp} \cap \varepsilon' \rangle$. The plane γ' is of type III while ε is of type IV. Clearly, the type of α' is III or IV. If it is III, then $x \in \alpha'_2$, so $x \in \alpha_3$, a contradiction. If α' is of type IV, then the setting for $\alpha', \varepsilon', \gamma', \alpha, \beta, \gamma$ is as for $\alpha, \beta, \gamma, \alpha', \varepsilon', \gamma'$ in the previous paragraph, so the conclusion obtained for ε' is also valid for β.

In particular, β has a line in $(A \cap B) \cup (A \cap C)$. Without loss of generality, we (may) assume that β_{AC} is a line. Clearly then α_{AC} is a line as well.

STEP 3. *Every plane on xy has a line in $A \cap C$.*

Suppose that ε is a plane on xy with $\varepsilon_{AC} = \varepsilon_{ABC}$. The previous step yields that ε_{AB} is a line. The plane $\zeta = \langle y'^{\perp} \cap \varepsilon, y' \rangle$ meets ε outside $\varepsilon_{ABC} = \beta_{ABC}$, so is of type III or IV. Suppose that ζ is of type IV. Now ζ also meets α outside α_{ABC}. Thus, $\zeta \cap \alpha$ contains a point of $A \cap C \setminus B$ and so ζ contains a line of $A \cap C$, whence $\zeta \cap \varepsilon$ contains a point of $A \cap C \setminus B$, contrary to the assumption $\varepsilon_{AC} = \varepsilon_{ABC}$. Hence ζ is of type III. By Lemma 8.4.20 and the fact that y' belongs to $\delta_2 \cap \zeta$, we obtain $y' \in \zeta_2$. As $x \in \beta_3$, Lemma 8.4.22 forces $x \in \zeta_2$, whence $xy' = \zeta_2$, contradicting $xy' \cap ((B \cap C) \cup (P \cap Q)) = \emptyset$. This establishes Step 3.

We observe that $x \notin A \cap C$, for otherwise $x\beta_{ABC} \subseteq A \cap C$, so $y \in A \cap C$, contradicting that $\gamma_{AC}, \gamma_{BC}, \gamma_{PQ}$ are non-collinear.

STEP 4. *Every plane on x adjacent to α is of type IV and contains a line in $A \cap C$.*

Let ε be a plane on x adjacent to α. If $\varepsilon \cap \alpha = \beta \cap \alpha$, then ε must be of type IV (otherwise the type is III and, by Lemma 8.4.22, x and x' are in ε_2, hence coincide), and $\varepsilon \cap \alpha$ has a point in $A \cap C \setminus B$, so we are done. Suppose therefore $x' \notin \varepsilon$. Since $\langle y^\perp \cap \varepsilon, y \rangle$ is a plane of type IV with a line in $A \cap C$ (see the previous step), meeting ε in a line having points z of $A \cap C \setminus B$ but no point of $(B \cap C) \cup (P \cap Q)$, the type of ε is III or IV, and in the latter case we are done. But ε cannot be of type III since z and $\varepsilon \cap \alpha_{AC}$ are distinct points in ε_{AC}. This ends Step 4.

STEP 5. *Every plane on x is of type IV and contains a line of $A \cap C$.*

Let ε be a plane on x nonadjacent to α and not containing xy. Consideration of $\zeta = \langle y, y^\perp \cap \varepsilon \rangle$ shows that ε is of type III or IV, and that there is a point $z \in (\zeta_{AC} \cap \varepsilon) \setminus B$. Take two lines l_1, l_2 in α on x such that $l_i^\perp \cap \varepsilon$ does not pass through z. By the previous step, $\zeta_i = \langle l_i, l_i^\perp \cap \varepsilon \rangle$ is of type IV and contains a line of $A \cap C$. For at least one $i \in \{1, 2\}$ say $i = 1$ we have $\varepsilon_{ABC} \nsubseteq \zeta_i$. Now $\zeta_1 \cap \varepsilon_{AC}$ and z are distinct points in ε_{AC}, so ε is of type IV and ε_{AC} is a line. Hence Step 5.

STEP 6. *The plane β is contained in A.*

Suppose not. Then $\beta_{AC} = \beta_A = \beta_C$ and $x \notin \beta_A$. If $z \in x^\perp \cap A$ then, $\varepsilon = \langle z, z^\perp \cap \beta \rangle$ is a plane on x for which $\varepsilon \cap \beta_A$ is a point, so the line ε_{AC} given by Step 5 is equal to ε_A, whence $z \in \varepsilon_C \subseteq x^\perp \cap C$. Thus $x^\perp \cap A \subseteq x^\perp \cap C$, and $x^\perp, A, C$ are collinear in $\mathcal{V}(Z)$. Similarly, arguing with x', y' instead of x, y we see that $x'^\perp, A, C$ are collinear in $\mathcal{V}(Z)$. As $x \neq x'$, this implies that $A = a^\perp$, $C = c^\perp$ for certain $a, c \in xx'$. But then $\beta \subseteq a^\perp = A$, a contradiction. Hence Step 6.

Since $\beta \subseteq A$, the set β_{AB} must be a line, too. Hence, by Step 5 applied to $A \cap B$ instead of $A \cap C$, every plane ρ on x is spanned by the two distinct lines ρ_{AB} and ρ_{AC}, so lies in A, showing $x^\perp \subseteq A$ and so $A = x^\perp$. Arguing with x', y' instead of x, y, we get $x^\perp = x'^\perp$, whence $x = x'$, the final contradiction for (i).

(ii) *β is of type IV, γ is of type I.* Then $x \in \beta \cap \gamma \cap \alpha$. Taking $\delta = \langle \alpha_{ABC}, \alpha_{ABC}^\perp \cap \gamma \rangle$, we obtain a plane adjacent to α, of type I, with $x \in \delta_2$ whence $x \in \alpha \cap \delta_2 \subseteq \alpha_3$.

(iii) *β is of type II, γ is of type I.* If $\alpha_{ABC} = \alpha \cap \beta \cap \gamma$ and $x \in \alpha \cap \beta \setminus \alpha_{ABC}$, we can finish verbatim as in Case (iii) of the proof of Lemma 8.4.22. If $\alpha_{ABC} \neq \gamma_{ABC}$ we can finish as in (ii). This settles the lemma. $\square$

Finally, we have enough material for the proof of the following theorem.

Theorem 8.4.25 *Suppose that Z is a thick nondegenerate polar space of rank three all of whose singular planes are Desarguesian and in which every line is on at least three planes. Then Z embeds in the projective space $\mathcal{V}(Z)$.*

Proof By Theorem 8.2.9, Z embeds into the Veldkamp space $\mathcal{V}(Z)$, so it remains to show that $\mathcal{V}(Z)$ is a projective space. As noticed before, Theorem 5.2.6 can be applied to show that $\mathcal{V}(Z)$ is a projective space. This requires that Pasch's axiom be established. This turned out to be easy by Theorem 8.2.8 if all lines have exactly three points, so we could focus on the case where all lines have at least four points. We constructed the subspace R of Z (see Lemma 8.4.8) from the configuration depicted in Fig. 8.2. As Proposition 8.4.11 shows that every line meets R, it remains to show that R is a proper subspace. To this end we fix a singular plane α of type III (cf. Lemma 8.4.13) and show that $\alpha \cap R = \alpha_2$ is a line. This statement follows from the following three claims concerning a singular plane β of Z.

(1) If β is of type I, then $\beta_n = \beta_1$ for all $n \geq 1$.
(2) If β is of type III, then $\beta_n = \beta_2$ for all $n \geq 2$.
(3) If β is of type II or IV, then $\beta_n = \beta_3$ for all $n \geq 3$.

The claims are obvious for $n = 1$ and $n = 2$, and follow from Lemma 8.4.18 if $n = 3$. So, suppose $n > 3$ and proceed by induction on n. Let $y \in \beta_n \setminus \beta_{n-1}$. Then β does not have type I and $y \in \beta \cap \gamma_{n-1}$ for some $\gamma \sim \beta$. By induction $\gamma_{n-1} = \gamma_j$, where $j = 1, 2, 3$ if the type of γ is equal to I, III, respectively II or IV, so $y \in \beta_{j+1}$. If β is of type II or IV, then we find $y \in \beta_4$ and so $y \in \beta_3$ by Lemmas 8.4.23 and 8.4.24. If β is of type III, then $y \in \beta_2$ by Lemmas 8.4.18 and 8.4.22. This establishes the validity of the claims and the statement. As a consequence, R is a proper subspace of Z, and therefore (by Proposition 8.4.11) a geometric hyperplane of Z. We conclude that Pasch's axiom holds in $\mathcal{V}(Z)$. $\square$

Combining Theorems 8.3.16 and 8.4.25, we find

Corollary 8.4.26 *Suppose that Z is a thick nondegenerate polar space of rank at least three all of whose singular planes are Desarguesian and in which every line is on at least three singular planes. The map $x \mapsto x^{\perp} : Z \to \mathcal{V}(Z)$ is an embedding of Z in the projective space $\mathcal{V}(Z)$.*

8.5 Automorphisms of Polar Spaces

In this section we construct automorphisms of nondegenerate polar spaces of rank at least three in a fairly general setting; they appear in Proposition 8.5.9. We begin by showing (in Theorem 8.5.5) that certain partial isomorphisms between polar spaces extend to isomorphisms.

Throughout this section, Z and Z' are thick nondegenerate polar spaces of rank at least three. Recall from Definition 2.5.8 that we occasionally use Z and Z' to indicate the point sets of these spaces.

Definition 8.5.1 If $A \subseteq Z$ and $A' \subseteq Z'$, a bijection $i : A \to A'$ is called an *isomorphism* from A to A' if both i and i^{-1} map every subset of a line of Z to a subset of a line of Z'.

Thus, for each line l of Z, there is a line l' of Z' such that $i(l \cap A) = l' \cap A'$. In particular, if i is an isomorphism and l is a line of Z in A, then $i(l)$ is a line of Z in A'. The above notion of isomorphism coincides with the notion of isomorphism of the subspaces induced on A and A' as given in Definition 2.5.8.

For the duration of this section, let p, q, and p', q', be non-collinear points of Z and Z', respectively. Assuming that there is an isomorphism $i : p^\perp \cup \{q\} \to p'^\perp \cup \{q'\}$, we first extend the isomorphism to an isomorphism $p^\perp \cup q^\perp \to (p')^\perp \cup (q')^\perp$, next to $\{p,q\} \cup Z \setminus \{p,q\}^{\perp\perp} \to \{p',q'\} \cup Z \setminus \{p',q'\}^{\perp\perp}$, and finally to $Z \to Z'$. The result is an isomorphism of polar spaces.

Exercise 8.6.4 shows how to extend i to an isomorphism i_1 on $p^\perp \cup q^\perp$. What remains is the question whether the intersection of $q'^\perp \setminus (p'^\perp \cup \{q'\})$ and $i(\{x, p\})^{\perp\perp}$ is non-empty. This is taken care of by Proposition 8.1.10 as will become clear in the next lemma.

Lemma 8.5.2 *Each isomorphism $i : p^\perp \cup \{q\} \to (p')^\perp \cup \{q'\}$ has a unique extension to an isomorphism $i_1 : p^\perp \cup q^\perp \to (p')^\perp \cup (q')^\perp$.*

Proof Clearly, $i(q) = q'$. As Z is nondegenerate, $p^\perp \cap p^{\perp\perp} = \{p\}$ and similarly for p' instead of p, so $i(p) = p'$. If $x \in p^\perp$, then $i_1(x) = i(x)$. Assume, therefore, $x \in q^\perp \setminus (p^\perp \cup \{q\})$. The line qx meets $p^\perp \cap q^\perp$ in a point $\overline{x}$ distinct from x. Now $i_1(x)$ must belong to the line $i(\overline{x})q'$. Moreover, $H_x := \{x, p\}^\perp$ is a geometric hyperplane of $p^\perp$ such that $H_x \cap q^\perp \subseteq \overline{x}^\perp$. As i is an isomorphism, $i(H_x)$ is a geometric hyperplane of $(p')^\perp$ such that $i(H_x) \cap q'^\perp \subseteq i(\overline{x})^\perp$. Proposition 8.1.10 provides a unique point x' on $i(\overline{x})q'$ such that $\{x', p'\}^\perp = i(H_x)$. The point $i_1(x)$ is necessarily x'. Thus, i_1 is uniquely determined.

In order to check that i_1 preserves lines, assume that $v, w \in q^\perp$ are distinct and collinear, and set $l = vw$. If $q \in l$, then $i_1(l)$ is obviously a line, so assume $q \notin l$. If $l \subseteq p^\perp$ there is nothing to show, so assume that $p^\perp \cap l$ is a singleton, say $\{r\}$. First, note that $i_1(v)$ and $i_1(w)$ both lie in the plane $\langle i(\overline{v}), i(\overline{w}), q' \rangle$ containing $i(r)$. There exists $u \in p^\perp \cap r^\perp \cap w^\perp$ such that $u^\perp \cap \langle \overline{v}, \overline{w}, q \rangle = l$. Then $u \in H_v \cap H_w \cap \{p, r\}^\perp$, so $i(u) \in i(H_v) \cap i(H_w) \cap \{p', i(r)\}^\perp$, whence $i_1(v), i_1(w), i(r) \in i(u)^\perp \cap \langle i(\overline{v}), i(\overline{w}), q' \rangle =: l'$. But then l' is a line satisfying $i_1(l) \subseteq l'$, as required.

Clearly i_1 is bijective, and, by symmetry, i_1^{-1} also maps lines onto lines. $\qquad\square$

For a line m and a point x of Z with $x \notin m^\perp$, write x_m to denote the unique point on m collinear with x.

Lemma 8.5.3 *Assume that $i : p^\perp \cup q^\perp \to p'^\perp \cup q'^\perp$ is an isomorphism. Put $A = \{p,q\} \cup (Z \setminus \{p,q\}^{\perp\perp})$ and $A' = \{p',q'\} \cup (Z' \setminus \{p',q'\}^{\perp\perp})$.*

(i) *If l is a line of $p^\perp$ and m a line of $q^\perp$ such that l, m are coplanar, then so are $i(l), i(m)$.*

(ii) *Let l be a line of Z disjoint from $\{p,q\}^\perp$. Then $l \subseteq A$ and $i_2[l] := i(p_l)i(q_l)$ is a line in A'.*

(iii) *If $x \in A \setminus (p^\perp \cup q^\perp)$, then there exists a line l on x disjoint from $\{p,q\}^\perp$. For each such line l, there is a unique point $i_2(x)$ on $i_2(l)$ such that $i_2(x)^\perp \cap i(p)^\perp = i(x^\perp \cap p^\perp)$. The point $i_2(x)$ does not depend on the choice of the line l on x disjoint from $\{p,q\}^\perp$. In particular, setting $i_2(x) = i(x)$ for $x \in p^\perp \cup q^\perp$, we obtain a well-defined bijective map $i_2 : A \to A'$.*

(iv) *If l is a line in A, then $i_2(l) = \bigcup_{x \in l}\{i_2(x)\}$ is a line in A' (equal to $i_2[l]$ if l does not meet $\{p, q\}^{\perp}$).*

(v) *If l is a line of Z with $l \not\subseteq A$, then there is a unique point $r \in l \setminus A$ such that $l = (l \cap A) \cup \{r\}$; moreover, there is a unique line l' in Z' with $i_2(l \cap A) \subseteq l'$.*

(vi) *The map i_2 is the unique extension of i to an isomorphism $A \to A'$.*

Proof (i) The lines $i(l)$ and $i(m)$ are coplanar because they meet in $i(l \cap m)$ and any pair of points in $i(l) \cup i(m)$ is collinear.

(ii) Notice that $l \subseteq p^{\perp}$ leads to $l \cap \{p, q\}^{\perp} \neq \emptyset$, a contradiction, so $l \not\subseteq p^{\perp}$. Similarly, $l \not\subseteq q^{\perp}$, and p_l and q_l are well defined.

We first prove $l \subseteq A$. Suppose the contrary, so there is a point x in $\{p, q\}^{\perp\perp} \cap l$. Let a be the unique point of $p^{\perp} \cap qq_l$. Then $a \perp x$. Observe that $a \neq q_l$, for otherwise $l \cap \{p, q\}^{\perp} \neq \emptyset$, against our assumptions. Hence $a \perp xq_l = l$ and $p_l \perp aq_l$, so $p_l \in \{p, q\}^{\perp}$, a contradiction with $l \cap \{p, q\}^{\perp} = \emptyset$.

By definition of isomorphism, Z' has a line l' on $i(p_l)$ and $i(q_l)$. Because $p' \not\perp q'_{l'}$, we have $i(p_l) = p'_{l'}$; similarly, $i(q_l) = q'_{l'}$, and l' lies in A'. Therefore, $i_2[l] = l'$.

(iii) Let $x \in A \setminus (p^{\perp} \cup q^{\perp})$. As $x \notin \{p, q\}^{\perp\perp}$, we can find a point r in $\{p, q\}^{\perp} \setminus x^{\perp}$. Clearly, $r \neq p$, and there is a point y on pr collinear with x. Now $l := xy$ is a line as required. For, if $l \cap \{p, q\}^{\perp} \neq \emptyset$, the point y would be collinear with q, so $p \in yr \subseteq q^{\perp}$, a contradiction.

Now let l be any line on x disjoint from $\{p, q\}^{\perp}$. The line $i_2[l]$ of (ii) is disjoint from $\{p', q'\}^{\perp}$ and satisfies $i(p^{\perp} \cap x^{\perp}) \cap i(p_l)^{\perp} = i_2[l]^{\perp} \cap p'^{\perp}$. Therefore, by Proposition 8.1.10, the line $i_2[l]$ has a unique point, $i_l(x)$ say, such that $i_l(x)^{\perp} \cap p'^{\perp} = i(p^{\perp} \cap x^{\perp})$.

Since $\{x, p\}^{\perp}$ is a geometric hyperplane of $p^{\perp}$ not on p and distinct from $\{p, q\}^{\perp}$, a line of it not in $\{p, q\}^{\perp}$ has at least two points outside of $\{p, q\}^{\perp}$; thus, x is on at least two lines l as in (ii). The images of two distinct such lines, say l and m, are distinct, so their intersection contains at most one point. We have to show $i_l(x) = i_m(x)$. If l and m are coplanar, then so are $i_2[l]$ and $i_2[m]$ (as they lie in the plane $\langle i(p_l) i(p_m), i(q_l) i(q_m)\rangle$), and their intersection is readily seen to be $i_l(x) = i_m(x)$. On the other hand, since $p^{\perp} \cap x^{\perp} \setminus q^{\perp}$ is connected, there is a series of lines $l_1 = l, l_2, \ldots, l_s = m$ with l_i, l_{i+1} coplanar for each i ($1 \leq i < s$), so, repetition of the earlier argument for each i leads to $i_l(x) = i_{l_2}(x) = \cdots = i_m(x)$. This establishes (iii).

(iv) If $l \subset p^{\perp}$ or $l \subset q^{\perp}$, the statement is trivial, so assume that this is not the case. If $l \cap \{p, q\}^{\perp} = \emptyset$, then l is as in (iii) and we are done.

Thus, we (may) assume that $l \cap \{p, q\}^{\perp} = \{u\}$ for some point u. Take $x \in l \setminus \{u\}$ and suppose that m is a line on x coplanar with l and disjoint from $\{p, q\}^{\perp}$ (as $x \in A$, such a line m can be found). The singular plane $\alpha' := \langle i_2[l], i_2[m]\rangle$ is the unique plane of Z' containing the union of all $i_2[m']$ for m' running over all lines on x lying in the plane $\alpha := \langle l, m\rangle$ and distinct from l. Since all points of l distinct from u are on lines meeting $p^{\perp}$ and $q^{\perp}$ in distinct points, it follows from (iii) that $i_2(\alpha) = \alpha'$. Now $i_2(l \setminus \{x\})$ consists of points of α' not on any line of α' containing $i_2(x)$ distinct from the line $i(u) i_2(x)$, and so must be contained in $i(u) i_2(x)$. The same arguments with i^{-1} instead of i then show that $i_2(l) = i(u) i_2(x)$. Hence (iv).

(v) We claim that $l \cap \{p, q\}^\perp$ is a singleton. For, as $l \not\subseteq A$, there exists $x \in l \cap \{p, q\}^{\perp\perp}$. By (ii), $l \cap \{p, q\}^\perp$ contains a point. If it would contain two points, then $l \subseteq \{p, q\}^\perp$, so $x \in \{p, q\}^\perp \cap \{p, q\}^{\perp\perp} = \mathrm{Rad}(\{p, q\}^\perp)$. As Z is nondegenerate, this conflicts with Lemma 7.4.8(ii). Hence the claim.

Now, let $u \in Z$ be such that $\{u\} = l \cap \{p, q\}^\perp$. If $q^\perp \cap l^\perp \not\subseteq p^\perp$, then a line m on the point x, defined by $\{x\} = l \cap \{p, q\}^{\perp\perp}$, in $l^\perp$ and disjoint from $\{p, q\}^\perp$ can be found; as before, this gives $m \subset A$, leading to the contradiction $x \in A$. Therefore, $q^\perp \cap l^\perp \subseteq p^\perp$. Take $y \in l \cap A \setminus \{u\}$. If $w \in \{u, p, q\}^\perp \setminus l^\perp$, then $y^\perp$ meets wq in a point of $q^\perp \cap l^\perp \setminus p^\perp$, which is absurd. Thus, $\{u, p, q\}^\perp \subseteq \{y, p, q\}^\perp$. But $y^\perp$ does not contain $\{p, q\}^\perp$ and $\{u, p, q\}^\perp$ is a geometric hyperplane of $\{p, q\}^\perp$, so Proposition 8.1.6 gives $\{u, p, q\}^\perp = \{y, p, q\}^\perp$, whence $\{u, p, q\}^\perp = l^\perp \cap \{p, q\}^\perp$.

Since Z is nondegenerate, there exists $v \in \{p, q\}^\perp \setminus u^\perp$. Then $v^\perp \cap l$ contains a single point, say r. Now $\{r, p, q\}^\perp$ strictly contains $\{u, p, q\}^\perp$, so coincides with $\{p, q\}^\perp$, showing that $r \in \{p, q\}^{\perp\perp}$. As $|\{p, q\}^{\perp\perp} \cap l| \leq 1$, we must have $l = (l \cap A) \cup \{r\}$.

Let α be a plane on l. Since each line on r contains a unique point of $\{p, q\}^{\perp\perp}$, we have $\{p, q\}^{\perp\perp} \cap \alpha = \{r\}$. Therefore, by the above, the image $i_2(m)$ of every line m in α not on r is a line of Z'. Thus there is a unique plane α' in Z' containing all of these lines, whence $i_2(l \setminus \{r\})$. Taking another plane β on l, we similarly find another plane β' containing $i_2(l \setminus \{r\})$, from which we conclude that $i_2(l \setminus \{r\})$ is contained in the line $\alpha' \cap \beta'$ of Z'. This proves (v).

(vi) The map i_2^{-1} exists and, by (iv), (v) also maps sets of collinear points to sets of collinear points. This ends the proof of the lemma. $\qquad\square$

Corollary 8.5.4 *If $i : \{p, q\} \cup (Z \setminus \{p, q\}^{\perp\perp}) \to \{p', q'\} \cup (Z' \setminus \{p', q'\}^{\perp\perp})$ is an isomorphism, then it extends uniquely to an isomorphism $i_3 : Z \to Z'$.*

Proof As before, set $A = \{p, q\} \cup (Z \setminus \{p, q\}^{\perp\perp})$ and $A' = \{p', q'\} \cup (Z' \setminus \{p', q'\}^{\perp\perp})$. Suppose that $x \in Z \setminus A$ and l is a line on x. Then l is as in Lemma 8.5.3(v), so $l = (l \cap A) \cup \{x\}$, and $i(l)$ is a line of Z' with a unique point, say $i_l(x)$, in $Z' \setminus A'$. We claim that $i_l(x)$ does not depend on the choice of l on x. For, if m is another line on x coplanar with l, then the proof of (v) above shows that $i_l(x) = i_m(x)$, and a connectedness argument does the rest. Therefore, $i_3(x) = i_l(x)$, which ensues uniqueness. Clearly, i_3 is a bijection.

Let n be a line of Z. We have to show that $i_3(n)$ is a line of Z'. Since $Z \setminus A$ is a coclique, n has at least two points in A. Hence there is a line $i(n)$ in Z with $i(n \cap A) = i(n) \cap A'$. But then either $n \subseteq A$ or there is $x \in n \setminus A$ for which $i_3(x) = i_n(x) \in i(n)$. This establishes that i_3 maps lines onto lines. The same holds for i_3^{-1}, whence the corollary. $\qquad\square$

Putting Lemmas 8.5.2, 8.5.3(vi), and Corollary 8.5.4 together, we obtain the extension of an isomorphism defined on $p^\perp \cup \{q\}$ to one on Z.

Theorem 8.5.5 *Let Z and Z' be nondegenerate polar spaces of rank at least three whose lines are thick. Suppose that p, q, respectively p', q' are non-collinear points* *of Z, respectively Z'. Each isomorphism from $p^\perp \cup \{q\}$ onto $p'^\perp \cup \{q'\}$ extends uniquely to an isomorphism from Z onto Z'.*

In order to apply Theorem 8.5.5 we need isomorphisms from $p^\perp$ to $p'^\perp$. Our present goal will be to construct nontrivial automorphisms of $p^\perp$, that is, we will deal with the important special case $p = p'$.

Proposition 8.5.6 *Let Z be a nondegenerate polar space of rank at least three all of whose lines are thick. For each point p of Z and distinct geometric hyperplanes H, H' of $p^\perp$ not on p, there is a unique automorphism α of $p^\perp$ leaving invariant each line on p, mapping H onto H' and fixing each point on every line of Z containing p and intersecting $H \cap H'$ nontrivially.*

Proof By Theorem 7.7.10 all planes of Z are Moufang. Every plane A on p meets both H and H' in a line, and $A \cap H \cap H'$ contains a point q. Since A is Moufang, it has a unique automorphism α_A fixing each line on p, each point on pq and mapping $A \cap H$ onto $A \cap H'$. Since necessarily $\alpha|_A = \alpha_A$ for an automorphism α as required, and each point of $p^\perp$ lies in some plane A, the automorphism α is unique.

To end the proof, it suffices to show that, for planes A, B on p meeting in a line l, the restrictions to l of the maps α_A and α_B coincide. This is trivial if $l \cap H \cap H'$ is non-empty. Assume the contrary, and let $\{z\} = l \cap H$, $\{z'\} = l \cap H'$. There are unique points q_A and q_B such that $\{q_A\} = A \cap H \cap H'$ and $\{q_B\} = B \cap H \cap H'$. By Corollary 7.7.7 there exists an isomorphism $\gamma : A \to B$ fixing l point-wise and mapping q_A onto some point u. The fact that B is a Moufang plane provides an automorphism $\varepsilon : B \to B$ fixing l point-wise and mapping u to q_B. Therefore, $\varepsilon\gamma : A \to B$ fixes l point-wise and maps q_A to q_B. Now the automorphism $\varepsilon\gamma\alpha_A\gamma^{-1}\varepsilon^{-1}$ of B fixes pq_B point-wise, fixes every line on p in B, and maps z to z', so coincides with α_B. Taking restrictions to l, we find $\alpha_A|_l = \alpha_B|_l$. $\qquad\square$

Theorem 8.5.7 *Let Z and Z' be nondegenerate polar spaces of rank at least three all of whose lines are thick. Suppose that p is a point of Z and p' is a point of Z' such that $p^\perp$ and $p'^\perp$ are isomorphic. Then Z and Z' are isomorphic.*

Proof Let $i : p^\perp \to p'^\perp$ be an isomorphism. Choose points $q \in Z \setminus p^\perp$ and $q' \in Z' \setminus p'^\perp$. As $H_1 = \{p, q\}^\perp$ and $H_2 = i^{-1}(\{p', q'\}^\perp)$ are geometric hyperplanes of $p^\perp$, Proposition 8.5.6 provides an automorphism j of $p^\perp$ mapping H_1 onto H_2. Thus, $i\,j$ is an isomorphism $p^\perp \to p'^\perp$ mapping $\{p, q\}^\perp$ onto $\{p', q'\}^\perp$. It obviously extends to an isomorphism $p^\perp \cup \{q\} \to p'^\perp \cup \{q'\}$, and so Theorem 8.5.5 applies yielding the required isomorphism. $\qquad\square$

Example 8.5.8 Theorem 8.5.7 states that an isomorphism between $p^\perp$ and $p'^\perp$ implies the existence of an isomorphism between Z and Z'. The stronger statement that every isomorphism $p^\perp \to p'^\perp$ can be extended to an isomorphism $Z \to Z'$ is incorrect: Take V to be a real vector space with infinite basis $(\varepsilon_i)_{i \in \mathbb{N} \setminus \{0\}}$ and equipped with the nondegenerate quadratic form κ given by

$$\kappa(x) = x_1 x_2 + x_3 x_4 + x_5 x_6 + \sum_{i \geq 7} x_i^2 \quad \left(x = \sum_{i \geq 1} \varepsilon_i x_i \right).$$

We consider the polar space $Z := \mathbb{P}(V)_\kappa$ of Example 7.8.1. The points $p := \langle \varepsilon_1 \rangle$ and $q := \langle \varepsilon_2 \rangle$ of Z are not collinear. Let T be the linear subspace $x_2 = 0$ of V, so T is the hyperplane of $\mathbb{P}(V)$ intersecting the point set of Z in $p^\perp$. Let $U = \{x \in T \mid \sum_{i \geq 1} x_i = 0\}$; it is well defined as each vector in T has only finitely many nonzero coordinates. The set U is a hyperplane of T not on p which is not of the form $\{x \in T \mid \sum_{i \geq 1} v_i x_i = 0\}$ for $v \in V$. It is readily verified that the set H' of points of Z in U is a geometric hyperplane of $p^\perp$ not on p and spanning U. By Proposition 8.5.6 there is an automorphism α of $p^\perp$ mapping the geometric hyperplane $H = p^\perp \cap q^\perp$ of $p^\perp$ onto H'. Suppose that α extends to an automorphism $\bar{\alpha}$ of Z. Putting $r = \bar{\alpha}(q)$, we find $p^\perp \cap r^\perp = \bar{\alpha}(H) = H'$, so $p^\perp \cap r^\perp$ spans U. Therefore, $U = \{x \in T \mid \sum_{i \geq 1} v_i x_i = 0\}$, where $v = \varepsilon_1 r_2 + \varepsilon_2 r_1 + \varepsilon_3 r_4 + \varepsilon_4 r_3 + \varepsilon_5 r_6 + \varepsilon_6 r_5 + \sum_{i \geq 7} \varepsilon_i r_i$ is a vector of V, a contradiction with the choice of U.

Here is another implication of Theorem 8.5.5.

Proposition 8.5.9 *Let Z be a nondegenerate polar space of rank at least three all of whose lines are thick.*

(i) *If both a, d and a, d' are pairs of non-collinear points, then there is an automorphism of Z fixing $\{a\} \cup \{a, d, d'\}^\perp$ point-wise and mapping d onto d'. In particular, $\mathrm{Aut}(Z)$ is transitive on the set of all ordered pairs of non-collinear points.*

(ii) *The automorphism group of Z is transitive on the set of all ordered four-tuples forming a quadrangle in Z.*

Proof (i) Set $H = \{a, d\}^\perp$ and $H' = \{a, d'\}^\perp$. Both H and H' are geometric hyperplanes of $a^\perp$ not containing a. By Proposition 8.5.6, there is an automorphism α of $a^\perp$ mapping H onto H' and fixing $H \cap H'$ point-wise. It follows that α extends to an isomorphism $\alpha' : a^\perp \cup \{d\} \to a^\perp \cup \{d'\}$ fixing $\{a, d, d'\}^\perp$ point-wise. By Theorem 8.5.5, α' extends uniquely to an automorphism of Z. This gives the first part of (i).

Clearly the stabilizer in $\mathrm{Aut}(Z)$ of a is transitive on $Z \setminus a^\perp$. Suppose that q and q' are distinct points of Z. By nondegeneracy of Z and the fact that lines have more than two points, a point p can be found that is non-collinear with each of q, q'. The previous paragraph then shows the existence of an automorphism of Z (fixing p) and mapping q to q'. Thus, $\mathrm{Aut}(Z)$ is transitive on Z, and, using the previous paragraph once more, we obtain the second part of assertion (i).

(ii) Let $\mathcal{Q}$ be the set of all quadrangles and define a graph structure on $\mathcal{Q}$ by letting $Q \sim Q'$ if they meet in a path of length two. The proof of (ii) can now be finished by establishing that $(\mathcal{Q}, \sim)$ is a connected graph. But we give a different proof.

Let a, b, c, d and a', b', c', d' be two 4-circuits forming quadrangles. We will establish the existence of an automorphism mapping one onto the other. By (i) there is an automorphism of Z mapping a' and c' onto a and c, respectively. Therefore, we (may) take $a' = a$ and $c' = c$ without loss of generality. Now similarly to the proof of (i), thickness of the lines of the polar space $\{a, c\}^{\perp}$ gives that there exists $p \in a^{\perp} \cap c^{\perp} \setminus (b^{\perp} \cup b'^{\perp})$. But then Part (i) shows the existence of an automorphism of Z fixing p, a, c and mapping b onto b'. This shows that we may assume $b = b'$. But then we must have $d \not\perp b \not\perp d'$ so that, again by (i), there is an automorphism mapping d to d' and fixing a, b, c. $\qquad\qquad\square$

8.6 Exercises

Section 8.1

Exercise 8.6.1 Suppose that Z is a polar space of rank at least 2 which is not a rosette and that all of its lines have at least four points. Show that, if A, B are geometric hyperplanes of Z, then $Z \setminus (A \cup B)$ generates Z.

Exercise 8.6.2 Suppose that Z is the thin line space $K_{3,2}$, the complete bipartite graph with cocliques of size 3 and 2; here, as in Example 1.5.5, a bipartite graph is called complete if each unordered pair of points from distinct parts is an edge. Prove that Z is a polar space of rank two, which is neither amply connected nor a rosette. Conclude that the thickness hypothesis in Theorem 8.1.4 is needed.

Exercise 8.6.3 Consider again the generalized hexagon of order $(2, 2)$ of Example 2.2.15. Prove that the 32 points at distance three from $\langle a \rangle$ partition into two sets of size 16 that are disconnected in the collinearity graph of the generalized hexagon. Conclude that the geometric hyperplane of the points at distance at most two from a is not a maximal subspace.

Exercise 8.6.4 Let Z be a thick nondegenerate polar space of rank at least three and let p and q be non-collinear points of Z. Prove that each point x of $q^{\perp} \setminus (p^{\perp} \cup \{q\})$ is uniquely determined by its collinearity with each point of $\{p, x\}^{\perp}$:

$$\{x\} = \left(q^{\perp} \setminus \left(p^{\perp} \cup \{q\}\right)\right) \cap \{x, p\}^{\perp\perp}.$$

Section 8.2

Exercise 8.6.5 Let Z be a finite generalized quadrangle of order (s, t).

(a) Let Z' be a subspace of Z that is a generalized quadrangle of order (s, t'). Show that Z' is a geometric hyperplane of Z if and only if $t' = t/s$.

(b) Let q be a prime power. Use a hermitian polarity to show the existence of a generalized quadrangle Z of order (q, q^2).
(c) Show that Lemma 8.2.1 need not hold if geometric hyperplanes are not assumed to be amply connected.
   (*Hint:* For Z as in (b), consider the geometric hyperplanes Z' as in (a).)

Exercise 8.6.6 Let Z be a nondegenerate polar space of rank at least three with the property that, for each pair $\{A, B\}$ of distinct geometric hyperplanes of Z and point p of Z outside $A \cup B$, there is a geometric hyperplane of Z containing $\{p\} \cup (A \cap B)$.

(a) Prove that the Veldkamp space $\mathcal{V}(Z)$ is a projective space.
(b) Show that $\mathcal{V}_x := \{H \in \mathcal{V}(Z) \mid x \in H\}$ is a hyperplane of $\mathcal{V}(Z)$ for each point x of Z.
(c) Verify that the collection $\{\mathcal{V}_x \mid x \in Z\}$ of hyperplanes of $\mathcal{V}(Z)$ satisfies Condition (2) but not (1) of Exercise 5.7.19.

Exercise 8.6.7 Give a line space whose Veldkamp space is not a linear space.

Exercise 8.6.8 Suppose that Z is a generalized quadrangle of order $(2, 4)$. Prove that Z embeds in a projective space of dimension 5.

Exercise 8.6.9 Suppose that Z is a generalized quadrangle embedded in a projective space $\mathbb{P}$. Show that the dimension of $\mathbb{P}$ is at least three.

Exercise 8.6.10 Determine the Veldkamp space of an $m \times n$-grid.

Exercise 8.6.11 Suppose that Z is a finite generalized quadrangle of order (s, t) with two disjoint ovoids O_1, O_2. Show that $|O_1 O_2| \leq s + 1$, with equality if and only if the members of the line $O_1 O_2$ in $\mathcal{V}(Z)$ partition the point set of Z.

Section 8.3

Exercise 8.6.12 In Proposition 8.3.5(iii), a part, namely R^3 of a geometric hyperplane R in a nondegenerate polar space Z of rank at least four is assumed to be non-empty. As this part needs considerable attention in Sect. 8.3, it will be a comfort to know that it may actually happen that it is non-empty. Here we provide such an example. We take Z to be the absolute in $\mathbb{P}(\mathbb{R}^7)$ with respect to the symmetric bilinear form f on $\mathbb{R}^7$ given by

$$f(x, y) = x_1 y_2 + x_2 y_1 + x_3 y_4 + x_4 y_3 + x_5 y_6 + x_6 y_5 + x_7 y_7$$

where $x = \sum_{i=1}^{7} \varepsilon_i x_i$ and $y = \sum_{i=1}^{7} \varepsilon_i y_i$ are in $\mathbb{R}^7$. Furthermore, we set

$$A = \{\langle x \rangle \in Z \mid x_5 = 0\}, \qquad B = \{\langle x \rangle \in Z \mid x_7 = x_6\},$$

$$C = \{\langle x \rangle \in Z \mid x_7 = -x_6\}, \qquad P = \{\langle x \rangle \in Z \mid x_6 = x_5 + x_7\},$$
$$Q = \{\langle x \rangle \in Z \mid x_6 = x_5 - x_7\}, \qquad R = \{\langle x \rangle \in Z \mid x_7 = 0\}.$$

Prove the following statements.

(a) The sets A, B, C, P, Q, and R are geometric hyperplanes of Z.
(b) The point A of the Veldkamp space $\mathcal{V}(Z)$ is the intersection point of BP and CQ, and R is the intersection point of BC and PQ. So they are the points of a Pasch configuration as in Fig. 8.2.
(c) Show that $\langle \varepsilon_6 \rangle$ lies in R^3.

Exercise 8.6.13 (This exercise is used in Example 8.3.17) Prove that the line space Z_p of Example 8.3.17 is indeed a generalized quadrangle.

♮ **Exercise 8.6.14** Let Z be a nondegenerate polar space of rank at least five. Suppose that A is a geometric hyperplane of Z and K is a geometric hyperplane of A. Fix $p \in Z \setminus A$. and consider the following sets of lines of Z.

(1) L_K: the lines of K;
(2) L_p: the lines joining p to a point of K;
(3) L_∞: the lines joining a point of K to a point of $Z \setminus (K \cup \{p\})$ on a member of L_p.

Take H to be the union of all lines in $L_K \cup L_p \cup L_\infty$. Prove that H is a geometric hyperplane of Z containing $K \cup \{p\}$. Use Exercise 8.6.6 or Theorem 8.2.7 to conclude that $\mathcal{V}(Z)$ is a projective space.
(*Hint:* The hard part is to prove that H is a subspace. To this end, distinguish cases and use the existence of singular planes on p having a line in K.)

Section 8.4

♮ **Exercise 8.6.15** Let Z be the Grassmannian of lines of $\mathbb{P}(\mathbb{D}^4)$ for a non-commutative division ring $\mathbb{D}$. Prove that Z does not embed in a projective space.

Exercise 8.6.16 Show that the Veldkamp space of a dualized projective space, discussed in Example 7.10.1(ii), is a projective space (some of whose lines are thin).

Section 8.5

Exercise 8.6.17 Consider the notion of isomorphism given in Definition 8.5.1. Show that there exist isomorphisms on the union of two lines in a projective plane of order at least three that do not extend to isomorphisms of the full plane.

Exercise 8.6.18 Let Z be a dualized projective space built by means of a non-Moufang projective plane; cf. Example 7.10.1. Show that Proposition 8.5.6 does not hold for Z.

8.7 Notes

Although, Tits' classification of nondegenerate polar spaces of rank at least three [286] proves their embeddability under the conditions stated in this chapter, Veldkamp [297] indicated a direct approach that was followed up in this chapter. We needed to fill various gaps in Veldkamp's proof. This led us to the long series of proofs of Sect. 8.4. It is conceivable that Teirlinck's [273] methods (cf. Exercise 5.7.19) as well as the results at the end of this chapter on the existence of automorphisms, lead to alternative proofs to those given in Sects. 8.3 and 8.4.

Section 8.1

The basic ingredient in Veldkamp's approach is the concept of a geometric hyperplane (sometimes also referred to as a projective hyperplane). Early occurrences of it can be found in [297] and [114]. In [52] it does not even have a name. Geometric hyperplanes also appeared in, for instance, [243] and [82]. Remark 8.1.11 stems from [107, Proposition 2.4]. The ideas involving ample connectedness and Veldkamp space have been found and developed during our revision of Veldkamp's work. For some further analysis of Veldkamp's embedding techniques, see [256]. A broader study of connectivity among hyperplane complements of shadow spaces of buildings can be found in [189].

Section 8.2

A generalization of the notion of embedding of a polar space in projective space, namely where lines are embedded in (but not necessarily equal to) projective lines, is dealt with in [108, 268, 269]. In the first paper, applications to the study of subgroups of Lie type are also given.

For $j \in [n]$, the Grassmannian of j-dimensional singular subspaces of the polar space of a nondegenerate quadratic form κ on $\mathbb{F}_q^{2n}$ of Witt index n is understood to be $\mathrm{ShSp}(\mathbb{P}(\mathbb{F}_q^{2n})_\kappa, j)$. (Notice that this is in accordance with Definition 7.9.1). In [1], its Veldkamp space is shown to be a projective space.

The Veldkamp space of some generalized quadrangles with lines of length three have been investigated in relation to quantum computation; see [298].

Section 8.3

It turned out that a relatively simple proof could be given for rank at least four (which resulted in Sect. 8.3) and that the difficult Veldkamp proof would become a little more transparent by restricting to the rank three case (as done in Sect. 8.4). It took some effort to complete a couple of arguments in our original proof of the case where the rank is at least four, and we are grateful to Pasini for his help.

An even simpler proof (due to Pasini) in the case where the rank is at least five is indicated in Exercise 8.6.14. Cuypers, Johnson, and Pasini [109] have given an(other) elegant proof of the case of rank at least four. Recently, Shult [260] gave a proof of this case using Teirlinck's characterization of a projective space that can be found in Exercise 5.7.19. The version of this section though provides a good introduction to the hard case of rank three.

Example 8.3.17 deals with a construction of generalized quadrangles given in [145]. Most of these generalized quadrangles have been known from before this publication; cf. [236].

Section 8.4

The proof of the fundamental Theorem 8.4.25 follows Veldkamp's original strategy developed in [297]. Our revision of Veldkamp's proof took place in 1986–1988. It was influenced by Teirlinck's theory [273].

Section 8.5

The material in this section is based on suggestions of Tits made to Buekenhout.

Section 8.6

Example 9.1.5(iv) provides a solution to Exercise 8.6.9.

Exercise 8.6.14 shows how the proof of Theorem 8.3.16 can be replaced by a much simpler one if the rank of Z is at least five. We owe the exercise to Antonio Pasini. It shows how Theorem 8.2.7 can be applied. It would be a challenge to establish that the construction suggested in the exercise also works for rank four.

The result appearing in Exercise 8.6.15 can be found with proof in [156, 157, Satz 3.6].

Chapter 9
Embedding Polar Spaces in Absolutes

In Theorems 8.3.16 and 8.4.25, we saw that the most common nondegenerate polar spaces (cf. Definition 7.4.1) of rank at least three are embeddable in a projective space. So, in the continued study of a nondegenerate polar space Z of rank at least three, it is a mild restriction to assume that Z is embedded in a projective space $\mathbb{P}$. Still, the methods used in this chapter only require that the rank of Z be at least two.

Our main goal is to show that such a space Z is a subspace of the polar space $\mathbb{P}_\pi$ related to a quasi-polarity (cf. Definition 7.1.9) π of $\mathbb{P}$. This goal is achieved in Theorem 9.5.8.

Before we reach it, we will construct a subspace $\pi(p)$ of $\mathbb{P}$ for each point p of $\mathbb{P}$ that will play the role of the image of p under a quasi-polarity of $\mathbb{P}$ whose absolute (cf. Definition 7.1.9) contains Z. It is easy to see that $\pi(p)$ is a hyperplane or coincides with $\mathbb{P}$ (see Lemma 9.2.3). It is harder to show that π is a quasi-polarity.

In Sect. 9.1, we discuss the notion of a projective embedding and some properties of spaces embedded in a projective space. The construction of π takes place in Sect. 9.2. We define a set, called the defect of Z in $\mathbb{P}$, that will turn out to be the kernel of π (cf. Definition 7.1.9) once this map has been established to be a quasi-polarity. In Sect. 9.3, we prove that the map π is a quasi-polarity if the defect of Z in $\mathbb{P}$ is empty and $\mathbb{P}$ does not belong to the image of π. We use this fact to provide a relatively easy proof that, if Z is nondegenerate of rank at least two and spans $\mathbb{P}$ and if $\dim(\mathbb{P}) \geq 4$, then π is a quasi-polarity and Z is a subspace of the absolute $\mathbb{P}_\pi$ with respect to π. In Theorem 9.3.3 of this section, we prove that the point set of Z is invariant under special perspectivities of $\mathbb{P}$. These perspectivities will play a role in the case where $\dim(\mathbb{P}) = 3$ (Sect. 9.5) and in Chap. 10, where we classify proper subspaces of absolutes.

In Sect. 9.4, we deploy some algebraic machinery underlying polarities. This is used in Sect. 9.5 to prove that π is a quasi-polarity and Z embeds in $\mathbb{P}_\pi$ even if $\dim(\mathbb{P}) = 3$.

Throughout this chapter, $\mathbb{P}$ will be a (possibly non-Desarguesian) projective space of dimension $n \geq 1$, possibly infinite.

F. Buekenhout, A.M. Cohen, *Diagram Geometry*,
Ergebnisse der Mathematik und ihrer Grenzgebiete. 3. Folge / A Series of
Modern Surveys in Mathematics 57, DOI 10.1007/978-3-642-34453-4_9,
© Springer-Verlag Berlin Heidelberg 2013

9.1 Embedded Spaces

We consider (line) spaces (cf. Definition 2.5.8) occurring in the projective space $\mathbb{P}$. Later, the choice of spaces will be narrowed down to polar spaces.

Definition 9.1.1 Let $Z = (P, L)$ be a line space. It is said to be *embedded* in $\mathbb{P}$ if

(1) P is a set of points of $\mathbb{P}$;
(2) each member of L (i.e., line of Z) is a line of $\mathbb{P}$.

This means that the identity map $P \to \mathbb{P}$ determines an embedding of the line space Z in $\mathbb{P}$ (cf. Definition 2.5.8). We will say that Z is *ruled* in $\mathbb{P}$ if it is embedded in $\mathbb{P}$ and

(3) P is the union of all members of L;
(4) P generates $\mathbb{P}$.

If Z is a polar space embedded in $\mathbb{P}$, we call it an *embedded polar space* in $\mathbb{P}$. If, in addition, Z is ruled in $\mathbb{P}$, we call it a *ruled polar space* in $\mathbb{P}$.

Notice that Z is embedded in $\mathbb{P}$ in the sense of Definition 9.1.1 if and only if the identity map $Z \to \mathbb{P}$ is an embedding of Z in $\mathbb{P}$ as in Definition 2.5.8.

For the next few generalities of embedded spaces, we need the notion of partial linear space introduced in Definition 2.5.13.

Lemma 9.1.2 *If Z is a line space embedded in $\mathbb{P}$, then Z is a partial linear space and every singular subspace of Z is a subspace of $\mathbb{P}$.*

Proof Let a and b be distinct points, and suppose that l and m are lines of Z on a and b. By Definition 9.1.1(2), the lines l and m are also lines of the linear space $\mathbb{P}$, so $l = m$, which proves that Z is a partial linear space.

Suppose that a and b belong to a singular subspace X of Z. The projective line ab coincides with the line of Z on a and b and so is also contained in X. Therefore, X is a subspace of $\mathbb{P}$. $\square$

Lemma 9.1.3 *If $Z = (P, L)$ is a connected line space embedded in $\mathbb{P}$ with $L \neq \emptyset$, then Z is ruled in $\langle P \rangle_{\mathbb{P}}$.*

Proof We only need to check Definition 9.1.1(3). Pick $l \in L$. If $a \in P$, then there is a path $a, a_1, \ldots, a_t = b$ in the collinearity graph of Z from a to a point b of l. In particular, a lies on a line in L containing a_1, which, by Lemma 9.1.2, coincides with the projective line aa_1. This shows that P is contained in the union of all members of L, as required. $\square$

In view of Lemma 9.1.3, the study of line spaces embedded in $\mathbb{P}$ can usually be reduced to the study of ruled spaces in a subspace of $\mathbb{P}$.

Example 9.1.4 Let Q be a set of points of $\mathbb{P}$ generating $\mathbb{P}$ such that every $p \in Q$ is on some line entirely contained in Q. Take M to be the set of lines of $\mathbb{P}$ entirely contained in Q. Then (Q, M) is a ruled line space in $\mathbb{P}$.

Example 9.1.5 We exhibit some embedded polar spaces. Notice that a ruled polar space has rank at least two.

(i) If Z is a nondegenerate polar space of rank at least three with 'sufficiently nice' singular planes and 'sufficiently many' singular planes on each line, then, by Theorem 8.4.25, the Veldkamp space $\mathcal{V}(Z)$ is a projective space in which Z embeds.

(ii) Let $\mathbb{P} = \mathbb{P}(\mathbb{D}^4)$ be the 3-dimensional projective space over a division ring $\mathbb{D}$. In Example 7.8.2 we saw that $\mathbb{P}$ gives rise to a polar space Z whose points are the lines of $\mathbb{P}$ and whose lines correspond to incident point-plane pairs of $\mathbb{P}$. If $\mathbb{D}$ is commutative, then, by the Klein correspondence (cf. Theorem 7.8.3), Z is embeddable in a projective space of dimension five over $\mathbb{D}$. If $\mathbb{D}$ is not commutative, then Z is not embeddable in a projective space (this is the content of Exercise 8.6.15).

(iii) Every quadric $\mathbb{P}_\kappa$ (cf. Example 7.8.1) is a polar space embedded in $\mathbb{P}$. A quadric having points but no lines is an example of an embedded polar space which is not ruled. All results so far regarding polar spaces are trivial for spaces without lines. Still, in Sect. 10.3, we study quadrics without lines by use of perspectivities.

(iv) If $\mathbb{P}$ is a projective plane and $Z = (P, L)$ is an embedded polar space in $\mathbb{P}$, then L is either empty, a singleton, the set of all lines of $\mathbb{P}$, or a collection of lines on a given point of $\mathbb{P}$; cf. Exercise 9.6.1. Therefore, interesting examples of embedded polar spaces occur only if $\dim(\mathbb{P}) \geq 3$.

Notation 9.1.6 In order to avoid confusion between subspaces of Z and those of $\mathbb{P}$ generated by a set X of points in Z, we will index brackets: $\langle X \rangle_Z$ as opposed to $\langle X \rangle_{\mathbb{P}}$, whenever appropriate. As in Definition 2.2.1, we will write $x \perp y$, or, if necessary, $x \perp_Z y$, to express that x and y are collinear in Z. Finally, $x^{\perp_Z}$ or, if no confusion arises, $x^\perp$ denotes the set of points of Z collinear with x in Z.

Confusion of $x \perp y$ with $x \perp_{\mathbb{P}} y$ is not likely as the latter expresses an assertion which is always true (namely, that x and y are collinear in $\mathbb{P}$). But $x \perp y$ might be confused with $x \perp_f y$ for a sesquilinear form f (as in Notation 7.2.8).

Given an embedding of Z in $\mathbb{P}$, we can sometimes factor out some points of $\mathbb{P}$ so as to get an embedding of Z in a smaller projective space, actually a quotient in the sense of Exercise 5.7.25.

Proposition 9.1.7 *If $Z = (P, L)$ is a line space embedded in the projective space $\mathbb{P}$, and K is a subspace of $\mathbb{P}$ such that $l \cap K = \emptyset$ for each projective line l with $|l \cap P| \geq 2$, then the restriction to P of the natural quotient map $\phi : \mathbb{P} \setminus K \to \mathbb{P}/K$ is an embedding of Z in $\mathbb{P}/K$.*

Proof Recall from Example 5.3.10(iii) (which is used in Exercise 5.7.25) that ϕ is determined by $\phi(x) = \langle x, K \rangle_\mathbb{P}$ for each point x of $\mathbb{P}$ off K. If $p, q \in P$, then $l := \langle p, q \rangle_\mathbb{P} = pq$ is a projective line with $|l \cap P| \geq 2$, so $l \cap K = \emptyset$, implying that $\langle p, K \rangle_\mathbb{P}$ and $\langle q, K \rangle_\mathbb{P}$ are distinct hyperplanes of $\langle l \cup K \rangle_\mathbb{P}$, so $\phi(p) \neq \phi(q)$. This proves that P embeds in $\mathbb{P}/K$. By the properties of the natural quotient map ϕ, lines of Z are mapped onto lines of $\mathbb{P}/K$. $\qquad\square$

Example 9.1.8 Let $\mathbb{F}$ be a field and consider the quadric $\mathbb{P}(\mathbb{F}^3)_\kappa$ of the quadratic form κ on $\mathbb{F}^3$ given by $\kappa(x_1, x_2, x_3) = x_1 x_2 + x_3^2$; cf. Example 7.8.1. The subspace $K = \langle \varepsilon_3 \rangle$ (where ε_3 stands for the third standard basis vector) satisfies the conditions of Proposition 9.1.7 if and only if no projective line on $\langle \varepsilon_3 \rangle$ meets the polar space in two distinct points. There are no lines of this kind passing through $\langle \varepsilon_1 \rangle$ and $\langle \varepsilon_2 \rangle$. As all other points of $\mathbb{P}(\mathbb{F}^3)_\kappa$ are of the form $\langle \varepsilon_1 x - \varepsilon_2 x^{-1} + \varepsilon_3 \rangle$, the conditions of Proposition 9.1.7 are satisfied for K if and only if there are no two distinct nonzero elements x, y in $\mathbb{F}$ such that

$$\det \begin{pmatrix} x & -1/x & 1 \\ y & -1/y & 1 \\ 0 & 0 & 1 \end{pmatrix} = 0.$$

As this equation is equivalent to $x^2 = y^2$, the quadric $\mathbb{P}(\mathbb{F}^3)_\kappa$ is embeddable in $\mathbb{P}(\mathbb{F}^2)$ if $\mathbb{F}$ has characteristic two.

Notation 9.1.9 Let $Z = (P, L)$ be a ruled space in the projective space $\mathbb{P}$. If U is a subspace of $\mathbb{P}$, then $U \cap L$ denotes the set $\{l \in L \mid l \subseteq U\}$, and $U \cap Z$ denotes the pair $(U \cap P, U \cap L)$.

In general, such a pair $U \cap Z$ need not be a ruled line space in $\mathbb{P}$. We now focus on polar spaces embedded in a projective space.

Lemma 9.1.10 *Let $Z = (P, L)$ be an embedded polar space in the projective space $\mathbb{P}$. If U is a linear subspace of $\mathbb{P}$, generated in $\mathbb{P}$ by a set of points of Z, then $U \cap P$ is a subspace of Z and $U \cap Z$ is an embedded polar space in U.*

Proof Straightforward. $\qquad\square$

The next result uses Corollary 8.1.8 which requires lines to be thick. Recall from Definition 7.5.1 that the nondegenerate rank of a polar space Z is the rank of $Z/\mathrm{Rad}(Z)$ (for $\mathrm{Rad}(Z)$, see Definition 7.4.3).

Proposition 9.1.11 *Let Z be a ruled polar space of nondegenerate rank ≥ 2 in a projective space $\mathbb{P}$ and let H be a hyperplane of $\mathbb{P}$. Then $H \cap Z$ is a geometric hyperplane of Z and, if $\mathbb{P}$ is thick, then $H \cap Z$ generates H in $\mathbb{P}$.*

Proof As the point set of Z generates $\mathbb{P}$ (by the definition of ruled space), there is a point $p \in Z \setminus H$. Using Lemma 9.1.10, we see that $H \cap Z$ is a proper subspace of Z. Let l be a line of Z on p. The intersection $l \cap H$ consists of a point, which belongs to $H \cap Z$, so $H \cap Z$ is a geometric hyperplane of Z.

Suppose next that $\mathbb{P}$ is thick. In order to prove $\langle H \cap Z \rangle_{\mathbb{P}} = H$, it suffices to show that $H \cap Z$ and p generate $\mathbb{P}$. By maximality of geometric hyperplanes as proper subspaces of Z (use the fact that Z is nondegenerate of rank at least two and Corollary 8.1.8), the subspace of Z generated by p and $H \cap Z$ coincides with Z. Hence, $\langle p, H \cap Z \rangle_{\mathbb{P}} = \langle \langle p, H \cap Z \rangle_Z \rangle_{\mathbb{P}} = \langle Z \rangle_{\mathbb{P}} = \mathbb{P}$. $\qquad\square$

The radical of an embedded polar space can be conveniently set aside, as we show now.

Definition 9.1.12 Given two spaces A and B embedded in the projective space $\mathbb{P}$, the *join* of A and B in $\mathbb{P}$ is the space whose point set is the union of all projective lines joining a point of A and a point of B, and whose line set consists of all joining lines as well as the lines of A and of B.

By Lemma 9.1.2, the radical of a polar space embedded in $\mathbb{P}$ is a subspace of $\mathbb{P}$. The following result, which is not needed for further developments, shows how the nondegenerate quotient (cf. Proposition 7.4.10) of a polar space embedded in a projective space can be embedded in a subspace of the projective space.

Proposition 9.1.13 *Suppose that Z is an embedded polar space in the projective space $\mathbb{P}$. Let U be a subspace of $\mathbb{P}$ such that $U \cap \mathrm{Rad}(Z) = \emptyset$ and suppose that U is maximal with this property.*

(i) $U \cap Z \cong Z/\mathrm{Rad}(Z)$.
(ii) *The space $U \cap Z$ is a nondegenerate polar space embedded in U and Z is the join of $\mathrm{Rad}(Z)$ and $U \cap Z$ in $\mathbb{P}$.*
(iii) *If, in addition, Z is ruled in $\mathbb{P}$ and has nondegenerate rank at least two, then $U \cap Z$ is ruled in U.*

Proof First of all, we may assume $\mathrm{Rad}(Z) \neq \emptyset$, for otherwise $U = \mathbb{P}$, so $U \cap Z = Z$ and we are done.

(i) The subspace U is a complementary subspace to $\mathrm{Rad}(Z)$ in $\mathbb{P}$; cf. Exercise 5.7.14. The map $x \mapsto \langle x, \mathrm{Rad}(Z) \rangle_{\mathbb{P}}$ establishes the isomorphism $U \cap Z \to Z/\mathrm{Rad}(Z)$.

(ii) Let p be a point of Z outside $\mathrm{Rad}(Z) \cup (U \cap Z)$. We claim that p is on a line joining a point in $\mathrm{Rad}(Z)$ to a point in $U \cap Z$. For, the maximality of U forces $\langle U, p \rangle_{\mathbb{P}}$ to intersect $\mathrm{Rad}(Z)$ in some point s. Thus, ps is a line of Z intersecting U in some point $u \in \langle \mathrm{Rad}(Z), p \rangle_Z$. Hence $u \in Z$ and so p is on a line joining a point in $\mathrm{Rad}(Z)$ to a point in $U \cap Z$. Consequently, Z is the join in $\mathbb{P}$ of $\mathrm{Rad}(Z)$ and $U \cap Z$. Now (1), (2), (3), (4) of Definition 9.1.1 are obvious for $U \cap Z$ in U. As $a \in \mathrm{Rad}(U \cap Z)$ implies $a \in \mathrm{Rad}(Z)$ by the above, $U \cap Z$ is nondegenerate. Hence (ii).

(iii) Put $Z = (P, L)$. By (i) and the assumption that Z has nondegenerate rank at least two, the collection of lines $U \cap L$ (cf. Notation 9.1.9) is non-empty.

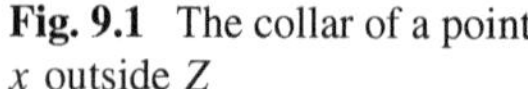

Fig. 9.1 The collar of a point x outside Z

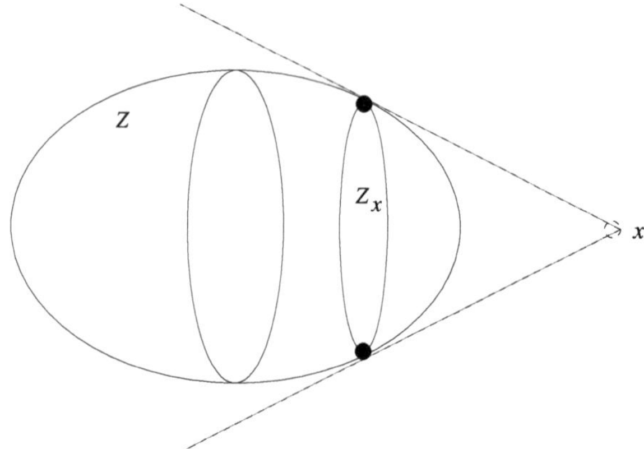

If $U \cap P$ generates a proper subspace T of U in $\mathbb{P}$, then $\langle \mathrm{Rad}(Z), T \rangle_{\mathbb{P}} = \langle \mathrm{Rad}(Z), U \cap P \rangle_{\mathbb{P}} = \langle P \rangle_{\mathbb{P}} = \mathbb{P}$ in view of (ii) and the fact that Z is a ruled space in $\mathbb{P}$. Consequently, $U \cap \mathrm{Rad}(Z) \neq \emptyset$, a contradiction. Hence $U \cap P$ generates U and Lemma 9.1.3 shows that $U \cap Z$ is ruled in U. $\qquad\square$

Example 9.1.14 Let U and V be disjoint subspaces of the projective space $\mathbb{P}$ such that $\mathbb{P} = \langle U, V \rangle_{\mathbb{P}}$. Let Z be a nondegenerate embedded polar space in V. The join of U and Z in $\mathbb{P}$ is an embedded polar space in $\mathbb{P}$ whose radical is U.

9.2 Collars and Tangent Hyperplanes

This section is devoted to the construction of a quasi-polarity for an embedded polar space. More precisely, given a polar space Z embedded in a projective space $\mathbb{P}$, we construct a map $\pi : \mathbb{P} \to \mathbb{P}^\star \cup \{\mathbb{P}\}$ which, in later sections, will turn out to be a quasi-polarity such that Z is a subspace of the absolute $\mathbb{P}_\pi$. The highlights of this section are the properties of π formulated in Theorem 9.2.5 and the transition to a quotient space of $\mathbb{P}$ so as to mod out a cumbersome subspace, called the defect, in Proposition 9.2.10.

Throughout this section, $Z = (P, L)$ is a nondegenerate polar space of rank at least two embedded in the projective space $\mathbb{P}$. For Definition 9.2.1 below, however, the restriction to polar spaces is not necessary.

Definition 9.2.1 Let $p \in P$. A line l of $\mathbb{P}$ is called a *tangent* to Z at p if either $l \cap P = \{p\}$ or $l \in L$. The *tangent set* $T(p)$ of Z at p is the union of $\{p\}$ and all tangents to Z at p.

Let $x \in \mathbb{P}$. The *collar* Z_x of x with respect to Z is the set of all points $q \in P$ such that either $q = x$ or qx is a tangent to Z at q (see Fig. 9.1). The *polar* of x with respect to Z is the projective subspace $\pi(x) := \langle Z_x \rangle_{\mathbb{P}}$ of $\mathbb{P}$.

For a tangent line l to Z not in L, the point p at which l is tangent is unique, so we can speak of l as a tangent to Z, without referring to p explicitly.

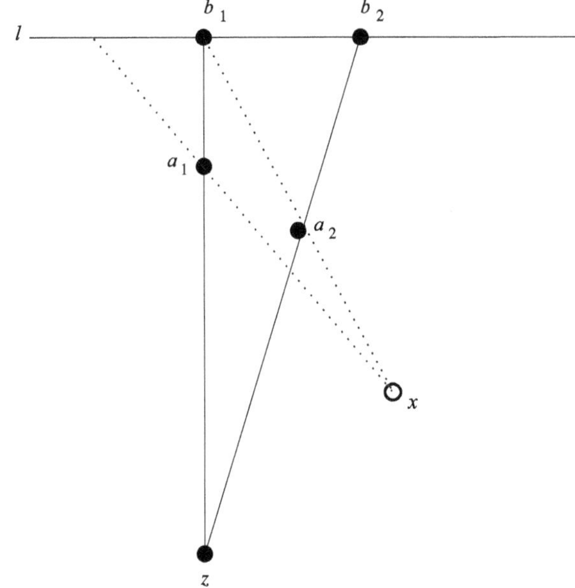

Fig. 9.2 The projective plane $\langle l, x \rangle_{\mathbb{P}}$. Points of Z are drawn *black* and lines of Z are drawn as *full lines*

In Theorem 9.2.5 below, we will prove that, if Z is a nondegenerate ruled polar space in $\mathbb{P}$, the tangent set $T(p)$ of Z at a point p of Z is a hyperplane of $\mathbb{P}$ and coincides with $\pi(p)$. In the course of this chapter, we show that, if Z is an embedded polar space in $\mathbb{P}$, the polar map π, defined on $\mathbb{P}$ by $\pi(x) = \langle Z_x \rangle_{\mathbb{P}}$, is a quasi-polarity of $\mathbb{P}$ with the property that Z is a subspace of its absolute $\mathbb{P}_\pi$ (cf. Definition 7.1.9).

Lemma 9.2.2 *For each point x of $\mathbb{P}$, the collar Z_x of x with respect to Z is either the entire point set of Z or a geometric hyperplane of Z.*

Proof Since $Z_x = x^\perp$ is a geometric hyperplane of Z when $x \in P$ (cf. Lemma 8.1.2(i)), we may assume $x \notin P$. Let $l \in L$ be such that $l \cap Z_x = \emptyset$. Consider a line in $\mathbb{P}$ on x meeting l. Beside its point on l it must have another point of Z, say a_1, for otherwise this line would be tangent to Z, contradicting $l \cap Z_x = \emptyset$. By the polar space axiom, there is a point b_1 on l collinear in Z with a_1 (cf. Fig. 9.2). Clearly, $a_1 \neq b_1$ as $a_1 \notin l$. The line $a_1 b_1$ of Z does not go through x, for otherwise x would belong to P. On the line $x b_1$ of $\mathbb{P}$, there must be a point $a_2 \in P$ for the same reason as for a_1. Similarly, there is point b_2 on l collinear in Z to a_2 such that $a_2 b_2$ does not contain x. Now, by Definition 7.4.1 of a polar space, there is a common point, z say, to $a_1 b_1$ and $a_2 b_2$, which belongs to P, since $a_1 b_1 \in L$. As z is not on l and there are two points (viz., b_1, b_2) on l collinear in Z to it, the whole plane $\langle z, l \rangle_{\mathbb{P}}$ is contained in P, a contradiction with $x \notin P$. This establishes that Z_x meets every line of Z.

Now suppose that $m \in L$ has two points, say c_1, c_2, in Z_x. Suppose that $c \in m$ does not belong to Z_x. Then the projective line cx contains a point, a say, of Z distinct from c. By the polar space axiom, there is a point $b \in m \cap a^\perp$, and, by

Pasch's Axiom, the line ab meets xc_i in a point of P, whence in c_i for $i = 1, 2$. But then $ab = c_1c_2$, contradicting that c is distinct from a. Hence every line on x meeting m is tangent to Z, so $m \subseteq Z_x$. This proves that Z_x is a subspace of Z. $\square$

The next result translates information on the collar of a point of $\mathbb{P}$ appearing in Lemma 9.2.2 to the polar of that point.

Lemma 9.2.3 *Let Z be ruled in the thick projective space $\mathbb{P}$. For $x \in \mathbb{P}$, the polar $\pi(x)$ of x with respect to Z is either equal to $\mathbb{P}$ or a hyperplane of $\mathbb{P}$.*

Proof By definition, $\pi(x)$ is a subspace of $\mathbb{P}$. Suppose that it does not coincide with $\mathbb{P}$. As Z is ruled in $\mathbb{P}$, Lemma 9.2.2 gives that Z_x is a geometric hyperplane of Z. The rank of Z is at least two and lines of Z, having the same cardinalities as lines of $\mathbb{P}$, are thick, so Corollary 8.1.8 shows that, for each $q \in Z \setminus Z_x$, we have $Z = \langle Z_x, q \rangle$. As Z is ruled, $\mathbb{P} = \langle P \rangle_{\mathbb{P}} = \langle \pi(x), q \rangle_{\mathbb{P}}$, proving that $\pi(x)$ contains a hyperplane of $\mathbb{P}$. $\square$

Lemma 9.2.4 *Suppose that Z is ruled in $\mathbb{P}$ and let $p \in Z$.*

(i) $Z_p = p^\perp = T(p) \cap Z$.

(ii) *If l is a projective line on p in a plane $\langle l_1, l_2 \rangle_{\mathbb{P}}$ generated by a line $l_1 \in L$ on p and a tangent line l_2 to Z at p, then $l \subseteq T(p)$.*

(iii) $\pi(p) \subseteq T(p)$.

(iv) *The subspace $\pi(p)$ is a hyperplane of $\mathbb{P}$.*

Proof The validity of Assertion (i) is immediate from the definitions.

(ii) Suppose that l has a point q with $q \neq p$. We show $q \in T(p)$. This suffices for the proof as $l \subseteq T(p)$ if there are no points of l in Z distinct from p. By the polar space axiom for Z (cf. Definition 7.4.1), there is a line m of Z on q intersecting l_1 in a point p_1 of Z. If m contains p, then $q \in T(p)$, and we are done. Therefore, we assume $p_1 \neq p$. By Pasch's Axiom, m intersects l_2 in a point, p_2 say, distinct from p; see Fig. 9.3. Now l_2 contains the two points p and p_2 of Z and so, being a tangent line, is a line of Z. Moreover, $p^\perp$ contains p_1 (as $p, p_1 \in l_1$) and p_2, whence m and q, proving $q \in T(p)$.

(iii) Let l be a line on p in $\pi(p)$. By the definition $\pi(p) = \langle Z_p \rangle_{\mathbb{P}}$ and Exercise 2.8.21, there is a finite number of lines of Z on p, say $l_1, l_2, \ldots, l_k$, such that $l \in \langle l_1, \ldots, l_k \rangle_{\mathbb{P}}$. We establish that l is tangent to Z at p by induction on k. For $k = 2$, this is done in (ii). Put $W = \langle l_1, \ldots, l_{k-1} \rangle_{\mathbb{P}}$ and examine $U = \langle W, l_k \rangle_{\mathbb{P}}$. Without loss of generality, we may assume that $l \neq l_k$. The plane $\alpha = \langle l, l_k \rangle_{\mathbb{P}}$ intersects W in a line l'. Thus, $\alpha = \langle l', l_k \rangle_{\mathbb{P}}$. By the induction hypothesis, l' is tangent to Z at p, so α is a plane as in (ii); therefore l is tangent to Z at p. This gives $\pi(p) = \langle Z_p \rangle_{\mathbb{P}} \subseteq T(p)$.

(iv) If $\pi(p) = \mathbb{P}$, then, by (i) and (iii), $p^\perp = T(p) \cap Z \supseteq \pi(p) \cap Z = Z$, so $p \in \mathrm{Rad}(Z)$, contradicting that Z is nondegenerate. This shows that $\pi(p)$ is contained in a hyperplane of $\mathbb{P}$. But, for $q \in Z \setminus p^\perp$, the subspace $\langle Z_p, q \rangle_{\mathbb{P}}$ of $\mathbb{P}$ contains $\langle p^\perp, q \rangle_Z$, which, by Corollary 8.1.8, coincides with Z. Therefore, the assumption

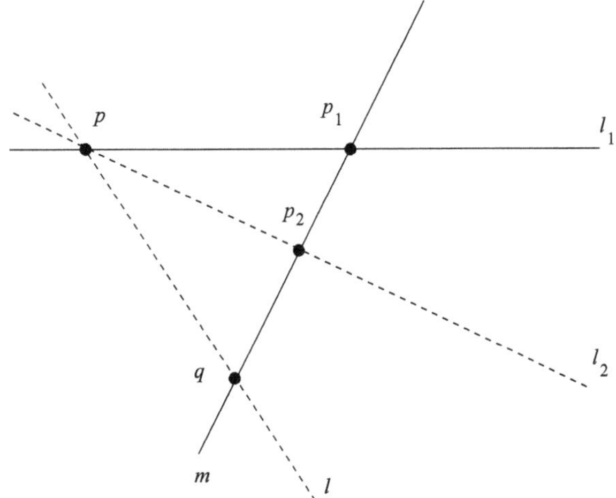

Fig. 9.3 The projective plane $\langle l_1, l_2 \rangle_{\mathbb{P}}$ of Lemma 9.2.4(ii). Lines of $\mathbb{P}$ that do not necessarily belong to Z are interrupted

that Z is ruled forces $\langle \pi(p), q \rangle_{\mathbb{P}} = \langle Z_p, q \rangle_{\mathbb{P}} = \langle Z \rangle_{\mathbb{P}} = \mathbb{P}$. From Proposition 5.2.9 we conclude that $\pi(p)$ is a hyperplane of $\mathbb{P}$.   □

We next show that, in the setting of Lemma 9.2.4, the tangent set $T(p)$ of Z at p coincides with the hyperplane of $\mathbb{P}$ generated by the subspace $p^{\perp}$ of Z, and that the points of this hyperplane lying in Z form the subspace $p^{\perp}$ of Z.

Theorem 9.2.5 *If Z is a nondegenerate ruled polar space in the thick projective space $\mathbb{P}$ and $p \in Z$, then $\pi(p)$ is a hyperplane of $\mathbb{P}$ satisfying*

$$\pi(p) = T(p) \quad and \quad Z \cap \pi(p) = p^{\perp}.$$

Moreover, for $q \in Z$ with $q \neq p$, we have $\pi(p) \neq \pi(q)$.

Proof By Lemma 9.2.4(i) and Theorem 7.4.13 (applied with $X = Z$), the collar $Z_p = p^{\perp}$ is a geometric hyperplane of Z, so, by Corollary 8.1.8 and the fact that Z is nondegenerate and (being ruled) of rank at least two, it is a maximal subspace of Z. But $Z \cap \pi(p)$ is a subspace of Z containing Z_p and distinct from Z (for otherwise, $\mathbb{P} = \langle Z \rangle_{\mathbb{P}} = \langle Z \cap \pi(p) \rangle_{\mathbb{P}} \subseteq \pi(p)$, contradicting Lemma 9.2.4(iv)). We conclude that Z_p coincides with $Z \cap \pi(p)$.

The inclusion $\pi(p) \subseteq T(p)$ is stated in Lemma 9.2.4(iii). Thus, for the proof of the first equality, it suffices to show that every tangent line to Z at p is contained in $\pi(p)$. Assume that l is a tangent line to Z at p that is not contained in $\pi(p)$. Let $q \in Z \setminus p^{\perp}$. By the above, $Z \cap \pi(q) = q^{\perp}$, so $p \notin \pi(q)$. Lemma 9.2.3 shows that l meets $\pi(q)$ in a single point, say x. Both p and q belong to Z_x.

Let m be a line of Z on p. Then $\alpha = \langle m, l \rangle_{\mathbb{P}}$ is a projective plane in which $\alpha \cap Z$ is the line m (for otherwise, by Lemma 9.2.4(ii), $\alpha \cap Z$ would contain more lines on p, contradicting $l \not\subseteq \pi(p)$). But this implies $m \subseteq Z_x$, and so $p^{\perp} \subseteq Z_x$. Now Z_x contains both q and $p^{\perp}$, so it is a subspace of Z (cf. Lemma 9.2.2) properly containing the maximal subspace $p^{\perp}$ of Z. This gives $Z_x = Z$.

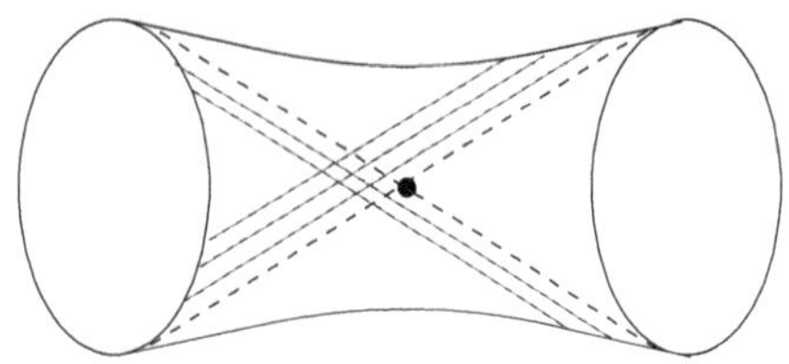

Fig. 9.4 A ruled surface, representing a quadric in $\mathbb{P}(\mathbb{R}^4)$. The two *dotted lines* are the only two lines on the *bold point* of the quadric that belong to the quadric

According to Proposition 9.1.7, there is a quotient map $\phi : \mathbb{P} \setminus \{x\} \to \mathbb{P}/\{x\}$ of projective spaces with kernel $\{x\}$ such that $\phi|_Z : Z \to \phi(Z)$ is an isomorphism of line spaces. In the quotient space $\mathbb{P}/\{x\}$, the embedded polar space $\phi(Z)$ has points sx for $s \in Z$, and lines $\langle l, x \rangle_{\mathbb{P}}$ for l a line of Z. Applying Lemma 9.2.4(iv) to the point px of $\phi(Z)$, we find that $(px)^{\perp \phi(Z)}$ spans a hyperplane of $\mathbb{P}/\{x\}$. As $\langle p^\perp \cup \{x\} \rangle_{\mathbb{P}/\{x\}} = \phi(\langle p^\perp \rangle_{\mathbb{P}}) = \langle (px)^{\perp \phi(Z)} \rangle_{\mathbb{P}/\{x\}}$, the subset $p^\perp \cup \{x\}$ of $\mathbb{P}$ spans a hyperplane of $\mathbb{P}$ containing the hyperplane $\pi(p)$ (cf. Lemma 9.2.4(i)). This forces $\langle p^\perp \cup \{x\} \rangle_{\mathbb{P}} = \pi(p)$ and so $x \in \pi(p)$. Consequently, $l \subseteq \pi(p)$. We have shown $T(p) = \pi(p)$.

Finally, let $q \in Z \setminus \{p\}$. If $\pi(p) = \pi(q)$, then $p^\perp = Z_p = Z \cap \pi(p) = Z \cap \pi(q) = Z_q = q^\perp$, contradicting Lemma 7.4.8(iv). $\qquad\square$

Corollary 9.2.6 *If $\mathbb{P}$ is a thick projective space, p is a point of a nondegenerate ruled polar space Z in $\mathbb{P}$, and if $q \in \mathbb{P}$, then $q \in \pi(p)$ if and only if $p \in Z_q$.*

Proof By Theorem 9.2.5, $q \in \pi(p)$ if and only if $q \in p^\perp$ or $pq \cap Z = \{p\}$, which is equivalent to $p \in Z_q$. $\qquad\square$

Our goal is of course to show that π is a quasi-polarity. This will suffice for the embedding of Z in the absolute of π, as the next lemma shows.

Lemma 9.2.7 *If π is a quasi-polarity, then Z is a subspace of $\mathbb{P}_\pi$.*

Proof Suppose that x and y are points of Z. They are collinear if and only if $x \in y^\perp$, which is equivalent to $x \in \pi(y)$ by Theorem 9.2.5. Taking $y = x$, we find $x \in \pi(x)$, proving that points of Z are points of $\mathbb{P}_\pi$. If x and y are distinct, we find, as $y \in \pi(y)$, that x and y are collinear if and only if xy is contained in $\pi(y)$, and similarly in $\pi(x)$. Consequently, x and y are on a line of Z if and only if they satisfy $xy \subseteq \pi(x) \cap \pi(y)$, which means they are on a line of $\mathbb{P}_\pi$. $\qquad\square$

Example 9.2.8 In the projective space $\mathbb{P}(\mathbb{R}^4)$, the quadric

$$Z = \left\{ \langle x \rangle \in \mathbb{P}(\mathbb{R}^4) \mid x_1 x_2 + x_3 x_4 = 0 \right\}$$

contains the point $p = \langle \varepsilon_1 \rangle$. The tangent set of Z at p is the plane defined by the equation $x_2 = 0$. It intersects Z along two lines, which generate the tangent plane at p; see Fig. 9.4.

We need the following notion to take care of a kind of degeneracy for embedded polar spaces.

Definition 9.2.9 Let Z be a ruled polar space in $\mathbb{P}$. A point $x \in \mathbb{P}$ is called *defective* for Z if x is in the polar $\pi(p)$ of each point p of Z. The set of all defective points for Z is called the *defect* of Z in $\mathbb{P}$.

Similarly to the last part of the proof of Theorem 9.2.5, it is useful to mod out points x of $\mathbb{P}$ whose collars coincide with Z. The prudent approach seems to go over to the quotient with respect to the defect only rather than the set $\{x \in \mathbb{P} \mid \pi(x) = \mathbb{P}\}$, which would be the kernel of π (cf. Definition 7.1.9) if it were a quasi-polarity. Later, in Theorem 9.5.8, once π has been established to be a quasi-polarity, there will appear to be no difference between the defect and $\mathrm{Ker}\,(\pi)$.

Proposition 9.2.10 *The defect D of a nondegenerate ruled polar space $Z = (P, L)$ in the thick projective space $\mathbb{P}$ has the following properties.*

 (i) *D is a subspace of $\mathbb{P}$ disjoint from P.*
 (ii) *$D = \{x \in \mathbb{P} \mid Z_x = P\} \subseteq \{x \in \mathbb{P} \mid \pi(x) = \mathbb{P}\}$.*
 (iii) *The restriction to P of the natural quotient map $\phi : \mathbb{P} \setminus D \to \mathbb{P}/D$ given by $\phi(x) = \langle x, D \rangle_{\mathbb{P}}$ is an embedding of Z in $\mathbb{P}/D$ whose image has empty defect.*
 (iv) *If $x, x' \in \mathbb{P} \setminus D$ satisfy $\phi(x) = \phi(x')$, then $\pi(x) = \pi(x')$.*
 (v) *For $x \in \mathbb{P} \setminus D$, the polar $\overline{\pi}(\phi(x))$ of $\phi(x)$ with respect to $\phi(Z)$ coincides with $\phi(\pi(x))$.*
 (vi) *If $\overline{\pi}$ as in (v) is a nondegenerate quasi-polarity of $\mathbb{P}/D$, then π is a quasipolarity of $\mathbb{P}$ with kernel D.*

Proof (i) The set $D = \bigcap_{p \in P} \pi(p)$ is a subspace as each $\pi(p)$ is a subspace of $\mathbb{P}$ by definition of $\pi(p)$. If $x \in P \cap D$, then, for each $y \in P$, by Theorem 9.2.5, $x \in \pi(y) \cap P = y^{\perp}$, so $x \in \mathrm{Rad}(Z)$, contradicting that Z is nondegenerate. Hence $P \cap D = \emptyset$.

(ii) Suppose that $x \in \mathbb{P}$ is defective. Then $x \in \pi(p)$ for each $p \in P$. This means that xp is tangent to Z at p, so $p \in Z_x$, proving $P = Z_x$. Moreover, as Z is ruled, $\pi(x) = \langle Z_x \rangle_{\mathbb{P}} = \langle P \rangle_{\mathbb{P}} = \mathbb{P}$, so $\pi(x) = \mathbb{P}$.

Conversely, suppose that $x \in \mathbb{P}$ satisfies $Z_x = P$. Theorem 9.2.5 gives $x \notin P$ and that, for each $p \in P$, the line xp is a tangent of Z at p. This implies $x \in \pi(p)$ for each $p \in P$, proving $x \in D$.

(iii) If l is a projective line with $|l \cap P| \geq 2$, then, by Definitions 9.2.1 and 9.2.9, we have $l \cap D = \emptyset$. Therefore Proposition 9.1.7 applies, proving that the restriction of ϕ to P is an isomorphism $Z \to \phi(Z)$, so $\phi(Z)$ is a nondegenerate ruled polar space in $\mathbb{P}/D$.

Now suppose that x is a point of $\mathbb{P} \setminus D$ such that $\langle x, D \rangle_{\mathbb{P}}$ is a point of the defect of $\phi(Z)$ in $\mathbb{P}/D$. Then, for each point $p \in P$, the line $\langle px, D \rangle_{\mathbb{P}}$ of $\mathbb{P}/D$ meets $\phi(P)$ in $\langle p, D \rangle_{\mathbb{P}}$; in particular, $px \cap P = \{p\}$, so px is a tangent line to Z, and $x \in \pi(p)$ by Theorem 9.2.5. But then $x \in D$, a contradiction. Hence the defect of $\phi(Z)$ in $\mathbb{P}/D$ is empty.

(iv) For x and x' as stated, the projective line xx' has a point d in D. If $y \in Z_x$, then $x, d \in \pi(y)$ so $x' \in xd \subseteq \pi(y)$ and Corollary 9.2.6 gives $y \in Z_{x'}$. This shows, by symmetry in the argument with respect to x and x', that $Z_x = Z_{x'}$, whence $\pi(x) = \pi(x')$.

(v) Let $y \in P$ be such that $\phi(y) \in (\phi(Z))_{\phi(x)}$. If $\phi(x) = \phi(y)$, then $x = y$ or $xy \cap D = \{d\}$ for some point d in D, and, as $Z = Z_d$ by (ii), the point y is the unique point of Z on the line yd through x, so $x \in T(y) = \pi(y)$ (cf. Theorem 9.2.5) and $y \in Z_x$ (cf. Corollary 9.2.6). Suppose now $\phi(x) \neq \phi(y)$. The projective line on $\phi(x)$ and $\phi(y)$ is tangent to $\phi(Z)$ at $\phi(y)$. If $\phi(x) \in \phi(P)$, then $\phi(x)\phi(y)$ belongs to $\phi(L)$, and, by (iv), we may take $x \in P$, so x and y are collinear in Z, whence $y \in Z_x$. Assume therefore $\phi(x) \notin \phi(P)$. As $\phi(x) \neq \phi(y)$, the projective line xy is disjoint from D. If $z \in xy \cap P$ is distinct from y, then $\phi(z)$ is in $\phi(x)\phi(y) \cap \phi(P)$ and distinct from $\phi(y)$, contradicting the fact that $\phi(x)\phi(y)$ is a tangent line to $\phi(Z)$ at $\phi(y)$. Therefore, xy is a tangent line to Z at y, so $y \in Z_x$. In all cases, we find $y \in Z_x$ (up to a suitable change of x within $\phi(x)$), so $\overline{\pi}(\phi(x)) = (\phi(Z))_{\phi(x)} \subseteq \phi(Z_x)$. As $\langle \phi(Z_x) \rangle_{\mathbb{P}/D} = \phi(\langle Z_x \rangle_{\mathbb{P}}) = \phi(\pi(x))$, it follows that $\overline{\pi}(\phi(x)) \subseteq \phi(\pi(x))$.

Conversely, let $y \in Z_x$. Suppose that $z \in P \setminus \{y\}$ satisfies $\phi(z) \in \phi(xy)$. Then $z \notin xy$, so $\langle x, y, z \rangle_{\mathbb{P}}$ is a projective plane meeting D in a point, say d. Now xd and yz are two lines in that plane, and so they meet in a point, say a. As $y \in Z_d \cap Z_x$, we have $a \in xd \subseteq \pi(y)$ (cf. Corollary 9.2.6), so $y \in Z_a$, and y is the unique point of Z on $ya = yz$, a contradiction with the choice of z. We conclude that $\phi(Z_x) \subseteq (\phi(Z))_{\phi(x)}$, and the inclusion $\phi(\pi(x)) \subseteq \langle (\phi(Z))_{\phi(x)} \rangle_{\mathbb{P}/D}$ follows.

(vi) Suppose $x \in \pi(y)$. Then $\phi(x) \in \phi(\pi(y)) = \overline{\pi}(\phi(y))$ by (iv). As $\overline{\pi}$ is a quasi-polarity, we also have $\phi(y) \in \overline{\pi}(\phi(x)) = \phi(\pi(x))$ (by (iv) again) and so $y \in \langle \pi(x), D \rangle$. As $\overline{\pi}(\phi(x))$ is a hyperplane of $\mathbb{P}/D$ by the hypothesis that $\overline{\pi}$ is non-degenerate, $D \subseteq \pi(x)$ and $\pi(x) = \langle D, \pi(x) \rangle_{\mathbb{P}}$ is a hyperplane of $\mathbb{P}$ containing y. This proves that π is a quasi-polarity.

Finally, suppose $x \in \mathrm{Ker}\,(\pi)$. Then $\pi(x) = \mathbb{P}$. As π is a quasi-polarity, x lies in $\bigcap_{y \in \mathbb{P}} \pi(y)$ and so in D, which settles $\mathrm{Ker}\,(\pi) \subseteq D$. The other inclusion is part of Assertion (ii). $\square$

Example 9.2.11 Let $\mathbb{F}$ be the field $\mathbb{F}_2(t)$ of rational functions over $\mathbb{F}_2$ in t. Consider the projective space $\mathbb{P} := \mathbb{P}(\mathbb{F}^6)$ and the quadric Z in $\mathbb{P}$ defined by the equation $x_1x_2 + x_3x_4 = x_5^2 + tx_6^2$. The quadric is a nondegenerate polar space. The defect of Z in $\mathbb{P}$ is the line $D := \langle \varepsilon_5, \varepsilon_6 \rangle$ where ε_i $(i \in [6])$ is the standard basis of $\mathbb{F}^6$. So $\mathbb{P}/D$ can be identified with the projective space on the first four coordinates of $\mathbb{F}^6$. By construction, each point of $\phi(Z)$ has a unique point of Z in its preimage under the natural quotient map $\phi : \mathbb{P} \setminus D \to \mathbb{P}/D$. We claim that the point sets of $\phi(P)$ and $\mathbb{P}(\mathbb{F}^4)$ coincide. Let $(x_1 : x_2 : x_3 : x_4)$ be an arbitrary point of $\mathbb{P}(\mathbb{F}^4)$, with all four x_i cleared of denominators, so $a := x_1x_2 + x_3x_4$ is a polynomial in $\mathbb{F}_2[t]$. Choosing x_5 and x_6 such that x_5^2 is the sum of all terms of a with even powers of t and x_6^2 is the multiple by $1/t$ of the sum of all terms of a with odd powers of t (observe that we can indeed take the square root of these expressions in $\mathbb{F}_2[t]$), we obtain a point $x = (x_1 : \cdots : x_6)$ of Z with quotient $\phi(x) = (x_1 : \cdots : x_4)$. This shows that each point of $\mathbb{P}(\mathbb{F}^4)$ belongs to $\phi(Z)$, and establishes the claim.

Let $f : \mathbb{F}^4 \times \mathbb{F}^4 \to \mathbb{F}$ be the bilinear alternating form given by $f(x, y) = x_1 y_2 + x_2 y_1 + x_3 y_4 + x_4 y_3$. Two points x, y of Z are collinear if and only if $f(\phi(x), \phi(y)) = 0$. In other words, $\phi(Z)$ is the absolute of the polarity δ_f introduced in Theorem 7.2.12. By Proposition 9.2.10(iii), Z is isomorphic to $\phi(Z)$. The defect of Z coincides with $\{x \in \mathbb{P} \mid \delta_f(x) = \mathbb{P}\}$.

Lemma 9.2.12 *Let a, b, c be distinct points of a line of $\mathbb{P}$. If Z is ruled in the thick projective space $\mathbb{P}$ with defect D, then*

(i) $Z_a \cap Z_b = Z_a \cap Z_c$;
(ii) $\langle a, b \rangle_{\mathbb{P}} \cap D \neq \emptyset$ *whenever* $Z_a = Z_b$.

Proof Denote by l the line of $\mathbb{P}$ on a, b, and c.

(i) Let $p \in Z_a \cap Z_b$. Corollary 9.2.6 gives $a, b \in \pi(p)$, whence $l \subseteq \pi(p)$. In particular, $c \in \pi(p)$, so $p \in Z_c$ by Corollary 9.2.6 again. Hence $Z_a \cap Z_b \subseteq Z_a \cap Z_c$. The other inclusion follows by symmetry.

(ii) Suppose $Z_a = Z_b$. Without loss of generality, we may assume that a is not defective, so by Proposition 9.2.10(ii), there is a point p of Z outside Z_a. The line l meets the hyperplane $\pi(p)$ (cf. Lemma 9.2.3(iv)) of $\mathbb{P}$ in a point, say d. By Corollary 9.2.6, $p \in Z_d \setminus Z_a$ and, by (i), $Z_a = Z_a \cap Z_b = Z_a \cap Z_d \subseteq Z_d$, so, by Lemma 9.2.2 and Corollary 8.1.8, $Z_d = Z$, proving that d is defective by Proposition 9.2.10(ii). $\qquad\square$

9.3 A Quasi-polarity

Let $\mathbb{P}$ be a thick projective space and Z a nondegenerate ruled polar space in $\mathbb{P}$. We continue to pursue the goal of establishing that the polar map π of Definition 9.2.1, which assigns a subspace of $\mathbb{P}$ to a point of $\mathbb{P}$, is a quasi-polarity. In view of Proposition 9.2.10, by going over to a quotient space of $\mathbb{P}$, we can (and often will) assume that the defect of Z in $\mathbb{P}$ is empty.

In Proposition 9.3.2, it will become clear that all work can be reduced to establishing that $\pi(x)$ is a hyperplane of $\mathbb{P}$ for all $x \in \mathbb{P}$. By Theorem 9.2.5 we know this is true when x is a point of Z. The presence of Z in $\mathbb{P}$ forces $\dim(\mathbb{P}) \geq 3$ (cf. Exercise 9.6.1). At the end of this section, in Theorem 9.3.7, we will achieve our goal for $\dim(\mathbb{P}) \geq 4$.

As a side effect of our analysis of ruled polar spaces in $\mathbb{P}$ we find that certain perspectivities of $\mathbb{P}$ leave Z invariant and induce automorphisms on Z; see Theorem 9.3.3.

Proposition 9.3.1 *Suppose that Z is a nondegenerate ruled polar space in $\mathbb{P}$ with empty defect. Let a, b, c be distinct points of Z.*

(i) *If $a \not\perp b$, then $\pi(a) \cap \pi(b) = \langle a^\perp \cap b^\perp \rangle_{\mathbb{P}}$.*
(ii) *If a, b, c are on a line of $\mathbb{P}$, then $\pi(a) \cap \pi(b) = \pi(a) \cap \pi(c)$.*
(iii) *If $x \in \mathbb{P}$ is such that $a, b \in Z_x$, then $\{a, b\}^{\perp\perp} \subseteq Z_x$.*
(iv) *$\{a, b\}^{\perp\perp} = \langle a, b \rangle_{\mathbb{P}} \cap Z$.*

Proof Let P be the point set of Z.

(i) Clearly, $\pi(a) \cap \pi(b) \supseteq \langle a^\perp \cap b^\perp \rangle_\mathbb{P}$. As for the converse, observe that $\langle a, a^\perp \cap b^\perp \rangle_Z = a^\perp$, whence $\langle a, a^\perp \cap b^\perp \rangle_\mathbb{P} = \langle a^\perp \rangle_\mathbb{P} = \pi(a)$ by use of Lemma 9.2.4. Proposition 5.2.9 shows that $\langle a^\perp \cap b^\perp \rangle_\mathbb{P}$ is a hyperplane of $\pi(a)$. As it is contained in $\pi(a) \cap \pi(b)$, we find the equality of (i).

(ii) In view of symmetry in the points a, b, c, it suffices to prove $\pi(a) \cap \pi(b) \subseteq \pi(c)$. First, assume that a, b, c are not on a line of Z. Let $s \in a^\perp \cap b^\perp$. Then $\pi(s)$ contains a and b, whence c, so $c \in \pi(s) \cap P = s^\perp = Z_s$ (cf. Theorem 9.2.5 and Lemma 9.2.4) and $s \in \pi(c)$ by Corollary 9.2.6. This establishes $a^\perp \cap b^\perp \subseteq \pi(c)$. By (i), $\pi(a) \cap \pi(b) = \langle a^\perp \cap b^\perp \rangle_\mathbb{P} \subseteq \pi(c)$.

Next, assume that a, b, c are on a line of Z. Let $p \in \pi(a) \cap \pi(b)$. We want to derive $p \in \pi(c)$. If $p \in P$, then Theorem 9.2.5 gives $p \in a^\perp \cap b^\perp \subseteq c^\perp$ so $p \in \pi(c)$ and we are done. Therefore, without loss of generality, we may assume $p \notin P$. Then, by Theorem 9.2.5 again, pa and pb intersect P in unique points, namely a and b, so $a, b \in Z_p$. By Lemma 9.2.2, the line ab is contained in Z_p, whence also c, so $p \in \pi(c)$, proving $\pi(a) \cap \pi(b) \subseteq \pi(c)$.

(iii) Let $x \in \mathbb{P}$ be such that $a, b \in Z_x$. If $a \perp b$, then Corollary 7.4.12 implies $\{a, b\}^{\perp\perp} = ab$ and Lemma 9.2.2 gives $ab \subseteq Z_x$, so we are done. Otherwise, by (i), $x \in \pi(a) \cap \pi(b) = \langle \{a, b\}^\perp \rangle_\mathbb{P}$, so if $c \in \{a, b\}^{\perp\perp}$, then (cf. Lemma 7.4.7) $\{a, b\}^\perp \subseteq c^\perp$, whence $x \in \langle \{a, b\}^\perp \rangle_\mathbb{P} \subseteq \langle c^\perp \rangle_\mathbb{P} = \pi(c)$, so Corollary 9.2.6 gives $c \in Z_x$, as required.

(iv) First suppose $x \in \langle a, b \rangle_\mathbb{P} \cap P$. For $y \in \{a, b\}^\perp$, we have $x \in \langle a, b \rangle_\mathbb{P} \cap P \subseteq \langle y^\perp \rangle_\mathbb{P} \cap P = y^\perp$ (cf. Theorem 9.2.5), so $\{a, b\}^\perp \subseteq x^\perp$, and hence $x \in x^{\perp\perp} \subseteq \{a, b\}^{\perp\perp}$ (cf. Lemma 7.4.7). This establishes one inclusion.

For the converse, suppose $x \in \{a, b\}^{\perp\perp}$. Then $\{a, b\}^\perp \subseteq x^\perp$. If $a \perp b$, then, by Corollary 7.4.12, $\{a, b\}^{\perp\perp} = ab = \langle a, b \rangle_\mathbb{P}$ and we are done. Suppose therefore $a \not\perp b$. Without loss of generality, we may assume $x \neq a$. Take z to be the point of $\mathbb{P}$ on the projective line xa lying in $\pi(b)$. By Corollary 9.2.6, we have $b \in Z_z$. For $y \in \{a, b\}^\perp$, we have $x \perp y$, so $z \in xa \subseteq \pi(y)$, whence $y \in Z_z$ by Corollary 9.2.6 again. This proves $\{a, b\}^\perp \subseteq Z_z$. Consequently, $b^\perp = \langle b, \{a, b\}^\perp \rangle_Z \subseteq Z_z$, whence $z = b$ (cf. Lemma 9.2.12(ii)). We conclude that x, being on the line az, belongs to the projective line $ab = \langle a, b \rangle_\mathbb{P}$. As $x \in P$, we have shown $\{a, b\}^{\perp\perp} \subseteq \langle a, b \rangle_\mathbb{P} \cap P$. $\square$

Part (ii) of the following result says that, once each polar $\pi(x)$ is a hyperplane of $\mathbb{P}$, the map π is a quasi-polarity.

Proposition 9.3.2 *Suppose that Z is a nondegenerate ruled polar space in a thick projective space $\mathbb{P}$. Let x and y be points of $\mathbb{P}$.*

(i) *If the polar $\pi(x)$ of x is a hyperplane of $\mathbb{P}$, then $Z_x = \pi(x) \cap Z$.*

(ii) *If Z has empty defect in $\mathbb{P}$ and $\pi(y)$ is a hyperplane of $\mathbb{P}$, then $x \in \pi(y)$ implies $y \in \pi(x)$. In particular, π is a nondegenerate injective quasi-polarity if no image under π coincides with $\mathbb{P}$.*

Proof Let P be the point set of Z.

(i) Assume that x is a counterexample. As $Z_x \subseteq P \cap \pi(x)$, there is a point $p \in P \cap \pi(x) \setminus Z_x$, so $\pi(x) = \langle Z_x \cup \{p\} \rangle_\mathbb{P}$. In particular, Z_x is a geometric hyperplane of Z; for otherwise, Lemma 9.2.2 gives $Z_x = P$ contradicting the existence of p. As Z is nondegenerate of rank at least two, Corollary 8.1.8 implies that Z_x is a maximal subspace of Z, so $\langle Z_x \cup \{p\} \rangle = P$. But then $\pi(x) = \langle Z_x \cup \{p\} \rangle_\mathbb{P} = \langle P \rangle_\mathbb{P} = \mathbb{P}$, contradicting the hypothesis that $\pi(x)$ is a hyperplane.

(ii) Assume that Z has empty defect, that $\pi(y)$ is a hyperplane of $\mathbb{P}$, and that $x \in \pi(y)$. If $\pi(x) = \mathbb{P}$, then clearly $y \in \pi(x)$, so, in view of Lemma 9.2.3, we may assume that $\pi(x)$ is a hyperplane of $\mathbb{P}$. Now both Z_x and Z_y are geometric hyperplanes of Z and hence (again by Corollary 8.1.8) maximal subspaces of Z. By Lemma 9.2.12(ii), they are distinct, and so there exists $p \in Z_y \setminus Z_x$. Since $p, x \in \pi(y)$, the projective line px lies in $\pi(y)$, but, as $p \notin Z_x$, it contains a point, q say, of Z distinct from p. By Lemma 9.2.12(i), $p^\perp \cap q^\perp = Z_p \cap Z_q \subseteq Z_x$. By (i), $q \in \pi(y) \cap Z = Z_y$, so $y \in \pi(q)$ (cf. Corollary 9.2.6). If $p \perp q$, then $x \in pq \subseteq Z_y$, so $y \in \pi(x)$ and we are done. If not, then Proposition 9.3.1(i) shows $\pi(p) \cap \pi(q) = \langle p^\perp \cap q^\perp \rangle_\mathbb{P}$, so $y \in \langle p^\perp \cap q^\perp \rangle_\mathbb{P} \subseteq \langle Z_x \rangle_\mathbb{P} = \pi(x)$. We have established the first assertion of (ii).

As for the second assertion of (ii), it follows from Lemma 9.2.3 that under the condition that no image of π coincides with $\mathbb{P}$, the subspace $\pi(x)$ is a hyperplane of $\mathbb{P}$, so π is a nondegenerate quasi-polarity. If $\pi(x) = \pi(y)$, then, by (i), $Z_x = Z_y$, so Lemma 9.2.12(ii) and the assumption that Z has empty defect imply $x = y$. This proves that π is injective. $\qquad\qquad\square$

An interesting consequence of the above proposition is the existence of automorphisms of Z induced from perspectivities of $\mathbb{P}$. As the dimension of the ambient projective space $\mathbb{P}$ of Z is at least three, it is Desarguesian and so it has perspectivities; cf. Example 6.1.2.

Theorem 9.3.3 *Let Z be a nondegenerate ruled polar space embedded in the thick projective space $\mathbb{P}$ with empty defect. Let $a \in \mathbb{P}$ be such that its polar $\pi(a)$ is a hyperplane of $\mathbb{P}$. For any two points b, c of Z on a projective line containing a, distinct from a and not in $\pi(a)$, the perspectivity σ of $\mathbb{P}$ with center a and axis $\pi(a)$ mapping b onto c, leaves Z invariant. Moreover, σ induces an automorphism on Z.*

Proof Put $Z = (P, L)$. We claim that, if $x \in P \setminus \pi(a)$ is mapped by σ to a point of Z, then every line $l \in L$ on x is mapped under σ onto a line of Z.

This claim suffices for the proof of the theorem as its condition is satisfied for $x = b$, and the collinarity graph of Z restricted to $P \setminus \pi(a)$ is connected (by Theorem 8.1.4), so that the conclusion of the claim can be shown to hold by induction on the distance of x to b. (There is nothing to show for lines of Z on a, nor for points in $\pi(a)$.)

In order to prove the claim, let z be the unique point in $\pi(a) \cap l$, so $z \in P$. By Proposition 9.3.2(ii), $\pi(z)$ contains a. But $\pi(z)$ also contains x and hence all points of xa, in particular $\sigma(x)$. As a result, the image $\sigma(x)$ belongs to $P \cap \pi(z) = Z_z$ (cf. Theorem 9.2.5). But then $\sigma(x) \perp z$, so $\sigma(l)$, being $z\sigma(x)$, is a line in L. $\qquad\square$

In view of Proposition 9.3.2, the remaining task for finding an embedding of Z in the absolute of a nondegenerate quasi-polarity of $\mathbb{P}$ is to prove that the polar $\pi(x)$ of a point x of $\mathbb{P}$ does not coincide with $\mathbb{P}$. In Theorem 9.3.7 below, we will prove this under the hypothesis that $\dim(\mathbb{P}) \geq 4$. The following lemma may clarify why the case $\dim(\mathbb{P}) = 3$ is hard: it is possible for the geometric hyperplane Z_x not to have lines.

Lemma 9.3.4 *Let Z be a nondegenerate ruled polar space in $\mathbb{P}$ with empty defect. If $x \in \mathbb{P}$ satisfies $\pi(x) = \mathbb{P}$, then Z_x contains no line of Z.*

Proof Suppose that x is a counterexample, so $\pi(x) = \mathbb{P}$ and Z_x contains a line of Z. The collar Z_x is contained in $p^\perp$ for no point p of Z, as $\pi(p) = \langle p^\perp \rangle_{\mathbb{P}}$ is a hyperplane of $\mathbb{P}$ by Theorem 9.2.5 and $Z_x \subseteq p^\perp$ would imply $\mathbb{P} = \pi(x) = \langle Z_x \rangle_{\mathbb{P}} \subseteq \langle p^\perp \rangle_{\mathbb{P}} = \pi(p)$. Thus (cf. Exercise 7.11.25) Z_x, being either the point set P of Z or a geometric hyperplane of Z by Lemma 9.2.2, is a nondegenerate polar space. As it contains a line by assumption, it is of rank at least two. Notice that Z_x is ruled by Lemma 9.1.3 as $\langle Z_x \rangle_{\mathbb{P}} = \pi(x) = \mathbb{P}$ by assumption. Therefore Theorem 9.2.5 applies to Z_x embedded in $\mathbb{P}$. For each point $y \in Z_x$, the tangent hyperplane $\pi(y)$ to Z at y coincides with the tangent hyperplane to Z_x at y (for both are hyperplanes of $\mathbb{P}$ by Theorem 9.2.5 and each line on y in $\pi(y)$ is clearly a tangent of Z_x). Let $y \in Z_x$ and $z \in P \setminus y^\perp$. Then $z \in (Z_x)_y$ would imply $z \in \langle (Z_x)_y \rangle_{\mathbb{P}} \cap P = \pi(y) \cap P = y^\perp$, a contradiction. Hence the projective line yz is not a tangent to Z_x at y and so contains a point $w \in Z_x$ with $w \neq y$. If $s \in y^\perp \cap w^\perp$, then $z \in P \cap yw \subseteq P \cap \langle s^\perp \rangle_{\mathbb{P}} = P \cap \pi(s) = s^\perp$, so $y^\perp \cap w^\perp \subseteq z^\perp$, whence $z \in \{y, w\}^{\perp\perp}$. According to Proposition 9.3.1(iii), this yields $z \in Z_x$. Therefore $P \setminus y^\perp \subseteq Z_x$. But then, by Corollary 8.1.7, also $P = \langle P \setminus y^\perp \rangle_Z \subseteq Z_x$, that is, $P = Z_x$, so, by Proposition 9.2.10(ii), x is defective with respect to Z, contrary to the assumption that the defect of Z is empty. $\square$

Lemma 9.3.5 *Let Z be a nondegenerate ruled polar space in $\mathbb{P}$ with empty defect. Suppose $\dim(\mathbb{P}) \geq 4$. If $x \in \mathbb{P}$ satisfies $\pi(x) = \mathbb{P}$ and p, q, r are points of Z_x, then $\langle p, q, r \rangle_{\mathbb{P}} \cap Z \subseteq Z_x$.*

Proof By Proposition 9.3.1(iii), (iv), it suffices to deal with the case where p, q, r are not on a single line of $\mathbb{P}$.

Since $\dim(\mathbb{P}) \geq 4$, the subspace $\pi(p) \cap \pi(q) \cap \pi(r)$ contains a line l on x. For $y \in l$, the points p, q, r are in Z_y.

We claim that there are distinct points y, z on l such that $\langle p, q, r \rangle_{\mathbb{P}} \cap Z \subseteq Z_y \cap Z_z$. Take $s \in \{p, q\}^\perp$. Then $s \notin Z_x$, because Z_x contains no lines of Z by Lemma 9.3.4. Hence, $\pi(s)$ intersects l at some point y distinct from x. Here, Z_y has rank ≥ 2 because it contains the line ps. Because y is not defective and Z_y is a proper subspace of Z (cf. Theorem 9.2.5), Z_y cannot contain $\{p, q\}^\perp$, p and q, so there is a point $t \neq s$ in $\{p, q\}^\perp$ such that $\pi(t)$ intersects l in a unique point z distinct from y and x. As Z_y has rank ≥ 2, Lemma 9.3.4 gives that $\pi(y)$ is a hyperplane of $\mathbb{P}$. Moreover, from $p, q, r \in Z_y$, we derive $\langle p, q, r \rangle_{\mathbb{P}} \cap Z \subseteq \langle Z_y \rangle_{\mathbb{P}} \cap Z$ which is equal to Z_y by Proposition 9.3.2(i). Consequently, $\langle p, q, r \rangle_\mathbb{P} \cap Z \subseteq Z_y$. The same argument applies with z instead of y, and so $\langle p, q, r \rangle_\mathbb{P} \cap Z \subseteq Z_y \cap Z_z$, as claimed.

By Lemma 9.2.12(i), for distinct points y, z on l, we have $Z_y \cap Z_z \subseteq Z_x$. Therefore, $\langle p, q, r \rangle_\mathbb{P} \cap Z \subseteq Z_x$, as required for the proof of the lemma. $\square$

Proposition 9.3.6 *Let* $\dim(\mathbb{P}) \geq 4$ *and let* Z *be a nondegenerate ruled polar space in* $\mathbb{P}$ *with empty defect. The polar* $\pi(x)$ *of each point* $x \in \mathbb{P}$ *is a hyperplane of* $\mathbb{P}$.

Proof Let P be the point set of Z. Suppose that $x \in \mathbb{P}$ is a counterexample. By Lemma 9.2.3, we have $\pi(x) = \mathbb{P}$ and by Theorem 9.2.5, $x \notin P$, whereas, by Lemma 9.3.4, Z_x contains no line of Z. Since x is not a defective point with respect to Z, there are distinct points p, q of Z such that p, q, and x are on a line of $\mathbb{P}$. Each $s \in \{p, q\}^\perp$ satisfies $x \in pq \in \pi(s)$, so $\{p, q\}^\perp \subseteq Z_x$. In particular, Z_x, being a geometric hyperplane of Z (cf. Lemma 9.2.2), also contains a point, r say, of $P \setminus \{p, q\}^\perp$. Now, by Lemma 9.3.1(i), $\langle \{p, q\}^\perp \rangle_\mathbb{P} = \pi(p) \cap \pi(q)$ and, together with r, this set generates a hyperplane H of $\mathbb{P}$ (as $r \notin \{p, q\}^\perp$, we have $r \notin \pi(p)$ or $r \notin \pi(q)$).

We claim that $H \cap P \subseteq Z_x$. To establish this, let s be an arbitrary point of $H \cap P$ other than r. (Obviously, r belongs to Z_x.) By construction of H, the subspace $\pi(p) \cap \pi(q)$ is a hyperplane of H, and so the line rs of H meets $\pi(p) \cap \pi(q)$ in some point y. Then Z_y contains p and q, so there is some point t in $\{p, q\}^\perp$ but not in Z_y (for otherwise Z_y would contain $\{p, q\}^\perp$, p, q and so all of P, contradicting that y is not defective). This means that there is a point $u \in P \cap yt$ distinct from t. As y and t belong to $\pi(p) \cap \pi(q)$, so does u. Consequently, the lines pu and qu are tangents to Z at p and q, respectively, and, as $u \in P$, this forces $u \in \{p, q\}^\perp$, whence $u \in Z_x$. Moreover, $s \in ry \subseteq \langle r, t, u \rangle_\mathbb{P} \cap P$. By Lemma 9.3.5, $\langle r, t, u \rangle_\mathbb{P} \cap P \subseteq Z_x$, so we find $s \in Z_x$, proving the claim $H \cap P \subseteq Z_x$.

As H is a hyperplane of $\mathbb{P}$, the intersection $H \cap P$ is a geometric hyperplane of Z (cf. Proposition 9.1.11). Since $\mathrm{rk}(Z) \geq 2$, it is a maximal subspace of Z (by Corollary 8.1.8). As $Z_x \neq P$ (for x is not defective), we find $H \cap P = Z_x$. But then $\mathbb{P} = \pi(x) = \langle Z_x \rangle_\mathbb{P} = \langle H \cap P \rangle_\mathbb{P} \subseteq H$, a contradiction with H being a hyperplane of $\mathbb{P}$. $\square$

Theorem 9.3.7 *Let* Z *be a nondegenerate ruled polar space in a thick projective space* $\mathbb{P}$ *with empty defect. If* $\dim(\mathbb{P}) \geq 4$*, then* π *is a nondegenerate injective quasi-polarity of* $\mathbb{P}$.

Proof Let x be a point of $\mathbb{P}$. By Proposition 9.3.6, its polar $\pi(x)$ is a hyperplane of $\mathbb{P}$. It follows from Proposition 9.3.2(ii) that π is a nondegenerate injective quasi-polarity. $\square$

Recall that Lemma 9.2.7 implies that, in the setting of Theorem 9.3.7, the space Z is a subspace of the polar space $\mathbb{P}_\pi$. In Exercise 9.6.5 the reader is asked to prove that π is the unique polarity with this property.

9.4 Technical Results on Division Rings

In the next section, it will be shown that even in a projective space $\mathbb{P}$ of dimension three, an embedded polar space is the subspace of the absolute of a polarity of $\mathbb{P}$. In other words, Theorem 9.3.7 also holds if $\dim(\mathbb{P}) = 3$. The proof of this fact is harder than the one given for higher dimensions. It will be given in Theorem 9.5.7 and needs the algebraic machinery of this section. Throughout this section, $\mathbb{D}$ is a division ring.

Lemma 9.4.1 *If α is a surjective map $\mathbb{D} \to \mathbb{D}$ satisfying*

  (i) $\alpha(x + y) = \alpha(x) + \alpha(y)$,
 (ii) $\alpha(x^{-1}) = \alpha(x)^{-1}$ *whenever x is nonzero,*
(iii) $\alpha(1) = 1$

for all $x, y \in \mathbb{D}$, then α is an automorphism or an anti-automorphism of $\mathbb{D}$.

Proof We provide a proof in five steps.

STEP 1. *All nonzero $a, b \in \mathbb{D}$ with $ab \neq 1$ satisfy the* Hua identity

$$a - \left(a^{-1} + \left(b^{-1} - a\right)^{-1}\right)^{-1} = aba.$$

Due to the assumptions, the expression $a^{-1} + (b^{-1} - a)^{-1}$ is well defined. Factor out a^{-1} to the left and $(b^{-1} - a)^{-1}$ to the right:

$$a^{-1} + \left(b^{-1} - a\right)^{-1} = a^{-1}\left(\left(b^{-1} - a\right) + a\right)\left(b^{-1} - a\right)^{-1} = a^{-1}b^{-1}\left(b^{-1} - a\right)^{-1}.$$

Inspection of the right hand side shows that we have a nonzero element of $\mathbb{D}$ at hand. Taking inverses, we find

$$\left(a^{-1} + \left(b^{-1} - a\right)^{-1}\right)^{-1} = \left(b^{-1} - a\right)ba = a - aba,$$

whence Step 1.

STEP 2. *For all $a, b \in \mathbb{D}$, we have $\alpha(aba) = \alpha(a)\alpha(b)\alpha(a)$.*

Let $x \in \mathbb{D}$. From (i) we find $\alpha(0) + \alpha(x) = \alpha(x)$ and $\alpha(0) = \alpha(x) + \alpha(-x)$, which directly imply $\alpha(0) = 0$ and $\alpha(-x) = -\alpha(x)$. When we apply α to the left hand side of the Hua identity in Step 1, Conditions (i) and (ii) and the above observation allow us to interchange α each time with the operations of subtracting, taking inverses, and addition. So we end up with an expression like the one we started with, but with a replaced by $\alpha(a)$ and b replaced by $\alpha(b)$. According to the Hua identity, the resulting left side equals $\alpha(a)\alpha(b)\alpha(a)$ and so $\alpha(aba) = \alpha(a)\alpha(b)\alpha(a)$. Observe that this equation is also true if a or b are 0. If $ab = 1$, then the left side of the identity of Step 2 is equal to $\alpha(a)$. As $\alpha(a)\alpha(b)\alpha(a) = \alpha(a)\alpha(a^{-1})\alpha(a) = \alpha(a)\alpha(a)^{-1}\alpha(a) = \alpha(a)$, the equation of Step 2 holds for all $a, b \in \mathbb{D}$.

STEP 3. *For all $a, b \in \mathbb{D}$, we have $\alpha(ab) + \alpha(ba) = \alpha(a)\alpha(b) + \alpha(b)\alpha(a)$.*

Setting $b = 1$ in the equation of Step 2, we find $\alpha(a^2) = \alpha(a)^2$. Replacing a by $a + b$, using (i), and subtracting quadratic terms, we derive

$$\alpha(ab + ba) = \alpha\big((a + b)^2 - a^2 - b^2\big) = \big(\alpha(a) + \alpha(b)\big)^2 - \alpha(a)^2 - \alpha(b)^2$$
$$= \alpha(a)\alpha(b) + \alpha(b)\alpha(a).$$

STEP 4. *All $a, b \in \mathbb{D}$ satisfy $\alpha(ab) = \alpha(a)\alpha(b)$ or $\alpha(b)\alpha(a)$.*

For $a = 0$ or $b = 0$, the assertion is trivially true, so let a and b be nonzero. Consider

$$\big(1 - \alpha(a)\alpha(b)\alpha\big((ab)^{-1}\big)\big)\big(\alpha(ab) - \alpha(b)\alpha(a)\big). \tag{9.1}$$

Expanding this expression yields

$$\alpha(ab) - \alpha(b)\alpha(a) - \alpha(a)\alpha(b) + \alpha(a)\alpha(b)\alpha\big((ab)^{-1}\big)\alpha(b)\alpha(a).$$

Applying Step 2 twice to the last summand, we find

$$\alpha(a)\big(\alpha(b)\alpha\big((ab)^{-1}\big)\alpha(b)\big)\alpha(a) = \alpha(a)\alpha\big(b(ab)^{-1}b\big)\alpha(a) = \alpha\big(a\big(b(ab)^{-1}b\big)a\big)$$
$$= \alpha(ba).$$

Therefore, Expression (9.1) is equal to $\alpha(ab) - \alpha(b)\alpha(a) - \alpha(a)\alpha(b) + \alpha(ba)$, which is zero by Step 3. As (9.1) is a product, one of the factors must vanish. This proves Step 4.

STEP 5. We conclude the proof by contradiction. Observe that α will be bijective if it is a homomorphism or an antihomomorphism, as it is assumed to be surjective and a nontrivial kernel would conflict with $\mathbb{D}$ being a division ring. So, if the lemma does not hold, then, in view of Step 4, there are four elements a, b, c, d in $\mathbb{D}$ such that

$$\alpha(ab) = \alpha(a)\alpha(b) \neq \alpha(b)\alpha(a) \quad \text{and} \quad \alpha(cd) = \alpha(d)\alpha(c) \neq \alpha(c)\alpha(d). \tag{9.2}$$

Let $x \in \mathbb{D}$. By Step 4 applied to a and $b + x$, we have

$$\alpha\big(a(b + x)\big) = \alpha(a)\alpha(b + x) = \alpha(a)\alpha(b) + \alpha(a)\alpha(x) \quad \text{or} \tag{9.3}$$
$$\alpha\big(a(b + x)\big) = \alpha(b + x)\alpha(a) = \alpha(b)\alpha(a) + \alpha(x)\alpha(a). \tag{9.4}$$

The left hand side in either case is equal to $\alpha(a)\alpha(b) + \alpha(ax)$. If (9.3) holds, we have $\alpha(ax) = \alpha(a)\alpha(x)$. If (9.4) holds, we must have $\alpha(ax) \neq \alpha(x)\alpha(a)$, as $\alpha(a)\alpha(b) \neq \alpha(b)\alpha(a)$. In view of Step 4, this means that $\alpha(ax) = \alpha(a)\alpha(x)$. So we always have $\alpha(ax) = \alpha(a)\alpha(x)$. The same method is used on the expressions $\alpha((a + x)b)$, $\alpha(c(d + x))$, and $\alpha((c + x)d)$. Collecting all four cases, we find:

$$\alpha(ax) = \alpha(a)\alpha(x), \tag{9.5}$$

$$\alpha(xb) = \alpha(x)\alpha(b), \tag{9.6}$$

$$\alpha(cx) = \alpha(x)\alpha(c), \tag{9.7}$$

$$\alpha(xd) = \alpha(d)\alpha(x). \tag{9.8}$$

Set $x = d$ in (9.5) and $x = a$ in (9.8); set also $x = c$ in (9.6) and $x = b$ in (9.7). We obtain

$$\alpha(a)\alpha(d) = \alpha(d)\alpha(a) \quad \text{and} \quad \alpha(c)\alpha(b) = \alpha(b)\alpha(c). \tag{9.9}$$

According to whether $\alpha((a + c)(b + d)) = \alpha(a + c)\alpha(b + d)$ or $\alpha(b + d)\alpha(a + c)$, we have

$$\alpha\big((a + c)(b + d)\big) = \alpha(a)\alpha(b) + \alpha(a)\alpha(d) + \alpha(c)\alpha(b) + \alpha(c)\alpha(d), \tag{9.10}$$

or

$$\alpha\big((a + c)(b + d)\big) = \alpha(b)\alpha(a) + \alpha(d)\alpha(a) + \alpha(b)\alpha(c) + \alpha(d)\alpha(c). \tag{9.11}$$

A direct computation of the common left hand side of (9.10) and (9.11) gives

$$\alpha(ab) + \alpha(ad) + \alpha(cb) + \alpha(cd) = \alpha(a)\alpha(b) + \alpha(a)\alpha(d) + \alpha(c)\alpha(b) + \alpha(d)\alpha(c).$$

Equation (9.10) would give $\alpha(c)\alpha(d) = \alpha(d)\alpha(c)$ which contradicts (9.2). In view of (9.9), Eq. (9.11) would give $\alpha(a)\alpha(b) = \alpha(b)\alpha(a)$, which contradicts (9.2) again. This establishes the lemma. $\qquad\square$

We recall from Example 1.8.16 that $Z(\mathbb{D})$ denotes the center of $\mathbb{D}$, that is, the subring (with identity element) of all elements commuting with each element of $\mathbb{D}$.

Theorem 9.4.2 *Let α be an automorphism of $\mathbb{D}$ such that $\alpha^2 = \mathrm{id}$. If $x\alpha(x)$, $x + \alpha(x) \in Z(\mathbb{D})$ for all $x \in \mathbb{D}$, then $\mathbb{D}$ is commutative.*

Proof For an automorphism γ of $\mathbb{D}$ and for $\varepsilon = \pm$, denote by $\mathbb{D}_\gamma^\varepsilon$ the subset $\{x \in \mathbb{D} \mid \gamma(x) = \varepsilon x\}$ of $\mathbb{D}$. We proceed in four steps.

STEP 1. *If γ is a non-identity automorphism of $\mathbb{D}$ with $\gamma^2 = \mathrm{id}$ and $x + \gamma(x) \in Z(\mathbb{D})$ for all $x \in \mathbb{D}$, then $\mathbb{D}_\gamma^+ \subseteq Z(\mathbb{D})$.*

Let $z \in \mathbb{D}_\gamma^+$. Notice that $2z = z + \gamma(z) \in Z(\mathbb{D})$. Hence, if $\mathrm{char}(\mathbb{D}) \neq 2$, then $z \in Z(\mathbb{D})$, as required.

Suppose, therefore, $\mathrm{char}(\mathbb{D}) = 2$. By assumption, there is $\lambda \in \mathbb{D}$ such that $\gamma(\lambda) \neq \lambda$. Now $z(\lambda + \gamma(\lambda)) = z\lambda + \gamma(z\lambda) \in Z(\mathbb{D})$ and $\lambda + \gamma(\lambda) \neq \lambda + \lambda = 0$, so $z = (z\lambda + \gamma(z\lambda))(\lambda + \gamma(\lambda))^{-1} \in Z(\mathbb{D})$.

STEP 2. *Assume that $\mathbb{D}$ is not commutative. If $\alpha = \mathrm{id}$, then there exists an automorphism β of $\mathbb{D}$, with $\beta \neq \mathrm{id}$, $\beta^2 = \mathrm{id}$, and $x + \beta(x) \in Z(\mathbb{D})$ for all $x \in \mathbb{D}$, such that $Z(\mathbb{D}) = \mathbb{D}_\beta^+$.*

As $\mathbb{D}$ is not commutative, there exists $\lambda \in \mathbb{D} \setminus Z(\mathbb{D})$. The conditions on α imply $2x \in Z(\mathbb{D})$ and $x^2 \in Z(\mathbb{D})$ for all $x \in \mathbb{D}$. In particular, the characteristic of $\mathbb{D}$ is equal to two and $\lambda^2 \in Z(\mathbb{D})$.

The inner automorphism $\beta : \mathbb{D} \to \mathbb{D}$ given by $\beta(x) = \lambda x \lambda^{-1}$ satisfies $\beta \neq \mathrm{id}$. As $\beta^2(x) = \lambda^2 x \lambda^{-2}$ and $\lambda^2 \in Z(\mathbb{D})$, we have $\beta^2 = \mathrm{id}$. Also, $(\lambda + x\lambda^{-1})^2 \in Z(\mathbb{D})$ and so $x + \beta(x) = (\lambda + x\lambda^{-1})^2 - \lambda^2 - (x\lambda^{-1})^2$ lies in $Z(\mathbb{D})$.

Clearly, $Z(\mathbb{D}) \subseteq \{x \in \mathbb{D} \mid \lambda x \lambda^{-1} = x\} = \mathbb{D}_\beta^+$ and, by Step 1, $\mathbb{D}_\beta^+ \subseteq Z(\mathbb{D})$.

STEP 3. *The theorem holds if $\mathbb{D}$ has characteristic two.*

Assume the contrary, so $\mathbb{D}$ is not commutative, and, in view of Step 2, there is a non-identity automorphism γ of $\mathbb{D}$ as in Step 1 (namely, α if $\alpha \neq \mathrm{id}$ and β otherwise). For $x \in \mathbb{D}$, write $x_\gamma = x + \gamma(x)$, so $x_\gamma \in \mathbb{D}_\gamma^+$.

By Step 1, $\mathbb{D}_\gamma^+$ is commutative. Fix $x \in \mathbb{D}$ with $\gamma(x) \neq x$. Then $x_\gamma \neq 0$ as $\mathrm{char}(\mathbb{D}) = 2$. We claim that every $y \in \mathbb{D} \setminus \mathbb{D}_\gamma^+$ belongs to the subfield $Z(\mathbb{D})(x)$. For,
$$xy + (xy)_\gamma = \gamma(xy) = \gamma(x)\gamma(y) = (x + x_\gamma)(y + y_\gamma) = xy + x_\gamma y + x y_\gamma + x_\gamma y_\gamma,$$
whence $x_\gamma y + x y_\gamma = (xy)_\gamma + x_\gamma y_\gamma \in \mathbb{D}_\gamma^+$, so $y = x_\gamma^{-1}(x_\gamma y + x y_\gamma) + x_\gamma^{-1} x y_\gamma \in \mathbb{D}_\gamma^+(x)$, and the claim follows from Step 1. We conclude that $\mathbb{D} = Z(\mathbb{D})(x)$ is commutative.

STEP 4. *The theorem holds if $\mathbb{D}$ has characteristic distinct from two.*

Assume the contrary. As before, by Step 2, there is a non-identity automorphism γ of $\mathbb{D}$ satisfying the properties of Step 1. As $2x = (x + \gamma(x)) + (x - \gamma(x))$, the division ring decomposes as $\mathbb{D} = \mathbb{D}_\gamma^+ + \mathbb{D}_\gamma^-$. But $(\mathbb{D}_\gamma^-)^2 \subseteq \mathbb{D}_\gamma^+$, so for $x \in \mathbb{D}_\gamma^- \setminus \{0\}$, we have $x^{-1} = x^{-2} \cdot x \in \mathbb{D}_\gamma^+ x$. If y is also in $\mathbb{D}_\gamma^-$, then $yx^{-1} \in \mathbb{D}_\gamma^+$ and so $y \in \mathbb{D}_\gamma^+ x$. Therefore $\mathbb{D}_\gamma^- = \mathbb{D}_\gamma^+ x$, which implies $\mathbb{D} = \mathbb{D}_\gamma^+ + \mathbb{D}_\gamma^+ x$. As $\mathbb{D}_\gamma^+ \subseteq Z(\mathbb{D})$ by Step 1, it follows that $\mathbb{D}$ is commutative, and we have reached a final contradiction. $\square$

9.5 Embedding in 3-Dimensional Space

Until the final theorem of this section, we let $\mathbb{P}$ be a thick projective space of dimension three. By Theorem 6.3.1, there is a division ring $\mathbb{D}$ such that $\mathbb{P} = \mathbb{P}(\mathbb{D}^4)$. Let $Z = (P, L)$ be a nondegenerate ruled polar space in $\mathbb{P}$ with empty defect (cf. Definition 9.2.9). Then Z has rank at least two and, according to Proposition 7.5.3(i) and Lemma 9.1.2, its rank is at most two, so Z is a generalized quadrangle. We want to show that π as constructed in Definition 9.2.1 is a polarity of $\mathbb{P}$. Lemma 9.2.7 then implies that Z is a subspace of $\mathbb{P}_\pi$.

By Proposition 9.3.2, this is equivalent to showing that $\pi(p)$ is a plane for each point p of $\mathbb{P}$. In fact, this is known to hold for $p \in P$ (cf. Theorem 9.2.5), so we only need verify that $\pi(p)$ is a plane for $p \in \mathbb{P} \setminus P$.

The proof occupies the greater part of this section and results in Theorem 9.5.7. The section ends with a combination of this result with those for the higher dimensional embeddings in Theorem 9.5.8. It starts with the action of π on lines of $\mathbb{P}$ that are secants of Z.

Definition 9.5.1 A line of $\mathbb{P}$ is called a *secant* of Z if it intersects P in at least two points and is not tangent to Z.

For a secant line l of Z, we define $\pi(l)$ to be the intersection of all planes $\pi(x)$ with $x \in l \cap P$. We call it the *polar line* of l.

A secant line of Z is a line of $\mathbb{P}$ but not a line of Z. By Proposition 9.3.1(ii), polar lines are lines of $\mathbb{P}$ of the form $\pi(a) \cap \pi(b)$ for non-collinear points a, b of Z.

Lemma 9.5.2 *If l is a secant line of Z, then so is $\pi(l)$, and $\pi(\pi(l)) = l$.*

Proof If a and b are distinct points in $l \cap P$, then $\{a, b\}^{\perp} \subseteq \pi(a) \cap \pi(b) \cap P \subseteq \pi(l) \cap P$, so $\pi(l)$ has at least two points of Z. As Z has rank two, $\pi(l)$ cannot be a line of Z, so $\pi(l)$ is again a secant line.

The last statement is immediate from $a, b \in \pi(\pi(l))$. $\qquad\qquad\square$

We first consider the case where all secants have exactly two points on Z. Recall that quadrics have been introduced in Definition 7.8.1.

Theorem 9.5.3 *Let $\mathbb{D}$ be a division ring. Suppose that Z is a nondegenerate ruled polar space in $\mathbb{P} = \mathbb{P}(\mathbb{D}^4)$ with empty defect. If all secants of Z have exactly two points on Z, then $\mathbb{D}$ is commutative, the polar map π is a polarity, and Z is a quadric which is a subspace of the absolute $\mathbb{P}_\pi$.*

Proof Write $Z = (P, L)$. For $p \in P$, Lemma 9.2.4 tells us that the polar $\pi(p)$ with respect to Z is a hyperplane of $\mathbb{P}$ containing p and hence of the form $\langle p, x, y \rangle$ for any two non-collinear points x and y in $p^{\perp}$. We proceed in four steps.

STEP 1. *Each point of Z is on exactly two lines of Z.*

If $p \in P$ is on more than two lines of Z, then the plane $\pi(p)$ contains at least three lines of Z through p and so the hypothesis of the theorem is contradicted by any line in $\pi(p)$ not on p (such a line does not belong to L as Z is a generalized quadrangle). Moreover, p cannot be on a single line l of Z, for otherwise $p^{\perp} \subseteq l^{\perp}$, contradicting that Z is nondegenerate (cf. Lemma 7.4.8(iii)).

STEP 2. *There is a partition of L in two families F_1, F_2 such that*

(1) *for each $i \in [2]$, the members of F_i partition P;*
(2) *each $l \in F_1$ and each $m \in F_2$ have a common point.*

Fix a line n of Z. Let F_1 be the set of lines of Z intersecting n in exactly one point. For each point $p \in n$, the unique line of Z on p distinct from n, is in F_1. If $p \in P \setminus n$, then $p^{\perp}$ intersects n in a unique point q and so pq is the only line on p in F_1. Therefore F_1 partitions P. Let F_2 be the set of lines not in F_1. It partitions the point set as well, in view of Step 1. So (1) holds.

As for (2), let $l \in F_1$, $m \in F_2$. Take $c \in m$. If $l \cap m \neq \emptyset$, we are done, so suppose $c \notin l$. Now $c^{\perp}$ contains a line k on c intersecting l. As $l \in F_1$, we must have $k \in$

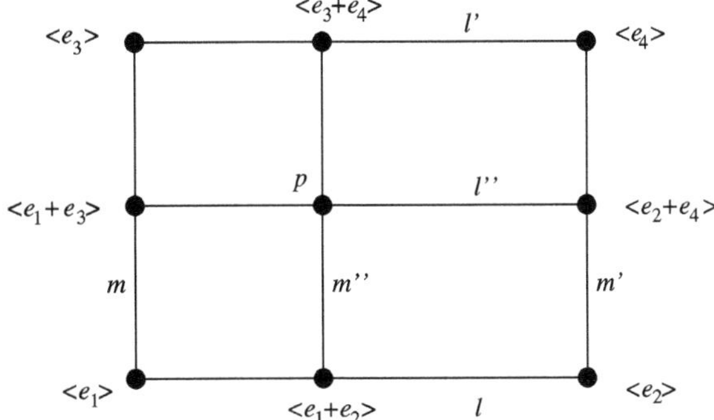

Fig. 9.5 A configuration of points and lines in Z

F_2, so both k and m belong to F_2 and contain c. Therefore, (1) forces $k = m$. In particular, $l \cap m$ is non-empty.

STEP 3. *The division ring* $\mathbb{D}$ *is commutative and the point set of the quadric* $\mathbb{P}_\kappa$ *of* $\mathbb{P}$ *corresponding to the quadratic form* κ *on* $\mathbb{D}^4$ *given by* $\kappa(x) = x_1 x_4 - x_2 x_3$ *is contained in* P.

It is not hard (cf. Exercise 9.6.11) to show that there is a basis e_1, e_2, e_3, e_4 of the right vector space $\mathbb{D}^4$ such that $l = \langle e_1, e_2\rangle \in F_1$, $l' = \langle e_3, e_4\rangle \in F_1$, $l'' = \langle e_1 + e_3, e_2 + e_4\rangle \in F_1$, $m = \langle e_1, e_3\rangle \in F_2$ and $m' = \langle e_2, e_4\rangle \in F_2$; cf. Fig. 9.5. For each scalar $\lambda \in \mathbb{D}$, consider the point

$$p := \langle e_1 + e_3 + (e_2 + e_4)\lambda\rangle \quad \text{on } l''.$$

By Step 2, the line m_p of F_2 containing p must be the intersection of the planes $\langle p, l\rangle$ and $\langle p, l'\rangle$ whose equations are, respectively,

$$\lambda x_3 = x_4 \quad \text{and} \quad \lambda x_1 = x_2.$$

Thus, for any choice of scalars $\lambda, \mu \in \mathbb{D}$, the point

$$(1 : \lambda : \mu : \lambda\mu) \tag{9.12}$$

lies on such a line m_p and so belongs to P. Taking $\lambda = 1$ in the expression for p, we find that m_p becomes the line $m'' = \langle e_1 + e_2, e_3 + e_4\rangle$. This shows that the triples l, l', l'' and m, m', m'' play symmetric roles. More precisely, it shows that we may fix e_1, e_4 and interchange e_2, e_3 in the arguments above. As a result, we find a line of Z as the intersection of the planes $\mu x_2 = x_4$ and $\mu x_1 = x_3$. This line contains the point

$$(1 : \lambda : \mu : \mu\lambda). \tag{9.13}$$

The planes $x_2 = \lambda x_1$ and $x_3 = \mu x_1$ intersect in a projective line t containing $\langle e_4\rangle$. The plane $\pi(\langle e_4\rangle)$ is given by the equation $x_1 = 0$, and so meets the line t in $\langle e_4\rangle$. In particular, t is not contained in $\pi(\langle e_4\rangle)$ and Theorem 9.2.5 shows that t is a secant

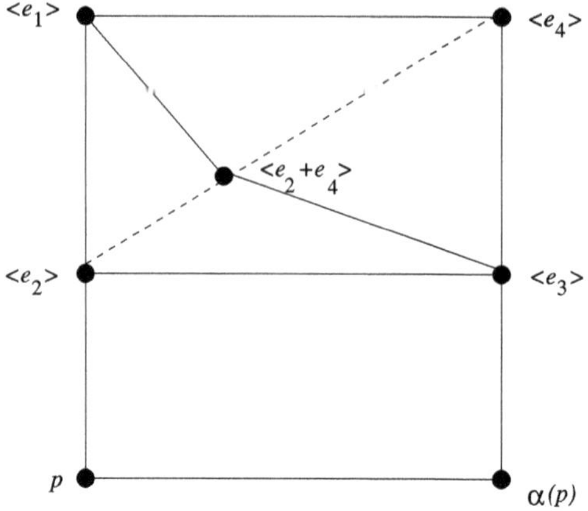

Fig. 9.6 A configuration of points and lines related to Z in the case of a secant with three points of Z. *Full lines* are lines of Z

of Z. Now the points given in (9.12) and (9.13) are distinct from $\langle e_4 \rangle$ and belong to t, so they must coincide. Inspection of the fourth coordinate shows $\lambda\mu = \mu\lambda$, from which we conclude that $\mathbb{D}$ is commutative.

Finally, suppose $\langle x \rangle \in \mathbb{P}_\kappa$. If $x_1 = 0$, then $\kappa(x) = 0$ implies $x_2 = 0$ or $x_3 = 0$, so $\langle x \rangle \in \langle e_3, e_4 \rangle \cup \langle e_2, e_4 \rangle = l' \cup m' \subseteq P$. Otherwise, we can take $x_1 = 1$, so $\kappa(x) = 0$ forces $x_4 = x_2 x_3$ and $\langle x \rangle = (1 : x_2 : x_3 : x_2 x_3) \in P$. This establishes that the point set of $\mathbb{P}_\kappa$ is contained in P.

$\textsc{Step}$ 4. *The quadric and the polarity.*

From Step 3 we know that the point set of $\mathbb{P}_\kappa$ is contained in P. The quadric $\mathbb{P}_\kappa$ has the same (two) lines on $\langle e_1 \rangle$ as Z, hence the same secants, and on each of these secants it has the same points as Z. Hence Z coincides with $\mathbb{P}_\kappa$. It is embedded in the absolute of the polarity δ_f determined by the form f given by

$$f(x, y) = x_1 y_4 + x_4 y_1 - x_2 y_3 - x_3 y_2.$$

On the other hand, the polar map π is readily checked to coincide with δ_f, and so is a polarity. The fact that Z is a subspace of $\mathbb{P}_\pi$ now follows from Lemma 9.2.7 (or from Example 7.8.1). $\square$

Remark 9.5.4 In view of Theorem 9.5.3, we may assume the existence of a secant intersecting P in at least three points. Therefore, there exists a basis e_1, e_2, e_3, e_4 of $\mathbb{D}^4$ such that $\langle e_1 \rangle, \langle e_2 \rangle, \langle e_3 \rangle, \langle e_4 \rangle, \langle e_2 + e_4 \rangle$ are points of Z, with (see Fig. 9.6)

$$\langle e_1 \rangle \perp \langle e_2 \rangle \perp \langle e_3 \rangle \perp \langle e_4 \rangle \perp \langle e_1 \rangle \quad \text{and} \quad \langle e_1 \rangle \perp \langle e_2 + e_4 \rangle \perp \langle e_3 \rangle. \tag{9.14}$$

By means of a coordinate transformation of $\mathbb{D}^4$, we can (and will) arrange for e_1, e_2, e_3, e_4 to be the standard basis. Consequently, each point p on $\langle e_1, e_2 \rangle$ other than $\langle e_1 \rangle$ can be written $p = (\lambda : 1 : 0 : 0)$ for some $\lambda \in \mathbb{D}$. Now $p^\perp$ meets $\langle e_3, e_4 \rangle$ in a unique point, denoted by $\alpha(p)$, with coordinates $\alpha(p) = (0 : 0 : 1 : \alpha(\lambda))$ for some $\alpha(\lambda) \in \mathbb{D}$. In particular,

$$(\lambda : 1 : 0 : 0) \perp \big(0 : 0 : 1 : \alpha(\lambda)\big). \tag{9.15}$$

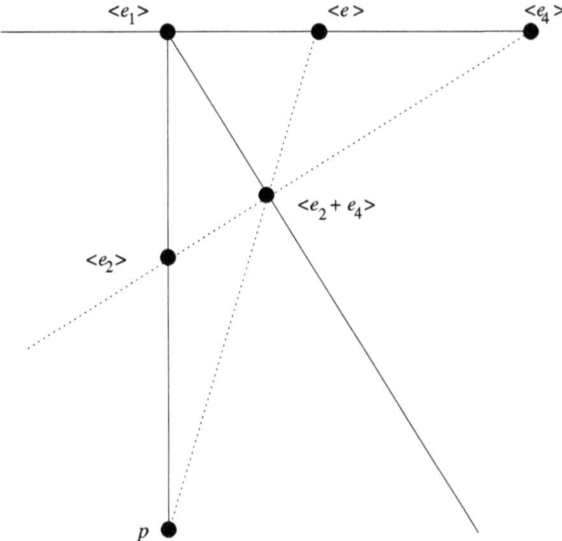

Fig. 9.7 A configuration of points in Z for σ_λ

As $\alpha(\langle e_2 \rangle) = \langle e_3 \rangle$, we have $\alpha(0) = 0$. Clearly, $\alpha : \langle e_1, e_2 \rangle \to \langle e_3, e_4 \rangle$ is a bijection and so is the map $\alpha : \mathbb{D} \to \mathbb{D}$. After replacing e_3 by a suitable scalar multiple, we can (and will) assume $\alpha(1) = 1$. We will maintain this set-up until the end of the proof of Lemma 9.5.6.

Proposition 9.5.5 *Suppose that Z is a nondegenerate ruled polar space in $\mathbb{P} = \mathbb{P}(\mathbb{D}^4)$ having a secant with at least three points on Z. The map α on $\mathbb{D}$ is an anti-automorphism of $\mathbb{D}$ with $\alpha^2 = \mathrm{id}$.*

Proof The proof is long and uses the perspectivity σ of $\mathbb{P}$ with center $\langle e_2 \rangle$ and axis $\pi(\langle e_2 \rangle)$ mapping $\langle e_4 \rangle$ to $\langle e_2 + e_4 \rangle$. By Theorem 9.3.3, it leaves Z invariant. Observe that $\pi(\langle e_2 \rangle) = \langle e_1, e_2, e_3 \rangle$. Here is an algebraic description of σ, where, as usual, $x = e_1 x_1 + e_2 x_2 + e_3 x_3 + e_4 x_4 \in \mathbb{D}^4$.

$$\sigma\big(\langle x \rangle\big) = (x_1 : x_2 + x_4 : x_3 : x_4). \tag{9.16}$$

Fix $\lambda \in \mathbb{D}$. For p as in Remark 9.5.4, the tangent plane $\pi(p)$ is given by

$$\pi(\lambda : 1 : 0 : 0) = \big\{ \langle x \rangle \mid x_4 = \alpha(\lambda) x_3 \big\}. \tag{9.17}$$

Let σ_λ be the perspectivity with center p, and axis $\pi(p)$ mapping $\langle e_2 + e_4 \rangle$ to a point $\langle e \rangle$ of $\langle e_1, e_4 \rangle$ (see Fig. 9.7). Again by Theorem 9.3.3, it leaves Z invariant. As $\langle e \rangle$ lies on $\langle e_1, e_4 \rangle$ and $\langle p, e_2 + e_4 \rangle$, we can take $e = e_4 - e_1 \lambda$. Consequently,

$$\sigma_\lambda\big(\langle x \rangle\big) = \big(x_1 + \lambda\alpha(\lambda)x_3 - \lambda x_4 : x_2 + \alpha(\lambda)x_3 - x_4 : x_3 : x_4\big). \tag{9.18}$$

Now $\pi(\langle e_2 + e_4 \rangle)$ is the plane $\langle e_1, e_3, e_2 + e_4 \rangle$, alternatively given by

$$\pi\big(\langle e_2 + e_4 \rangle\big) = \big\{ \langle x \rangle \mid x_2 = x_4 \big\}. \tag{9.19}$$

The image of this plane under σ_λ is the plane $\langle e_1, e_4, e_2\alpha(\lambda) + e_3 \rangle$, which has equation $x_2 = \alpha(\lambda)x_3$. For this plane, $\sigma_\lambda \pi(\langle e_2 + e_4 \rangle) = \pi(\sigma_\lambda \langle e_2 + e_4 \rangle) = \pi(\langle e \rangle)$. Hence, the tangent plane at $\langle e \rangle$ is

$$\pi(-\lambda : 0 : 0 : 1) = \big\{ \langle x \rangle \mid x_2 = \alpha(\lambda)x_3 \big\}. \tag{9.20}$$

We will also use $\sigma_\lambda \sigma$. In its action on Z, it leaves $\langle e_1, e_4 \rangle$ invariant. It is determined by the equation

$$\sigma_\lambda \sigma \big(\langle x \rangle \big) = \big(x_1 + \lambda\alpha(\lambda)x_3 - \lambda x_4 : x_2 + \alpha(\lambda)x_3 : x_3 : x_4 \big). \tag{9.21}$$

We need seven steps to finish the proof.

STEP 1. $\alpha(\lambda + \mu) = \alpha(\lambda) + \alpha(\mu)$ *for all* $\lambda, \mu \in \mathbb{D}$.

Using (9.21) for both λ and μ in $\mathbb{D}$, we obtain that $\sigma_\mu \sigma \sigma_\lambda \sigma(\langle x \rangle)$ is equal to

$$\big(x_1 + \big(\lambda\alpha(\lambda) + \mu\alpha(\mu)\big)x_3 - (\lambda + \mu)x_4 : x_2 + \big(\alpha(\lambda) + \alpha(\mu)\big)x_3 : x_3 : x_4 \big).$$

In particular, $\sigma_\mu \sigma \sigma_\lambda \sigma(\langle e_4 \rangle) = \langle e_4 - e_1(\lambda + \mu) \rangle$. As $\pi(\langle e_4 \rangle) = \langle e_1, e_3, e_4 \rangle$, its transform under $\sigma_\mu \sigma \sigma_\lambda \sigma$ is

$$\sigma_\mu \sigma \sigma_\lambda \sigma \pi \big(\langle e_4 \rangle \big) = \big\langle e_1, e_4, e_2\big(\alpha(\lambda) + \alpha(\mu)\big) + e_3 \big\rangle,$$

whose equation is $x_2 = (\alpha(\lambda) + \alpha(\mu))x_3$. On the other hand, it must also coincide with $\pi(\sigma_\mu \sigma \sigma_\lambda \sigma \langle e_4 \rangle) = \pi(\langle e_4 - e_1(\lambda + \mu) \rangle)$, which, by (9.20), is given by $x_2 = \alpha(\lambda + \mu)x_3$. Hence Step 1.

STEP 2. $\alpha^2 = \mathrm{id}$.

Taking $\lambda = 1$ in (9.20), we see that $\pi(-1 : 0 : 0 : 1)$ is given by $x_2 = x_3$. This plane intersects the line $\langle e_2, e_3 \rangle$ of Z in $(0 : 1 : 1 : 0)$ and so

$$(-1 : 0 : 0 : 1) \perp (0 : 1 : 1 : 0).$$

Let l be the line on these two points, so l is a line of Z. It contains the point

$$u = \big(-\alpha(\lambda) : 1 : 1 : \alpha(\lambda) \big).$$

Now consider the points $v = (-\alpha(\lambda) : 1 : 0 : 0)$ and $w = (0 : 0 : 1 : \alpha(\lambda)) = \alpha(p)$. Clearly, u, v, and w are on a projective line. Moreover, v and w, lying on the lines $\langle e_1, e_2 \rangle$ and $\langle e_3, e_4 \rangle$, respectively, of Z, are points of Z. Therefore, according to Proposition 9.3.1(ii), the tangent planes $\pi(u)$, $\pi(v)$, $\pi(w)$ have the line $\pi(v) \cap \pi(w)$ in common. Here, $\pi(v)$ is given by $x_4 = \alpha(-\alpha(\lambda))x_3$ (cf. (9.17)) and $\pi(w)$ by $x_1 = \lambda x_2$ (for, $w \perp \langle e_3, e_4, e_1\lambda + e_2 \rangle$ by (9.15)). Notice that $\alpha(-\alpha(\lambda)) = -\alpha^2(\lambda)$ in view of Step 1. We have shown

$$\pi(v) \cap \pi(w) = \big\langle e_1\lambda + e_2, e_4 - e_3\alpha^2(\lambda) \big\rangle.$$

Both l and the projective line $\pi(v) \cap \pi(w)$ are in $\pi(u)$, so they intersect in a point $(-a, b, b, a) = (\lambda c, c, d, -\alpha^2(\lambda)d)$ for certain scalars a, b, c, $d \in \mathbb{D}$. This gives $a = -\alpha^2(\lambda)d = -\lambda c$ and $b = c = d$, whence $\alpha^2(\lambda) = \lambda$ for all $\lambda \in \mathbb{D}$. This settles Step 2.

STEP 3. $\alpha(\lambda^{-1}) = (\alpha(\lambda))^{-1}$ *for all* $\lambda \in \mathbb{D} \setminus \{0\}$.

Applying $\sigma_\lambda \sigma$ as in (9.21) with $\lambda = 1$ to $\langle e_2, e_3 \rangle$, we obtain the line $\langle e_2, e_1 + e_2 + e_3 \rangle$ of Z. On this line lies the point $\langle e_1 + e_3 \rangle$, which must belong to Z. By (9.19), the plane $\pi(\langle e_2 + e_4 \rangle)$ has equation $x_2 = x_4$, so $\langle e_1 + e_3 \rangle \perp \langle e_2 + e_4 \rangle$ and $m = \langle e_1 + e_3, e_2 + e_4 \rangle$ is a line of Z.

The points $q = (\lambda : 1 : \lambda : 1)$, $p = (\lambda : 1 : 0 : 0)$, and $s = (0 : 0 : \lambda : 1)$ are on a line meeting $\langle e_1 + e_3, e_2 + e_4 \rangle$ in q. Since q, p, and s are on a projective line, their tangent planes have a common line in $\pi(q)$, which meets $\langle e_1 + e_3, e_2 + e_4 \rangle$ in some point $\langle z \rangle$. Using the fact that $\pi(p)$ is given by $x_4 = \alpha(\lambda)x_3$ (see (9.17)) and $\pi(s)$ by $x_1 = \alpha^{-1}(\lambda^{-1})x_2$ (as for w above), we see that there are scalars a, b, c, d in $\mathbb{D}$ such that

$$z = (a, b, a, b) = \left(\alpha^{-1}(\lambda^{-1})c, c, d, \alpha(\lambda)d\right).$$

Therefore $a = d = \alpha^{-1}(\lambda^{-1})c$ and $b = c = \alpha(\lambda)d$, so $d = \alpha^{-1}(\lambda^{-1})\alpha(\lambda)d$. If $d = 0$, then $a = b = 0$, so $z = 0$, a contradiction. Consequently, $d \neq 0$, and $\alpha^{-1}(\lambda^{-1})\alpha(\lambda) = 1$, so, by Step 2, $\alpha(\lambda)^{-1} = \alpha^{-1}(\lambda^{-1}) = \alpha(\lambda^{-1})$.

STEP 4. α *is an automorphism or an anti-automorphism.*

In view of Steps 1 and 3, this is immediate from Lemma 9.4.1.

If α is an anti-automorphism, we are done. So, from now on, we may assume that α is an automorphism of $\mathbb{D}$. We must show that $\mathbb{D}$ is commutative.

STEP 5. $\alpha(\mu)\alpha(\lambda)\lambda = \lambda\alpha(\lambda)\alpha(\mu)$ *for all* λ, μ *in* $\mathbb{D}$.

The perspectivity $\sigma_\lambda \sigma$ of (9.21) maps the line $\langle e_2, e_3 \rangle$ of Z to the line $\langle e_2, (\lambda\alpha(\lambda), \alpha(\lambda), 1, 0) \rangle$ of Z; this line meets the projective line $\langle e_1, e_3 \rangle$ in the point $\langle f \rangle$ with $f = e_1 \lambda\alpha(\lambda) + e_3$ of Z.

By Theorem 9.3.3, the perspectivity ρ_λ with center $\langle e_1 \rangle$ and axis $\pi(\langle e_1 \rangle)$ mapping $\langle e_3 \rangle$ to f leaves Z invariant. The polar plane $\pi\langle e_1 \rangle$ is spanned by e_1, e_2, e_4, so ρ_λ is determined by

$$\rho_\lambda(\langle x \rangle) = (x_1 + \lambda\alpha(\lambda)x_3 : x_2 : x_3 : x_4).$$

We point out yet another perspectivity. For each μ in $\mathbb{D}$, the vectors f as above, $g = e_3 - e_2\mu\lambda\alpha(\lambda)$, and $e_1 + e_2\mu$ represent three points of Z on a projective line, as $e_1 + e_2\mu = (f - g)(\lambda\alpha(\lambda))^{-1}$. The point $\langle e_1 + e_2\mu \rangle$ coincides with p when λ is replaced by μ^{-1} and so (9.17) shows that its polar plane is given by $x_4 = \alpha(\mu^{-1})x_3$. Hence (cf. Theorem 9.3.3), the perspectivity $\sigma_{\lambda,\mu}$ with center $\langle e_1 + e_2\mu \rangle$, axis $\pi\langle e_1 + e_2\mu \rangle$, mapping $\langle f \rangle$ to $\langle g \rangle$, leaves Z invariant. In coordinates, $\sigma_{\lambda,\mu}(\langle x \rangle)$ is given by

$$\left(x_1 - \lambda\alpha(\lambda)x_3 + \lambda\alpha(\lambda)\alpha(\mu)x_4 : x_2 - \mu\lambda\alpha(\lambda)x_3 + \mu\lambda\alpha(\lambda)\alpha(\mu)x_4 : x_3 : x_4\right).$$

Composing $\sigma_{\lambda,\mu}$ with ρ_λ, we find

$$\sigma_{\lambda,\mu}\rho_\lambda(\langle x\rangle) = \big(x_1 + \lambda\alpha(\lambda)\alpha(\mu)x_4 : x_2 - \mu\lambda\alpha(\lambda)x_3 + \mu\lambda\alpha(\lambda)\alpha(\mu)x_4 : x_3 : x_4\big).$$

The composed map fixes $\langle e_2\rangle$ and maps $\langle e_3\rangle$ to $\langle e_3 - e_2\mu\lambda\alpha(\lambda)\rangle$, and $\langle e_4\rangle$ to $\langle e_1\lambda\alpha(\lambda)\alpha(\mu) + e_2\mu\lambda\alpha(\lambda)\alpha(\mu) + e_4\rangle$. As $\pi(\langle e_3\rangle) = \langle e_2, e_3, e_4\rangle$, we find

$$\sigma_{\lambda,\mu}\rho_\lambda\pi\big(\langle e_3\rangle\big) = \big\{\langle x\rangle \mid x_1 = \lambda\alpha(\lambda)\alpha(\mu)x_4\big\}. \tag{9.22}$$

By use of (9.20), we see that the polar $\pi(\langle e_4 - e_1\lambda\rangle)$ meets $\langle e_2, e_3\rangle$ at $\langle e_2\alpha(\lambda) + e_3\rangle$. Therefore, the polar $\pi(\langle e_2\alpha(\lambda) + e_3\rangle)$ is given by the equation $x_1 = -\lambda x_4$. Replacing $\alpha(\lambda)$ by $-\mu\lambda\alpha(\lambda)$ and using (9.22), we find that $\pi(\sigma_{\lambda,\mu}\rho_\lambda\langle e_3\rangle)$ is given by $x_1 = -\alpha^{-1}(-\mu\lambda\alpha(\lambda))x_4$, which rewrites to $x_1 = \alpha(\mu)\alpha(\lambda)\lambda x_4$ as $\alpha^2 = \mathrm{id}$ by Step 2 and α is an automorphism by assumption. Comparing this with (9.22), we obtain $\alpha(\mu)\alpha(\lambda)\lambda = \lambda\alpha(\lambda)\alpha(\mu)$, as required.

STEP 6. *Each $\lambda \in \mathbb{D}$ satisfies $\lambda\alpha(\lambda) \in Z(\mathbb{D})$.*

By Step 5 with $\mu = 1$, we have $\alpha(\lambda)\lambda = \lambda\alpha(\lambda)$. Hence, again by Step 5,

$$\lambda\alpha(\lambda)\alpha(\mu) = \alpha(\mu)\alpha(\lambda)\lambda = \alpha(\mu)\lambda\alpha(\lambda),$$

so $\lambda\alpha(\lambda)$ centralizes $\alpha(\mu)$ for each $\mu \in \mathbb{D}$, whence all of $\mathbb{D}$.

STEP 7. *For each $\lambda \in \mathbb{D}$, we have $\lambda + \alpha(\lambda) \in Z(\mathbb{D})$.*

Replacing λ by $\lambda + 1$ in the equality of Step 6, using the automorphism assumption on α, and the additivity of α established in Step 1, we derive $\alpha(\lambda)\lambda\mu + \alpha(\lambda)\mu + \lambda\mu = \mu\alpha(\lambda)\lambda + \mu\alpha(\lambda) + \mu\lambda$. As $\alpha(\lambda)\lambda\mu = \mu\alpha(\lambda)\lambda$ by Step 6, we are left with $(\alpha(\lambda) + \lambda)\mu = \mu(\alpha(\lambda) + \lambda)$, as required.

In view of Steps 1, 2, 6, 7, the hypotheses of Theorem 9.4.2 are satisfied, so $\mathbb{D}$ is commutative. As α is assumed to be an automorphism, it is also an anti-automorphism of $\mathbb{D}$. $\qquad\square$

In order to establish that π is a nondegenerate quasi-polarity (cf. Definition 7.1.9), we need one more technical result, which uses Proposition 9.5.5.

Lemma 9.5.6 *Suppose that Z is a nondegenerate ruled polar space in $\mathbb{P} = \mathbb{P}(\mathbb{D}^4)$ having a secant with at least three points on Z. If l, l' are distinct lines of $\mathbb{P}$ which are secants of Z with a common point in $\mathbb{P}$, then $\pi(l)$ and $\pi(l')$ have a common point as well.*

Proof Suppose that p lies in $l \cap l'$. We may assume that $p \notin P$, for otherwise $\pi(l)$ and $\pi(l')$ are contained in the plane $\pi(p)$ and so they meet. Fix distinct points $\langle a\rangle$, $\langle b\rangle$ in $l \cap P$, and points $\langle r\rangle$, $\langle r'\rangle$ in $l' \cap P$. Take $\langle c\rangle$, $\langle d\rangle$ in $\{\langle a\rangle, \langle b\rangle\}^\perp$. By Proposition 9.3.1(i), $\pi(l) = \pi(\langle a\rangle) \cap \pi(\langle b\rangle) = \langle\{\langle a\rangle, \langle b\rangle\}^\perp\rangle_\mathbb{P} = \langle c, d\rangle_\mathbb{P}$; see Fig. 9.8.

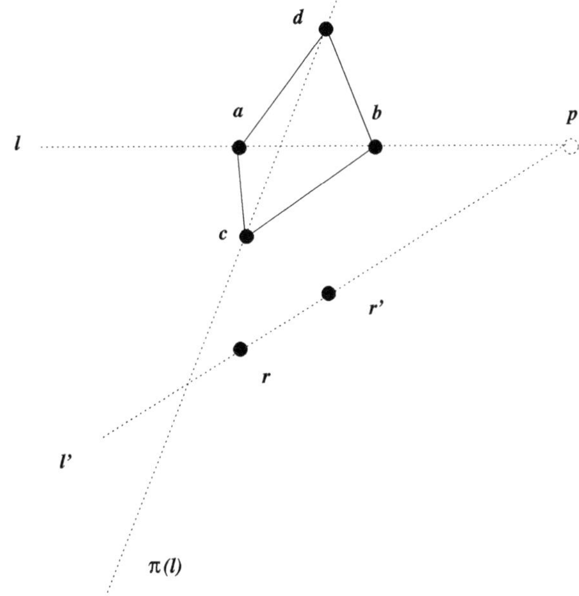

Fig. 9.8 A configuration of points and lines in $\mathbb{P}(\mathbb{D}^4)$ related to Z

The line $\pi(l)$ intersects the plane $\langle l, l' \rangle$ in a point $\langle h \rangle$, where $h = c\xi + d\theta$ for certain ξ, $\theta \in \mathbb{D}$. The plane $\langle l, l' \rangle$ is generated by the points $\langle a \rangle$, $\langle b \rangle$, and $\langle h \rangle$, so there are scalars λ, λ', μ, μ', η, η' in $\mathbb{D}$ such that

$$r = a\lambda + b\mu + h\eta \quad \text{and} \quad r' = a\lambda' + b\mu' + h\eta'.$$

Referring to Remark 9.5.4, we identify a, b, c, and d with e_1, e_3, e_2, and e_4, respectively, and use the map α introduced there. Hence $r = (\lambda, \xi\eta, \mu, \theta\eta)$ and $r' = (\lambda', \xi\eta', \mu', \theta\eta')$.

Suppose first $\mu = 0$. Then $\langle r \rangle \in \langle a, h \rangle \subseteq \pi(\langle a \rangle)$, so $\langle a \rangle \perp \langle r \rangle$ and $\langle a, r \rangle$ is a line of Z. As Z is a polar space, there is a nonzero vector c such that $\langle c \rangle = \langle b \rangle^{\perp} \cap \langle a, r \rangle$. The point satisfies $\pi(\langle c \rangle) = \langle a, b, c \rangle = \langle l, l' \rangle$, so $\langle l, l' \rangle$ is a tangent plane to Z at $\langle c \rangle$. Consequently, $\langle c, r' \rangle$ is a line of Z, so $c \in \{\langle r \rangle, \langle r' \rangle\}^{\perp} = \pi(l)$. We conclude that $\langle c \rangle$ belongs to $\pi(l) \cap \pi(l')$.

By interchanging the roles of r and r', or those of a and b, we can also deal with the cases $\mu' = 0$, $\lambda = 0$, and $\lambda' = 0$.

It remains, therefore, to consider the case where $\mu, \mu', \lambda, \lambda' \neq 0$. As in (9.17), and using $\alpha^2 = \mathrm{id}$ (cf. Proposition 9.5.5), we see

$$\pi\left(\alpha\left(\theta\eta\mu^{-1}\right) : 1 : 0 : 0\right) = \left\{\langle x \rangle \mid x_4 = \theta\eta\mu^{-1}x_3\right\},$$

and so $\langle r \rangle$ is in it. Similarly, by (9.20),

$$\pi\left(-\alpha\left(\xi\eta\mu^{-1}\right) : 0 : 0 : 1\right) = \left\{\langle x \rangle \mid x_2 = \xi\eta\mu^{-1}x_3\right\},$$

and so $\langle r \rangle$ is also in this plane. The points $\langle r \rangle$, $\langle e_1 \alpha(\theta \eta \mu^{-1}) + e_2 \rangle$, and $\langle e_4 - e_1 \alpha(\xi \eta \mu^{-1}) \rangle$ of Z, being linearly independent, generate the plane $\pi(\langle r \rangle)$, and

$$\pi(\langle r \rangle) = \left\{ \langle x \rangle \mid x_1 - \alpha(\theta \eta \mu^{-1}) x_2 + \zeta x_3 + \alpha(\xi \eta \mu^{-1}) x_4 = 0 \right\},$$

for some $\zeta \in \mathbb{D}$ (which can be determined from the equation for $\pi(\langle r \rangle)$ by substitution of the coordinates of r as $\mu \neq 0$). Now $\pi(r) \cap \pi(l) = \pi(r) \cap \langle c, d \rangle$ is given by the equations $x_1 = x_3 = 0$. Together with the fact that α is an anti-automorphism (cf. Proposition 9.5.5), this gives $\alpha(\theta) x_2 = \alpha(\xi) x_4$, so $\pi(r) \cap \pi(l)$ is the point $q = \langle c\alpha(\theta^{-1})\alpha(\xi) + d \rangle$ if $\theta \neq 0$ and $q = \langle c + d\alpha(\xi^{-1})\alpha(\theta) \rangle$ if $\xi \neq 0$ (observe that θ and ξ cannot simultaneously be zero). This shows that q does not depend on λ, λ', μ, μ', η, η'; so, if we replace r by r' (which is allowed as $\mu' \neq 0$), the result must be the same: $\pi(r') \cap \pi(l) = \{q\}$. In this case $\pi(r') \cap \pi(r) \cap \pi(l) = \{q\}$ and so $\pi(l') = \pi(r') \cap \pi(r)$ meets $\pi(l)$ in $\{q\}$, as required. $\square$

Theorem 9.5.7 *Suppose that $\mathbb{P}$ is a thick projective space of dimension three and Z is a nondegenerate ruled polar space in $\mathbb{P}$ with empty defect. The map π is a polarity on $\mathbb{P}$ such that Z is a subspace of $\mathbb{P}_\pi$.*

Proof Theorem 9.5.3 takes care of the case where each secant has exactly two points of Z, so assume this is not the case. This means that the setting of Remark 9.5.4 applies. By Propositions 9.3.2(ii) and 7.1.10, π is a polarity if (and only if), for each $p \in \mathbb{P} \setminus P$, the subspace $\pi(p)$ does not coincide with $\mathbb{P}$.

Assume by way of contradiction, that there is $p \in \mathbb{P} \setminus P$ satisfying $\pi(p) = \mathbb{P}$. By Lemmas 9.2.2 and 9.3.4, every line of Z intersects the collar Z_p of p in exactly one point. Therefore, the projective line on any two points of Z_p is a secant. As $\langle Z_p \rangle_{\mathbb{P}} = \pi(p) = \mathbb{P}$, there are two secants, say l and m, with at least two points in Z_p that are disjoint (for, otherwise any two lines in $\langle Z_p \rangle_{\mathbb{P}}$ would meet, so $\dim(\mathbb{P}) = 3 \neq 4$; cf. Exercise 5.7.11). Now $\pi(l) \cap \pi(m)$ contains p (as $p \in \pi(a)$ for each $a \in (l \cap Z) \cup (m \cap Z)$ according to Corollary 9.2.6). By Lemma 9.5.2, $l = \pi(\pi(l))$ and $m = \pi(\pi(m))$ are secants and, by Lemma 9.5.6, they also have a common point, contradicting that l and m are disjoint. We conclude that π is a polarity.

The last part of the theorem follows from Lemma 9.2.7. $\square$

The main goal stated at the beginning of the chapter can now be formulated by use of the polar map π of Definition 9.2.1, which uses the collar Z_x of a point x of the ambient space $\mathbb{P}$ of a ruled polar space.

Theorem 9.5.8 *Let Z be a nondegenerate ruled polar space in a thick projective space $\mathbb{P}$. The polar map π is the unique quasi-polarity whose kernel is the defect D of Z and whose absolute $\mathbb{P}_\pi$ contains Z as a subspace. Moreover, the restriction to Z of the natural quotient map $\phi : \mathbb{P} \setminus D \to \mathbb{P}/D$ embeds Z as a ruled polar space in $\mathbb{P}/D$ in such a way that the corresponding polar map $\overline{\pi}$ is a nondegenerate injective quasi-polarity of $\mathbb{P}/D$. In particular, $\phi(Z)$ is a subspace of $(\mathbb{P}/D)_{\overline{\pi}}$ isomorphic to Z.*

Proof Let $\phi : \mathbb{P} \setminus D \to \mathbb{P}/D$ be as in Proposition 9.2.10(iii). By the results of this proposition, the map $\overline{\pi} : \mathbb{P}/D \to (\mathbb{P}/D)^{\star}$ given by $\overline{\pi}(\langle x, D \rangle_{\mathbb{P}}) = \phi(\pi(x))$ is the polar of the nondegenerate ruled polar space $\phi(Z)$ in $\mathbb{P}/D$. The proposition also gives that the defect of $\phi(Z)$ in $\mathbb{P}/D$ is empty.

By Theorems 9.3.7 (if $\dim(\mathbb{P}/D) \geq 4$) and 9.5.7 (if $\dim(\mathbb{P}/D) = 3$), the map $\overline{\pi}$ is a nondegenerate injective quasi-polarity of $\mathbb{P}/D$. By Proposition 9.2.10(vi), the map π is a quasi-polarity with kernel D. Now, according to Proposition 7.4.4, the absolutes $\mathbb{P}_{\pi}$ and $(\mathbb{P}/D)_{\overline{\pi}}$ are embedded polar spaces. By Lemma 9.2.7, Z is a subspace of $\mathbb{P}_{\pi}$ and $\phi(Z)$ is a subspace of $(\mathbb{P}/D)_{\overline{\pi}}$.

The proof that π is unique is left to the reader; cf. Exercise 9.6.5. $\square$

A comparison of an arbitrary embedding of a polar space Z in a projective space with the Veldkamp space embedding (cf. Definition 8.2.3 and Corollary 8.4.26) gives the following universal property in case Z has rank at least three.

Corollary 9.5.9 *Suppose that Z is a ruled nondegenerate polar space, of rank at least three with thick lines and empty defect, in the projective space $\mathbb{P}$. The map $\psi : \mathbb{P} \to \mathcal{V}(Z)$ given by $\psi(x) = Z_x$ is an injective homomorphism of projective spaces whose image is the subspace of $\mathcal{V}(Z)$ spanned by the geometric hyperplanes $p^{\perp}$ of Z for p a point of Z.*

Proof Put $Z = (P, L)$ and denote by U the subspace of $\mathcal{V}(Z)$ generated by the image of P in $\mathcal{V}(Z)$ under the natural embedding (cf. Theorem 8.2.9), that is, by all $p^{\perp}$ for $p \in P$. As Z has empty defect in $\mathbb{P}$, Theorems 9.5.8 and 9.2.5 give that π is an embedding $\mathbb{P} \to \mathbb{P}^{\star}$ such that $P \cap \pi(x) = x^{\perp}$ for $x \in P$. As Z is a ruled subspace of $\mathbb{P}$, the map $\chi : \mathbb{P}^{\star} \to \mathcal{V}(Z)$ given by $\chi(H) = P \cap H$ is well defined. Now $\psi = \chi\pi$ is readily seen to be a homomorphism: if z is a point on the line xy for distinct points x, y of $\mathbb{P}$, then $\pi(z)$ contains $\pi(x) \cap \pi(y)$, so Z_z contains $(P \cap \pi(x)) \cap (P \cap \pi(y)) = Z_x \cap Z_y$. According to Theorem 8.2.4, this implies that $\psi(z)$ is on the Veldkamp line spanned by Z_x and Z_y. In view of the empty defect, Lemma 9.2.12(ii) implies that the map ψ is injective, so it is an embedding. Its restriction to P coincides with the natural embedding of Z in U. But U is spanned by $\psi(P)$ in $\mathcal{V}(Z)$, and $\mathbb{P}$ is spanned by P, so $\psi(\mathbb{P})$ is spanned by $\psi(P)$. Therefore, $U = \psi(\mathbb{P})$. $\square$

As a consequence, if Z is a ruled polar space in an infinite-dimensional projective space without defect, then it does not embed in a finite-dimensional projective space; see Exercise 9.6.13 for an example.

The nondegenerate ruled polar spaces $\mathbb{P}_{\pi}$ of this chapter appear as maximal nondegenerate polar spaces embedded in a projective space. The major question now is whether they are minimal also. While this is often the case, it is not true in general. Actually, there are embedded polar spaces that we have not yet come across, namely pseudo-quadrics. We will deal with this final part of the classification of polar spaces with Desarguesian planes in the next chapter.

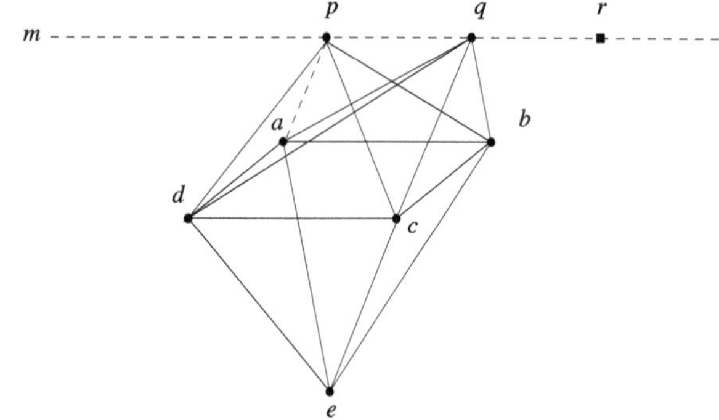

Fig. 9.9 An embedded polar space with *thin lines*. The *interrupted line* is a thick projective line

9.6 Exercises

Section 9.1

Exercise 9.6.1 Let Z be a polar space of rank at least two embedded in the projective plane $\mathbb{P}$. Prove that Z is either a subspace of $\mathbb{P}$ or a rosette (cf. Definition 8.1.3).

Exercise 9.6.2 Let Z be a nondegenerate ruled polar space of rank ≥ 2 in a thick projective space $\mathbb{P}$. Recall from Exercise 5.7.12 that $\mathbb{P}^\star$ denotes the hyperplane dual of $\mathbb{P}$.

(a) Prove that the map $\phi : \mathbb{P}^\star \to \mathcal{V}(Z)$ given by $H \mapsto H \cap Z$ is an injective homomorphism of spaces.
(b) Suppose that Z has rank at least three and each geometric hyperplane of Z spans a hyperplane of $\mathbb{P}$. Prove that ϕ maps lines onto lines, i.e., the image $\phi(l)$ of each line l of $\mathbb{P}^\star$ is a line of $\mathcal{V}(Z)$.

Exercise 9.6.3 Verify by means of the four steps below, that the condition in Proposition 9.1.11 that $\mathbb{P}$ be thick is necessary by use of the following example, which is also depicted in Fig. 9.9. The projective space $\mathbb{P}$ is the direct sum (cf. Definition 6.3.6) of the thin projective space on the five points a, b, c, d, e, and the projective line m consisting of the three points p, q, r. The embedded polar space Z is the direct sum of the quadrangle on a, b, c, d and the space on $\{p, q, e\}$ without lines.

(a) The space Z is a nondegenerate ruled polar space of rank three in $\mathbb{P}$.
(b) The set $H := \{a, b, c, d, e, r\}$ is a hyperplane of $\mathbb{P}$.
(c) The set $H \cap Z = \{a, b, c, d, e\}$ is a geometric hyperplane of Z.
(d) The span $\langle H \cap Z \rangle_{\mathbb{P}}$ is strictly contained in H.

Section 9.2

Exercise 9.6.4 Let Q be a set of points in the projective space $\mathbb{P}$. For $x \in Q$, define $T(x)$ as in Definition 9.2.1, so $T(x)$ is the union of $\{x\}$, all projective lines on x

entirely contained in Q, and all projective lines l with $Q \cap l = \{x\}$. The set Q is called *quadratic* if

(1) each line l of $\mathbb{P}$ has either at most two or all points in Q;
(2) for each point $x \in Q$, the tangent set $T(x)$ is either a hyperplane of $\mathbb{P}$ or equal to $\mathbb{P}$.

Prove that the pair (Q, L), where L is the collection of lines of $\mathbb{P}$ that are contained in Q, is a polar space embedded in $\mathbb{P}$. Verify that the set of points on a quadric is a quadratic set.

Exercise 9.6.5 (This exercise is used in Theorems 9.5.3 and 9.5.8) Let Z be a nondegenerate ruled polar space in $\mathbb{P}$ with empty defect. Prove that there is at most one nondegenerate quasi-polarity π of $\mathbb{P}$ such that Z embeds in $\mathbb{P}_\pi$.

Exercise 9.6.6 Suppose that Z is a ruled polar space in $\mathbb{P}$. Prove the following assertions.

(a) The subspace $R := \mathrm{Rad}(Z)$ of Z is also a subspace of $\mathbb{P}$.
(b) The restriction to the point set of Z of the natural quotient map $\mathbb{P} \setminus R \to \mathbb{P}/R$ is the natural homomorphism $Z \to Z/R$ onto the nondegenerate quotient of Z described in Proposition 7.4.10.
(c) Let V be a complementary subspace to R in $\mathbb{P}$ (cf. Exercise 5.7.14). Then $V \cap Z$ is isomorphic to Z/R.
(d) The defect of Z in $\mathbb{P}$ is generated by the defect of $V \cap Z$ in V and R.

Section 9.3

Exercise 9.6.7 A *semi-ovoid* of the projective space $\mathbb{P}$ is a non-empty set X of points of $\mathbb{P}$ such that, for each $p \in X$, the union of all lines of $\mathbb{P}$ intersecting X exactly in p, is a hyperplane of $\mathbb{P}$. An ovoid in $\mathbb{P}$, as defined in Example 5.5.10, is a semi-ovoid X such that no line of $\mathbb{P}$ intersects X in more than two points. Prove that the absolute of the polarity of a hermitian form of trace type without lines is a semi-ovoid that is not an ovoid.

Exercise 9.6.8 Let $\mathbb{P}$ be the Fano plane, whose points are labelled as in Fig. 1.21. We will use these labels as names for the points. We explore the notions of ovoid and semi-ovoid of Exercise 9.6.7 for $\mathbb{P}$.

(a) Verify that $X = \{1, 3, 4\}$, or any other triple forming a basis of $\mathbb{P}$, is a semi-ovoid but not an ovoid.
(b) Show that the map sending the point 5 to $\mathbb{P}$ and each other point p to the line on $p5$ is a quasi-polarity π of $\mathbb{P}$ such that $\pi(p)$ for $p \in X$ is the unique line on p meeting X in p only.
(c) Show that $\{1, 3, 4, 5\}$ is not a semi-ovoid of X. Conclude that $\mathbb{P}$ has no ovoids.

Exercise 9.6.9 Prove that there are ovoids, as defined in Exercise 9.6.7, in the real projective plane that cannot be embedded in a polarity.
(*Hint.* Take halves of two distinct quadrics and join them smoothly to an ovoid; more explicitly, take all affine points $\varepsilon_1 x_1 + \varepsilon_2 x_2 \in \mathbb{R}^2$ with $x_1^2 + x_2^2 = 1$ for $x_1 \leq 0$ and $\frac{1}{2}x_1^2 + x_2^2 = 1$ for $x_1 \geq 0$.)

Exercise 9.6.10 Show, by means of an example, that the condition that Z has empty defect is essential in Lemma 9.3.4.

Section 9.4

Exercise 9.6.11 (This exercise is used in Theorem 9.5.3) Suppose that Z is a nondegenerate generalized quadrangle embedded in $\mathbb{P}(\mathbb{D}^4)$. Show that Z is ruled in $\mathbb{P}$ and show (as stated in Step 3 of the proof of Theorem 9.5.3) that there is a basis e_1, e_2, e_3, e_4 of $\mathbb{D}^4$ such that $l = \langle e_1, e_2 \rangle$, $l' = \langle e_3, e_4 \rangle$, $l'' = \langle e_1 + e_3, e_2 + e_4 \rangle$, $m = \langle e_1, e_3 \rangle$, and $m' = \langle e_2, e_4 \rangle$ are lines of Z.

Exercise 9.6.12 Show that the condition that α be surjective is needed in Lemma 9.4.1.
(*Hint:* Take $\mathbb{D}$ to be a field of characteristic two possessing elements that are no squares.)

Section 9.5

Exercise 9.6.13 Consider the embedded polar space $Z = \mathbb{P}(V)_f$ of Exercise 7.11.9. Prove that the kernel of δ_f is empty and show that, for $n \geq 3$, there is no embedding of Z in a projective space of finite dimension.

9.7 Notes

The construction of polarities whose absolutes contain embedded polar spaces was initiated by Buekenhout-Lefèvre [53, 54] and completed by Dienst [118]. Improvements, notably to Sect. 9.3, are due to Johnson [179].

There are many generalizations of the notion of embeddings of a line space. Some carry the name weak embeddings; cf. [104].

Section 9.1

Proposition 9.1.7 stems from [307], where embeddings in projective spaces of d-shadow spaces of geometries of type A_n over a field (Grassmannians), shadow spaces on n of geometries of type D_n (half dual polar spaces), and shadow spaces on n of geometries of type C_n (dual polar space) are also studied.

Section 9.2

Much of this section originates from [53], where notions like collar, tangent set, and polar appear. The notion of defect is also present in [286].

Section 9.3

The existence of perspectivities leaving the embedded polar space invariant, Theorem 9.3.3, stems from [54]. It will be used in Proposition 9.5.5, which in turn is needed for the case where $\mathbb{P}$ is 3-dimensional, Theorem 9.5.7.

The proof of Theorem 9.3.7 differs from the one given in [118], where a reduction to the case $\dim(\mathbb{P}) = 3$ is carried out.

Section 9.4

The proof of Lemma 9.4.1 follows [4, Theorem 1.15]. The Hua identity appearing in Step 1 of the proof of this lemma stems from [165], where more complicated versions of the identity are found. The proof of Lemma 9.4.1 is as in [4, Theorem 1.15]. The identity is also used in analysis of additive subgroups of fields that are closed under inversion; see [211] and [132].

Step 2 raises the question whether additive maps ϕ of division rings with $\phi(aba) = \phi(a)\phi(b)\phi(a)$ for all elements a and b in the division ring are automorphisms or anti-automorphisms. The affirmative answer is given by Hua [166] if ϕ preserves the identity element of the division ring. Herstein [159] studied semi-homomorphisms of groups and showed that there are various other possibilities in this more general setting.

Step 4 of the lemma states that the map α on the multiplicative group of the division ring is what Scott [250] calls a half-homomorphism. He shows that a half-homomorphism of groups is also either a homomorphism or an anti-homomorphism. The proof of his result, which is stronger than Step 4, is more elaborate.

Section 9.5

The proof as given in this section is from [118].

The question of characterizing generalized hexagons embeddable in a projective space does not have such a concise answer as for generalized quadrangles; see [59, 78, 278].

Chapter 10
Classical Polar Spaces

In this chapter, we conclude the study of polar spaces. In Chap. 7 they made their appearance as line spaces connected with diagram geometries of type B_n. In Chap. 8 they were shown to embed in projective spaces under some mild conditions, like the rank n being at least three and every line being on at least three maximal singular subspaces. Grassmannians of lines of a thick projective space over a noncommutative division ring are examples of nondegenerate polar spaces of rank three that do not satisfy these conditions. In Chap. 9, the polar spaces of rank at least two embedded in a projective space were shown to be subspaces of absolutes of quasi-polarities of the ambient projective space. In this chapter, we completely determine these polar spaces. The main result is Theorem 10.3.13 and Sect. 10.3 is devoted to its proof. Proposition 10.3.11 points out which nondegenerate polar spaces amongst those embedded in absolutes of quasi-polarities on projective spaces are proper subspaces of the absolutes. The new examples are generalizations of quadrics, called pseudo-quadrics, which are introduced in Sect. 10.2. They are characterized as the minimal polar spaces embeddable in an absolute that are invariant under perspectivities. The same property was exploited successfully in the proof of Theorem 9.5.7 (via Proposition 9.5.5), where the ambient projective space is 3-dimensional. Table 10.1 of Remark 10.3.15 surveys the relations between polar spaces (of finite rank and embeddable in projective spaces) and polar geometries, similarly to Table 6.1 for the projective case.

In Sect. 10.4, we study thin substructures, called apartments, of the polar spaces of Theorem 10.3.13 that behave as analogues of apartments of projective spaces. Next, in Sect. 10.5, we use the perspectivities found earlier as well as the apartments to study the automorphism groups of polar spaces. A remarkable feature is that they are transitive on the set of apartments. Finally, in Sect. 10.6, we focus on the finite examples. The information gathered regarding the action of their automorphism groups on the polar spaces is of use in determining their orders.

Throughout this chapter, $\mathbb{D}$ is a division ring and (σ, ε) is an admissible pair for $\mathbb{D}$ (see Definition 7.3.10). Moreover, V is a right vector space over $\mathbb{D}$.

We begin with a section on algebraic aspects of (σ, ε)-hermitian forms related to quasi-polarities. Such forms were introduced in Definition 7.3.10 in order to de-

F. Buekenhout, A.M. Cohen, *Diagram Geometry*,
Ergebnisse der Mathematik und ihrer Grenzgebiete. 3. Folge / A Series of
Modern Surveys in Mathematics 57, DOI 10.1007/978-3-642-34453-4_10,
© Springer-Verlag Berlin Heidelberg 2013

scribe reflexivity. These technicalities are needed for the introduction of pseudo-quadrics in Sect. 10.2.

10.1 Trace Valued Forms

Let $\mathbb{P} = \mathbb{P}(V)$ and let π be a quasi-polarity of $\mathbb{P}$. By Proposition 7.4.4, its absolute space $\mathbb{P}_\pi$ (cf. Definition 7.1.9) is a polar space embedded in $\mathbb{P}$.

We will study the algebraic counterpart of quasi-polarities, certain kinds of sesquilinear forms, satisfying conditions involving an admissible pair (σ, ε) for $\mathbb{D}$ as introduced in Definition 7.3.10. This means that σ is an anti-automorphism of $\mathbb{D}$ and ε is an element of $\mathbb{D}$ such that, for each $t \in \mathbb{D}$,

$$\sigma^2(t) = \varepsilon t \varepsilon^{-1} \quad \text{and} \quad \sigma(\varepsilon) = \varepsilon^{-1}.$$

By Theorem 7.2.14(i) and Corollary 7.3.9, the quasi-polarity π is determined by a (σ, ε)-hermitian form f on V in the sense that $\pi = \delta_f$ (see Theorem 7.2.12). In Theorem 10.1.3 and Proposition 10.1.8, we find algebraic criteria for $\mathbb{P}(V)_\pi$ to be ruled. In Corollary 10.1.9, we classify the forms into three relevant types and further on, we introduce the stabilizer subgroups of $\mathrm{GL}(V)$ of these forms. The section ends with a description of perspectivities of $\mathbb{P}(V)$ in these stabilizers by means of linear transformations of V.

By Theorem 7.3.11, the embedded polar space defined by a quasi-polarity depends only on the proportionality class of f and so, in view of Corollary 7.3.16, there is no harm in imposing the additional requirement that $\varepsilon = \pm 1$. Then the admissibility conditions for the pair (σ, ε) become

$$\sigma^2 = \mathrm{id} \quad \text{and} \quad \varepsilon = \pm 1. \tag{10.1}$$

However, we will not always exploit this restriction, as working with the entire proportionality class may give more insight.

Notation 10.1.1 For an anti-automorphism σ of $\mathbb{D}$ and an element ε of $\mathbb{D}$, we write

(1) $\mathbb{D}_{\sigma,\varepsilon} = \{x - \sigma(x)\varepsilon \mid x \in \mathbb{D}\}$,
(2) $\mathbb{D}^{\sigma,\varepsilon} = \{x \in \mathbb{D} \mid \sigma(x)\varepsilon = -x\}$.

According to Definition 7.3.10, every (σ, ε)-hermitian form on V satisfies $f(y, x) = \sigma(f(x, y))\varepsilon$ for all $x, y \in V$, which implies $f(x, x) \in \mathbb{D}^{\sigma,-\varepsilon}$. This motivates part of the technicalities of the following lemma.

Lemma 10.1.2 *Each admissible pair (σ, ε) for $\mathbb{D}$ satisfies the following properties.*

(i) *Both $\mathbb{D}_{\sigma,\varepsilon}$ and $\mathbb{D}^{\sigma,\varepsilon}$ are additive subgroups of $\mathbb{D}$. If $\varepsilon = \pm 1$, then they are σ-invariant.*
(ii) $\mathbb{D}_{\sigma,\varepsilon} \subseteq \mathbb{D}^{\sigma,\varepsilon}$.
(iii) *For each $\lambda \in \mathbb{D}$ we have $\sigma(\lambda)\mathbb{D}_{\sigma,\varepsilon}\lambda \subseteq \mathbb{D}_{\sigma,\varepsilon}$ and $\sigma(\lambda)\mathbb{D}^{\sigma,\varepsilon}\lambda \subseteq \mathbb{D}^{\sigma,\varepsilon}$.*

(iv) *The equality $\mathbb{D}_{\sigma,\varepsilon} = \mathbb{D}$ holds if and only if $\sigma = \mathrm{id}$, $\varepsilon = -1$, and the characteristic of $\mathbb{D}$ is distinct from two.*

(v) *If the characteristic of $\mathbb{D}$ is distinct from two, then $\mathbb{D}_{\sigma,\varepsilon} = \mathbb{D}^{\sigma,\varepsilon}$.*

Proof (i), (ii), and (iii) are straightforward.

(iv) Suppose $\mathbb{D}_{\sigma,\varepsilon} = \mathbb{D}$. Since $1 \in \mathbb{D}_{\sigma,\varepsilon}$, assertion (ii) gives $\sigma(1)\varepsilon = -1$, whence $\varepsilon = -1$. Therefore, $\sigma^2 = \mathrm{id}$. If $x \in \mathbb{D}$, then $x \in \mathbb{D}_{\sigma,\varepsilon}$, so $x = t + \sigma(t)$ for some $t \in \mathbb{D}$. Now $\sigma(x) = \sigma(t) + \sigma^2(t) = \sigma(t) + t = x$ and $\sigma = \mathrm{id}$. Taking $x = 1$, we obtain $1 = 2t$, so $\mathbb{D}$ is of characteristic distinct from two.

For the converse, assume $(\sigma, \varepsilon) = (\mathrm{id}, -1)$ and $\mathrm{char}(\mathbb{D}) \neq 2$. Then each $x \in \mathbb{D}$ can be written as $(x/2) - \sigma(x/2)\varepsilon$, and so belongs to $\mathbb{D}_{\sigma,\varepsilon}$.

(v) If $\mathbb{D}$ has characteristic distinct from two, then, for each $x \in \mathbb{D}^{\sigma,\varepsilon}$, the equality $\sigma(x)\varepsilon = -x$ implies $x = (x/2) - \sigma(x/2)\varepsilon \in \mathbb{D}_{\sigma,\varepsilon}$, so $\mathbb{D}^{\sigma,\varepsilon} \subseteq \mathbb{D}_{\sigma,\varepsilon}$, and the result follows from (ii). $\qquad\qquad\square$

The use of $\mathbb{D}_{\sigma,\varepsilon}$ will be clear from the following result.

Theorem 10.1.3 *Let f be a nondegenerate (σ, ε)-hermitian form on the vector space V. Let $\pi = \delta_f$ be the quasi-polarity determined by f on $\mathbb{P} := \mathbb{P}(V)$. Assume that the absolute space $\mathbb{P}_\pi$ is non-empty.*

(i) *The absolute points of π span $\mathbb{P}$ if and only if $f(x,x) \in \mathbb{D}_{\sigma,-\varepsilon}$ for all $x \in V$.*

(ii) *If $\mathbb{P}_\pi$ spans $\mathbb{P}$, then, for each point d of $\mathbb{P}_\pi$ and each line l on d not in $\pi(d)$, the line l meets $\mathbb{P}_\pi$ in at least one point distinct from d.*

(iii) *The set $\{x \in V \mid f(x,x) \in \mathbb{D}_{\sigma,-\varepsilon}\}$ is a linear subspace of V.*

Proof (i) '$\Rightarrow$' Suppose that $\mathbb{P}_\pi$ spans $\mathbb{P}$. Then there are points $\langle e_i \rangle \in \mathbb{P}_\pi$ ($i \in I$) such that the e_i span V for some index set I. Now $\langle e_i \rangle \in \mathbb{P}_\pi$ means $f(e_i, e_i) = 0$. For $x \in V$, there are scalars $\lambda_i \in \mathbb{D}$ ($i \in I$), of which only a finite number are nonzero, such that $x = \sum_i e_i \lambda_i$. Expansion of $f(x,x)$ gives

$$f(x,x) = \sum_{i<j} \big(\sigma(\lambda_i) f(e_i, e_j)\lambda_j + \sigma(\lambda_j) f(e_j, e_i)\lambda_i\big)$$

$$= \sum_{i<j} \big(\sigma(\lambda_i) f(e_i, e_j)\lambda_j + \sigma\big(\sigma(\lambda_i) f(e_i, e_j)\lambda_j\big)\varepsilon\big).$$

Each summand is in $\mathbb{D}_{\sigma,-\varepsilon}$, and as this set is additively closed (see Lemma 10.1.2(i)), it follows that $f(x,x) \in \mathbb{D}_{\sigma,-\varepsilon}$.

(i) '$\Leftarrow$' Assume $f(x,x) \in \mathbb{D}_{\sigma,-\varepsilon}$ for all $x \in V$. Let $d = \langle v \rangle$ be a point of $\mathbb{P}_\pi$, and let $b = \langle w \rangle$ be a point of $\mathbb{P}$ with $f(d,b) \neq 0$; it exists as f is nondegenerate. Let l be the projective line on b and d and $p \in l \setminus \{d\}$, so $p = \langle v\lambda + w \rangle$ for some scalar λ. We claim that there is $\lambda \in \mathbb{D}$ such that $f(p,p) = 0$, i.e., such that $p \in \mathbb{P}_\pi$. Choosing $t \in \mathbb{D}$ such that $f(w,w) = t + \sigma(t)\varepsilon$, we have

$$f(v\lambda + w, v\lambda + w) = f(v\lambda, v\lambda) + f(v\lambda, w) + f(w, v\lambda) + f(w, w)$$

$$= \sigma(\lambda) f(v, w) + \sigma\big(\sigma(\lambda) f(v, w)\big)\varepsilon + t + \sigma(t)\varepsilon$$

$$= \big(\sigma(\lambda) f(v, w) + t\big) + \sigma\big(\sigma(\lambda) f(v, w) + t\big)\varepsilon.$$

Consequently, $f(p, p) = 0$ for each λ satisfying $\sigma(\lambda)f(v, w) + t = 0$. As $f(v, w) \neq 0$, this equation has a solution λ, which settles the claim. Now $b \in l = \langle d, p \rangle_{\mathbb{P}}$, and so each line on a point of $\mathbb{P}_\pi$ is contained in the span of $\mathbb{P}_\pi$. This implies (i).

(ii) is a direct consequence of the above, the point p being the required point on l distinct from d.

(iii) Write $U = \{x \in V \mid f(x, x) \in \mathbb{D}_{\sigma, -\varepsilon}\}$. Clearly, $0 \in U$. Suppose, $v, w \in U$ and $\lambda \in \mathbb{D}$. Then $f(v + w\lambda, v + w\lambda) = f(v, v) + \sigma(\lambda)f(w, w)\lambda + f(v, w)\lambda + \sigma(\lambda)f(w, v) = f(v, v) + \sigma(\lambda)f(w, w)\lambda + (f(v, w)\lambda + \sigma(f(v, w)\lambda)\varepsilon) \in \mathbb{D}_{\sigma, -\varepsilon}$ in view of Lemma 10.1.2(i), (iii). $\qquad\square$

We take a closer look at the condition $f(x, x) \in \mathbb{D}_{\sigma, -\varepsilon}$ appearing in Theorem 10.1.3(i).

Lemma 10.1.4 *For a (σ, ε)-hermitian form f on the vector space V over $\mathbb{D}$, the following properties are equivalent.*

(i) $f(x, y) = g(x, y) + \sigma(g(y, x))\varepsilon$ *for some σ-sesquilinear form g on V and all $x, y \in V$.*

(ii) $f(x, x) \in \mathbb{D}_{\sigma, -\varepsilon}$ *for all $x \in V$.*

(iii) *There is a basis $(e_i)_{i \in I}$ of V such that $f(e_i, e_i) \in \mathbb{D}_{\sigma, -\varepsilon}$ for all $i \in I$.*

Proof (i) $\Rightarrow$ (ii). Indeed, $f(x, x) = g(x, x) + \sigma(g(x, x))\varepsilon$.

(ii) $\Rightarrow$ (iii) is obvious. By the way, (iii)$\Rightarrow$(ii) is a direct consequence of Theorem 10.1.3(iii).

(iii) $\Rightarrow$ (i). Choose $g_i \in \mathbb{D}$ such that $f(e_i, e_i) = g_i + \sigma(g_i)\varepsilon$ for $i \in I$. Fix an ordering $<$ on I and put

$$g_{ij} = \begin{cases} f(e_i, e_j) & \text{if } i < j, \\ g_i & \text{if } i = j, \\ 0 & \text{if } i > j. \end{cases}$$

The σ-sesquilinear form g defined by $g(x, y) = \sum_{i,j} \sigma(x_i)g_{ij}y_j$ satisfies (i) because

$$g(x, y) + \sigma(g(y, x))\varepsilon$$
$$= \sum_{i,j} \sigma(x_i)g_{ij}y_j + \sum_{i,j} \sigma(\sigma(y_i)g_{ij}x_j)\varepsilon = \sum_{i,j} \sigma(x_i)(g_{ij} + \sigma(g_{ji})\varepsilon)y_j$$
$$= \sum_{i<j} \sigma(x_i)f(e_i, e_j)y_j + \sum_{i>j} \sigma(x_i)\sigma(f(e_j, e_i))\varepsilon y_j$$
$$+ \sum_i \sigma(x_i)(g_i + \sigma(g_i)\varepsilon)y_i$$
$$= f(x, y). \qquad\square$$

Definition 10.1.5 A reflexive σ-sesquilinear form f is called *trace valued* if it has the equivalent properties of the lemma. We will see in a while that most reflexive forms are trace valued. For that purpose we also say that an admissible pair (σ, ε) is a *pair of trace type* if every (σ, ε)-hermitian form on a vector space over $\mathbb{D}$ is trace valued. Finally, we say that a division ring $\mathbb{D}$ is of *trace type* if all admissible pairs for $\mathbb{D}$ are of trace type.

Remark 10.1.6 Let f be a nondegenerate (σ, ε)-hermitian form on a vector space V. Consider the set U of all vectors x such that $f(x, x) \in \mathbb{D}_{\sigma, -\varepsilon}$. By Theorem 10.1.3(iii), U is a subspace of V. Moreover, the restriction of f to $U \times U$ is trace valued. So, forms that are not necessarily trace valued have a canonical trace valued restriction. Also, the points of $\mathbb{P}(V)_f$ (cf. Notation 7.2.13) lie in $\mathbb{P}(U)$. In view of these observations, for the study of embedded polar spaces, we can restrict attention to trace valued forms.

Example 10.1.7 Let $\mathbb{F}$ be a field of characteristic two, $\sigma = \mathrm{id}$, and let f be a symmetric form on a vector space V over $\mathbb{F}$ such that $f(x, x) \neq 0$ for some $x \in V$. Then $\mathbb{F}_{\sigma, -\varepsilon} = \mathbb{F}_{\sigma, \varepsilon} = 0$, so f cannot be trace valued. An example is the form f on $\mathbb{F}^3$ defined by $f(x, y) = x_1 y_1 + x_2 y_3 + x_3 y_2$. By Theorem 10.1.3(iii), $\{x \in V \mid f(x, x) = 0\}$ is a linear subspace of $\mathbb{F}^3$; in fact it is given by the equation $x_1 = 0$. By Remark 10.1.6, $\mathbb{P}_f$ is embedded in $\mathbb{P}(U)$ and the restriction of f to $U \times U$ is trace valued; indeed, on U, we have $f(x, y) = x_2 y_3 + x_3 y_2$.

The next result shows that many reflexive sesquilinear forms are trace valued.

Proposition 10.1.8 *Let (σ, ε) be admissible pair for $\mathbb{D}$.*

(i) *The pair (σ, ε) is of trace type if and only if $\mathbb{D}_{\sigma, -\varepsilon} = \mathbb{D}^{\sigma, -\varepsilon}$.*
(ii) *If the characteristic of $\mathbb{D}$ is distinct from two, then $\mathbb{D}$ is of trace type.*
(iii) *If $\mathbb{D}$ is of characteristic two and if σ does not fix all elements of the center $\mathrm{Z}(\mathbb{D})$ of $\mathbb{D}$, then $\mathbb{D}$ is of trace type.*

Proof If (σ, ε) is an admissible pair, then so is $(\sigma, -\varepsilon)$. Recall from Lemma 10.1.2(i) that $\mathbb{D}_{\sigma, -\varepsilon} \subseteq \mathbb{D}^{\sigma, -\varepsilon} = \{u \in \mathbb{D} \mid \sigma(u)\varepsilon = u\}$.

(i) Assume that (σ, ε) is of trace type. Let $u \in \mathbb{D}$ with $\sigma(u)\varepsilon = u$. Consider the form $f : \mathbb{D} \times \mathbb{D} \to \mathbb{D}$ defined by $f(x, y) = \sigma(x)uy$. It is (σ, ε)-hermitian. Thus, f is trace valued, whence $u = f(1, 1) \in \mathbb{D}_{\sigma, -\varepsilon}$.

Conversely, let $\mathbb{D}_{\sigma, -\varepsilon} = \mathbb{D}^{\sigma, -\varepsilon}$ and let f be a (σ, ε)-hermitian form. Then $f(x, x) = \sigma(f(x, x))\varepsilon$ and so condition (ii) of Lemma 10.1.4 holds. Hence f is trace valued.

(ii) This is direct from (i) and Lemma 10.1.2(v).

(iii) Let $a \in \mathrm{Z}(\mathbb{D})$ such that $\sigma(a) \neq a$. Then $b = a + \sigma(a) \neq 0$ and $\sigma(b) = \sigma^2(a) + \sigma(a) = \varepsilon a \varepsilon^{-1} + \sigma(a) = a + \sigma(a) = b$. If $u \in \mathbb{D}$ and $\sigma(u)\varepsilon = u$, then $u = b(b^{-1}u) = (a + \sigma(a))b^{-1}u = ab^{-1}u + \sigma(a)b^{-1}u$ and $\sigma(a)b^{-1}u = \sigma(ab^{-1}u)\varepsilon$ because $\sigma(ab^{-1}u)\varepsilon = \sigma(u)\sigma(b^{-1})\sigma(a)\varepsilon = \sigma(a)b^{-1}\sigma(u)\varepsilon = \sigma(a)b^{-1}u$. Therefore $u \in \mathbb{D}_{\sigma, -\varepsilon}$ and so (i) applies. $\qquad\square$

Quasi-polarities of a projective space $\mathbb{P}$ with absolutes spanning $\mathbb{P}$ obey the following classification. It uses the notions alternating, symmetric, antisymmetric, and σ-hermitian introduced in Definition 7.3.10.

Corollary 10.1.9 *Let* $\dim(V) \geq 1$. *Every quasi-polarity on the projective space* $\mathbb{P} := \mathbb{P}(V)$ *whose absolute spans* $\mathbb{P}$ *falls into one of the three following mutually exclusive cases.*

(1) *A quasi-polarity determined by a nonzero alternating form over a field.*
(2) *A quasi-polarity determined by a symmetric bilinear form over a field of characteristic distinct from two.*
(3) *A quasi-polarity determined by a nonzero σ-hermitian form such that $\sigma \neq$ id.*

Proof According to Theorem 7.2.14(i), a quasi-polarity as in the hypothesis is of the form δ_f for f a nondegenerate reflexive σ-sesquilinear form on V. Indeed, $\mathrm{Rad}(f) = \{0\}$ as $\pi(\langle x \rangle)$ is a hyperplane of $\mathbb{P}(V)$ for each nonzero vector x in V. By Corollary 7.3.9, such an f is (σ, ε)-hermitian, and by Corollary 7.3.16, we may assume, by changing f within its proportionality class, that f is either antisymmetric or non-alternating and hermitian. In the latter case, f is as in (2) or (3) according to whether $\sigma = $ id or $\sigma \neq$ id. So assume that f is antisymmetric and nonzero. Then $\sigma = $ id and $\varepsilon = -1$, so $\mathbb{D}_{\sigma,-\varepsilon} = \{0\}$. As $\mathbb{P}_\pi$ spans $\mathbb{P}$, Theorem 10.1.3(i) implies $f(x, x) = 0$ for all $x \in V$, so we are in case (1). By the assumption that f is nonzero in this case, there are x, y in V with $f(x, y) = 1$ and, hence, $f(y, x) = -1$, so that $f(x, y) \neq f(y, x)$ if the characteristic of the field is not equal to two. In particular, f is not as in (2). $\qquad\square$

The following definitions are consistent with those of Exercise 1.9.31.

Definition 10.1.10 Quasi-polarities as in (1), (2), (3) of Corollary 10.1.9 are called *symplectic, orthogonal, unitary,* respectively. The corresponding linear groups

$$\mathrm{GL}(V)_f = \big\{ g \in \mathrm{GL}(V) \mid \forall_{x,y \in V}\ f\big(g(x), g(y)\big) = f(x, y) \big\}$$

are also given these names. More precisely, in the respective cases of the corollary, we call $\mathrm{GL}(V)_f$

(1) the *symplectic group* and denote it by $\mathrm{Sp}(V, f)$;
(2) the *orthogonal group* and denote it by $\mathrm{O}(V, \kappa)$, where κ is the quadratic form on V given by $\kappa(x) = f(x, x)/2$;
(3) the *unitary group* and denote it by $\mathrm{U}(V, f)$.

The images of these groups in $\mathrm{PGL}(V, f)$ (introduced in Example 1.8.16) under the natural quotient map are denoted $\mathrm{PSp}(V, f)$, $\mathrm{PO}(V, \kappa)$, and $\mathrm{PU}(V, f)$, respectively.

In the second case, the characteristic is distinct from two and the form f can be recovered from κ by

$$f(x, y) = \frac{1}{2}\big(\kappa(x + y) - \kappa(x) - \kappa(y)\big). \tag{10.2}$$

Later (in Definition 10.2.16) we will define a wider class of orthogonal groups.

Remark 10.1.11 The group $GL(V)$ acts on the linear space of (σ, ε)-hermitian forms on V and on the linear space of σ-quadratic forms on V by $(gf)(x, y) = f(g^{-1}x, g^{-1}y)$ and $(g\kappa)(x) = \kappa(g^{-1}x)$ for $x, y \in V$, respectively. The groups that have just been introduced can be seen as stabilizers of κ and f, respectively, under these actions.

Example 10.1.12 In special cases, orthogonal and symplectic groups are isomorphic. Let V be a vector space of odd dimension n over a field $\mathbb{F}$ of characteristic two. Let κ be a quadratic form on V for which $\mathbb{P}(V)_\kappa$ is nondegenerate and spans $\mathbb{P}(V)$ (cf. Exercise 10.7.12 for a simple criterion for this property). Then κ has a unique kernel, that is, a point $\langle v \rangle$ of $\mathbb{P}(V)$ with $\kappa(v) \neq 0$ and $f(v, V) = 0$, where f is the hermitian form of κ. Modding out $\langle v \rangle$ leads to a linear map $\sigma : V \to \overline{V}$, where $\overline{V} = V/\langle v \rangle$. As $\langle v \rangle$ lies in $\mathrm{Rad}(f)$, the bilinear form f induces a unique symmetric bilinear form $\overline{f}$ on $\overline{V}$. Its stabilizer in $GL(\overline{V})$ is $\mathrm{Sp}(\overline{V}, \overline{f})$. Each $g \in O(V, \kappa)$ fixes $\langle v \rangle$ and so induces a unique linear transformation $\overline{g} \in \mathrm{Sp}(\overline{V}, \overline{f})$. This map is injective: if $\overline{g}$ is the identity on $\overline{V}$, then, for each $w \in V$, there is a scalar $\lambda_g \in \mathbb{F}$ such that $gw = w + v\lambda_g$; now $\kappa(w) = \kappa(gw) = \kappa(w) + \lambda_g f(w, v) + \lambda_g^2 \kappa(v)$. As $f(w, v) = 0 \neq \kappa(v)$, this implies $\lambda_g = 0$, so g is the identity map. Moreover, if $\mathbb{F}$ is perfect (that is, every element of $\mathbb{F}$ is a square), then the map is surjective: if $h \in \mathrm{Sp}(\overline{V}, \overline{f})$, fix a complement U of $\langle v \rangle$ in V and let h act on U as on $\overline{V}$. Now, for each $u \in U$, let λ_u be chosen so that $\lambda_u^2 \kappa(v) = \kappa(hu) + \kappa(u)$ and define $g(v\lambda + u) = v(\lambda_u + \lambda) + hu$. Then $\kappa(v\lambda + u) = \kappa(v\lambda) + \kappa(u) = \lambda_u^2 \kappa(v) + \kappa(v\lambda) + \kappa(hu) = \kappa(v(\lambda_u + \lambda) + hu) = \kappa(g(v\lambda + u))$, so $g \in O(V, \kappa)$ is in the inverse image of h. We conclude that $O(V, \kappa) \cong \mathrm{Sp}(\overline{V}, \overline{f})$ if $\mathbb{F}$ is perfect.

Conversely, if f is a nondegenerate bilinear alternating form on a vector space U over $\mathbb{F}$, extend U by one dimension to a vector space $V = U \oplus \langle v \rangle$, and let κ be a nondegenerate symmetric bilinear form on V with $\kappa(v) \neq 0$ and hermitian form f. Then $O(V, \kappa) \cong \mathrm{Sp}(U, f)$.

Clearly, $U(V, f)$ leaves invariant the absolute $\mathbb{P}(V)_f$ and acts on the corresponding geometry $\mathrm{Abs}(\mathrm{PG}(V), \delta_f)$. If $\mathbb{P}(V)_f$ is ruled in $\mathbb{P}(V)$, then the group induced on $\mathbb{P}(V)_f$ by $U(V, f)$ coincides with $\mathrm{PU}(V, f)$.

In Theorem 9.3.3, the existence of perspectivities leaving invariant a ruled polar space in $\mathbb{P}(V)$ has been derived. Here we exhibit the explicit form of these perspectivities for $U(V, f)$, using the transformations $r_{a,\phi}$ of V defined in Exercise 1.9.30.

Proposition 10.1.13 *Let f be a (σ, ε)-hermitian form on V. Suppose that $c \in V \setminus \mathrm{Rad}(f)$ and $a, b \in V \setminus \delta_f(c)$ are such that $\langle c \rangle$, $\langle a \rangle$, and $\langle b \rangle$ are collinear points in $\mathbb{P}(V)$, scaled in such a way that $c = a - b$ and $f(a, a) = f(b, b)$. The map $r_{c,\phi} : V \to V$, where $\phi : V \to \mathbb{D}$ is given by $\phi(z) = f(c, a)^{-1} f(c, z)$, belongs to $U(V, f)$. It induces a perspectivity in $\mathbb{P}(V)$ with center $\langle c \rangle$, axis $\delta_f(c)$, mapping $\langle a \rangle$ to $\langle b \rangle$.*

Proof The assignment $x \mapsto f(c, x)$ $(x \in V)$ defines a linear form on V, so we may apply the observation made in Example 6.1.2. Therefore we need only show that $f(r_{c,\phi}(x), r_{c,\phi}(y)) = f(x, y)$ for any two vectors $x, y \in V$; but this follows from a straightforward computation. $\square$

Later, in Sect. 10.5, we will pursue a further study of the unitary and orthogonal groups.

10.2 Pseudo-quadrics

In this section we introduce certain embeddable polar spaces that have not yet been introduced. They are called pseudo-quadrics. As the name suggests, they are generalizations of the quadrics appearing in Example 7.8.1. It will turn out that novelties arise only over certain division rings of characteristic two (not of trace type); see Theorem 10.2.15 and Example 7.8.1, which describes the familiar situation of quadrics.

We now extend the results of Example 7.8.1 to the case of a division algebra $\mathbb{D}$. We will make use of the additive quotient group $\mathbb{D}/\mathbb{D}_{\sigma,\varepsilon}$ of $\mathbb{D}$. The key idea is to introduce a map $\kappa : V \to \mathbb{D}/\mathbb{D}_{\sigma,\varepsilon}$ generalizing the notion of a quadratic form κ and to consider the set of points $\langle x \rangle$ of $\mathbb{P}(V)$ such that $\kappa(x)$ is equal to zero in $\mathbb{D}/\mathbb{D}_{\sigma,\varepsilon}$.

Notation 10.2.1 For $a \in \mathbb{D}$, write $\overline{a} = a + \mathbb{D}_{\sigma,\varepsilon}$ and consider the product

$$\mathbb{D}/\mathbb{D}_{\sigma,\varepsilon} \times \mathbb{D} \to \mathbb{D}/\mathbb{D}_{\sigma,\varepsilon} \ \text{ given by } \ \overline{a} * b = \overline{\sigma(b)ab} \qquad (a, b \in \mathbb{D}).$$

In view of Lemma 10.1.2(iii), $\overline{a'} * b$ does not depend on the particular choice of $a' \in \overline{a}$.

Lemma 10.2.2 *Let (σ, ε) be an admissible pair for $\mathbb{D}$. Then $\mathbb{D}^{\sigma,\varepsilon}/\mathbb{D}_{\sigma,\varepsilon}$, supplied with right scalar multiplication $*$, is a right vector space over $\mathbb{D}$.*

Proof By Lemma 10.1.2(i), (ii), $\mathbb{D}_{\sigma,\varepsilon}$ and $\mathbb{D}^{\sigma,\varepsilon}$ are additive groups such that the former is contained in the latter. By Lemma 10.1.2(iii), the scalar multiplication $*$ on $\mathbb{D}$ leaves $\mathbb{D}_{\sigma,\varepsilon}$ and $\mathbb{D}^{\sigma,\varepsilon}$ invariant, and so is well defined on the quotient $\mathbb{D}^{\sigma,\varepsilon}/\mathbb{D}_{\sigma,\varepsilon}$. For $a \in \mathbb{D}^{\sigma,\varepsilon}$ and $b, c \in \mathbb{D}$, we have

$$\sigma\big(\sigma(b)ac\big)\varepsilon + \sigma(c)ab = \sigma(c)\sigma(a)\varepsilon b + \sigma(c)ab = \sigma(c)\big(\sigma(a)\varepsilon + a\big)b = 0,$$

so

$$\overline{a} * (b + c) = \overline{\sigma(b + c)a(b + c)} = \overline{\sigma(b)ab + \sigma(b)ac + \sigma(c)ab + \sigma(c)ac}$$

$$= \overline{\sigma(b)ab} + \overline{\sigma(b)ac - \sigma\big(\sigma(b)ac\big)\varepsilon} + \overline{\sigma\big(\sigma(b)ac\big)\varepsilon + \sigma(c)ab} + \overline{\sigma(c)ac}$$

$$= \overline{\sigma(b)ab} + \overline{0} + \overline{0} + \overline{\sigma(c)ac}$$

$$= \overline{a} * b + \overline{a} * c,$$

proving that the scalar multiplication is distributive. The other axioms of a vector space are straightforward to verify. $\square$

Example 10.2.3 In the orthogonal case, where $(\sigma, \varepsilon) = (\mathrm{id}, 1)$, we have $\mathbb{D}_{\sigma,\varepsilon} = \{0\}$ and $\bar{a} * b = \overline{b^2 a}$. As $\mathbb{D}^{\sigma,\varepsilon} = \{x \in \mathbb{D} \mid 2x = 0\}$, the vector space on $\mathbb{D}^{\sigma,\varepsilon}/\mathbb{D}_{\sigma,\varepsilon}$ with scalar multiplication $*$ is trivial if $\mathbb{D}$ has characteristic distinct from two and is a twisted version of $\mathbb{D}$ otherwise.

In the symplectic case, $(\sigma, \varepsilon) = (\mathrm{id}, -1)$, with $\mathbb{D}$ of characteristic $\neq 2$, we have $\mathbb{D}_{\sigma,\varepsilon} = \mathbb{D}$ (see Lemma 10.1.2(iv)) and $\mathbb{D}^{\sigma,\varepsilon}/\mathbb{D}_{\sigma,\varepsilon} = \{0\}$ (see Lemma 10.1.2(v)).

Proposition 10.2.4 *The following properties concerning a map* $\kappa : V \to \mathbb{D}/\mathbb{D}_{\sigma,\varepsilon}$ *are equivalent.*

(i) *There exists a σ-sesquilinear form* $g : V \times V \to \mathbb{D}$ *such that*

$$\kappa(x) = g(x, x) + \mathbb{D}_{\sigma,\varepsilon} \quad \text{for all } x \in V.$$

(ii) *$\kappa(x\lambda) = \kappa(x) * \lambda$ for all $x \in V$, $\lambda \in \mathbb{D}$, and there exists a trace valued (σ, ε)-hermitian form* $f : V \times V \to \mathbb{D}$ *such that*

$$\kappa(x + y) = \kappa(x) + \kappa(y) + \big(f(x, y) + \mathbb{D}_{\sigma,\varepsilon}\big)$$

for all $x, y \in V$.

Moreover, if g is as in (i), *then the form f of* (ii) *can be chosen in such a way that* $f(x, y) = g(x, y) + \sigma(g(y, x))\varepsilon$.

Proof (i) $\Rightarrow$ (ii) Assume (i). Then $\kappa(x\lambda) = g(x\lambda, x\lambda) + \mathbb{D}_{\sigma,\varepsilon} = \sigma(\lambda)g(x, x)\lambda + \mathbb{D}_{\sigma,\varepsilon} = \kappa(x) * \lambda$. Moreover, if we put $f(x, y) = g(x, y) + \sigma(g(y, x))\varepsilon$, then f is trace valued by definition and

$$\kappa(x + y) = g(x + y, x + y) + \mathbb{D}_{\sigma,\varepsilon}$$
$$= g(x, x) + g(y, y) + g(x, y) + g(y, x) + \mathbb{D}_{\sigma,\varepsilon}$$
$$= \kappa(x) + \kappa(y) + g(x, y) + \sigma\big(g(y, x)\big)\varepsilon + g(y, x) - \sigma\big(g(y, x)\big)\varepsilon + \mathbb{D}_{\sigma,\varepsilon}$$
$$= \kappa(x) + \kappa(y) + \big(f(x, y) + \mathbb{D}_{\sigma,\varepsilon}\big).$$

We have to check that f is a (σ, ε)-hermitian form on V. Clearly, f is biadditive. Next,

$$f(x\lambda, y\mu) = \sigma(\lambda)g(x, y)\mu + \sigma\big(\sigma(\mu)g(y, x)\lambda\big)\varepsilon$$
$$= \sigma(\lambda)g(x, y)\mu + \sigma(\lambda)\sigma\big(g(y, x)\big)\sigma^2(\mu)\varepsilon$$
$$= \sigma(\lambda)g(x, y)\mu + \sigma(\lambda)\sigma\big(g(y, x)\big)\varepsilon\mu$$
$$= \sigma(\lambda)f(x, y)\mu.$$

Also,

$$f(y, x) = g(y, x) + \sigma\big(g(x, y)\big)\varepsilon = \sigma\big(\sigma\big(g(y, x)\big)\varepsilon + g(x, y)\big)\varepsilon = \sigma\big(f(x, y)\big)\varepsilon.$$

(ii) $\Rightarrow$ (i) Assume (ii). Let $(e_i)_{i\in I}$ be a basis of V, which we provide with a total ordering $<$. Take $g_i \in \mathbb{D}$ such that $\kappa(e_i) = g_i + \mathbb{D}_{\sigma,\varepsilon}$ and define the σ-sesquilinear form g by

$$g(e_i, e_j) = \begin{cases} f(e_i, e_j) & \text{if } i < j, \\ g_i & \text{if } i = j, \\ 0 & \text{if } i > j. \end{cases}$$

Then (i) holds. Indeed, writing $x = \sum_i e_i x_i$ with $x_i \in \mathbb{D}$, nonzero for a finite number of $i \in I$, we have $g(x, x) = g(\sum_i e_i x_i, \sum_j e_j x_j) = \sum_{i,j} \sigma(x_i) g(e_i, e_j) x_j$ while

$$\kappa(x) = \kappa\left(\sum_i e_i x_i\right) = \sum_i \kappa(e_i x_i) + \sum_{i<j} f(e_i x_i, e_j x_j) + \mathbb{D}_{\sigma,\varepsilon}$$

$$= \sum_i \kappa(e_i) * x_i + \sum_{i<j} \sigma(x_i) f(e_i, e_j) x_j + \mathbb{D}_{\sigma,\varepsilon}$$

$$= \sum_i \sigma(x_i) g_i x_i + \sum_{i<j} \sigma(x_i) g(e_i, e_j) x_j + \mathbb{D}_{\sigma,\varepsilon}$$

$$= g(x, x) + \mathbb{D}_{\sigma,\varepsilon}.$$

The final assertion is immediate from the proof of the first implication. $\square$

Corollary 10.2.5 *Let κ be as in Proposition* 10.2.4. *If $\mathbb{D}_{\sigma,\varepsilon} \neq \mathbb{D}$, then the trace valued (σ, ε)-hermitian form f of Proposition* 10.2.4 *is unique.*

Proof Suppose that f_1 is another such form on V. The form $h = f - f_1$ is (σ, ε)-hermitian with $h(x, y) \in \mathbb{D}_{\sigma,\varepsilon}$ for all $x, y \in V$ and satisfies $h(u, v) \neq 0$ for certain $u, v \in V$. Now $\mathbb{D} = h(u, v)\mathbb{D} = h(u, v\mathbb{D}) \subseteq \mathbb{D}_{\sigma,\varepsilon}$, whence $\mathbb{D} = \mathbb{D}_{\sigma,\varepsilon}$, contradicting the hypotheses. $\square$

If $(\sigma, \varepsilon) = (\mathrm{id}, -1)$, then either the characteristic of $\mathbb{D}$ equals two and so $(\sigma, \varepsilon) = (\mathrm{id}, 1)$, or the characteristic is distinct from two and each function κ as in Proposition 10.2.4 is trivial by Lemma 10.1.2(iv). So there is little lost in excluding the latter case. Besides, Corollary 10.2.5 shows that, by excluding it, we obtain uniqueness of the form f pertaining to κ.

Definition 10.2.6 Suppose that (σ, ε) is an admissible pair distinct from $(\mathrm{id}, -1)$ if $\mathrm{char}(\mathbb{D}) \neq 2$. A function κ with the properties (i) and (ii) of Proposition 10.2.4 is called a (σ, ε)-*quadratic form* or *pseudo-quadratic form* relative to σ and ε. We call the trace valued (σ, ε)-hermitian form f as in Proposition 10.2.4(ii) the *hermitian form* of κ. As earlier, by a σ-*quadratic form* we mean a $(\sigma, 1)$-quadratic and *quadratic* stands for id-quadratic. A form g as in (i) of the proposition is called a *facilitating form* for κ.

The proof of Proposition 10.2.4 actually shows that the facilitating form for κ can be chosen to be upper triangular with respect to any ordered basis of V.

In view of the restrictions on (σ, ε), there is no need to define an *antisymmetric quadratic form*: any $(\mathrm{id}, -1)$-quadratic form is also a symmetric quadratic form if the characteristic of $\mathbb{D}$ equals two and is not defined (and would be trivial if it were) otherwise.

Definition 10.2.7 Let κ be a (σ, ε)-quadratic form on the right vector space V over $\mathbb{D}$. The *pseudo-quadric* $\mathbb{P}(V)_\kappa$ defined by κ is the space whose points are those points $\langle v \rangle$ of $\mathbb{P}(V)$ on which κ vanishes, i.e., $\kappa(v) \in \mathbb{D}_{\sigma, \varepsilon}$, and whose lines are the lines of $\mathbb{P}(V)$ fully contained in the point set of $\mathbb{P}(V)_\kappa$.

Example 10.2.8 Let $V = \mathbb{D}^2$ and fix an admissible pair (σ, ε). For $x = \varepsilon_1 x_1 + \varepsilon_2 x_2 \in V$, put $\kappa(x) = \sigma(x_1)x_2 + \mathbb{D}_{\sigma, \varepsilon}$. Then κ is a pseudo-quadratic form relative to σ and ε such that $\mathbb{P}(V)_\kappa$ contains $(1 : 0)$ and $(0 : 1)$. As a matter of fact, every nonzero such pseudo-quadratic form on V for which $\mathbb{P}(V)_\kappa$ is non-empty can be transformed into κ after a coordinate transformation fixing these two points of $\mathbb{P}(V)$ (cf. Exercise 10.7.9).

Naturally, for a given pseudo-quadratic form κ on V, we are interested in the relation between the space $\mathbb{P}(V)_\kappa$ and the polar space $\mathbb{P}(V)_f$ where f is the hermitian form of κ. Recall from Notation 9.1.9 that we also use the $\sqcap$ symbol when one of its arguments is a line space.

Theorem 10.2.9 *Suppose that κ is a (σ, ε)-quadratic form on a right vector space V over $\mathbb{D}$ and that f is the hermitian form of κ. Put $\mathbb{P} = \mathbb{P}(V)$.*

(i) *The space $\mathbb{P}_\kappa$ is a subspace of $\mathbb{P}_f$. In particular, $\mathbb{P}_\kappa$ is a polar space.*
(ii) *If the points of $\mathbb{P}_\kappa$ span $\mathbb{P}$, then $\mathrm{Rad}(\mathbb{P}_\kappa) = \mathbb{P}_\kappa \sqcap \mathrm{Rad}(\mathbb{P}_f) = \mathbb{P}_\kappa \sqcap \mathrm{Rad}(f)$.*
(iii) *If $\mathbb{D}_{\sigma, \varepsilon} = \mathbb{D}^{\sigma, \varepsilon}$, then $\mathbb{P}_\kappa = \mathbb{P}_f$.*

Proof If f is a nonzero alternating form and the characteristic of $\mathbb{D}$ is distinct from two, then no pseudo-quadratic form κ is defined whose hermitian form is f. For, if it were, then $\kappa(V) = \{\bar{0}\}$, so the space $\mathbb{P}_\kappa$ coincides with $\mathbb{P}$ and the theorem holds trivially. Therefore, we exclude admissible pairs $(\mathrm{id}, -1)$ if the characteristic of $\mathbb{D}$ is not two.

For a facilitating form g of κ (as in Proposition 10.2.4(i)), $\langle x \rangle \in \mathbb{P}_\kappa$ implies $g(x, x) \in \mathbb{D}_{\sigma, \varepsilon}$, so, by Lemma 10.1.2(ii), $f(x, x) = \sigma(g(x, x))\varepsilon + g(x, x) = 0$, yielding $\langle x \rangle \in \mathbb{P}_f$, so the points of $\mathbb{P}_\kappa$ belong to $\mathbb{P}_f$.

(i) Let x and y be points of $\mathbb{P}_\kappa$. If xy is a line of $\mathbb{P}_\kappa$, then $\kappa(x + y\lambda) = \mathbb{D}_{\sigma, \varepsilon}$ for all $\lambda \in \mathbb{D}$. By Proposition 10.2.4(ii), this gives $f(x, y)\lambda \in \mathbb{D}_{\sigma, \varepsilon}$ for all $\lambda \in \mathbb{D}$. Suppose $f(x, y) \neq 0$. Then $\mathbb{D} = f(x, y)\mathbb{D} \subseteq \mathbb{D}_{\sigma, \varepsilon}$, so, by Lemma 10.1.2(iv), $\sigma = \mathrm{id}$, $\varepsilon = -1$ and $\mathbb{D}$ is a field of characteristic distinct from two, which is an excluded case. We conclude that $f(x, y) = 0$, establishing that xy is a line of $\mathbb{P}_f$.

Conversely, if xy is a line of $\mathbb{P}_f$, then $f(x, y) = 0$, so $\kappa(x + y\lambda) = 0$ for all $\lambda \in \mathbb{D}$, proving that xy is a line of $\mathbb{P}_f$. Hence $\mathbb{P}_\kappa$ is a subspace of $\mathbb{P}_f$.

The final assertion of (i) is immediate from the first thanks to Lemma 7.4.8(i).

(ii) Clearly $\mathbb{P}_\kappa \cap \mathrm{Rad}(\mathbb{P}_f) \subseteq \mathrm{Rad}(\mathbb{P}_\kappa)$. Suppose $\langle x \rangle \in \mathrm{Rad}(\mathbb{P}_\kappa)$ for some $x \in V \setminus \{0\}$. Then $f(x, y) = 0$ whenever $y \in V \setminus \{0\}$ satisfies $\langle y \rangle \in \mathbb{P}_\kappa$. By the assumption that $\mathbb{P}_\kappa$ spans $\mathbb{P}$, this implies $f(x, z) = 0$ for all $z \in V$, whence $\langle x \rangle \in \mathrm{Rad}(\mathbb{P}_f)$. This settles the first equality. As the points of $\mathbb{P}_\kappa$ are in $\mathbb{P}_f$ and span $\mathbb{P}$, Lemma 10.1.4 gives that the form f is trace valued, so the second equality follows from Proposition 7.4.4.

(iii) Thanks to the assumption $\mathbb{D}_{\sigma,\varepsilon} = \mathbb{D}^{\sigma,\varepsilon}$, Lemma 10.1.2(ii) can be used the other way around and so the argument in the second paragraph of this proof can be reversed. This implies that the point sets of $\mathbb{P}_f$ and $\mathbb{P}_\kappa$ coincide, so that $\mathbb{P}_\kappa = \mathbb{P}_f$. $\square$

By Theorem 10.2.9(i), the space $\mathbb{P}(V)_\kappa$ for a pseudo-quadric κ on V is a polar space which, in case it is nondegenerate, we can view as a subspace of a unique polar space of the form $\mathbb{P}(V)_f$ for a given hermitian form f of κ. We can then apply Proposition 9.1.7 to the pseudo-quadric $\mathbb{P}(V)_\kappa$, using $K = \mathrm{Rad}(f)$ as the kernel, for, by Proposition 10.2.9(ii), $\mathbb{P}(V)_\kappa \cap K = \mathbb{P}(V)_\kappa \cap \mathrm{Rad}(\mathbb{P}(V)_f) = \mathrm{Rad}(\mathbb{P}(V)_\kappa) = \emptyset$. Thus, $\mathbb{P}(V)_\kappa$ embeds in the nondegenerate polar space $\mathbb{P}(V)_{\overline{f}}$ of the form $\overline{f}$ induced by f on $\mathbb{P}(V)/\mathrm{Rad}(f)$. We conclude that there is no loss of generality in the study of nondegenerate $\mathbb{P}(V)_\kappa$ to assume that f is nondegenerate.

Here are two ways of finding pseudo-quadrics corresponding to a given (σ, ε)-hermitian form.

Proposition 10.2.10 *Suppose that f is a (σ, ε)-hermitian form on the right $\mathbb{D}$-vector space V.*

(i) *If $\mathbb{D}$ is of characteristic distinct from two, consider the map $\kappa : V \to \mathbb{D}/\mathbb{D}_{\sigma,\varepsilon}$ given by*

$$\kappa(x) = \frac{1}{2} f(x, x) + \mathbb{D}_{\sigma,\varepsilon}.$$

(ii) *If f is trace valued, let g be as in Lemma 10.1.4(i), and consider the map $\kappa : V \to \mathbb{D}/\mathbb{D}_{\sigma,\varepsilon}$ given by $\kappa(x) = g(x, x) + \mathbb{D}_{\sigma,\varepsilon}$ $(x \in V)$.*

In each case, κ is a (σ, ε)-quadratic form and f is a hermitian form of κ. In case (ii), g is a facilitating form for κ.

Proof In each case, we verify (ii) of Proposition 10.2.4. Let $x, y \in V$ and $\lambda \in \mathbb{D}$.

(i) Now $\kappa(x\lambda) = \frac{1}{2} f(x\lambda, x\lambda) + \mathbb{D}_{\sigma,\varepsilon} = \sigma(\lambda) \frac{1}{2} f(x, x) \lambda + \mathbb{D}_{\sigma,\varepsilon} = \sigma(\lambda)(\frac{1}{2} f(x, x) + \mathbb{D}_{\sigma,\varepsilon})\lambda = \kappa(x) * \lambda$ in view of Lemma 10.1.2(iii). Moreover,

$$\kappa(x + y) = \frac{1}{2} f(x + y, x + y) + \mathbb{D}_{\sigma,\varepsilon}$$

$$= \frac{1}{2} f(x, x) + \frac{1}{2} f(y, y) + f(x, y) - \frac{1}{2}\big(f(x, y) - \sigma\big(f(x, y)\varepsilon\big)\big) + \mathbb{D}_{\sigma,\varepsilon}$$

$$= \kappa(x) + \kappa(y) + \big(f(x, y) + \mathbb{D}_{\sigma,\varepsilon}\big).$$

(ii) The identity $\kappa(x\lambda) = \kappa(x) * \lambda$ follows as in (i) (with g instead of f). Furthermore,

$$\kappa(x+y) = g(x,x) + g(y,y) + g(x,y) + g(y,x) + \mathbb{D}_{\sigma,\varepsilon}$$
$$= \kappa(x) + \kappa(y) + f(x,y) + g(y,x) - \sigma\big(g(y,x)\big)\varepsilon + \mathbb{D}_{\sigma,\varepsilon}$$
$$= \kappa(x) + \kappa(y) + \big(f(x,y) + \mathbb{D}_{\sigma,\varepsilon}\big). \qquad \square$$

Example 10.2.11 Put $\mathbb{P} = \mathbb{P}(V)$. We are interested in pseudo-quadrics $\mathbb{P}_\kappa$ of $\mathbb{P}$ that are strictly contained in $\mathbb{P}_f$. Theorem 10.2.9(iii) shows that $\mathbb{D}_{\sigma,\varepsilon}$ should not coincide with $\mathbb{D}^{\sigma,\varepsilon}$. Proposition 10.1.8(i) shows that we require an admissible pair (σ,ε) such that $(\sigma,-\varepsilon)$ is not of trace type. By Proposition 10.1.8(ii), (iii), this forces $\mathbb{D}$ to be of characteristic two while σ fixes all elements of the center $Z(\mathbb{D})$ of $\mathbb{D}$.

(i) Consider the case where $\mathbb{D}$ is a field. Then $\sigma = \mathrm{id}$, whence $\varepsilon = 1$ and κ is a quadratic form. Now $\mathbb{P}_\kappa$ is a quadric which can be strictly contained in $\mathbb{P}_f$; but this case already appeared in Example 7.8.1.

(ii) Let $V = \mathbb{D}^2$. For $x = \varepsilon_1 x_1 + \varepsilon_2 x_2 \in V$, let $\kappa(x) = \sigma(x_1)x_2 + \mathbb{D}_{\sigma,\varepsilon}$, as in Example 10.2.8. Then κ is a pseudo-quadratic form relative to σ and ε. The form f given by $f(x,y) = \sigma(x_1)y_2 + \sigma(x_2)\varepsilon y_1$ $(x, y \in V)$ is a hermitian form of κ. The point $(0{:}1)$ belongs to $\mathbb{P}_\kappa$. All other points of $\mathbb{P}_\kappa$ are of the form $(1{:}a)$ with $a \in \mathbb{D}_{\sigma,\varepsilon}$. The points of $\mathbb{P}_f$ are $(0{:}1)$ and those of the form $(1{:}a)$ with $a \in \mathbb{D}^{\sigma,\varepsilon}$. Thus, we see again from Lemma 10.1.2(ii) that $\mathbb{P}_\kappa$ is contained in $\mathbb{P}_f$. If we take $\sigma = \mathrm{id}$, $\varepsilon = 1$, and $\mathbb{D}$ of characteristic two, then $\mathbb{D}^{\sigma,\varepsilon} = \mathbb{D}$ and, as we have seen in Example 10.2.3, $\mathbb{D}_{\sigma,\varepsilon} = \{0\}$, so $\mathbb{P}_\kappa$ is a proper subspace of $\mathbb{P}_f$.

(iii) Let $\mathbb{D}$ and σ be as in Exercise 10.7.5(d). This implies that $\mathbb{F} = Z(\mathbb{D})$ is a field of characteristic two and that $\mathbb{D}_{\sigma,1} = \mathbb{F}$ and $\mathbb{D}^{\sigma,1} = \mathbb{F} + e_2\mathbb{F} + e_3\mathbb{F}$. Consider the admissible pair $(\sigma,1)$. Take $V = \mathbb{D}^2$ and put $g(v,w) = \sigma(v_1)w_2$ for $v = (v_1, v_2)$, $w = (w_1, w_2) \in V$. The form f defined by $f(v,w) = g(v,w) + \sigma(g(w,v)) = \sigma(v_1)w_2 + \sigma(v_2)w_1$ is nondegenerate and trace valued. The vectors v in V with $f(v,v) = 0$ are those for which $\sigma(v_1)v_2 = \sigma(\sigma(v_1)v_2)$, that is, $\sigma(v_1)v_2 \in \mathbb{D}^{\sigma,1}$. Since $\mathbb{F} = \mathbb{D}_{\sigma,1}$ is strictly contained in $\mathbb{D}^{\sigma,1}$, there are elements v of V such that $f(v,v) = 0$ while $g(v,v) = \sigma(v_1)v_2 \notin \mathbb{F}$. Let $\kappa : V \to \mathbb{D}/\mathbb{D}_{\sigma,\varepsilon}$ be a pseudo-quadratic form with facilitating form g as in Proposition 10.2.4(i). Then $\mathbb{P}_\kappa$ spans $\mathbb{P}(V)$ and the point set of $\mathbb{P}_\kappa$ is properly contained in the point set of $\mathbb{P}_f$.

We need some more algebraic developments. First of all, just as for sesquilinear forms, the geometric interpretation of pseudo-quadratic forms requires a concept of proportionality which can be used for reduction purposes.

Remark 10.2.12 Fix $\lambda \in \mathbb{D} \setminus \{0\}$ and consider the pair (ρ,δ), where

$$\rho(t) = \lambda\sigma(t)\lambda^{-1} \quad \text{and} \quad \delta = \lambda\sigma(\lambda)^{-1}\varepsilon.$$

Then $\lambda\mathbb{D}_{\sigma,\varepsilon} = \mathbb{D}_{\rho,\delta}$ (this is Exercise 10.7.3), and $\lambda\kappa : V \to \mathbb{D}/\mathbb{D}_{\rho,\delta}$, given by $(\lambda\kappa)(x) = \lambda(\kappa(x))$, is (well defined and) a (ρ,δ)-quadratic form. If $\lambda \neq 0$, we say that $\lambda\kappa$ is *proportional* to κ. Clearly, $\mathbb{P}_\kappa = \mathbb{P}_{\lambda\kappa}$.

Proportionality helps to classify pseudo-quadrics in two particular types, much like Theorem 7.3.15 for reflexive sesquilinear forms.

Proposition 10.2.13 *Let (σ, ε) be an admissible pair for $\mathbb{D}$ with $\mathbb{D}_{\sigma,\varepsilon} \neq \mathbb{D}$ and suppose that κ is a (σ, ε)-quadratic form on V.*

(i) *If $(\sigma, \varepsilon) \neq (\mathrm{id}, -1)$, then κ is proportional to a ρ-quadratic form for some anti-automorphism ρ of $\mathbb{D}$ with $\rho^2 = \mathrm{id}$.*

(ii) *If κ is not a σ-quadratic form, then it is proportional to a $(\rho, -1)$-quadratic form where ρ can be chosen in such a way that $1 \in \mathbb{D}_{\rho,-1}$.*

Proof (i) Since $(\sigma, \varepsilon) \neq (\mathrm{id}, -1)$, there is some $\lambda \in \mathbb{D}$, $\lambda \neq 0$, such that $\sigma(\lambda)\lambda^{-1} = \varepsilon$ (see the proof of Theorem 7.3.15). Then $\delta := \lambda\sigma(\lambda^{-1})\varepsilon = 1$, so $\lambda\kappa$ is a ρ-quadratic form for ρ as in Remark 10.2.12.

(ii) Since κ is not σ-quadratic, $\lambda := 1 - \sigma(\varepsilon) \neq 0$. Then $\sigma(\lambda) = 1 - \varepsilon$, whence $\sigma(\lambda) = -\varepsilon\lambda$ and so $\lambda\kappa$ is (ρ, δ)-quadratic with $\delta = \lambda\sigma(\lambda)^{-1}\varepsilon = -1$ and ρ given by $\rho(x) = \lambda\sigma(x)\lambda^{-1}$. Moreover $\lambda^{-1} + \rho(\lambda^{-1}) \in \mathbb{D}_{\rho,-1}$ by definition and $\lambda^{-1} + \rho(\lambda^{-1}) = \lambda^{-1} + \lambda\sigma(\lambda^{-1})\lambda^{-1} = \lambda^{-1} - \lambda(\varepsilon\lambda)^{-1}\lambda^{-1} = (1 - \varepsilon^{-1})\lambda^{-1} = \lambda\lambda^{-1} = 1$. Hence $1 \in \mathbb{D}_{\rho,-1}$.					$\square$

Recall from Lemma 10.2.2 that the quotient $\mathbb{D}^{\sigma,\varepsilon}/\mathbb{D}_{\sigma,\varepsilon}$ is a right $\mathbb{D}$-vector space with scalar multiplication given by $(v + \mathbb{D}_{\sigma,\varepsilon}) * \lambda = \sigma(\lambda)v\lambda + \mathbb{D}_{\sigma,\varepsilon}$ ($v \in \mathbb{D}^{\sigma,\varepsilon}$, $\lambda \in \mathbb{D}$). It will be used in Theorem 10.2.15(ii) to 'lift' the ambient projective space $\mathbb{P}$ of the absolute of a non-symplectic quasi-polarity π to a projective space $\mathbb{P}'$ in such a way that, for some pseudo-quadratic form κ on $\mathbb{P}'$, we have $\mathbb{P}_\pi \cong \mathbb{P}'_\kappa$. Although the statement below holds for arbitrary characteristic, it only gives useful information if the characteristic of $\mathbb{D}$ is two; see Proposition 10.1.8.

Lemma 10.2.14 *Suppose that σ is an involutory anti-automorphism of $\mathbb{D}$. The inclusion map $\kappa_\sigma : \mathbb{D}^{\sigma,1}/\mathbb{D}_{\sigma,1} \to \mathbb{D}/\mathbb{D}_{\sigma,1}$ is a σ-pseudo-quadratic form on the right vector space $\mathbb{D}^{\sigma,1}/\mathbb{D}_{\sigma,1}$ over $\mathbb{D}$. Its hermitian form is trivial.*

Proof By straightforward verification of the properties in Proposition 10.2.4(ii) of a pseudo-quadratic form with hermitian form $f = 0$.					$\square$

The map κ_σ of this lemma is used in the lifting result below, which, in turn, will be used in the proof of Theorem 10.3.13 to describe an embedded polar space as the absolute of a reflexive sesquilinear form or as a pseudo-quadric.

Theorem 10.2.15 *Let (σ, ε) be an admissible pair for $\mathbb{D}$ with $\varepsilon \in \{\pm1\}$. Suppose that f is a trace valued (σ, ε)-hermitian form on V. There exists a (σ, ε)-quadratic form κ with hermitian form f. If, for such a form κ, the point set of $\mathbb{P}(V)_\kappa$ does not coincide with the point set of the ambient polar space $\mathbb{P}(V)_f$, then the following statements hold.*

(i) *The characteristic of $\mathbb{D}$ is two, $\varepsilon = 1$, and $(\sigma, 1)$ is not of trace type.*

(ii) *Set $V' = V \oplus \mathbb{D}^{\sigma,1}/\mathbb{D}_{\sigma,1}$ and $\kappa' = \kappa \oplus \kappa_\sigma$. The map κ' is a pseudo-quadratic form on V' and the canonical projection $V' \to V$ onto the first factor induces an isomorphism $\mathbb{P}(V')_{\kappa'} \to \mathbb{P}(V)_f$ of polar spaces.*

(iii) *Let $\mathbb{P}' := \mathbb{P}(V')$ and κ' be as in (ii). If f' is a hermitian form of κ' and f is nondegenerate, then $\mathrm{Rad}(\mathbb{P}'_{f'}) = \mathbb{P}(\mathbb{D}^{\sigma,1}/\mathbb{D}_{\sigma,1})$ and the quotient space $\mathbb{P}'_{f'}/\mathrm{Rad}(\mathbb{P}'_{f'})$ is isomorphic to $\mathbb{P}(V)_f$.*

Proof The existence of κ follows from Proposition 10.2.10(ii).

(i) By Theorem 10.2.9(iii), we have $\mathbb{D}_{\sigma,\varepsilon} \neq \mathbb{D}^{\sigma,\varepsilon}$, so, by Proposition 10.1.8(i), (ii), the pair (σ, ε) is not of trace type and the characteristic of $\mathbb{D}$ equals two. In particular, $\varepsilon = 1$.

(ii) Thanks to Lemma 10.2.14 the form κ_σ is (σ, ε)-quadratic. It easily follows that the sum κ' of κ and κ_σ has the same property.

Let g be the σ-sesquilinear facilitating form $V \times V \to \mathbb{D}$ for κ (as in Proposition 10.2.4(ii)) so $\kappa(x) = g(x, x) + \mathbb{D}_{\sigma,1}$ for all $x \in V$. Consider an arbitrary vector $(x, \overline{a}) \in V'$, where $x \in V \setminus \{0\}$ and $a \in \mathbb{D}^{\sigma,1}$. The corresponding projective point belongs to $\mathbb{P}(V')_{\kappa'}$ if and only if $\kappa(x) + \overline{a} = \overline{0} = \mathbb{D}_{\sigma,1}$. This implies $\overline{g(x, x)} = \overline{a} \subseteq \mathbb{D}^{\sigma,1}$, and so $f(x, x) = g(x, x) + \sigma(g(x, x)) = 0$, whence $\langle x \rangle \in \mathbb{P}(V)_f$. Thus, indeed, the projection $V' \to V$ onto the first factor induces a map $\mathbb{P}(V')_{\kappa'} \to \mathbb{P}(V)_f$.

Suppose $\langle x \rangle \in \mathbb{P}(V)_f$ for some nonzero vector x of V. Then $g(x, x) + \sigma(g(x, x)) = f(x, x) = 0$, so $g(x, x) \in \mathbb{D}^{\sigma,1}$. Now $\langle (x, \overline{g(x, x)}) \rangle$ is easily seen to be the one and only preimage of $\langle x \rangle$ under the map $\mathbb{P}(V')_{\kappa'} \to \mathbb{P}(V)_f$.

The check that the projection and its inverse preserve lines comes down to verifying that $f'((x, \overline{a}), (y, \overline{b})) = f(x, y)$ for all $x, y \in V$ and $a, b \in \mathbb{D}^{\sigma,1}$, which is immediate.

(iii) The equality of Proposition 7.4.4 for f' in terms of f shows that, if f is nondegenerate, then $\mathrm{Rad}(f') = \mathbb{D}^{\sigma,1}/\mathbb{D}_{\sigma,1}$. This proves the statement about the radical. $\qquad\square$

In terms of Definition 9.2.9, the space $\mathrm{Rad}(\mathbb{P}'_{f'})$ is the defect of the polar space $\mathbb{P}'_{\kappa'}$ embedded in $\mathbb{P}'$. By joining it to $\mathbb{P}'_{\kappa'}$, we obtain $\mathbb{P}'_{f'}$.

In Definition 10.1.10, the orthogonal group $\mathrm{O}(V, \kappa)$ was defined for a quadratic form on V. Here is the natural extension of that definition to pseudo-quadratic forms.

Definition 10.2.16 The group of all linear transformations g of V preserving the pseudo-quadratic form κ on V (in the sense that $\kappa(gx) = \kappa(x)$ for each $x \in V$) is called the *orthogonal group* related to κ and denoted by $\mathrm{O}(V, \kappa)$.

Thus, we have defined orthogonal groups for pseudo-quadratic forms, unitary groups for sesquilinear forms, and symplectic groups for alternating bilinear forms (Definition 10.1.10).

Proposition 10.2.17 *If $\mathbb{D}_{\sigma,\varepsilon} \neq \mathbb{D}$ and f is the hermitian form of a pseudo-quadratic form κ, then $\mathrm{O}(V, \kappa)$ is a subgroup of $\mathrm{U}(V, f)$.*

Proof This follows directly from the uniqueness of f as established in Corollary 10.2.5. □

For the analogue of Proposition 10.1.13 regarding $O(V, \kappa)$, see Exercise 10.7.13. The orthogonal groups will be studied further in Sect. 10.5.

10.3 Perspective Sets

According to Theorems 9.5.8 and 10.2.15 every nondegenerate polar space of rank at least two with thick lines that can be embedded in a projective space embeds in a subspace of the absolute of a quasi-polarity. Every such polar space that we have encountered so far can be described as a pseudo-quadric or the absolute of a quasi-polarity. One might ask whether there are more polar spaces embedded in a projective space. In this section, we show that the answer is negative. In proving this fact, we obtain a complete classification of these spaces; see Theorem 10.3.13. As a consequence, in Corollary 10.3.14, we are able to classify the polar geometries of type B_n for $n \in \mathbb{N}$, $n \geq 4$.

For the proof we will use a remarkable invariance property of polar spaces embedded in a projective space $\mathbb{P}$, namely invariance under many perspectivities of $\mathbb{P}$.

Definition 10.3.1 Let $\mathbb{P}$ be a projective space with quasi-polarity π. A set X of points of $\mathbb{P}_\pi$ is called *perspective* (with respect to π) if, for all $p \in \mathbb{P}$ and all $a, b \in X \setminus (\{p\} \cup \pi(p))$ such that a, b, and p are on a projective line, the perspectivity with center p and axis $\pi(p)$ mapping a to b leaves X invariant.

In Definition 6.2.1 a projective line was defined by means of a set of perspectivities. This gives Definition 10.3.1 a meaning even if the projective space $\mathbb{P}$ is a line.

Example 10.3.2 It is a direct consequence of Theorem 9.3.3 that, if f is a nondegenerate (σ, ε)-hermitian form on a right vector space V over $\mathbb{D}$, then, under some mild restrictions, $\mathbb{P}(V)_f$ is perspective with respect to δ_f. Similarly, for $\mathbb{P}(V)_\kappa$ if κ is a pseudo-quadratic form on V (see Exercise 10.7.13 for an explicit description of the perspectivity).

The perspectivities of Example 10.3.2 all come from the corresponding unitary and orthogonal groups.

Lemma 10.3.3 *Let f be a (σ, ε)-hermitian form f on a right vector space V over $\mathbb{D}$ and write $\mathbb{P} = \mathbb{P}(V)$.*

(i) *The perspectivities with respect to δ_f leaving $\mathbb{P}_f$ invariant belong to $\mathrm{PU}(V, f)$.*
(ii) *Let κ be a (σ, ε)-quadratic form with hermitian form f such that $\mathbb{P}_\kappa$ is nonempty. The perspectivities with respect to δ_f leaving $\mathbb{P}_\kappa$ invariant belong to $\mathrm{PO}(V, \kappa)$.*

Proof Each perspectivity with respect to δ_f is of the form $r_{c,\phi}$ as in Exercise 1.9.30, where $c \in V \setminus \{0\}$ and ϕ is a linear form on V such that $\phi(x) = \lambda f(c, x)$ for some $\lambda \in \mathbb{D}$. If $a, b \in V$ represent distinct points of $\mathbb{P}_f$ such that $\langle c \rangle$ is in $\langle a, b \rangle \setminus \{\langle a \rangle, \langle b \rangle\}$ then, without loss of generality, we may assume $c = a + b$. If, moreover, $\langle a \rangle$ and $\langle b \rangle$ are not on $\delta_f(c)$, the perspectivity $r_{c,\phi}$ maps $\langle a \rangle$ to $\langle b \rangle$ if and only if $r_{c,\phi}(a) = b\mu$ for some $\mu \in \mathbb{D}$, which is equivalent to $\lambda = f(b, a)^{-1}$ and $\mu = -1$, and hence to $\phi(x) = f(b, a)^{-1} f(c, x)$ for all $x \in V$. This, in turn, is equivalent to $r_{c,\phi} \in \mathrm{U}(V, f)$. Hence (i). The proof of (ii) is part of Exercise 10.7.13. $\square$

The radical of a subset of a polar space is introduced in Definition 7.4.3.

Proposition 10.3.4 *Let π be a quasi-polarity of a thick projective space $\mathbb{P}$.*

  (i) *Each subspace of the polar space $\mathbb{P}_\pi$ containing a line is perspective with respect to π.*
 (ii) *The intersection of any collection of perspective subsets of $\mathbb{P}_\pi$ is again perspective.*
(iii) *If X is a perspective subset of $\mathbb{P}_\pi$ with empty radical all of whose induced lines have at least three points, then the group generated by all perspectivities of $\mathbb{P}$ with respect to π leaving X invariant is transitive on X.*

Proof (i) By Theorem 9.3.3, $\mathbb{P}_\pi$ itself is perspective. Suppose now that Z is a subspace of $\mathbb{P}_\pi$ containing the points a, b, c, and that $p \in \mathbb{P}$ is on the projective line ab, while $a, b \notin \{p\} \cup \pi(p)$. Let α be the perspectivity with center p and axis $\pi(p)$ mapping a onto b. We show that $\alpha(c)$ belongs to Z.

First, assume that a and c are collinear in Z. The line ac lies in Z and $a \notin \pi(p)$, so $ac \cap \pi(p)$ contains a single point d belonging to Z. Under α, the line ac of $\mathbb{P}_\pi$ is mapped onto the line bd of $\mathbb{P}_\pi$ containing the points b and d of Z. As Z is a subspace of $\mathbb{P}_\pi$, the line bd is also a line of Z. In particular, the point $\alpha(c)$ belonging to it lies in Z.

Remains the case where a and c are non collinear. Then, as Z itself is a polar space containing a line, there is a point h of Z collinear with both a and c. By applying the previous argument first to h, we find that $\alpha(h)$ belongs to Z. Replacing a and b by h and $\alpha(h)$, we can apply the previous argument once more so as to obtain the required conclusion.

(ii) is obvious.

(iii) Suppose $x, y \in X$. If they are not collinear in $\mathbb{P}_\pi$, choose a point $p \in xy \setminus \{x, y\}$. We have $x, y \notin \pi(p)$ as $\pi(x) \cap xy = \{x\}$ and $\pi(y) \cap xy = \{y\}$. The perspectivity with center p and axis $\pi(p)$ mapping x to y leaves X invariant. The result now follows because the non-collinearity graph of X is connected; cf. Exercise 7.11.16. $\square$

Remark 10.3.5 Suppose that Z is a nondegenerate ruled polar space in $\mathbb{P}(V)$ with empty defect. By Theorem 9.5.8, there is a (unique) quasi-polarity π such that Z is a subspace of $\mathbb{P}(V)_\pi$. Therefore, every perspectivity leaving invariant Z will be a perspectivity with respect to π.

In view of Proposition 10.3.4(ii), we can talk about perspective subsets generated by a subset of $\mathbb{P}_\pi$.

Notation 10.3.6 Let π be a quasi-polarity of $\mathbb{P}$. For a set X of points of $\mathbb{P}_\pi$, we denote by $\langle X \rangle_\pi$ the perspective subset of $\mathbb{P}_\pi$ generated by X. As usual, we also write $\langle x \rangle_\pi$ instead of $\langle \{x\} \rangle_\pi$ and $\langle x, y \rangle_\pi$ instead of $\langle \{x, y\} \rangle_\pi$, etc.

Remark 10.3.7 The essential and motivating example of a perspective subset is a pseudo-quadric in the projective line. For further algebraic analysis, we identify the point set of the projective line $\mathbb{P}^1(\mathbb{D})$ (cf. Example 6.2.7) with the set $\mathbb{D} \cup \{\infty\}$ via the correspondence

$$(a : 1) \leftrightarrow a \quad (a \in \mathbb{D}), \qquad (1 : 0) \leftrightarrow \infty.$$

In the sequel, we will write $\mathbb{P}(\mathbb{D}^2)$ rather than $\mathbb{P}^1(\mathbb{D})$.

Lemma 10.3.8 *Let π be the polarity of the Desarguesian projective line $\mathbb{P} := \mathbb{P}(\mathbb{D}^2)$ over $\mathbb{D}$ corresponding to a nondegenerate (σ, ε)-hermitian form f. Suppose that 0 and ∞ belong to $\mathbb{P}_\pi$. There is a unique (σ, ε)-quadratic form κ on $\mathbb{D}^2$ with hermitian form f such that 0 and ∞ are points of $\mathbb{P}_\kappa$. Moreover, $\mathbb{P}_\kappa$ coincides with both the perspective subset $\langle 0, \infty \rangle_\pi$ of $\mathbb{P}_\pi$ and $\{\infty\} \cup \sigma^{-1}(\mathbb{D}_{\sigma,\varepsilon})$.*

Proof Since 0 and ∞ belong to $\mathbb{P}_\pi = \mathbb{P}_f$, the form f looks like

$$f(x, y) = \sigma(x_1)\alpha y_2 + \sigma(x_2)\sigma(\alpha)\varepsilon y_1,$$

for some $\alpha \in \mathbb{D}$, $\alpha \neq 0$. After a change of coordinates by means of the diagonal matrix with entries $1, \alpha^{-1}$ (which fixes both 0 and ∞), we may even take f to be of the form

$$f(x, y) = \sigma(x_1)y_2 + \sigma(x_2)\varepsilon y_1.$$

Suppose now that κ is a (σ, ε)-quadratic form on $\mathbb{D}^2$ with hermitian form f such that 0 and ∞ belong to $\mathbb{P}_\kappa$. It follows that $f(\varepsilon_1, \varepsilon_2) = 1$, where $\varepsilon_1, \varepsilon_2$ is the standard basis of $\mathbb{D}^2$. As

$$\kappa(\varepsilon_1 x_1 + \varepsilon_2 x_2) = \kappa(\varepsilon_1 x_1) + \kappa(\varepsilon_2 x_2) + \overline{f(\varepsilon_1 x_1, \varepsilon_2 x_2)} = \overline{\sigma(x_1)x_2},$$

κ is uniquely determined. Moreover,

$$\mathbb{P}_\kappa = \left\{ (x_1 : x_2) \mid \sigma(x_1)x_2 \in \mathbb{D}_{\sigma,\varepsilon} \right\} = \{\infty\} \cup \left\{ x \in \mathbb{D} \mid \sigma(x) \in \mathbb{D}_{\sigma,\varepsilon} \right\}$$
$$= \{\infty\} \cup \sigma^{-1}(\mathbb{D}_{\sigma,\varepsilon}).$$

By applying Theorem 10.2.9(i) and Proposition 10.3.4(i) to a higher dimensional pseudo-quadric Q whose restriction to $\mathbb{D}^2$ is κ, and intersecting Q with $\mathbb{P}(\mathbb{D}^2)$, we find from Proposition 10.3.4(ii) that $\mathbb{P}_\kappa$ is perspective. (Of course, this can also be verified by direct computation or by use of Exercise 10.7.13.)

It remains to prove that each element of $\sigma^{-1}(\mathbb{D}_{\sigma,\varepsilon})$ is in $\langle 0, \infty \rangle_\pi$. To this end, take $a \in \mathbb{D} \setminus \{0\}$ and put $d = \sigma^{-1}(a) = \varepsilon^{-1}\sigma(a)\varepsilon$. The perspectivity α with center d and axis $\pi(d)$ mapping 0 to ∞ is given by

$$\alpha(x) = x - (\varepsilon_1 d + \varepsilon_2)\sigma(d)^{-1}f(\varepsilon_1 d + \varepsilon_2, x) \quad (x \in \mathbb{D}^2).$$

Observe that $0 \notin \pi(d)$ as $f(\varepsilon_1 d + \varepsilon_2, \varepsilon_2) = \sigma(d) = a \neq 0$. Moreover, $\infty \notin \pi(d)$ as $f(\varepsilon_1 d + \varepsilon_2, \varepsilon_1) = \varepsilon \neq 0$. Now $\alpha(\infty) = d - \varepsilon^{-1}\sigma(d)$, for

$$\alpha(\varepsilon_1) = \varepsilon_1 - (\varepsilon_1 d + \varepsilon_2)\sigma(d)^{-1}\varepsilon = \left(\varepsilon_1\left(d - \varepsilon^{-1}\sigma(d)\right) + \varepsilon_2\right)\left(-\sigma(d)^{-1}\varepsilon\right).$$

This implies $\sigma^{-1}(a - \sigma(a)\varepsilon) = \sigma^{-1}(a) - \varepsilon^{-1}a = d - \varepsilon^{-1}\sigma(d) = \alpha(\infty)$, showing $\sigma^{-1}(\mathbb{D}_{\sigma,\varepsilon}) \subseteq \langle 0, \infty \rangle_\pi$. $\qquad\square$

Example 10.3.9 By Lemma 10.3.8, the perspectively closed sets generated by two points of $\mathbb{P}_f$ are known. In view of Theorem 10.2.15 we may expect other perspective sets than those found in Lemma 10.3.8 only if $\mathrm{char}(\mathbb{D}) = 2$. Consider Example 10.2.11(ii), where $\mathbb{D}$ is a field of characteristic two, $\mathbb{P} = \mathbb{P}(\mathbb{D}^2)$, and $(\sigma, \varepsilon) = (\mathrm{id}, 1)$, while $f(x, y) = x_1 y_2 + x_2 y_1$ for $x, y \in \mathbb{D}^2$. We have $\mathbb{D}_{\sigma,\varepsilon} = \{0\}$ and $\mathbb{D}^{\sigma,\varepsilon} = \mathbb{D}$. In particular, with $\pi = \delta_f$, we find $\langle 0, \infty \rangle_\pi = \{0, \infty\}$ and, with a little more work,

$$\langle 0, 1, \infty \rangle_\pi = \{\infty\} \cup Q,$$

where Q is the set of squares in $\mathbb{D}$ (proving this equality is the content of Exercise 10.7.17). Since there are fields of characteristic two for which Q does not coincide with $\mathbb{D}$, there are proper perspective subsets of $\mathbb{P}_f$ which properly contain the pseudo-quadric $\mathbb{P}_\kappa$ for κ as in Lemma 10.3.8. A specific example is the field $\mathbb{D} = \mathbb{F}_2(X)$ of rational functions in the indeterminate X with coefficients in $\mathbb{F}_2$, for which we have $X \notin Q$. In this case, $\{\infty\} \cup Q$ is not of the form $\mathbb{P}_\kappa$ for a (σ, ε)-quadratic form κ with hermitian form f, for, by Lemma 10.3.8, κ is uniquely determined by the requirements that 0 and ∞ are points of $\mathbb{P}_\kappa$.

We translate Lemma 10.3.8 into a coordinate-free statement and make a case distinction for the type of form f.

Proposition 10.3.10 *Let f be a nondegenerate (σ, ε)-sesquilinear form on a 2-dimensional vector space V over $\mathbb{D}$ and write $\pi = \delta_f$ and $\mathbb{P} = \mathbb{P}(V)$.*

(i) *If κ is a non-trivial (σ, ε)-quadratic form on V with hermitian form f, and if p, p' are two points of $\mathbb{P}_\kappa$, then the point set of the pseudo-quadric $\mathbb{P}_\kappa$ is the smallest perspective subset of $\mathbb{P}_f$ containing p and p'.*

(ii) *If f is an alternating form on V and $\mathbb{D}$ has characteristic distinct from two, then, for any two points p, p' of $\mathbb{P}_f$, the point set of the absolute space $\mathbb{P}_f$ coincides with $\langle p, p' \rangle_\pi$.*

Proof Without loss of generality, we choose a basis of V so that $p = (1 : 0)$ and $p' = (0 : 1)$ in both (i) and (ii). Now (i) is immediate from Lemma 10.3.8. As for (ii), in this case, $(\sigma, \varepsilon) = (\mathrm{id}, -1)$, so, as the characteristic of $\mathbb{D}$ differs from two, $\mathbb{D}_{\sigma,\varepsilon} = \{2a \mid a \in \mathbb{D}\} = \mathbb{D}$. Thus, by Lemma 10.3.8, $\langle p, p' \rangle_\pi$ contains all points of $\mathbb{P}_f$. $\qquad\square$

Notice the subtlety that, in (ii), f is allowed to be degenerate. By taking intersections with appropriate lines, the more general case can be reduced to the above proposition, with the following result.

Proposition 10.3.11 *Let $\mathbb{P} = \mathbb{P}(V)$ for some vector space V over $\mathbb{D}$. Suppose that Z is a nondegenerate polar space of rank at least two that is a subspace of $\mathbb{P}_f$, for a (σ, ε)-hermitian form f, that is ruled in $\mathbb{P}$.*

(i) *If f is proportional to a nondegenerate alternating form and $\mathbb{D}$ has characteristic distinct from two, then $Z = \mathbb{P}_f$.*

(ii) *If Z is a subspace of a nondegenerate pseudo-quadric $\mathbb{P}_\kappa$ for a pseudo-quadratic form κ with hermitian form f, then $Z = \mathbb{P}_\kappa$.*

Proof Let P be the set of points of Z. Put $R = \mathbb{P}_f$ in case (i) and $R = \mathbb{P}_\kappa$ in case (ii), so that $P \subseteq R$ and R is a nondegenerate polar space. We need to show that equality holds. Suppose not, so there is a point, x say, of R (whence also of $\mathbb{P}_f$) that does not belong to P. If there is a projective line l on x meeting P in at least two points, say y and z, then y and z are not collinear in $\mathbb{P}_f$ (for otherwise yz, whence x, would belong to P). Hence f induces a nondegenerate quasi-polarity, π say, on l. Take a complementary subspace U of $\mathrm{Rad}(\mathbb{P}_f)$ in V spanned by elements of P, including x and y (this is possible as Z is ruled in $\mathbb{P}$; see Exercise 5.7.14). Now the restriction of f to U is nondegenerate, so it induces a nondegenerate quasi-polarity, ρ say, on U. By Proposition 10.3.4(i), $U \cap P$ (which contains l) is perspective with respect to ρ, so $l \cap P$ is perspective with respect to π. Therefore, Proposition 10.3.10 implies that $\langle y, z \rangle_\pi = R \cap l$ is contained in P, leading again to the contradiction $x \in P$.

Thus, every line of $\mathbb{P}$ on x meets P in at most one point, that is, in the notation of Definition 9.2.1, $P \subseteq Z_x$. Since P spans $\mathbb{P}$, this implies $\mathbb{P} \subseteq \delta_f(x)$, and so $x \in R^\perp$, contradicting the fact that R is nondegenerate. Hence (i) and (ii). $\qquad\square$

Definition 10.3.12 A polar space is called *classical* if it is isomorphic to at least one of

(1) the Grassmannian of lines of a thick projective space of dimension three,
(2) the absolute $\mathbb{P}(V)_f$ for a (σ, ε)-hermitian form $f : V \times V \to \mathbb{D}$, or
(3) the pseudo-quadric $\mathbb{P}(V)_\kappa$ of a pseudo-quadratic form κ on V,

for some right vector space V over a division ring $\mathbb{D}$.

Corollary 10.1.9 further divides the second case of Definition 10.3.12 into absolutes of symplectic forms and of σ-hermitian forms for non-identity anti-involutions σ.

In Example 7.5.6, we have seen classical polar spaces of infinite rank. These do not lead to polar geometries of finite rank. On the other hand, there are classical polar spaces of finite rank that embed in projective spaces, but not in finite-dimensional projective spaces; cf. Exercise 10.7.19.

We collect our results in the following classification. In the proof of this result, for an embedded polar space Z, we first mod out the defect, and next put in place a suitable new defect.

Theorem 10.3.13 *Let Z be a nondegenerate polar space of finite rank all of whose lines are thick. If Z has a singular subspace that is a Desarguesian plane, or has* *rank at least two and is embedded in a projective space, then Z is a classical polar space.*

Proof If Z has rank three and some line is on precisely two singular planes, then Corollary 7.8.10 gives that Z is isomorphic to the Grassmannian of lines of a projective space of dimension three over a division ring. (In this case we did not need to require that planes are Desarguesian.)

If Z has rank four or if Z has rank at least three and each line is on at least three singular planes, then, by Theorems 8.3.16 and 8.4.25 (which uses that the singular planes are Desarguesian), Z embeds in a projective space.

Therefore, in each case, Z embeds in a thick projective space $\mathbb{P}$. By going over to the span in $\mathbb{P}$ of the points of Z, we may assume that Z is a ruled polar space in $\mathbb{P}$; cf. Lemma 9.1.3. Let D be the defect of Z in $\mathbb{P}$ and denote by ϕ the natural quotient map $\mathbb{P} \setminus D \to \mathbb{P}/D$ introduced in Exercise 5.7.25. According to Theorem 9.5.8, there is a unique quasi-polarity π of $\mathbb{P}/D$ such that $\phi(Z)$ is a ruled polar space in $\mathbb{P}/D$ with empty defect and such that $\phi(Z)$ is a subspace of the polar space $(\mathbb{P}/D)_\pi$ isomorphic to Z. After a transition from $\mathbb{P}$ to $\mathbb{P}/D$, we may assume that Z is a ruled polar space with empty defect in $\mathbb{P}$, and a subspace of $\mathbb{P}_\pi$.

In view of Example 9.1.5(iv), $\mathbb{P}$ has dimension at least three and so, by Corollary 6.3.2, there is a right vector space V over a division ring $\mathbb{D}$ such that $\mathbb{P} = \mathbb{P}(V)$.

By Theorem 7.3.11, there is an admissible pair (σ, ε) for $\mathbb{D}$ and a nondegenerate (σ, ε)-hermitian form f on V such that $\pi = \delta_f$ for some admissible pair (σ, ε) for $\mathbb{D}$. As Z is ruled in $\mathbb{P}$, Lemma 10.1.4 tells us that f is trace valued. According to Corollary 7.3.16, we may assume $\varepsilon \in \{\pm 1\}$.

If f is antisymmetric and $\mathrm{char}(\mathbb{D}) \neq 2$, then $Z = \mathbb{P}_f$ by Proposition 10.3.11(i), and we are done. So suppose that this is not the case. By Theorem 10.2.15, there is a (σ, ε)-quadratic form κ' on a vector space V' over $\mathbb{D}$ containing V as a subspace such that $\mathbb{P}(V)_f$ is isomorphic to $\mathbb{P}(V')_{\kappa'}$. Consequently, Z can be viewed as a subspace of the pseudo-quadric $\mathbb{P}(V')_{\kappa'}$. After replacing V' and κ' by the subspace U of V' generated by the points of Z and the restriction of κ' to U, respectively, we may assume that Z is a subspace of $\mathbb{P}(V')_{\kappa'}$ that is ruled in $\mathbb{P}(V')$. Now Proposition 10.3.11(ii) applies and shows that $Z = \mathbb{P}(V')_{\kappa'}$, as required. $\square$

The following corollary summarizes our findings in terms of diagram geometries. Recall that a polar geometry of rank t is a geometry of type B_t with some additional properties specified in Definition 7.6.3.

Corollary 10.3.14 *Suppose that Γ is a polar geometry of finite rank $t \geq 3$ with point order at least two. If $t = 3$, assume also that its singular planes are Desarguesian. There is a classical polar space Z of rank t such that $\Gamma \cong \Gamma(Z)$.*

Proof The existence of a nondegenerate polar space Z of rank t with $\Gamma(Z) \cong \Gamma$ has been established in Corollary 7.6.8. The assumptions on Γ imply that Z satisfies the hypotheses of Theorem 10.3.13, which then gives that Z is classical of rank t. $\square$

Table 10.1 Dependencies among polar geometry constructs

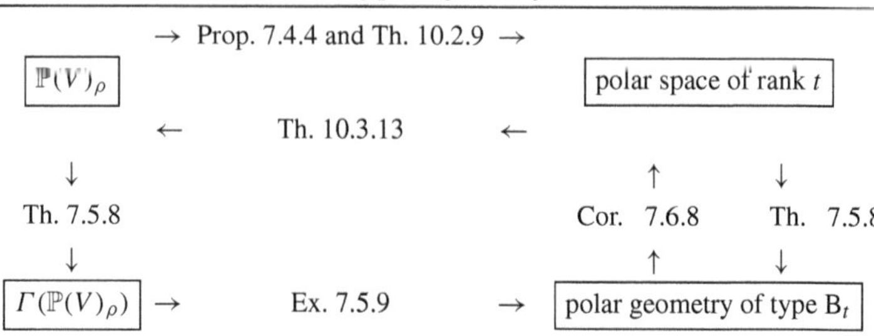

Remark 10.3.15 We summarize the polar results by means of Table 10.1, in the same guise as Table 6.1 for the projective case. Here, V is a vector space and ρ stands for either a quasi-polarity δ_f determined by a sesquilinear form f on V or a pseudo-quadratic form κ on V (cf. Definition 10.2.6). In the latter case, $\mathrm{PG}(\mathbb{P}(V), \rho)$, which is defined in Notation 7.3.2 if ρ is a quasi-polarity, should be read as the subgeometry of $\mathrm{PG}(\mathrm{PG}(V), \delta_f)$ induced on the subspaces of V on which κ vanishes if $\rho = \kappa$ is a pseudo-quadratic form with hermitian form f. The left hand side of Table 10.1 only deals with polar spaces embedded in a projective space, so the Grassmannians of lines (cf. Definition 10.3.12(1)) of a 3-dimensional projective space and polar spaces (of rank at most three) with singular planes that are non-Desarguesian are not mentioned (cf. Theorem 8.4.25).

By Proposition 7.4.4 and Theorem 10.2.9, the line spaces $\mathbb{P}(V)_\rho$ with $\rho = \delta_f$ or κ are polar spaces. If t is the rank of the polar space $\mathbb{P}(V)_{\delta_f}$, then Example 7.5.9 states that $\mathrm{Abs}(\mathrm{PG}(V), \delta_f)$ is isomorphic to the polar geometry $\Gamma(\mathbb{P}(V)_{\delta_f})$ of type B_t. For $\rho = \kappa$, the polar geometry $\Gamma(\mathbb{P}(V)_\kappa)$ is a subgeometry of $\Gamma(\mathbb{P}(V)_{\delta_f})$, where f is the hermitian form of κ.

A nondegenerate polar space of rank $t \geq 3$ whose singular planes are Desarguesian is classical by Theorem 10.3.13. By the exclusion of Grassmannians of projective lines, this means (cf. Definition 10.3.12) that it is of the form $\mathbb{P}(V)_{\delta_f}$ or $\mathbb{P}(V)_\kappa$.

A nondegenerate polar space of rank t corresponds to a residually connected geometry of type B_t that is a polar geometry by Theorem 7.5.8 and Corollary 7.6.8.

If the t-order of the polar geometry of type B_t is one, then the geometry also gives rise to a geometry of type D_t, an oriflamme geometry (cf. Theorem 7.8.6). For $t = 3$, the Grassmannians of lines of a projective space of dimension three arise (cf. Corollary 7.8.10), which is in accordance with the isomorphism between the diagrams D_3 and A_3 (see also the Klein Correspondence 7.8.3). For $t \geq 4$, the corresponding polar space is of the form $\mathbb{P}(V)_\kappa$ with V a $2t$-dimensional vector space over a field and κ a quadratic form on V of Witt index t (see Corollary 10.4.10).

10.4 Apartments

According to Theorem 6.5.5, projective geometries of finite rank are equipped with apartments: subgeometries whose induced structure is that of a thin projective geometry of the same rank as the ambient geometry. In the terminology of Example 4.1.13(i), these geometries are hypertetrahedra and in Definition 6.5.3, the corresponding subsets of points of the projective space are called frames. Here we prove that a similar phenomenon occurs in polar geometries. The subgeometries that will be called apartments again, have the induced structure of a thin polar geometry of the same rank as the ambient geometry. These thin geometries occurred as hyperoctahedra in Example 4.1.13(iii) and will reappear after the introduction of buildings in Chap. 11.

Most of the time, we will be working with the polar spaces that are shadow spaces of polar geometries of finite rank t. The set of points of the polar space belonging to an apartment will again be called a frame. In Theorem 10.4.6 we establish the existence of apartments and in Exercise 10.7.23 we find that there is only one orbit of apartments under the automorphism group of the polar space if its rank is at least three. In the last part of the section, we use apartments to find convenient bases of vector spaces in whose projective spaces the polar spaces embed.

Definition 10.4.1 Let Z be a polar space. An *octahedral set* of Z is a set O of points of Z such that, for each $x \in O$, the set $x^\perp$ contains all but a single point of O which we denote $O(x)$. A *frame* of Z is a maximal octahedral set of Z. For a subset X of an octahedral set O, we write $O(X) := \{O(x) \mid x \in X\}$.

By Zorn's lemma, frames exist in polar spaces of arbitrary rank. In Proposition 10.4.3 below we will see that, in nondegenerate polar spaces of finite rank t, frames have size $2t$.

Lemma 10.4.2 *Let O be an octahedral set of a polar space Z.*

(i) *If $x \in O$, then $O(O(x)) = x$, so $O \setminus \{x, O(x)\}$ is an octahedral set.*
(ii) *If $X \subseteq O$, then $O(O(X)) = X$.*
(iii) *If $X, Y \subseteq O$, then $X \subseteq Y$ implies $O(X) \subseteq O(Y)$.*
(iv) *Every finite clique X of O of size m generates a singular subspace of dimension $m - 1$.*
(v) *If X and Y are finite disjoint cliques of O, then $\langle X \rangle$ and $\langle Y \rangle$ are disjoint.*

Proof Properties (i), (ii), and (iii) are obvious.

(iv) Suppose that there is a point $x \in X$ such that $x \in \langle X \setminus \{x\} \rangle$. Then $O(x)^\perp$ contains $X \setminus \{x\}$, hence x, a contradiction. Therefore, the rank of $\langle X \rangle$ is equal to $m - 1$.

(v) It suffices to prove the assertion after $O(X)$ has been joined to Y and subsequently $O(Y)$ to X. But then $O(X) \subseteq Y$ and $O(Y) \subseteq X$, whence $O(X) = Y$ (for, by (ii) and (iii), the second inclusion implies $Y \subseteq O(X)$).

For $m = 1$, the assertion is clear as $x \in \langle X \rangle \cap \langle Y \rangle$ implies $X = \{x\} \subseteq \langle Y \rangle$, giving the contradiction $x \in O(x)^\perp$ as $\langle Y \rangle \subseteq O(x)^\perp$. We proceed by induction on m. Let $m > 1$ and assume $p \in \langle X \rangle \cap \langle Y \rangle$. For $x \in X$, the subspace $x^\perp \cap \langle Y \rangle$ of Z contains p and $Y \setminus \{O(x)\}$. But, by (i), $\langle Y \setminus \{O(x)\} \rangle$ is a geometric hyperplane of $\langle Y \rangle$. It is contained in $x^\perp \cap \langle Y \rangle$, and so $\langle Y \setminus \{O(x)\} \rangle = x^\perp \cap \langle Y \rangle$, which implies $p \in \langle Y \setminus$

$\{O(x)\}\rangle$. Similarly $p \in \langle X \setminus \{x\}\rangle$; this contradicts the induction hypothesis applied to $X \setminus \{x\}$ and $Y \setminus \{O(x)\}$. $\qquad\square$

In the proposition below, we use frames of projective spaces to study frames of polar spaces. Recall from Theorems 7.4.13 and 7.5.8 that maximal singular subspaces of a nondegenerate polar space of rank t are t-dimensional projective spaces.

Proposition 10.4.3 *Every nondegenerate polar space Z of finite rank t satisfies the following assertions.*

  (i) *Given disjoint maximal singular subspaces T, T' of Z and a frame B of t points in T, there is a unique octahedral set $O = B \cup O(B)$ with $O(B) \subseteq T'$.*
 (ii) *If $O = X \cup O(X)$ is an octahedral set of Z and X is a clique of points in Z, then $\langle X \rangle$ is a maximal singular subspace of Z if and only if O is a frame of Z.*
(iii) *All frames of Z have cardinality $2t$.*

Proof (i) For $b \in B$, consider the set $b^* = T' \cap (B \setminus \{b\})^{\perp}$. As $\dim(T) = \dim(T') = t - 1$ (cf. Theorem 7.5.5(ii)), there are $t - 1$ points y in $B \setminus \{b\}$ and since $U \cap y^{\perp}$, for a subspace U of T', has dimension at least $\dim(U) - 1$, the subspace b^* of T' contains at least one point. It cannot contain more than one point, for otherwise there would be a point, say a, in $b^{\perp} \cap b^*$, so that $a^{\perp}$ contains T, contradicting the maximality of T. Therefore b^* is a singleton disjoint from $b^{\perp}$. Now $O := B \cup \bigcup_{b \in B} b^*$ is an octahedral set as required.

(ii) If $\langle X \rangle$ is a maximal singular subspace of Z, then O is maximal by Lemma 10.4.2(iv). Conversely, if O is a frame and if $\langle X \rangle$ is not maximal, then, by Proposition 7.5.3, there are disjoint maximal singular subspaces $M \supseteq \langle X \rangle$ and $M' \supseteq \langle O(X) \rangle$. Furthermore, by Lemma 10.4.2(iv), X can be extended to a frame B of M. Applying (i) to B we obtain an octahedral set strictly containing O and B, which contradicts the maximality of O.

(iii) is obvious from (ii) and Theorem 7.4.13. $\qquad\square$

Definition 10.4.4 Let $t \in \mathbb{N}$, $t > 0$. The *apartment* of a polar space Z (or of $\Gamma(Z)$) of rank t corresponding to a frame O is the subgeometry of $\Gamma(Z)$ whose i-elements, for $i \in [t]$, are the singular subspaces $\langle T \rangle$ for T a clique in O of size i.

Remark 10.4.5 Let Z be a nondegenerate polar space of finite rank t, so its polar geometry $\Gamma(Z)$ is a geometry of type B_n by Theorem 7.5.8. In view of Lemma 10.4.2(iv), the subspace $\langle T \rangle$ of Z generated by a clique T of size i in a frame O is an i-element of $\Gamma(Z)$ for each $i \in [t]$. This justifies the introduction in Definition 10.4.4 of an apartment of Z as a subgeometry of $\Gamma(Z)$.

If O is a frame of Z, then the line space induced on O (cf. Definition 2.5.8) is a nondegenerate polar space of rank t all of whose lines are thin. The apartment corresponding to O is isomorphic to the polar geometry $\Gamma(O)$ of rank t (cf. Definition 7.5.7) whose point shadow space is the line space induced on O. It is also isomorphic to the hyperoctahedron introduced in Example 4.1.13(iii) and to the geometry $\Gamma(\mathcal{C}(\mathrm{B}_t))$ of the thin chamber system of type B_t introduced in Definition 4.2.5.

We next derive a property of apartments of Z that is characteristic of the building structure to be treated in Chap. 11.

Theorem 10.4.6 *If Z is a nondegenerate polar space of finite rank, then, for any pair of chambers c, d of $\Gamma(Z)$, there is an apartment A of Z such that c and d are chambers of A.*

Proof Write $t = \mathrm{rk}(Z)$. For $t = 1$ the theorem is obvious as frames are pairs of points. We proceed by induction on t and assume $t > 1$. Write $c = \{c_1, \ldots, c_t\}$ and $d = \{d_1, \ldots, d_t\}$, where $\dim(c_i) = \dim(d_i) = i - 1$.

Suppose that there exists $z_1 \in d_t \setminus c_1^{\perp}$. If $d_1 \not\perp c_1$, pick $z_1 \in d_1$. Set $\overline{P} = c_1^{\perp} \cap z_1^{\perp}$. Then $\overline{P}$ is a subspace of Z, which, by Lemma 7.4.8(ii) and Theorem 7.4.13(ii), is a nondegenerate polar space of rank $t - 1$. Now c and d induce chambers $\overline{c} = \{c_2 \cap \overline{P}, \ldots, c_t \cap \overline{P}\}$ and $\overline{d} = \{d_1 \cap \overline{P}, \ldots, d_t \cap \overline{P}\}$ in $\overline{P}$. Notice that, if $d_1 \not\perp c_1$, the intersection $d_1 \cap \overline{P}$ is empty and should be discarded and that, otherwise, two members of $\overline{d}$ will coincide. By the induction hypothesis, there is an apartment $\overline{A}$ of $\overline{P}$ such that $\overline{c}$ and $\overline{d}$ are chambers of $\overline{A}$. Then the geometry A on the union of the points of $\overline{A}$ and $c_1 \cup \{z_1\}$ is an apartment of Z such that c and d are chambers of it.

Therefore, we may assume that we cannot find such a point z_1, that is, $d_t \subseteq c_1^{\perp}$. By maximality of d_t as a singular subspace of Z, we have $c_1 \subseteq d_t$. Interchanging the roles of c and d, we may also assume $d_1 \subseteq c_t$.

If $c_1 = d_1$, a construction as above with z_1 a point of Z such that $z_1 \not\perp c_1$ also gives an apartment as required. So, without loss of generality, we may assume that c_1 and d_1 are singletons of distinct and collinear points. Then $\langle c_1 \cup d_1 \rangle$ is a line of Z contained in $c_t \cap d_t$. As Z is nondegenerate, we can find points h_1 and k_1 of Z such that $c_1 \perp d_1 \perp h_1 \perp k_1 \perp c_1 \not\perp h_1$ and $d_1 \not\perp k_1$. In other words the four points involved form a quadrangle.

If $t = 2$, then the quadrangle is an apartment as required, so we are left with the case where $t > 2$. Now $\widetilde{P} := (c_1 \cup d_1 \cup \{h_1, k_1\})^{\perp}$ is a nondegenerate polar space of rank $t - 2$. As $c_t \cap \widetilde{P}$ and $d_t \cap \widetilde{P}$ are maximal singular subspaces of $\widetilde{P}$, the set $\widetilde{c} = \{c_2 \cap \widetilde{P}, \ldots, c_t \cap \widetilde{P}\}$ has size $t - 2$ (for exactly one $j \in [t - 1]$, we will have $c_j \cap \widetilde{P} = c_{j+1} \cap \widetilde{P}$) and so is a chamber of $\widetilde{P}$. The same statement holds for $\widetilde{d} = \{d_2 \cap \widetilde{P}, \ldots, d_t \cap \widetilde{P}\}$ and so we can apply the induction hypothesis to $\widetilde{c}$ and $\widetilde{d}$ to find an apartment $\widetilde{A}$ containing them. Now the thin geometry corresponding to the union of $c_1 \cup d_1 \cup \{h_1, k_1\}$ and the points of $\widetilde{A}$ is an apartment as required. $\square$

Exercise 10.7.23 shows that the transitivity proved in Proposition 8.5.9 extends to finite octahedral sets of arbitrary finite size. In the remainder of this section, we will use the frames of polar spaces embedded in $\mathbb{P}(V)$, the projective space of a vector space V, to find canonical bases of V for alternating and quadratic forms.

Lemma 10.4.7 *Let Z be a nondegenerate polar space of finite rank that is ruled in $\mathbb{P}(V)$. If O is an octahedral set of Z of size $2m$, then the points of O are linearly independent in V. In particular, $\dim(V) \geq 2m$.*

Proof Put $\mathbb{P} = \mathbb{P}(V)$. If $m = 1$, then O consists of two non-collinear points, so the required property is obvious. We proceed by induction on m. Let $m > 1$ and assume that $a \in O$ is linearly dependent on $O \setminus \{a\}$. In Notation 9.1.6, this is expressed as $a \in \langle O \setminus \{a\} \rangle_\mathbb{P}$. As $O \setminus \{a\} \subseteq O(a)^\perp$, we have $a \in \langle O \setminus \{a\} \rangle_\mathbb{P} \cap Z \subseteq \langle O(a)^\perp \rangle_\mathbb{P} \cap Z = O(a)^\perp$. Here, the equality follows from Theorem 9.2.5 as Z is nondegenerate and ruled in $\mathbb{P}$, and the convention for $\cap$ is as in Notation 9.1.9. We conclude that $a \in O(a)^\perp$, contradicting that a and $O(a)$ are non-collinear. Hence a is linearly independent of $O \setminus \{a\}$. $\qquad\square$

The following notions for a pseudo-quadratic form are already known from Example 7.8.1 in the case of quadratic forms.

Definition 10.4.8 The *Witt index* of a pseudo-quadratic form κ on a vector space V is the maximum dimension of a linear subspace of V on which κ vanishes. A pseudo-quadratic form κ is called *anisotropic* if its Witt index is zero.

Thus, the Witt index of κ is equal to the rank (as in Definition 7.5.1) of the polar space $\mathbb{P}(V)_\kappa$. In the anisotropic case, this polar space has no points.

We now describe frames explicitly for the types (2) and (3) of embeddable classical polar spaces introduced in Definition 10.3.12. Type (1), the Grassmannian of lines of a projective space of dimension three, pertains to a geometry of type A_3, and its apartments coincide with those for a projective geometry by Exercise 10.7.24.

Proposition 10.4.9 *Let V be a right vector space over $\mathbb{D}$. Put $\mathbb{P} = \mathbb{P}(V)$.*

(i) *Suppose that f is a nondegenerate alternating form on V such that the absolute $\mathbb{P}_f$ has finite rank t. Then $\dim(V) = 2t$ and, for each frame O of $\mathbb{P}_f$, there is a basis $e_1, \ldots, e_{2t}$ of V with $O = \{\langle e_1 \rangle, \ldots, \langle e_{2t} \rangle\}$ such that*

$$f(x, y) = (x_1 y_2 - x_2 y_1) + \cdots + (x_{2t-1} y_{2t} - x_{2t} y_{2t-1})$$
$$\text{for } x = \sum_i e_i x_i \text{ and } y = \sum_i e_i y_i. \tag{10.3}$$

(ii) *Suppose that (σ, ε) is an admissible pair for $\mathbb{D}$ and κ is a (σ, ε)-quadratic form on V such that the corresponding pseudo-quadric $\mathbb{P}_\kappa$ is nondegenerate of finite rank t. Then, for each frame O of the pseudo-quadric $\mathbb{P}_\kappa$, there is a linearly independent set of vectors $e_1, \ldots, e_{2t}$ of V with $O = \{\langle e_1 \rangle, \ldots, \langle e_{2t} \rangle\}$ and a complement $U = \langle e_1, \ldots, e_{2t} \rangle^{\perp_f}$ of the linear span of O in V such that*

$$\kappa(x) = \sigma(x_1)x_2 + \cdots + \sigma(x_{2t-1})x_{2t} + \kappa_0(x_U)$$
$$\text{for } x = \sum_{i=1}^{2t} e_i x_i + x_U \quad \text{with } x_U \in U,$$

where κ_0 is an anisotropic (σ, ε)-quadratic form on U.

Proof Let $e_1, \ldots, e_{2t}$ be vectors of V such that the corresponding projective points form a frame O of the polar space involved (so t equals the rank of the polar space). Without loss of generality we can assume that the points are ordered in such a way that $\langle e_{2i-1} \rangle = O(\langle e_{2i} \rangle)$ for $i \in [t]$. Let T denote the subspace of V spanned by O. By Lemma 10.4.7, its dimension is $2t$.

(i) The value $f(e_i, e_j)$ for $i, j \in [2t]$ is nonzero only if $\{i, j\} = \{2k - 1, 2k\}$ for some $k \in [t]$. By rescaling e_i if necessary, we can therefore achieve that the restriction of f to $T \times T$ is as asserted.

It remains to show that $T = V$. Observe that $T^{\perp_f} \cap T = 0$ because the restriction of f to $T \times T$ is nondegenerate. If $T \neq V$, then there is a nonzero vector $e \in V \cap T^{\perp_f}$. But any projective point belongs to $\mathbb{P}_f$, so $\langle e \rangle \in \mathbb{P}_f \cap \{e_1, e_3, \ldots, e_{2t-1}\}^\perp$ and $\langle e, e_1, e_3, \ldots, e_{2t-1} \rangle$ is a singular subspace of $\mathbb{P}_f$ of dimension t, contradicting that maximal singular subspaces of $\mathbb{P}_f$ have dimension $t - 1$. Therefore, $T = V$.

(ii) Let f be the hermitian form of κ. Since $\langle e_i \rangle \in \mathbb{P}_\kappa$, we have $\kappa(e_i) = \overline{0} = \mathbb{D}_{\sigma,\varepsilon}$ for $1 \in [t]$. Similarly, since $\langle e_i \rangle$ and $\langle e_j \rangle$ are non-collinear only if $\{i, j\} = \{2k - 1, 2k\}$ for some $k \in [t]$, the value $f(e_i, e_j)$ for $i, j \in [2t]$ is nonzero in $\mathbb{D}_{\sigma,\varepsilon}$ only if $\{i, j\} = \{2k - 1, 2k\}$ for some k. Let g be a facilitating form for $\kappa|_T$ (as in Proposition 10.2.4) that is upper triangular with respect to the basis $(e_i)_{i \in [2t]}$. We find, by an argument as in the proof of '(ii)$\Rightarrow$(i)' of that proposition, that

$$\kappa(x) = \sum_{i=1}^{t} \sigma(x_{2i-1}) g(e_{2i-1}, e_{2i}) x_{2i} + \mathbb{D}_{\sigma,\varepsilon} \quad \left(x = \sum_{i=1}^{2t} e_i x_i \right).$$

By rescaling the e_{2i} appropriately, we obtain the required form of the restriction of κ to T.

Again, $T^{\perp_f} \cap T - \{0\}$ because the restriction of f to $T \times T$ is nondegenerate. Consequently, $\kappa = \kappa|_T + \kappa_0$ (point-wise sum), where κ_0 is the restriction of κ to $U := T^{\perp_f}$. Finally, reasoning similarly to the proof of $T = V$ in (i), we find that κ_0 is anisotropic on U. $\square$

The following result states that the oriflamme geometry of Definition 7.8.5 only arises in the cases described in Example 7.8.7. The notion of a dual polar space is introduced in Definition 7.5.4.

Corollary 10.4.10 *Let V be a right vector space of finite dimension $n = 2t$ over the division ring $\mathbb{D}$ and κ a (σ, ε)-quadratic form on V of Witt index t. The dual polar space of $\mathbb{P}(V)_\kappa$ has thin lines if and only if $\mathbb{D}$ is a field and $(\sigma, \varepsilon) = (\mathrm{id}, 1)$. In this case, the polar space $\mathbb{P}(V)_\kappa$ is the point shadow space of an oriflamme geometry with diagram D_t.*

Proof By Proposition 10.4.9 we may assume that κ is of the form

$$\kappa(x) = \sigma(x_1)x_2 + \cdots + \sigma(x_{2t-1})x_{2t} + \mathbb{D}_{\sigma,\varepsilon} \quad \left(x = \sum_{i=1}^{t} e_i x_i \right)$$

with respect to some basis $(e_i)_{i \in [2t]}$ of V. Consider the submaximal singular subspace $H = \langle e_1, e_3 \ldots, e_{2t-3} \rangle$ of $\mathbb{P}(V)_\kappa$. We need to establish the conditions under which there are exactly two maximal singular subspaces of $\mathbb{P}(V)_\kappa$ containing H. These maximal singular subspaces are exactly the points of the polar space $H^\perp / H$, which can be identified with the quadric induced by κ on the subspace of $\mathbb{P}(V)$ spanned by e_{2t-1} and e_{2t}. This means that the maximal singular subspaces containing H correspond to the projective points in $\langle e_{2t-1}x_{2t-1} + e_{2t}x_{2t} \rangle$ with $\sigma(x_{2t-1})x_{2t} \in \mathbb{D}_{\sigma, \varepsilon}$. This setting can be found in the proof of Lemma 10.3.8, where it is shown that their number is $1 + |\mathbb{D}_{\sigma, \varepsilon}|$. Thus, the dual polar space has thin lines if and only if $\mathbb{D}_{\sigma, \varepsilon} = \{0\}$, which in turn is equivalent to $x = \sigma(x)\varepsilon$ for each $x \in \mathbb{D}$. This implies $(\sigma, \varepsilon) = (\mathrm{id}, 1)$. The first assertion of the corollary is a direct consequence. The second assertion, about the oriflamme geometry, follows from Theorem 7.8.6. $\qquad\square$

The explicit description of quadratic forms of Proposition 10.4.9 allows us to conclude that a non-empty quadric determines the proportionality class of the defining quadratic form uniquely.

Corollary 10.4.11 *Let V be a right vector space over the field $\mathbb{F}$ and κ a quadratic form on V. If $\mathbb{P}(V)_\kappa$ is a non-empty nondegenerate quadric of finite rank in $\mathbb{P}(V)$, then it determines the quadratic form κ uniquely up to proportionality.*

Proof Suppose that $\mathbb{P}(V)_\kappa = \mathbb{P}(V)_{\kappa'}$ for some quadratic form κ' on V. Let t denote the rank of $\mathbb{P}(V)_\kappa$. As this polar space is non-empty, we have $t \geq 1$. Fix a frame O of $\mathbb{P}(V)_\kappa$ and let $e_1, \ldots, e_{2t}$ be a basis of the linear span T of O in V as in Proposition 10.4.9(ii) for κ with $(\sigma, \varepsilon) = (\mathrm{id}, 1)$. Now, for some facilitating form g' for $\kappa'|_T$ on T, we must have, for $x \in T$,

$$\kappa'(x) = x_1 g'(e_1, e_2)x_2 + \cdots + x_{2t-1} g'(e_{2t-1}, e_{2t})x_{2t}.$$

We first verify that the restrictions of κ and κ' to T are proportional. This is obvious for $t = 1$, so assume $t \geq 2$. Since $\kappa(e_1 + e_2 - e_{2i-1} - e_{2i}) = 0$ for any $i \in [t]$, we must also have $\kappa'(e_1 + e_2 - e_{2i-1} - e_{2i}) = 0$, whence $g'(e_1, e_2) = g'(e_{2i-1}, e_{2i})$. We find that κ' and $g'(e_1, e_2)\kappa$ agree on T. Now replace κ' by the appropriate scalar multiple so that κ and κ' are identical on T.

Let f and f' be the hermitian forms of κ and κ', respectively. Using $\perp_f$ from Notation 7.2.8, we verify that $U := T^{\perp_f}$ and $T^{\perp_{f'}}$ coincide. Let $u \in U \setminus \{0\}$. Then $\kappa(e_i + u) = \kappa(u) \neq 0$ because κ is anisotropic on U, and so $f'(e_i, u) + \kappa'(u) = \kappa'(e_i + u) \neq 0$. Replacing u by $u\lambda$ for $\lambda \in \mathbb{F} \setminus \{0\}$, we find $f'(e_i, u) + \kappa'(u)\lambda \neq 0$. If $f'(e_i, u) \neq 0$, then $\lambda = -f'(e_i, u)/\kappa'(u)$ would be a counterexample, so $f'(e_i, u) = 0$. It follows that U is contained in $\bigcap_{i \in [2t]} e_i^{\perp_{f'}} = T^{\perp_{f'}}$. As U and $T^{\perp_{f'}}$ have the same codimensions, they must be equal.

At this point, we need only show that the restrictions of κ and κ' to U are equal. For each $u \in U$, the point $\langle e_1 - e_2\kappa(u) + u \rangle$ lies on $\mathbb{P}(V)_\kappa$, so κ' vanishes also on it, from which we easily derive that $\kappa(u) = \kappa'(u)$, as required. $\qquad\square$

10.5 Automorphisms

In this section, we will be concerned with the group of linear transformations of V leaving invariant a polar space embedded in $\mathbb{P}(V)$. We will mostly work with the orthogonal groups $O(V, \kappa)$ and unitary groups $U(V, f)$ of Definitions 10.1.10 and 10.2.16. In particular, in Theorem 10.5.6 we will prove that these groups have high degrees of transitivity. This suffices to conclude that they have Tits systems as will be explained in Chap. 11.

The first result concerns transitivity on sets of collinear pairs and uses perspectivities that are known to act on the polar space by Theorem 9.3.3.

Proposition 10.5.1 *Let κ be a (σ, ε)-quadratic form on the vector space V over $\mathbb{D}$ such that $\mathbb{P}(V)_\kappa$ is a nondegenerate pseudo-quadric.*

(i) *The subgroup G of $O(V, \kappa)$ generated by the perspectivities of $\mathbb{P}(V)$ leaving invariant $\mathbb{P}(V)_\kappa$, is transitive on the set of ordered pairs of distinct collinear points of $\mathbb{P}(V)_\kappa$.*

(ii) *If κ is a quadratic form, then G is even transitive on the set of ordered triples of distinct collinear points of $\mathbb{P}(V)_\kappa$.*

Proof Write $(P, L) := \mathbb{P}(V)_\kappa$. As all perspectivities of Theorem 9.3.3 belong to $O(V, \kappa)$, Proposition 10.3.4(iii) gives that G acts transitively on P. Fix $a \in P$.

(i) Let b, $b' \in P$ be such that (a, b) and (a, b') are distinct pairs of points on lines in L. If $b' \not\perp b$, then the form κ does not vanish on the projective line bb', so $\mathbb{D}_{\sigma,\varepsilon} \neq \mathbb{D}$. Therefore, there is a point d of $\mathbb{P}(V)$ that is neither a point of $\mathbb{P}(V)_\kappa$ nor defective. The perspectivity with center d and axis $\delta_f(d)$, where f is the hermitian form of κ, mapping b to b' leaves P invariant and fixes a.

Suppose, therefore, that $b' \perp b$. Without loss of generality, we may assume $b \neq b'$. Let l be the line bb' and pick a point $c \in l \setminus \{b, b'\}$ such that $c = a$ if $a \in l$. As $\mathbb{P}(V)_\kappa$ is nondegenerate, so is $a^\perp/\{a\}$ (cf. Proposition 7.4.10), so, by Exercise 7.11.20, there is $d \in a^\perp \setminus \{a\}$ such that $d^\perp \cap l = \{c\}$. In particular, $d \not\perp b$ and $d \not\perp b'$. Applying the previous paragraph to the pairs (a, b) and (a, d), and subsequently to (a, d) and (a, b'), we obtain a product of perspectivities fixing a and mapping b to b'.

(ii) By assumption, $(\sigma, \varepsilon) = (\mathrm{id}, 1)$. Let (a, b, c) and (a, b', c') be distinct triples of points on lines in L. If $b' \not\perp b$, then the projective lines bb' and cc' intersect in a point d which is neither a point of $\mathbb{P}(V)_\kappa$ (as, by Corollary 10.4.10, there are no more than two points of $\mathbb{P}(V)_\kappa$ on bb') nor defective. The perspectivity with center d and axis $\delta_f(d)$, where f is the hermitian form of κ, mapping b to b' leaves P invariant, fixes a, and maps c to c'.

If $b' \perp b$, then it suffices to apply the preceding argument twice to a pair (a, b''), respectively a triple (a, b'', c''), of points on a line of $\mathbb{P}(V)_\kappa$ with $b'' \not\perp b$ and $b'' \not\perp b'$. Such a pair, respectively, triple, exists by arguments as in (i). (Alternatively, use Proposition 7.5.3 to find a maximal singular subspace M of the polar space $\mathbb{P}(V)_\kappa$ intersecting the plane $\langle a, b, b' \rangle$ only in a. The spaces $b^\perp$ and $b'^\perp$ intersect M in hyperplanes of M and there is a point $b'' \in M \setminus (b^\perp \cup b'^\perp)$.) $\qquad\square$

The linear transformations introduced in the next lemma are important in a bijective correspondence between the lines of a polar space Z and certain subgroups of $\mathrm{Aut}(Z)$ isomorphic to the additive group of $\mathbb{D}$.

Lemma 10.5.2 *Let κ be a nontrivial (σ, ε)-quadratic form and f a (σ, ε)-hermitian form on V. Suppose that $\langle a \rangle$ and $\langle b \rangle$ are distinct collinear points of $\mathbb{P}(V)_f$ and $\lambda \in \mathbb{D}$. Denote by $t_{a,b;\lambda}$ the linear transformation $V \to V$ given by*

$$t_{a,b;\lambda}(x) = x + a\lambda f(b, x) - b\varepsilon^{-1}\sigma(\lambda)f(a, x) \quad (x \in V).$$

The following statements hold.

(i) $t_{a,b;\lambda} \in \mathrm{U}(V, f)$.
(ii) *If $\kappa(a) = \kappa(b) = 0$ and f is the hermitian form of κ, then $t_{a,b;\lambda} \in \mathrm{O}(V, \kappa)$.*

Moreover, $t_{a,b;\lambda}$ fixes each point on the line ab.

Proof Notice that $t_{a,b;\lambda}$ is invertible, as its inverse is $t_{a,b;-\lambda}$.

(i) Let $x, y \in V$. By Definition 7.3.10, we have $\sigma(f(x, y)) = f(y, x)\varepsilon^{-1}$. Using this and $f(a, a) = f(b, b) = f(a, b) = 0$, we find

$$f\big(t_{a,b;\lambda}(x), t_{a,b;\lambda}(y)\big) = f(x, y) + f(x, a)\lambda f(b, y) - f(x, b)\varepsilon^{-1}\sigma(\lambda)f(a, y)$$
$$+ f(x, b)\varepsilon^{-1}\sigma(\lambda)f(a, y) - f(x, a)\lambda f(b, y)$$
$$= f(x, y).$$

(ii) For $x \in V$, the definition of $\mathbb{D}_{\sigma,\varepsilon}$ (applied at the last equality below) gives

$$\kappa\big(t_{a,b;\lambda}(x)\big) = \kappa\big(x + a\lambda f(b, x) - b\varepsilon^{-1}\sigma(\lambda)f(a, x)\big)$$
$$= \kappa(x) + f(x, a)\lambda f(b, x) - f(x, b)\varepsilon^{-1}\sigma(\lambda)f(a, x)$$
$$= \kappa(x) + f(x, a)\lambda f(b, x) - \sigma\big(f(x, a)\lambda f(b, x)\big)\varepsilon$$
$$= \kappa(x).$$

The final assertion of the lemma is easily verified. $\qquad\square$

Definition 10.5.3 The transformation $t_{a,b;\lambda}$ of Lemma 10.5.2 is called a *Siegel transvection* with respect to f. The group generated by all Siegel transvections with fixed projective points $\langle a \rangle$ and $\langle b \rangle$ is denoted $T_{a,b}$. The group generated by all $T_{a,b}$ with $f(a, b) = 0$ and $\kappa(a) = \kappa(b) = 0$ is denoted $\Omega(V, \kappa)$.

It is easy to see that $T_{a,b}$ is isomorphic to the additive group of $\mathbb{D}$ and that it only depends on the line ab (rather than a and b). It is the subgroup of $\mathrm{Aut}(Z)$ alluded to before, and helps to identify the points of the root filtration space built from Z as in Proposition 7.9.2 with a class of subgroups of $\mathrm{Aut}(Z)$. We next show that the transitivity proved in Proposition 8.5.9(i) even holds for the subgroup $\Omega(V, \kappa)$ of $\mathrm{O}(V, \kappa)$.

Lemma 10.5.4 *Suppose that κ is a nondegenerate pseudo-quadratic form on V, and that f is a nondegenerate (σ, ε)-hermitian form on V. Let Z be $\mathbb{P}(V)_\kappa$ or*

$\mathbb{P}(V)_f$ *and let G be the group $\Omega(V, \kappa)$ in the former and $\mathrm{U}(V, f)$ in the latter case. The group G is transitive on the set of ordered pairs of non-collinear points from Z.*

Proof Suppose that $\langle a \rangle$ and $\langle b \rangle$ are collinear points of Z. There exists a point $\langle c \rangle$ of Z in $\langle a \rangle^\perp$ but not in $\langle b \rangle^\perp$ (cf. Exercise 7.11.20). Lemma 10.5.2 gives that the subgroup $T_{a,c}$ of G fixes $\langle a \rangle$, stabilizes the line $\langle a, b \rangle$ of Z, and acts transitively on the points of the line $\langle a, b \rangle$ distinct from $\langle a \rangle$. By interchanging the roles of a and b, we find that the stabilizer in G of ab is doubly transitive on this line.

Next suppose that $\langle a \rangle$ and $\langle d \rangle$ are non-collinear points of Z. Choose a point $\langle x \rangle$ in Z collinear with $\langle a \rangle$ and $\langle d \rangle$. The composition of an element interchanging $\langle a \rangle$ and $\langle x \rangle$, and an element interchanging $\langle x \rangle$ and $\langle d \rangle$, both chosen from G as proven to exist above, will then move $\langle a \rangle$ to $\langle d \rangle$. We conclude that G is transitive on the point set P of Z.

Now fix a. In order to show that G is transitive on the set of pairs of non-collinear points, it suffices to show that the stabilizer of $\langle a \rangle$ in this group is transitive on $P \setminus \langle a \rangle^\perp$. Suppose that $\langle b \rangle$ and $\langle c \rangle$ are distinct points of $P \setminus \langle a \rangle^\perp$.

If $\langle b \rangle$ and $\langle c \rangle$ are collinear, let $\langle d \rangle$ be the unique point of Z in $\langle a \rangle^\perp \cap \langle b, c \rangle$. As we have seen above, a suitable member of $T_{a,d}$ will move $\langle b \rangle$ to $\langle c \rangle$.

If $\langle b \rangle$ and $\langle c \rangle$ are not collinear in $P \setminus a^\perp$, then, as Z is amply connected by Theorem 8.1.4, there is a path in the collinearity of $P \setminus \langle a \rangle^\perp$ from $\langle b \rangle$ to $\langle c \rangle$ and application of the previous paragraph gives us a product of elements of G fixing $\langle a \rangle$, and mapping $\langle b \rangle$ to $\langle c \rangle$. We have shown that, in all cases, the stabilizer of $\langle a \rangle$ in G is transitive on $P \setminus \langle a \rangle^\perp$, as required. $\qquad\square$

Next we show that the automorphism group of a non-empty nondegenerate quadric acts transitively on the collection of frames of Z.

Proposition 10.5.5 *If κ is a pseudo-quadratic form on V such that $Z = \mathbb{P}(V)_\kappa$ is a nondegenerate and non-empty pseudo-quadric of finite Witt index and $n = \dim(V)$ is finite, then $\mathrm{O}(V, \kappa)$ acts transitively on the set of all frames of Z.*

Proof We proceed by induction on the rank t of Z, which equals the Witt index of κ. As Z is non-empty, $t \geq 1$. Let $O = \{e_1, \ldots, e_{2t}\}$ and $O' = \{e'_1, \ldots, e'_{2t}\}$ be octahedral sets of Z of size $2t$ with $O(e_{2i-1}) = e_{2i}$ and $O(e'_{2i-1}) = e'_{2i}$ for $i \in [t]$. We want to show that $g(O) = O'$ for some $g \in \mathrm{O}(V, \kappa)$. By Lemma 10.5.4, we may assume that $e_1 = e'_1$ and $e_2 = e'_2$. For $t = 1$, there is nothing left to prove. For $t > 1$, consider the subspace $W = \{e_1, e_2\}^{\perp_f}$ of V and the restriction κ_1 of κ to W. Then $\mathbb{P}(W)_{\kappa_1}$ is a nondegenerate polar subspace of $\mathbb{P}(V)_\kappa$ of rank $t - 1 > 0$, and $e_3, \ldots, e_{2t}, e'_3, \ldots, e'_{2t} \in \mathbb{P}(W)_{\kappa_1}$. Induction applies and so there is α in $\mathrm{O}(W, \kappa_1)$ (leaving $\mathbb{P}_{\kappa_1}$ invariant and) mapping $e_3, \ldots, e_{2t}$, respectively, to $e'_3, \ldots, e'_{2t}$.

Extending α to a linear map (with the same name) on V by decreeing that it fixes e_1 and e_2, we find $\kappa(\alpha(x)) = \kappa(e_1 x_1 + e_2 x_2) + \kappa_1(\alpha(e_3 x_3 + \cdots + e_n x_n)) = \kappa(e_1 x_1 + e_2 x_2) + \kappa_1(e_3 x_3 + \cdots + e_n x_n) = \kappa(x)$, so $\alpha \in \mathrm{O}(V, \kappa)$, as required. $\qquad\square$

The transitivity properties of the orthogonal and unitary groups can be pushed further as follows.

Theorem 10.5.6 *Let f be a nondegenerate (σ, ε)-hermitian form on V and let κ be a nondegenerate (σ, ε)-quadratic form on V. Put $\mathbb{P} = \mathbb{P}(V)$.*

(i) *The group $\mathrm{U}(V, f)$ is transitive on the set of pairs consisting of an apartment of the polar geometry of $\mathbb{P}_f$ and a chamber of the apartment.*
(ii) *The same holds for $\mathrm{O}(V, \kappa)$ with respect to the polar geometry of $\mathbb{P}_\kappa$.*
(iii) *If κ is a quadratic form and n is even, then $\mathrm{SO}(V, \kappa)$, the subgroup of linear transformations of $\mathrm{O}(V, \kappa)$ having determinant 1, is also transitive on the set of pairs consisting of an apartment of the oriflamme geometry of $\mathbb{P}_\kappa$ and a chamber of the apartment.*

Proof Let $G = \mathrm{U}(V, f)$ in (i) and $G = \mathrm{O}(V, \kappa)$ in (ii). In view of Proposition 10.5.5 it suffices for (i) and (ii) to show that the set-wise stabilizer $N := N_G(O)$ (also called normalizer) of a fixed maximal octahedral set O of $\mathbb{P}(V)_f$ or $\mathbb{P}(V)_\kappa$, respectively, is transitive on the set of chambers of O.

By Exercise 4.9.26 (see Table 4.4), the automorphism group of the thin geometry O, being isomorphic to $W(\mathrm{B}_t)$, has size $2^t (t!)$, where $t = |O|/2$. This group consists of all permutations of O preserving the collection of subsets $\{p, O(p)\}$ for $p \in O$. Therefore, transitivity of N on the set of chambers of O will follow once we establish that N contains elements inducing the following permutations of points of O.

(1) The transposition $(p, O(p))$ interchanging a point p of O with $O(p)$
(2) The product $(p, q)(O(p), O(q))$ of the transposition (p, q) of two distinct collinear points p, q of O and the transposition $(O(p), O(q))$ of their opposites

Given $p \in O$, select a basis $e_1, \ldots, e_n$ of V whose first $2t$ elements represent the points of O in such a way that $p = \langle e_1 \rangle$ and $O(p) = \langle e_2 \rangle$. Then $f(e_1 x_1 + e_2 x_2, e_1 y_1 + e_2 y_2) = \sigma(x_1) y_2 + \sigma(x_2)\varepsilon y_1$ for $x_1, x_2, y_1, y_2 \in \mathbb{D}$.

(1) Consider the linear transformation M of V that fixes $e_3, \ldots, e_n$ with $Me_1 = e_2\varepsilon$ and $Me_2 = e_1$. It belongs to $\mathrm{U}(V, f)$ interchanges p and $O(p)$ and fixes all other points of O, so $M \in N$ is as required.

Moreover, in (ii), without loss of generality, $\kappa(e_1 x_1 + e_2 x_2) = \sigma(x_1) x_2$ for $x_1, x_2 \in \mathbb{D}$. Now $\kappa(M(e_1 x_1 + e_2 x_2)) = \kappa(e_1 x_2 + e_2\varepsilon x_1) = \sigma(x_2)\varepsilon x_1 + \mathbb{D}_{\sigma,\varepsilon} = \sigma(\sigma(x_1) x_2)\varepsilon + \mathbb{D}_{\sigma,\varepsilon} = \kappa(e_1 x_1 + e_2 x_2)$, so $M \in N$ also in this case.

(2) As for $(p, q)(O(p), O(q))$, assume $p = \langle e_1 \rangle$, $O(p) = \langle e_2 \rangle$, $q = \langle e_3 \rangle$, $O(q) = \langle e_4 \rangle$, and consider the linear transformation M of V fixing $e_5, \ldots, e_n$ with $M(e_1 x_1 + \cdots + e_4 x_4) = e_1 x_3 + e_2 x_4 - e_3 x_1 - e_4 x_2$. We easily verify that $M \in N$ and that M induces the required permutation $(p, q)(O(p), O(q))$ on O.

Thus far, we have shown (i) and (ii).

(iii) In case κ is a quadratic form, Exercise 4.9.23 gives that the order of $W(\mathrm{D}_t)$ is $2^{t-1}(t!)$ and that it is a subgroup of $W(\mathrm{B}_t)$ of index two. Viewed as permutations on O, the group $W(\mathrm{D}_t)$ consists of those elements of $W(\mathrm{B}_t)$, the group induced on

O by N (see (ii)), that preserve the two classes of maximal singular subspaces of O. According to Exercise 10.7.10, the elements of $\mathrm{O}(V, \kappa)$ have determinant ± 1. A reflection has determinant -1 and interchanges the two classes of maximal singular subspaces (see also Exercise 10.7.27), so the subgroup $N \cap \mathrm{SO}(V, \kappa)$ of N has index 2 and preserves the two classes. Therefore, the permutation group on O induced by $N \cap \mathrm{SO}(V, \kappa)$ coincides with $W(\mathrm{D}_t)$. This establishes the required transitivity. $\square$

Corollary 10.5.7 *Let $t \geq 3$. The oriflamme geometry of a nondegenerate quadratic form κ on a vector space V of dimension $2t$ is a geometry of type D_t having a duality.*

Proof This is direct from the proof of Theorem 10.5.6 as the duality is realized by any reflection. $\square$

10.6 Finite Classical Polar Spaces

A full classification of finite embedded polar spaces depends on induction with respect to the dimension n of the vector space V underlying the ambient projective space. A first requirement is the study of the cases where $n \leq 3$; this is taken care of in Propositions 10.6.6 and 10.6.7. In Theorem 10.6.2, the general classification in Theorem 10.3.13 leads to three kinds of polar spaces: the absolute of a nondegenerate alternating or unitary form f on V, or the quadric of a quadratic form κ on V. In the first two cases, a single form f suffices; this is known for alternating forms and proved in Theorem 10.6.15 for unitary forms. In the third case, a single form κ suffices if n is odd. If n is even, there are two essentially different cases: elliptic and hyperbolic quadrics; see Theorem 10.6.9.

In Theorems 10.6.4, 10.6.13, and 10.6.18, we determine the orders of the corresponding subgroups of $\mathrm{GL}(V)$.

Throughout this section, we consider a finite field $\mathbb{F}_q$, where q is a power of a prime p, and a vector space V over $\mathbb{F}_q$ of dimension n. We also write $\mathbb{F} = \mathbb{F}_q$ and $\mathbb{P} = \mathbb{P}(V)$.

The following technical lemma helps to analyze the finite pseudo-quadrics.

Lemma 10.6.1 *Let σ be an automorphism of $\mathbb{F} = \mathbb{F}_q$ of order two and $\varepsilon \in \{\pm 1\}$. Then $q = r^2$ is the square of a power r of p and $\sigma(x) = x^r$ $(x \in \mathbb{F}_q)$. Furthermore, the following holds.*

(i) *$\mathbb{F}^{\sigma,\varepsilon} = \mathbb{F}_{\sigma,\varepsilon}$. In particular, the admissible pairs $(\sigma, 1)$ and $(\sigma, -1)$ for $\mathbb{F}$ are of trace type.*
(ii) *$\{a \in \mathbb{F} \mid \sigma(a)a = 1\} = \{\sigma(x)x^{-1} \mid x \in \mathbb{F} \setminus \{0\}\}$.*
(iii) *For each $a \in \mathbb{F}$, the equation*

$$\sigma(x) + x = a$$

has no solutions $x \in \mathbb{F}$ if $a \notin \mathbb{F}_r$, and precisely r solutions if $a \in \mathbb{F}_r$.

(iv) *For each $a \in \mathbb{F}$, the equation*

$$\sigma(x)x = a$$

has no solutions $x \in \mathbb{F}$ if $a \notin \mathbb{F}_r$, a single solution if $a = 0$, and $r + 1$ solutions if $a \in \mathbb{F}_r \setminus \{0\}$.

Proof Let $s \in \mathbb{N}$ be such that $q = p^s$. The fact that σ is the unique automorphism $x \mapsto x^r$ follows from a standard property of finite fields, which states that the automorphism group of $\mathbb{F}$ is cyclic of order s.

(i) Take $c \in \mathbb{F}$ such that $\sigma(c) + \varepsilon c \neq 0$ (if $\sigma(y) + \varepsilon y = 0$ for all $y \in \mathbb{F}$, then the case $y = 1$ shows $\varepsilon = -1$, which implies $\sigma = \mathrm{id}$, a contradiction, so such a c exists). If $a \in \mathbb{F}^{\sigma,\varepsilon}$, then $\sigma(a) + \varepsilon a = 0$, so $x := (\sigma(c) + \varepsilon c)^{-1} a \sigma(c)$ satisfies $x - \varepsilon \sigma(x) = a$. This shows $\mathbb{F}^{\sigma,\varepsilon} \subseteq \mathbb{F}_{\sigma,\varepsilon}$, and equality follows from Lemma 10.1.2(ii). The second assertion is direct from Proposition 10.1.8(i).

(ii) For $a \in \mathbb{F}$ with $\sigma(a)a = 1$, take $c \in \mathbb{F}$ such that $\sigma(c)a + c \neq 0$ (such a c exists by an argument similar to the previous case). Then $x := (\sigma(c)a + c)^{-1}$ satisfies $\sigma(x)x^{-1} = a$. Hence $\{a \in \mathbb{F} \mid \sigma(a)a = 1\} \subseteq \{\sigma(x)x^{-1} \mid x \in \mathbb{F} \setminus \{0\}\}$. The other inclusion is straightforward to verify.

(iii) The left hand side of the equation belongs to $\mathbb{F}_r$. View $x \mapsto \sigma(x) + x$ as a linear map $\mathbb{F} \to \mathbb{F}_r$ from a 2-dimensional to a 1-dimensional vector space over $\mathbb{F}_r$. By (i), applied with $\varepsilon = -1$, it is surjective, so its kernel is one-dimensional. This implies that there are precisely r solutions for $a \in \mathbb{F}_r$.

(iv) The left hand side belongs to $\mathbb{F}_r$. If $a = 0$, then $\sigma(x) = 0$ or $x = 0$, which always implies $x = 0$.

If x is a nonzero solution to the equation, then so is λx whenever $\lambda \sigma(\lambda) = 1$. Hence the number of solutions for $a \in \mathbb{F}_r \setminus \{0\}$ is a constant. As there are $q - 1$ nonzero elements in $\mathbb{F}_q$, each of which is a solution to one of these equations, and $|\mathbb{F}_r \setminus \{0\}| = r - 1$, there are $(q - 1)/(r - 1) = r + 1$ solutions for $a \in \mathbb{F}_r \setminus \{0\}$. $\square$

Part (i) of Lemma 10.6.1 is of relevance to Case (iii) below.

Theorem 10.6.2 *Let Z be a finite nondegenerate classical polar space all of whose lines are thick, of rank at least two. If Z is ruled in $\mathbb{P}$, then it is one of the following embedded spaces.*

(i) *The absolute $\mathbb{P}_f$ of a nondegenerate alternating bilinear form f on V.*
(ii) *The quadric $\mathbb{P}_\kappa$ of a quadratic form κ on V.*
(iii) *The absolute $\mathbb{P}_f$ of a σ-hermitian form f on V, where $\sigma(x) = x^r$ $(x \in \mathbb{F}_q)$ with $r = \sqrt{q} \in \mathbb{N}$.*

Proof By Theorem 10.3.13, we may assume that there is an admissible pair (σ, ε) for $\mathbb{F}_q$ such that either $Z = \mathbb{P}_\kappa$ for some pseudo-quadratic form κ on V, or p is odd, $(\sigma, \varepsilon) = (\mathrm{id}, -1)$, and $Z = \mathbb{P}_f$ for a nondegenerate alternating bilinear form f on V. Since the latter case corresponds to Conclusion (i), let us assume the former.

By Proposition 10.2.13(i), changing κ by a proportional form if necessary, we may assume that κ is one of

(1) a quadratic form $(\sigma, \varepsilon) = (\mathrm{id}, 1)$,
(2) a σ-quadratic form for some involutory automorphism σ of $\mathbb{F}_q$.

These cases correspond to Conclusions (ii) and (iii), respectively. Let σ be an automorphism of $\mathbb{F}_q$ of order two. Then $q = r^2$ for a power r of p, and $\mathbb{F}_r$ is the fixed field in $\mathbb{F}_q$ of σ. By Lemma 10.6.1(i) the pairs $(\sigma, 1)$ and $(\sigma, -1)$ are of trace type, so, for a σ-quadratic form κ, we have $\mathbb{P}_\kappa = \mathbb{P}_f$, where f is the hermitian form of κ. $\square$

We study in greater detail the different cases appearing in Theorem 10.6.2, comparing them with Corollary 10.1.9 and using the terminology introduced in Definition 10.1.10. In particular, the polar space $\mathbb{P}_f$ is called symplectic in Case (i), orthogonal in Case (ii), and unitary in Case (iii).

In Tables 10.3 and 10.4, we collect some parameters for the finite polar spaces, like the sizes of their point sets and the group orders of the corresponding classical groups. We begin with Case (i) of Theorem 10.6.2, which is concerned with a symplectic quasi-polarity determined by a nondegenerate alternating bilinear form f, Case (1) of Corollary 10.1.9.

Notation 10.6.3 Recall that the corresponding symplectic group is denoted $\mathrm{Sp}(V, f)$. In the finite case, with f a nondegenerate alternating bilinear form, it is also denoted $\mathrm{Sp}(n, q)$. Similarly, we write $\mathrm{SL}(n, q)$ for $\mathrm{SL}(\mathbb{F}_q^n)$ and $\mathrm{PSL}(n, q)$ for $\mathrm{PSL}(\mathbb{F}_q^n)$ (cf. Example 1.8.16).

The omission of f from the notation of the group is justified by Proposition 10.4.9, which tells us they are all in the same $\mathrm{GL}(V)$-orbit.

Theorem 10.6.4 *If $n = 2t$ is even, then the order of the finite symplectic group* $\mathrm{Sp}(n, q)$ *is*

$$\left| \mathrm{Sp}(2t, q) \right| = q^{t^2} \prod_{j=1}^{t} (q^{2j} - 1).$$

Proof By Proposition 10.4.9(i), for each nondegenerate alternating bilinear form on V, there is a basis $e_1, \ldots, e_n$ on which it takes the form (10.3) (of a frame). Thus, a coordinate transformation moves every nondegenerate alternating form on V into f. In other words, each nondegenerate alternating form on V is in the $\mathrm{GL}(V)$-orbit of f.

If there are two bases of V, say d and e, with respect to which f takes the form (10.3), then the coordinate transformation $e_i \mapsto d_i$ $(i \in [n])$ belongs to $\mathrm{Sp}(V, f)$. We conclude that the order of the group $\mathrm{Sp}(V, f)$ equals the number of bases $(e_i)_{i \in [n]}$ on which f takes the form (10.3). We call such a basis a *symplectic basis* and will count how many there are. Denote this number by K_n, to show its dependence on n, the dimension of V.

If $n = 2$, then $\mathrm{Sp}(V, f)$ coincides with $\mathrm{SL}(V)$ (see Exercise 10.7.28), so $K_2 = q(q^2 - 1)$.

Table 10.2 Quadrics in $\mathbb{P}(\mathbb{F}_q^2)$

Type	Quadric size	Number
Hyperbolic	2	$q(q+1)/2$
Parabolic	1	$q+1$
Elliptic	0	$q(q-1)/2$

Suppose $n > 2$. There are $(q^n - 1)/(q - 1)$ points in $\mathbb{P}_f$. Each point is collinear with $(q^{n-1} - 1)/(q - 1)$ points, whence non-collinear with $(q^n - q^{n-1})/(q - 1) = q^{n-1}$ points. An ordered pair of non-collinear points corresponds to $(q - 1)$ pairs of vectors occurring as the first two elements of a symplectic basis. Each of the $(q^n - 1)q^{n-1}$ such pairs, can be extended in K_{n-2} ways to a symplectic basis of V. We conclude that

$$\left|\mathrm{Sp}(V, f)\right| = K_n = \left(q^n - 1\right)q^{n-1}K_{n-2}.$$

The required closed form for $|\mathrm{Sp}(V, f)|$ follows by induction on n. $\qquad\square$

We proceed with quadrics, starting with the projective line. In Case (ii) of Theorem 10.6.2, each hermitian form of κ is a symmetric bilinear form. This form may be degenerate. If it is not, it determines an orthogonal quasi-polarity, Case (2) of Corollary 10.1.9.

Definition 10.6.5 Let $V = \mathbb{F}_q^2$ (that is, $n = 2$). A quadratic form κ on V is called *hyperbolic, parabolic, elliptic* according to whether $\mathbb{P}(V)_\kappa$ has exactly two, one, or no points.

In the elliptic case there is little geometric information. Therefore, we need some more algebraic insight.

Proposition 10.6.6 *Let $n = 2$, so $V \cong \mathbb{F}_q^2$.*

(i) *A non-zero quadratic form on V is hyperbolic, parabolic, or elliptic.*
(ii) *The number of proportionality classes for each case equals the number of quadrics of the indicated size and is as given in the last column of Table* 10.2.
(iii) *If κ is an elliptic form on $\mathbb{F}_q^2$, then κ is hyperbolic over the quadratic extension field $\mathbb{F}_{q^2}$.*
(iv) *The group $\mathrm{PGL}(\mathbb{F}_q^2)$ acts transitively on each set of proportionality classes of hyperbolic, parabolic, and elliptic forms.*

Proof By Lemma 10.3.8, a non-zero quadric in $\mathbb{P}(V)$ does not have more than two points, hence (i).

(ii) There are q^3 distinct quadratic forms on V since the parameters $a, b, c \in \mathbb{F}_q$ of the polynomial $ax_1^2 + bx_1x_2 + cx_2^2$ in x_1, x_2 take that many values. The constant map 0 on V is the unique form vanishing everywhere. We now count the hyperbolic and parabolic forms. Given the points $\infty = (1 : 0)$ and $0 = (0 : 1)$ in $\mathbb{P}(V)$, the hyperbolic forms vanishing on these points are the $q - 1$ forms bx_1x_2 with $b \in \mathbb{F}_q \setminus \{0\}$. Since there are $q(q + 1)/2$ distinct choices for unordered pairs of points in $\mathbb{P}(\mathbb{F}_q^3)$, there are precisely $(q + 1)q(q - 1)/2$ hyperbolic forms. Given the point $(1 : 0)$, the parabolic forms vanishing on it are the $q - 1$ forms cx_2^2 with $c \in \mathbb{F}_q \setminus \{0\}$. Hence there are $(q + 1)(q - 1)$ parabolic forms. Consequently, the number of elliptic forms is $q^3 - 1 - (q + 1)(q - 1)q/2 - (q^2 - 1) = (q - 1)^2 q/2$ and since each proportionality class contains $q - 1$ forms, (ii) follows.

(iii) Let κ be elliptic on $\mathbb{F}_q^2$ and consider κ as a form on the 2-dimensional vector space $\mathbb{F}_{q^2}$ over $\mathbb{F}_q$. Then $\kappa(x) = 0$ is a homogeneous quadratic equation in x_1, x_2 with coefficients in $\mathbb{F}_q$, and so has at most two projective solutions $\mathbb{F}x = (x_1 : x_2)$. It clearly has a solution over $\mathbb{F}_{q^2}$. If κ is not hyperbolic over $\mathbb{F}_{q^2}$, it is parabolic and $\kappa(x) = 0$ has only one projective solution $(u_1 : u_2)$ in $\mathbb{F}_{q^2}^2$, so $\kappa(x) = a(x_1u_2 - x_2u_1)^2 = a(x_1^2u_2^2 - 2x_1x_2u_1u_2 + x_2^2u_1^2)$ for some $a \in \mathbb{F}_q$. Thus, u_1^2 and u_2^2 belong to $\mathbb{F}_q$. Now q is odd, for otherwise $a^{-1}\kappa(x) = x_1^2u_2^2 + x_2^2u_1^2$ is a square in $\mathbb{F}_q$ which forces κ to be parabolic on $\mathbb{F}_q^2$, contradicting the ellipticity assumption on κ. Hence $u_1/u_2 = u_1u_2/u_2^2 \in \mathbb{F}_q$ giving $(u_1 : u_2) \in \mathbb{P}(\mathbb{F}_q^2)$, a contradiction.

(iv) The observation is obvious for hyperbolic and parabolic forms, so assume that κ is elliptic. The stabilizer H in $\mathrm{PGL}(\mathbb{F}_q^2)$ of the proportionality class of κ is contained in $\mathrm{PGL}(\mathbb{F}_{q^2}^2)$. It stabilizes the unordered pair of points of $\mathbb{P}(\mathbb{F}_{q^2}^2)_\kappa$. Since the orbit of H on $\mathbb{P}(V)$ has at most $q + 1$ points and, by Exercise 5.7.10, the stabilizer in $\mathrm{PGL}(2, q^2)$ of three points of $\mathbb{P}(\mathbb{F}_{q^2}^2)$ is trivial, this yields $|H| \leq 2(q + 1)$. Therefore the $\mathrm{PGL}(\mathbb{F}_q^2)$-orbit of the proportionality class of κ has at least $q(q^2 - 1)/(2(q + 1)) = q(q - 1)/2$ members. Now (ii) shows that there is a unique such orbit, and that it is of length $q(q - 1)/2$. It follows that $\mathrm{PGL}(\mathbb{F}_q^2)$ is transitive as required (and $|H| = 2(q + 1)$). $\qquad\square$

Next we examine quadrics in $\mathbb{P}(\mathbb{F}_q^3)$.

Proposition 10.6.7 *Let $n = 3$, so $V = \mathbb{F}_q^3$. In the projective plane $\mathbb{P}(V)$, every quadric is non-empty. Moreover, if it is not contained in a union of two lines, then it has precisely $q + 1$ points.*

Proof A degenerate quadric is clearly contained in the union of two lines. By Proposition 10.4.9, a nondegenerate quadric having at least two points is of the form $\mathbb{P}_\kappa$ where $\kappa(x) = x_1x_2 + x_3^2$ on a suitable basis. It readily follows that $|\mathbb{P}_\kappa| = q + 1$.

Assume that the quadratic form κ determines an empty quadric. Fix a nonzero linear form $\phi \in V^\vee$ and consider the family of quadratic forms $\kappa_\lambda = \lambda\kappa + \phi^2$ where $\lambda \in \mathbb{F}_q$; denote by $Z_\lambda = \mathbb{P}_{\kappa_\lambda}$ the corresponding quadrics. Each point $y \in \mathbb{P}(\mathbb{F}_q^3)$ determines a unique $\lambda \in \mathbb{F}_q$ such that $\kappa_\lambda(y) = 0$ (for $\kappa(y) \neq 0$). Hence, the cardinality of the set $X = \{(x, \lambda) \in \mathbb{P} \times \mathbb{F}_q \mid x \in Z_\lambda\}$ equals $q^2 + q + 1$. The quadric Z_0 is the line $\mathrm{Ker}(\phi)$. If $\lambda \neq 0$, then Z_λ does not contain a line, so either has exactly $q + 1$ points, is a singleton, or is empty (cf. Exercise 10.7.11). Let α, β, γ be the numbers of nonzero λ for which Z_λ is a singleton, of size $q + 1$, and empty, respectively. By counting $\lambda \in \mathbb{F}_q$ and X each in two ways, we find

$$\alpha + \beta + \gamma + 1 = q,$$
$$\alpha + \beta(q + 1) + (q + 1) = q^2 + q + 1,$$

so $\alpha = -1/q - \gamma(1 + 1/q)$, a contradiction with $\alpha \in \mathbb{N}$. We conclude that every quadric in $\mathbb{P}(V)$ is non-empty. $\qquad\square$

Corollary 10.6.8 *For $n \geq 3$, each quadric in $\mathbb{P}(\mathbb{F}_q^n)$ is non-empty.*

Proof For $n \geq 3$, the space $\mathbb{P}(\mathbb{F}_q^n)$ contains a projective plane $\mathbb{P}(\mathbb{F}_q^3)$ and, by Proposition 10.6.7, this plane contains a point of the quadric. $\qquad\square$

The field $\mathbb{F}_q$ possesses non-squares (elements that are not the square of another element of $\mathbb{F}_q$) if and only if p is odd. If so, denote by g a specific non-square in $\mathbb{F}_q$ and by γ an elliptic quadratic form on $\mathbb{F}_q^2$.

Theorem 10.6.9 *Let κ be a quadratic form on the n-dimensional vector space V over $\mathbb{F}_q$ such that $\mathbb{P}_\kappa$ is a nondegenerate quadric in $\mathbb{P} = \mathbb{P}(V)$. Then precisely one of the following cases occurs, where t is the rank of the polar space $\mathbb{P}_\kappa$.*

(1) $n = 2t + 1$ and there is a basis of V on which κ has the form

$$\kappa(x) = x_1 x_2 + x_3 x_4 + \cdots + x_{2t-1} x_{2t} + \alpha x_n^2,$$

where $\alpha = 1$ or (if $p > 2$) $\alpha = g$;
(2) $n = 2t$ and there is a basis of V on which κ has the form

$$\kappa(x) = x_1 x_2 + \cdots + x_{2t-1} x_{2t},$$

(3) $n = 2t + 2$ and there is a basis of V on which κ has the form

$$\kappa(x) = x_1 x_2 + \cdots + x_{2t-1} x_{2t} + \gamma(x_{2t+1}, x_{2t+2}).$$

Proof Without loss of generality, we take $V = \mathbb{F}_q^n$. By Proposition 10.4.9, κ is of the form $x_1 x_2 + \cdots + x_{2t-1} x_{2t} + \kappa_0(x_{2t+1}, \ldots, x_n)$ where t is the Witt index of κ and κ_0 is nondegenerate of Witt index 0 on $\mathbb{F}_q^{n-2t}$. By Corollary 10.6.8, this implies $n - 2t \leq 2$, whence $t \geq \frac{1}{2}n - 1$. Since, obviously, $t \leq n/2$, we find $t = (n - 1)/2$ if $n = 2t + 1$, and $t \in \{m - 1, m\}$ if $n = 2m$.

Assume $n = 2t + 1$. Then $\kappa_0(x) = \lambda x_n^2$ for some nonzero $\lambda \in \mathbb{F}_q$. We can easily achieve $\lambda \in \{1, g\}$ if we replace e_n by $e_n \mu$, where $\mu^2 \in \{\lambda^{-1}, g\lambda^{-1}\}$. This brings us to Case (1).

Assume $n = 2m$ with $t \in \{m - 1, m\}$. If $t = m$, Case (2) is at hand. If $t = m - 1$, Proposition 10.4.9 provides a basis of $\mathbb{F}_q^2$ such that $\kappa_0 = \gamma$, bringing us to Case (3). $\qquad\square$

Table 10.3 Finite classical polar spaces

Name	Type	$\lvert P \rvert$	$\lvert P \setminus a^{\perp} \rvert$
$\mathrm{Sp}(2m, q)$	Symplectic	$(q^{2m} - 1)/(q - 1)$	q^{2m-1}
$\mathrm{O}(2m + 1, q)$	Odd dim. quadric	$(q^{2m} - 1)/(q - 1)$	q^{2m-1}
$\mathrm{O}^{+}(2m, q)$	Hyperbolic quadric	$(q^{m-1} + 1)(q^{m} - 1)/(q - 1)$	q^{2m-2}
$\mathrm{O}^{-}(2m, q)$	Elliptic quadric	$(q^{m-1} - 1)(q^{m} + 1)/(q - 1)$	q^{2m-2}
$\mathrm{U}(n, r)$	Unitary	$(r^{n} - (-1)^{n})(r^{n-1} - (-1)^{n-1})/(r - 1)$	r^{2n-3}

Remark 10.6.10 Observe that the quadratic forms in (1) for $\alpha = 1$ and $\alpha = g$ are in distinct $\mathrm{GL}(V)$-orbits. However, forms from one orbit are proportional to forms from the other orbit, so the corresponding orthogonal groups are essentially the same.

Definition 10.6.11 A nondegenerate finite quadric $\mathbb{P}_{\kappa}$ as in Theorem 10.6.9 is called *odd-dimensional, hyperbolic*, or *elliptic*, and the group $\mathrm{O}(V, \kappa)$ is also denoted by $\mathrm{O}(n, q)$, $\mathrm{O}^{+}(n, q)$, $\mathrm{O}^{-}(n, q)$, according to whether it belongs to Cases (1), (2), or (3) of Theorem 10.6.9.

These definitions are in accordance with the projective lines case, cf. Definition 10.6.5.

Lemma 10.6.12 *Suppose that* (P, L) *is a nondegenerate quadric in* $\mathbb{P}(\mathbb{F}_{q}^{n})$ *and* $a \in P$.

(i) *The sizes of* P *and* $P \setminus a^{\perp}$ *are as indicated in Table* 10.3 *for the odd-dimensional* $(n = 2m + 1)$, *hyperbolic* $(n = 2m)$, *and elliptic* $(n = 2m)$ *quadrics.*

(ii) *If* $n = 2m$ *and* (P, L) *is a hyperbolic quadric, then its dual polar space is thin and so it defines an oriflamme geometry with diagram* D_{m}.

(iii) *The quadric* (P, L) *has a defective point if and only if* q *is even and* n *is odd. If so, there is a unique defective point.*

Proof As before, denote the rank of (P, L) by t.

(i) The values of the numbers $N_{n,t} := \lvert P \rvert$ of points are obtained by induction on n. If $n = 2$, then $N_{2,1} = 2$ and $N_{2,0} = 0$ by Proposition 10.6.6. If $n = 3$, then $N_{3,1} = q + 1$ by Proposition 10.6.7.

Suppose now $n \geq 4$, so $t \geq 1$. Consider a basis of V as in Proposition 10.4.9. Take $a = \langle e_{1} \rangle$. Each point of P that is not perpendicular to a with respect to the hermitian form of κ can be represented by a vector $x = \sum_{i} e_{i} x_{i}$ with $x_{2} = 1$. For such a point $\langle x \rangle$ to belong to P, it is necessary and sufficient that $x_{1} = \kappa(\sum_{i \geq 3} e_{i} x_{i})$. Thus there are q^{n-2} points in $P \setminus a^{\perp}$.

Next we count the number of points in $a^{\perp}$. Take $b \in P \setminus a^{\perp}$. By the transitivity result of Lemma 10.5.4, we may assume $b = \langle e_{2} \rangle$. The line space $a^{\perp} / \mathrm{Rad}(a^{\perp}) =$

$a^\perp/\{a\}$ is isomorphic to $a^\perp \cap b^\perp$. This is a nondegenerate quadric in $\mathbb{P}(\mathbb{F}_q^{n-1})$ with rank $t-1$ of the same type (odd-dimensional, hyperbolic, or elliptic) as (P, L), and so has $N_{n-2,t-1}$ points. Consequently,

$$N_{n,t} = 1 + q N_{n-2,t-1} + q^{n-2}.$$

By induction on n, the formulae for $N_{n,t}$ in the three quadric cases can now be verified to be as in the third column of Table 10.3.

(ii) The same induction as in (i) shows that, in the case where $(n, t) = (2m, m)$, there are exactly two maximal singular subspaces containing a singular subspace of (projective) dimension $m - 1$. (See also Example 7.8.7.) Now we are in the setting of Corollary 10.4.10, so we are done.

(iii) follows easily from inspection of the reduced forms obtained in Theorem 10.6.9. (If $p = 2$ and $n = 2t + 1$, then $\langle e_n \rangle$ is the unique defective point.) $\square$

As an application of our last results we compute the order of $O(V, \kappa)$ for the various cases. The approach is similar to the one for symplectic forms in Theorem 10.6.4.

Theorem 10.6.13 *The orders of* $O(2m + 1, q)$, $O^+(2m, q)$, *and* $O^-(2m, q)$ *are as described in Table* 10.4.

Proof Let κ be the form in question, which we take as in Theorem 10.6.9, and denote by G the group $O(V, \kappa)$. We count the number $K_{n,t}$ of bases $(e_i)_{1 \leq i \leq n}$ of V such that κ is as in Theorem 10.6.9 with respect to $(e_i)_{1 \leq i \leq n}$. This number is equal to $|G|$ by the same arguments as for Theorem 10.6.4.

By counting the non-collinear points of $\mathbb{P}_\kappa$ by use of Lemma 10.6.12, we have, whenever $t > 0$,

$$K_{n,t} = N_{n,t} q^{n-2}(q - 1) K_{n-2,t-1},$$

where $N_{n,t}$ is as in the proof of Lemma 10.6.12. Therefore, it remains to determine $K_{n,t}$ in the cases $(n, t) = (0, 0)$, $(1, 0)$, and $(2, 0)$. A look at the automorphism group of a zero-dimensional space leads to $K_{0,0} = 1$. The second one, $K_{1,0}$, is the number of $\lambda \in \mathbb{F}_q$ such that $\lambda^2 = 1$, so $K_{1,0} = \gcd(2, q - 1)$. Finally, Proposition 10.6.6(iv) (see also its proof) gives $K_{2,0} = 2(q + 1)$. $\square$

It remains to treat the unitary case, Case (iii) of Theorem 10.6.2. We let σ be an automorphism of $\mathbb{F}_q$ of order two. Then $q = r^2$ for a power r of p, and $\mathbb{F}_r$ is the fixed field in $\mathbb{F}_q$ of σ. In Definition 10.1.10 the corresponding group $U(V, f)$ is called the unitary group.

Lemma 10.6.14 *Let* σ *be an involutory automorphism of* $\mathbb{F}_q$ *and let* f *be a nonde-generate* σ*-hermitian form on* V.

(i) *If* $n = 1$, *then there is a basis vector* $d_1 \in V$ *such that* f *is of the form*

$$f(x, y) = \sigma(x_1) y_1 \quad (x = d_1 x_1, \ y = d_1 y_1 \in V).$$

So $\mathbb{P}_f$ *is empty and* $U(V, f)$ *is a cyclic group of order* $r + 1$.

(ii) *If $n = 2$, then there are bases e_1, e_2 and d_1, d_2 of V such that f is one of the forms*

$$f(x, y) = \sigma(x_1)y_2 + \sigma(x_2)y_1 \quad (x = e_1x_1 + e_2x_2,\, y = e_1y_1 + e_2y_2 \in V),$$
$$f(x, y) = \sigma(x_1)y_1 + \sigma(x_2)y_2 \quad (x = d_1x_1 + d_2x_2,\, y = d_1y_1 + d_2y_2 \in V).$$

In particular, $\mathbb{P}_f$ has exactly $r + 1$ points.

Proof (i) Clearly, there is a projective point $\langle d_1 \rangle$ that is not a point of $\mathbb{P}_f$ (for otherwise we would have $f = 0$, contradicting nondegeneracy of f). Observe that $a = f(d_1, d_1)$ belongs to $\mathbb{F}_r \setminus \{0\}$ since it is fixed by σ and $a = 0$ would imply $\langle d_1 \rangle$ to be a point of $\mathbb{P}_f$. Replacing d_1 by its scalar multiple αd_1, where $\alpha \in \mathbb{F}_q$ satisfies $\sigma(\alpha)\alpha = a^{-1}$ (cf. Lemma 10.6.1(iv)), we obtain $f(d_1, d_1) = 1$. Now (i) follows.

(ii) Assume $n = 2$. As the restriction of f to $V \cap d_1^\perp$ must be nondegenerate again, we can apply (i) to it, thus finding a vector d_2 with $f(d_1, d_2) = 0$ and $f(d_2, d_2) = 1$. This establishes the existence of the basis d_1, d_2.

As for the basis e_1, e_2, if we can find a point of $\mathbb{P}_f$, then nondegeneracy and scaling will do the rest. Consider $y = d_1y_1 + d_2y_2$ with $y_i \in \mathbb{F}$. It satisfies $f(y, y) = 0$ if and only if $\sigma(y_1)y_1 = -\sigma(y_2)y_2$. Fixing a nonzero y_2, we find from Lemma 10.6.1(iv) that there are exactly $r + 1$ solutions y_1 to this equation. This shows that $\mathbb{P}_f$ has exactly $r + 1$ points and so settles the proof of the lemma. $\square$

Theorem 10.6.15 *Suppose $r = \sqrt{q} \in \mathbb{N}$. Let σ be the automorphism $x \mapsto x^r$ of $\mathbb{F}_q$ and consider a nondegenerate σ-hermitian form f on the n-dimensional vector space V over $\mathbb{F}_q$. The polar space $\mathbb{P}_f$ has rank $t = \lfloor n/2 \rfloor$, and there is a basis $e_1, \ldots, e_n$ such that*

$$f(x, y) = \big(\sigma(x_1)y_2 + \sigma(x_2)y_1\big) + \cdots + \big(\sigma(x_{2t-1})y_{2t} + \sigma(x_{2t})y_{2t-1}\big)$$
$$\text{for } x = \sum_i e_ix_i,\ \text{and } y = \sum_i e_iy_i$$

if $n = 2t$ and

$$f(x, y) = \big(\sigma(x_1)y_2 + \sigma(x_2)y_1\big) + \cdots + \big(\sigma(x_{2t-1})y_{2t} + \sigma(x_{2t})y_{2t-1}\big) + \sigma(x_n)y_n$$
$$\text{for } x = \sum_i e_ix_i,\ \text{and } y = \sum_i e_iy_i$$

if $n = 2t + 1$. The $2t$ basis elements $e_1, \ldots, e_{2t}$ correspond to a frame of $\mathbb{P}_f$. Also, there is a basis $d_1, \ldots, d_n$ of V such that

$$f(x, y) = \sigma(x_1)y_1 + \cdots + \sigma(x_n)y_n$$
$$\text{for } x = \sum_i d_ix_i,\ \text{and } y = \sum_i d_iy_i.$$

The basis $d_1, \ldots, d_n$ is a unitary basis in the sense that $f(d_i, d_j) = 0$ if $i \neq j$ and $f(d_i, d_i) = \sigma(d_i)d_i$ for all $i, j \in [n]$.

Proof By Proposition 10.4.9 and Lemma 10.6.14 (in particular, the fact that a unitary pseudo-quadric with respect to σ in dimension at least two is non-empty), we find the basis $e_1, \ldots, e_n$ as in the theorem. By Lemma 10.6.14, the pairs e_{2i-1}, e_{2i} for $i \in [t]$ can be replaced by pairs d_{2i-1}, d_{2i} so as to obtain a basis $(d_i)_i$ as required (with $d_n = e_n$ if $n = 2t + 1$). $\qquad\square$

Lemma 10.6.16 *Suppose that f is a nondegenerate σ-hermitian form on $V = \mathbb{F}_q^n$ and a is a point of the polar space $\mathbb{P}_f$. Then the sizes of the point set P of $\mathbb{P}_f$ and of $P \setminus a^\perp$ are as indicated in Table* 10.3.

Proof The values of the numbers $N_n = |P|$ of points of $\mathbb{P}_f$ are obtained by induction on n. Clearly, $N_1 = 0$. Moreover, $N_2 = r + 1$, where $r = \sqrt{q}$, by Lemma 10.6.14.

Suppose now $n \geq 3$. Consider a basis of V as in Proposition 10.4.9. Take $a = \langle e_1 \rangle$, so $a \in P$. Each point of P that is not perpendicular to a with respect to f can be represented by a vector $x = \sum_i e_i x_i$ with $x_2 = 1$. For such a point $\langle x \rangle$ to belong to P, it is necessary and sufficient that

$$\sigma(x_1) + x_1 = -f\left(\sum_{i \geq 3} e_i x_i, \sum_{i \geq 3} e_i x_i \right). \tag{10.4}$$

There are q^{n-2} choices for $x_3, \ldots, x_n \in \mathbb{F}_q$; for each choice, by Lemma 10.6.1(iii), there are r solutions x_1 to (10.4). Hence $|P \setminus a^\perp| = q^{n-2}r = r^{2n-3}$, as required.

Next we count the number of points in $a^\perp$. Take $b \in P \setminus a^\perp$ (by transitivity, we may assume $b = \langle e_2 \rangle$). The line space $a^\perp / \mathrm{Rad}(a^\perp) = a^\perp / \{a\}$ is $a^\perp \cap b^\perp$. The latter is the absolute in $\mathbb{P}(\mathbb{F}_q^{n-1})$ of a nondegenerate σ-hermitian form, and so has N_{n-2} points. Consequently,

$$N_n = 1 + q N_{n-2} + r^{2n-3}.$$

By induction on n, the formula for N_n can easily be verified to be as in the third column, last row, of Table 10.3. $\qquad\square$

Notation 10.6.17 For f as in Theorem 10.6.15, we also write $\mathrm{U}(n, r)$ instead of $\mathrm{U}(\mathbb{F}_q^n, f)$, where $r = \sqrt{q}$.

Theorem 10.6.18 *The order of the unitary group* $\mathrm{U}(n, r)$ *is*

$$\left| \mathrm{U}(n, r) \right| = r^{\binom{n}{2}} \prod_{j=1}^{n} \left(r^j - (-1)^j \right).$$

Proof As before, $|\mathrm{U}(V, f)|$ equals the number of bases $(e_i)_i$ of V as described in Theorem 10.6.15. By (i) the number does not depend on the choice of f, so we denote it by K_n. If $n = 0$, then $\mathrm{U}(V, f)$ is the trivial group. If $n = 1$, then $\mathrm{U}(V, f)$ coincides with $\{\lambda \in \mathbb{F}_q \mid \sigma(\lambda)\lambda = 1\}$; a group of order $r + 1$.

Table 10.4 Classical groups

Name	Lie name	Order	Center order
$\mathrm{SL}(n, q)$	$\mathrm{A}_{n-1}(q)$	$q^{\binom{n}{2}} \prod_{j=2}^{n}(q^j - 1)$	$\gcd(n, q - 1)$
$\mathrm{Sp}(2m, q)$	$\mathrm{C}_m(q)$	$q^{m^2} \prod_{j=1}^{m}(q^{2j} - 1)$	$\gcd(2, q - 1)$
$\mathrm{O}(2m + 1, q)$	$\mathrm{B}_m(q)$	$\gcd(2, q - 1)q^{m^2} \prod_{j=1}^{m}(q^{2j} - 1)$	$\gcd(2, q - 1)$
$\mathrm{O}^+(2m, q)$	$\mathrm{D}_m(q)$	$q^{m(m-1)}(q^m - 1) \prod_{j=1}^{m-1}(q^{2j} - 1)$	$\gcd(4, q^m - 1)$
$\mathrm{O}^-(2m, q)$	${}^2\mathrm{D}_m(q)$	$q^{m(m-1)}(q^m + 1) \prod_{j=1}^{m-1}(q^{2j} - 1)$	$\gcd(4, q^m + 1)$
$\mathrm{U}(n, r)$	${}^2\mathrm{A}_{n-1}(r)$	$r^{\binom{n}{2}} \prod_{j=1}^{n}(r^j - (-1)^j)$	$\gcd(n, r + 1)$

If $n \geq 2$, then, by Lemma 10.6.14, there are points in $\mathbb{P}_f$, so, by counting pairs of non-collinear points and going over to the case of dimension $n - 2$, we find

$$K_n = \left(r^n - (-1)^n\right)\left(r^{n-1} - (-1)^{n-1}\right)r^{2n-3}K_{n-2}.$$

The required closed form expression for K_n follows by induction on n. $\qquad\square$

Remark 10.6.19 The totality of finite groups found so far are summarized in Table 10.4. We have added the special linear groups, because the resulting set of groups are the finite species of the so-called *classical groups*. A way to make sense of this choice is to view the linear groups as groups of automorphisms of the degenerate polar space $\mathbb{P}(\mathbb{F}_q^n)$. A deeper reason is that these groups cover all series of finite Chevalley groups defined for arbitrary finite rank, a quantity directly related to the (singular or polar) rank of the line space.

Also, some more information is added, like the order of the center. Most of the classical groups are *almost simple*, that is, they lie between a simple group and their automorphism group. There are some exceptions to simplicity for low values of n, e.g., $\mathrm{O}^+(4, q)$ is not simple; but also for $n \leq 2$ or small fields some of these groups are not almost simple. In the second column we have listed the name of the group that is common in the context of Lie theory.

10.7 Exercises

Section 10.1

Exercise 10.7.1 Let (σ, ε) be an admissible pair for $\mathbb{D}$. Prove that $x \mapsto -\sigma(x)\varepsilon$ is an involution and that $\mathbb{D}^{\sigma,\varepsilon}$ is its set of fixed elements in $\mathbb{D}$.

Exercise 10.7.2 Let (σ, ε) be an admissible pair for $\mathbb{D}$. Prove that $\mathbb{D}_{\sigma,\varepsilon}$ and $\mathbb{D}^{\sigma,\varepsilon}$ of Lemma 10.1.2(i) are σ-invariant if and only if $\varepsilon \in \{\pm 1\}$.
(*Hint:* Consider the element $1 - \varepsilon$ of $\mathbb{D}$.)

Exercise 10.7.3 (This exercise is used in Remark 10.2.12) Let (σ, ε) be an admissible pair for the division ring $\mathbb{D}$ and $\lambda \in \mathbb{D} \setminus \{0\}$. Show that $\lambda \mathbb{D}_{\sigma,\varepsilon} = \mathbb{D}_{\rho,\delta}$, where $\rho(x) = \sigma(\lambda)x\lambda^{-1}$ for all $x \in \mathbb{D}$ and $\delta = \lambda\sigma(\lambda)^{-1}\varepsilon$.

Exercise 10.7.4 Let $\mathbb{F}$ be a field of characteristic two and consider the admissible pair $(\sigma, \varepsilon) = (\mathrm{id}, 1)$ for $\mathbb{F}$. Now $\mathbb{F}^{\sigma,\varepsilon}/\mathbb{F}_{\sigma,\varepsilon} \cong \mathbb{F}$.

(a) Prove that $*$ gives $\mathbb{F}$ the structure of a vector space over the subfield $\mathbb{F}^{(2)} = \{x^2 \in \mathbb{F} \mid x \in \mathbb{F}\}$.
(b) Give an example showing that the dimension of this vector space can be infinite.

Exercise 10.7.5 (This exercise is used in Example 10.2.11(iii)) Let $\mathbb{D}$ be a 4-dimensional associative algebra over a field $\mathbb{F}$ with identity element 1 and basis 1, e_1, e_2, e_3, whose multiplication table is given below.

	e_1	e_2	e_3
e_1	$e_1 + a$	e_3	$e_3 + e_2 a$
e_2	$e_2 - e_3$	b	$b - e_1 b$
e_3	$-e_2 a$	$e_1 b$	$-ab$

for certain $a, b \in \mathbb{F}$. Supply $\mathbb{D}$ with the quadratic form $\kappa : \mathbb{D} \to \mathbb{F}$ given by

$$x_0^2 + x_0 x_1 - a x_1^2 - \left(x_2^2 + x_2 x_3 - a x_3^2\right)b$$

for $x = x_0 + e_1 x_1 + e_2 x_2 + e_3 x_3$. Prove the following assertions.

(a) $\kappa(xy) = \kappa(x)\kappa(y)$ for all $x, y \in \mathbb{D}$.
(b) The map $\sigma : x \mapsto x_1 + 2x_0 - x$ defines an anti-automorphism of $\mathbb{D}$.
(c) If $\kappa(x) \neq 0$ for all $x \neq 0$, then $\mathbb{D}$ is a division ring.
(d) For $\mathbb{F} = \mathbb{R}$, the ring $\mathbb{H}$ discussed in Exercise 1.9.11(b) is a special case.
(e) The condition in (c) is satisfied for $\mathbb{F} = \mathbb{F}_2(X)$, the field of rational functions over $\mathbb{F}_2$ in the indeterminate X, and $a = 1, b = X$.
(f) Assume now that $\mathbb{D}$ is a division ring and that $\mathbb{F}$ has characteristic two. The pair $(\sigma, 1)$ is admissible with $\mathbb{D}_{\sigma,1} = \mathbb{F}$ and $\mathbb{D}^{\sigma,1} = \mathbb{F} + e_2\mathbb{F} + e_3\mathbb{F}$.

Exercise 10.7.6 Determine the kernels of the homomorphisms $\mathrm{GL}(V)_f \to \mathrm{Aut}(Z)$ for f as in (1), (2), and (3) of Definition 10.1.10.

Exercise 10.7.7 Let $\mathbb{F}$ be a field of characteristic two and set $V = \mathbb{F}^{2n+1}$, with standard basis $\varepsilon_1, \ldots, \varepsilon_{2n+1}$. Consider the quadratic from κ on V given by

$$\kappa(x) = x_1 x_2 + x_3 x_4 + \cdots + x_{2n-1} x_{2n} + x_{2n+1}^2 \quad (x \in V).$$

(a) Let W be the subspace of V spanned by $\varepsilon_1, \ldots, \varepsilon_{2n}$ and let f be the alternating bilinear form on W given by

$$f(x, y) = \kappa(x + y) - \kappa(x) - \kappa(y) \quad (x, y \in W).$$

Show that $\mathbb{P}(V)_\kappa$ and $\mathbb{P}(W)_f$ are isomorphic polar spaces.
(*Hint:* Look ahead to Theorem 10.2.15.)
(b) Prove that $\mathrm{O}(V, \kappa)$ and $\mathrm{Sp}(W, f)$ are isomorphic groups.

Section 10.2

Exercise 10.7.8 Use Example 10.2.11(iii) to establish that not every $(\sigma, 1)$-quadratic form is proportional to a quadratic form.

Exercise 10.7.9 (This exercise is used in Example 10.2.8) Let (σ, ε) be an admissible pair for $\mathbb{D}$. Suppose that κ is a nonzero (σ, ε)-quadratic form on the vector space $\mathbb{D}^2$.

(a) Show that, if κ vanishes at the standard basis elements ε_1 and ε_2, then it is of the form

$$\kappa(x) = \sigma(x_1)ax_2 + \sigma(x_2)\sigma(a)\varepsilon x_1 + \mathbb{D}_{\sigma,\varepsilon} \quad \left(x = \varepsilon_1 x_1 + \varepsilon_2 x_2 \in \mathbb{D}^2\right)$$

for some nonzero $a \in \mathbb{D}$.
(b) Let A be the linear transformation on $\mathbb{D}^2$ with diagonal matrix relative to the standard basis of $\mathbb{D}^2$ and with diagonal entries $1, a^{-1}$ for $a \in \mathbb{D} \setminus \{0\}$. Show that the composition $\kappa \circ A$ is the pseudo-quadratic form

$$\kappa(x) = \sigma(x_1)x_2 + \sigma(x_2)\varepsilon x_1 + \mathbb{D}_{\sigma,\varepsilon} \quad \left(x = \varepsilon_1 x_1 + \varepsilon_2 x_2 \in \mathbb{D}^2\right).$$

(c) Conclude that the statement in Example 10.2.8 about the choice of κ up to a coordinate transformation is correct.

Exercise 10.7.10 Let κ be a nondegenerate quadratic form and f a nondegenerate symplectic form on the vector space V over the field $\mathbb{F}$. Prove the following properties of the determinant.

(a) $\det(\mathrm{O}(V, \kappa)) = \{\pm 1\}$.
(b) $\det(\mathrm{Sp}(V, f)) = \{1\}$.

Exercise 10.7.11 (This exercise is used in Proposition 10.6.7) Quadrics of a projective plane $\mathbb{P} = \mathbb{P}(\mathbb{F}^3)$ over a field are usually called *conics*. Prove that if Z is a conic in $\mathbb{P}$, then it is empty, a singleton, a line, the union of two distinct lines, the whole plane, or Z has rank 1 and each point of Z lies on a unique projective line intersecting Z only in that point. In the latter case, we call Z an *oval*.

Exercise 10.7.12 Let κ be a pseudo-quadratic form on V such that $\mathbb{P}(V)_\kappa$ is a nonempty nondegenerate pseudo-quadric. Prove that $\mathbb{P}(V)_\kappa$ spans $\mathbb{P}(V)$.

Section 10.3

Exercise 10.7.13 (This exercise is used in Example 10.3.2) Let κ be a (σ, ε)-quadratic form with hermitian form f on the right vector space V over $\mathbb{D}$ and write $\mathbb{P} = \mathbb{P}(V)$. We show the analogue of Proposition 10.1.13 for $\mathbb{P}_\kappa$ instead of $\mathbb{P}_f$. Let

$p \in V \setminus \{0\}$ and $\lambda \in \mathbb{D}$. Write ϕ for the map $V \to \mathbb{D}$ given by $\phi(x) = \lambda f(p, x)$, so the map $r_{p,\phi}$ of Exercise 1.9.30 is given by

$$r_{p,\phi}(x) = x - p\lambda f(p, x) \quad (x \in V)$$

and induces a perspectivity on $\mathbb{P}(V)$ with center $\langle p \rangle$ and axis $\langle p^{\perp} \rangle_{\mathbb{P}}$.

(a) Show that $r_{p,\phi}$ belongs to $O(V, \kappa)$ if and only if $\lambda = 0$ or $\kappa(p) = \lambda^{-1} + \mathbb{D}_{\sigma,\varepsilon}$.
(b) Suppose that $\langle a \rangle$ and $\langle b \rangle$ are non-collinear points of $\mathbb{P}_{\kappa}$ and that $p = a + b$ represents a point $\langle p \rangle$ of $\mathbb{P}(V)$ such that $\langle a \rangle, \langle b \rangle \notin p^{\perp}$. Then $f(a, b) \neq 0$. Set $\lambda = f(b, a)^{-1}$. Prove that $r_{p,\phi}$ is well defined and that it maps $\langle a \rangle$ to $\langle b \rangle$. Conclude that, for each collinear triple of projective points $\langle p \rangle$, $\langle a \rangle$, $\langle b \rangle$ with the latter two outside $\langle p \rangle \cup \langle p \rangle^{\perp}$, the perspectivity with center $\langle p \rangle$ and axis $\langle p \rangle^{\perp}$ mapping $\langle a \rangle$ to $\langle b \rangle$ is induced by an element of $O(V, \kappa)$.
(c) Show that $r_{p,\phi}$ is an *involution* (that is, an element of order 2) if and only if $2\lambda^{-1} = f(p, p)$. Compare the result with Exercise 1.9.30(e).

Exercise 10.7.14 Consider perspective sets in $\mathbb{P}(\mathbb{D}^2)$. Let α be a linear transformation of $\mathbb{D}^2$. If α is a perspectivity with center $(a : 1)$ and axis $(b : 1)$, for $a, b \in \mathbb{D}$, then the matrix of α on the standard basis is

$$\begin{pmatrix} a\lambda + 1 & -a\lambda b \\ \lambda & 1 - \lambda b \end{pmatrix}$$

for some $\lambda \in \mathbb{D}$ distinct from $(b - a)^{-1}$.

Exercise 10.7.15 Suppose that $\mathbb{D}$ has characteristic two. We are concerned with a polarity π on the projective line $\mathbb{P} := \mathbb{P}(\mathbb{D}^2)$. Let $0, 1, \infty \in \mathbb{P}_{\pi}$ (cf. Remark 10.3.7). The map π induces an involutory anti-automorphism of $\mathbb{D}$ and $\mathbb{P}_{\pi}$ can be identified with the set $\{x \in \mathbb{D} \mid \pi(x) = x\} \cup \{\infty\}$. Show that $\mathbb{D}_{\sigma,-1} \setminus \{0\}$ is closed under inversion.
(*Hint:* For $t \in \mathbb{D}$ with $t + \pi(t) \neq 0$, write $u = (t + \pi(t))^{-1}t(t + \pi(t))^{-1}$, and show $(t + \pi(t))^{-1} = u + \pi(u)$.)

Exercise 10.7.16 Use the perspectivities occurring in the proof of Lemma 10.3.8 to establish that the group of perspectivities leaving invariant $\mathbb{D}_{\sigma,-1}$ is 2-transitive on $\mathbb{D}_{\sigma,-1}$.
(*Hint:* The linear transformation with matrix

$$\begin{pmatrix} 0 & 1 \\ -1 & 0 \end{pmatrix}$$

interchanges 0 and ∞, while the one with matrix

$$\begin{pmatrix} 1 & \lambda \\ 0 & 1 \end{pmatrix},$$

moves 0 to any $\lambda \in \mathbb{D}_{\sigma,-1} \setminus \{\infty\}$.)

Exercise 10.7.17 (This exercise is used in Example 10.3.9) Let $\mathbb{F}$ be a field of characteristic two and Q the set of squares in $\mathbb{F}$. Denote by f the symmetric bilinear form $(x, y) \mapsto x_1 y_2 + x_2 y_1$ on $\mathbb{F}^2$. Prove that $\{\infty\} \cup Q$ is a perspective subset of $\mathbb{P}(\mathbb{F}^2)_f$.

Exercise 10.7.18 Let π be an injective nondegenerate quasi-polarity of a thick projective space $\mathbb{P}$. Let X be a perspective subset of $\mathbb{P}_\pi$ with empty radical and with at least three points on every line of the induced line space on X (cf. Definition 2.5.8). Prove that the group generated by all perspectivities of $\mathbb{P}$ leaving X invariant is transitive on X.

(*Hint:* Suppose $x, y \in X$. If they are not collinear, choose a point $p \in xy \setminus \{x, y\}$ and argue with the perspectivity with center p and axis $\pi(p)$. The result now follows because the non-collinearity graph of X is connected (cf. Exercise 7.11.16).)

Exercise 10.7.19 Prove that the polar space of Exercise 8.5.8 does not embed in a finite-dimensional projective space.

Section 10.4

Exercise 10.7.20 Consider the quadric forms κ_1 and κ_2 on $\mathbb{R}^2$ given by $\kappa_1(x) = x_1^2 + x_2^2$ and $\kappa_2(x) = x_1^2 + \frac{1}{2} x_2^2$ for $x = \varepsilon_1 x_1 + \varepsilon_2 x_2 \in \mathbb{R}^2$, where ε_1, ε_2 is the standard basis of $\mathbb{R}^2$.

(a) Prove that both quadratic forms are anisotropic.
(a) Verify that κ_1 and κ_2 are not proportional to each other. Conclude that the requirement that $\mathbb{P}_\kappa$ be non-empty is necessary in Corollary 10.4.11.

Exercise 10.7.21 Let κ be a nondegenerate (σ, ε)-quadratic form on a vector space V of dimension at least three over a division ring $\mathbb{D}$ and let f be its hermitian form. Suppose that $a, b, c \in V$ are nonzero vectors spanning points of $\mathbb{P}(V)_\kappa$ such that $\langle a \rangle$ and $\langle b \rangle$ are distinct points not collinear with $\langle c \rangle$. Define the map $\alpha : V \to V$ by

$$\alpha(x) = x + c\big(f(a, c)^{-1} f(a, x) - f(b, c)^{-1} f(b, x)$$
$$+ f(b, c)^{-1} f(b, a) f(c, a)^{-1} f(c, x)\big)$$
$$- a f(c, a)^{-1} f(c, x) + b f(c, b)^{-1} f(c, x).$$

Prove the following properties of α.

(a) $\alpha \in \mathrm{U}(V, f) \cap \mathrm{O}(V, \kappa)$.
(b) $\alpha(\langle a \rangle) = \langle b \rangle$.
(c) $\alpha(c) = c$.
(d) If $f(x, c) = 0$, then $\langle x \rangle$, $\langle c \rangle$, $\langle \alpha(x) \rangle$ are collinear in $\mathbb{P}(V)_\kappa$.

Exercise 10.7.22   Suppose, in the setting of Exercise 10.7.21, that $\mathbb{P}(V)_\kappa$ is a polar space of rank one. Prove that this point set admits the structure of a Moufang set (cf. Remark 6.2.3) on which $O(V, \kappa)$ acts nontrivially.

Exercise 10.7.23   Let S be a nondegenerate polar space of rank $t \geq 3$ all of whose lines are thick. Prove the following two assertions.

(i) For each $m \in \mathbb{N}$ with $m \leq t$, the automorphism group of S is transitive on the collection of all octahedral sets of S of size $2m$.
(*Hint:* Use Proposition 8.5.9.)
(ii) If t is finite, then the stabilizer in $\mathrm{Aut}(S)$ of a frame O of S induces the symmetric group on every clique in O.

Exercise 10.7.24   Prove that the notion of an apartment for a polar space as in Definition 10.3.12(1) coincides with the notion of an apartment of the corresponding projective geometry of type A_3.

Section 10.5

Exercise 10.7.25   In the three cases of Definition 10.1.10, determine the subgroup of $\mathrm{GL}(V)_f$ leaving an ordered frame invariant.

Exercise 10.7.26   Let κ be a nonzero quadratic form on a vector space V over $\mathbb{F}$ whose hermitian form is nondegenerate. Define $\mathrm{GO}(V, \kappa) = \{g \in \mathrm{GL}(V) \mid \exists \lambda_g \in \mathbb{F} \ \forall x \in V \ \kappa(gx) = \lambda_g \kappa(x)\}$. Show that, if $\mathbb{P}(V)_\kappa$ is non-empty and $\dim(V) > 1$, the map $g \mapsto \lambda_g$ is a surjective homomorphism $\mathrm{GO}(V, \kappa) \to \mathbb{F}^*$ with kernel $O(V, \kappa)$.

Exercise 10.7.27   Prove the following two statements in the setting of the proof of Theorem 10.5.6.

(a) If the symplectic case prevails (that is, $(\sigma, \varepsilon) = (\mathrm{id}, -1)$ and $\mathbb{D}$ is of characteristic distinct from two), then the linear transformation M of V is not a perspectivity with respect to δ_f.
(b) More generally, there is a perspectivity in N inducing the permutation $(p, O(p))$ on O if and only if there is $\mu \in \mathbb{D}$ such that $\varepsilon = \sigma(\mu)^{-1}\mu$.

Section 10.6

Exercise 10.7.28   Prove that the groups $\mathrm{SL}(2, q)$ and $\mathrm{Sp}(2, q)$ are isomorphic.
(*Hint:* The group $\mathrm{SL}(2, q)$ preserves the nondegenerate alternating form $\det(v_1 \ v_2)$, where $v_1, v_2 \in \mathbb{F}_q^2$ are column vectors forming the 2×2 matrix $(v_1 \ v_2)$ and det is the usual determinant of a square matrix.)

Table 10.5 Classical generalized quadrangles

Name	$Sp(4, q)$	$O(5, q)$	$O^+(4, q)$	$O^-(6, q)$	$U(4, q)$	$U(5, q)$
Order	(q, q)	(q, q)	$(q, 1)$	(q, q^2)	(q^2, q)	(q^2, q^3)

Exercise 10.7.29 Let q be an odd prime power. Prove that the groups $PSL(2, q)$ and $O(3, q)/Z(O(3, q))$ are isomorphic.

(*Hint:* Construct a homomorphism of groups $SL(\mathbb{F}_q^2) \to O(\mathbb{F}_q^3, f)$ for a nondegenerate symmetric bilinear form f on $\mathbb{F}_q^3$, as in Exercise 4.9.12(a)–(d). Verify that $Z(SL(\mathbb{F}_q^2))$ is the kernel of the composition of this homomorphism with the natural quotient map $O(3, q) \to O(3, q)/Z(O(3, q))$, and compare orders of domain and codomain.)

Exercise 10.7.30 Consider the group W of Exercise 4.9.30.

(a) Verify that the matrices of the representation of W on $\mathbb{C}^3$ with respect to the standard basis can be written over the ring $R := \mathbb{Z}[\sigma, 1/2]$.
(b) Show that taking the matrix entries modulo 7 gives a group homomorphism $W \to O(\mathbb{F}_7^3, f)$, where f is the bilinear form obtained from the restriction of the standard hermitian inner product on $\mathbb{C}^3$ to R^3 by reduction modulo 7.
(c) Establish that the map $W \to O(\mathbb{F}_7^3, f)$ is an isomorphism.
(d) Conclude that $W \cong C_2 \times PSL(2, 7)$.
   (*Hint:* Use Exercise 10.7.29 and the group homomorphisms det on W and on $O(\mathbb{F}_7^3, f)$ to isolate the kernels as complements of $\langle -\,\mathrm{id}\rangle$.)

Exercise 10.7.31 Let $\mathbb{F}$ be a finite field of order q and $V = \mathbb{F}^n$ for some natural number n. Consider the quadratic form

$$\kappa(x) = x_1^2 + \cdots + x_n^2.$$

In which form of Theorem 10.6.9 can κ be written?
(*Hint:* The answer depends on the parity of n and q.)

Exercise 10.7.32 Show that the finite classical generalized quadrangles are as indicated in Table 10.5, as well as the corresponding point and line orders. Deduce from this information that the dual line space of the generalized quadrangle associated with $U(5, q)$ is not embeddable.

Exercise 10.7.33 Let κ be a nondegenerate quadratic form on a finite-dimensional vector space V over $\mathbb{F}$ and let f be its hermitian form. For $a \in V$ with $\kappa(a) \neq 0$, define $\phi : V \to \mathbb{D}$ by $\phi(x) = \kappa(a)^{-1} f(a, x)$. We consider $s_a := r_{a,\phi}$ of Exercise 1.9.30. The transformation s_a is a reflection and induces a perspectivity on $\mathbb{P}(V)_\kappa$ as introduced in Theorem 9.3.3. By Exercise 10.7.13(a), it belongs to $O(V, \kappa)$. Prove the following properties of s_a.

(a) s_a and s_b commute if $b \in V$ with $\kappa(b) \neq 0$ and $f(a, b) = 0$.
(b) The group $O(V, \kappa)$ is generated by the set of all s_a.

(c) Suppose $\dim(V) = 2m$ and κ has Witt index m. Choose a maximal singular subspace U of V, so $\dim(U) = m$. Take $a \in V$ with $\kappa(a) \neq 0$ and $a^{\perp} \cap U \neq U$. Show that $\dim(U \cap s_a(U)) = m - 1$. Conclude that s_a induces a duality of the oriflamme geometry of $\mathbb{P}_\kappa$ that is not an isomorphism (cf. Definition 1.3.1).

10.8 Notes

In our completion of the classification of nondegenerate polar spaces whose singular planes are Desarguesian and of nondegenerate generalized quadrangles embedded in a projective space, we employed the ideas on perspective sets written down in [42]. The remainder of the chapter covers material on groups and polar geometries that is fairly general knowledge.

Section 10.1

In [143], the set $\mathbb{D}^{\sigma, -\varepsilon}$ of Notation 10.1.1 is called the set of symmetric elements of $\mathbb{D}$ and $\mathbb{D}_{\sigma, -\varepsilon}$ its subgroup of traces.

The anti-automorphisms used in this section are usually referred to as 'involutions of the second kind'. See [195] for more on involutory anti-automorphisms.

Section 10.2

For general fields, there are many more different forms. The determination of the related absolute spaces and their automorphism groups [286] and the normal structure of these groups [3, 120, 143, 175, 249] have been studied quite intensively, and still present some open problems.

Section 10.3

The case in which secants (that is, projective lines that are not lines of Z but meet Z in at least two points) meet Z in precisely two points, is dealt with by the following result: If Z is a nondegenerate generalized quadrangle embedded in a projective space, and every secant meets Z in precisely two points, then Z is a quadric. This is proved in [40].

Section 10.4

Apartments within the context of buildings appear in Definition 11.2.1.

Section 10.5

In [286] it is shown that each nondegenerate polar space of rank at least three with thick lines has enough automorphisms for the corresponding building (cf. Proposition 11.1.9) to be Moufang.

Good references for the classical groups are [7, 62, 120, 272, 313]. These groups have Tits systems, defined in Chap. 11, which leads to an elegant proof that they are quasi-simple; see [29].

Often, the full group of automorphisms of the polar space is somewhat larger than the classical groups described: diagonal automorphisms (corresponding to linear transformations having determinant different from 1) and field automorphisms (corresponding to Galois automorphisms of the underlying field $\mathbb{D}$) are usually the only extra ones needed to generate the full group.

Section 10.6

The classification of quadratic forms over infinite fields is quite delicate. See [238] for a good introduction, but also [164], where the question whether the quadric defined by the form is empty or not receives special attention.

Substructures such as ovoids and spreads of polar spaces have been the study of intense investigations and yet several outstanding problems remain; see [277].

Chapter 11
Buildings

The most important geometries of this book are of Coxeter type (cf. Definition 2.4.2). Their 'building blocks', that is, their rank two residues, are the generalized polygons (cf. Definition 2.2.7), which are precisely the rank two geometries of Coxeter type. In Chap. 4, we studied thin chamber systems of Coxeter type M and found that these are quotients of the very nice and regular universal chamber system $C(M)$ for Coxeter systems of type M. In this chapter, we study special chamber systems of Coxeter type, called buildings, in which $C(M)$ frequently occurs as a subsystem, which is called apartment.

There are many possible definitions of buildings. In Sect. 11.1 we take a topological approach: closed galleries cannot be of types that are minimal expressions of nontrivial Coxeter group elements. In Corollary 11.2.6, we give the more classical definition in terms of apartments (but still in terms of chamber systems rather than geometries).

In Sect. 11.2 we explore several useful properties of buildings, such as, in Corollary 11.2.12, their residual connectedness. By Theorem 3.4.6, this implies that buildings correspond to certain residually connected geometries over a Coxeter diagram. Accordingly, the term building will also be applied to geometries. In Proposition 11.1.9, we find that all of the projective geometries (Coxeter type A_n), studied in Chaps. 5 and 6, and all of the polar geometries (Coxeter type B_n), studied in Chaps. 7–10, are buildings. The geometries of type $\widetilde{A}_{n-1}$ of Theorem 2.7.14 are also buildings (cf. Exercise 11.8.2). We will be mainly concerned with spherical buildings, which means that these have a spherical Coxeter type (cf. Definition 4.6.8).

In Sect. 11.3, we study groups acting highly transitively (the technical term is 'strongly transitively') on buildings and arrive at the famous Tits systems in groups. In Sect. 11.4, we focus on shadow spaces of buildings. We derive various properties of these line spaces that have been used to characterize some of these spaces as shadow spaces of buildings. Often, these spaces have a family of convex subspaces isomorphic to polar spaces. Such spaces are called parapolar spaces and are studied in Sect. 11.5. In Sect. 11.6, we pay special attention to a particular kind of shadow spaces of buildings, called root shadow spaces, and show that these are root filtration spaces (cf. Definition 6.7.2). Root shadow spaces exist for each Weyl type, that is,

F. Buekenhout, A.M. Cohen, *Diagram Geometry*,
Ergebnisse der Mathematik und ihrer Grenzgebiete. 3. Folge / A Series of
Modern Surveys in Mathematics 57, DOI 10.1007/978-3-642-34453-4_11,
© Springer-Verlag Berlin Heidelberg 2013

a spherical irreducible Coxeter type not isomorphic to one of H_3, H_4, $I_2^{(m)}$ ($m \in \{5, 7, 8, 9, \ldots\}$). We are motivated by the fact (not proved in this book) that each finite simple group of Lie type and rank at least three acts faithfully on a building of Weyl type.

Finally, in Sect. 11.7 we explore to what extent line spaces can be recognized as shadow spaces of buildings by means of properties found in previous sections. The proofs in this section are sketched rather than given in full.

Throughout this chapter, (W, S) is a Coxeter system of type $M = (m_{s,r})_{s,r \in S}$ and S will serve as the set of types of the chamber systems to be considered. We put $n = |S|$ and assume $n < \infty$. For notational convenience, we will often identify S and $[n]$.

By $\mathcal{C}$ we denote a chamber system of type M (cf. Definition 3.5.1) and by Γ its incidence system $\Gamma(\mathcal{C})$, as in Definition 3.3.1. We recall from Definition 4.2.7 that $\mathcal{C}$ is non-empty and connected and that, for distinct $i, j \in S$, each $\{i, j\}$-cell of $\mathcal{C}$ is isomorphic to the chamber system of a generalized $m_{i,j}$-gon.

11.1 Building Axioms

In Definition 11.1.4 of this section, buildings will be introduced as chamber systems. The defining property is phrased by means of galleries. After verifications that the most important geometries of Coxeter type that we have seen are buildings (Propositions 11.1.9 and 11.1.10), we provide an alternative building definition in Theorem 11.1.13.

We will use the Coxeter system (W, S) of type M to collect information about chamber systems $\mathcal{C}$ of type M. Recall from Definition 4.2.12 the notion of a minimal expression and from Remark 4.2.15 the free monoid S^* on the alphabet S, the homomorphism of monoids $\zeta : S^* \to W$ with

$$\zeta(r_1 \cdots r_q) = r_1 \cdots r_q \quad (r_1, \ldots, r_q \in S),$$

and the length function l on W with respect to S. For $w \in W$, an expression for w is an element of the pre-image $\zeta^{-1}(w)$; it is minimal if and only if its length is $l(w)$.

In Definition 3.2.2, a gallery of $\mathcal{C}$ was introduced as a path in the graph of the chamber system $\mathcal{C}$ with adjacency given by $\sim = \bigcup_{r \in S} \sim_r$. In Exercise 3.7.4, the notion of a simple gallery appeared: a gallery without repeated chambers.

Definition 11.1.1 Suppose that γ is a simple gallery $c_0, c_1, c_2, \ldots, c_q$ in $\mathcal{C}$. We say that $r = r_1 r_2 \cdots r_q \in S^*$ is the *type* of γ if, for each $j \in [q]$, the chamber c_j is r_j-adjacent to c_{j-1}.

For subsets $J_1, \ldots, J_q$ of S and chambers c, d of $\mathcal{C}$, there is a simple gallery from c to d whose type lies in $J_1^* \cdots J_q^*$ (viewed as a set of words in the free monoid S^*) if and only if $d \in c J_1^* \cdots J_q^*$ (cf. Notation 3.4.8).

Remark 11.1.2 Each pair (c, d) of distinct chambers in $\mathcal{C}$ with $c \sim d$ has a unique adjacency type; for, $c \sim_r d$ and $c \sim_s d$ would imply that c and d belong to the chamber system of a generalized $m_{r,s}$-gon, which, according to Example 3.1.16, leads to $c = d$, a contradiction. Consequently, the type of a simple gallery is uniquely determined by the gallery.

Recall that a minimal gallery (cf. Definitions 1.6.1 and 3.2.2) in $\mathcal{C}$ is a geodesic path in the graph of $\mathcal{C}$. We will relate minimal galleries to elements of the Coxeter group W. The type r of a gallery is an element of S^*; it will be called minimal if it is a minimal expression for $\zeta(r)$.

Lemma 11.1.3 *For any two chambers c, d of $\mathcal{C}$, the following assertions hold.*

(i) *The type of a minimal gallery in $\mathcal{C}$ from c to d is a minimal expression in S^*.*
(ii) *If $r \in S^*$ is a minimal expression and is the type of a simple gallery from c to d, then, for each $r' \in S^*$ with $\zeta(r') = \zeta(r)$, there is a simple gallery from c to d of type r'.*

Proof (i) Suppose that γ is a simple gallery $c = c_0, c_1, \ldots, c_q = d$ from c to d in $\mathcal{C}$ of type r. If $r = tst \cdots$ ($m_{t,s}$ factors) for certain $t, s \in S$, then, by the axiom for chamber systems of type M (cf. Exercise 3.7.4), there is another simple gallery from c to d of type $sts \cdots$ ($m_{t,s}$ factors). Now, applying this observation to subgalleries of γ as well, we obtain from Corollary 4.5.11 that, if r is not minimal, we may assume, without loss of generality, that γ contains a simple subgallery $c_{i-1} \sim_u c_i \sim_u c_{i+1}$ for some $u \in S$ and some $i \in [q-1]$. But then $c_{i-1} \sim_u c_{i+1}$, so $c = c_0, c_1, \ldots, c_{i-1}, c_{i+1}, \ldots, c_q = d$ is a gallery from c to d which is strictly shorter than γ. Hence (i).

(ii) In view of Corollary 4.5.11, and induction on the length of r', it suffices to show that, if the word $r_1 r_2 \cdots r_q$ in S^* is the type of a simple gallery from c to d, then, for each $i \in [q-1]$ and each $s \in S$, there is a simple gallery from c to d of type $r_1 r_2 \cdots r_i s s r_{i+1} \cdots r_q$. But this is immediate from the definition of chamber system. $\qquad\square$

See Exercise 11.8.1 for an alternative proof of Lemma 11.1.3, which uses the monoid homomorphism of Lemma 4.5.9.

Lemma 11.1.3 shows that, given two chambers c and d of $\mathcal{C}$, there is an element w of W such that, for each word r in $\zeta^{-1}(w)$, we can find a simple gallery from c to d of type r. For thin chamber systems, the universal object $\mathcal{C}(M)$ differs from its quotients in that the element w is uniquely determined by c and d. The following definition is inspired by this observation.

Definition 11.1.4 The chamber system $\mathcal{C}$ of type M is called a *building of type M* if every simple closed gallery of $\mathcal{C}$ of minimal type is trivial (i.e., consists of a single chamber).

If M is spherical, then the building is said to be *spherical*.

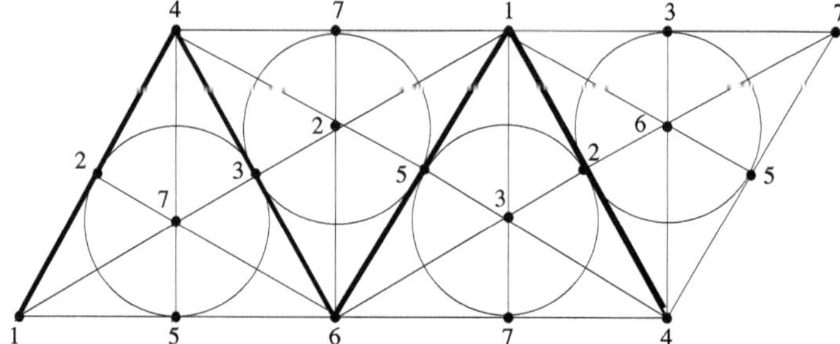

Fig. 11.1 The closed gallery of type 123123123 in the Neumaier geometry. Four Fano planes are drawn. Call the first three, in order of appearance from left to right, π, π', and π''. The fourth is equal to π again. The gallery is indicated by the *thick lines*. More precisely, the gallery is $(1, 124, \pi)$, $(4, 124, \pi)$, $(4, 346, \pi)$, $(4, 346, \pi')$, $(6, 346, \pi')$, $(6, 156, \pi')$, $(6, 156, \pi'')$, $(1, 156, \pi'')$, $(1, 124, \pi'')$, $(1, 124, \pi)$

Example 11.1.5 (i) A thin chamber system $\mathcal{C}(M)/A$ of Coxeter type M (cf. Definition 4.2.5) is a building if and only if $A = 1$.

(ii) The chamber system of the Neumaier geometry of Example 2.4.11 is residually connected of type B_3 but is not a building. This follows from the existence of a closed simple gallery with type 123123123 as indicated in Fig. 11.1.

(iii) By definition, a chamber system over [2] of type M is the chamber system of a generalized m-gon, where $m = m_{1,2}$. The condition that there are no nontrivial closed simple galleries of minimal types is satisfied in each such chamber system, so a building of rank two is the chamber system of a generalized polygon (and conversely).

Recall the notion of a J-cell from Definition 3.2.2 and the notion of the direct sum of chamber systems from Definition 3.5.3. The direct sum of two diagrams was introduced in Exercise 3.7.11.

Proposition 11.1.6 *Each building $\mathcal{C}$ of type M satisfies the following properties.*

(i) *For every pair (r, r') of minimal expressions in S^*, the existence of two simple galleries in $\mathcal{C}$ with common origin and common end point, and types r and r', respectively, implies $\zeta(r) = \zeta(r')$.*

(ii) *For each subset J of S and each chamber c of $\mathcal{C}$, the J-cell cJ^* is a building of type $M|_J$.*

(iii) *If $\mathcal{C}'$ is a building of type M', where M' is a Coxeter matrix over an index set S' disjoint from S, then the direct sum $\mathcal{C} \oplus \mathcal{C}'$ is a building of type $M \oplus M'$ over $S \cup S'$.*

Proof (i) Denote the two simple galleries of minimal types r, r' by γ, γ', respectively, and let c, d be their common origin and end point, respectively. We proceed by induction on $l(r)$. If $l(r) = 0$, then $c = d$ (and $\zeta(r) = 1$) so $\zeta(r') = 1$ by Definition 11.1.4, whence $\zeta(r) = \zeta(r')$ as required. Assume $l(r) > 0$.

Thus, $r = sr''$ with $s \in S$ and $r'' \in S^*$, so $l(s\zeta(r'')) > l(r'')$. Let e be the chamber of γ following c and denote by γ'' the tail of γ starting at e. Clearly, γ'' is a simple gallery from e to d of minimal type r'' and of length $l(r) - 1$. If $l(s\zeta(r')) > l(r')$, then e, γ' is a simple gallery from e to d of minimal type sr', so by the induction hypothesis applied to γ'' and e, γ', we have $\zeta(r'') = s\zeta(r')$, leading to $\zeta(r) = \zeta(r')$, as required.

Suppose, therefore, that $l(s\zeta(r')) < l(r')$. Then, in view of Theorem 4.5.8 and Lemma 11.1.3(ii), changing γ' if necessary, we (may) assume that the first element of r' is s. Thus, $r' = sr'''$ where r''' is minimal. Denote by e' the chamber of γ' following c and by γ''' the tail of γ' with origin e'. If $e \neq e'$, then e, γ''' is a simple gallery from e to d of minimal type $sr''' = r'$, so by the induction hypothesis $\zeta(r'') = s\zeta(r''') = \zeta(r')$. But then $l(r) = l(s\zeta(r'')) = l(s\zeta(r')) < l(r') = l(r'')$, which contradicts that r'' is part of r. Hence, $e = e'$ and, again by the induction hypothesis, $\zeta(r'') = \zeta(r''')$, so that $\zeta(r) = \zeta(sr'') = \zeta(sr''') = \zeta(r')$. This establishes (i).

(ii) and (iii) are immediate from the definition of a chamber system. $\square$

Remark 11.1.7 Let c and d be chambers of a building C of type M. The converse of Lemma 11.1.3(i) holds in the sense that every simple gallery of minimal type from c to d is minimal. For, if γ is a minimal gallery of type r and γ' is a gallery of minimal type r', then according to Lemma 11.1.3(i) the type r is minimal and so Proposition 11.1.6(i) gives $\zeta(r) = \zeta(r')$. By minimality of r and r', this implies that r and r' have the same length. Therefore, γ and γ' have the same length, proving that γ is minimal.

The proposition below shows examples of buildings coming from projective and polar geometries (of types A_n and B_n, respectively).

Lemma 11.1.8 *Suppose that C is a residually connected chamber system of type M over S. There is no closed simple gallery in C whose type contains a given element of S exactly once.*

Proof Suppose that γ is a closed simple gallery in C of type $r = r'sr'' \in S^*$ for some $s \in S$ and $r', r'' \in (S \setminus \{s\})^*$ and with origin c. Let d, e be the s-adjacent chambers of γ (occurring in this order). Then $d, e \in c(S \setminus \{s\})^*$ so $d \in e(S \setminus \{s\})^* \cap es^*$. Since C is residually connected, Lemma 3.4.9 forces $d = e$, which contradicts the simplicity of γ. $\square$

The projective geometries introduced in Example 1.4.9 and the polar geometries introduced in Definition 7.6.3 are buildings. By Proposition 2.4.7 and Definition 7.6.3, these are $[n]$-geometries of type A_n and B_n, respectively.

Proposition 11.1.9 *The following geometries are buildings of type M.*

(i) *Each $[n]$-geometry of type $M = A_n$.*

(ii) *Each polar geometry of rank n, which is of type $M = B_n$.*

Proof Let C be the chamber system of the geometry. In view of the Chamber System Correspondence Theorem 3.4.6, Definition 1.6.7 for Case (i), and Theorem 7.5.8 and Corollary 7.6.8 for Case (ii), the chamber system C is residually connected of type $M = A_n$, B_n, respectively.

If $n \leq 1$, there is nothing to show and if $n = 2$, the polar geometry is a generalized quadrangle, so the result follows from the definition, cf. Example 11.1.5(iii). We proceed by induction on n and assume $n \geq 3$. Let c be a chamber of C and let γ be a simple gallery in C of minimal type $r \in S^*$ starting and ending at c. We will show $\gamma = c$. Assume that γ is nontrivial. By residual connectedness and induction on n (cf. Lemma 7.6.5 for $M = B_n$), we may assume $r_1 \in S_{\zeta(r)}$. We will reach a contradiction in each case.

(i) By Example 4.5.12, we can choose $w_1, w_2 \in \langle S \setminus \{r_1\}\rangle$ in such a way that $l(r) = l(w_1) + 1 + l(w_2)$, so that, by Lemma 11.1.3, we may assume, without loss of generality, that γ is of type $r'r_1r''$ with r', $r'' \in (S \setminus \{r_1\})^*$. This contradicts Lemma 11.1.8.

(ii) If r_1 occurs only once in r, the argument goes as for (i). By Exercise 4.9.20, it remains to consider the case where r_1 occurs twice in r, i.e.,

$$r = r'r_1r''r_1r'''.$$

By moving from c to a neighbor and using reduction, we may assume that r''' is the empty word, so $r = r'r_1r''r_1$. Consider the chambers c', c'', c''' on γ occurring at the end of the subgallery of γ starting at c and of type r', $r'r_1$, and $r'r_1r''$, respectively. If x and m denote the point and line of c, then c' has the point x in common with c, while c''' has the line m in common with c. Let m' be the line of c' and x' be the point of c'''. Then x' and m' must be the point and line of c''. Thus, m' is a line on both x and x', but so is m. But $x \neq x'$ by minimality of the type, so Condition (2) of Definition 7.6.3 implies $m = m'$. By residual connectedness, Lemma 3.4.9(iv) applies with $L = \emptyset$, $J = S \setminus \{r_1\}$ and $K = S \setminus \{r_2\}$, giving $c''' \in c''(S \setminus \{r_1\})^* \cap c''(S \setminus \{r_1\})^* = c''(S \setminus \{r_1, r_2\})^*$. Therefore, we (may) assume that r_1 and r_2 do not occur in r''. But then $\zeta(r''r_1) = \zeta(r_1r'')$, and so, $\zeta(r) = \zeta(r'r_1r_1r'')$, a contradiction with the minimality of r. $\qquad\square$

The oriflamme geometry was introduced in Definition 7.8.5. For each integer $n \geq 4$, the oriflamme geometry $\Delta(X)$ of a polar space X of rank n whose n-order (of the corresponding polar geometry) is equal to 1, is a residually connected geometry of type D_n (by Theorem 7.8.6), and so also represents a residually connected chamber system of this type (cf. the Chamber System Correspondence Theorem 3.4.6).

Proposition 11.1.10 *For $n \geq 3$, each residually connected chamber system of type D_n is a building.*

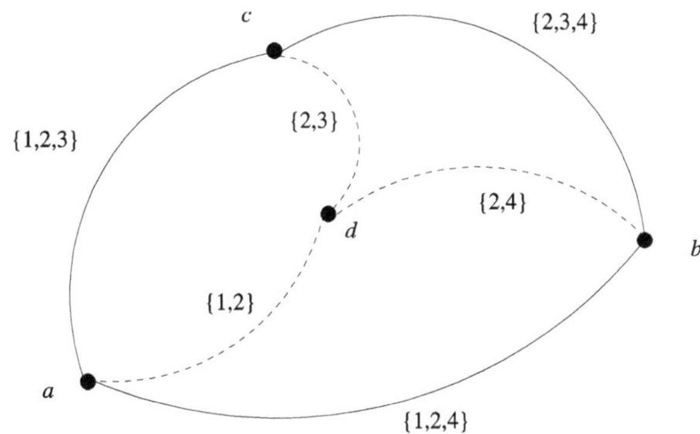

Fig. 11.2 This picture illustrates the proof of Proposition 11.1.10 for $n = 4$. A closed simple gallery of type in $\{1, 2, 3\}^*\{1, 2, 4\}^*\{2, 3, 4\}^*$ through the chambers a, b, c in a building of type D_4

Proof Let $\mathcal{C}$ be a residually connected chamber system of type D_n. We proceed by induction on n, the case $n = 3$ being clear from Proposition 11.1.9 (as $D_3 \cong A_3$). We verify Definition 11.1.4. Suppose that there is a simple closed gallery γ in $\mathcal{C}$ with end point c whose type is a minimal expression. By consideration of the longest element of the Coxeter group of type D_n (see Table 4.5) and the fact that each element of W is less than or equal to it in the Bruhat order (cf. Exercise 11.8.6), we know that the type of γ is in $J^*K^*L^*$, where $J = \{1, \ldots, n - 1\}$, $K = \{1, \ldots, n - 2, n\}$, $L = \{2, \ldots, n\}$.

The chamber c lies at the crossover of γ from the L-part to the J-part. Let a be the crossover chamber of γ from the J-part to the K-part, and let b be the crossover chamber of γ from the K-part to the L-part. Since $a \in aJ^* \cap bK^*$, $b \in bK^* \cap cL^*$, and $c \in aJ^* \cap cL^*$, residual connectedness of $\mathcal{C}$ implies that there is a chamber d of $\mathcal{C}$ such that $aJ^* \cap bK^* \cap cL^* = d(J \cap K \cap L)^*$; see Lemma 3.4.9. The chambers and galleries are depicted in Fig. 11.2 for $n = 4$. Also, by residual connectedness of $\mathcal{C}$, minimal galleries from a to d have type in $(J \cap K)^*$, those from b to d have type in $(K \cap L)^*$, those from c to d have type in $(J \cap L)^*$. All J-cells and all K-cells are residually connected chamber systems of type A_{n-1} and hence, by Proposition 11.1.9, buildings. Each L-cell is a building of type D_{n-1} by the induction hypothesis, because it is a residually connected chamber system of type D_{n-1}. According to Proposition 11.1.6(i), this means that all of the minimal galleries from a to b have the same image under ζ. Since there is a simple gallery through d, its type must be in $(J \cap K)^*(K \cap L)^*$. Similarly, the type of a minimal gallery from b to c must be in $(K \cap L)^*(J \cap L)^*$, and the type of a minimal gallery from c to a must be in $(J \cap L)^*(J \cap K)^*$. Consequently, the type of γ lies in $(J \cap L)^*(J \cap K)^*(K \cap L)^*(J \cap L)^*$. Of the three intersections, only $(J \cap K)^*$ contains the node 1. Since the restriction of D_n to $(J \cap K) \times (J \cap K)$ is of type A_{n-2}, the type of this part of γ can be written with at most one occurrence of 1 (this follows from Example 4.5.12 but can also be seen, for instance, by noticing that n occurs only once in the longest element of $W(A_n)$ given in Table 4.5, and using the nontrivial diagram automorphism). But then, by Lemma 11.1.8, there is no occurrence of 1, and so γ lies in the L-cell cL^*, a building of type D_{n-1}, and we finish by invoking the induction hypothesis. $\square$

Proposition 11.1.11 *Each residually connected chamber system of type* E_6 *is a building.*

Proof We verify Definition 11.1.4. For the longest element w_{E_6} of the Coxeter group of type E_6, we have

$$
\begin{aligned}
w_{E_6} &= 12314231435423143542(654231435426)5431 \\
&= 12314231435423143542(654234561342)5431 \\
&\in \{1, 2, 3, 4, 5\}^* 65423456\{1, 2, 3, 4, 5\}^* \\
&\subseteq \{1, 2, 3, 4, 5\}^* \{2, 3, 4, 5, 6\}^* \{1, 2, 3, 4, 5\}^*.
\end{aligned}
$$

Hence, if γ is a closed simple gallery of minimal type, then it contains chambers c, d such that one part of γ is a minimal gallery from c to d of type in $[5]^*$ and the other is a minimal gallery from d to c of type in $\{2, 3, 4, 5, 6\}^*$. In particular, $d \in c\{1, 2, 3, 4, 5\}^* \cap c\{2, 3, 4, 5, 6\}^*$, which, by Lemma 3.4.9, implies $d \in c\{2, 3, 4, 5\}^*$. The [5]-cell containing c and d is a building by Proposition 11.1.10, and so the type of the first part of γ lies in $\{2, 3, 4, 5\}^*$. The $\{2, 3, 4, 5, 6\}$-cell containing c and d is a building by Proposition 11.1.9, and so the type of the second part of γ lies in $\{2, 3, 4, 5\}^*$. We have argued that the whole closed gallery γ lies in a cell of type D_4. But then, again by Proposition 11.1.10, it must be trivial. Hence $\mathcal{C}$ is a building. $\square$

Theorem 11.1.13 below shows that the axiom of buildings can be weakened. As a consequence, we will obtain some fundamental properties of buildings. We need the following converse of Proposition 11.1.6(i).

Lemma 11.1.12 *For every chamber system $\mathcal{C}$ of type M, the following two properties hold.*

(i) *Let c be a chamber of $\mathcal{C}$. Suppose that, for every pair (r, r') of minimal expressions in S^*, the existence of two simple galleries with origin c and common end point, and types r and r', respectively, implies $\zeta(r) = \zeta(r')$. All simple galleries starting at c having the same endpoint and the same minimal type, are the same.*
(ii) *If, for every pair (r, r') of minimal expressions in S^*, the existence of two simple galleries with common origin and common endpoint, and types r and r', respectively, implies $\zeta(r) = \zeta(r')$, then $\mathcal{C}$ is a building.*

Proof (i) Let δ, δ' be simple galleries in $\mathcal{C}$, from the chamber c to a chamber, e say, of minimal type r. We prove by induction on $l(r)$ that δ and δ' coincide. The case $l(r) = 0$ being trivial, assume $l(r) > 0$. In particular, there are chambers h, h' of $\mathcal{C}$ which are s-adjacent to e for some $s \in S$ such that δ has tail h, e and δ' has tail h', e. Let δ'', δ''' be the head part of δ, δ' ending at h, h', respectively. These galleries have minimal type r' such that $r = r's$. If $h \neq h'$, then δ'' and δ''', h are simple galleries starting at c and ending at h of minimal types r' and r, respectively. According to the hypothesis, this implies $\zeta(r') = \zeta(r)$. This is absurd as $\zeta(r) = \zeta(r')s$. Therefore, $h = h'$, and we can finish by invoking the induction hypothesis.

(ii) Let γ be a closed simple gallery containing a chamber, say c, of minimal type r. We need to show $\gamma = c$. But this follows immediately by applying the hypothesis to the galleries γ and c. $\qquad\square$

The following characterization uses a property that holds for buildings according to Proposition 11.1.6(i). It is remarkable in that the fact that the image in W under ζ of the type of γ is independent of the choice of gallery γ of minimal type with given origin and given endpoint, is only required for a single origin.

Theorem 11.1.13 *Let C be a chamber system of type M. Suppose that C has a chamber c such that, for every pair r, r', of minimal expressions in S^*, the existence of simple galleries with common origin c and a common endpoint, and types r, r', respectively, implies $\zeta(r) = \zeta(r')$. Then C is a building.*

Proof Let d be a chamber of C, and let γ, γ' be two distinct simple galleries starting at d, with the same endpoint, and with respective minimal types r and r'. We first show that $\zeta(r) = \zeta(r')$. By induction on the length of a gallery from c to d (and connectedness of C) it suffices to establish $\zeta(r) = \zeta(r')$ in the case where $c \sim d$. To this end, suppose $c \in ds^*$ for some $s \in S$. The galleries c, γ and c, γ' have types sr, and sr', respectively. If these types are both minimal, then $\zeta(sr) = \zeta(sr')$ by the hypothesis, whence $s\zeta(r) = s\zeta(r')$, and $\zeta(r) = \zeta(r')$, as required.

Assume that neither sr nor sr' is a minimal expression. Then, by Theorem 4.5.8, there are minimal expressions sr'', sr''' in S^* such that $\zeta(r) = \zeta(sr'')$ and $\zeta(r') = \zeta(sr''')$. According to Lemma 11.1.3(ii), there are minimal galleries with the same origin and endpoint as γ (and γ') of types sr'' and sr''', respectively. Let the first chamber following d on these galleries be e, e', respectively. Denote the tail of these galleries from e, e' respectively (to the end), by γ'', γ''', respectively.

If $e' = c \neq e$, then the simple galleries c, γ'', and γ''' both start at c and have types sr'' and r''', respectively; both are minimal and have the same extremities, so $\zeta(sr'') = \zeta(r''')$. Hence, $l(r'') + 1 = l(sr'') = l(r''')$ in view of minimality. But also $\zeta(r'') = \zeta(sr''')$, whence $l(r'') = l(sr''') = l(r''') + 1$, a contradiction. Similarly, we can rule out $e = c \neq e'$. If $c \neq e, e'$, then the galleries c, γ'' and c, γ''' are simple of type sr'' and sr''', respectively, starting at c with the same endpoint, and if $c = e = e'$, then γ'' and γ''' are simple galleries of type r'' and r''', respectively, starting at c with the same endpoint. In both cases the desired conclusion follows from the hypotheses.

Finally, assume that sr' is a minimal expression, but sr is not (the other remaining case being the same up to a change of roles for r and r'). Again, replace r by sr'' such that $\zeta(r) = \zeta(sr'')$, denote by e the first member following d on a simple gallery with same extremities as γ and type sr'', and by γ'' the tail end of this gallery starting at e. If $c = e$, then c, γ' and γ'' are simple galleries starting at c with the same endpoints, of types sr' and r'', respectively. Thus, by the hypothesis, $\zeta(sr') = \zeta(r'')$, whence $\zeta(r') = \zeta(sr'') = \zeta(r)$, as required. Therefore, we may restrict attention to the case where $c \neq e$. Consideration of the galleries c, γ' and

c, γ'' leads to $\zeta(sr') = \zeta(sr'')$, whence $\zeta(r') = \zeta(r'')$. Upon replacing γ' by a suitable gallery (cf. Lemma 11.1.3(ii)), we may assume $r'' = r'$. Now the galleries c, γ' and c, γ'' start at c and have the same endpoint and the same type sr'. Since $d \neq e$ (due to the simplicity of d, γ''), we have a contradiction with Lemma 11.1.12(i). $\square$

Recall from Example 11.1.5(i) that the chamber system $\mathcal{C}(M)$ is a building. The simple galleries from the chamber 1 to a chamber $w \in W$ correspond to the elements of $\zeta^{-1}(w)$ in S^*. In this light, the first statement below connects $\mathcal{C}$ with $\mathcal{C}(M)$.

Corollary 11.1.14 *For chambers c, d, e of the building $\mathcal{C}$ of type M, the following holds.*

  (i) *There is a unique $w \in W$, denoted by $w = \delta_{\mathcal{C}}(c, d)$, such that the type map on galleries induces a bijective correspondence*

$$\{\text{minimal galleries from } c \text{ to } d\} \to \{\text{minimal expressions for } w\}.$$

  (ii) *If γ is a simple gallery from c to d of minimal type, then γ is minimal, and $\delta_{\mathcal{C}}(c, d)$ is the image under ζ of the type of γ.*

 (iii) *If $c \in ds^*$, where $s \in S$, then $\delta_{\mathcal{C}}(c, e) \in \langle s \rangle \delta_{\mathcal{C}}(d, e)$.*

 (iv) *The map $x \mapsto \delta_{\mathcal{C}}(d, x)$ is a homomorphism $\mathcal{C} \to \mathcal{C}(M)$ of chamber systems over S.*

  (v) $l(\delta_{\mathcal{C}}(c, d)\delta_{\mathcal{C}}(d, e)) \leq l(\delta_{\mathcal{C}}(c, e))$.

Proof (i) Let γ and γ' be minimal galleries in $\mathcal{C}$ from c to d. Then the types of γ and of γ' are minimal by Lemma 11.1.3(i), so, by Proposition 11.1.6(i), there is $w \in W$ such that these types belong to $\zeta^{-1}(w)$. If the types of γ and γ' are equal, then $\gamma = \gamma'$ by Lemma 11.1.12(i). Finally, the restriction to minimal galleries of the type is surjective onto the set of minimal expressions by Lemma 11.1.3(ii).

  (ii) This is immediate from (i) and Proposition 11.1.6(i).

  (iii) Let γ be a minimal gallery from d to e of type, say, r. Then $\zeta(r) = \delta_{\mathcal{C}}(d, e)$. Consider the gallery c, γ. If $c = d$, there is nothing to show, so we may assume that the gallery is simple. If sr is a minimal expression, then, by (ii), the gallery c, γ is minimal, and $\delta_{\mathcal{C}}(c, e) = s\zeta(r) \in \langle s \rangle \delta_{\mathcal{C}}(d, e)$. Otherwise, we may assume, without loss of generality (cf. Theorem 4.5.8 and Lemma 11.1.3), that $r = sr'$. Let d' be the first chamber of γ following d. Then $d' \in cs^*$, so either $d' = c$ and there is a simple gallery from c to e of type r', or $d' \neq c$ and there is a simple gallery from c (via d') to e of type sr'. Since both types are minimal, we have, again by (ii), that

$$\delta_{\mathcal{C}}(c, e) \in \{\zeta(r'), \zeta(sr')\} = \{\zeta(r), s\zeta(r)\} = \langle s \rangle \delta_{\mathcal{C}}(d, e)$$

in all cases. Hence (iii).

  (iv) Let x, x' be two s-adjacent chambers of $\mathcal{C}$, where $s \in S$. Then, by (iii), $\delta_{\mathcal{C}}(x, d) \in \langle s \rangle \delta_{\mathcal{C}}(x', d)$. Since, obviously, $\delta_{\mathcal{C}}(c, e) = \delta_{\mathcal{C}}(e, c)^{-1}$, it follows that $\delta_{\mathcal{C}}(d, x) \in \delta_{\mathcal{C}}(d, x')\langle s \rangle$, which is equivalent to saying that $\delta_{\mathcal{C}}(d, x)$ and $\delta_{\mathcal{C}}(d, x')$ are s-adjacent in $\mathcal{C}(M)$ (cf. Definition 4.2.5). This establishes (iv).

  (v) Since distances decrease under homomorphisms, it follows from (i) and (iv) that

$$d_{\mathcal{C}(M)}\bigl(\delta_{\mathcal{C}}(d,c),\delta_{\mathcal{C}}(d,e)\bigr)\le d_{\mathcal{C}}(c,e).$$

But the left hand side equals $l(\delta_{\mathcal{C}}(d,c)^{-1}\delta_{\mathcal{C}}(d,e))$, and, in view of (i), we have $\delta_{\mathcal{C}}(c,e)=l(d_{\mathcal{C}}(c,e))$. Hence the corollary. $\qquad\square$

Definition 11.1.15 The element $\delta_{\mathcal{C}}(c,d)\in W$ of Corollary 11.1.14(i) is called the *Weyl distance* between c and d in $\mathcal{C}$.

11.2 Properties of Buildings

The study of thin chamber systems of Coxeter type is of good use to buildings. Recall the notion of an isomorphism of chamber systems from Definition 3.1.7 and the notion of a chamber subsystem from Definition 3.1.13. Fix a Coxeter type M over $S=[n]$, and a Coxeter system (W,S) of this type. The thin chamber system $\mathcal{C}(M)$ is introduced in Definition 4.2.5. A characterization of buildings in terms of apartments appears in Corollary 11.2.6.

Definition 11.2.1 An *apartment* of a chamber system $\mathcal{C}$ of type M is a chamber subsystem of $\mathcal{C}$ that is isomorphic to $\mathcal{C}(M)$ as a chamber system over S.

Observe that if $\alpha:\mathcal{C}(M)\to\mathcal{C}$ is an injective homomorphism of chamber systems, then $d(\alpha(x),\alpha(y))=d(x,y)$. In order to find an apartment in $\mathcal{C}$ containing a given chamber c, we have to find a set A of chambers of $\mathcal{C}$ with the property that the restriction to A of the homomorphism $x\mapsto\delta_{\mathcal{C}}(c,x)$ is an isomorphism. In particular, for $d,e\in A$ we will have equality in Corollary 11.1.14(v).

Definition 11.2.2 Let $\mathcal{C}$ be a building of type M and let W be the Coxeter group of type M. For $X\subseteq W$, a map $\alpha:X\to\mathcal{C}$ is called a *strong isometry* if $\delta_{\mathcal{C}}(\alpha(x),\alpha(y))=x^{-1}y=\delta_{\mathcal{C}}(x,y)\in\mathcal{C}(M)$ for all $x,y\in X$.

The existence of apartments is an easy consequence of the following result.

Lemma 11.2.3 *Let $X\subseteq W$. If $\alpha:X\to\mathcal{C}$ is a strong isometry, then α can be extended to a strong isometry $\alpha:W\to\mathcal{C}$.*

Proof By Zorn's lemma, it suffices to show that if $X\subset W$ and $\alpha:X\to\mathcal{C}$ is a strong isometry, then there is $w\in W\setminus X$ such that α can be extended to a strong isometry on $X\cup\{w\}$. If $X=\emptyset$, we can take any $w\in W$, so assume $X\ne\emptyset$. By connectedness of $\mathcal{C}(M)$, there must be an s-adjacent pair of chambers for some $s\in S$, such that only one of the two is in X. Applying an automorphism of $\mathcal{C}(M)$ if necessary, we may assume this pair to be $\{1,s\}$, so that $1\in X$, and $s\notin X$.

If $l(sx)>l(x)$ for all $x\in X$, then let $\alpha(s)$ denote any chamber s-adjacent to $\alpha(1)$. Every minimal gallery γ from $\alpha(1)$ to $\alpha(x)$ extends to a minimal gallery $\alpha(s),\gamma$ from $\alpha(s)$ to $\alpha(x)$, so $\delta_{\mathcal{C}}(\alpha(s),\alpha(x))=sx$ for all $x\in X$. This shows that α extends to a strong isometry on $X\cup\{s\}$.

Assume $l(sy) < l(y)$ for some $y \in X$. Let $\alpha(s)$ be the second chamber of a minimal gallery from $\alpha(1)$ to $\alpha(y)$ whose type begins with s (such a gallery exists in view of Corollary 11.1.14(i)). Now $\delta_C(\alpha(s), \alpha(y)) = sy$ by construction of $\alpha(s)$. We have to show that $\delta_C(\alpha(s), \alpha(x)) = sx$ for all $x \in X$. Take $x \in X$. Since $\delta_C(\alpha(1), \alpha(x)) = x$ by the hypothesis on α, we obtain from Corollary 11.1.14(iii) that $\delta_C(\alpha(s), \alpha(x)) \in \langle s \rangle x$. Suppose that $\delta_C(\alpha(s), \alpha(x)) = x$. Clearly, $l(sx) < l(x)$. Moreover, by Corollary 11.1.14(v), $l(\delta_C(\alpha(x), \alpha(s))\delta_C(\alpha(s), \alpha(y))) \le l(\delta_C(\alpha(x), \alpha(y)))$, whence $l(x^{-1}sy) \le l(x^{-1}y)$, a contradiction with Corollary 4.5.22. Therefore, $\delta_C(\alpha(s), \alpha(x)) \in \langle s \rangle x \setminus \{x\} = \{sx\}$, and $\delta_C(\alpha(1), \alpha(x)) = x$ as required for α to be a strong isometry on $X \cup \{s\}$. This establishes the lemma. $\qquad\square$

The map $\alpha : W \to C$ of Lemma 11.2.3 can be viewed as an injective chamber system homomorphism $\alpha : C(M) \to C$.

Example 11.2.4 By Proposition 11.1.9, projective geometries of finite rank n are buildings of type A_n. The notion of an apartment in such a geometry appears already in Definition 6.5.3. Indeed, an apartment as defined there coincides with an apartment of a building of type A_n when we take into account the Chamber System Correspondence 3.4.6. Theorem 6.5.5 shows how apartments can be found in geometries of type A_n. Theorem 10.4.6 shows the analogue for polar geometries of type B_n. The next theorem shows that this pattern carries over to the general case.

Recall from Definition 7.9.13 that a set X of chambers of C is called convex if every minimal gallery between chambers in X is entirely contained in X.

Theorem 11.2.5 *Each building C satisfies the following properties.*

(i) *Every pair of chambers of C is contained in an apartment.*
(ii) *Every apartment of C is convex.*

Proof (i) follows from Lemma 11.2.3 by taking $X = \{1, \delta_C(c, d)\}$, $\alpha(1) = c$, and $\alpha(\delta_C(c, d)) = d$ for a pair (c, d) of chambers of C.

(ii) is a direct consequence of Corollary 11.1.14(i) as $C(M)$ contains a path from c to d for each minimal expression of $\delta_C(c, d)$. $\qquad\square$

The following corollary characterizes a building in terms of apartments.

Corollary 11.2.6 *Let C be a chamber system of type M. It is a building if and only if there is a collection Σ of injective homomorphisms $C(M) \to C$ and a collection of homomorphisms $\rho_c : C \to C(M)$, one for each chamber c of C, with the following two properties.*

(1) *For each $c, d \in C$ there is a member σ of Σ such that $c, d \in \sigma(C(M))$.*
(2) *For $\sigma \in \Sigma$ and $c \in \sigma(C(M))$, the composite map $\rho_c \circ \sigma$ is an automorphism of $C(M)$.*

Proof Suppose that $\mathcal{C}$ satisfies (1) and (2). The image $\sigma(\mathcal{C}(M))$ of each $\sigma \in \Sigma$ is an apartment of $\mathcal{C}$ as $\sigma^{-1} = (\rho_c\sigma)^{-1} \circ \rho_c|_A$ (well defined in view of (2)), being the composition of two homomorphisms, is a homomorphism.

Let γ and γ' be simple galleries in $\mathcal{C}$ of minimal types r and r', respectively, with origin c and end point d. By (1), there is $\sigma \in \Sigma$ such that $c, d \in \sigma(W)$. If both γ and γ' lie in A, then $\sigma^{-1}(\gamma)$ and $\sigma^{-1}(\gamma')$ are galleries in $\mathcal{C}(M)$ of types r and r', respectively, with same origin and end point, and so $\zeta(r) = \zeta(r')$. This shows that Theorem 11.1.13 applies, so $\mathcal{C}$ is a building.

Therefore, it suffices to prove that every simple gallery of minimal type with origin c and end point d lies in every apartment of $\mathcal{C}$ containing c and d that is an image of $\mathcal{C}(M)$ under a member of Σ. Let γ be a simple gallery of minimal type r from c to d and suppose $\sigma \in \Sigma$ is such that $A := \sigma(\mathcal{C}(M))$ contains both c and d. If γ has length at most one, then obviously γ lies in A. We proceed by induction on the length q of γ and assume $q \geq 2$. Denote by e and f the chambers on γ adjacent to c and d, respectively, and let $s, t \in S$ be such that $c \sim_s e$ and $f \sim_t d$. Thus, there is $u \in S^*$ of length $q - 2$ with $r = sut$.

If γ does not lie in A, then we (may) assume that e is not contained in A, for otherwise we may shorten γ by replacing d by its adjacent chamber in γ and invoke the induction hypothesis to conclude that γ is contained in A. Similarly, we may assume $f \notin A$.

Use (1) to find $\tau \in \Sigma$ with $c, f \in \tau(W)$. The part γ' of γ starting at c and ending at f has length less than q and so, by induction, it lies in $\tau(W)$.

For any two adjacent chambers x and y of $\tau(W)$, we have $\rho_c(x) \neq \rho_c(y)$. For, otherwise $\tau^{-1}(x) = \tau^{-1}(y)$ (as $\rho_c\tau$ is an automorphism of $\mathcal{C}(M)$ by (2)), which would imply the contradiction $x = y$. This implies that $\sigma\rho_c$ embeds γ' isometrically in A. After composing ρ_c with a suitable automorphism of $\mathrm{Aut}(\mathcal{C}(M))$, we (may) assume $\sigma(\rho_c(c)) = c$. This implies that $\sigma \circ \rho_c|_A$ is the identity on A, so $\sigma\rho_c(d) = d$. Put $e' = \sigma\rho_c(e)$ and $f' = \sigma\rho_c(f)$. We have $e' \sim_s c \sim_s e$, so $e' \sim_s e$ and, similarly $f' \sim_s f$. In particular, there are two galleries from e to f', one of type su via c' and one of type ut via f. Now consider $\sigma' \in \Sigma$ such that $e, f' \in \sigma'(W)$. As the length of each of the two galleries from e to f' is $q - 1$, both galleries lie in $\sigma'(W)$ thanks to the induction hypothesis. This implies $\zeta(su) = \zeta(ut)$. In particular, $\zeta(r) = \zeta(sut) = \zeta(u)$, contradicting that r is of minimal type. The conclusion is that γ lies in A, as required.

Conversely, suppose that $\mathcal{C}$ is a building. Take Σ to be the set of all strong isometries $\mathcal{C}(M) \to \mathcal{C}$ and, for $c \in \mathcal{C}$, take ρ_c to be the map $d \mapsto \delta_{\mathcal{C}}(c, d)$. Clearly, (1) is a consequence of Theorem 11.2.5(i). As for (2), let $\sigma \in \Sigma$ and $c \in \sigma(\mathcal{C}(M))$. For $x \in W$, we have $\rho_c \circ \sigma(x) = \delta_{\mathcal{C}}(c, \sigma(x)) = \delta_{\mathcal{C}(M)}(\sigma^{-1}c, x) = (\sigma^{-1}c)^{-1}x$, so $\rho_c \circ \sigma$ is an automorphism of $\mathcal{C}(M)$ (equal to $\alpha_{(\sigma^{-1}c)^{-1}}$ as in Notation 4.2.9), which settles (2). This ends the proof of the corollary. $\square$

We proceed to derive properties of buildings that are of significance to the related geometry and shadow spaces.

Lemma 11.2.7 *Each cell of a building is convex.*

Proof Suppose that c and d are chambers of the same J-cell X for some subset J of types of a building $\mathcal{C}$. Repeated application of Corollary 11.1.14(iii) yields $\delta_{\mathcal{C}}(c,d) \in \langle J \rangle$, so each chamber on a minimal gallery from c to d lies in cJ^*. $\sqcup$

The following useful result needs the set ${}^J W^K$ of (J, K)-reduced elements of Definition 4.5.14.

Theorem 11.2.8 *Suppose that c, d are chambers of a building $\mathcal{C}$ over S and that $J, K \subseteq S$. Let w be the (J, K)-reduced element in W of $\langle J \rangle \delta_{\mathcal{C}}(c,d) \langle K \rangle$. There is a unique $(J \cap {}^w K)$-cell in cJ^* each of whose chambers has Weyl distance w to some chamber of dK^*.*

Proof By Proposition 4.5.15(iv), there is a minimal expression $w_1 w w_2$ of $\delta_{\mathcal{C}}(c,d)$ with $w_1 \in \langle J \rangle$ and $w_2 \in \langle K \rangle$. As $\mathcal{C}$ is a chamber system of type M, there are $a \in cJ^*$ and $b \in dK^*$ such that $\delta_{\mathcal{C}}(c,a) = w_1$, $\delta_{\mathcal{C}}(a,b) = w$, and $\delta_{\mathcal{C}}(b,d) = w_2$. Suppose $a' \in a(J \cap {}^w K)^*$, say $z := \delta_{\mathcal{C}}(a',a) \in \langle J \cap {}^w K \rangle$. Then there is $y \in \langle K \rangle$ such that $z = wyw^{-1}$, so zw and wy are two minimal expressions of the same element. Moreover $\delta_{\mathcal{C}}(a',b) = zw$, so there is an element $b' \in bK^*$ with $\delta_{\mathcal{C}}(a',b') = w$ and $\delta_{\mathcal{C}}(b',b) = y$. In particular, a' is at Weyl distance w to a chamber of dK^*.

In order to prove uniqueness, suppose that $a'' \in cJ^*$ has distance w to a chamber b'' of dK^*. Then $\delta_{\mathcal{C}}(a,b'') = z'w = wy'$ for certain $y' \in \langle K \rangle$ and $z' \in \langle J \rangle$ (as there are simple galleries via a'' and via b, respectively). Now $\delta_{\mathcal{C}}(a'',a) = z' \in \langle J \rangle \cap w \langle K \rangle w^{-1} = \langle J \cap {}^w K \rangle$, so $a'' \in a(J \cap {}^w K)^*$. $\square$

The following consequence uses the set ${}^J W$ of left J-reduced elements of the Coxeter group W; see Definition 4.2.18.

Corollary 11.2.9 (Gate Property) *Let $\mathcal{C}$ be a building over S. For each $J \subseteq S$ and any two chambers c, d of $\mathcal{C}$, there is a unique chamber $e \in cJ^*$ such that the distance from e to d is minimal among all distances of chambers from cJ^* to d. The chamber e lies on a minimal gallery from c to d. The Weyl distance $\delta_{\mathcal{C}}(e,d)$ is left J-reduced.*

Proof Immediate from Theorem 11.2.8 with $K = \emptyset$. $\square$

Remark 11.2.10 The chamber e closest to d in cJ^* of Corollary 11.2.9 is called the *projection* of d on cJ^*.

We have the following converse to Corollary 11.2.9. Suppose that c, d, and e are chambers of a building $\mathcal{C}$ over S with $e \in cJ^*$ and $\delta_{\mathcal{C}}(e,d) \in {}^J W$. Then e is the unique chamber of cJ^* with smallest distance to d. For, if x is a chamber in cJ^*, then Lemma 11.2.7 gives $\delta_{\mathcal{C}}(x,e) \in \langle J \rangle$, so $l(\delta_{\mathcal{C}}(x,e)\delta_{\mathcal{C}}(e,d)) = l(\delta_{\mathcal{C}}(x,e)) + l(\delta_{\mathcal{C}}(e,d))$. In particular, there is a simple gallery from x to d via e of minimal type, so $\delta_{\mathcal{C}}(x,d) = \delta_{\mathcal{C}}(x,e)\delta_{\mathcal{C}}(e,d)$ and $l(\delta_{\mathcal{C}}(x,d)) = l(\delta_{\mathcal{C}}(x,e)) + l(\delta_{\mathcal{C}}(e,d))$, which shows that e is as claimed.

Theorem 11.2.11 (Intersection Property) *Let $\mathcal{C}$ be a building over S. If c, d are chambers of $\mathcal{C}$ and $J, K, L \subseteq S$ are such that $cJ^*L^* \cap dK^*L^* \neq \emptyset$, then there is*

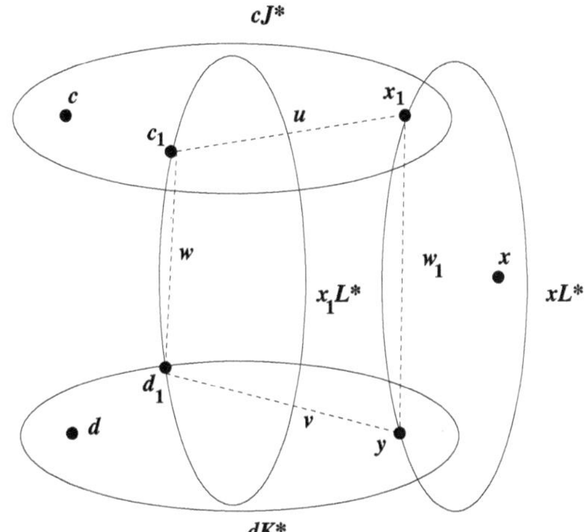

Fig. 11.3 The chambers and cells of the proof of Theorem 11.2.11

a chamber a in cJ^* such that $cJ^*L^* \cap dK^*L^* = a(J \cap {}^w K)^*L^*$, where w is the shortest element of $\langle J \rangle \delta_{\mathcal{C}}(c, d)\langle K \rangle$ (which belongs to $\langle L \rangle$).

Proof By Proposition 4.5.15 and Corollary 11.1.14(i) we can find $c_1 \in cJ^*$ and $d_1 \in dK^*$ with $\delta_{\mathcal{C}}(c_1, d_1) = w$. As $d \in cJ^*L^*K^*$, we have $\delta_{\mathcal{C}}(c, d) \in \langle J \rangle \langle L \rangle \langle K \rangle$, and so, by Exercise 4.9.16, $w \in {}^J W^K \cap \langle J \rangle \langle L \rangle \langle K \rangle \subseteq \langle L \rangle$.

Suppose $x \in cJ^*L^* \cap dK^*L^*$. Let x_1 be the projection of c on xL^* and take $y \in xL^* \cap dK^*$. Put $u = \delta_{\mathcal{C}}(c_1, x_1)$, $v = \delta_{\mathcal{C}}(d_1, y)$ and $w_1 = \delta_{\mathcal{C}}(x_1, y)$ (see Fig. 11.3). Lemma 11.2.7 shows that $u \in \langle J \rangle$, $v \in \langle K \rangle$ and $w_1 \in \langle L \rangle$. By Corollary 11.2.9, $u^{-1} = \delta_{\mathcal{C}}(x_1, c_1) \in {}^L W$, so $u \in W^L$. Furthermore, $w \in {}^J W^K \subseteq W^K$, and so, by Remark 11.2.10, we obtain $uw_1 = \delta_{\mathcal{C}}(c_1, y) = wv$, whence $w_1 \in \langle J \rangle w \langle K \rangle$. Therefore, there are $u_1 \in \langle J \rangle$ and $v_1 \in \langle K \rangle$ such that $w_1 = u_1 w v_1$ and $l(w_1) = l(u_1) + l(w) + l(v_1)$ (cf. Proposition 4.5.15). It follows from Corollary 4.5.13(i) that $u_1 \in \langle J \cap L \rangle$ and $v_1 \in \langle K \cap L \rangle$. Consequently, by Lemma 4.5.19, $\delta_{\mathcal{C}}(c_1, x_1)u_1 = uu_1 = wvv_1^{-1}w^{-1} \in \langle J \rangle \cap {}^w \langle K \rangle = \langle J \cap {}^w K \rangle$, so $\delta_{\mathcal{C}}(c_1, x_1) \in \langle J \cap {}^w K \rangle \langle L \rangle$, and $x \in x_1 L^* \subseteq c_1 (J \cap {}^w K)^* L^*$. This shows that $a := c_1$ satisfies $cJ^*L^* \cap dK^*L^* \subseteq c_1(J \cap {}^w K)^* L^*$. Since the other inclusion is trivial, this establishes the theorem. $\square$

Corollary 11.2.12 *Buildings are residually connected.*

Proof Let c, d, e be chambers and J, K, L subsets of S such that cJ^*, dK^*, eL^* are cells that pairwise meet in a non-empty set. We show that $cJ^* \cap dK^* \cap eL^*$ is non-empty. Let w be the shortest element in $\langle J \rangle \delta_{\mathcal{C}}(c, d)\langle K \rangle$. Since $cJ^* \cap dK^* \neq \emptyset$, we have $w = 1$. As $e \in cJ^*L^* \cap dK^*L^*$, by Theorem 11.2.11, this shows the existence of a chamber $a \in cJ^*$ such that $cJ^*L^* \cap dK^*L^* = a(J \cap {}^w K)^* L^*$. Replacing a by the projection of e on $a(J \cap K)^*$, we find $a \in eL^* \cap cJ^* \cap dK^*$. Now

$$eL^* \cap cJ^* \cap dK^* = aL^* \cap aJ^* \cap aK^* = a(L \cap J \cap K)^*$$

is an $L \cap J \cap K$-cell, whence, by Lemma 3.4.9, $\mathcal{C}$ is residually connected. $\square$

Remark 11.2.13 By use of Corollary 11.2.12, Theorem 3.4.6, and Exercise 3.7.10, buildings of rank n can be viewed as $[n]$-geometries of Coxeter type. As a consequence, we can translate the classification results on geometries of Coxeter type into classification results on buildings.

More specifically, each building of type A_n ($n \geq 2$), is the chamber system of a projective space of dimension n (by Corollary 5.4.4), which is of the form $\mathbb{P}(\mathbb{D}^{n+1})$ if the building is thick and $n \geq 3$ (by Corollary 6.3.2).

Each building of type B_n ($n \geq 2$) is the chamber system of the polar geometry of a nondegenerate polar space of rank n (by Example 11.4.8 and Corollary 7.6.8), which is classical if the building is thick and $n \geq 4$ (by Theorem 10.3.13).

Each building of type D_n ($n \geq 4$) is the chamber system of the oriflamme geometry of a polar space of rank n (by Proposition 7.8.9 and Corollary 7.6.8), which is of the form $\mathbb{P}(\mathbb{F}^{2n})_\kappa$ if the building is thick (by Theorem 10.3.13 and Corollary 10.4.10).

11.3 Tits Systems

Fix a Coxeter matrix M. Let $\mathcal{C}$ be a building of type M. Assuming a sufficiently transitive action of a group G on $\mathcal{C}$, we will provide another variation of Theorem 1.7.5. The translation of the transitive action of a group G on a building in group data takes place in Proposition 11.3.3, which by Definition 11.3.7 leads to a Tits system in the group G. The converse is stated in Corollary 11.3.11.

The remainder of the section deals with examples of type A_n, B_n, and D_n. A final proposition shows how, in these cases, the shadow spaces can be fully described in terms of subgroups of G.

Definition 11.3.1 A group G acting on a building $\mathcal{C}$ is called *strongly transitive* if it is transitive on the set of all pairs (c', A') with c' a chamber of $\mathcal{C}$ and A' an apartment of $\mathcal{C}$ containing c'.

Remark 11.3.2 Fix a chamber c of $\mathcal{C}$ and an apartment A on c, and suppose that G acts chamber transitively on $\mathcal{C}$. By Theorem 1.7.5, writing $B = G_c$, we can identify G/B with the set of all chambers of $\mathcal{C}$. If d is a chamber of $\mathcal{C}$, then, by Theorem 11.2.5, there is an apartment containing c and d. Therefore, assuming strong transitivity of G on $\mathcal{C}$, we obtain that $g(d) \in A$ for some $g \in B$; furthermore, by the same assumption, there is $n \in N_G(A) := \{g \in G \mid g(A) = A\}$ with $n(gd) = c$. If d represents the coset yB, the latter equality is equivalent to $ngyB = B$, and hence to $y \in g^{-1}n^{-1}B$. Consequently, $G = BN_G(A)B$.

Fix a Coxeter system (W, S) of type M. The transitivity properties of G imply that

(1) B is transitive on the set of apartments containing c;
(2) $N := N_G(A)$ is transitive on A;
(3) for each $w \in W$, the group G is transitive on the set of pairs (x, y) of chambers with $\delta_{\mathcal{C}}(x, y) = w$.

The last property follows as every pair of chambers is contained in an apartment (cf. Theorem 11.2.5) and as in A there is a unique chamber d such that $\delta_{\mathcal{C}}(c, d) = w$.

Writing $H = C_G(A) := \{g \in G \mid \forall a \in A \ g(a) = a\}$, we have that N/H is isomorphic to a subgroup of the chamber system $\mathrm{Aut}(A)$ over S which is transitive on A. By Proposition 4.2.10, we find $N/H \cong \mathrm{Aut}(A) \cong W$ and $B \cap N = H$.

Putting all this together, we see that every chamber corresponds to a coset gB with $g \in G$, which can be written in the form bwB where $b \in B$ and $w = nH$ (with $n \in N$) is an element of $W = N/H$. Notice that wB is well defined as $H \subseteq B$. The chambers r-adjacent to B are of the form brB $(b \in B)$, so $N_G(cr^*) = B \cup BrB$. In particular, $r \notin B$, $Br^{-1}B = BrB$, and $BrBrB \subseteq B \cup BrB$. Writing $G^{(r)} = N_G(cr^*)$ as in Definition 3.6.3, we obtain that $\mathcal{C} \cong \mathcal{C}(G, B, (G^{(s)})_{s \in S})$. The chamber c corresponds to B and the apartment A to the set $\{nB \mid n \in N\} = \{wB \mid w \in W\}$.

Suppose that $d = gB$ is the endpoint of a minimal gallery of type $r_1 \cdots r_t$ starting at c. Then the minimal gallery may be described as follows for certain $n_i \in r_i$ $(i \in [t])$.

$$c = B \sim_{r_1} b_1 n_1 B \sim_{r_2} b_1 n_1 b_2 n_2 B \sim_{r_3} \cdots \sim_{r_t} b_1 n_1 b_2 n_2 \cdots b_t n_t B = d.$$

On the other hand, by what we have seen above, there are $b \in B$ and $w \in W$ such that $g \in bwB$. But then $r_1 \cdots r_t = \delta_{\mathcal{C}}(c, d) = \delta_{\mathcal{C}}(B, bwB) = w$.

More generally, whenever $w = r_1 \cdots r_t$ is a minimal expression for w, we have

$$Br_1 Br_2 B \cdots r_t B = BwB.$$

For, letting $b_i \in B$ and $n_i \in r_i$ be arbitrary, we can argue as before for $d = gB$ where $g = b_1 n_1 \cdots b_t n_t$ and obtain $g \in Br_1 \cdots r_t B$, which proves the non-trivial inclusion.

The above remark leads to the following proposition. Recall from Definition 3.1.1 that a building is thick if all of its panels have at least three chambers.

Proposition 11.3.3 *Let $\mathcal{C}$ be a thick building over S of type M admitting an action of a group G. Let (c, A) be a pair consisting of a chamber c contained in an apartment A. If G is strongly transitive on $\mathcal{C}$, then the following assertions hold with $B = G_c$, $N = N_G(A)$, $H = C_G(A)$, $W = N/H$.*

(i) *B and N are subgroups of G generating the full group G.*
(ii) *$H = B \cap N \trianglelefteq N$.*
(iii) *The set S embeds in W as a generating set of W satisfying the following relations for $w \in W$, $r \in S$.*

$$BrBwB \subseteq BwB \cup BrwB$$

(iv) *For each $r \in S$, we have $rBr^{-1} \not\subseteq B$.*

As before, for $w \in W$, the subset wB of G is well defined as for $n \in w$, we have $wB = nHB = nB$. Similar observations hold for Bw^{-1} and wBw^{-1}.

Proof (i) This follows from the relation $G = BNB$.

(ii) is obvious from the above.

(iii) If $l(rw) = l(w) + 1$, this follows from the above. Otherwise, as $Br\,Br\,B \subseteq B \cup Br\,B$, there is a minimal expression $w = rr_2 \cdots r_t$ such that

$$Br\,Bw\,B \subseteq Br\,Brr_2 \cdots r_t\,B = (Br\,Br\,B)(Br_2\,B \cdots Br_t\,B)$$

$$\subseteq (B \cup Br\,B)(Br_2 \cdots r_t\,B) = Brw\,B \cup Br\,Br_2 \cdots r_t\,B$$

$$= Brw\,B \cup Bw\,B.$$

(iv) Observe that $Br = Br^{-1}$ as $r^2 = 1 \in W$. If $rBr^{-1} \subseteq B$, then $Br = rrBr^{-1} \subseteq rB$, so $brB = rB$ for each $b \in B$. This means that cr^*, where $c = B$, consists of c and cr only; therefore $\mathcal{C}$ is thin, contradicting the hypotheses. $\qquad\square$

Remark 11.3.4 Proposition 6.5.6 shows that the automorphism group of PG(V), where V is a vector space of dimension $n + 1$, is a strongly transitive group on the corresponding building with $M = \mathrm{A}_n$, and Theorem 10.5.6 shows that U(V, f) and O(V, κ), where V is a vector space of dimension n admitting a pseudo-quadratic form κ and a nondegenerate (σ, ε)-hermitian form f, are examples with $M = \mathrm{B}_n$. The last part of Theorem 10.5.6 shows that, if κ is a nondegenerate quadratic form of Witt index n on a vector space of dimension $2n$, the group SO(V, κ) is strongly transitive on the corresponding building of type $M = \mathrm{D}_n$.

The above properties of a group G are now abstracted from the chamber system setting. We drop the assumption that (W, S) be a Coxeter system.

Definition 11.3.5 Let G be a group. A *Tits system* in G is a quadruple (B, N, W, S) for which Conditions (i)–(iv) of the above proposition hold.

We will work towards Corollary 11.3.11, which is a partial converse to Proposition 11.3.3. We begin by deriving some convenient properties of Tits systems.

Theorem 11.3.6 *Each Tits system (B, N, W, S) in a group G satisfies the following properties.*

(i) *The pair (W, S) is a Coxeter system, and, for each $r \in S$ and $w \in W$, we have $l(rw) > l(w)$ if and only if $Br\,Bw\,B = Brw\,B$.*

(ii) *If $J, K \subseteq S$, then $B\langle J\rangle B\langle K\rangle B = B\langle J\rangle\langle K\rangle B$. In particular, $G^{(J)} := B\langle J\rangle B$ is a subgroup of G. Moreover, $G^{(S)} = G$, $G^{(\emptyset)} = B$.*

(iii) *If $w_1, w_2 \in W$ satisfy $w_1 \neq w_2$, then $Bw_1 B \neq Bw_2 B$.*

(iv) *If $J, K, L \subseteq S$, then $G^{(J)} \cap (G^{(K)}G^{(L)}) = (G^{(J)} \cap G^{(K)})(G^{(J)} \cap G^{(L)}) = G^{(J\cap K)}G^{(J\cap L)}$.*

Proof We first show that S consists of involutions in W. Let $r \in S$. Applying (iii) with $w = r^{-1}$ yields $Br\,Br^{-1}B \subseteq Br^{-1}B \cup B$. In view of (iv) and $B \subseteq Br\,Br^{-1}B$, this implies

$$Br\,Br^{-1}B = Br^{-1}B \cup B. \tag{11.1}$$

Inverting the sets at both sides of the equation, we get $Br\,Br^{-1}B = Br\,B \cup B$, which, again by use of (iv), together with (11.1) leads to

$$Br\,B = Br^{-1}B. \tag{11.2}$$

Applying (iii) with $w = r$ shows $Br\,Br\,B \subseteq Br\,B \cup Br^2 B$. On the other hand, (11.1) and (11.2) give

$$Br\,Br\,B = Br\,Br^{-1}B = Br\,B \cup B. \tag{11.3}$$

Thus, we must have $B = Br^2 B$, i.e., $r^2 \subseteq B$. Since $r^2 \subseteq N$, by definition, we derive $r^2 = H$, so $r^2 = 1 \in W$. Since $r = 1$ would contradict (iv), it follows that r is an involution of W. An immediate consequence (inversion of (iii)) is

$$wBr \subseteq BwB \cup Bwr\,B \quad \text{for all } r \in S \text{ and } w \in W. \tag{11.4}$$

(ii) Obviously, $B\langle J\rangle B\langle K\rangle B \supseteq B\langle J\rangle\langle K\rangle B$. We next show $B\langle J\rangle B\langle J\rangle\langle K\rangle B \subseteq B\langle J\rangle\langle K\rangle B$. Let $g \in B\langle J\rangle B\langle J\rangle\langle K\rangle B$. Then there are $r_1, \ldots, r_q \in J$ such that $g \in Br_1 \cdots r_q B\langle J\rangle\langle K\rangle B$. If $q = 0$, then $g \in B\langle J\rangle\langle K\rangle B$ and there is nothing to prove. If $q = 1$, then $g \in Br_1 B\langle J\rangle\langle K\rangle B$, so there are $u \in \langle J\rangle$ and $v \in \langle K\rangle$ such that $g \in Br_1 BuvB$. By Axiom (iii), this gives $g \in BuvB \cup Br_1 uvB \subseteq B\langle J\rangle\langle K\rangle B$, and we are done. Otherwise, the case $q = 1$ gives

$$Br_1 \cdots r_q B\langle J\rangle\langle K\rangle B \subseteq Br_1 \cdots r_{q-1}Br_q B\langle J\rangle\langle K\rangle B$$

$$\subseteq Br_1 \cdots r_{q-1}B\langle J\rangle\langle K\rangle B,$$

whence $g \subseteq Br_1 \cdots r_{q-1}B\langle J\rangle\langle K\rangle B$. By induction on q, it follows that $g \in B\langle J\rangle\langle K\rangle B$, so $B\langle J\rangle B\langle J\rangle\langle K\rangle B \subseteq B\langle J\rangle\langle K\rangle B$ indeed.

As a consequence, $B\langle J\rangle B\langle K\rangle B \subseteq B\langle J\rangle B\langle J\rangle\langle K\rangle B \subseteq B\langle J\rangle\langle K\rangle B$, and the first statement of (ii) is proved.

The subset $G^{(J)}$ of G is clearly non-empty and closed under taking inverses. From what we have just seen, it is also closed under multiplication, so it is a subgroup of G. Finally, due to Condition (i) of a Tits system, we have $G^{(\emptyset)} = B1B = B$, and $G^{(S)} = BSB = BNB = \langle B, N\rangle = G$, which establishes (ii).

(iii) Suppose that w_1 and w_2 are distinct elements of W. Without harming generality for the proof of (iii), we may assume $l(w_1) \geq l(w_2)$. If $l(w_2) = 0$, then $Bw_1 B = Bw_2 B$ would imply $w_1 H \subseteq B \cap N = H$, whence $w_1 H = H = w_2 H$, a contradiction. Thus $Bw_1 B \neq Bw_2 B$ and we are done. Let $l(w_2) \geq 1$. Then there is an involution $r \in S$ such that $l(rw_2) < l(w_2)$. By induction on $l(w_2)$ we have $Brw_2 B \neq Bw_1 B, Brw_1 B$, so $Brw_2 B \cap Br\,Bw_1 B = \emptyset$. Now $Bw_1 B = Bw_2 B$ would imply $Brw_2 B \cap Br\,Bw_2 B = \emptyset$, which contradicts that $Brw_2 B$ is contained in this intersection. We conclude $Bw_1 B \neq Bw_2 B$, which proves (iii).

(i) For $r \in S$, set $C_r = \{w \in W \mid Br\,Bw\,B = Br\,w\,B\}$. We first prove two claims on these C_r , stated in (11.5) and (11.6).

$$C_r \cap rC_r = \emptyset. \tag{11.5}$$

Suppose $w \in C_r$. Then $Br\,Br\,w\,B = Br\,Br\,Bw\,B = Bw\,B \cup Br\,w\,B$, so $rw \notin C_r$, and $w \notin rC_r$, settling (11.5).

$$\text{If } w \in C_r \text{ and } s \in S \text{ with } ws \notin C_r, \text{ then } rw = ws. \tag{11.6}$$

For,

$$
\begin{aligned}
Bw\,B &\subseteq Bws\,Bs\,B \\
&\subseteq Br\,Bws\,Bs\,B \quad (\text{as } ws \notin C_r) \\
&\subseteq Br\,Bw\,Bs\,Bs\,B = Br\,w\,Bs\,Bs\,B \quad (\text{as } w \in C_r) \\
&= Br\,w\,B \cup Br\,w\,Bs\,B \quad \big(\text{by } (11.3)\big) \\
&= Br\,w\,B \cup Br\,ws\,B \quad \big(\text{by } (11.4)\big)
\end{aligned}
$$

so $w \in \{rw, rws\}$ by (iii). But $w = rw$ conflicts $r \neq 1$, so $w = rws$. This yields $rw = ws$ as wanted.

Now we verify the exchange condition (cf. Definition 4.5.7) at the same time establish that $w \in C_r$ is equivalent to $l(rw) > l(w)$. This suffices for the proof of (i). Thus, let $w \in W$. Suppose $w \notin C_r$. Let $r_1 \cdots r_q$ be a minimal expression of w as a product of elements from S. Write $w_j = r_1 \cdots r_j$ for $j = 0, 1, \ldots, q$. Since $w_0 = 1 \in C_r$ and $w_q = w \notin C_r$, there is an element $j \in \{0, 1, \ldots, q-1\}$ such that $w_j \in C_r$ and $w_j r_{j+1} = w_{j+1} \notin C_r$. Applying (11.6), we obtain $rw_j = w_j r_{j+1} = w_{j+1}$. This means $rr_1 \cdots r_j = r_1 \cdots r_{j+1}$, and implies $l(rw) < l(w)$. Next, suppose $w \in C_r$. Then by (11.5), $rw \notin C_r$, so by what we have just seen $l(w) = l(r(rw)) < l(rw)$. Consequently, $w \in C_r$ if and only if $l(rw) < l(w)$, and the exchange condition holds. In view of Theorem 4.5.8, this ends the proof of (i).

(iv) We have

$$
\begin{aligned}
G^{(J)} \cap \big(G^{(K)}G^{(L)}\big) &= \big(B\langle J\rangle B\big) \cap \big(B\langle K\rangle B\langle L\rangle B\big) \\
&= \big(B\langle J\rangle B\big) \cap \big(B\langle K\rangle\langle L\rangle B\big) \quad (\text{by (ii)}) \\
&= B\big(\langle J\rangle \cap \langle K\rangle\langle L\rangle\big)B \quad (\text{by (iii)}) \\
&= B\big(\langle J\rangle \cap \langle K\rangle\big)\big(\langle J\rangle \cap \langle L\rangle\big)B \\
&\qquad (\text{by (i) and Proposition 4.5.15(ii)}) \\
&= B\big(\langle J\rangle \cap \langle K\rangle\big)BB\big(\langle J\rangle \cap \langle L\rangle\big)B \quad (\text{by (ii)}) \\
&= \big((B\langle J\rangle B) \cap (B\langle K\rangle B)\big)\big((B\langle J\rangle B) \cap (B\langle L\rangle B)\big) \quad (\text{by (iii)}) \\
&= \big(G^{(J)} \cap G^{(K)}\big)\big(G^{(J)} \cap G^{(L)}\big).
\end{aligned}
$$

But also

$$\begin{aligned}
G^{(J)} \cap G^{(K)} &= B\langle J\rangle B \cap B\langle K\rangle B \\
&= B\big(\langle J\rangle \cap \langle K\rangle\big)B \quad \text{(by (iii))} \\
&= B\big(\langle J \cap K\rangle\big)B \quad \text{(by Corollary 4.5.13(ii))} \\
&= G^{(J\cap K)}
\end{aligned}$$

and similarly $G^{(J)} \cap G^{(L)} = G^{(J\cap L)}$. This ends the proof of (iv) and hence the theorem. $\square$

The notation $G^{(J)}$ coincides with the one of Notation 3.6.8. This is justified since by Theorem 11.3.6(ii) $G^{(J)} = \langle G^{(j)} \mid j \in J\rangle$. The notation is also in accordance with Theorem 3.6.9, where a chamber system is constructed from such data. We come back to this construction in Definition 11.3.7 below.

Definition 11.3.7 Let (B, N, W, S) be a Tits system in a group G. Then, according to Theorem 11.3.6(i), (W, S) is a Coxeter system, so there exists a Coxeter matrix $M = (m_{r,s})_{r,s\in S}$ such that (W, S) is of type M. We will also refer to M as the *type* of the Tits system and to $|S|$ as the *rank* of the Tits system. The *chamber system associated with the Tits system* (B, N, W, S) in G, denoted by $\mathcal{C}(B, N, W, S)$, is the chamber system over S whose chambers are the cosets gB for $g \in G$ and in which, for each $r \in S$, the chambers gB and hB are r-adjacent if and only if $Bh^{-1}gB \subseteq B\langle r\rangle B$.

Observe that this is indeed a chamber system because if gB, hB and hB, kB are r-adjacent pairs, then, by Theorem 11.3.6(ii),

$$Bk^{-1}gB \subseteq Bk^{-1}hBh^{-1}gB \subseteq B\langle r\rangle B\langle r\rangle B = B\langle r\rangle B,$$

so gB and kB are also r-adjacent, proving that r-adjacency is an equivalence relation.

Remark 11.3.8 For a Tits system (B, N, W, S) in a group G, the chamber system $\mathcal{C}(B, N, W, S)$ coincides with the chamber system $\mathcal{C}(G, B, (G^{(s)})_{s\in S})$ determined by G on B with respect to $(G^{(s)})_{s\in S}$ of Definition 3.6.3. For, if $g, h \in G$ and $r \in S$, then gB and hB are r-adjacent in $\mathcal{C}(G, B, (G^{(s)})_{s\in S})$ if and only if $gG^{(r)} = hG^{(r)}$, which holds if and only if $gB \cup gBrB = hB \cup hBrB$; the latter is easily seen to be equivalent to $Bh^{-1}gB \subseteq B\langle r\rangle B$, which is the definition of r-adjacency in $\mathcal{C}(B, N, W, S)$.

Example 11.3.9 Let G be a group with a Tits system (B, N, W, S) of rank one. Write $S = \{r\}$. By Theorem 11.3.6(ii), $G = G^{\{r\}} = B\langle r\rangle B = B \cup BrB$, so G acts 2-transitively on G/B by left multiplication (cf. Exercise 2.8.12(a)). The subgroup B of G is the stabilizer of the point B in this permutation representation and r comes from an element n of G such that the $\langle n\rangle$-orbit of B in G/B has length two.

Conversely, suppose G acts 2-transitively on a set X (of size at least three). Take two distinct points $x, y \in X$ and set $B = G_x$. Choose an element $n \in G$ moving the pair (x, y) to (y, x). This element satisfies $G = B \cup BnB$ and $n^2 \in B$. Define N as the subgroup of G generated by n and set $H = B \cap N$. The group G, being generated by B and n, is also generated by B and N, so (i) of Proposition 11.3.3 is satisfied. As H has index two in N, it is a normal subgroup of N and the quotient N/H is a cyclic group of order two, that is, isomorphic to $W(A_1)$. We have verified (ii) of Proposition 11.3.3. Condition (iii) of Proposition 11.3.3 needs to be checked only for $r = n$ and $w \in \{1, n\}$. It is trivial for $w = 1$ and follows from $G = BnB \cup Bn^2B$ in case $w = n$. Now, as X has size at least three, there is an element $b \in B$ mapping y to an element $z \neq y$. In particular $nbn^{-1}x = nby = nz \neq ny = x$, showing that $nbn^{-1} \in nBn^{-1} \setminus B$. Hence (iv) of Proposition 11.3.3.

The conclusion is that a Tits system of rank one is equivalent to a doubly transitive permutation group on a set of size at least three.

Corollary 11.3.10 *Let $C = C(B, N, W, S)$ be the chamber system associated with the Tits system (B, N, W, S) in a group G. For each $J \subseteq S$, the following assertions hold.*

  (i) *The quadruple $(B, \langle J\rangle H, \langle J\rangle, J)$ is a Tits system in $G^{(J)} = B\langle J\rangle B$.*
 (ii) *All J-cells of C are mutually isomorphic as chamber systems over J.*
(iii) *The J-cell of C containing B coincides with $C(B, \langle J\rangle H, \langle J\rangle, J)$.*

Proof Easy. $\qquad\square$

Corollary 11.3.11 *Let C be the chamber system associated with the Tits system (B, N, W, S) in a group G.*

  (i) *The chamber system C is a thick building of type M.*
 (ii) *The group G acts on C by left multiplication; for each $w \in W$, it is transitive on the set of pairs of chambers (c, d) with $\delta_C(c, d) = w$.*
(iii) *If M is of spherical Coxeter type, then G is strongly transitive on C.*

Proof (i) Let $r \in S$. The chambers rB and B are distinct (as $r \notin B$) and r-adjacent. If the two of them would constitute an r-cell, then $B \cup BrB = B \cup rB$, so $rB \subseteq Br$, which contradicts (iv) of the definition of Tits systems (recall $r^{-1} = r$). Hence C is a thick chamber system.

As each element of G lies in $Br_1 Br_2 B \cdots Br_q B = G^{(r_1)} \cdots G^{(r_q)}$ for some minimal type $r_1 \cdots r_q \in S^*$, the group G is generated by the $G^{(r)}$ ($r \in S$), and so C is connected in view of Lemma 3.6.7.

We next prove that C is of type M. In view of Corollary 11.3.10 it suffices to show that, in case $S = \{r, s\}$ is of size two, the chamber system C is a generalized m-gon, where $m = m_{r,s}$. We will use Exercise 3.7.4 for this purpose. Let $S = \{r, s\}$ be of size two. As we have seen above, C is thick and connected. Also, as $G = BWB$, every chamber is of the form bwB with $b \in B$ and $w = srs \cdots$ or $rsr \cdots$ of length at most m, and so it is at distance at most m from B. Besides, distance m occurs if $m < \infty$ and all finite distances occur if $m = \infty$. Since G is transitive on the set of chambers of $\mathcal{C}$, this shows that the graph of $\mathcal{C}$ has diameter m. Now Conditions (a) and (b) of Exercise 3.7.4 are verified.

Suppose that there is a closed simple gallery of length $n < 2m$ and of type $rsrs \cdots \in S^*$. We can choose chambers gB and hB on this gallery such that there are two simple galleries from gB to hB, one of type $u = rsr \cdots \in S^*$ and of length at most m, the other of type $v = srs \cdots$, and of length less than m. Since u and v are minimal expressions, we obtain from $h^{-1}g \in (Br\,Bs\,Br\,B \cdots B) \cap (Bs\,Br\,Bs\,B \cdots B)$ that $h^{-1}g \in (B\zeta(u)B) \cap (B\zeta(v)B)$. Now the lengths of u and v imply that $\zeta(u) \neq \zeta(v)$, contradicting Theorem 11.3.6(iii). This establishes Condition (c) of Exercise 3.7.4.

If $m = \infty$, this suffices, so assume next that $m < \infty$. The proof of the final condition is similar: if there is a simple gallery from gB to hB of type $rsr \cdots$ (length m), then $g^{-1}h \in Br\,Bs\,Br\,B \cdots B = Brsr \cdots B$ (since the type is minimal). But $rsr \cdots = srs \cdots$ (both sides of length m), so $g^{-1}h \in Bsrs \cdots B = Bs\,Br\,Bs\,B \cdots B$, proving that there is a simple gallery of type $srs \cdots$ (length m) from gB to hB. This establishes Condition (d). All conditions of Exercise 3.7.4 are satisfied, so we conclude that $\mathcal{C}$ is the chamber system of a generalized m-gon.

We have derived that $\mathcal{C}$ is a chamber system of type M. In order to finish the proof that $\mathcal{C}$ (no longer assumed to be of rank two) is a building, we suppose that

$$B, bn_1 B, b_1 n_1 b_2 n_2 B, \ldots, b_1 n_1 b_2 n_2 \cdots b_t n_t B = B,$$

with $n_i \in r_i$ ($i \in [t]$) is a closed simple gallery of minimal type $r_1 \cdots r_t$, and show that it is trivial. By Theorem 11.3.6(ii), $B = Bn_1 b_2 n_2 \cdots b_t n_t B = Br_1 r_2 \cdots r_t B$, so that, by Theorem 11.3.6(iii), $r_1 r_2 \cdots r_t - 1$, proving that the only closed simple gallery starting at B of minimal type is the trivial gallery. As G is transitive on $\mathcal{C}$, this establishes (i).

(ii) Clearly, G is transitive on the set of chambers of $\mathcal{C}$. Let c be the chamber B. Then for the stabilizer G_c of the chamber B, we have $G_c = \{g \subset G \mid gB = B\} = B$. If d is a chamber with $\delta_{\mathcal{C}}(c, d) = w$, then $d \in BwB/B$, so there exists $b \in B$ such that $d = bwB$, implying $b(c, wB) = (c, d)$.

(iii) The standard apartment $A = WB/B$ contains the chamber $c = B$. Suppose that A' is another apartment containing the chamber c. We want to show that there exists $b \in B$ such that $A' = bA$. Let w_0 be the longest element of W (cf. Theorem 4.6.6). By (ii), there is $b \in B = G_c$ such that $bw_0 B$ is the chamber of A' at distance w_0 from c. But then, as A' is convex (cf. Theorem 11.2.5), we must have $A' = bWB/B = bA$. As G acts transitively on the set of chambers of $\mathcal{C}$, it follows that G is strongly transitive on $\mathcal{C}$. $\qquad\square$

In the general case, there may be more apartments on c than present in the B-orbit of A. See Exercise 11.8.12 for an example.

Definition 11.3.12 A Tits system is called *split* if there is a normal subgroup U of B such that B is the semidirect product of U and H.

Example 11.3.13 The group $GL(\mathbb{F}^{n+1})$ has Tits system (B, N, W, S), where B is the subgroup of upper triangular matrices, N is the group of monomial matrices (that is, with a single nonzero entry in each row and in each column), and (W, S) is the Coxeter system of type A_n. By strong transitivity of $GL(\mathbb{F}^{n+1})$ on the building $\mathcal{C}(PG(\mathbb{F}^{n+1}))$ (cf. Remark 11.3.4) and inspection of Proposition 11.3.3, it will be clear that the building $\mathcal{C}$ associated with the Tits system is isomorphic to $\mathcal{C}(PG(\mathbb{F}^{n+1}))$. Conversely, by Corollary 11.3.11, the existence of the Tits system in $GL(\mathbb{F}^{n+1})$ implies that the group acts strongly transitively on $\mathcal{C}$.

The group B is the semidirect product of the group U of all matrices in B with ones on the main diagonal and the subgroup H of diagonal matrices in B, so the Tits system is split. In fact, all of the chamber systems of the geometries of Propositions 11.1.9 and 11.1.10 of rank at least three have split Tits systems.

Proposition 11.3.14 *Each thick* $[n]$*-geometry of type* A_n *for* $n \geq 3$*, each thick polar geometry of rank* $n \geq 3$ *whose singular planes are Desarguesian, and each thick* $[n]$*-geometry of type* D_n *for* $n \geq 4$ *has a Tits system.*

Proof By Propositions 11.1.9 and 11.1.10, these geometries are buildings. In view of the classification results mentioned in Remark 11.2.13, these buildings are as in Remark 11.3.4, where it was observed that their automorphism groups are strongly transitive. Therefore, a Tits system as required exists by Proposition 11.3.3. $\square$

Particular examples of buildings, and in fact all thick spherical buildings of rank at least three, are obtained from Tits systems. Exercise 11.8.14 shows how to view shadows in terms of subgroups of the Tits system.

11.4 Shadow Spaces

In Definition 2.5.1 shadow spaces of a geometry were introduced in order to arrive at the more classical viewpoints of the geometry. In this section, we study the shadow spaces of buildings and derive some remarkable properties.

Fix a Coxeter type M and a Coxeter system (W, S) of type M. Let $\mathcal{C}$ be a building of type M and put $\Gamma = \Gamma(\mathcal{C})$. As $\mathcal{C}$ is residually connected (cf. Corollary 11.2.12), so is the geometry Γ (cf. Proposition 3.4.5). We assume $n = |S| < \infty$ and often identify S with $[n]$. For the duration of the section, we let T and J be subsets of S with $T \subseteq J$, and write $L = S \setminus J$.

Recall from Definition 2.5.1 that the J-shadow of a flag F of Γ is the set $Sh_J(F)$ of all flags of type J which are incident with F. In terms of the chamber system $\mathcal{C}$, if c is a chamber of $\mathcal{C}$ on F, then F is represented by the K-cell cK^*, where K is the cotype of F. Each chamber of the J-shadow of F is an element of cK^*L^*. In particular, $Sh_J(F) = \{dL^* \mid d \in cK^*L^*\}$.

We will study the shadow space $Z = ShSp(\Gamma, J, T)$ of Γ on J. Its points are the sets cL^* for c a chamber of $\mathcal{C}$ and its lines are the sets of points in cr^*L^* for c a chamber of $\mathcal{C}$ and $r \in T$. So a line in Z corresponds to cr^*L^*, the union of L-cells of the shadow on J of the flag of Γ of type $S \setminus \{r\}$ whose r-cell in $\mathcal{C}$ is cr^*.

The Cartesian product $Z \times Z$ is partitioned into relations Z_w ($w \in {}^L W^L$) defined by

$$Z_w = \left\{ (xL^*, yL^*) \mid x, y \text{ chambers of } \mathcal{C} \text{ with } \delta_{\mathcal{C}}(x, y) \in \langle L \rangle w \langle L \rangle \right\}.$$

This idea will be at the heart of the analysis of Z in this section and the next two. We will be mainly concerned with those relations Z_w that represent pairs of points at mutual distance at most two.

Definition 11.4.1 A *shadow space of type* (M, J) is the shadow space $\mathrm{ShSp}(\Gamma(\mathcal{C}), J)$ of a building $\mathcal{C}$ of type M. If $M = Y_n$ and $J = \{j\}$, we also write $Y_{n,j}$ instead of $(Y_n, \{j\})$.

Since apartments tell us a lot about buildings, it is no surprise that shadow spaces of apartments will be a great help in analyzing shadow spaces of buildings. Therefore, we frequently look into the thin examples, that is, Coxeter groups. The following result is such an instance. Here, we write $r^\perp = \{s \in S \mid m_{r,s} \le 2\}$ and $J^\perp = \bigcap_{r \in J} r^\perp$ as in Notation 4.5.18.

Proposition 11.4.2 *Let* L, K_1, K_2 *be subsets of* S. *If* $w \in W$ *satisfies*

$$\langle K_1 \rangle \langle L \rangle = w \langle K_2 \rangle \langle L \rangle,$$

then the union C *of all components of* K_1 *containing an element of* $K_1 \setminus L$ *coincides with the same union for* K_2, *and* $w \in \langle K_1 \rangle \langle L \cap C^\perp \rangle \langle K_2 \rangle$.

Proof Without loss of generality, we (can) take $w \in {}^{K_1} W^{K_2}$ (see Definition 4.5.14), so $w \in \langle L \rangle$. If $K_2 \subseteq L$, then $\langle K_1 \rangle \langle L \rangle = \langle L \rangle$, whence $K_1 \subseteq L$ and $C = \emptyset$, from which the assertion follows immediately. Let $r \in K_2 \setminus L$ and suppose t lies in the connected component of K_2 containing r. Choose a shortest path $r = r_1, \ldots, r_m = t$ in K_2. All r_i are distinct, and r_i and r_{i+1} do not commute. Let S_w be as in Theorem 4.5.10, that is, the set of all members of S occurring in a minimal expression of w.

We claim $r_i \in K_1 \cap S_w^\perp$. We prove this by induction on m. For $m = 1$, we have $t = r \in K_2 \setminus L$ and $wr \in \langle K_1 \rangle \langle L \rangle$. This forces $r \in K_1$, and $w \in \langle K_1 \rangle \langle L \rangle r$, which means that there are $u \in \langle L \rangle$ and $k \in \langle K_1 \rangle$ such that $w = kur$. We must have $l(wr) > l(w)$, so by the Exchange Condition 4.5.8 applied to wr upon right multiplication with r, we have $w = kur = k'u$ with $k' \in \langle K_1 \rangle$ obtained by removal of one member of S from a minimal expression of k. As $w \in {}^{K_1} W^{K_2}$, we must have $k' = 1$ and so $w = k'u = u$. This gives $S_w = S_u$. As $r \notin L$, we have $\{r\} \cup S_w = S_{wr} = S_{ku}$, which implies $k = r$ and $rwr = u = w$. In particular, $r \in S_w^\perp$ and so $r_1 = r \in K_1 \cap S_w^\perp$.

By the induction hypothesis, $r_i \in K_1 \cap S_w^\perp$ for $i < m$. Consider $z := wr_m \cdots r_2 r_1 \in w \langle K_2 \rangle \subseteq \langle K_1 \rangle \langle L \rangle$. As $w \in {}^{K_1} W^{K_2}$, the above expression of z is minimal. It contains precisely one factor r, and so z does not belong to $\langle L \rangle$. Hence, there is $s \in K_1$ such that $l(sz) < l(z)$. By the Exchange Condition 4.5.8, and minimality of

w, there is $j \in [m]$ such that $sz = wr_m \cdots r_{j+1}r_{j-1}\cdots r_1$, whence $swr_m \cdots r_{j+1} = wr_m \cdots r_{j+1}r_j$. Now $w^{-1}sw = r_m \cdots r_{j+1}r_j r_{j+1}\cdots r_m \in \langle S_w \cup \{s\}\rangle \cap \langle \{r_m\} \cup K_2 \setminus S_w\rangle$. We conclude $s \in K_1 \cap {}^w K_2$. For, if $s \in K_2$, then $w^{-1}sw \in \langle (S_w \cup \{s\}) \cap (\{r_m\} \cup K_2 \setminus S_w)\rangle \subseteq \langle s, r_m\rangle \subseteq \langle K_2\rangle$, so $s \in \langle K_1\rangle \cap {}^w\langle K_2\rangle = \langle K_1 \cap {}^w K_2\rangle$, which forces the conclusion, and otherwise $w^{-1}sw \in \langle (S_w \cup \{s\}) \cap (\{r_m\} \cup K_2 \setminus S_w)\rangle \subseteq \langle \{s\} \cap \{r_m\}\rangle$, so $s = r_m = t$ and $sw = wt$, giving $s \in K_1 \cap {}^w K_2$, again. Now Lemma 4.5.19(ii) implies $s \in S_w \cup S_w^{\perp}$. If $s \in S_w$, then $s \in r_{m-1}^{\perp}$, contradicting that r_m and r_{m-1} are adjacent in K_2. Therefore, $r_m = s \in S_w^{\perp}$, whence the claim.

The claim gives that the connected component C_2 of K_2 containing r is contained in K_1 and $S_w \subseteq C_2^{\perp}$, so $w \in C_2^{\perp}$. By symmetry, we also have that C is contained in K_2 and $w \in C^{\perp}$. It follows that $C = C_2$, whence the proposition. $\qquad\square$

Corollary 11.4.3 *Suppose that (W, S) is a Coxeter system of irreducible type, and that J, K are subsets of S. If $W = \langle J\rangle\langle K\rangle$ is a factorization of W, then $J = S$ or $K = S$.*

Proof Suppose that both J and K are proper subsets of S. Then there are $k \in S \setminus J$ and $j \in S \setminus K$. By Corollary 4.5.13(i), $\{j\} = S_j \subseteq J \cup K$, so $j \in J$ and, similarly, $k \in K$. As $kj \in \langle J\rangle\langle K\rangle$, Proposition 4.5.15(i) forces $jk = kj$, so $m_{j,k} = 2$.

Let A be the set of nodes of the shortest path in M connecting j and k. The diagram on A is linear, with end nodes j and k, so the component C of $A \setminus \{k\}$ containing j is $A \setminus \{k\}$. By Proposition 4.5.15(ii), $\langle A\rangle = \langle A \cap J\rangle\langle A \cap K\rangle$, whence $\langle A\rangle = \langle A \setminus \{k\}\rangle\langle A \setminus \{j\}\rangle$, and $\langle A \setminus \{k\}\rangle\langle A \setminus \{j\}\rangle = \langle A\rangle = k\langle A\rangle = k\langle A \setminus \{k\}\rangle\langle A \setminus \{j\}\rangle$, so Proposition 11.4.2 applies with $C = A \setminus \{k\}$, showing that k lies in $(A \setminus \{k\})^{\perp}$, a contradiction with the choice of A. $\qquad\square$

Corollary 11.4.4 *Suppose that M is irreducible, and let L, K_1, K_2 be subsets of S. Write $J = S \setminus L$ and assume that K_1 and K_2 do not have connected components disjoint from J.*

(i) If $w \in \langle L\rangle$ satisfies ${}^w\langle K_1\rangle = \langle K_2\rangle$, then $K_1 = K_2$ and $w \in \langle K_1\rangle\langle L \cap K_1^{\perp}\rangle$.

(ii) $N_{\langle L\rangle}(\langle K_1\rangle) = \langle L \cap K_1\rangle\langle L \cap K_1^{\perp}\rangle$.

Proof (i) For w as stated, we have $w\langle K_1\rangle\langle L\rangle = \langle K_2\rangle w\langle L\rangle = \langle K_2\rangle\langle L\rangle$, so Proposition 11.4.2 applies. This gives $K_1 = K_2$ and $w \in \langle K_1\rangle\langle L \cap K_1^{\perp}\rangle\langle K_1\rangle = \langle K_1\rangle\langle L \cap K_1^{\perp}\rangle$.

(ii) Apply (i) to $w \in N_{\langle L\rangle}(\langle K_1\rangle)$ with $K_1 = K_2$. This shows $N_{\langle L\rangle}(\langle K_1\rangle) \subseteq \langle K_1\rangle\langle L \cap K_1^{\perp}\rangle$. By intersecting both sides with $\langle L\rangle$ and using Proposition 4.5.15(ii), we find that the normalizer is contained in $\langle L \cap K_1\rangle\langle L \cap K_1^{\perp}\rangle$. Since this is clearly contained in $N_{\langle L\rangle}(\langle K_1\rangle)$, we are done. $\qquad\square$

We first address the question whether the flag with respect to which the shadow was taken, can be reconstructed from the shadow alone. In general, when the J-reduction of a set of types $B \subset S$ is not equal to B, this is not the case, as we have seen in Lemma 2.5.5. For a chamber c and subsets K and L of S, the shadow

cK^*L^* only depends on the union K_0 of the connected components of K containing elements of $J = S \setminus L$. For, $K \setminus K_0$ lies in L and commutes with K_0, so $cK^*L^* = cK_0^*(K \setminus K_0)^*L^* = cK_0^*L^*$. Therefore, in studying shadows on J of flags of cotype K, we may assume that each connected component of K meets J nontrivially.

Lemma 11.4.5 *Let c, d be chambers of C and let K_1, K_2, L be subsets of S such that, for $i = 1, 2$, each connected component of K_i meets $J = S \setminus L$. The two shadows $cK_1^*L^*$ and $dK_2^*L^*$ on the shadow space of $\Gamma(C)$ on J coincide if and only if $K_1 = K_2$ and the Weyl distance $\delta_C(c, d)$ lies in $\langle K_1 \rangle \langle L \cap K_1^\perp \rangle$.*

Proof Suppose that $cK_1^*L^* = dK_2^*L^*$. Since c and d are connected by a simple gallery in $K_1^*L^*K_2^*$, after replacing c and d by suitable chambers in their K_i-cells, we may assume $w = \delta_{C(M)}(c, d) \in {}^{K_1}W^{K_2} \cap \langle L \rangle$. By the Intersection Property 11.2.11, there is a chamber a in cK_1^* such that $cK_1^*L^* = cK_1^*L^* \cap dK_2^*L^* = a(K_1 \cap {}^wK_2)^*L^*$. As $cK_1^* = aK_1^*$, it follows that $aK_1^*L^* = a(K_1 \cap {}^wK_2)^*L^*$. Considering the types of galleries from a to chambers in both sides of the equation and using Corollary 11.1.14, we find $\langle K_1 \rangle \langle L \rangle = \langle K_1 \cap {}^wK_2 \rangle \langle L \rangle$. By Proposition 4.5.15(ii), intersecting with $\langle D \rangle$, where D is a connected component of K_1, we find $\langle D \rangle = \langle D \rangle \langle L \cap D \rangle = \langle D \cap {}^wK_2 \rangle \langle L \cap D \rangle$. Corollary 11.4.3 implies that one of the two factors is equal to $\langle D \rangle$. But $D \setminus L \neq \emptyset$ by assumption, so we cannot have $L \cap D = D$. Therefore, $D \cap {}^wK_2 = D$, whence $D \subseteq {}^wK_2$. Since this holds for all components D of K_1, we conclude $K_1 \subseteq {}^wK_2$. By symmetry, also $K_2 \subseteq {}^{w^{-1}}K_1$, whence $K_1 = {}^wK_2$. As $w \in \langle L \rangle$, we obtain $\langle K_1 \rangle \langle L \rangle = \langle {}^wK_2 \rangle \langle L \rangle = w \langle K_2 \rangle w^{-1} \langle L \rangle = w \langle K_2 \rangle \langle L \rangle$. Recalling $w \in {}^{K_1}W^{K_2}$, we derive from Corollary 11.4.4(i) that $K_1 = K_2$ and $w \in \langle L \cap K_1^\perp \rangle$. $\square$

Lemma 11.4.6 *Shadow spaces of buildings are partial linear spaces.*

Proof Consider the shadow space $Z = \mathrm{ShSp}(\Gamma, J, T)$. Suppose that a, b, and c are chambers of C such that the point cL^* of Z is on each of the lines $ar_1^*L^*$ and $br_2^*L^*$ for some $r_1, r_2 \in T$. Then, according to Theorem 11.2.11, there is a chamber $e \in ar_1^*$ such that, for w the shortest element of $\langle r_1 \rangle \delta_C(a, b) \langle r_2 \rangle$, we have $ar_1^*L^* \cap br_2^*L^* = e(\{r_1\} \cap {}^w\{r_2\})^*L^*$. If ${}^w\{r_2\} = \{r_1\}$, then $ar_1^*L^* \cap br_2^*L^* = er_1^*L^* = ar_1^*L^* = br_2^*L^*$, so the two lines coincide, and, by Lemma 11.4.5, $r_1 = r_2$. If ${}^w\{r_2\} \neq \{r_1\}$, then $ar_1^*L^* \cap br_2^*L^* = eL^*$, and so $cL^* = eL^*$, that is, the two lines of Z corresponding to $ar_1^*L^*$ and $br_2^*L^*$, respectively, meet in a single point. $\square$

Notation 11.4.7 For $m \in \mathbb{N}$, a line space Z, and a point x of Z, write $Z_m(x)$ for the set of points of Z at distance exactly m from x, and $Z_{\leq m}(x)$ for the set of points at distance at most m from x in the collinearity graph of Z.

Now consider $Z = \mathrm{ShSp}(\Gamma, J, T)$. We have the following recursive description of $Z_m(cL^*)$ for the L-cell cL^* containing a chamber c of C.

$$Z_0(cL^*) = \{cL^*\} \quad \text{and} \quad Z_m(cL^*) = \{dL^* \mid d \in cL^*(TL^*)^m\} \setminus Z_{\leq m-1}(cL^*).$$

Example 11.4.8 Let Z be a shadow space of type $B_{n,1}$ for some $n \in \mathbb{N}$, $n \geq 2$. By Lemma 11.4.6, it is a partial linear space. We will verify the polar space axiom, Definition 7.4.1, for Z. Let c and d be chambers of the underlying building of Z of type B_n. Write $S = \{s_1, \ldots, s_n\}$, with s_i corresponding to the i-th node of B_n in Table 4.2. Now cL^* and $ds_1^*L^*$ represent an arbitrary point and line, respectively, of Z. The long element w_S of $W(B_n)$ (cf. Corollary 4.6.7), as represented in Table 4.5, contains s_1 (written there as 1) only twice, so $Z = Z_{\leq 2}(cL^*)$. In fact, it is easily shown by induction on n that, apart from the identity, and s_1, the (L, L)-reduced element, v say, in $\langle L \rangle w_S \langle L \rangle$, is the only other member of $^L W^L$. Corollary 4.6.10(iv) gives $w_S S w_S = S$. As conjugation by w_S induces an automorphism of B_n, which is necessarily the identity, this implies $w_S s_i w_S = s_i$ for all $i \in [n]$. We conclude that w_S, and hence also v, normalizes $\langle L \rangle$, so $W(B_n) = \langle L \rangle \cup \langle L \rangle s_1 \langle L \rangle \cup v \langle L \rangle$. As v has a minimal expression with only two occurrences of s_1, it can be written as $s_1 u s_1$ for some $u \in \langle L \rangle$. It follows that $cL^* \subseteq dS^* = dL^* \cup dL^* s_1 L^* \cup ds_1 L^* s_1 L^*$. In terms of the space Z, this means that the point cL^* is either equal to dL^*, collinear with the point dL^*, or collinear with a point $d_1 L^*$ for d_1 a chamber on the s_1-panel of d, which is therefore on the line $ds_1^* L^*$. This gives the polar space axiom.

We will prove the convexity of shadows in shadow spaces of buildings, beginning with apartments. The notion of a minimal expression for an element of a Coxeter group was introduced in Definition 4.2.12.

Lemma 11.4.9 *Write* $Y = \mathrm{ShSp}(\Gamma(\mathcal{C}(M)), J, T)$ *and let* $K \subseteq L \cup T$ *and* $w \in Y_m(\langle L \rangle) \cap \langle K \rangle \cap W^L$. *Suppose that* $w_0, w_1, \ldots, w_{m-1} \in \langle L \rangle$ *and* $r_1, \ldots, r_m \in T$ *are such that* $w_0 r_1 w_1 r_2 \cdots r_{m-1} w_{m-1} r_m$ *is a minimal expression for* w. *For* $j < m$, *we have*

$$w_0 r_1 w_1 r_2 \cdots r_{j-1} w_{j-1} r_j w_j \in Y_j(\langle L \rangle) \cap \langle K \rangle.$$

Proof If $m = 0$, there is nothing to show. We proceed by induction on pairs (m, j), lexicographically ordered. Assume $m \geq 1$.

First, suppose $j = 0$. Put $K_0 := K \setminus T$. As $Y_0(\langle L \rangle) \cap \langle K \rangle = \langle L \rangle \cap \langle K \rangle = \langle L \cap K \rangle = \langle K_0 \rangle$, we need to show $w_0 \in \langle K_0 \rangle$. To this end, we may assume, without loss of generality, $w \in {}^{K_0} W$. If $s \in L$ satisfies $l(sw) < l(w)$, then $s \in S_w \subseteq K$, so $s \in K \cap L = K_0$. Hence, $w \in {}^L W \cap W^L = {}^L W^L$ (cf. Definition 4.5.14). It follows that $w_0 = 1 \in \langle K_0 \rangle$, as required.

Suppose, therefore, $j \in [m-1]$. By the case $j = 0$, we have $w_0 \in \langle K \setminus T \rangle$. The element $r_1 w_0^{-1} w$ has minimal expression $w_1 r_2 \cdots r_{m-1} w_{m-1} r_m w_m$, so it belongs to $Y_{m-1}(\langle L \rangle) \cap \langle K \rangle \cap W^L$, and the induction hypothesis on m gives

$$w_0 r_1 w_1 r_2 \cdots r_{j-1} w_{j-1} r_j w_j \in w_0 r_1 \left(Y_{j-1}(\langle L \rangle) \cap \langle K \rangle \right) \subseteq Y_j(\langle L \rangle) \cap \langle K \rangle.$$

This proves the lemma.   □

By the Exchange Condition 4.5.8, minimal expressions for w as assumed in the hypothesis of Lemma 11.4.9 exist. Here is the convexity result for an arbitrary building $\mathcal{C}$.

Proposition 11.4.10 *Suppose that $X = dK^*L^*$ is the shadow on J in $Z = \mathrm{ShSp}(\Gamma, J, T)$ of the flag corresponding to dK^* for some chamber d of $\mathcal{C}$ and subset K of S. Then X is a convex subspace of Z. As a space, X is isomorphic to $\mathrm{ShSp}(\Gamma(dK^*), J \cap K, T \cap K)$, the shadow space of type $(J \cap K, T \cap K)$ of the building dK^*.*

Proof We first prove that X is a subspace of Z. Suppose that l is a line of Z with $|l \cap X| \geq 1$. Then there is a chamber c such that the point cL^* lies in X and on the line $l = cr^*L^*$, where $r \in T$. By Theorem 11.2.11, we have $l \cap X = cr^*L^* \cap dK^*L^* = e(\{r\} \cap {}^w K)^* L^*$ for some $e \in cr^*$ and some $w \in {}^{\{r\}} W^K$ such that $\delta_{\mathcal{C}}(c, d) = xwy$ for certain $x \in \langle r \rangle$ and $y \in \langle K \rangle$. If $r \in {}^w K$ then $l = cr^*L^* \subseteq e(\{r\} \cap {}^w K)^* L^* = l \cap X$ and we are done. Otherwise, $\{r\} \cap {}^w K = \emptyset$, so $cL^* \subseteq eL^*$, whence $cL^* = eL^*$, proving that $l \cap X$ is the single point cL^*. We conclude that $|l \cap X| \geq 2$ implies $l \subseteq X$, so X is a subspace of S.

In order to prove that X is convex, we assume that $a = a_0, a_1, \ldots, a_m$ are chambers of $\mathcal{C}$ and $r_1, \ldots, r_m$ are members of T such that

$$aL^* \sim_{r_1} a_1 L^* \sim_{r_2} \cdots \sim_{r_{m-1}} a_{m-1} L^* \sim_{r_m} a_m L^*$$

is a geodesic in Z whose extremities aL^* and $a_m L^*$ belong to X. We need to show that each point $a_i L^*$ also belongs to X.

As aL^* and $a_m L^*$ belong to X, we can choose $a, a_m \in dK^*$ in such a way that $\delta_{\mathcal{C}}(a, a_m) \in \langle K \rangle \cap W^L$. On the other hand, $\delta_{\mathcal{C}}(a, a_m) \in Y_m(\langle L \rangle)$, where $Y = \mathrm{ShSp}(\Gamma(\mathcal{C}(M)), J, T)$ is the shadow space of the thin geometry of the Coxeter group W. Let $w_0 r_1 w_1 r_2 \cdots r_{m-1} w_{m-1} r_m w_m$ be a minimal expression for $\delta_{\mathcal{C}}(a, a_m)$ with $w_i \in \langle L \rangle$ for each i. Put $K_0 = K \setminus T$. By Lemma 11.4.9 applied with $K \cap S_w$ instead of K (so $K \cap S_w \subseteq L \cup T$), for each $j \in \{0, \ldots, m-1\}$,

$$w_0 r_1 w_1 r_2 \cdots r_{j-1} w_{j-1} r_j w_j \in Y_j\big(\langle L \rangle\big) \cap \langle K \cap S_w \rangle.$$

In particular, $a_1 \in a w_0 r_1 L^* \subseteq a K_0^* r_1 L^*$, so the point $a_1 L^*$ belongs to the shadow $X = dK^*L^*$. By induction on m, it follows that all $a_i L^*$ do. This establishes that X is convex.

It remains to show that X is isomorphic to a shadow space of the building dK^* of type $M|_K$ (cf. Proposition 11.1.6(ii)). Set $U = \mathrm{ShSp}(\Gamma(dK^*), J \cap K, T \cap K)$ and consider the map $\varepsilon : U \to Z$ given by $x(K \cap L)^* \mapsto xL^*$ for $x \in dK^*$. As $K \cap L \subseteq L$, this map is well defined. For $H \subseteq K$, the set of chambers $x H^*(K \cap L)^*$ maps onto $x H^* L^*$, which is the shadow in Z of a flag of cotype H.

We show that ε is injective. Suppose that a and b are chambers in dK^* with $\varepsilon(a(K \cap L)^*) = \varepsilon(b(K \cap L)^*)$. Then $aL^* = bL^*$. On the other hand $a, b \in dK^*$ implies $\delta_{\mathcal{C}}(a, b) \in \langle K \rangle$, so $\delta_{\mathcal{C}}(a, b) \in \langle L \rangle \cap \langle K \rangle = \langle K \cap L \rangle$ (by Corollary 4.5.13), whence $a(K \cap L)^* = b(K \cap L)^*$.

Now we show that ε maps shadows into shadows. A shadow in U is of the form $x G^*(K \cap L)^*$ for some subset G of K, and so clearly maps onto $x G^* L^*$, a shadow of a flag of cotype G in Z.

Conversely, assume that, for some $G \subseteq K$, we have a shadow of a flag of Γ of cotype G entirely contained in dK^*L^*. Without loss of generality, we (may) assume that each connected component of G contains an element of J. This shadow is of the form xG^*L^* for some chamber x of C. Fix x and let w be the shortest element of $\langle G\rangle\delta_C(x,d)\langle K\rangle$. As $\delta_C(x,d) \in \langle G\rangle\langle L\rangle\langle K\rangle$, we have $w \in \langle L\rangle$. By the Intersection Property 11.2.11, there is a chamber $e \in dK^*$ such that $xG^*L^* = dK^*L^* \cap xG^*L^* = e(K \cap {}^wG)^*L^*$. Lemma 11.4.5 gives $G \subseteq K \cap {}^wG$, so $G \subseteq K$. Comparing sizes, we also find $G = {}^wG$, which implies $w \in N_{\langle L\rangle}(\langle G\rangle)$. In view of Corollary 11.4.4(ii), the shadow xG^*L^* is the image of $eG^*(K \cap L)^*$ in X. $\square$

If T is a proper subset of J, then Z is not connected and, by Proposition 11.4.10, the connected component of cL^* for a chamber c is isomorphic to $\mathrm{ShSp}(\Gamma(c(L \cup T)^*), T, T)$. So, when studying connected line spaces, we can restrict ourselves to shadow spaces of the form $Z = \mathrm{ShSp}(\Gamma(C), J)$. Here are some subspaces of shadow spaces. In line with Definition 1.6.1, we let d_Z denote distance in the collinearity graph of Z. The Bruhat order is introduced in Exercise 11.8.6.

Lemma 11.4.11 *The following sets of points of $Z = \mathrm{ShSp}(\Gamma, J)$ are subspaces of Z.*

(i) $Z_{\leq w}(cL^*) := \{dL^* \mid d \in C, \delta_C(c,d) \leq w\}$, *where $\leq$ stands for the Bruhat order, c is a chamber of C, and $w \in {}^LW^L$.*

(ii) $Z_{\leq r}(cL^*) \cap Z_{\leq v^{-1}}(dL^*)$, *where c, d are chambers and $r \in J$ is such that rv is a minimal expression of the shortest element of $\langle L\rangle\delta_C(c,d)\langle L\rangle$.*

(iii) *For each shadow X and each point p of Z,*

$$\pi_p(X) := \left\{ q \in X \mid d_Z(p,q) = \min_{u \in X} d_Z(p,u) \right\}.$$

Proof (i) Take a line dr^*L^* with two points in $Z_{\leq w}(cL^*)$. Consider $v = \delta_C(c,d)$. Each chamber e in dr^* satisfies $\delta_C(c,e) \in v\langle r\rangle$, and by the Gate Property 11.2.9 there is a unique chamber $a \in dr^*$ with $\delta_C(c,a)$ the shortest element of $\{v, vr\}$. Since $dr^*L^* \cap Z_{\leq w}(cL^*)$ contains two points, at least one of them is of the form bL^* with $\delta_C(c,b)$ the longest element of $\{v, vr\}$. Therefore, both $v \leq w$ and $vr \leq w$, so each point of dr^*L^* belongs to $Z_{\leq w}(cL^*)$.

(ii) This follows from (i), as the subset is the intersection of the two subspaces $Z_{\leq r}(cL^*)$ and $Z_{\leq v^{-1}}(dL^*)$.

(iii) We have $p = xL^*$ for some chamber x of C. Suppose that y_1 and y_2 are chambers such that y_1L^* and y_2L^* are distinct collinear points of $\pi_p(X)$. Without loss of generality, we may take the line on these two points to be $y_1r^*L^*$ for some $r \in J$ with $y_2 \in y_1r^*$. Consequently, $\delta_C(x,y_2) \in \delta_C(x,y_1)\langle r\rangle$. Without loss of generality, we assume $l(\delta_C(x,y_1)) > l(\delta_C(x,y_2)) = l(\delta_C(x,y_1)r)$. By the Gate Property 11.2.9, there is a unique chamber $z \in y_1r^*L^*$ with $l(\delta_C(x,z)) = \min\{l(\delta_C(x,y)) \mid y \in y_1r^*L^*\}$. In particular, $l(\delta_C(x,z)) \leq l(\delta_C(x,y_2))$. Write m for $d_Z(p,y_1L^*)$, so zL^* belongs to $Z_m(p)$. By the length assumption, y_1L^* is distinct from zL^* and also belongs to $Z_m(p)$. This implies that both $\delta_C(x,y_1)$ and $\delta_C(x,y_1)r$

belong to $Y_m(\langle L \rangle)$, where $Y = \mathrm{ShSp}(\Gamma(\mathcal{C}(M)), J)$ is the shadow space of the thin geometry of the Coxeter group W. But then, for each chamber $v \in y_1 r^*$, we have $\delta_{\mathcal{C}}(x, v) \in \{\delta_{\mathcal{C}}(x, y_1), \delta_{\mathcal{C}}(x, y_1)r\} \subseteq Y_m(\langle L \rangle)$, whence $d_Z(p, vL^*) = m$. This shows that each element vL^* of the line $y_1 r^* L^*$ belongs to $\pi_p(X)$. $\square$

Our next goal is to analyze the set of common neighbors of two points at mutual distance two.

As a consequence of Proposition 11.4.10, the common neighbor set of two points cL^* and dL^* is part of the shadow space $cS_w^* L^*$, where w is the minimal element of $\langle L \rangle \delta_{\mathcal{C}}(c, d)\langle L \rangle$. Lemma 11.4.11(ii) shows that this set is a subspace of $cS_w^* L^*$. We will determine more of the structure of this subspace by use of Coxeter group properties.

Lemma 11.4.12 *Let cL^* and dL^* be two points at distance two in the shadow space $Z = \mathrm{ShSp}(\Gamma, J)$. Assume further that $w = \delta_{\mathcal{C}}(c, d) \in {}^L W^L$. Then $w = r_1 y r_2$, with $r_1, r_2 \in J$ and $y \in \langle L \rangle$. If x is a chamber with xL^* collinear with both cL^* and dL^* in Z, then either $x \in c(L \cap {}^w L)^* r_1 L^*$ or $r_1 \neq r_2$, $w = r_1 r_2 = r_2 r_1$, and $x \in cr_2 L^*$.*

Proof Let x be a chamber such that xL^* is collinear in Z with both cL^* and dL^*. Without loss of generality, we (can) choose $x \in cL^* J$. Let c_1 be the chamber in cL^* nearest to x (cf. the Gate Property 11.2.9), so $w_0 := \delta_{\mathcal{C}}(c, c_1) \in \langle L \rangle$ and $\delta_{\mathcal{C}}(c_1, x) = r_3$ for some $r_3 \in J$. There are $w_1, w_2 \in \langle L \rangle$ and $r_4 \in J$ such that $w_1 r_4 w_2$ is a minimal expression for $\delta_{\mathcal{C}}(d, x)$. As $w \in {}^L W^L$ and $w_2 \in {}^J W^J$, the expressions $w_0^{-1} w w_1$ and $r_3 w_2^{-1} r_4$, where d_1 is a chamber in $dL^* \cap xL^* r_4$, are minimal for $\delta_{\mathcal{C}}(c_1, d_1)$. Hence

$$w_0^{-1} r_1 y r_2 w_1 = w_0^{-1} w w_1 = r_3 w_2^{-1} r_4. \tag{11.7}$$

As $S_w = S_y \cup \{r_1, r_2\} \subseteq L \cup \{r_3, r_4\}$, we have either $r_1 = r_3$ and $r_2 = r_4$, or $r_1 = r_4$ and $r_2 = r_3$.

Suppose that $r_1 \neq r_2$ and we are in the latter case, that is, $w_0^{-1} r_1 y r_2 w_1 = r_2 w_2^{-1} r_1$. Then $r_1 y r_2$, which is (L, L)-reduced, can also be written as a minimal expression of the form $r_2 v r_1$ with $v \in \langle L \rangle$. Now $y r_2 = r_1 r_2 v r_1 \in \langle L \cup \{r_2\} \rangle \cap {}^{r_1}\langle L \cup \{r_2\} \rangle \subseteq \langle r_1^{\perp} \rangle$, whence $y = 1$ and $w = r_1 r_2$. Also, $v = r_2 w r_1 = r_2 r_1 r_2 r_1 \in \langle r_1, r_2 \rangle \cap \langle L \rangle = \langle \emptyset \rangle$ implies $v = 1$ and $r_1 r_2 = r_2 r_1$. It follows from Lemma 4.5.19(iii) that $w_0 = r_1 r_2 w_1 r_1 w_2 r_2 = r_2 (r_1 w_1 r_1 w_2) r_2$, which lies in $\langle L \cup \{r_1\} \rangle \cap {}^{r_2}\langle L \cup \{r_1\} \rangle$, belongs to $\langle r_2^{\perp} \rangle$, so commutes with r_2. This gives $\delta_{\mathcal{C}}(x, c) = w_0 r_3 = w_0 r_2 = r_2 w_0 \in r_2 \langle L \rangle$, so $x \in cr_2 L^*$, as required.

It remains to consider the case where $r_1 = r_3$ and $r_2 = r_4$. Rewrite (11.7) to $r_2 y^{-1} r_1 w_0 = w_1 r_2 w_2 r_1$. As $r_2 y^{-1} r_1 \in {}^L W^L$ and $w_0 \in \langle L \rangle$, the left hand side is minimal. Also $w_1 r_2 w_2$ is minimal. If right multiplication of it by r_1 would reduce the expression, it would lead to an element in $\langle L \rangle \langle r_2 \rangle \langle L \rangle$, a contradiction with cL^* and dL^* having mutual distance two. This implies that the right hand side, $w_1 r_2 w_2 r_1$, is also minimal. Hence

$$l(y) + l(w_0) = l(w_1) + l(w_2).$$

Rewrite (11.7) once more, but now as $r_1 yr_2 w_1 = w_0 r_1 w_2^{-1} r_2$. Again, the left hand side is minimal. As $w_0 \in W^{r_1^{\perp}}$, the expression $w_0 r_1 w_2^{-1}$ is minimal and, as before, we infer that $w_0 r_1 w_2^{-1} r_2$, the right hand side, is also minimal. Consequently,

$$l(y) + l(w_1) = l(w_0) + l(w_2).$$

Comparing these two length equations, we see $l(w_0) = l(w_1)$ and $l(y) = l(w_2)$. As $r_1 yr_2$ and $r_1 w_2^{-1} r_2$ are both the unique shortest element of $\langle L \rangle r_1 yr_2 \langle L \rangle = \langle L \rangle r_1 w_2^{-1} r_2 \langle L \rangle$ (see Proposition 4.5.15) this means $y = w_2^{-1}$. Now the equation reads $w_0 = (r_1 yr_2) w_1 (r_1 yr_2)^{-1} = w w_1 w^{-1}$, so $w_0 \in \langle L \rangle \cap {}^w \langle L \rangle = \langle L \cap {}^w L \rangle$. Therefore, $x \in c w_0 r_3 \subseteq c \langle L \cap {}^w L \rangle r_1 L^*$, as required. $\qquad\square$

Lemma 11.4.12 shows the need to study $L \cap {}^w L$, where, we recall, $L = S \setminus J$. For $r_1 \in J$ and $i \geq 2$, we will use the notation of Lemma 4.5.19 and write $L_i = \{s \in L \mid m_{r_1,s} = i\}$.

Proposition 11.4.13 *Suppose that c, $d \in W$ are such that $c\langle L \rangle$ and $d\langle L \rangle$ are points at mutual distance two in $Y := \mathrm{ShSp}(\Gamma(\mathcal{C}(M)), J)$. The (L, L)-reduced element w of $\langle L \rangle \delta_{\mathcal{C}(M)}(c, d)\langle L \rangle$ is of the form $w = r_1 yr_2$ with $r_1, r_2 \in J$ and y an $(L \cap r_1^{\perp}, L \cap r_2^{\perp})$-reduced element of $\langle L \rangle$. Put $E = L \cap {}^w L$ and $F = (L \cap r_1^{\perp}) \cap {}^y(L \cap r_2^{\perp})$. The connected component K of $E \cup \{r_1\}$ containing r_1 in M is the diagram A_m $(m \geq 1)$, B_m $(m \geq 2)$ or D_m $(m \geq 3)$ with r_1 appearing as the end node labelled 1 in the labelling of Table 4.6.*

Moreover, $E \cap r_1^{\perp} = F$ and one of the following five assertions holds.

  (i) *$y \notin L_4 \cup L_3 \langle L_2 \rangle L_3$ or $r_1 \neq r_2$. Furthermore, $E = F$ and $K \cong \mathrm{A}_1$.*

 (ii) *$y \in L_4$. Furthermore, $E = \{y\} \cup (L_2 \cap F \cap y^{\perp})$, and $K \cong \mathrm{B}_2$.*

(iii) *$y = st = ts$ with $s, t \in L_3, s \neq t$. Furthermore, $E = \{s, t\} \cup (L_2 \cap \{s, t\}^{\perp})$, and $K \cong \mathrm{D}_3$.*

(iv) *$y = svt$ with $s, t \in L_3, s \neq t$, and $v \in \langle F \rangle$, where $st \neq ts$ or $v \neq 1$. Furthermore, $E = \{s\} \cup (L_2 \cap s^{\perp})$, and $K \cong \mathrm{A}_2$.*

 (v) *$y = svs$ with $s \in L_3$ and $v \in \langle F \rangle \setminus \{1\}$. Furthermore, $E = \{s\} \cup F$, and $K \cong \mathrm{A}_m$ $(m \geq 2)$, B_m $(m \geq 3)$, or D_m $(m \geq 4)$.*

Table 11.1 illustrates the possibilities encountered in the above proposition. Here, $\mathrm{K}_{m,n}$ is as defined in Example 1.5.5 and $\mathrm{K}_{m \times n}$ is a complete multipartite graph with m parts of size n. This means that $\mathrm{K}_{m \times n}$ is a multipartite graph all of whose m parts have size n and all of whose unordered pairs of nodes from distinct parts are edges.

Proof Let $s \in E$. Then $sr_1 yr_2 = r_1 yr_2 t$ for some $t \in L$. If $s \in L_2$, then $syr_2 = r_1 sr_1 yr_2 = yr_2 t$, so, by Proposition 4.5.19(iii), $y^{-1} sy = r_2 tr_2 \in \langle L \cap {}^{r_2} L \rangle \subseteq \langle r_2^{\perp} \rangle$, whence $y^{-1} sy = t \in L \cap r_2^{\perp}$ and $s \in F$, proving $E \cap L_2 \subseteq F$. Conversely, if $t_1 \in F$, then $t_1 \in L_2$ and $t_1 = yuy^{-1}$ for some $u \in L \cap r_2^{\perp}$, so $w^{-1} t_1 w = r_2 y^{-1} r_1 t_1 r_1 yr_2 = r_2 y^{-1} t_1 yr_2 = r_2 ur_2 = u$, whence $t_1 \in L \cap {}^w L = E$ and $t_1 \in E \cap L_2$. We conclude $E \cap L_2 = F$. This proves $E \cap r_1^{\perp} = F$.

Table 11.1 Points wL^* at distance two from L^* in $\mathrm{ShSp}(\Gamma(\mathcal{C}(M)), J)$

$w = ryr$	$E \setminus r_1^\perp$	K	$\{L^*, wL^*\}^\perp$
Special	$\emptyset$	A_1	Singleton
$y \in L_4$	$\{y\}$	B_2	$K_{1,1}$
$y = st = ts \in \langle L_3 \rangle$	$\{s, t\}$	D_3	$K_{2,2}$
$y = svt, s, t \in L_3, v \in \langle F \rangle$	$\{s\}$	A_2	K_2
$y = svs, s \in L_3, v \in \langle L \cap r_1^\perp \rangle$	$\{s\}$	$A_m \ (m \geq 2)$	K_m
$y = svs, s \in L_3, v \in \langle L \cap r_1^\perp \rangle$	$\{s\}$	$B_m \ (m \geq 3)$	$K_{(m-1)\times 2}$
$y = svs, s \in L_3, v \in \langle L \cap r_1^\perp \rangle$	$\{s\}$	$D_m \ (m \geq 4)$	$K_{(m-1)\times 2}$

We claim that, if $E \not\subseteq L_2$, then $r_1 = r_2$ and $y \in L_4 \cup L_3 \langle L_2 \rangle L_3$. To establish this claim, assume $s \in E \setminus L_2$. As above, $sr_1 yr_2 = r_1 yr_2 t$ for some $t \in L$. Rewrite this to

$$yr_2 t = r_1 sr_1 yr_2. \tag{11.8}$$

Now $sr_1 yr_2$ is a minimal expression, but $r_1 sr_1 yr_2$ is not. Since $s \notin L_2$, the Exchange Condition 4.5.8 implies that $r := r_1 = r_2$ and that either

(a) there is an element $\widehat{y} \in \langle L \rangle$, obtained from a minimal expression of y by deleting a member of S, such that $yrt = sr\widehat{y}r$; or

(b) $yrt = sry$.

In Case (a), $rsryr = sr\widehat{y}r$ gives $\widehat{y}y^{-1} = rsrsr \in \langle \{r, s\} \rangle \cap \langle L \rangle = \langle s \rangle$, whence $y = s\widehat{y}$ and $m_{r,s} = 3$. Now $yrt = rsryr$ implies $\widehat{y} = r\widehat{y}rtr$, so by Lemma 4.5.19(iv) $\widehat{y} \in \langle L_2 \rangle L_3 \langle L_2 \rangle$. Since $y \in {}^{L_2}W^{L_2}$ and $l(\widehat{y}) = l(y) - 1$, we have $\widehat{y} \in W^{L_2}$ and so $\widehat{y} \in \langle L_2 \rangle L_3$. It follows that $y = s\widehat{y} \in L_3 \langle L_2 \rangle L_3$.

In Case (b), we have $sy = rytr \in \langle L \rangle \cap r\langle L \rangle r = \langle L_2 \rangle$ (by Lemma 4.5.19(iii)), so $y \in s\langle L_2 \rangle \cap {}^{L_2}W^{L_2} = \{s\}$, that is, $y = s$. The equation in (a) gives $t = s$, and (11.8) gives $rsrsr = srs$, so $m_{r,s} = 4$ and $y \in L_4$. This settles the claim.

(i) Assume $y \notin L_4 \cup L_3 \langle L_2 \rangle L_3$ or $r_1 \neq r_2$. By the claim, $E \subseteq L_2$. Consequently, $E = E \cap L_2 = F$, and so $\langle E \rangle r_1 \langle L \rangle = \langle F \rangle r_1 \langle L \rangle = r_1 \langle L \rangle$. This implies that the common neighbor set of $\langle L \rangle$ and $w\langle L \rangle$ consists of a single point in U. Therefore, $K = \{r_1\}$. This establishes (i).

For the other cases, we (can) assume $y \in L_4 \cup L_3 \langle L_2 \rangle L_3$ and $r := r_1 = r_2$.

(ii) Assume $y \in L_4$. Then $y = {}^w y \in L \cap {}^w L = E$. By the above analysis for Case (b), we find $E = \{y\} \cup F$, whence $K = \{y, r\}$, of type B_2. Hence (ii).

For the remainder of this proof, we (may) assume $y = svt$ with $s, t \in L_3$ and $v \in \langle L_2 \rangle$. Observe that $s \in E$ as $s = rsvrv^{-1}sr = rsvtrtrtv^{-1}sr = wtw^{-1} \in E$.

(iii) If $|E \setminus L_2| > 1$, then, as in Case (a) above, this expression is not the unique one in $L_3 \langle L_2 \rangle L_3$, and, as $S_y \subseteq \{s, t\} \cup L_2$, we must also have $y = tus$ for some $u \in \langle L_2 \rangle$. This forces $vt = stus \in \langle L_2 \cup \{t\} \rangle \cap {}^s \langle L_2 \cup \{t\} \rangle$, so by Lemma 4.5.19(iii), $v, t \in \langle r^\perp \rangle$. In particular, $y = svt = vst$. But also $y \in {}^{L_2}W^{L_2}$, so $y = st = ts$ and $m_{s,t} = 2$. Lemma 4.5.19(ii) gives $E = \{s, t\} \cup (L_2 \cap {}^{st}L_2) \subseteq \{s, t\} \cup \{r, s, t\}^\perp$, so

$K = \{r, s, t\}$ is of type $A_3 \cong D_3$ and the node r is labelled 1 in the D_3 labelling. This gives (iii).

From now on, we (may) assume $E \setminus L_2 = \{s\}$, that is, $E = \{s\} \cup F$.

(iv) Assume $t \neq s$. If $x \in L_2 \cap {}^y L_2$, then there is $z \in L_2$ with $x = svtztv^{-1}s$. Now $sxsv = vtzt \in \langle L_2 \cup \{s\} \rangle \cap \langle L_2 \cup \{t\} \rangle = \langle L_2 \rangle$, so $sxs = x$, proving $L_2 \cap {}^y L_2 \subseteq s^\perp$. In particular, $E = \{s\} \cup (E \cap s^\perp)$, and $K = \{r, s\}$ is of type A_2. This proves (iv).

We are left with the case where $t = s$.

(v) This assertion follows by induction on the length of y. Here s, v, $L_2 \cap {}^y L_2$, L_2 replace the roles of r, y, E, L, respectively, in the induction step. $\qquad \square$

Definition 11.4.14 A Coxeter diagram is said to be of *Weyl type* if it is one of A_n $(n \geq 1)$, B_n $(n \geq 2)$, D_n $(n \geq 4)$, E_n $(n = 6, 7, 8)$, F_4, G_2.

Remark 11.4.15 Comparison with Theorem 4.7.3 shows that $I_2^{(m)}$ for $m \notin \{2, 3, 4, 6\}$, H_3, and H_4 are the only spherical irreducible Coxeter types that are not Weyl types. The Weyl types distinguish themselves from other spherical irreducible Coxeter types in the following aspects.

(i) The real vector space underlying the reflection representation of a Coxeter group of Weyl type has a lattice (over $\mathbb{Z}$) that is invariant under the Coxeter group action.

(ii) If M is a spherical irreducible Coxeter diagram of rank at least three such that there is a thick building over M, then M is a Weyl type.

Definition 11.4.16 The *convex closure* of a subspace X of a line space Z is the intersection of all convex subspaces containing X.

As in Proposition 11.4.13, we will write $L_i = \{s \in L \mid m_{r_1, s} = i\}$.

Theorem 11.4.17 *Let M be a Weyl type. Suppose that c and d are chambers of $\mathcal{C}$ such that cL^* and dL^* are points of $\mathrm{ShSp}(\Gamma, J)$ at mutual distance two. Suppose, furthermore, that c and d have been chosen in such a way that $w := \delta_\mathcal{C}(c, d) \in {}^L W^L$. There are $r_1, r_2 \in J$ and $y \in \langle L \rangle \cap {}^{L \cap r_1^\perp} W^{L \cap r_2^\perp}$ with $w = r_1 y r_2$. Moreover, the set N of common neighbors of cL^* and dL^* satisfies exactly one of the following properties.*

(1) *If $r_1 = r_2$, then N is a singleton if and only if $y \notin L_4 \cup L_3 \langle L_2 \rangle L_3$. If $r_1 \neq r_2$, then N is a singleton if and only if $y \neq 1$ or $r_1 r_2 \neq r_2 r_1$.*

(2) *N is a coclique of size at least two if and only if either $r_1 = r_2$ and $y \in L_4$, or $r_1 \neq r_2$, $y = 1$, and $r_1 r_2 = r_2 r_1$. In this case, the convex closure of cL^* and dL^* is a generalized quadrangle.*

(3) *For each $k \geq 2$, the set N is a polar space of rank k if and only if $r_1 = r_2$ and $((L \cap {}^w L) \cup \{r_1\}, \{r_1\})$ is isomorphic to $(D_k, \{1\})$ or $(B_k, \{1\})$. In this case, the convex closure of $\{cL^*, dL^*\}$ is a nondegenerate polar space of rank $k + 1$.*

Exactly one of the three cases occurs.

Proof The points cL^* and dL^* belong to the shadow space $cS_w^* L^*$. By Proposition 11.4.10, this shadow is a convex subspace, so N is also contained in it. By use of Proposition 11.4.13, we can recognize the type S_w and hence the type of the shadow space $cS_w^* L^*$. Clearly, in the shadow on J of an apartment containing c and d (cf. Theorem 11.2.5(i)), the points $c\langle L\rangle$ and $d\langle L\rangle$ also have distance two, so we can apply that proposition.

(1) Suppose that either $y \notin L_4 \cup L_3\langle L \cap r_1^\perp\rangle L_3$, or $y \neq 1$ and $r_1 \neq r_2$, or $r_1 r_2 \neq r_2 r_1$. Then, in view of Lemma 11.4.12 and Proposition 11.4.13, $L \cap {}^w L \subseteq L \cap r_1^\perp$, and $N \subseteq c(L \cap r_1^\perp)^* r_1 L^* = c r_1 L^*$. In particular, $xL^* \in N$ implies $x \in c r_1^* L^*$. Choose $x \in c r_1^*$ such that $xL^* \in N$. Then, by the Gate Property 11.2.9, x is the unique chamber of $c r_1^*$ contained in $dL^* r_2^* L^*$. In particular, xL^* is the unique common neighbor of cL^* and dL^*.

Conversely, if $N = xL^*$ is a singleton, then so is the common neighbor set of the points cL^* and dL^* in the shadow space of an apartment containing these two (which exists by Theorem 11.2.5). If $r_1 = r_2$, then, by Proposition 11.4.13, $y \notin L_4 \cup L_3\langle L \cap r_1^\perp\rangle L_3$, as required. The case where $r_1 \neq r_2$ follows directly from the proof of Lemma 11.4.12.

(2) Suppose N is a coclique of size at least two. By (1), we either have $r_1 = r_2$ and $y \in L_4 \cup L_3\langle L_2\rangle L_3$, or $r_1 \neq r_2$, $y = 1$, and $r_2 \in L_2$. In the latter case, as we have seen in the first paragraph of the proof, cL^* and dL^* belong to $c\{r_1, r_2\}^* L^*$, which by Proposition 11.4.10 is a generalized quadrangle. Suppose, therefore, $r_1 = r_2$. If $y \in L_4$, then cL^* and dL^* belong to $c\{r_1, y\}^* L^*$, which is also a generalized quadrangle. It remains to consider $y = svt$, with $s, t \in L_3$ and $v \in \langle L \cap r_1^\perp\rangle$. Then, in the shadow of an apartment containing c and d, we can find a pair of distinct collinear points (viz., $c s r_1\langle L\rangle$ and $c r_1\langle L\rangle$), contradicting that N is a coclique.

The converse is clear from Proposition 11.4.10.

(3) Suppose that N contains a line. In view of (1) and (2), we have $r_1 = r_2 = r$ and $y = svt$, with $s, t \in L_3$ and $v \in \langle L_2 \cap {}^w L\rangle$. In view of Example 11.4.8, we only need to rule out Cases (iv) and (v) of Proposition 11.4.13 with $K \cong A_m$ for some $m \geq 2$. In Case (iv), the diagram M has a circuit $r_1, t, \ldots, s, r_1$, so cannot be of Weyl type. Suppose, therefore, that we are in Case (v), that is, $s = t$ and $v \in \langle L \cap {}^w L \cap r_1^\perp\rangle$, and $K \cong A_m$ for some $m \in \{2, \ldots, n\}$.

As $w \in \langle \{r\} \cup E\rangle = \langle \{r, s\} \cup F\rangle$, we have $S_w \subseteq \{r, s\} \cup F$, so $N \subseteq c S_w L^* = c(\{r\} \cup F)^* L^*$. The complement H of the connected component K of $\{r\} \cup E$ is contained in L_2 and commutes with K, so $c(\{r\} \cup E)^* L^* \subseteq c(K \cup H)^* L^* = c K^* H^* L^* = c K^* L^*$ (cf. Lemma 3.5.2). According to Proposition 11.4.10, this is the shadow on 1 of a building of type A_m. As cL^* and dL^* are contained in this shadow, they have distance at most one, contradicting that the distance between cL^* and dL^* is two. Therefore, $K \cong A_m$ is ruled out. $\qquad\square$

11.5 Parapolar Spaces

In this section we introduce and study parapolar spaces, a class of line spaces encompassing most shadow spaces of buildings (as will become clear from Theorem 11.5.12). According to Theorem 11.5.24, nondegenerate root filtration spaces, with the exclusion of those appearing in Theorem 7.9.19, are parapolar spaces. The result is actually stronger than that: it indicates precisely what kind of parapolar spaces the root filtration spaces are.

Our first goal, however, is Theorem 11.5.9, a characterization of parapolar spaces by descriptions of the space on the common neighbor set of two points at mutual distance two. Deviating from the introduction of this chapter, we sometimes use S to denote a certain subspace.

Definition 11.5.1 A pair of non-collinear points in a space Z is called *special* if they have a single common neighbor in the collinearity graph of Z. A pair of non-collinear points of Z is called *polar* if they have at least two non-adjacent neighbors in the collinearity graph of Z.

It is a consequence of Theorem 11.4.17 that, if M is a Weyl type, then two points at mutual distance two are either polar or special. In the thin case, this implies that they either have a unique common neighbor or they are contained in a graph isomorphic to $K_{m \times 2}$ for some $m \geq 2$, which is a convex subgraph. It is essential here that M is a Weyl type: in the shadow space of type $H_{3,1}$ (cf. Definition 11.4.1), two points at mutual distance two exist whose common neighbors form a clique on two points; see the left hand side of Fig. 1.2.

Most of the interesting shadow spaces of buildings are spaces of the following kind.

Definition 11.5.2 A *parapolar* space is a connected partial linear gamma space possessing a collection of convex subspaces, called *symplecta* (singular: *symplecton*), isomorphic to nondegenerate polar spaces of rank at least two, with the following two properties.

(1) Each pair of points at mutual distance two that is not special, is contained in a symplecton.
(2) Each line is contained in a symplecton.

Similarly, a space is called *weakly parapolar* if it satisfies the same conditions except possibly (2). A space is called *strongly parapolar* if it is parapolar and has no special pairs.

If all symplecta have rank k, or rank at least k, the space is said to have *polar rank k*, or *polar rank at least k*, respectively.

Remark 11.5.3 Definition 11.5.2 implies that any pair of points at mutual distance two in a weakly parapolar space is either special or polar. For, if p and q are points of a parapolar space at mutual distance two that do not form a special pair, then they are contained in a symplecton, which is a nondegenerate polar space of rank at least two containing $\{p, q\}^{\perp}$. By Lemma 7.4.8(ii), the latter is a nondegenerate polar space of rank at least one, and so contains two non-collinear points. This implies that (p, q) is a polar pair according to Definition 11.5.1.

Example 11.5.4 Linear spaces and generalized n-gons with $n > 4$ are examples of weakly parapolar spaces (with an empty set of symplecta) that are not parapolar.

By Theorem 11.4.17, shadow spaces of buildings of Weyl type are weakly parapolar (see also Theorem 11.5.12). Those of type $A_{5,3}$, $E_{6,1}$, and $E_{7,7}$ are strongly parapolar. To see this for $A_{5,3}$, take $L = \{1, 2, 4, 5\}$, so $L_4 = \emptyset$ and $L_3 = \{2, 4\}$ in the notation of Theorem 11.4.17. Now ${}^L W^L$, where $W = W(A_5)$, consists of the elements (written in $[5]^*$)

$$\emptyset, 3, 3243, 321432543.$$

Two chambers whose Weyl distance is the last word correspond to points at mutual distance three in the shadow space of a building of type A_5 on 3. The first two elements correspond to the identity and the collinearity relation. The third element, $w = 3243$, satisfies $L \cap {}^w L = \{2, 4\}$ and so, Theorem 11.4.17 gives that the convex closure of two points that are shadows of chambers at this Weyl distance is a polar space of rank three isomorphic to the Grassmannian of lines of a projective space of dimension three.

Also, the direct product space $\Pi \times Z$ of any projective space Π and any nondegenerate polar space Z is strongly parapolar.

As the intersection of convex subspaces of a line space is again a convex subspace, and the line space itself is convex, the notion of convex closure is well defined.

Lemma 11.5.5 *Suppose that Z is a weakly parapolar space.*

(i) *If (p, q) is a polar pair of Z, then the convex closure $S(p, q)$ of the subspace $\{p, q\}$ of Z is a symplecton of Z. Each symplecton of Z is obtained in this way.*
(ii) *The intersection of two distinct symplecta is a singular subspace of Z.*

Proof (i) Suppose that Y is a symplecton containing two points p and q, say, at mutual distance two in Z. Then Y is the convex closure of p and q. For, by definition, Y is convex, and so it contains $S(p, q)$. Conversely, let x be a point of Y. Take a quadrangle Q on p, u, q, v in Y for two common neighbors u, v of p and q. Such a quadrangle exists because Y is a nondegenerate polar space of rank at least two (cf. Lemma 7.4.8(ii)). Clearly, Q is contained in the convex closure of p and q and hence in Y. If x belongs to Q, there is nothing to show. Otherwise, as Y is a polar space, x has at least two non-collinear neighbors in Q, say a and b. But then x lies on a geodesic from a to b and hence belongs to the convex closure of Q. It follows that x lies in $S(p, q)$.

As for the final assertion of (i), since each symplecton is nondegenerate, it contains a polar pair. Hence (i).

(ii) Suppose that p and q are non-collinear points of Z belonging to a symplecton. As we have seen in (i), this symplecton must be $S(p, q)$. Therefore, the collinearity graph of the intersection of two distinct symplecta is a clique. Being the intersection of two subspaces, it is a subspace itself.   □

The lemma tells us that we need not specify the set $\mathcal{S}$ of symplecta of a weakly parapolar space: it is uniquely determined as the collection of all $S(p, q)$ for (p, q) a polar pair.

For a parapolar space of polar rank at least three, the existence of symplecta can be derived from some seemingly weaker axioms concerning the common neighbors of two points at mutual distance two. This is the content of Theorem 11.5.9 below.

Definition 11.5.6 A *preparapolar space* is a partial linear gamma space with the property that, for each pair (x, y) of points at mutual distance two, the subspace $\{x, y\}^\perp$ is either a singular space or a nondegenerate polar space of rank at least two.

Grassmannians of lines of polar spaces of rank three are examples of parapolar spaces that are not preparapolar.

By Theorem 7.9.5, nondegenerate root filtration spaces in which every line is on at least two maximal singular subspaces are preparapolar.

Under some mild conditions, preparapolar spaces will be shown to be weakly parapolar. The collections of subspaces defined in (11.9) below will be the symplecta needed for this conclusion. For any two points x, y of a preparapolar space Z at mutual distance two, write

$$S(x, y) = \left\{ z \in Z \mid z^\perp \cap l \neq \emptyset \text{ for each line } l \subseteq \{x, y\}^\perp \right\}. \qquad (11.9)$$

This set is of interest to us only if (x, y) is a polar pair. Clearly, for such a pair, $\{x, y\} \cup \{x, y\}^\perp \subseteq S(x, y)$.

Lemma 11.5.7 *For a preparapolar space Z, the following statements hold.*

(i) *If l, m are two lines and if there are two points of m collinear with a point of l, then each point of m is collinear with some point of l.*

(ii) *For each polar pair (x, y) of points from Z, the subset $S(x, y)$ is a subspace of Z.*

Proof (i) Let q and q' be points of m collinear with points p and p' of l, respectively. If $p = p'$ we need only apply the gamma space property of Z, so assume $p \neq p'$ and $p \notin q'^\perp$ as well as $p' \notin q^\perp$. Then $\{p', q\}^\perp$ is a polar space of rank at least two containing both p and q' and so $\{p, p', q, q'\}^\perp$ must contain at least two non-collinear points, say u and v. Now $\{u, v\}^\perp$ is a polar space containing l and m, and (i) follows.

(ii) Let m be a line intersecting $S(x, y)$ in distinct points c, d. Furthermore, let l be a line in $\{x, y\}^\perp$. We must show that each point of m is collinear with some point of l. Since c and d belong to $S(x, y)$, there are points in $l \cap c^\perp$ and in $l \cap d^\perp$. Hence, by (i), each point of m is collinear with some point of l, as required.   □

Proposition 11.5.8 *For each polar pair (x, y) of a preparapolar space, the subset $S(x, y)$ is a nondegenerate polar space.*

Proof Suppose that Z is a preparapolar space and let (x, y) be a polar pair of Z. We proceed in seven steps.

STEP 1. *In $S(x, y)$ every line on x has some point in $y^{\perp}$.*

Let l be a line on x in $S(x, y)$. Choose a point $z \in l \setminus \{x\}$. If $z^{\perp}$ contains $\{x, y\}^{\perp}$, then, for non-collinear $u, v \in \{x, y\}^{\perp}$, the polar space $A = \{u, v\}^{\perp}$ contains x, y, z, so $\{x, y\}^{\perp} \cap A \subseteq z^{\perp} \cap A$, and $\{x, y, z\}^{\perp} \cap A = \{x, y\}^{\perp} \cap A$. Lemma 7.4.8(iii), applied to the subspace $\{u, v, y\}^{\perp}$, gives that there is a point on l collinear with y.

Therefore, we may assume that there is a point $w \in \{x, y\}^{\perp}$ that is not collinear with z. Let m, n be lines on w in $\{x, y\}^{\perp}$ with $m \not\subseteq n^{\perp}$. Then, by the definition of $S(x, y)$, there are points $u \in m \cap z^{\perp}$ and $v \in n \cap z^{\perp}$. Observe that u and v differ from w as z is not collinear with w. Now u, v is a polar pair as they are not collinear and $\{u, v\}^{\perp}$ contains $\{x, y\}$. We conclude that the line l and the point y lie in the polar space $\{u, v\}^{\perp}$, whence $l \cap y^{\perp} \neq \emptyset$.

STEP 2. *If l is a line of $S(x, y)$ on x and $a \in S(x, y)$, then $l \cap a^{\perp} \neq \emptyset$.*

By Step 1 there is a point $b \in l \cap y^{\perp}$. If $a \in b^{\perp}$ we are done, so assume the contrary. Let m, n be lines of $\{x, y\}^{\perp}$ on b with $m \not\subseteq n^{\perp}$. By the definition of $S(x, y)$, there are points $u \in m \cap a^{\perp}$ and $v \in n \cap a^{\perp}$. Notice that u and v are non-collinear as they are distinct from a (for otherwise $b \perp a$). Since $x, y \in \{u, v\}^{\perp}$, the pair (u, v) is polar. The polar space $\{u, v\}^{\perp}$ contains l and a, and so $l \cap a^{\perp} \neq \emptyset$.

STEP 3. *If l is a line of $S(x, y)$ and $a \in \{x, y\}^{\perp}$, then $l \cap a^{\perp} \neq \emptyset$.*

Choose two points, b, c say, on l. Let m, n be lines on a in $\{x, y\}^{\perp}$ with $m \not\subseteq n^{\perp}$. By the definition of $S(x, y)$, there are points $b' \in m \cap b^{\perp}$ and $c' \in m \cap c^{\perp}$. If $b' \neq c'$, then by Lemma 11.5.7(i) every point of m, in particular a, is collinear with a point of l, so we are done. So we may assume $b' = c'$. This gives $l \subseteq b'^{\perp}$. Similarly, we may assume that there is a point $b'' \in n \cap l^{\perp}$. If b' and b'' are collinear, then they must coincide with a, in which case the assertion is proved. So we assume that they are non-collinear. As $a, b, c \in \{b', b''\}^{\perp}$, the pair (b', b'') is polar and contains both l and a, and so $l \cap a^{\perp} \neq \emptyset$.

STEP 4. *If l is a line of $S(x, y)$ then $l \cap x^{\perp} \neq \emptyset$.*

Let $u \in \{x, y\}^{\perp}$. If xu has two distinct points each of which is collinear with a point of l, then we are done by Lemma 11.5.7(i). So assume the contrary. By Step 3, u is collinear with a point of l and so it is the unique point on xu with this property. Hence, in view of Step 2, every point of l is collinear with the point u of xu.

Now take $v \in \{x, y\}^{\perp} \setminus u^{\perp}$, so that the pair (u, v) is polar (observe that $\{u, v\}^{\perp}$ contains x and y). By the same argument as for u, we may assume $v \in l^{\perp}$, and so both l and x belong to the polar space $\{u, v\}^{\perp}$, whence $l \cap x^{\perp} \neq \emptyset$.

STEP 5. *If l is a line of $S(x, y)$ and $a \in S(x, y) \cap x^{\perp}$, then $l \cap a^{\perp} \neq \emptyset$.*

By Step 1, there is a point $z \in xa \cap y^{\perp}$. According to Step 4, x is collinear with a point of l and according to Step 3, z is collinear with a point of l. Hence, by Lemma 11.5.7(i), each point of ax, in particular a, is collinear with a point of l.

STEP 6. *The subspace $S(x, y)$ of Z is a polar space.*

Let l be a line and a be a point of $S(x, y)$. We prove $l \cap a^{\perp} \neq \emptyset$. Suppose that we can find a point $z \in \{x, y\}^{\perp}$ that is not collinear with a. By Step 2, there is a point $c \in xz \cap a^{\perp}$. By Lemma 11.5.7(ii), the line ac lies in $S(x, y)$ and, by Step 4, there is a point $d \in ac \cap y^{\perp}$. According to Step 5, both c and d are collinear with a point of l, and hence, by Lemma 11.5.7(i), so is a (which lies on cd). Therefore, we (may) assume that $\{x, y\}^{\perp}$ is contained in $a^{\perp}$.

By Step 4, there are points $x' \in l \cap x^{\perp}$ and $y' \in l \cap y^{\perp}$. If $x' = y'$, then $x' \in l \cap \{x, y\}^{\perp} \subseteq l \cap a^{\perp}$ and we are done, so we may assume $x' \neq y'$. By Step 1, there exist $x'' \in xx' \cap y^{\perp}$ and $y'' \in yy' \cap x^{\perp}$. Clearly, $x'' \neq x$. If $x'' = x'$ or $x'' = y'$, then $x'' \in l \cap \{x, y\}^{\perp} \subseteq l \cap a^{\perp}$, so we may also assume $x'' \neq x, x', y'$. Furthermore, $x'' \not\perp y''$ for otherwise x'' would be collinear with both y and y'' and so with y', implying that y' is collinear with $x'x'' = xx''$, whence in $l \cap \{x, y\}^{\perp}$, giving a point of $l \cap a^{\perp}$. Thus, $\{x'', y''\}^{\perp}$ is a nondegenerate polar space of rank at least two, and so $\{x, y, x'', y''\}^{\perp}$ has at least two non-collinear points u, v. By the assumption that $\{x, y\}^{\perp}$ is contained in $a^{\perp}$, we have $a \in \{u, v\}^{\perp}$. Since also $l \subseteq \{u, v\}^{\perp}$, we can finish by invoking the polar space axiom for $\{u, v\}^{\perp}$.

STEP 7. *The subspace $S(x, y)$ of Z is a nondegenerate polar space.*

By Step 6, it is a polar space, so we only need to establish it is nondegenerate. If $z \in \mathrm{Rad}(S(x, y))$, then $z \in \{x, y\}^{\perp}$, and so $z \in \mathrm{Rad}(\{x, y\}^{\perp}) = \emptyset$. This shows that $S(x, y)$ is nondegenerate. $\qquad\square$

The rank of the polar space $S(x, y)$ of Proposition 11.5.8 is one more than the rank of $\{x, y\}^{\perp}$. The collection of $S(x, y)$ for x, y a polar pair of a preparapolar space Z will be used to show that Z is a weakly parapolar space.

Theorem 11.5.9 *Suppose that Z is a preparapolar space. If each pair of points of Z at mutual distance two is either polar or special, then Z is weakly parapolar. If, moreover, each line of Z lies on at least two maximal singular subspaces, then Z is parapolar.*

Proof We proceed in five steps, working with the symplecta $S(x, y)$ as in (11.9) for polar pairs (x, y) of Z. We will be using Proposition 11.5.8 many times to invoke polar space properties of these symplecta.

STEP 1. $S(x, y) = S(x, z)$ *whenever* $z \in S(x, y)$ *is non-collinear with* x.

Let l be a line of $\{x, y\}^{\perp}$ and $p \in S(x, z)$. We show $p^{\perp} \cap l \neq \emptyset$. If $l \subseteq z^{\perp}$, then $l \subseteq \{x, z\}^{\perp} \subseteq S(x, z)$, and the fact that $S(x, z)$ is a polar space implies $p^{\perp} \cap l \neq \emptyset$.

Suppose, therefore, $l \not\subseteq z^{\perp}$. There is a unique point w such that $\{w\} = l \cap z^{\perp}$. Take $u \in l \setminus \{w\}$ and $v \in xu \cap z^{\perp}$. Then u and v are distinct points. Observe that

$v, w \in \{x, z\}^\perp \subseteq S(x, z)$. In particular, $xu = xv \subseteq S(x, z)$ (as, by Lemma 11.5.7(ii), $S(x, z)$ is a subspace) and hence $u \in S(x, z)$. But then also $l = uw \subseteq S(x, z)$. As $S(x, z)$ is a polar space, this implies that p is collinear with a point of l.

We have shown $S(x, z) \subseteq S(x, y)$. By symmetry of the roles of y and z, we conclude $S(x, z) = S(x, y)$.

STEP 2. *$S(x, y) = S(a, b)$ whenever $a, b \in S(x, y)$ are non-collinear.*

If $u \in S(x, y) \setminus (a^\perp \cup x^\perp)$, then $S(x, y) = S(x, u) = S(u, x) = S(u, a) = S(a, u) = S(a, b)$ by repeated application of Step 1.

Apparently, we are done if $S(x, y) \nsubseteq a^\perp \cup x^\perp$. Hence we may assume $x \perp b$ and $a \perp y$. But, in view of interchangeability of the roles of a and b, we may also assume $x \perp a$ and $b \perp y$, that is x, a, y, b form a quadrangle.

We claim that $\{x, y\}^\perp$ is contained in $S(a, b)$. In order to prove the claim, suppose $z \in \{x, y\}^\perp$. If $z \in b^\perp \setminus a^\perp$, then z, a, b are in the polar space $S(x, y)$, and so there is a point $c \in a^\perp \cap zb$ distinct from z and b. Then $S(a, b)$ contains cb, whence z.

So, without loss of generality, we may assume $z \notin b^\perp \cup a^\perp$. Let m be a line in $\{a, b\}^\perp$. We must show that $z^\perp \cap m \neq \emptyset$. In the polar space $\{a, b\}^\perp$, we can find $u \in x^\perp \cap m$ and $v \in y^\perp \cap m$. If $x = u$, then $z \perp x = u \in m$, so $u \in z^\perp \cap m$, as required. So we may assume $x \neq u$ and, similarly, $y \neq v$.

Suppose that $u = v$. Then, taking $a' \in z^\perp \cap au$ and $b' \in z^\perp \cap bu$ (which exist since $z, a, b\ u$ all four belong to $\{x, y\}^\perp$), we find that both m and z belong to $\{a', b'\}^\perp$, whence $z^\perp \cap m \neq \emptyset$.

Assume, therefore, $u \neq v$. In the polar space $\{a, b\}^\perp$, the line yv meets $x^\perp$ in a point v' distinct from y and v. Now $v' \in \{x, y\}^\perp$, whence $yv = yv' \subseteq S(x, y)$, so $v \in S(x, y)$. Similarly, $u \in S(x, y)$, so $m = uv \subseteq S(x, y)$, whence indeed $z^\perp \cap m \neq \emptyset$. By definition of $S(a, b)$, we have found that z belongs to $S(a, b)$. This proves the claim.

To finish Step 2, assume now that $p \in S(a, b)$ and let l be a line of $\{x, y\}^\perp$. Then, by the claim, l is contained in $S(a, b)$. Since this is a polar space and $p \in S(a, b)$, we find $p^\perp \cap l \neq \emptyset$, proving that p belongs to $S(x, y)$. We have shown $S(x, y) \subseteq S(a, b)$. But by symmetry of the roles of x, y and those of a, b, we conclude $S(x, y) = S(a, b)$.

STEP 3. *If $x, y \in Z$ are polar, then $S(x, y)$ is the convex closure of $\{x, y\}$.*

Let $a, b \in S(x, y)$. If they are collinear, all points on the line ab belong to $S(x, y)$ because of Lemma 11.5.7(ii). Otherwise they are at mutual distance two and polar, with $S(a, b) = S(x, y)$ by Step 2. In particular, all points of $\{a, b\}^\perp$ belong to $S(x, y)$. Hence $S(x, y)$ is convex.

As for the converse, denote by K the convex closure of $\{x, y\}$. Clearly, $\{x, y\}^\perp$ is contained in K. Let $z \in S(x, y) \setminus \{x, y\}^\perp$. Choose $a, b \in \{x, y\}^\perp$ with $a \not\perp b$. Then $a, b \in K$. If $z \in \{a, b\}^\perp$, then z is in the convex closure of a and b, and hence in K. So suppose, after interchanging a and b if necessary, that $a \not\perp z$. Then, for $u \in xa \cap z^\perp$ and $v \in ya \cap z^\perp$, we find that $u \not\perp v$ and that z lies in the convex closure of u and v, and, since $u, v \in K$, also $z \in K$. This proves $S(x, y) \subseteq K$.

STEP 4. *Z is a weakly parapolar space.*

We need to verify (1) of Definition 11.5.2. By Step 3, each pair of points at mutual distance two that is not special, is contained in a symplecton. This proves (1).

STEP 5. *If each line is on at least two maximal singular subspaces, then Z is parapolar.*

We need to verify (2) of Definition 11.5.2. If l is a line of Z, then, by the hypothesis, there are non-collinear points x and y in $l^\perp$. As $l \subseteq \{x, y\}^\perp$, the pair (x, y) is polar and l is contained in the symplecton $S(x, y)$. Hence (2). $\square$

Corollary 11.5.10 *Every nondegenerate root filtration space is either a generalized hexagon or a parapolar space such that, for each symplecton X and each point x off X, the set $x^\perp \cap X$ is not a singleton. If, moreover, each line of this space is on at least two maximal singular subspaces, then it has polar rank at least three.*

Proof Let $Z = (E, F)$ be a nondegenerate root filtration space with respect to the quintuple $(E_i)_{-2 \le i \le 2}$. Definition 6.7.2 implies that Z is a partial linear gamma space.

Suppose that p and q are non-collinear points of Z having at least two common neighbors. By Remark 6.7.17, the pair (p, q) is polar (in the sense of Definition 6.7.2).

If each line in F is on at least two maximal singular subspaces, then, by Theorem 7.9.5, $E_{-1}(p, q) = \{p, q\}^\perp$ is a nondegenerate polar space of rank at least two. This proves that Z is a preparapolar space in which each symplecton is a polar space of rank at least three. We just saw that each pair of points at mutual distance two is either special or polar. As a consequence, Theorem 11.5.9 applies and gives that Z is a parapolar space. Otherwise, that is, if there is a line of Z lying on a single maximal singular subspace, Theorem 7.9.19 applies, and gives that Z is either a generalized hexagon or a parapolar space.

Assume that X is a symplecton of Z containing a point z collinear with a point x outside X. It remains to show that $x^\perp \cap X$ is strictly bigger than $\{z\}$. As X is nondegenerate, it has a point $y \in X \setminus z^\perp$. The triangle condition gives $x \in E_{\le 1}(y)$. If $x \in E_1(y)$, then, by the filtration around z, the point $[x, y]$ is collinear with z, so $[x, y] \in \{x, y, z\}^\perp \subseteq x^\perp \cap X$, and the line on $[x, y]$ and z lies in $x^\perp \cap X$. Therefore, we may assume $x \in E_{\le 0}(y)$. By Definition 6.7.2(5), $E_{\le 0}(y)$ is a subspace of Z, so the line l on x and z lies in it. If $u \in l \cap y^\perp$, then $u \in \{y, z\}^\perp \subset X$, so l, being the line on x and u, lies in X, implying $x \in X$, a contradiction. We conclude $l \subseteq E_0(y)$, so Proposition 7.9.4 shows there is a point $v \in y^\perp \cap l^\perp = \{x, y, z\}^\perp \subseteq x^\perp \cap X$, necessarily distinct from z, so the line on z and v lies in $x^\perp \cap X$, as required. $\square$

The critical property of Corollary 11.5.10 is attached to the following name.

Definition 11.5.11 A parapolar space Z is called a *root parapolar space* if, for each point x and symplecton Y of Z, we have $|x^\perp \cap Y| \ne 1$.

By Exercise 11.8.22, the diameter of a root parapolar space of rank at least three is at most three.

In Example 11.5.4 we already mentioned that all shadow spaces of buildings of Weyl type are weakly parapolar. Later, in Proposition 11.6.5 we will indicate the shadow spaces of buildings that are root parapolar spaces. Here, we indicate which shadow spaces are parapolar.

Theorem 11.5.12 *For nonempty $J \subseteq S$, the shadow space Z on J of a building of type M satisfies the following properties, where $L = S \setminus J$.*

(i) *All singular subspaces of Z are projective.*

(ii) *If M is of Weyl type, then Z is a weakly parapolar space. If, moreover, for no $r \in J$ the connected component of $L \cup \{r\}$ containing r has end node r and type A_m for some $m \geq 1$, then Z is a parapolar space.*

(iii) *If $(M, J) = (B_m, \{m - 1\})$ for some $m \geq 3$, $(F_4, \{2\})$, or $(F_4, \{3\})$, then each line of Z lies in a unique maximal singular subspace.*

Proof (i) It suffices to prove Pasch's Axiom 5.2.4. To this end, we show that every triple of pairwise collinear points can be embedded in a subspace isomorphic to a projective plane.

Let $r_1, r_2, r_3 \in J$. Suppose c, d, e are chambers of $\mathcal{C}$ with cL^*, dL^*, eL^* pairwise collinear points of Z, but not all three on the same line. By suitably adapting c, d, and e within their L-cells, we may assume that $\delta_{\mathcal{C}}(c, d) = r_1 w_d \in r_1 \langle L \rangle$, $\delta_{\mathcal{C}}(d, e) = r_2 w_e \in r_2 \langle L \rangle$, and $\delta_{\mathcal{C}}(e, c) = r_3 w_c \in r_3 \langle L \rangle$. Let c_1 be the chamber of cL^* nearest to e (cf. Gate Property 11.2.9), so $\delta_{\mathcal{C}}(e, c_1) = r_3$. Clearly, $r_2 w_e r_3$ is a minimal expression, as it corresponds to the type of a minimal gallery from d to c_1. By further adapting c and d, we may assume that $w_c r_1 w_d$ is a minimal expression for $\delta_{\mathcal{C}}(c_1, d)$. Now $w_c r_1 w_d = \delta_{\mathcal{C}}(c_1, d) = r_3 w_e^{-1} r_2$. This implies $r_1 \in \{r_2, r_3\}$. It readily follows that $r_1 = r_2 = r_3$. Put $r = r_1$. Rewriting the above equation, we find $w_c^{-1} = r w_d r w_e r$, so, by Lemma 4.5.19(iv), $w_c \in \langle L \cap r^\perp \rangle s \langle L \cap r^\perp \rangle$, where $s \in L$ satisfies $m_{r,s} = 3$. We claim that the three points belong to the shadow of $c_1\{r, s\}^*$ for a suitable chamber $c_1 \in cL^*$. Without loss of generality, we may assume that $w_d r w_e$ is minimal, so $w_d = w_d' s$ and $w_e = s w_e'$ for certain $w_d', w_e' \in \langle L \cap r^\perp \rangle$. This implies that we can find chambers $c_1 \in cL^*$ and $d_1 \in ds^* \cap c_1 r^*$, as well as $e_1 \in eL^* \cap dr^* s^*$. But then d_1 and e_1 are both in the shadow of $c_1\{r, s\}^*$ on J whereas c_1, d_1, and e_1 represent the same points of Z as c, d, e, respectively. Hence the claim. It follows from Proposition 11.4.10 that, up to isomorphism, cL^*, dL^*, eL^* are three points, pairwise r-collinear, in the shadow space of the building of type A_2 (from $\{r, s\}$) over r, whence a projective plane. This establishes (i).

(ii) By Lemmas 11.4.6 and 11.4.11(i), Z is a partial linear gamma space. If c and d are chambers of the underlying building, then the number of elements of J appearing in $\delta_{\mathcal{C}}(c, d)$ is an upper bound for the length of a path from cL^* to dL^* in Z, so Z is connected.

Suppose that cL^* and dL^* are at mutual distance two in Z. If they have more than one common neighbor in the collinearity graph, then, Case (2) or (3) of Theorem 11.4.17 prevails, and the geodesic closure of $\{cL^*, dL^*\}$ is a nondegenerate polar space of rank at least two. We conclude that Z is weakly parapolar.

Suppose that cr^*L^* is a line of type $r \in J$. By the additional assumption on J, there is a subset K of $L \cup \{r\}$ with $K \cap J = \{r\}$ such that (K, r) is of type $(D_m, 1)$ for some $m \geq 3$ or $(D_m, 1)$ for some $m \geq 2$. This gives that the shadow cK^*L^* is a symplecton containing the line cr^*L^*. Thus, Condition (2) of Definition 11.5.2 is satisfied, so Z is a parapolar space.

(iii) Suppose that there are two maximal singular subspaces, say X and Y, containing a line l. Then l consists of common neighbors of a point in $X \setminus Y$ and a point in $Y \setminus X$. According to the definition of parapolar space, the line is contained in a symplecton. However, application of Proposition 11.4.13 to the various cases shows that there is no subdiagram corresponding to such a symplecton. $\qquad\square$

The special case $(M, J) = (B_3, \{2\})$ of Theorem 11.5.12(iii) was dealt with in Theorem 7.9.19.

According to Exercise 11.8.18, two shadows $cK_1^*L^*$ and $dK_2^*L^*$ have an empty intersection if and only if the shortest element of $\langle K_1 \rangle \delta_C(c, d) \langle K_2 \rangle$ does not belong to $\langle L \rangle$.

Example 11.5.13 The Intersection Property 11.2.11 gives us information about the intersections of two shadows of flags. The strategy to determine the possible ways in which two shadows of cotype K_1 and K_2, respectively, may intersect is to determine first the set ${}^{K_1}W^{K_2}$ of (K_1, K_2)-reduced elements (cf. Definition 4.5.14), and next, for each $w \in {}^{K_1}W^{K_2}$, the intersection $K_1 \cap {}^w K_2$. The latter index set informs us about the feasible types of intersection.

For example, consider the shadow space Z on 1 of a building of type E_6. The relative positions of cells of type $K_1 = K_2 = \{1, 2, 3, 4, 5\}$ are indexed by the members w of ${}^{K_1}W^{K_2} = \{\emptyset, \; 6, \; 65423456\}$. In the respective cases we find $K_1 \cap {}^w K_2 = \{1, 2, 3, 4, 5\}, \{1, 2, 3, 4\}$, and $\{2, 3, 4, 5\}$. Now $L = S \setminus \{1\} = \{2, 3, 4, 5, 6\}$ and any two symplecta (cf. Definition 11.5.2) in Z, being shadows on 1 of flags of type 6, meet in, respectively, the whole symplecton, a maximal singular subspace of projective dimension four, and a point.

If K_1 and K_2 are as before but $J = \{2\}$, then Z is a shadow space of type $E_{6,2}$. From the previous paragraph, we know ${}^{K_1}W^{K_2}$ and the three possibilities for $K_1 \cap {}^w K_2$. The result is that two distinct shadows on 2 of flags of the type 6 either meet in the Grassmannian of lines of a projective space of dimension four (isomorphic to a shadow space of type $A_{4,2}$) or in a symplecton (isomorphic to a shadow space of type $D_{4,1}$).

Proposition 11.5.14 *Let $k \in \mathbb{N}$, $k \geq 0$ and $r \in [n]$. All singular subspaces of dimension k of the shadow space on r of a building of Weyl type over $S = [n]$ are shadow spaces of type $A_{k,1}$.*

Proof Let X be a singular subspace of dimension k of a shadow space Z of a building of Weyl type M as in the hypothesis. If $k \leq 1$, then the proposition is obvious. We proceed by induction on (n, k), and assume $k > 1$. Let Y be a singular subspace of X of dimension $k - 1$. By the induction hypothesis, there are nodes $r_1, \ldots, r_{k-1}$

of M such that $r_1 = r$ and $r_i \in L = S \setminus \{r\}$ with $H := \{r_1, \ldots, r_{k-1}\} \cong A_{k-1}$ and $Y = cH^*L^*$ for some chamber c. Suppose d is a chamber such that dL^* is in $X \setminus Y$.

We will re-use the notation $L_i = \{s \in L \mid m_{r,s} = i\}$ for $i \in \mathbb{N}$. By suitably adapting c within cH^* and d within dL^*, we may assume $w := \delta_C(c, d) \in {}^H W^L \cap \langle L \rangle r$. In particular, there exists $v \in {}^H W^{L_2} \cap \langle L \rangle$ such that $w = vr$. For each $a \in \langle H \setminus \{r\} \rangle$, we can find a chamber $x \in cH^*$ with $\delta_C(x, c) = ra$. The point $aL^* \in Y$ is collinear with dL^*, so $ravr = \delta_C(x, d) \in \langle L \rangle r \langle L \rangle$. By use of Lemma 4.5.19(iv), we find $av \in r \langle L \rangle r \langle L \rangle r \cap \langle L \rangle = \langle L_2 \rangle L_3 \langle L_2 \rangle$. On the one hand, taking $a = 1$, we see that there exists $s \in L_3$ such that $v \in \langle L_2 \rangle s \langle L_2 \rangle$ and, similarly, taking $a = r_2$, we find $r_2 v \in \langle L_2 \rangle r_2 \langle L_2 \rangle$. On the other hand, using the notation of Theorem 4.5.10(ii), we have $s \in S_{r_2 v} \setminus L_2 = \{r_2\}$, so $s = r_2$. Consequently, $w = vr = ur_2 r_1$ for some $u \in \langle H \rangle$. Fix a chamber e in the r-panel of d on a geodesic from c to d. Consideration of the point $e(L \setminus \{r_2\})^*$ and the singular subspace $c(K \setminus \{r\})^*(L \setminus \{r_2\})^*$ of the shadow space on r_2 of the $(S \setminus \{r\})$-cell of the building containing c and induction on the rank n, shows that $v = \delta_C(c, e) = r_k r_{k-1} \cdots r_2$ for some $r_k \in L \cap \{r_1, \ldots, r_{k-2}\}^\perp$ with $m_{r_k, r_{k-1}} = 3$. Now $w = r_k r_{k-1} \cdots r_1$, so $d \in cK^*L^*$, where $K := H \cup \{r_k\}$ is a subset of S on which M induces the Coxeter diagram A_k. As cK^*L^* is obviously a singular subspace of Z of type $A_{k,1}$ generated by Y and dL^*, this proves $X = cK^*L^*$, which finishes the proof of the proposition. $\square$

To finish this section, we show that there is a close connection between root parapolar spaces of polar rank at least three and nondegenerate root filtration spaces all of whose lines are on at least two maximal singular subspaces. Our goal is Theorem 11.5.24. We begin by deriving some properties from the characteristic root parapolar space condition; these are stated in Proposition 11.5.16.

Lemma 11.5.15 *Suppose that Z is a parapolar space of polar rank at least three. Each singular plane is contained in a symplecton of Z.*

Proof Let π be a singular plane. Take a line m on π. By Definition 11.5.2, m lies on a symplecton Y. Pick a plane ρ on m inside Y (this is possible as the polar rank is at least three). If $\rho = \pi$, we have embedded π in a symplecton. Suppose $\rho \neq \pi$, so $\rho \cap \pi = m$. Pick points $u \in \pi \setminus m$ and $v \in \rho \setminus m$. As $m \subseteq \{u, v\}^\perp$, the pair (u, v) is polar, and $\pi = \langle u, m \rangle \subseteq S(u, v)$. We conclude that, in each case, π is contained in a symplecton. $\square$

Proposition 11.5.16 *Suppose that Z is a thick root parapolar space.*

(i) *If Y is a symplecton of Z and x a point of Z off Y such that $l := x^\perp \cap Y$ is not empty, then either (x, y) is polar for each $y \in Y \setminus x^\perp$ or l is a line and (x, y) is special for each $y \in Y \setminus l^\perp$.*

(ii) *For each point x of Z, the set $C_Z(x)$ of points z of Z for which there is a symplecton containing both x and z, is a subspace of Z.*

(iii) *If Z has polar rank at least three, then, for each singular plane π and each line l meeting π at a point x of Z, either one or all lines m on x in π have the property that $l \cup m$ is contained in a symplecton.*

Proof (i) Let $y \in Y \setminus x^\perp$. As $Y \cap x^\perp$ is a singular subspace of the nondegenerate polar space Y containing a line, the set $Y \cap x^\perp \cap y^\perp$ is nonempty, so each point of Y is at distance at most two from x. Assume that (x, y) is special. Then l must be a line and there is a point u such that $y^\perp \cap l = \{u\}$.

Let $z \in Y \setminus l^\perp$. We need to show that (z, x) is special. Let v be the unique point of $z^\perp \cap l$. The line zv contains a point, w say, in $y^\perp$. Suppose that (x, z) is a polar pair. Then the symplecton, R say, on x and z contains the point w of $y^\perp$, so there is a line m in $R \cap y^\perp$. Now $x^\perp \cap m$ contains a point, which is necessarily collinear with x and y, so it must be equal to u, the only common neighbor of x and y. Notice that w is collinear with both v and u. If $u \neq v$, then $z \in wv \subseteq l^\perp$, a contradiction. As $u \in l$, we have $l^\perp \subseteq u^\perp$, so we have shown that each pair (z, x) with $z \in Y \setminus u^\perp$ is special. Repeating the argument for an arbitrary element of $z \in Y \setminus u^\perp$ instead of y, we find that all pairs (z, x) with $z \in (Y \cap u^\perp) \setminus l^\perp$ are also special.

(ii) Fix a point x, suppose that l is a line of Z having distinct points x_1 and x_2 in $C_Z(x)$, and assume $z \in l$. We need to show $z \in C_Z(x)$. This is clearly true if $z \in \{x_1, x_2\}$ or if $x^\perp \cap l \neq \emptyset$. Assume therefore that this is not the case. Now x_1 and x_2 have distance two to x, so there are symplecta $S_1 = S(x, x_i)$ for $i = 1, 2$. Observe that $x_1 \notin S_2$ and $x_2 \in x_1^\perp \cap S_2$, so by the hypothesis of the proposition, there is a line l_2 in $x_1^\perp \cap S_2$, and similarly, a line l_1 in $x_2^\perp \cap S_1$. As S_i is a polar space containing x and l_i containing the point x_i non-collinear with x, there is a unique point u_i in $x^\perp \cap l_i$, for each $i \in [2]$. As u_i lies in $\{x, x_1, x_2\}^\perp$, it belongs to $S_1 \cap S_2$ by the convex closure of S_1 and S_2. But this intersection is a singular space, so u_1 and u_2 are collinear. If $u_1 \neq u_2$, then the line $u_1 u_2$ is contained in $\{x, z\}^\perp$ and so (x, z) is a polar pair. Therefore, we may assume $u_1 = u_2$. As this must hold for any other choice of line l_i inside $x_{3-i}^\perp \cap S_i$, we may also assume $x_{3-i}^\perp \cap S_i = l_i$, and $S_1 \cap S_2 = xu_1$.

By (i) and the fact that $x \in S_2 \setminus l_2^\perp$ is polar with x_1, the pair (x_1, y) is polar for every point $y \in S_2 \setminus x_1^\perp$. Pick $y \in x_2^\perp \cap S_2 \setminus u^\perp$ (this is possible in view of Lemma 7.4.8(iii)). Then z belongs to $S(x_1, y)$ and, as z is not collinear with y (for otherwise $y \in l^\perp \cap S_2 = l_2$), it is polar with y. Now x_1 and x_2 belong to $S(y, z) \setminus x^\perp$ and are polar with x. If they are both collinear with $x^\perp \cap S(y, z)$, then so is each point of l, so $\{z, x\}^\perp$ contains a line and (x, z) is polar. Otherwise, (i) gives that each point of $S(y, z) \setminus x^\perp$ is polar with x, so (z, x) is polar, again.

(iii) Let l, π, and x be as assumed in (iii). By Lemma 11.5.15 there is a symplecton, Y say, on π. If $l \subset Y$, then we are done, so assume this is not the case. By Definition 11.5.11, $l^\perp \cap Y$ contains a line, n say, on x. By the polar space properties, $n^\perp \cap \pi$ contains a line, m, on x. As $n \subseteq l^\perp \cap m^\perp$, there is a symplecton containing l and the line m of π. As a consequence of (ii), the set of points of π polar with a fixed point of $l \setminus \{x\}$, is a subspace of π containing m, and so is either equal to m or to all of π. $\qquad\square$

Definition 11.5.17 Since a parapolar space Z is a gamma space, so is $x^\perp$ for every point x. By Proposition 7.4.10, the nondegenerate quotient space $x^\perp/\{x\}$, defined in Definition 7.4.9 in case Z is nondegenerate, is also a gamma space. It is called the *local space* of Z at x.

Proposition 11.5.18 *Let Z be a thick parapolar space of polar rank at least three containing a point p and denote the local space $p^\perp/\{p\}$ of Z at p by V.*

(i) *The space V is strongly parapolar and its symplecta are of the form $(p^\perp \cap Y)/\{p\}$ for Y a symplecton of Z containing p.*

(ii) *Each pair (x, y) of points in $p^\perp \setminus \{p\}$ is special if and only if the points xp and yp of V are at mutual distance at least three.*

(iii) *If, in addition, Z is root parapolar, then, for each point x of V, the set $x^\perp \cap Q$ is nonempty whenever Q is a symplecton of V, and the set $V_{\leq 2}(x)$ of points of V at distance at most two from x, is a subspace of V.*

Proof Straightforward. For instance, Condition (2) of Definition 11.5.2 for V follows from Lemma 11.5.15. $\qquad\square$

Example 11.5.19 The dual polar space of a nondegenerate polar space of rank three is a strongly parapolar space satisfying the conclusion of Proposition 11.5.18(iii). Other examples are direct products of a line and a nondegenerate polar space; see Exercise 11.8.23.

Proposition 11.5.20 *Let Z be a thick strongly parapolar space such that for each point x of Z, the set $x^\perp \cap S$ is nonempty whenever S is a symplecton, and the set $Z_{\leq 2}(x)$ of points at distance at most two from x, is a subspace of Z.*

(i) *For each point x of Z, the set $Z_{\leq 2}(x)$ of points at distance at most two from x, is a geometric hyperplane, or all of Z.*

(ii) *If S is a symplecton and x is a point of Z such that $S \not\subseteq Z_{\leq 2}(x)$, then there is a unique point x_S of S such that $x^\perp \cap S = \{x_S\}$ and each point of $S \setminus x_S^\perp$ has distance three to x.*

(iii) *If S is a symplecton and x is a point of Z such that $S \not\subseteq Z_{\leq 2}(x)$, then for each symplecton T containing x, either $S \cap T$ is a line on x_S, or $S \cap T = \emptyset$. Moreover, both cases occur.*

(iv) *For symplecta S and T with $S \cap T = \emptyset$, write $T_S := \bigcap_{x \in T} S \cap Z_{\leq 2}(x)$. This is a subspace of S. If, in this setting, $y \in S \setminus T_S$, then $y^\perp \cap T_S \subseteq y_T^\perp$.*

(v) *If S and T are symplecta with $S \cap T = \emptyset$, then S_T is either empty, a singular subspace of T containing a line, or all of T. Moreover, if $T_S \neq \emptyset$ and $y \in S \setminus T_S$, then $y_T \in S_T$.*

Proof (i) Let l be a line of Z. As $Z_{\leq 2}(x)$ is a subspace by assumption, it suffices to show that l has a point in common with $Z_{\leq 2}(x)$. By Condition (2) for parapolar spaces, there is a symplecton Y containing l. If $x \in Y$, then clearly, $l \subseteq Z_2(x)$, so we may assume $x \notin Y$. By the hypotheses, there is $y \in x^\perp \cap Y$, and by the polar axiom for Y, there is $z \in y^\perp \cap l$, so $z \in l \cap Z_{\leq 2}(x)$, as required.

(ii) By assumption, there is a point $y \in S$ at distance three from x. If $x^\perp \cap S$ would contain a line, then, by the polar axiom for S, the set $y^\perp \cap x^\perp \cap S$ would be nonempty, contradicting that x and y are at mutual distance three. So, by assumption, $x^\perp \cap S$ is a singleton, $\{x_S\}$. Now $x_S^\perp \cap S$ is a geometric hyperplane of S contained in the proper subspace $Z_{\leq 2}(x) \cap S$ of S. By Corollary 8.1.8, $x_S^\perp \cap S = Z_{\leq 2}(x) \cap S$, so each point of $S \setminus x_S^\perp$ is at distance three from x.

(iii) Let $y \subset S$ be a point at distance three from x. If $S \cap T$ contains a line, this line contains the points x_S and y_T which must be distinct, and so the line coincides with $x_S y_T$, in which case we must have $S \cap T = x_S y_T$.

Suppose that $S \cap T = \{p\}$ for a point p. As p and x are in T, they have distance at most two, so $p \in S \cap Z_{\leq 2}(x) \subseteq x_S^\perp$, and p is collinear with x_S. If p and x are non-collinear, then T, being convex, contains their common neighbor x_S, so $p x_S$ is a line in $S \cap T$, a contradiction. If p and x are collinear and distinct, then $p x_S$ is a line in $x^\perp \cap S$ on which $y^\perp$ has a point, contradicting that the distance between x and y is three. Therefore, $p = x_S$, and similarly, $p = y_T$. But then p is a common neighbor of x and y, contradicting that x and y are at mutual distance three. We conclude that $S \cap T$ is either empty or equal to $x_S y_T$.

Let Q be a symplecton on $x x_S$, so $Q \cap S$ is a line on x_S. Pick a line n on x inside Q but distinct from $x x_S$. By the polar pace axioms, there is a unique point r such that $y_Q^\perp \cap n = \{r\}$. The pair (r, y) is polar and so $\{r, y\}^\perp$ contains a point $z \notin y_Q^\perp$. The line rz does not lie in Q as $Q \cap S(r, y)$ contains y_Q, which is not collinear with z, so the symplecton $S(x, z)$ (notice that z and x are not collinear as $z \perp y$) is distinct from Q. It meets Q at n and so does not contain x_S, so $S(x, z)$ is a symplecton on x with $x_S \notin S(x, z) \cap S$.

(iv) By the hypotheses, each $Z_{\leq 2}(x)$ is a subspace of Z, so S_T, being the intersection of such subspaces and T, is a subspace.

Let $z \in y^\perp \cap T_S$. Since $y \notin T_S$, we have $y \neq z$. As yz is a thick line, there is $w \in yz \setminus \{y, z\}$. Because T_S is a subspace containing z but not y, we have $w \notin T_S$. If $w_T = y_T$, then $z \in wy \subseteq y_T^\perp$, and we are done. Assume, therefore, $w_T \neq y_T$. Now $y \perp w \perp w_T$ shows that y and w_T have distance two, which, by (ii), implies $w_T \perp y_T$. Suppose $x \in T \setminus y_T^\perp$. As $Z_{\leq 2}(x)$ is a subspace of Z containing z but not y, the points x and w are at mutual distance three, and so $x \in T \setminus w_T^\perp$. The same argument works with w and y interchanged and shows $T \setminus w_T^\perp = T \setminus y_T^\perp$. Nondegeneracy of T forces the contradiction $w_T = y_T$.

(v) We first show that if S_T is non-singular, it coincides with T. Suppose that u and v are non-collinear points of S_T. If $h \in \{u, v\}^\perp \setminus S_T$, then $h_S \perp \{u, v\}$ by (iv), and so $h_S \in S(u, v) = T$, contradicting $S \cap T = \emptyset$. Therefore, $\{u, v\}^\perp \subseteq S_T$. By repetition of this argument and using that S_T is a subspace of T (see (iv)), we find that S_T is the convex closure of u and v, and so coincides with T.

Let $y \in S \setminus T_S$. If $x \in S_T$, then $d_Z(x, y) \leq 2$, so $x \perp y_T$ by (ii). This proves $y_T \in S_T^\perp$, so $S_T \cap y_T^\perp = S_T$. Suppose $y_T \notin S_T$. Then $(y_T)_S = y$ and by (iv), $S_T = S_T \cap y_T^\perp \subseteq y^\perp$. As $y^\perp \cap T = \{y_T\}$, we find $S_T \subseteq y^\perp \cap T \setminus \{y_T\} = \emptyset$. Therefore, either $S_T = \emptyset$ or $y_T \in S_T$. We show that the former does not occur if $T_S \neq \emptyset$, by selecting $z \in S \setminus T_S^\perp$. Such an element exists as T_S, being distinct from S, must be a singular subspace by the above paragraph and S is a nondegenerate polar space. Arguing as for y, we see that $z_T \notin S_T$ would imply $z = (z_T)_S \perp T_S$, contrary to the choice of z. We conclude that $z_T \in S_T$, so $S_T \neq \emptyset$, and hence $y_T \in S_T$, which proves the last assertion of (v).

Fig. 11.4 The collinearity
graph on eight points in Z

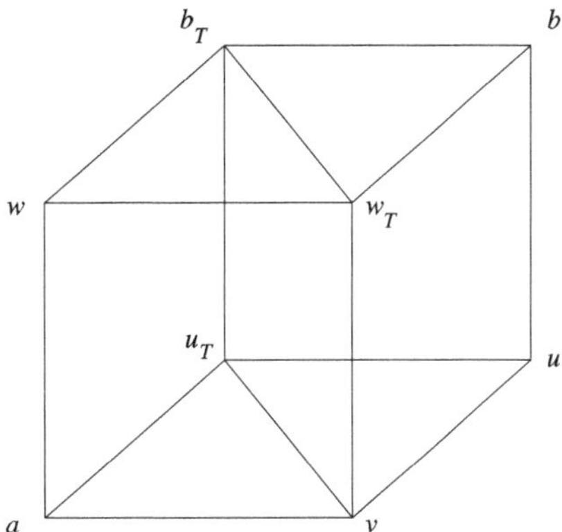

We finish by showing that S_T is not a singleton. If it were equal to $\{v\}$ for some
point v of T, then, for each $y \in S \setminus T_S$, we have $y_T = v$ by the above. This would
imply $S \setminus T_S \subseteq v^\perp \cap S$, which is a singular subspace of S. As $T \setminus S_T = T \setminus \{v\}$ is
not empty, neither is $S \setminus T_S$. This implies that T_S is a singular subspace of S whose
complement in S is a clique, which is a contradiction with Theorem 7.4.13. $\square$

Proposition 11.5.21 *Let Z be a thick strongly parapolar space such that for each
point x of Z, the set $x^\perp \cap S$ is nonempty whenever S is a symplecton, and $Z_{\leq 2}(x)$
is a subspace of Z. If Z has diameter three, then each path of length two in Z can
be extended to a path of length three.*

Proof We proceed in two steps.

STEP 1. *For each point of Z, there is a point at distance three from it.*

By the hypothesis, there are points a, b at mutual distance three in Z. Let v be
an arbitrary neighbor of a in the collinearity graph of Z. We claim that v has a
point in Z at distance three from it. Suppose not. Then $d_Z(b, v) = 2$, so, by the
strongly parapolar property, there is a symplecton $S := S(b, v)$ on b and v and, by
Proposition 11.5.20(iii), there is a symplecton T on a disjoint from S. As $v \in S$ and
$Z_{\leq 2}(v) = Z$, we have $v \in T_S$.

Observe that $a \in T \setminus S_T$ and $b \in S \setminus T_S$. It follows from Proposition 11.5.20(v)
that T_S is a singular subspace of Z, and so in the nondegenerate polar space S,
there is a point $u \in \{b, v\}^\perp \setminus T_S$ (cf. Lemma 7.4.8(ii)). By Proposition 11.5.20(ii),
the point u_T exists. As $d_Z(u, a) = 2$, and $a \in T$, we must have $a \sim u_T$; similarly,
$d_Z(b, u_T) = 2$ implies $b_T \sim u_T$.

As T is a nondegenerate polar space, there is $w \in \{a, b_T\}^\perp \setminus u_T^\perp$ (cf. Lemma 7.4.8(ii)). Because S_T is a singular subspace of T containing u_T (cf. Proposition 11.5.20(v)), we have $w \in T \setminus S_T$, so w_T exists. Now, $d_Z(w, v) = 2 = d_Z(w, b)$
gives $v \sim w_T \sim b$. The collinearities established between the eight points a, v, u,
u_T, w, b_T, b, w_T are as in a cube; see Fig. 11.4. But there are (at least) two more:
by Proposition 11.5.20(iv) we also have $u_T \sim v$ and $b_T \sim w_T$.

We have seen $d(a, u) = d(b, w) = 2$. Put $X = S(a, u)$ and $Y = S(b, w)$. We claim $X \cap Y = \emptyset$. For, suppose $x \in X \cap Y$. Clearly, $a \notin Y$ (for otherwise, $bb_T \cap a^\perp$ would contain a common neighbor of u and b). As $b \in Y$ and $d(a, b) = 3$, the point a_Y exists and coincides with w. But $d(x, a) \leq 2$ (as they are both in X), so $x \perp a_Y = w$. Now $w \in \{a, x\}^\perp$. If $a \not\perp x$, then $w \in X$, contradicting $d(w, u) = 3$, so $a \perp x$. Recall that $a = x$ is impossible as $a \notin Y$. Now there is a point in $ax \cap u^\perp$ collinear with w, again contradicting $d(w, u) = 3$. Hence $X \cap Y = \emptyset$.

Notice that X_Y and Y_X are non-empty. Indeed, as $v \in X$ and $Z_{\leq 2}(v) = Z$, we have $v \in Y_X$, which implies $X_Y \neq \emptyset$. As $d(u, w) = 3$ whereas $u \in X$ and $w \in Y$, the subspaces X_Y and Y_X are singular and $w \in Y \setminus X_Y$. But Proposition 11.5.20(v) implies $w = a_Y \in X_Y$, which leads to the final contradiction, showing that there is a point in Z at distance three from v.

STEP 2. *Each path of length two in Z can be extended to a path of length three.*

Let $x \in Z$. Put $A = \{z \in Z_2(x) \mid z^\perp \subseteq Z_{\leq 2}(x)\}$ and $B = Z_2(x) \setminus A$. We need to show that A is empty. Suppose $a \in A$. By Step 1, $B \neq \emptyset$. Choose $b \in B$, and $f \in b^\perp \cap Z_3(x)$. Suppose $a \perp b$. By definition of A, we have $a \not\perp f$, so there is a symplecton $S(a, f)$. As $d(x, f) = 3$, the point x_S exists. Now $a^\perp \cap S \subseteq Z_{\leq 2}(x) \cap S = x_S^\perp \cap S$, so, as S is a nondegenerate polar space, $a = x_S \perp x$, contradicting $a \in Z_2(x)$.

Suppose $d_Z(a, b) = 2$. Consider $T := S(a, b)$. If T contains a point of $Z_3(x)$, then x_T exists and $Z_{\leq 2}(x) \cap T = x_T^\perp \cap T$. It follows that $x_T \in \{a, b\}^\perp$. As T is a nondegenerate polar space, there is a point $u \in \{a, b\}^\perp \setminus x_T^\perp$, which necessarily belongs to $Z_3(x)$, contradicting $a \in A$. Therefore, $T \subseteq Z_{\leq 2}(x)$. If $x \in T$, then $T \setminus x^\perp$ is connected by Theorem 8.1.4 (we use here that lines are thick); if $x \notin T$, then $x^\perp \cap T$ is a singular subspace and so $T \setminus x^\perp$ is again connected. Applying the argument of the previous paragraph, we find $T \setminus x^\perp \subseteq B$, contradicting $a \in A$. We conclude $d_Z(a, b) = 3$.

Take a symplecton Q on b. Now a_Q exists and belongs to B. But then a, being collinear with a_Q also belongs to B by the first paragraph. This contradiction ends the proof of the proposition. $\qquad\square$

Corollary 11.5.22 *Let Z be a thick root parapolar space of diameter three and of polar rank at least three. If Z contains a special pair, then, for each pair (b, c) of distinct collinear points of Z, there is a point a collinear with c such that (a, b) is special.*

Proof Let c be the common neighbor of a special pair of points of Z. Assume that v is a point collinear with but distinct from c. We claim that v is also the common neighbor of a special pair of points. To see this, take a symplecton Q on the line vc. Pick a point $p \in c^\perp \cap Q \setminus v^\perp$. By Proposition 11.5.18(i), the local space $c^\perp/\{c\}$ is a strongly parapolar space of polar rank at least two, and by Proposition 11.5.18(iii), it satisfies the conditions of Proposition 11.5.20. Pick a path of length two from pc to vc in $c^\perp/\{c\}$. By Proposition 11.5.21, there is a point $y \in c^\perp \setminus \{c\}$ such that yc

is collinear with vc and at distance three from pc in $c^{\perp}/\{c\}$. In view of Proposition 11.5.16(i) we conclude that $y^{\perp} \cap Q$ is a line and each point of $Q \setminus (vc)^{\perp}$ is at distance three from y. Taking a point $z \in Q \cap v^{\perp} \setminus c^{\perp}$, we find that v is the common neighbor of the special pair (y, z), which proves the claim. As Z is connected, we conclude that each point is the common neighbor of a special pair.

To finish, suppose that c and b are distinct collinear points, so bc is a point of $c^{\perp}/\{c\}$. By the claim, the latter space has diameter three, and so Proposition 11.5.21 gives that any of its points has a point at distance three from it. In particular, by Proposition 11.5.18(ii), there is a point a collinear with c such that (a, b) is special, as required. $\qquad\qquad\square$

Lemma 11.5.23 *Every root parapolar space of polar rank at least three satisfies the following properties.*

(i) *Each line is contained in a singular plane.*
(ii) *If x, y, z are distinct points with $x \sim y$ and (y, z) polar, then there is a point collinear with x, y, and z.*
(iii) *All non-collinear pairs of points of a pentagon are polar.*

Proof Let Z be a root parapolar space of polar rank at least three.

(i) Let l be a line of Z. By Definition 11.5.2(2), l is contained in a symplecton, which, by the hypotheses, has rank at least three. This implies that the symplecton has a singular plane containing l.

(ii) Let S be the symplecton containing both y and z. Then x is collinear with the point y of S, whence, by the root parapolarity of Z, with a line m of S containing y. Since S is a polar space containing m and z, there is a point on m collinear with z. This point is as required.

(iii) Let a, b, c, d, e be a pentagon. We first show that at least one non-collinear pair must be polar. Suppose the contrary, that is, all non-collinear pairs are special. By (i), there is a plane π on de. By Proposition 11.5.16(iii) there is a point y on $\pi \setminus \{d\}$ and a symplecton S on y and c. By (ii), there is a point z collinear with b, c, and y. Proposition 11.5.16(iii) applied to the point a and the plane bcz, shows that there is a point f on cz and a symplecton T containing f and a. Both f and the line yd lie in the polar space S, and so there is a point g on yd collinear with f. Now consider the pentagon P consisting of a, e, g, f, b. Observe that it is a pentagon indeed (e.g., if $a{\sim}g$, then (a, d) is a polar pair, a contradiction). In P the pair (a, f) is polar.

Suppose that (e, f) is a special pair, so $g = [e, f]$. The point a is collinear with e and polar with f, so, by (ii), it must be collinear with g, contradicting that P is a pentagon. Therefore, (e, f) is a polar pair. As b is collinear with f, it follows again from (ii) that f is collinear with $[b, e] = a$. This contradicts again that P is a pentagon. The conclusion is that there are no pentagons all of whose non-collinear pairs are special.

Suppose now that a, b, c, d, e are a pentagon in which (a, c) is a polar pair. The symplecton S on a and c contains b as well as a point f collinear with a, c and d. In particular, f, being collinear with c, is distinct from e, so the pair a, d, having two distinct neighbors (viz. e and f), must be polar. Going round the pentagon, we see that all non-collinear pairs are in fact polar. $\qquad\square$

Theorem 11.5.24 *Let Z be a thick line space. Then Z is a nondegenerate root filtration space each of whose lines is on at least two maximal singular subspaces if and only if it is a root parapolar space of polar rank at least three and of diameter three.*

Proof Let Z be a thick nondegenerate root filtration space each of whose lines is on at least two maximal singular subspaces. Corollary 11.5.10 gives that Z is a root parapolar space of polar rank at least three. Furthermore, it follows from Remark 6.7.17 that its diameter is three. This proves one implication.

As for the 'if' part, suppose that $Z = (E, F)$ is a thick root parapolar space of polar rank at least three and of diameter three. By definition, Z is a partial linear gamma space. We verify Conditions (1)–(8) of Definition 6.7.2 for Z to be a nondegenerate root filtration space with respect to the following partition $(E_i)_{-2 \le i \le 2}$ of $E \times E$.

(-2) E_{-2} is the equality relation.
(-1) $E_{\le -1}$ is the collinearity relation.
(0) E_0 is the polar relation.
(1) E_1 is the special relation.
(2) E_2 is the complement in $E \times E$ of $\bigcup_{-2 \le i \le 1} E_i$.

Conditions (1) and (2) follow directly from the definitions of E_{-2} and E_{-1}.

(3) Let $x, u, v \in E$ and $i, j \in \{-2, \ldots, 2\}$. Suppose $(u, v) \in E_1$ and assume $u \in E_i(x)$ and $v \in E_j(x)$. We need to show that $[u, v] \in E_{\le i+j}(x)$. Without loss of generality, we may take $-2 \le i \le j \le 2$ with $i + j < 2$.

Suppose $i = -2$. Then $u = x$ and $j = 1$ so $[u, v] = [x, v]$ is collinear with x, as required.

Next, suppose $i = -1$. If $j = -1$, then $x = [u, v]$ by the definition of special pair, so there is nothing to prove. If $j = 0$, then by Lemma 11.5.23(ii) there is a point collinear with x, u and v, which must be $[u, v]$, and so $[u, v] \in E_{\le -1}(x)$, as required. If $j = 1$, then due to Lemma 11.5.23(iii) the 5-circuit $x, u, [u, v], v, [x, v]$ cannot be a pentagon. Since no other pair from the five points can be collinear, $[u, v]$ and $[x, v]$ must be collinear. If they are equal, then $[u, v]$ is collinear with x, and there is nothing left to prove. Otherwise, there are at least two common neighbors to x and $[u, v]$, so $[u, v] \in E_{\le 0}(x)$, as required. If $j = 2$, then $[u, v]$, being a neighbor of u, is at distance at most two from x, whence in $E_{\le 1}(x)$.

Now suppose $i = 0$. If $j = 0$, take u_1 to be a point collinear with x, u and $[u, v]$ (which exists by Lemma 11.5.23(ii)) and, likewise, v_1 to be a point collinear with x, v and $[u, v]$. If u_1 and v_1 are distinct, then $[u, v]$ must be in $E_{\le 0}(x)$, so assume they coincide. But then u_1 is a common neighbor of u and v and so coincides with $[u, v]$, whence $[u, v] = u_1 \in E_{-1}(x)$. This shows that $[u, v]$ belongs to $E_{\le 0}(x)$.

If $j = 1$, then by Lemma 11.5.23(ii) there is a point collinear with x, u, and $[u, v]$. Consequently, $[u, v]$ and x are at mutual distance at most two, whence $[u, v] \in E_{\le 1}(x)$.

This ends the verification of the filtration axiom (3).

(4) Suppose $(x, y) \in E_2$. If a were to belong to $E_{\leq -1}(x) \cap E_{\leq 0}(y)$, then, by Lemma 11.5.23(ii), there is a point b collinear with x, y (and a), so $y \in E_{\leq 1}(x)$, contradicting the hypothesis $(x, y) \in E_2$. Hence $E_{\leq -1}(x) \cap E_{\leq 0}(y)$ is empty.

(5) Let $x \in E$. Since Z is a gamma space, $E_{\leq -1}(x)$ is a subspace. Suppose that $y, z \in E_{\leq 0}(x)$ are distinct and collinear. In view of the previous case, we may assume that at least one of these, say z, is not collinear with x. Thus, (x, z) is a polar pair. If some point of yz is collinear with x, then yz is contained in the symplecton determined by x and z, and so lies in $E_{\leq 0}(x)$. Therefore, we may assume that (x, y) is also a polar pair. By Lemma 11.5.23(ii), there is a point u in E such that u is collinear with x, y, and z. Applying Proposition 11.5.16(iii) to the line xu and the plane uyz, we conclude that all points of yz are polar with x. Hence $E_{\leq 0}(x)$ is a subspace of Z.

(6) Let $x \in E$. We show that $E_{\leq 1}(x)$ is a geometric hyperplane of Z. Suppose that S is a symplecton. We claim that there is a symplecton T containing x and meeting S nontrivially. To see this, we argue by length of a path in the collinearity graph of Z (which is connected as it is a parapolar space) from x to S. If this length is 0, we can take $T = S$ and if it is 1, then the claim follows from the existence of a symplecton on a line joining x to a point of S (a parapolar space property). If the length of this path is two, let y be a point collinear with x and to a point of S. By the root parapolar space property there is a line of S collinear with y, and by Proposition 11.5.16(iii), at least one point, say z on this line is polar with x. Now the symplecton on x and z is as required. Suppose therefore, that the distance of x to S is at least three. By induction on the length, there is a symplecton U containing a point v collinear with x and point w of S. By Lemma 11.5.23(ii) there is a point c collinear with x, v, and w. But then x has distance at most two to S, a contradiction, so we have established the claim.

Suppose that $l \in F$ has two points in $E_2(x)$. Take a symplecton S containing l. By the above claim and Condition (5), there is a point, say p, of S polar with x. By the polar space property of S, there is a point q on l collinear with p. By Lemma 11.5.23(ii), there is a point a collinear with p, q and x. But then x has distance at most two to the point q of l, so l has a point in $E_{\leq 1}(x)$.

In order to establish that $E_{\leq 1}(x)$ is a geometric hyperplane of Z, it remains to show that $E_{\leq 1}(x)$ is a subspace. Suppose that y and z are distinct collinear points of $E_{\leq 1}(x)$. In view of Condition (5), we may assume that at least one of them, say z, is special to x. If y and x lie in a symplecton S, then by Lemma 11.5.23(ii) there is a point u collinear with x, y, z, in which case all points of yz are collinear with u whence in $E_{\leq 1}(x)$. Therefore, we may assume that both y and z are special to x. But then, by Lemma 11.5.23(iii), the 5-circuit x, $[x, y]$, y, z, $[z, x]$ is not a pentagon. This implies that $[x, y]$ and $[x, z]$ are collinear (observe that y cannot be collinear with $[x, z]$ because of $(x, y) \in E_1$). If $[x, y] = [x, z]$, then $\langle y, z, [x, y] \rangle$ is a singular plane, and $yz \subseteq [x, y]^{\perp}$, so each point of yz has distance at most two to x, proving $yz \subseteq E_{\leq 1}(x)$. Otherwise, each point of yz is collinear with a point of the line joining $[x, y]$ and $[x, z]$ (as all these points belong to a single symplecton), and so has distance at most two to x, and again belongs to $E_{\leq 1}(x)$. This proves that $E_{\leq 1}(x)$ is a subspace, whence (6).

Table 11.2 Root types (M, J), root orbit lengths l, and ranks of symplecta (srk)

M	J	l	srk	M	J	l	srk
A_n	$\{1, n\}$	$\binom{n+1}{2}$	2	E_7	$\{1\}$	126	5
B_n	$\{1\}$	$2n$	n	E_8	$\{8\}$	240	7
B_n, D_n	$\{2\}$	$2n(n-1)$	$3, n-1$	F_4	$\{1\}, \{4\}$	24	3
E_6	$\{2\}$	72	4	G_2	$\{1\}, \{2\}$	6	–

Condition (7) follows from Corollary 11.5.22 and Lemma 6.7.13 (recall that this lemma was proved from Conditions (1)–(6) only), while Condition (8) is a property of parapolar spaces and therefore also of Z. $\qquad\square$

11.6 Root Shadow Spaces

In this section we introduce and study root shadow spaces of buildings of Weyl type. In Corollary 11.6.6 we find that root shadow spaces are root filtration spaces. Each building of Weyl type, has at least one root shadow space that is a nondegenerate root filtration space. The converse will be discussed in the next section (Theorem 11.7.11).

In the presence of a Coxeter type M, we will write Φ for the root system of type M, as constructed in Definition 4.5.1.

Definition 11.6.1 A *root type* is a pair (M, J) consisting of a Weyl type M and a subset J of nodes of M as appearing in Table 11.2. A *root shadow space* is a shadow space of root type.

In Table 11.2, we list, for each root type (M, J), the length l of its root orbit $|W\alpha|$ under the Coxeter group $W = W(M)$, where α is the unique root in Φ^+ (cf. Proposition 11.6.2) orthogonal to all α_i for $i \in S \setminus J$, and, under the header srk, the ranks of symplecta in $\mathrm{ShSp}(\Gamma(\mathcal{C}), J)$ for a building $\mathcal{C}$ of type M.

In the lemma below, we write W_v for a vector v in the reflection representation of the Coxeter group W, to denote the stabilizer of v in W. We also use the roots e_i, the root system Φ, and the set Φ^+ of positive roots as in Definition 4.5.1.

Proposition 11.6.2 *If (M, J) is a root type, then there is a unique root $\alpha \in \Phi^+$ such that $(e_i, \alpha) = 0$ for all $i \in L$. This root satisfies $W_\alpha = \langle L \rangle$.*

Proof Take $M = A_n$. Then $L = \{2, \ldots, n-1\}$ and so $\{e_i \mid i \in L\}^\perp = \alpha\mathbb{R}$ with $\alpha = e_1 + \cdots + e_n \in \Phi^+$. It follows that α is the unique positive root orthogonal to all e_i for $i \in L$.

In all other cases, the size of L is equal to $n-1$, so there is at most one positive root in $\{e_i \mid i \in L\}^\perp$. It is readily checked (case-by-case) that such a root $\alpha \in \Phi^+$ exists.

Clearly, $\langle L \rangle$ fixes α, so $\langle L \rangle \subseteq W_\alpha$. As $\langle L \rangle$ is the Coxeter group of type $M|_L$, its size is easily determined. By Theorem 1.7.5, the size of W_α is equal to $|W|/|W\alpha|$. Verifying, for each root type (M, J), that the orbit $W\alpha$ has length $|W/\langle L \rangle|$ (see Table 11.2), we find that the sizes of W_α and $\langle L \rangle$ are equal, so they coincide. $\square$

Example 11.6.3 As a consequence of Proposition 11.6.2, there is a map $\mu : {}^L W^L \to \{-2, -1, 0, 1, 2\}$ given by $\mu(w) = i$ if and only if the root $\alpha \in \Phi^+$ of Proposition 11.6.2 satisfies $(\alpha, w\alpha) = -i$. The map μ is surjective in all cases but $M = G_2$, A_1, B_2. For root type $F_{4,1}$, it is even bijective:

$w \in {}^L W^L$	i	$w\alpha$
$\emptyset$	-2	$2e_1 + 3e_2 + 4e_3 + 2e_4$
1	-1	$e_1 + 3e_2 + 4e_3 + 2e_4$
12321	0	$e_2 + 2e_3 + 2e_4$
12324321	1	$-e_1$
123214321324321	2	$-2e_1 - 3e_2 - 4e_3 - 2e_4$

But for root type $B_{5,2}$, the set

$${}^L W^L = \{\emptyset, 2, 2132, 2345432, 21345432, 213243543215432\}$$

of (L, L)-reduced elements of W has size six, and so two of its elements map onto the same inner product. In fact, there are two distinct $\langle L \rangle$-orbits of roots in $W\alpha$ orthogonal to α. In the associated polar space, this reflects the two ways in which a pair of lines can be polar in the Grassmannian of lines: they can be disjoint inside a singular subspace and they can meet in a point but not lie in a common singular subspace. The convex closure of the former polar pair is a symplecton of rank three whereas the convex closure of the latter pair is a symplecton of rank 4.

Lemma 11.6.4 *Let (W, S) be a Coxeter system of Weyl type M, let α be the unique root of $\Phi^\perp$ as in Proposition 11.6.2, and put $\Psi := W\alpha$. If $\beta, \gamma \in \Psi$ satisfy $(\alpha, \beta) = 0$ and $(\alpha, \gamma) = 1$ but $(\beta, \gamma) \neq 1$, then there is a root δ in Ψ with $(\alpha, \delta) = (\beta, \delta) = (\gamma, \delta) = 1$.*

Proof Because Ψ is a single W-orbit of roots and M is a Weyl type, we only need consider pairs of roots in the systems A_n, D_n, E_n. For diagrams distinct from D_n, there is only one W-orbit of orthogonal pairs (α, β) from Φ and for D_n, the number of orbits is at most three if $n = 4$ and two if $n > 4$. Therefore, the proof can be reduced to a series of straightforward verifications. If, for instance, the type is D_n and $\alpha = \varepsilon_1 - \varepsilon_2$, $\beta = \varepsilon_1 + \varepsilon_2$ (with Φ as in Exercise 4.9.23), then, without loss of generality we have $\gamma = \varepsilon_3 - \varepsilon_2$, and $\delta = \varepsilon_1 + \varepsilon_3$ is as required. $\square$

Theorem 11.6.5 *Every root shadow space of a thick building of Weyl type whose type is distinct from $A_{n,\{1,n\}}$ and $B_{3,2}$ is either a generalized hexagon, a nondegenerate polar space, or a root parapolar space of polar rank at least three and of diameter three.*

Proof Let C be the chamber system of a thick building of Weyl type Y_n, let $Y_{n,J}$ be the type of its root shadow space Z, and put $L = S \setminus J$. A look at Table 11.2 shows that, without loss of generality, we may assume that Y_n is distinct from A_n and B_3 (by assumption), G_2 (the generalized hexagon case) and that $Y_{n,J}$ is distinct from $B_{n,1}$ (in which case Z would be a nondegenerate polar space; see Example 11.4.8). In particular, $J = \{r\}$ for some $r \in S$.

By Theorem 11.5.12, Z is a parapolar space. The element r occurs at most three times in the longest element of W and hence in any (L, L)-reduced element of W; this shows that the diameter of Z is three. Inspection of the diagrams shows that all symplecta have rank at least three. For instance, if $Y_{n,J} = B_{n,2}$, then the symplecta are the shadows on 2 of flags of type $[n] \setminus [3]$, which are polar spaces of rank three, and the shadows on 2 of flags of type $\{1\}$, which are polar spaces of rank $n - 1$.

Suppose that Y is a symplecton of Z and that x is a point outside Y collinear with a point y of Y. There are chambers c and d such that $y = cL^* = dL^*$, $Y = cK^*L^*$ for a suitable subset $K \subseteq S$ (on the complement of which M induces the diagram D_m or B_m for some $m \geq 3$), and $xy = dr^*L^*$. Embedding c and d in an apartment and identifying the image with roots, we find a root α corresponding to y, a root β with $(\alpha, \beta) = 0$ corresponding to a point z of Y non-collinear with y, and a root γ with $(\alpha, \gamma) = 1$ representing a point v of xy. By Lemma 11.6.4, there is a root $\delta \in W\alpha$ representing a point u collinear with v, y, and z. As Z, being a parapolar space, is a gamma space, the line yu lies in Y and consists of points collinear with x. This proves that Z is a root parapolar space. $\square$

Corollary 11.6.6 *Every root shadow space of a thick building of Weyl type M is a root filtration space with respect to $(E_i)_{-2 \leq i \leq 2}$. If its type is distinct from $B_{n,1}$, then it is nondegenerate.*

Proof It is known that the shadow spaces of type $B_{n,1}$ are nondegenerate polar spaces. The root spaces of type $A_{n,\{1,n\}}$ and $G_{2,1}$ are shown to be nondegenerate root filtration spaces in Theorem 6.7.7 and Example 6.7.10, respectively. According to Theorem 11.6.5, the remaining root shadow spaces are thick root parapolar spaces to which we can apply Theorem 11.5.24. This gives that these are nondegenerate root filtration spaces, as required. $\square$

11.7 Recognizing Shadow Spaces of Buildings

Let $j \in [n]$ and let Y_n be a Weyl type over $[n]$. In the previous section, we found various properties of line spaces that are typical of shadow spaces of the buildings of type Y_n. Here we overview axioms for partial linear spaces that characterize the shadow spaces of type $Y_{n,j}$ among all partial linear spaces. We omit most proofs, but give some rough outlines, and, as usual, provide some pointers to the literature in Sect. 11.9.

We begin with type A_n. We will work with parapolar spaces and strongly parapolar spaces introduced in Definition 11.5.2. In Theorem 11.5.12(ii), we saw that most shadow spaces of buildings, including those of type $A_{n,j}$ for $j \notin \{1, n\}$, are parapolar. Proposition 11.4.13 shows that the latter are actually strongly parapolar (see also Example 11.5.4).

For $j \leq n$, the shadow spaces of type $A_{n,j}$ are of course isomorphic to those of type $A_{n,n+1-j}$, so we may restrict ourselves to types $A_{n,j}$ with $j \leq (n+1)/2$. The case $j = 1$ corresponds to projective spaces of dimension n and is characterized by Theorem 6.3.1 and Proposition 11.1.9. For $j > 1$, we have the following result, based on the property stated in Proposition 6.6.2(iii).

Theorem 11.7.1 *Let Z be a thick strongly parapolar space of finite singular dimension that is not complete. If,*

for each point x and line l of Z such that $x^{\perp} \cap l = \emptyset$, the set $x^{\perp} \cap l^{\perp}$ is either empty or a line of Z,

then Z is either a polar space, a shadow space of type $A_{n,j}$ with $n \geq 4$ and $1 < j \leq (n+1)/2$, or the quotient space of an infinite shadow space of type $A_{2j-1,j}$ with $j \geq 5$ by a polarity of Witt index at most $j - 5$.

Notice that, in the latter case, the polarity preserves j as it is the middle node of A_{2j-1}, so it induces an automorphism on the shadow space of type $A_{2j-1,j}$ and the quotient space is as defined in Exercise 2.8.24.

Sketch of Proof The proof starts with a space Z satisfying the hypotheses that is not a polar space. Its maximal singular subspaces will have to be identified as the elements of types $j - 1$ or $j + 1$ of the geometry. The members of the second class have dimension j, and those of the first class have dimension $n + 1 - j$. If $j \neq (n+1)/2$, these two classes can be separated by dimension. Otherwise more subtle arguments are needed to separate the classes; in fact, in the quotient case, the two classes are fused and the effort lies in constructing the covering space, in which the two classes can be separated. The proof runs by induction on j: a new strongly parapolar space is constructed whose point set is the class of maximal singular subspaces of dimension $n + 1 - j$; if $j > 2$, the new space is shown to satisfy the same properties as Z but with a class of maximal singular subspaces of dimension $j - 1$. $\square$

We continue with shadow spaces of buildings of type B_n. Those of type $B_{n,1}$ are polar spaces, and the converse holds as well in view of Theorem 7.5.8 and Proposition 11.1.9.

In view of Proposition 6.4.2, the analogue of an affine projective space for a polar space would be a polar space from which a geometric hyperplane is removed. The following result characterizes such spaces. It uses the notion of an induced subspace of a line space of Definition 2.5.8.

Theorem 11.7.2 *If Z is a connected non-complete partial linear space such that, supplied with the class of planes, a geometry of type*

$$\underset{1}{\circ} \overset{\text{Af}}{\underline{\qquad}} \underset{2}{\circ} \overset{\text{Polar}_n}{\underline{\qquad}} \underset{3}{\circ}$$

results, where Polar_n *stands for the class of nondegenerate polar spaces of rank n, then Z is isomorphic to the subspace induced on the complement of a geometric hyperplane of a nondegenerate polar space of rank $n + 1$.*

Sketch of Proof The proof consists of reconstructing the points and lines missing from Z to turn it into a polar space. The new elements are reconstructed as may be expected from the setting of a missing geometric hyperplane, in a way similar to the construction of the projective space from an affine space as in the proof of Proposition 6.4.2. The complication with respect to the projective space case lies in the existence of different kinds of geometric hyperplanes in a nondegenerate polar space. $\qquad\qquad\square$

Definition 11.7.3 Extending Definition 5.6.8, we call a connected partial linear space Z a *near-polygon* if, for every point p and every line l of Z, there is a unique point on l nearest to p.

A typical example of a near-polygon is the hypercube, the shadow space on n of the thin building of type B_n.

Theorem 11.7.4 *If Z is a thick near-polygon such that any two points at mutual distance two have at least two common neighbors, then any two points of Z at mutual distance i are contained in a unique convex subnear $2i$-gon.*

The proof uses an intricate induction on i, starting with the case $i = 2$. The resulting convex subspaces of diameter two are the symplecta of the parapolar space Z. The complete collection of subspaces does not suffice to conclude that the space is a shadow space of type $\mathrm{B}_{n,n}$, as the counterexample of Corollary 5.6.7 shows.

Theorem 11.7.5 *Suppose that Z is a thick near n-gon such that every pair of points at mutual distance two have at least two common neighbors, so X is a parapolar space. If, for any pair of a point z and a symplecton of Z, the symplecton contains a unique point nearest z, then Z is a shadow space of type $\mathrm{B}_{n,n}$.*

In Proposition 7.9.2, the Grassmannian of lines of a polar space is shown to be a root filtration space. So a general result recognizing root filtration spaces as shadows of buildings will take care of a recognition theorem for these Grassmannians. This is the content of Theorem 11.7.10.

Now let us consider the 'thin from above' case for polar spaces of rank n. Definition 7.8.5 gave a way to construct a building of type D_n from these, next to the usual ones of type B_n, which have thin n-panels. Taking one class of maximal singular subspaces of the polar space gives a shadow space of type $D_{n,n}$, called a *half dual polar space*, that can be characterized as follows.

Theorem 11.7.6 *Let Z be a parapolar space of polar rank at least three having a collection $\mathcal{M}$ of maximal singular subspaces such that each line of Z is contained in a member of $\mathcal{M}$. Suppose that d is a positive integer such that, for each point x of Z and each $M \in \mathcal{M}$ with $x \notin M$, the set $x^{\perp} \cap M$ is either empty or a projective space of dimension d. If $d = 1$, assume also that some line of Z is contained in at least two members of $\mathcal{M}$. If Z is not a polar space, then either $d = 1$ and Z is a shadow space of type $A_{n,k}$ for some $k \geq 2$, or $d = 2$ and Z is a homomorphic image of a shadow space of type $D_{n,n}$.*

Shadow spaces of type $B_{d+2,1}$, that is, polar spaces of rank $d + 2$, are examples satisfying the hypotheses for arbitrary $d > 0$.

Remark 11.7.7 We can push the reduction from building to space a little further in restricting ourselves to collinearity graphs of shadow spaces. This does not work for $A_{n,1}$, since the collinearity graphs of these shadow spaces are complete. Although some other complications arise (due to the fact that, for two collinear points x and y, the set $\{x, y\}^{\perp\perp}$ strictly contains the line on x and y in cases described in Theorem 11.5.12(iii)), these are essentially the only counterexamples of Weyl type. Thus, apart from $A_{n,1}$ and its isomorph $A_{n,n}$, a shadow space of type $Y_{n,j}$ for any Weyl type Y_n is uniquely determined by its collinearity graph.

As we have seen, there is at least one shadow space of each building of Weyl type that is a root filtration space. This calls for a converse. Theorem 7.9.19 made a step in this direction and has the following generalization.

Theorem 11.7.8 *Suppose that Z is a thick nondegenerate root filtration space having a maximal singular subspace of dimension two. Then Z is either the Grassmannian of lines of a nondegenerate polar space or the root shadow space of type $F_{4,1}$ or $A_{3,\{1,3\}}$.*

Sketch of Proof If Z has a line that is contained in a unique maximal singular subspace, then Theorem 7.9.19 and the hypothesis that some maximal singular subspace has dimension two imply that Z is a shadow space of type $B_{3,2}$ (in which case Z is the Grassmannian of lines of a nondegenerate polar space) or of type $A_{3,\{1,3\}}$. Therefore, we (may) assume that every line is contained in at least two maximal singular subspaces. Corollary 11.5.10 shows that Z is a root parapolar space (cf. Definition 11.5.11) of polar rank at least three.

Denote by $\mathcal{S}$ the collection of all symplecta and by $\mathcal{D}$ the collection of oriflamme symplecta of rank three, that is, those isomorphic to the Grassmannian of lines of a projective space of dimension three (alternatively, an oriflamme geometry of rank three).

Put $Z = (E, F)$. Assume first $\mathcal{D} \neq \emptyset$. On each line there is at most one member of $\mathcal{T} := \mathcal{S} \setminus \mathcal{D}$. If there is such a symplecton, then the triple $(\mathcal{T}, E, *)$, where $T * x$ for $(T, x) \subset \mathcal{T} \times E$ stands for $x \subset T$, is the rank two geometry of points and lines of a nondegenerate polar space of rank at least four, and Z is isomorphic to the Grassmannian of lines of this polar space. If there is no such a symplecton, then $\mathcal{D}$ can be partitioned into precisely three classes $\mathcal{D}_1$, $\mathcal{D}_2$, $\mathcal{D}_3$ such that distinct symplecta D, D' belong to the same class if they meet in a single point of Z. In this case, consider the incidence system $(\mathcal{D}_1, E, \mathcal{D}_2, \mathcal{D}_3, *)$, where

(1) $D * D'$ for distinct symplecta D, D' if and only if $D \cap D'$ is a maximal singular subspace of Z, and
(2) $D * x$ for a symplecton D and a point x if and only if $x \in D$.

It is a residually connected geometry of type D_4 and hence a building of type D_4 (cf. Proposition 11.1.10 and the Chamber System Correspondence 3.4.6). The space Z is then easily seen to be isomorphic to the Grassmannian of lines of each of the three polar spaces that are shadow spaces of this building.

Assume next $\mathcal{D} = \emptyset$. In this case all symplecta necessarily have the same rank, which must be three. The incidence system $(E, F, \mathcal{M}, \mathcal{S}, *)$, where $*$ stands for symmetrized containment, is a residually connected geometry of type F_4, which can be shown to be a building. Clearly, Z is the point shadow of this geometry. $\square$

Theorem 11.7.9 *Suppose that Z is a thick strongly parapolar space of finite singular dimension at least two and diameter three satisfying the following properties.*

(1) *If x is a point and Q is a symplecton with $x \notin Q$, then $x^{\perp} \cap Q$ is not empty.*
(2) *For each point x of Z, the set $Z_{\leq 2}(x)$ is a subspace of Z.*

The space Z is a shadow space type of $A_{5,3}$, $D_{6,6}$, or $E_{7,7}$.

Sketch of Proof Besides singular subspaces and symplecta, certain convex spaces subspaces of diameter two need to be constructed, in order to construct the geometries of diagrams of the types appearing in the conclusion. These subspaces are of type $A_{4,2}$ for $A_{5,3}$, of type $D_{5,5}$ for $D_{6,6}$. In the case $E_{7,7}$, no such subspaces are needed. $\square$

Theorem 11.7.10 *Every thick nondegenerate root filtration space of finite singular dimension all of whose lines are on at least two maximal singular subspaces is a shadow space of type $B_{n,2}$ $(n \geq 4)$, $E_{6,2}$, $E_{7,1}$, $E_{8,8}$, or $F_{4,1}$.*

Sketch of Proof If the root filtration space, Z say, has maximal singular subspaces of dimension two, then Theorem 11.7.8 applies, giving a shadow space of type $B_{n,2}$ $(n \geq 4)$ or $F_{4,1}$ (the type $A_{3,\{1,3\}}$ is excluded by the requirement that there are at least two maximal singular subspaces on a line; see Example 6.7.20). Suppose, therefore, that Z has singular dimension at least three.

By Theorem 11.5.24, Z is a thick root parapolar space of polar rank at least three and of diameter three. Pick $x \in Z$. As Z contains a special pair, Corollary 11.5.22 and Proposition 11.5.18(ii) force that the local space $x^{\perp}/\{x\}$ has diameter three. The singular dimension of Z is at least three, so Proposition 11.5.18(iii) gives that $x^{\perp}/\{x\}$ satisfies the hypotheses of Theorem 11.7.9. Therefore, $x^{\perp}/\{x\}$ is a shadow space of type $A_{5,3}$, $D_{6,6}$, or $E_{7,7}$. It is fairly standard to show that all local subspaces are isomorphic. In particular, maximal singular spaces have dimensions in $\{4\}$, $\{4, 6\}$, $\{6, 7\}$, in the respective cases. With this information, subspaces of Z can be constructed corresponding to each node of the diagram of type E_n, where $n = 6$, 7, 8. The construction of residually connected geometries of these types is relatively straightforward. In view of Proposition 11.1.11, this geometry is a building if $n = 6$. For $n = 7$, 8, some extra care is needed to derive this conclusion. The final step, establishing that Z is isomorphic to the root shadow space of this building, is easy. $\qquad\square$

Notice that, although they are not explicitly listed, the conclusion includes the shadow spaces of type $D_{n,2}$ $(n \geq 4)$, as these are special cases of those of type $B_{n,2}$ $(n \geq 4)$.

All spaces in the conclusion are root shadow spaces but not polar spaces. Therefore, Theorem 11.7.10 is a partial converse of Corollary 11.6.6. Here is a more complete converse.

Theorem 11.7.11 *Let Z be a thick nondegenerate root filtration space. If the singular dimension of Z is finite, then Z is isomorphic to a shadow space of type $A_{n,\{1,n\}}$ $(n \geq 2)$, $B_{n,2}$ $(n \geq 3)$, $E_{6,2}$, $E_{7,1}$, $E_{8,8}$, $F_{4,1}$, or $G_{2,1}$.*

Sketch of Proof One of the following two cases prevails.

(1) Each line is contained in a unique maximal singular subspace and Z satisfies the hypotheses of Theorem 7.9.19.
(2) The space Z satisfies the hypotheses of Theorem 11.7.10.

Case (1) leads to $A_{n,\{1,n\}}$ $(n \geq 2)$, $B_{3,2}$, and $G_{2,1}$. If (1) does not hold, then, by Corollary 11.5.10, Z is a parapolar space as in (2), and application of Theorem 11.7.10 leads to the remaining types of the conclusion. $\qquad\square$

11.8 Exercises

Section 11.1

Exercise 11.8.1 Lemma 11.1.3 can also be proven by an application of Lemma 4.5.9. To this end, consider the monoid of all relations on $\mathcal{C}$ in which the product AB of two relations A and B is given by $(c, d) \in AB$ if and only if there is $b \in \mathcal{C}$ such that $(c, b) \in A$ and $(b, d) \in B$. We are interested in its submonoid F generated by the

relations G_r, for $r \in S^*$, given by $(c, d) \in G_r$ if there is a simple gallery from c to d of type r. Show that there is a uniquely determined map $G : W \to F$ such that $G(w) = G_r$ if r is a minimal expression of w.
(*Hint:* Use Exercise 3.7.4(d) to verify the hypotheses of Lemma 4.5.9.)

♮ **Exercise 11.8.2** Show that the geometries $(\mathcal{L}, *, \tau)$ of type $\widetilde{A}_{n-1}$ of Theorem 2.7.14 are buildings for each $n \in \mathbb{N}$, $n > 2$.

Exercise 11.8.3 Let (W, S) be a Coxeter system of type M and let $\mathcal{C}$ be a building of type M.

(a) Denote by C the set of chambers of $\mathcal{C}$. Show that $\delta_{\mathcal{C}}$ as defined in Corollary 11.1.14 is a map satisfying the following three properties for all $x, y \in C$, where $w = \delta_{\mathcal{C}}(x, y)$.
   (i) $w = 1$ if and only if $x = y$.
   (ii) For each $z \in C$ with $\delta_{\mathcal{C}}(y, z) = r \in S$, we have $\delta_{\mathcal{C}}(x, z) \in \{w, wr\}$; if, moreover, $l(wr) > l(w)$, then $\delta_{\mathcal{C}}(x, z) = wr$.
   (iii) For each $r \in S$, there exists $z \in C$ such that $\delta_{\mathcal{C}}(y, z) = r$ and $\delta_{\mathcal{C}}(x, z) = wr$.
(b) Suppose that C is a set and $\delta : C \times C \to W$ is a map satisfying (i), (ii), (iii) as stated for $\delta_{\mathcal{C}}$. Prove that C, supplied with the relations $\sim_r$ on C for $r \in S$ given by $x \sim_r y$ whenever $\delta_{\mathcal{C}}(x, y) \in \{1, r\}$, is a building of type M.

Exercise 11.8.4 Let $\mathcal{C}$ be a chamber system of Coxeter type M. Show that the following statements are equivalent. Here, the map $G : W \to F$ is as in Exercise 11.8.1, and id stands for the identity (diagonal) relation on $\mathcal{C}$.

(a) $\mathcal{C}$ is a building.
(b) For each $w \in W$ we have $G(w) \cap \mathrm{id} \neq \emptyset$ if and only if $w = 1$.
(c) For each pair (c, d) of chambers of $\mathcal{C}$, there is exactly one $w \in W$ with $(c, d) \in G(w)$.

Conclude that, if $\mathcal{C}$ is a building of type M, we have $(c, d) \in G(w)$ if and only if $\delta_{\mathcal{C}}(c, d) = w$.

Exercise 11.8.5 Combine Lemma 3.4.9, Proposition 4.5.15, and Corollary 4.5.13(iii) to prove that a building of type M is residually connected. (In the text, this is proved later, see Corollary 11.2.12.)

Exercise 11.8.6 (Cited in Proposition 11.1.10) Let (W, S) be a Coxeter system. Define the relation $\leq$ on W by $w_1 \leq w_2$ if and only if there is a minimal expression $r_1 \cdots r_q$ for w_2 with $r_1, \ldots, r_q \in S$ such that $w_1 = r_{i_1} \cdots r_{i_t}$, where $1 \leq i_1 < i_2 < \cdots < i_t \leq q$. Prove that $\leq$ is a partial ordering on W. This order is known as the *Bruhat order* on W. Show that, if W is finite, it has a unique maximum element with respect to the Bruhat order.

Section 11.2

Exercise 11.8.7 Show that, in each spherical building, any two opposite chambers are in a unique apartment. What about two chambers that are not opposite?

Exercise 11.8.8 (This exercise is used in Corollary 11.2.6) Take a building $\mathcal{C}$ over S and let (c, A) be a pair consisting of a chamber c of $\mathcal{C}$ and an apartment A of $\mathcal{C}$ containing it. As usual, C is the set of chambers of $\mathcal{C}$.

(a) Show that, for each $x \in C$, there is a unique chamber $\rho_{c,A}(x)$ in A with the property that $\delta_{\mathcal{C}}(c, x) = \delta_{\mathcal{C}}(c, \rho_{c,A}(x))$. This chamber is called *the retract of x onto A with center c*.
   (*Hint:* Since A is an apartment, there is a strong isometry $\alpha : W \to C$ with $\alpha(W) = A$ and $\alpha(1) = c$. Now $\rho_{c,A}$ is the composite of the homomorphism of chamber systems $x \mapsto \delta_{\mathcal{C}}(c, x)$ (cf. Corollary 11.1.14) and α.)
(b) Derive that $\rho_{c,A}$ is a homomorphism of chamber systems over S which is the identity on A. This justifies the name 'retract'.
(c) Verify that $\rho_{c,A}$ maps minimal galleries starting at c onto minimal galleries of the same type.

Exercise 11.8.9 Let $\mathcal{C}$ be a building over S. Suppose that X is a J-cell and Y is a K-cell of $\mathcal{C}$ for certain $J, K \subseteq S$. Prove the following assertions.

(a) There is a single element $w \in {}^{J}W^{K}$ such that, for each $x \in X$ and $y \in Y$, we have $\delta_{\mathcal{C}}(x, y) \in \langle J \rangle w \langle K \rangle$.
(b) The shortest distance between X and Y is equal to $l(w)$.
(c) The shortest distance is assumed as follows: take a point $x \in X$. Use the Gate Property 11.2.9 twice to determine the chamber d of Y nearest to x and, next, to determine the chamber c of X nearest to d. Then $w = \delta_{\mathcal{C}}(c, d)$.
(d) The shortest distance (as indicated in the previous part) is not necessarily assumed by a unique pair of chambers $(c, d) \in X \times Y$.

Section 11.3

Exercise 11.8.10 Prove that (B, N, W, S) is a Tits system if and only if it satisfies Conditions (i), (ii) of a Tits system (Proposition 11.3.3) and (W, S) is a Coxeter system such that for every $w \in W, r \in S$ with $l(wr) > l(w)$ we have

$$Bw\,Br\,B = Bwr\,B \quad \text{and} \quad Br\,Br\,B = B \cup Br\,B.$$

(*Hint:* Since (W, S) is a Coxeter system, we have $r = r^{-1}$ so $rBr^{-1} \subseteq B$ would imply $BrB \cup B = Br\,Br\,B \subseteq B$ and hence, $r \in B$, a contradiction with (ii). This establishes (iv). In order to prove (iii), suppose $l(wr) < l(w)$ (otherwise we are done). Then $l(wrr) > l(wr)$ so $Bw\,Br\,B = Bwrr\,Br\,B = Bwr\,Br\,Br\,B = Bwr\,B \cup Bwr\,Br\,B = Bwr\,B \cup Bw\,B$, whence (iii).)

Exercise 11.8.11 Recall from Definition 4.5.3 that R denotes the set of reflections of (W, S). For $w \in W$, write $T_w = \{r_\beta \in R \mid \beta \in \Phi_w\}$, where r_β is the reflection in R with root β. Use Corollary 4.5.5 and Exercise 4.9.15((b)) to prove the following properties of T_w.

(a) If $w = r_1 \cdots r_q$ is a minimal expression for w, then

$$T_w = \{r_q \cdots r_{i+1} r_i r_{i+1} \cdots r_q \mid i \in [q]\}.$$

(b) T_w is a finite set of reflections in W of cardinality $l(w)$.
(c) If $w = sv$ for $s \in S$ and $v \in W$ with $l(v) = l(w) - 1$, then $T_w = T_v \cup \{v^{-1}sv\}$.
(d) If (G, B, W, S) is a Tits system, then $BT_w B \subseteq Bw^{-1}BwB$.

Exercise 11.8.12 For a spherical Coxeter type M, the existence of a Tits system in a group G implies that G acts strongly transitively on the associated building; see Corollary 11.3.11(iii). This is not necessarily true if M is not spherical, as can be seen already from the following rank two example. Fix two integers p and q. Let $\mathcal{T}$ be an infinite tree whose vertices partition into two cocliques, $\mathcal{T}_1$ and $\mathcal{T}_2$, such that each vertex in $\mathcal{T}_1$ has all of its p neighbors in $\mathcal{T}_2$ and each vertex in $\mathcal{T}_2$ has all its q neighbors in $\mathcal{T}_1$. This leads to a geometry over $[2]$ in which incidence is given by adjacency in $\mathcal{T}$. So an edge is a chamber and an i-panel consists of all edges sharing a given vertex of type $3 - i$.

(a) Show that a subset of $\mathcal{T}$ is an apartment if and only if it is a 2-sided infinite path in $\mathcal{T}$.
(b) Take an edge $c = \{a, b\}$ of $\mathcal{T}$. Let r_a be an involutory automorphism fixing a and interchanging b with another neighbor b' of a. Show that there is an apartment A containing a, b, b' that is left invariant by r_a.
(c) Let r_b be an involutory automorphism of $\mathcal{T}$ fixing b, leaving invariant A and interchanging the two neighbors of b in A. Put $N = \langle r_a, r_b \rangle$. The group $H :=$ N_c fixes a and b and leaves invariant A and so induces the identity on A and $W := N/H$ is the infinite dihedral group. Take $p = q$ and label every edge of $\mathcal{T}$ with an integer from $[p - 1]$ in such a way that
 (1) edges from the same N-orbit have the same label (in particular, edges in A have the same label);
 (2) all $p - 1$ edges sharing a vertex x of $\mathcal{T}$ but not containing a second vertex on the unique geodesic from x to A, have distinct labels.
 For each $k \in \mathbb{N}$, let g_k be an automorphism of $\mathcal{T}$ that fixes all vertices at distance at most k from a, and, for each vertex x at distance k to a, permutes the $p - 1$ vertices adjacent to x and at distance $k + 1$ to a and distinct from b in such a way that the permutation $(1, 2, \ldots, p - 1)$ is induced on the edges containing x but not on the geodesic to a. Moreover, let h_k be an automorphism of $\mathcal{T}$ defined similarly with the roles of a and b interchanged. Finally, let B be the subgroup of the automorphism group of $\mathcal{T}$ generated by all g_k and h_k for $k \in \mathbb{N}$. Establish

that (B, N, W, S) is a Tits system in $G := \langle B, N \rangle$ for a Coxeter system (W, S) of type $\tilde{A}_1$.

(d) Prove that G is not transitive on the set of all apartments of $\mathcal{T}$.

Exercise 11.8.13 Let $\mathcal{C}$ be a thick spherical building of rank at least two. A *root* of $\mathcal{C}$ is a root of an apartment in $\mathcal{C}$ (cf. Exercise 4.9.22). The building $\mathcal{C}$ is called *Moufang* if, for each root α of $\mathcal{C}$, the root group U_α of all elements of $g \in \mathrm{Aut}(\mathcal{C})$ fixing each chamber in a panel of $\mathcal{C}$ having two chambers in α is transitive on the set of apartments of $\mathcal{C}$ containing α. Prove the following assertions.

(a) A thick projective plane is Moufang (according to Definition 7.7.8) if and only if the corresponding building (that is, the chamber system of the rank two geometry of the plane) is Moufang.
(b) If $\mathcal{C}$ is Moufang, then $\mathrm{Aut}(\mathcal{C})$ is strongly transitive on $\mathcal{C}$.
(c) If $\mathcal{C}$ is Moufang, then so is every J-cell of $\mathcal{C}$ for $J \subset S$ with $|J| \geq 2$.

Exercise 11.8.14 Let G be a group with Tits system (B, N, W, S) and let $J \subseteq S$. Prove that the shadow space $\mathrm{ShSp}(\Gamma(\mathcal{B}), J)$ on J of the building $\mathcal{B}$ corresponding to this Tits system is

$$\left(G/G^{(L)}, \left\{ gG^{(j)}G^{(L)}/G^{(L)} \mid g \in G,\ j \in J \right\} \right) \quad \text{where } L = S \setminus J.$$

Section 11.4

Exercise 11.8.15 In the case where A is an irreducible linear spherical Coxeter diagram with end nodes j and k, the proof of nonexistence of a factorization $W = \langle A \setminus \{k\} \rangle \langle A \setminus \{j\} \rangle$ can also be given by means of order computations. Finish the proof of Corollary 11.4.3 in this way.

Exercise 11.8.16 For $L = S$, the conclusion of Corollary 11.4.4 reads $N_W(\langle K \rangle) = \langle K \rangle \langle K^\perp \rangle$. This is false in general: take $w_S \in W$ to be the longest element of W for a Coxeter group of spherical type M where $w_S \in Z(W)$ (see Theorem 4.6.6). Then w_S normalizes $\langle K \rangle$ for every $K \subseteq S$, but w_S cannot be written as a product of an element of $\langle K \rangle$ and an element of $\langle K^\perp \rangle$ for most nontrivial choices of K. Prove this statement for at least one choice of K. Conclude that the hypothesis about K_1 and K_2 not having connected components disjoint from J in the corollary is needed.

Exercise 11.8.17 Let j, k, and l be elements of S such that $\{k\}$ separates $\{j\}$ from $\{l\}$ in M (cf. Definition 2.5.4). Consider the shadow space $\mathrm{ShSp}(\Gamma, j)$ on j of the geometry Γ of a building of type M. Prove, for elements u and v of Γ of type k and l, respectively, that u and v are incident in Γ if and only if $\mathrm{Sh}_j(u) \subseteq \mathrm{Sh}_j(v)$.

Exercise 11.8.18 Let K_1 and K_2 be subsets of S. Prove that two shadows $cK_1^*L^*$ and $dK_2^*L^*$ have an empty intersection if and only if $w \notin \langle L \rangle$, where w is the shortest element in $\langle K_1 \rangle \delta_{\mathcal{C}}(c, d) \langle K_2 \rangle$.

Section 11.5

Exercise 11.8.19 Show that the direct product of two parapolar spaces is a parapolar space.

Exercise 11.8.20 Recall the setting of Exercises 2.8.24, and 7.11.11. Suppose that A is a group of automorphisms of a parapolar space Z such that $d(x, \sigma(x)) \geq 5$ for each point x of Z and $\sigma \in A \setminus \{1\}$. Prove that Z/A is a parapolar space.

Exercise 11.8.21 Let (x, y) be a special pair of points of a parapolar space Z, with common neighbor z. Prove that each pair (x', y') of points with $x' \in xz \setminus \{z\}$ and $y' \in yz \setminus \{z\}$ is special.

Exercise 11.8.22 Prove that the diameter of a root parapolar space of polar rank at least three is at most three.

Exercise 11.8.23 Let $V := \Pi \times Z$ be the direct product of a thick projective space Π and a thick nondegenerate polar space Z of rank at least two, as in Example 11.5.4. Prove that the strongly parapolar space V satisfies the conclusion of Proposition 11.5.18(iii) if and only if Π is a line.

Section 11.6

Exercise 11.8.24 Verify that Proposition 11.6.2 also holds for types $(M, J) = (\mathrm{H}_3, \{2\})$ and $(\mathrm{H}_4, \{1\})$, where the nodes of the diagrams as in Table 4.2 are labelled 1, 2, 3 and 1, 2, 3, 4, respectively, from left to right.
(*Hint:* For H_3, take $e_1 = \varepsilon_1\sqrt{2}$, $e_2 = (-\varepsilon_1\frac{1}{2} + \varepsilon_2\cos(3\pi/5) - \varepsilon_1\cos(\pi/5))\sqrt{2}$, $e_3 = \varepsilon_3\sqrt{2}$, where ε_1, ε_2, ε_3 represent an orthogonal basis of $\mathbb{R}^3$. The set Φ consists of all vectors obtained from e_1 and e_2 by even permutations of the coordinates and arbitrary sign changes in each coordinate. Similarly for H_4 starting from $e_1 = (\varepsilon_4\cos(3\pi/5) - \varepsilon_1\frac{1}{2} + \varepsilon_2\cos(\pi/5))\sqrt{2}$ and e_i as e_{i-1} above for $i = 2, 3, 4$, and Φ obtained by the same operations as for H_3 from e_1, e_2, and $(\varepsilon_1 + \varepsilon_2 + \varepsilon_3 + \varepsilon_4)/\sqrt{2}$.)

Exercise 11.8.25 Show that each root shadow space is uniquely determined by its collinearity graph.
(*Hint:* In all cases except $\mathrm{B}_{3,2}$, the line on the two collinear points x and y coincides with $\{x, y\}^{\perp\perp}$. In the case of $\mathrm{B}_{3,2}$, a line is the intersection of a maximal singular subspace and a symplecton.)

Section 11.7

Exercise 11.8.26 Consider a line space Z as in Theorem 11.7.9. Prove that, for each point x of Z, the subspace $Z_{\leq 2}(x)$ is a geometric hyperplane of Z.

11.9 Notes

The theory of buildings was developed by Tits in the 1960s. The standard reference for buildings is [286]; the founding paper is [284]. The book [286] contains the classification of thick spherical buildings of rank at least three. Many classifications of groups, such as Timmesfeld's results [279] on abstract root subgroups of a group, depend on it. The introduction to buildings in this chapter follows the ideas of [287]. Easier introductions to the topic are [2, 37, 130, 247]. See [245] for a rather abstract way of constructing buildings. The recent book [305] gives a very lucid treatment of the most difficult (and crucial) parts (for the classification) of [286].

Section 11.1

The current definition of a building is a slight variation of Tits' definition in [287]. It differs entirely from the first approach, which is in [286]. There are many others, of which we only mention a definition using the metric properties, closely linked to the Gate Property 11.2.9, in [247].

A beautiful way of discovering buildings inside buildings rests on a theory of admissible partitions of the nodes of a Coxeter diagram. It is a geometric counterpart of Galois theory and makes it possible to construct the polar geometries from projective geometries; it is described in Mühlherr's thesis [225] and the construction in [224] of rank three polar spaces whose singular planes are projective planes defined over Cayley division rings (cf. Example 2.3.4) from buildings of type E_7 is an application.

Proposition 11.1.11 does not extend to type E_7: it is no longer true that every residually connected chamber system of type E_7 is a building; see [33]. This implies that there is no division of a closed gallery in three parts belonging to cells of proper subtypes, as was used in the proof of Proposition 11.1.11 for $M = E_6$.

Section 11.2

The chamber e closest to d in cJ^* of Remark 11.2.10 is named the projection of d on cJ^* in [286], and the gate for d in cJ^* in [247].

Corollary 11.2.6 gives the connection between the given definition of a building, in terms of homotopy, and the more classical definition, in terms of apartments. It occurs in [287].

Section 11.3

A Tits system (cf. Definition 11.3.5) is called a B, N-pair in [284, 286] and denoted by (G, B, N, S). Observe that $W = N/(B \cap N)$. They appear in the theory of algebraic groups, Lie groups, p-adic Lie groups, and Kac-Moody groups.

The classification of spherical buildings whose types have no components of rank one or two in [286] implies that all these buildings have a Tits system. The classification of spherical buildings of rank two is an entirely different story. There are many (wild) examples of generalized triangles (projective planes, cf. Example 2.3.4) and generalized quadrangles (cf. [274, 275] for the Fraïssé construction, and Example 8.3.17). But no finite generalized hexagons or generalized octagons are known that do not come from a Tits system in (twisted versions of) algebraic groups.

Section 11.4

Various parts of the proofs of Sect. 11.4 come from [36, 73]. For properties of shadow spaces of buildings, see also [74, 93]. The diameters of all shadow spaces of type $Y_{n,j}$ with Y_n spherical are known; see [36]. The number equals the maximum over all minimal numbers of occurrences of j in expressions for $([n] \setminus \{j\}, [n] \setminus \{j\})$-reduced elements of $W(Y_n)$.

An attempt at an explanation for the list of types in Definition 11.4.14 is given in Remark 11.4.15. The characterizations of the remark are connected with their appearance in normalizers of maximal split tori in algebraic groups, topics that we do not touch upon in this book. The proof of Remark 11.4.15(i) is easy and can be found in, for instance, [169] and [29]. The proof of Remark 11.4.15(ii) is more involved and can be found in [286, 291]. Usually (cf. [29]), Property (i) of Remark 11.4.15 is used to define Weyl groups.

Section 11.5

The material at the beginning of this section, especially Theorem 11.5.9, is known as Cooperstein theory; it originates from [94] and is also treated in [48, 70, 260]. An alternative name for symplecta (coined in [128] for the particular case of a building of type F_4) is *hyperline*, and for parapolar space (which, to our knowledge, first appeared in [77]) is polarized space, although in the latter case the precise conditions (regarding thickness of lines, connectedness of the space, and the common neighbors of two points at mutual distance two being allowed to be a singular space of rank bigger than one) differ slightly.

The nondegeneracy assumptions in Definition 11.5.2 are slightly stronger than those found in the literature regarding parapolar spaces [48, 77, 152, 190]. This does not greatly harm the strength of the results to be discussed, and simplifies their presentation.

The content of Theorem 11.5.9 originates from [93, 94]. It is somewhat extended in [48, 70].

The list of shadow spaces of diameter two is rather restricted. However, if the condition 'singular' in Proposition 11.5.14 is replaced by 'diameter two', the conclusion can no longer be that the subspace is the shadow of a flag. Counterexamples are given in [95].

The notion of a root filtration space and the main results in this direction, like Theorem 11.5.24, can be found in [79, 80]. It connects very well with root parapolar spaces, which were studied in [190], without use of the terminology, in order to characterize root shadow spaces. The results from Proposition 11.5.16 to Corollary 11.5.22 are due to [259].

Section 11.6

The sources for this section are [36, 79, 80].

New results on the connection between subspaces in shadow spaces and elements in buildings are to be found in [187, 188, 228].

Timmesfeld [279] showed that the root filtration spaces (E, F) that are sets of abstract root subgroups for which $E_{-1} = E_1 = \emptyset$ (which he calls degenerate) arise from polar spaces (cf. Exercise 7.11.35). Cuypers [105] gave a geometric interpretation of Timmesfeld's result, which also generalizes it. Cuypers' result is a characterization of nondegenerate polar spaces by means of hyperbolic lines (that is, subsets of the form $\{x, y\}^{\perp\perp}$, where x and y are non-collinear points of the polar space).

Timmesfeld's theory of abstract root groups was an initiative to view the beautiful results on 3-transpositions by Fischer [127] as a special case of a more general setup in which the original setting would be the case where the parameterizing field has two elements. A comprehensive treatment can be found in [8]. A geometric approach was initiated by Buekenhout [40] and brought further by Cuypers, Hall, Weiss, and Zara [102, 106, 149, 302, 303]. The basic notion is that of a *Fischer space*, a partial linear space in which each plane is isomorphic to a dual affine plane of order two or to an affine plane of order three.

Section 11.7

The general pattern of the section is that, from line spaces satisfying suitable axioms, subspaces can be found that correspond to shadows of elements of various types, from which the full geometry of the building can be reconstructed. There is a more general technique of constructing a chamber system of a given (suitable) type from a space. It is based on the idea that, for a chamber system, all that is needed to reconstruct it from a shadow space is the local picture of the flag, that is, of the subspaces related to the flag. Thus, we need not always construct the whole geometry from global knowledge of a space Z, but rather can construct chambers of the form $(x_i)_{i \in S}$ where x_1 is a point, x_2 a line containing the point x_1, and x_7, say, a subspace of the residue $x_1^\perp/\{x_1\}$ of x_1, rather than a globally defined subspace of Z. This generalization has been made possible by Tits' chamber system approach to buildings in [287].

Theorem 11.7.1 is proved in [72], which extends results by Cooperstein [94]. The proof in [72] uses a technical lemma involving a small part of Tits' classification [286] of polar spaces of rank 3. Shult and Thas [261] gave a shorter proof of that lemma using the fact that the generalized quadrangles involved are Moufang. See [152, 257] for other characterizations of the Grassmannians $A_{n,d}$ and [126] for affine versions.

Theorem 11.7.2 is proved in [82].

Theorem 11.7.4 is due to Brouwer and Wilbrink, [35], but the start of the induction, the case $i = 2$, establishing the existence of quads, stems from the initial work by Shult and Yanushka [262]. The existence of such subspaces was known in the case of classical dual polar spaces from work of Cameron [57].

Theorem 11.7.5 rests on Cameron's [57] characterization, which is improved by work of Shad and Shult [253]. See [110] for details on near polygons. A remarkable near-octagon is connected with the second Janko group J_2; see [83]. It shows that not every thick near-octagon satisfies the hypotheses of Theorem 11.7.4.

Theorem 11.7.6 is due to Shult [258].

Remark 11.7.7 is based on [73, 93].

According to [190], Theorem 11.7.9 can be generalized so as to allow for singular dimension at least one, at the cost of adding products of lines and nondegenerate polar spaces and dual polar spaces to the conclusion (cf. Exercise 11.8.23).

The proof of Theorem 11.7.10 is a composition of various results from the literature. First, Shult [259] showed how more properties can be derived for a root parapolar space of polar rank at least three, such as Proposition 11.5.16(iii). Next, Shult and Kasikova [190] proved that conditions of Cohen and Cooperstein [77] are satisfied, which suffice for the existence of all subspaces needed to recover the geometry of the building. The proof in [190] differs from the one sketched here in that the spaces of type $D_{4,2}$ are treated as spaces of type $F_{4,1}$ (cf. Exercise 7.11.34).

The only root shadow spaces missing from Theorem 11.7.11 are the polar spaces of type $C_{n,1}$. Indeed, polar spaces are root shadow spaces of type $C_{n,1}$. The shadow spaces of type $D_{n,1}$ and $A_{3,2}$ are polar spaces, but the corresponding nodes of D_n and A_3 are not root nodes. It is well known that a building of type D_n is a non-thick building of type B_n; since the types A_3 and D_3 are isomorphic, a building of type A_3 can also be seen as a non-thick building of type B_3. In this way, polar spaces of type $D_{n,1}$ and $A_{3,2}$ can be viewed as root shadow spaces of type $C_{n,1}$ (with $n = 3$ in the latter case).

The characterization of root filtration spaces plays a role in both the classification of simple modular Lie algebras (see [79, 86]) and the classification of groups generated by 'abstract root subgroups' developed by Timmesfeld [279].

A somewhat more complicated set of axioms for a characterization of shadows spaces of many different types is given by Hanssens [152].

Section 11.8

The definition of the Moufang property in Exercise 11.8.13 can easily be generalized to an arbitrary group G acting on a spherical building $\mathcal{C}$ instead of $\mathrm{Aut}(\mathcal{C})$. For such a group G and each root α of $\mathcal{C}$, the root subgroup U_α is a normal subgroup of G_α acting regularly on the set of apartments of $\mathcal{C}$ containing α (cf. [2, 248]). If $\mathcal{C}$ has rank one, then the definition of Exercise 11.8.13 for $\mathcal{C}$ to be Moufang does not work properly for three reasons. First, $\mathrm{Aut}(\mathcal{C})$ is the full symmetry group on the set C of chambers of $\mathcal{C}$. Second, a root consists of a single chamber and so there are no panels having two chambers in a root. Third, a root group U_α need not be a normal subgroup of $\mathrm{Aut}(\mathcal{C})$ acting regularly on the set of apartments containing α. Consider the following adjusted definition: A group G acting on $\mathcal{C}$ is called *Moufang* if, for each root α of $\mathcal{C}$, there is a normal subgroup U_α of G_α consisting of elements of G_α fixing each chamber of α and each chamber in a panel of $\mathcal{C}$ having two chambers in α, acting regularly on the set of apartments of $\mathcal{C}$ containing α. (The subgroup U_α is called the root subgroup of G corresponding to α.) This definition with $G = \mathrm{Aut}(\mathcal{C})$ coincides with the one of Exercise 11.8.13 if $\mathcal{C}$ has rank at least two. Moreover, it is equivalent to $\mathcal{C}$ being a Moufang set as defined in Remark 6.2.3.

By [286], every spherical building whose Coxeter type has no components of rank one or two is Moufang, and so all of its rank two residues are Moufang generalized polygons. As mentioned in Sect. 2.9, the classification of all Moufang generalized polygons is given in [291]. In particular, the finite Moufang generalized octagons are coset geometries of the finite groups ${}^2\mathrm{F}_4(q)$ of Lie type. But generalized octagons are neither of Weyl type nor nontrivial root filtration spaces (cf. Exercise 6.8.18). Accordingly, the groups ${}^2\mathrm{F}_4(q)$ are the only finite groups of Lie type and of rank at least two that are not covered by the theory of root filtration spaces.

References

1. A.O. Abou Zayda, The Veldkamp space of point-line geometry of type $D_{n,k}$. J. Rajasthan Acad. Phys. Sci. **2**, 89–96 (2003) (p. 408)
2. P. Abramenko, K.S. Brown, Theory and applications, in *Buildings*. Graduate Texts in Mathematics, vol. 248 (Springer, New York, 2008) (pp. 198, 565, 569)
3. E. Artin, Coordinates in affine geometry. Rep. Math. Colloq. (2) **2**, 15–20 (1940) (pp. 301, 496)
4. E. Artin, *Geometric Algebra* (Interscience, New York, 1957) (pp. 302, 303, 445)
5. M. Aschbacher, Flag structures on Tits geometries. Geom. Dedic. **14**, 21–32 (1983) (p. 47)
6. M. Aschbacher, Finite geometries of type C_3 with flag-transitive groups. Geom. Dedic. **16**, 195–200 (1984) (p. 100)
7. M. Aschbacher, On the maximal subgroups of the finite classical groups. Invent. Math. **76**, 469–514 (1984) (p. 497)
8. M. Aschbacher, *3-Transposition Groups*. Cambridge Tracts in Mathematics, vol. 124 (Cambridge University Press, Cambridge, 1997) (p. 567)
9. E.F. Assmus Jr., H.F. Mattson, On the number of inequivalent Steiner triple systems. J. Comb. Theory **1**, 301–305 (1966) (p. 248)
10. R. Baer, Homogeneity of projective planes. Am. J. Math. **64**, 137–152 (1942) (pp. 248, 301)
11. R. Baer, *Linear Algebra and Projective Geometry* (Academic Press, New York, 1952) (pp. 247, 301)
12. W.W.R. Ball, H.S.M. Coxeter, *Mathematical Recreations and Essays*, 3th edn. (Dover, New York, 1987) (p. 197)
13. S. Bang, J.H. Koolen, V. Moulton, There are only finitely many regular near polygons and geodetic distance-regular graphs with fixed valency. J. Reine Angew. Math. **635**, 213–235 (2009) (p. 368)
14. L.M. Batten, *Combinatorics of Finite Geometries*, 2nd edn. (Cambridge University Press, Cambridge, 1997) (p. 366)
15. B. Baumeister, A computer-free construction of the third group of Janko. J. Algebra **192**, 780–809 (1997) (pp. 249, 250)
16. B. Baumeister, A.A. Ivanov, D.V. Pasechnik, A characterization of the Petersen-type geometry of the McLaughlin group. Math. Proc. Camb. Philos. Soc. **128**, 21–44 (2000) (p. 250)
17. C.T. Benson, N.E. Losey, On a graph of Hoffman and Singleton. J. Comb. Theory, Ser. B **11**, 67–79 (1971) (p. 100)
18. T. Beth, D. Jungnickel, H. Lenz, *Design Theory* (Cambridge University Press, Cambridge, 1986) (p. 248)
19. N. Biggs, *Algebraic Graph Theory*, 2nd edn. Cambridge Mathematical Library (Cambridge University Press, Cambridge, 1993) (p. 47)

F. Buekenhout, A.M. Cohen, *Diagram Geometry*,
Ergebnisse der Mathematik und ihrer Grenzgebiete. 3. Folge / A Series of
Modern Surveys in Mathematics 57, DOI 10.1007/978-3-642-34453-4,
© Springer-Verlag Berlin Heidelberg 2013

20. G. Birkhoff, Combinatorial relations in projective geometries. Ann. Math. (2) **36**, 743–748 (1935) (pp. 248, 302)
21. G. Birkhoff, *Lattice Theory*, 3rd edn. American Mathematical Society Colloquium Publications, vol. XXV (American Mathematical Society, Providence, 1967) (p. 247)
22. A. Björner, F. Brenti, *Combinatorics of Coxeter Groups*. Graduate Texts in Mathematics, vol. 231 (Springer, New York, 2005) (p. 198)
23. A. Blanchard, *Les corps non commutatifs*. Collection Sup: Le Mathématicien, vol. 9 (Presses Universitaires de France, Vendôme, 1972) (p. 366)
24. A. Blass, F. Harary, Z. Miller, Which trees are link graphs? J. Comb. Theory, Ser. B **29**, 277–292 (1980) (p. 100)
25. A. Blokhuis, A.E. Brouwer, D. Buset, A.M. Cohen, The locally icosahedral graphs, in *Finite Geometries*, Winnipeg, Man., 1984. Lecture Notes in Pure and Appl. Math., vol. 103 (Dekker, New York, 1985), pp. 19–22 (p. 100)
26. A. Blokhuis, T. Kloks, H. Wilbrink, A class of graphs containing the polar spaces. Eur. J. Comb. **7**, 105–114 (1986) (p. 367)
27. G. Bol, Gewebe und gruppen. Math. Ann. **114**, 414–431 (1937) (p. 304)
28. A. Borel, *Linear Algebraic Groups*, 2nd edn. Graduate Texts in Mathematics, vol. 126 (Springer, New York, 1991) (p. 46)
29. N. Bourbaki, *Lie Groups and Lie Algebras*. Elements of Mathematics (Berlin) (Springer, Berlin, 2002). Chaps. 4–6. Translated from the 1968 French original by Andrew Pressley (pp. 198, 199, 497, 566)
30. M.R. Bridson, A. Haefliger, *Metric Spaces of Non-positive Curvature*. Grundlehren der Mathematischen Wissenschaften [Fundamental Principles of Mathematical Sciences], vol. 319 (Springer, Berlin, 1999) (p. 131)
31. A.E. Brouwer, The complement of a geometric hyperplane in a generalized polygon is usually connected, in *Finite Geometry and Combinatorics*, Proc. Deinze, 1992. London Math. Soc. Lect. Notes, vol. 191 (Cambridge University Press, Cambridge, 1993), pp. 53–57 (p. 251)
32. A.E. Brouwer, Block designs, in *Handbook of Combinatorics*, vol. 1 (Elsevier, Amsterdam, 1995), pp. 693–745 (p. 248)
33. A.E. Brouwer, A.M. Cohen, Some remarks on Tits geometries. Ned. Akad. Wet. Indag. Math. **45**, 393–402 (1983). With an appendix by J. Tits (p. 565)
34. A.E. Brouwer, A.M. Cohen, Local recognition of Tits geometries of classical type. Geom. Dedic. **20**, 181–199 (1986) (p. 367)
35. A.E. Brouwer, H.A. Wilbrink, The structure of near polygons with quads. Geom. Dedic. **14**, 145–176 (1983) (p. 568)
36. A.E. Brouwer, A.M. Cohen, A. Neumaier, *Distance-Regular Graphs*. Ergebnisse der Mathematik und ihrer Grenzgebiete (3) [Results in Mathematics and Related Areas (3)], vol. 18 (Springer, Berlin, 1989) (pp. 46, 47, 99, 100, 101, 102, 249, 368, 566, 567)
37. K.S. Brown, *Buildings*. Springer Monographs in Mathematics (Springer, New York, 1998). Reprint of the 1989 original (p. 565)
38. R.H. Bruck, *A Survey of Binary Systems*. Ergebnisse der Mathematik und ihrer Grenzgebiete. Neue Folge, Heft 20. Reihe: Gruppentheorie (Springer, Berlin, 1958) (p. 303)
39. F. Bruhat, J. Tits, Groupes réductifs sur un corps local. Publ. Math. Inst. Hautes Études Sci. 5–251 (1972) (p. 102)
40. F. Buekenhout, Une caractérisation des espaces affins basée sur la notion de droite. Math. Z. **111**, 367–371 (1969) (pp. 303, 496, 567)
41. F. Buekenhout, Foundations of one dimensional projective geometry based on perspectivities. Abh. Math. Semin. Univ. Hamb. **43**, 21–29 (1975) (p. 302)
42. F. Buekenhout, On weakly perspective subsets of Desarguesian projective lines. Abh. Math. Semin. Univ. Hamb. **43**, 30–37 (1975) (p. 496)
43. F. Buekenhout, Extensions of polar spaces and the doubly transitive symplectic groups. Geom. Dedic. **6**, 13–21 (1977) (p. 367)

44. F. Buekenhout, Diagrams for geometries and groups. J. Comb. Theory, Ser. A **27**, 121–151 (1979) (pp. 45, 99, 247, 249, 250)
45. F. Buekenhout, The basic diagram of a geometry, in *Geometries and Groups*, Berlin, 1981. Lecture Notes in Math., vol. 893 (Springer, Berlin, 1981), pp. 1–29 (pp. 45, 99)
46. F. Buekenhout, (g, d^*, d)-gons, in *Finite Geometries*, Pullman, Wash., 1981. Lecture Notes in Pure and Appl. Math., vol. 82 (Dekker, New York, 1981), pp. 93–111 (p. 99)
47. F. Buekenhout, Diagram geometries for sporadic groups, in *Finite Groups—Coming of Age*, Montreal, Que., 1982. Contemp. Math., vol. 45 (American Mathematical Society, Providence, 1982), pp. 1–32 (pp. 249, 250)
48. F. Buekenhout, Cooperstein's theory. Simon Stevin **57**, 125–140 (1983) (pp. 566, 567)
49. F. Buekenhout, On the foundations of polar geometry. II. Geom. Dedic. **33**, 21–26 (1990) (p. 367)
50. F. Buekenhout, Foundations of incidence geometry, in *Handbook of Incidence Geometry* (North-Holland, Amsterdam, 1995), pp. 63–105 (p. 47)
51. F. Buekenhout, An introduction to incidence geometry, in *Handbook of Incidence Geometry* (North-Holland, Amsterdam, 1995), pp. 1–25 (pp. 131, 367)
52. F. Buekenhout, X. Hubaut, Locally polar spaces and related rank 3 groups. J. Algebra **45**, 391–434 (1977) (pp. 100, 408)
53. F. Buekenhout, C. Lefèvre, Generalized quadrangles in projective spaces. Arch. Math. (Basel) **25**, 540–552 (1974) (pp. 444, 445)
54. F. Buekenhout, C. Lefèvre, Semi-quadratic sets in projective spaces. J. Geom. **7**, 17–42 (1976) (pp. 444, 445)
55. F. Buekenhout, E. Shult, On the foundations of polar geometry. Geom. Dedic. **3**, 155–170 (1974) (pp. 366, 367)
56. F. Buekenhout, A. Sprague, Polar spaces having some line of cardinality two. J. Comb. Theory, Ser. A **33**, 223–228 (1982) (p. 368)
57. P.J. Cameron, Dual polar spaces. Geom. Dedic. **12**, 75–85 (1982) (pp. 367, 568)
58. P.J. Cameron, There are only finitely many finite distance-transitive graphs of given valency greater than two. Combinatorica **2**, 9–13 (1982) (p. 368)
59. P.J. Cameron, W.M. Kantor, 2-transitive and antiflag transitive collineation groups of finite projective spaces. J. Algebra **60**, 384–422 (1979) (p. 445)
60. R.D. Carmichael, Tactical configurations of rank two. Am. J. Math. **53**, 217–240 (1931) (p. 248)
61. É. Cartan, *Œuvres complètes. Partie I*, 2nd edn. (Éditions du Centre National de la Recherche Scientifique (CNRS), Paris, 1984). Groups de Lie. [Lie groups] (p. 197)
62. R.W. Carter, *Simple Groups of Lie Type*. Wiley Classics Library (Wiley, New York, 1989). Reprint of the 1972 original, A Wiley–Interscience Publication (pp. 302, 368, 497)
63. B. Casselman, Computation in Coxeter groups. I. Multiplication. Electron. J. Comb. **9**, 25 (2002). 22 pp. (electronic) (p. 198)
64. B. Casselman, Computation in Coxeter groups. II. Constructing minimal roots. Represent. Theory **12**, 260–293 (2008) (p. 198)
65. O. Chein, H.O. Pflugfelder, J.D.H. Smith (eds.), *Quasigroups and Loops: Theory and Applications*. Sigma Series in Pure Mathematics, vol. 8 (Heldermann, Berlin, 1990) (p. 303)
66. W. Cherowitzo, Hyperovals in Desarguesian planes: an update. Discrete Math. **155**, 31–38 (1996). Combinatorics (Acireale, 1992) (p. 249)
67. C. Chevalley, Sur certains groupes simples. Tôhoku Math. J. (2) **7**, 14–66 (1955) (p. v)
68. W.-L. Chow, On the geometry of algebraic homogeneous spaces. Ann. Math. (2) **50**, 32–67 (1949) (p. 303)
69. A.M. Cohen, Finite complex reflection groups. Ann. Sci. Éc. Norm. Super. (4) **9**, 379–436 (1976) (pp. 102, 200)
70. A.M. Cohen, An axiom system for metasymplectic spaces. Geom. Dedic. **12**, 417–433 (1982) (pp. 566, 567)
71. A.M. Cohen, Exceptional presentations of three generalized hexagons of order 2. J. Comb. Theory, Ser. A **35**, 79–88 (1983) (p. 100)

72. A.M. Cohen, On a theorem of Cooperstein. Eur. J. Comb. **4**, 107–126 (1983) (pp. 48, 568)
73. A.M. Cohen, Point-line characterizations of buildings, in *Buildings and the Geometry of Diagrams, Como, 1984. Lecture Notes in Math.*, vol. 1181 (Springer, Berlin, 1986), pp. 191–206 (pp. 566, 568)
74. A.M. Cohen, Point-line spaces related to buildings, in *Handbook of Incidence Geometry* (North-Holland, Amsterdam, 1995), pp. 647–737 (p. 566)
75. A.M. Cohen, *Diagram Geometry and Groups of Lie Type* (Springer, Berlin, 2014, forthcoming) (p. 367)
76. A.M. Cohen, The geometry of extremal elements in Lie algebras, in *Groups, Finite Geometries and Buildings, Springer Conference Proceedings* (Springer, Berlin, 2012), pp. 15–35 (p. 368)
77. A.M. Cohen, B.N. Cooperstein, A characterization of some geometries of Lie type. Geom. Dedic. **15**, 73–105 (1983) (pp. 566, 568)
78. A.M. Cohen, B.N. Cooperstein, On the local recognition of finite metasymplectic spaces. J. Algebra **125**, 348–366 (1989) (pp. 100, 445)
79. A.M. Cohen, G. Ivanyos, Root filtration spaces from Lie algebras and abstract root groups. J. Algebra **300**, 433–454 (2006) (pp. 303, 368, 567, 568)
80. A.M. Cohen, G. Ivanyos, Root shadow spaces. Eur. J. Comb. **28**, 1419–1441 (2007) (pp. 303, 368, 567)
81. A.M. Cohen, E.J. Postma, Covers of point-hyperplane graphs. J. Algebr. Comb. **22**, 317–329 (2005) (p. 100)
82. A.M. Cohen, E.E. Shult, Affine polar spaces. Geom. Dedic. **35**, 43–76 (1990) (pp. 408, 568)
83. A.M. Cohen, J. Tits, On generalized hexagons and a near octagon whose lines have three points. Eur. J. Comb. **6**, 13–27 (1985) (pp. 100, 568)
84. A.M. Cohen, H. Cuypers, H. Sterk (eds.), *Some Tapas of Computer Algebra*. Algorithms and Computation in Mathematics, vol. 4 (Springer, Berlin, 1999) (pp. 101, 198)
85. A.M. Cohen, H. Cuypers, R. Gramlich, Local recognition of non-incident point-hyperplane graphs. Combinatorica **25**, 271–296 (2005) (p. 100)
86. A.M. Cohen, G. Ivanyos, D. Roozemond, Simple Lie algebras having extremal elements. Indag. Math. (N.S.) **19**, 177–188 (2008) (p. 568)
87. P.M. Cohn, *Algebra*, vol. 1, 2nd edn. (Wiley, Chichester, 1982) (p. 366)
88. P.M. Cohn, *Algebra*, vol. 2, 2nd edn. (Wiley, Chichester, 1989) (p. 247)
89. C.J. Colbourn, R. Mathon, Steiner systems, in *Handbook of Combinatorial Designs* 2nd edn. (Elsevier, Amsterdam, 2007), pp. 102–110 (p. 248)
90. C.J. Colbourn, A. Rosa, *Triple Systems*. Oxford Mathematical Monographs (Clarendon, Oxford, 1999) (p. 249)
91. J.H. Conway, R.T. Curtis, R.A. Wilson, A brief history of the ATLAS, in *The Atlas of Finite Groups: Ten Years on*, Birmingham, 1995. London Math. Soc. Lecture Note Ser., vol. 249 (Cambridge University Press, Cambridge, 1998), pp. 288–293 (p. 249)
92. K. Coolsaet, A construction of the simple group of Rudvalis from the group $U_3(5){:}2$. J. Group Theory **1**, 143–163 (1998) (p. 100)
93. B.N. Cooperstein, Some geometries associated with parabolic representations of groups of Lie type. Can. J. Math. **28**, 1021–1031 (1976) (pp. 566, 567, 568)
94. B.N. Cooperstein, A characterization of some Lie incidence structures. Geom. Dedic. **6**, 205–258 (1977) (pp. 566, 567, 568)
95. B.N. Cooperstein, E.E. Shult, Frames and bases of Lie incidence geometries. J. Geom. **60**, 17–46 (1997) (p. 567)
96. H. Coxeter, Finite groups generated by reflections, and their subgroups generated by reflections. Proc. Camb. Philos. Soc. **30**, 466–482 (1934) (p. 198)
97. H.S.M. Coxeter, Discrete groups generated by reflections. Ann. Math. (2) **35**, 588–621 (1934) (p. 198)
98. H. Coxeter, The complete enumeration of finite groups of the form $R_i^2 = (R_i R_j)^{k_{ij}} = 1$. J. Lond. Math. Soc. **10**, 21–25 (1935) (p. 198)
99. H.S.M. Coxeter, *Regular Polytopes*, 3rd edn. (Dover, New York, 1973) (pp. 197, 198)

100. H.S.M. Coxeter, *Regular Complex Polytopes*, 2nd edn. (Cambridge University Press, Cambridge, 1991) (p. 197)
101. H.S.M. Coxeter, W.O.J. Moser, *Generators and Relations for Discrete Groups*, 3rd edn. Ergebnisse der Mathematik und ihrer Grenzgebiete, vol. 14 (Springer, New York, 1972) (pp. 197, 198)
102. H. Cuypers, On a generalization of Fischer spaces. Geom. Dedic. **34**, 67–87 (1990) (p. 567)
103. H. Cuypers, Affine Grassmannians. J. Comb. Theory, Ser. A **70**, 289–304 (1995) (p. 304)
104. H. Cuypers, The geometry of secants in embedded polar spaces. Eur. J. Comb. **28**, 1455–1472 (2007) (p. 444)
105. H. Cuypers, The geometry of hyperbolic lines in polar spaces. Preprint (2009), pp. 1–23. (p. 567)
106. H. Cuypers, J.I. Hall, The 3-transposition groups with trivial center. J. Algebra **178**, 149–193 (1995) (p. 567)
107. H. Cuypers, A. Pasini, Locally polar geometries with affine planes. Eur. J. Comb. **13**, 39–57 (1992) (pp. 102, 408)
108. H. Cuypers, A. Steinbach, Linear transvection groups and embedded polar spaces. Invent. Math. **137**, 169–198 (1999) (p. 408)
109. H. Cuypers, P. Johnson, A. Pasini, On the embeddability of polar spaces. Geom. Dedic. **44**, 349–358 (1992) (p. 409)
110. B. De Bruyn, *Near Polygons*. Frontiers in Mathematics (Birkhäuser, Basel, 2006) (pp. 368, 568)
111. T. De Medts, Y. Segev, A course on Moufang sets. Innov. Incid. Geom. **9**, 79–122 (2009) (p. 302)
112. A. Delgado, D. Goldschmidt, B. Stellmacher, *Groups and Graphs: New Results and Methods*. DMV Seminar, vol. 6 (Birkhäuser, Basel, 1985). With a preface by the authors and Bernd Fischer (p. 47)
113. P. Dembowski, *Finite Geometries*. Classics in Mathematics (Springer, Berlin, 1997). Reprint of the 1968 original (pp. 247, 248)
114. P. Dembowski, A. Wagner, Some characterizations of finite projective spaces. Arch. Math. **11**, 465–469 (1960) (pp. 251, 408)
115. V.V. Deodhar, On the root system of a Coxeter group. Commun. Algebra **10**, 611–630 (1982) (p. 199)
116. V.V. Deodhar, Some characterizations of Coxeter groups. Enseign. Math. (2) **32**, 111–120 (1986) (p. 199)
117. W. Dicks, M.J. Dunwoody, *Groups Acting on Graphs*. Cambridge Studies in Advanced Mathematics, vol. 17 (Cambridge University Press, Cambridge, 1989) (p. 47)
118. K.J. Dienst, Verallgemeinerte Vierecke in projektiven Räumen. Arch. Math. (Basel) **35**, 177–186 (1980) (pp. 444, 445)
119. J. Dieudonné, Sur les groupes classiques, in *Actualités Sci. Ind. No. 1040 = Publ. Inst. Math. Univ. Strasbourg (N.S.) No. 1 (1945)* (Hermann, Paris, 1945) (p. v)
120. J. Dieudonné, *La géométrie des groupes classiques*. Ergebnisse der Mathematik und ihrer Grenzgebiete (N.F.), vol. 5 (Springer, Berlin, 1955) (pp. v, 496, 497)
121. J. Doyen, X. Hubaut, Finite regular locally projective spaces. Math. Z. **119**, 83–88 (1971) (pp. 100, 303)
122. J. Doyen, M. Vandensavel, Non isomorphic Steiner quadruple systems. Bull. Soc. Math. Belg. **23**, 393–410 (1971). Collection of articles honoring P. Burniat, Th. Lepage and P. Libois (p. 249)
123. P. Du Val, *Homographies, Quaternions and Rotations*. Oxford Mathematical Monographs (Clarendon, Oxford, 1964) (p. 197)
124. A. Errera, Sur les polyèdres réguliers de l'analysis situs. Belg. Acad. Mém. **7**, 1–17 (1922) (p. 197)
125. G. Fano, Sui postulati fondamentali della geometria proiettiva. G. Mat. **30**, 114–124 (1892) (p. 47)

126. E. Ferrara Dentice, P.M. Lo Re, Affine Tallini sets and Grassmannians. Adv. Geom. **10**, 659–682 (2010) (p. 568)

127. B. Fischer, Finite groups generated by 3-transpositions. I. Invent. Math. **13**, 232–246 (1971) (p. 567)

128. H. Freudenthal, H. de Vries, *Linear Lie Groups*. Pure and Applied Mathematics, vol. 35 (Academic Press, New York, 1969) (p. 566)

129. W. Fulton, *Algebraic Topology*. Graduate Texts in Mathematics, vol. 153 (Springer, New York, 1995). A first course (p. 200)

130. P. Garrett, *Buildings and Classical Groups* (Chapman & Hall, London, 1997) (p. 565)

131. Gergonne, Démonstration d'un théorème relatif aux lignes du second ordre circonscrites à un même quadrilatère, renfermant la solution du premier des trois problèmes de géométrie énoncés à la page 284 du précédent volume. Ann. Math. Pures Appl. [Ann. Gergonne] **18**, 100–110 (1827/1828) (p. 366)

132. D. Goldstein, R.M. Guralnick, L. Small, E. Zelmanov, Inversion invariant additive subgroups of division rings. Pac. J. Math. **227**, 287–294 (2006) (p. 445)

133. D. Gorenstein, R. Lyons, R. Solomon, General group theory, in *The Classification of the Finite Simple Groups. Number 2. Part I. Chap. G*. Mathematical Surveys and Monographs, vol. 40 (American Mathematical Society, Providence, 1996) (pp. 249, 250, 302)

134. D. Gorenstein, R. Lyons, R. Solomon, Almost simple K-groups, in *The Classification of the Finite Simple Groups. Number 3. Part I. Chap. A*. Mathematical Surveys and Monographs, vol. 40 (American Mathematical Society, Providence, 1998). (p. 249)

135. D. Gorenstein, R. Lyons, R. Solomon, Uniqueness theorems, in *The Classification of the Finite Simple Groups. Number 4. Part II, Chaps. 1–4*. Mathematical Surveys and Monographs Chaps., vol. 40 (American Mathematical Society, Providence, 1999). With errata: *The Classification of the Finite Simple Groups. Number 3. Part I. Chap. A* (Am. Math. Soc., Providence, 1998) (p. 249)

136. D. Gorenstein, R. Lyons, R. Solomon, The generic case, stages 1–3a, in *The Classification of the Finite Simple Groups. Number 5. Part III. Chaps. 1–6*. Mathematical Surveys and Monographs, vol. 40 (American Mathematical Society, Providence, 2002) (p. 249)

137. D. Gorenstein, R. Lyons, R. Solomon, The special odd case, in *The Classification of the Finite Simple Groups. Number 6. Part IV*. Mathematical Surveys and Monographs, vol. 40 (American Mathematical Society, Providence, 2005) (p. 249)

138. H. Gottschalk, D. Leemans, The residually weakly primitive geometries of the Janko group J_1, in *Groups and Geometries*, Siena, 1996. Trends Math. (Birkhäuser, Basel, 1998), pp. 65–79 (p. 249)

139. R. Gramlich, H. van Maldeghem, Intransitive geometries. Proc. Lond. Math. Soc. (3) **93**, 666–692 (2006) (pp. 47, 131)

140. R. Gramlich, J.I. Hall, A. Straub, The local recognition of reflection graphs of spherical Coxeter groups. J. Algebr. Comb. **32**, 1–14 (2010) (p. 100)

141. J. Gray, A course in the history of geometry in the 19th century, in *Worlds Out of Nothing*. Springer Undergraduate Mathematics Series (Springer, London, 2007) (p. 301)

142. R.L. Griess Jr., *Twelve Sporadic Groups*. Springer Monographs in Mathematics (Springer, Berlin, 1998) (p. 249)

143. H. Gross, *Quadratic Forms in Infinite-Dimensional Vector Spaces*. Progress in Mathematics, vol. 1 (Birkhäuser Boston, Cambridge, 1979) (pp. 366, 496)

144. B. Grünbaum, G.C. Shephard, An introduction, in *Tilings and Patterns. A Series of Books in the Mathematical Sciences* (Freeman, New York, 1989) (p. 45)

145. T. Grundhöfer, M. Joswig, M. Stroppel, Slanted symplectic quadrangles. Geom. Dedic. **49**, 143–154 (1994) (p. 409)

146. M. Hall, Projective planes. Trans. Am. Math. Soc. **54**, 229–277 (1943) (p. 100)

147. M. Hall Jr., Automorphisms of Steiner triple systems, in *Proc. Sympos. Pure Math.*, vol. VI (American Mathematical Society, Providence, 1962), pp. 47–66 (pp. 247, 248)

148. J.I. Hall, Locally Petersen graphs. J. Graph Theory **4**, 173–187 (1980) (p. 100)

149. J.I. Hall, Graphs, geometry, 3-transpositions, and symplectic $\mathbf{F}_2$-transvection groups. Proc. Lond. Math. Soc. (3) **58**, 89–111 (1989) (p. 567)

150. J.I. Hall, E.E. Shult, Locally cotriangular graphs. Geom. Dedic. **18**, 113–159 (1985) (p. 100)

151. H. Hanani, On quadruple systems. Can. J. Math. **12**, 145–157 (1960) (p. 249)

152. G. Hanssens, A characterization of point-line geometries for finite buildings. Geom. Dedic. **25**, 297–315 (1988). Geometries and groups (Noordwijkerhout, 1986) (pp. 566, 568, 569)

153. M.I. Hartley, An atlas of small regular abstract polytopes. Period. Math. Hung. **53**, 149–156 (2006) (p. 197)

154. A. Hartman, Counting quadruple systems, in *Proceedings of the Twelfth Southeastern Conference on Combinatorics, Graph Theory and Computing, Vol. II*, vol. 33, Baton Rouge, La., 1981 (1981), pp. 45–54 (p. 249)

155. R. Hartshorne, *Foundations of Projective Geometry*. Lecture Notes, Harvard University, vol. 1966/67 (Benjamin, New York, 1967) (p. 46)

156. H. Havlicek, Zur Theorie linearer Abbildungen. I. J. Geom. **16**, 152–167 (1981). 168–180 (p. 409)

157. H. Havlicek, Zur Theorie linearer Abbildungen. II. J. Geom. **16**, 168–180 (1981) (p. 409)

158. S. Heiss, Two sporadic geometries related to the Hoffman-Singleton graph. J. Comb. Theory, Ser. A **53**, 68–80 (1990) (p. 101)

159. I.N. Herstein, Semi-homomorphisms of groups. Can. J. Math. **20**, 384–388 (1968) (p. 445)

160. D.G. Higman, C.C. Sims, A simple group of order $44,352,000$. Math. Z. **105**, 110–113 (1968) (pp. 101, 250)

161. D. Hilbert, *Grundlagen der Geometrie*, 14th edn. Teubner-Archiv zur Mathematik. Supplement [Teubner Archive on Mathematics. Supplement], vol. 6 (Teubner, Stuttgart, 1999). With supplementary material by Paul Bernays, With contributions by Michael Toepell, Hubert Kiechle, Alexander Kreuzer and Heinrich Wefelscheid, Edited and with appendices by Toepell (pp. 45, 251)

162. J.W.P. Hirschfeld, *Projective Geometries over Finite Fields*, 2nd edn. Oxford Mathematical Monographs (Clarendon/Oxford University Press, New York, 1998) (pp. 248, 249, 301)

163. A.J. Hoffman, R.R. Singleton, On Moore graphs with diameters 2 and 3. IBM J. Res. Dev. **4**, 497–504 (1960) (p. 100)

164. D.W. Hoffmann, Isotropy of quadratic forms and field invariants, in *Quadratic Forms and Their Applications*, Dublin, 1999. Contemp. Math., vol. 272 (American Mathematical Society, Providence, 2000), pp. 73–102 (p. 497)

165. L.-K. Hua, Some properties of a sfield. Proc. Natl. Acad. Sci. USA **35**, 533–537 (1949) (p. 445)

166. L.K. Hua, On semi-homomorphisms of rings and their applications in projective geometry. Usp. Mat. Nauk (N.S.) **8**, 143–148 (1953) (p. 445)

167. D.R. Hughes, F.C. Piper, *Projective Planes*. Graduate Texts in Mathematics, vol. 6 (Springer, New York, 1973) (pp. 99, 100, 247, 302, 367)

168. J.E. Humphreys, *Linear Algebraic Groups*. Graduate Texts in Mathematics, vol. 21 (Springer, New York, 1975) (p. 46)

169. J.E. Humphreys, *Reflection Groups and Coxeter Groups*. Cambridge Studies in Advanced Mathematics, vol. 29 (Cambridge University Press, Cambridge, 1990) (pp. 198, 199, 566)

170. A.A. Ivanov, On geometries of the Fischer groups. Eur. J. Comb. **16**, 163–183 (1995) (p. 249)

171. A.A. Ivanov, Petersen and tilde geometries, in *Geometry of Sporadic Groups. I*. Encyclopedia of Mathematics and Its Applications, vol. 76 (Cambridge University Press, Cambridge, 1999) (pp. 101, 102, 250)

172. A.A. Ivanov, S.V. Shpectorov, A geometry for the O'Nan group connected with the Petersen graph. Russ. Math. Surv. **41**, 211–212 (1986) (pp. 249, 250)

173. A.A. Ivanov, S.V. Shpectorov, Representations and amalgams, in *Geometry of Sporadic Groups. II*. Encyclopedia of Mathematics and Its Applications, vol. 91 (Cambridge University Press, Cambridge, 2002) (pp. 101, 102, 250)

174. L.O. James, A combinatorial proof that the Moore (7, 2) graph is unique. Util. Math. **5**, 79–84 (1974) (p. 100)

175. D.G. James, Unitary geometry over local rings. J. Algebra **56**, 221–234 (1979) (p. 496)

176. T. Januszkiewicz, J. Świątkowski, Simplicial nonpositive curvature. Publ. Math. Inst. Hautes Études Sci. 1–85 (2006) (p. 100)

177. P.M. Johnson, Polar spaces of arbitrary rank. Geom. Dedic. **35**, 229–250 (1990) (p. 367)

178. D.L. Johnson, *Presentations of Groups*, 2nd edn. London Mathematical Society Student Texts, vol. 15 (Cambridge University Press, Cambridge, 1997) (p. 198)

179. P.M. Johnson, Semiquadratic sets and embedded polar spaces. J. Geom. **64**, 102–127 (1999) (p. 444)

180. P. Johnson, E. Shult, Local characterizations of polar spaces. Geom. Dedic. **28**, 127–151 (1988) (p. 367)

181. W.M. Kantor, Dimension and embedding theorems for geometric lattices. J. Comb. Theory, Ser. A **17**, 173–195 (1974) (p. 248)

182. W.M. Kantor, Some geometries that are almost buildings. Eur. J. Comb. **2**, 239–247 (1981) (pp. 102, 250)

183. W.M. Kantor, Some exceptional 2-adic buildings. J. Algebra **92**, 208–223 (1985) (p. 102)

184. W.M. Kantor, Generalized polygons, scabs and gabs, in *Buildings and the Geometry of Diagrams*, Proc. C.I.M.E. Session, Como 1984. Springer Lecture Notes, vol. 1181 (Springer, Berlin, 1986), pp. 79–158 (pp. 48, 102)

185. W.M. Kantor, R. Liebler, J. Tits, On discrete chamber-transitive automorphism groups of affine buildings. Bull. Am. Math. Soc. **16**, 129–133 (1987) (p. 102)

186. W.M. Kantor, T. Meixner, M. Wester, Two exceptional 3-adic affine buildings. Geom. Dedic. **33**, 1–11 (1990) (p. 102)

187. A. Kasikova, Characterization of some subgraphs of point-collinearity graphs of building geometries. Eur. J. Comb. **28**, 1493–1529 (2007) (p. 567)

188. A. Kasikova, Characterization of some subgraphs of point-collinearity graphs of building geometries. II. Adv. Geom. **9**, 45–84 (2009) (p. 567)

189. A. Kasikova, Simple connectedness of hyperplane complements in some geometries related to buildings. J. Comb. Theory, Ser. A **118**, 641–671 (2011) (p. 408)

190. A. Kasikova, E. Shult, Point-line characterizations of Lie geometries. Adv. Geom. **2**, 147–188 (2002) (pp. 566, 567, 568)

191. P. Kaski, P.R.J. Östergård, The Steiner triple systems of order 19. Math. Comput. **73**, 2075–2092 (2004) (electronic) (p. 249)

192. P. Kaski, P.R.J. Östergård, O. Pottonen, The Steiner quadruple systems of order 16. J. Comb. Theory, Ser. A **113**, 1764–1770 (2006) (p. 249)

193. A.B. Kempe, A memoir on the theory of mathematical form. Philos. Trans. R. Soc. Lond. **177**, 1–70 (1886) (p. 46)

194. G. Kist, H.-J. Kroll, K. Sörensen, Beispiel eines Körpers, der keinen Antiautomorphismus zuläßt. Mitt. Math. Ges. Hamb. **10**, 345–348 (1977) (p. 366)

195. M.-A. Knus, A. Merkurjev, M. Rost, J.-P. Tignol, *The Book of Involutions*. American Mathematical Society Colloquium Publications, vol. 44 (American Mathematical Society, Providence, 1998). With a preface in French by J. Tits (p. 496)

196. P. Köhler, T. Meixner, M. Wester, The affine building of type $\tilde{A}_2$ over a local field of characteristic two. Arch. Math. (Basel) **42**, 400–407 (1984) (p. 102)

197. T.Y. Lam, *A First Course in Noncommutative Rings*, 2nd edn. Graduate Texts in Mathematics, vol. 131 (Springer, New York, 2001) (pp. 46, 366)

198. C.W.H. Lam, L. Thiel, S. Swiercz, The nonexistence of finite projective planes of order 10. Can. J. Math. **41**, 1117–1123 (1989) (p. 248)

199. D. Leemans, Two rank six geometries for the Higman-Sims sporadic group. Discrete Math. **294**, 123–132 (2005) (p. 250)

200. D. Leemans, On the geometry of O'N. J. Geom. **97**, 83–97 (2010) (p. 47)

201. D. Leemans, E. Schulte, Groups of type $L_2(q)$ acting on polytopes. Adv. Geom. **7**, 529–539 (2007) (p. 200)

202. D. Leemans, E. Schulte, Polytopes with groups of type $PGL_2(q)$. Ars Math. Contemp. **2**, 163–171 (2009) (p. 200)

203. G.I. Lehrer, D.E. Taylor, *Unitary Reflection Groups*. Australian Mathematical Society Lecture Series, vol. 20 (Cambridge University Press, Cambridge, 2009) (p. 200)
204. H. Lenz, Über die Einführung einer absoluten Polarität in die projektive und affine Geometrie des Raumes. Math. Ann. **128**, 363–372 (1954) (p. 366)
205. H. Lenz, Zur Begründung der analytischen Geometrie. S.-B. Math. Nat. Kl. Bayer. Akad. Wiss. **1954**, 17–72 (1954) (p. 247)
206. P. Libois, Quelques espaces linéaires. Bull. Soc. Math. Belg. **16**, 13–22 (1964) (p. 101)
207. H. Lüneburg, Ein neuer Beweis eines Hauptsatzes der projektiven Geometrie. Math. Z. **87**, 32–36 (1965) (pp. 248, 302)
208. H. Lüneburg, Über die Struktursätze der projektiven Geometrie. Arch. Math. (Basel) **17**, 206–209 (1966) (p. 302)
209. H. Lüneburg, *Transitive Erweiterungen Endlicher Permutationsgruppen*. Lecture Notes in Mathematics, vol. 84 (Springer, Berlin, 1969) (p. 249)
210. H. Matsumoto, Générateurs et relations des groupes de Weyl généralisés. C. R. Acad. Sci. Paris **258**, 3419–3422 (1964) (p. 198)
211. S. Mattarei, Inverse-closed additive subgroups of fields. Isr. J. Math. **159**, 343–347 (2007) (p. 445)
212. P. McMullen, Four-dimensional regular polyhedra. Discrete Comput. Geom. **38**, 355–387 (2007) (p. 197)
213. P. McMullen, E. Schulte, *Abstract Regular Polytopes*. Encyclopedia of Mathematics and Its Applications, vol. 92 (Cambridge University Press, Cambridge, 2002) (pp. 197, 200)
214. T. Meixner, A note on three rank four geometries for $P\Omega_6^-(3)$. J. Geom. **54**, 84–90 (1995) (p. 100)
215. T. Meixner, Buekenhout geometries of rank 3 which involve the Petersen graph. J. Math. Soc. Jpn. **48**, 467–491 (1996) (p. 101)
216. T. Meixner, On Weiss' geometric characterization of the Rudvalis simple group. Forum Math. **10**, 135–146 (1998) (p. 250)
217. T. Meixner, F. Timmesfeld, Chamber systems with string diagrams. Geom. Dedic. **15**, 115–123 (1983) (pp. 101, 131)
218. E.H. Moore, Tactical Memoranda I–III (continued). Am. J. Math. **18**, 291–303 (1896) (p. 45)
219. P.J. Morandi, B.A. Sethuraman, J.-P. Tignol, Division algebras with an anti-automorphism but with no involution. Adv. Geom. **5**, 485–495 (2005) (p. 366)
220. P. Moszkowski, Involutions and reflection subgroups of finite Coxeter groups. Eur. J. Comb. **30**, 1902–1912 (2009) (p. 199)
221. P. Moszkowski, Addendum to: "Involutions and reflection subgroups of finite Coxeter groups" [Eur. J. Comb.] [mr2552672]. Eur. J. Comb. **30**, 1913–1918 (2009) (p. 199)
222. R. Moufang, Zur Struktur von Alternativkörpern. Math. Ann. **110**, 416–430 (1935) (p. 303)
223. F.R. Moulton, A simple non-Desarguesian plane geometry. Trans. Am. Math. Soc. **3**, 192–195 (1902) (p. 304)
224. B. Mühlherr, A geometric approach to non-embeddable polar spaces of rank 3. Bull. Soc. Math. Belg., Sér. A **42**, 577–594 (1990). Algebra, groups and geometry (pp. 367, 565)
225. B. Mühlherr, Some contributions to the theory of buildings based on the gate property. Ph.D. Thesis. Tübingen University, Tübingen (1994) (p. 565)
226. B. Mühlherr, On isomorphisms between Coxeter groups. Des. Codes Cryptogr. **21**, 189 (2000). Special issue dedicated to Dr. Jaap Seidel on the occasion of his 80th birthday (Oisterwijk, 1999) (p. 200)
227. B. Mühlherr, A. Nguyen, A combinatorial approach to Coxeter groups. Bull. Belg. Math. Soc. Simon Stevin **12**, 859–869 (2005) (p. 199)
228. B. Mühlherr, J. Tits, The center conjecture for non-exceptional buildings. J. Algebra **300**, 687–706 (2006) (p. 567)
229. G.P. Nagy, P. Vojtěchovský, The Moufang loops of order 64 and 81. J. Symb. Comput. **42**, 871–883 (2007) (p. 304)
230. A. Neumaier, Some sporadic geometries related to PG(3, 2). Arch. Math. (Basel) **42**, 89–96 (1984) (p. 100)

231. M. Pankov, Isometric embeddings of Johnson graphs in Grassmann graphs. J. Algebr. Comb. **33**, 555–570 (2011) (p. 303)

232. P. Papi, A characterization of a special ordering in a root system. Proc. Am. Math. Soc. **120**, 661–665 (1994) (p. 199)

233. A. Pasini, *Diagram Geometries*. Oxford Science Publications (Clarendon/Oxford University Press, New York, 1994) (pp. 45, 46, 47, 131)

234. A. Pasini, H. van Maldeghem, Some constructions and embeddings of the tilde geometry. Note Mat. **21**, 1–33 (2002/2003) (p. 102)

235. D. Passman, *Permutation Groups* (Benjamin, New York, 1968) (p. 47)

236. S.E. Payne, J.A. Thas, *Finite Generalized Quadrangles*, 2nd edn. EMS Series of Lectures in Mathematics (European Mathematical Society (EMS), Zürich, 2009) (pp. 99, 409)

237. N. Percsy, On the Buekenhout-Shult theorem on polar spaces. Bull. Soc. Math. Belg., Sér. B **41**, 283–294 (1989) (p. 367)

238. A. Pfister, *Quadratic Forms with Applications to Algebraic Geometry and Topology*. London Mathematical Society Lecture Note Series, vol. 217 (Cambridge University Press, Cambridge, 1995) (p. 497)

239. H.O. Pflugfelder, *Quasigroups and Loops: Introduction*. Sigma Series in Pure Mathematics, vol. 7 (Heldermann, Berlin, 1990) (p. 303)

240. G. Pickert, *Projektive Ebenen*. Die Grundlehren der mathematischen Wissenschaften in Einzeldarstellungen mit besonderer Berücksichtigung der Anwendungsgebiete, vol. LXXX (Springer, Berlin, 1955) (pp. 99, 247, 303)

241. Poncelet, Note sur divers articles du bulletin des sciences de 1826 et de 1827, relatifs à la théorie des polaires réciproques, à la dualité des propriétés de situation de l'étendue, etc. Ann. Math. Pures Appl. [Ann. Gergonne] **18**, 125–142 (1827/1828) (p. 366)

242. M.A. Ronan, Triangle geometries. J. Comb. Theory **37**, 294–319 (1984) (p. 102)

243. M.A. Ronan, Embeddings and hyperplanes of discrete geometries. Eur. J. Comb. **8**, 179–185 (1987) (p. 408)

244. M. Ronan, *Lectures on Buildings* (University of Chicago Press, Chicago, 2009). Updated and revised (p. 198)

245. M.A. Ronan, J. Tits, Building buildings. Math. Ann. **278**, 291–306 (1987) (p. 565)

246. R.D. Schafer, *An Introduction to Nonassociative Algebras*. Pure and Applied Mathematics, vol. 22 (Academic Press, New York, 1966) (p. 100)

247. R. Scharlau, A characterization of Tits buildings by metrical properties. J. Lond. Math. Soc. (2) **32**, 317–327 (1985) (p. 565)

248. R. Scharlau, Buildings, in *Handbook of Incidence Geometry* (North-Holland, Amsterdam, 1985), pp. 477–645 (p. 569)

249. W. Scharlau, *Quadratic and Hermitian Forms*. Grundlehren der Mathematischen Wissenschaften [Fundamental Principles of Mathematical Sciences], vol. 270 (Springer, Berlin, 1985) (p. 496)

250. W.R. Scott, Half-homomorphisms of groups. Proc. Am. Math. Soc. **8**, 1141–1144 (1957) (p. 445)

251. B. Segre, *Lectures on Modern Geometry*. With an Appendix by Lucio Lombardo-Radice. Consiglio Nazionale delle Ricerche Monografie Matematiche, vol. 7 (Edizioni Cremonese, Rome, 1961) (p. 302)

252. J.-P. Serre, *Trees*. Springer Monographs in Mathematics (Springer, Berlin, 2003). Translated from the French original by John Stillwell, Corrected 2nd printing of the 1980 English translation (p. 47)

253. S. Shad, E. Shult, Near n-gons, in *The Santa Cruz Conference on Finite Groups*. Univ. California, Santa Cruz, Calif., 1979. Proc. Sympos. Pure Math, vol. 37 (American Mathematical Society, Providence, 1980), pp. 461–463 (p. 568)

254. G.C. Shephard, J.A. Todd, Finite unitary reflection groups. Can. J. Math. **6**, 274–304 (1954) (p. 200)

255. E. Shult, Geometric hyperplanes of embeddable Grassmannians. J. Algebra **145**, 55–82 (1992) (p. 303)

256. E. Shult, On Veldkamp lines. Bull. Belg. Math. Soc. Simon Stevin **4**, 299–316 (1997) (p. 408)
257. E.E. Shult, Characterization of Grassmannians by one class of singular subspaces. Adv. Geom. **3**, 227–250 (2003) (p. 568)
258. E. Shult, Characterizing the half-spin geometries by a class of singular subspaces. Bull. Belg. Math. Soc. Simon Stevin **12**, 883–894 (2005) (p. 568)
259. E. Shult, On characterizing the long-root geometries. Adv. Geom. **10**, 353–370 (2010) (pp. 567, 568)
260. E.E. Shult, *Points and Lines*. Universitext. (Springer, Heidelberg, 2011). Characterizing the classical geometries (pp. 45, 48, 303, 366, 409, 566)
261. E.E. Shult, K. Thas, A theorem of Cohen on parapolar spaces. Combinatorica **30**, 435–444 (2010) (p. 568)
262. E. Shult, A. Yanushka, Near n-gons and line systems. Geom. Dedic. **9**, 1–72 (1980) (pp. 249, 367, 568)
263. A.P. Sprague, Rank 3 incidence structures admitting dual-linear, linear diagram. J. Comb. Theory, Ser. A **38**, 254–259 (1985) (p. 304)
264. T.A. Springer, *Linear Algebraic Groups*, 2nd edn. Modern Birkhäuser Classics (Birkhäuser Boston, Boston, 2009) (p. 46)
265. T.A. Springer, F.D. Veldkamp, *Octonions, Jordan Algebras and Exceptional Groups*. Springer Monographs in Mathematics (Springer, Berlin, 2000) (p. 100)
266. R.P. Stanley, On the number of reduced decompositions of elements of Coxeter groups. Eur. J. Comb. **5**, 359–372 (1984) (p. 198)
267. R.P. Stanley, *Enumerative Combinatorics. Volume 1*, 2nd edn. Cambridge Studies in Advanced Mathematics, vol. 49 (Cambridge University Press, Cambridge, 2012) (p. 99)
268. A. Steinbach, H. van Maldeghem, Generalized quadrangles weakly embedded of degree >2 in projective space. Forum Math. **11**, 139–176 (1999) (p. 408)
269. A. Steinbach, H. van Maldeghem, Generalized quadrangles weakly embedded of degree 2 in projective space. Pac. J. Math. **193**, 227–248 (2000) (p. 408)
270. J. Steiner, Combinatorische aufgabe. Pac. J. Math. **45**, 181–182 (1853) (p. 248)
271. H. Stichtenoth, *Algebraic Function Fields and Codes*, 2nd edn. Graduate Texts in Mathematics, vol. 254 (Springer, Berlin, 2009) (p. 366)
272. D.E. Taylor, *The Geometry of the Classical Groups*. Sigma Series in Pure Mathematics, vol. 9 (Heldermann, Berlin, 1992) (pp. 368, 497)
273. L. Teirlinck, On projective and affine hyperplanes. J. Comb. Theory, Ser. A **28**, 290–306 (1980) (pp. 251, 408, 409)
274. K. Tent, Very homogeneous generalized n-gons of finite Morley rank. J. Lond. Math. Soc. (2) **62**, 1–15 (2000) (p. 566)
275. K. Tent, Free polygons, twin trees and cat(1)-spaces. Pure Appl. Math. Q. **7**, 1037–1052 (2011) (pp. 100, 566)
276. J.A. Thas, Extensions of finite generalized quadrangles, in *Symposia Mathematica*, Rome, 1983. Sympos. Math., vol. XXVIII (Academic Press, London, 1986), pp. 127–143 (p. 367)
277. J.A. Thas, Projective geometry over a finite field, in *Handbook of Incidence Geometry* (North-Holland, Amsterdam, 1995), pp. 295–347 (p. 497)
278. J.A. Thas, H. van Maldeghem, Embedded thick finite generalized hexagons in projective space. J. Lond. Math. Soc. (2) **54**, 566–580 (1996) (p. 445)
279. F.G. Timmesfeld, *Abstract Root Subgroups and Simple Groups of Lie Type*. Monographs in Mathematics, vol. 95 (Birkhäuser, Basel, 2001) (pp. 304, 565, 567, 568)
280. J. Tits, Les groupes de Lie exceptionnels et leur interprétation géométrique. Bull. Soc. Math. Belg. **8**, 48–81 (1956) (pp. v, 45, 46, 99, 247)
281. J. Tits, Sur la trialité et certains groupes qui s'en déduisent. Publ. Math. Inst. Hautes Études Sci. 13–60 (1959) (pp. 99, 100, 101, 102, 368)
282. J. Tits, Groupes algébriques semi-simples et géométries associées, in *Algebraical and Topological Foundations of Geometry*, Proc. Colloq., Utrecht, 1959 (Pergamon, Oxford, 1962), pp. 175–192 (p. 45)
283. J. Tits, Ovoïdes et groupes de Suzuki. Arch. Math. **13**, 187–198 (1962) (p. 248)

284. J. Tits, Géométries polyédriques et groupes simples. Atti della II Riunione del Groupement de Mathématiciens d'Expression Latine, Firenze-Bologna 1961, pp. 66–88 (1963) (pp. 46, 47, 197, 565, 566)

285. J. Tits, Le problème des mots dans les groupes de Coxeter, in *Symposia Mathematica*, vol. 1, INDAM, Rome, 1967/1968 (Academic Press, London, 1969), pp. 175–185 (p. 198)

286. J. Tits, *Buildings of Spherical Type and Finite BN-Pairs*. Lecture Notes in Mathematics, vol. 386 (Springer, Berlin, 1974) (pp. v, vi, 47, 101, 197, 366, 367, 368, 408, 445, 496, 497, 565, 566, 568, 569)

287. J. Tits, A local approach to buildings, in *The Geometric Vein* (Springer, New York, 1981), pp. 519–547 (pp. 46, 47, 48, 131, 198, 367, 565, 568)

288. J. Tits, Buildings and group amalgamations, in *Proceedings of Groups—St. Andrews 1985*. London Math. Soc. Lecture Note Ser., vol. 121 (Cambridge University Press, Cambridge, 1986), pp. 110–127 (p. 131)

289. J. Tits, Ensembles ordonnés, immeubles et sommes amalgamées. Bull. Soc. Math. Belg., Sér. A **38**, 367–387 (1986) (p. 131)

290. J. Tits, Immeubles de type affine, in *Buildings and the Geometry of Diagrams*, Como, 1984. Lecture Notes in Math., vol. 1181 (Springer, Berlin, 1986), pp. 159–190 (p. v)

291. J. Tits, R.M. Weiss, *Moufang Polygons*. Springer Monographs in Mathematics (Springer, Berlin, 2002) (pp. 99, 303, 566, 569)

292. J. Überberg, *Foundations of Incidence Geometry: Projective and Polar Spaces*. Springer Monographs in Mathematics (Springer, Berlin, 2011) (pp. 45, 301, 366)

293. H. van Maldeghem, A geometric characterization of the perfect Suzuki-Tits ovoids. J. Geom. **58**, 192–202 (1997) (p. 248)

294. H. van Maldeghem, *Generalized Polygons*. Monographs in Mathematics, vol. 93 (Birkhäuser, Basel, 1998) (p. 99)

295. H. van Maldeghem, A geometric characterization of the perfect Ree-Tits generalized octagons. Proc. Lond. Math. Soc. (3) **76**, 203–256 (1998) (p. 100)

296. O. Veblen, J.W. Young, *Projective Geometry. Vols. 1, 2* (Blaisdell/Ginn, New York, 1965) (pp. 247, 301)

297. F.D. Veldkamp, Polar geometry. I, II, III, IV, V. Ned. Akad. Wet. Proc., Ser. A 62; 63 = Indag. Math. 21, 512–551 (1959) **22**, 207–2012 (1959) (pp. vi, 366, 367, 408, 409)

298. P. Vrana, P. Lévay, The Veldkamp space of multiple qubits. J. Phys. A **43**, 125303 (2010) (p. 408)

299. T.R.S. Walsh, Characterizing the vertex neighbourhoods of semi-regular polyhedra. Geom. Dedic. **1**, 117–123 (1972) (p. 197)

300. G.M. Weetman, A construction of locally homogeneous graphs. J. Lond. Math. Soc. (2) **50**, 68–86 (1994) (p. 100)

301. G.M. Weetman, Diameter bounds for graph extensions. J. Lond. Math. Soc. (2) **50**, 209–221 (1994) (p. 100)

302. R. Weiss, On Fischer's characterization of $Sp_{2n}(2)$ and $U_n(2)$. Commun. Algebra **11**, 2527–2554 (1983) (p. 567)

303. R. Weiss, 3-transpositions in infinite groups. Math. Proc. Camb. Philos. Soc. **96**, 371–377 (1984) (p. 567)

304. R. Weiss, A geometric characterization of the groups M_{12}, He and Ru. J. Math. Soc. Jpn. **43**, 795–814 (1991) (p. 250)

305. R.M. Weiss, *The Structure of Spherical Buildings* (Princeton University Press, Princeton, 2003) (pp. 198, 565)

306. R.M. Weiss, *The Structure of Affine Buildings*. Annals of Mathematics Studies, vol. 168 (Princeton University Press, Princeton, 2009) (pp. v, 102)

307. A.L. Wells Jr., Universal projective embeddings of the Grassmannian, half spinor, and dual orthogonal geometries. Q. J. Math. Oxford Ser. (2) **34**, 375–386 (1983) (p. 444)

308. D.J.A. Welsh, *Matroid Theory*. L.M.S. Monographs, vol. 8 (Academic Press/Harcourt Brace Jovanovich, London, 1976) (p. 247)

309. M. Wester, Endliche fahnentransitive Tits-Geometrien und ihre universellen Überlagerungen. Mitt. Math. Sem. Giess. **170**, 143 (1985) (p. 100)
310. H. Whitney, On the abstract properties of linear dependence. Am. J. Math. **57**, 509–533 (1935) (p. 247)
311. H. Wielandt, *Finite Permutation Groups* (Academic Press, New York, 1964). Translated from the German by R. Bercov (p. 47)
312. R. Wilson, Nonisomorphic Steiner triple systems. Math. Z. **135**, 303–313 (1974) (p. 249)
313. R.A. Wilson, *The Finite Simple Groups*. Graduate Texts in Mathematics, vol. 251 (Springer, London, 2009) (p. 497)
314. E. Witt, Die 5-fach transitiven Gruppen von Mathieu. Abh. Math. Semin. Hans. Univ. **12**, 256–264 (1938) (p. 248)
315. S. Yoshiara, The flag-transitive C_3-geometries of finite order. J. Algebr. Comb. **5**, 251–284 (1996) (p. 100)
316. G.M. Ziegler, *Lectures on Polytopes*. Graduate Texts in Mathematics, vol. 152 (Springer, New York, 1995) (p. 197)
317. P.-H. Zieschang, *An Algebraic Approach to Association Schemes*. Lecture Notes in Mathematics, vol. 1628 (Springer, Berlin, 1996) (p. 131)

Index

F. Buekenhout, A.M. Cohen, *Diagram Geometry*,
Ergebnisse der Mathematik und ihrer Grenzgebiete. 3. Folge / A Series of
Modern Surveys in Mathematics 57, DOI 10.1007/978-3-642-34453-4,
© Springer-Verlag Berlin Heidelberg 2013

9 783642 344527